International Conference on Applied Sciences (ICAS2018)

IOP Conference Series: Materials Science and Engineering
Volume 477

Banja Luke, Bosnia and Herzegovina
9 - 11 May 2018

Editor:

Ludovic Dan Lemle

ISBN: 978-1-5108-8216-4
ISSN: 1757-8981

Printed from e-media with permission by:

Curran Associates, Inc.
57 Morehouse Lane
Red Hook, NY 12571

Some format issues inherent in the e-media version may also appear in this print version.

Printed by Curran Associates, Inc. (2019)

For permission requests, please contact the Institute of Physics at the address below.

Institute of Physics
Dirac House, Temple Back
Bristol BS1 6BE UK

Phone: 44 1 17 929 7481
Fax: 44 1 17 920 0979

techtracking@iop.org

Additional copies of this publication are available from:

Curran Associates, Inc.
57 Morehouse Lane
Red Hook, NY 12571 USA
Phone: 845-758-0400
Fax: 845-758-2633
Email: curran@proceedings.com
Web: www.proceedings.com

Table of contents

Volume 477

International Conference on Applied Sciences

9–11 May 2018, Banja Luke, Bosnia and Herzegovina

Accepted papers received: 28 December 2018
Published online: 18 February 2019

Preface

Papers

IOP Conf. Series: Materials Science and Engineering **477** (2019) 011001 doi:10.1088/1757-899X/477/1/011001

International Conference on Applied Sciences ICAS2018

Banja Luka, Bosnia and Herzegovina, May 9-11, 2018

Organizers:

University Politehnica Timisoara **and** **University of Banja Luka**

in cooperation with

Ministry of Science and Technology of the Republic of Srpska, Bosnia and Herzegovina
Ministry of Trade and Tourism of the Republic of Srpska, Bosnia and Herzegovina
Academy of Romanian Scientists
Academy of Sciences of the Republic of Srpska, Bosnia and Herzegovina
Academy of Technical Sciences of Romania - Timisoara Branch
and
General Association of Romanian Engineers - Hunedoara Branch

Event sponsored by:

IOP Conf. Series: Materials Science and Engineering **477** (2019) 011001 doi:10.1088/1757-899X/477/1/011001

The **International Conference on Applied Sciences ICAS2018** took place in Banja Luka, Republic of Srpska, Bosnia and Herzegovina, in the period May 9–11, 2018, at the University of Banja Luka, with the aims to serve as a platform for exchange of information between various areas of applied sciences, and to promote the communication between the scientists of different nations, countries and continents.

The conference has been focused on several fields of application, operation and influence of the applied sciences and technologies on industry.

Topics of the conference covers a comprehensive spectrum of issues from:
1. Materials Science: Metallic Materials, Composite Materials, Metal Alloys, Metallurgy, Heat Transfer, and others...
2. Mechatronics: Mechanical Engineering, Robotic Systems Engineering, Control Engineering, Reliability, and others...
3. Electrical Engineering: Circuits and Systems, Signal Processing, Electric Motors, and others...
4. Computers Engineering: Modeling and Simulation, Computational Methods in Engineering, Software Engineering, Data Bases, and others...
5. Fundamental Sciences: Numerical approximation and analysis, Interdisciplinary applications of mathematics and physics, Chemistry, and others...

The conference gathered qualified researchers whose expertise can be used to develop new engineering knowledge that has applicability potential in Industry. The number of participants was 312 from 21 countries (Algeria, Austria, Bosnia and Herzegovina, Brazil, France, Germany, Croatia, Iran, Italy, Macedonia, Hungary, Malaysia, Morocco, Moldova, Norway, Poland, Romania, Serbia, Slovenia, Turkey, U.S.A.).

During the three days of the conference, 6 invited speakers and 144 oral talks were delivered. Based on the works presented at the conference, 60 papers have been selected to be included in this volume of *IOP Conference Series: Materials Science and Engineering.*

Invited speakers

Mohsen Razzaghi, Mississippi State University
Gabor Szeibig, UiT Arctic University of Norway
Branislav Borovac, University of Novi Sad
Ioan Ilca, Academy of Technical Sciences of Romania
Zora Konjevic, University of Novi Sad, University Singidunum
Nataša Trišović, University of Belgrade

Honorary Committee

Dana Manuela Constantinescu, Ambassador of Romania in Bosnia and Herzegovina
Jasmin Komić, Ministry of Science and Technology of the Republic of Srpsca, B&H
Predrag Gluhaković, Ministry of Trade and Tourism of the Republic of Srpska, B&H
Popoviciu Octavian Mircea, Romanian Academy of Scientists
Ilca Ioan, Academy of Technical Sciences in Romania
Harau Carmen, Parliament of Romania - Senate
Dragoljub Mirjanić, Academy of Sciences and Arts of the Republic of Srpska, B&H
Predrag Damjanović, Director of the Pedagogical Institute of the Republic of Srpska, B&H
Igor Radojičić, Mayor of Banja Luka
Doru Ion Petrescu, CEO Arcelor Mittal SA, Romania
Borko Đurić, President of the Chamber of Commerce of the Republic of Srpska, B&H
Boboutanu Dan, Mayor of Hunedoara, Romania

IOP Conf. Series: Materials Science and Engineering **477** (2019) 011001 doi:10.1088/1757-899X/477/1/011001

Organizing Committee

Co-Presidents
Strain Posavljak, University of Banja Luka
Serban Viorel-Aurel, University Politehnica Timisoara

Vice-Presidents
Knezevic Darko, University of Banja Luka
Panoiu Caius, University Politehnica Timisoara

Coordinators
Latinovic Tihomir, University of Banja Luka
Lemle Ludovic Dan, University Politehnica Timisoara

Program Chairmen
Rogić Miroslav, University of Banja Luka
Kiss Imre, University Politehnica Timisoara

Members
Abrudean Cristian, University Politehnica Timisoara
Alexa Vasile, University Politehnica Timisoara
Ardelean Marius, University Politehnica Timisoara
Dinis Corina, University Politehnica Timisoara
Đorđić Dijana, University of Banja Luka
Deaconu Sorin, University Politehnica Timisoara
Gherman Lucian, University Politehnica Timisoara
Kardaš Danijela, University of Banja Luka
Maglov Sanja, University of Banja Luka
Prochaska Biljana, University of Banja Luka
Ratiu Sorin, University Politehnica Timisoara
Savković Ivana, University of Banja Luka
Škundrić Jovan, University of Banja Luka
Socalici Ana, University Politehnica Timisoara
Sredanović Branislav, University of Banja Luka
Tirian Ovidiu, University Politehnica Timisoara
Tošić Gordana, University of Banja Luka

Secretariat
Berdie Adela, University Politehnica Timisoara
Bojinovic Mirko, University of Banja Luka
Josan Ana, University Politehnica Timisoara
Novakovic Bojana, University of Banja Luka
Petrovic Slavica, University of Banja Luka
Rob Raluca, University Politehnica Timisoara

IOP Conf. Series: Materials Science and Engineering **477** (2019) 011001 doi:10.1088/1757-899X/477/1/011001

Scientific Committee

President
Lemle Ludovic Dan, University Politehnica Timisoara

Vice-Presidents
Barz Cristian, Technical University of Cluj Napoca
Gherman Laurian, "Henri Coanda" Air Force Academy of Brasov
Jiang Yiwen, Military Economics Academy of Wuhan
Lakić-Globočki Gordana, University of Banja Luka

Chairmen of Materials Science Session
Ardelean Erika, University Politehnica Timisoara
Dragoslav Dobraš, University of Banja Luka

Chairmen of Mechatronics Session
Bordeasu Ilare, University Politehnica Timisoara
Rogic Miroslav, University of Banja Luka

Chairmen of Electrical Engineering Session
Blanusa Branko, University of Banja Luka
Popa Gabriel, University Politehnica Timisoara

Chairmen of Computers Engineering Session
Latinovic Tihomir, University of Banja Luka
Panoiu Manuela, University Politehnica Timisoara

Chairmen of Fundamental Sciences Session
Bistrian Diana, University Politehnica Timisoara
Romano Daniel, University of Banja Luka

Chairmen of Video-conference Session
Knezevic Bojan, University of Banja Luka
Muscalagiu Ionel, University Politehnica Timisoara

Members
Borojević Stevo, University of Banja Luka
Cioaca Catalin, "Henri Coanda" Air Force Academy of Brasov
Cioata Vasile, University Politehnica Timisoara
Constantinescu George, "Henri Coanda" Air Force Academy of Brasov
Cosma Marius, Technical University of Cluj Napoca
Cuntan Corina, University Politehnica Timisoara
Dobrnjac Mirko, University of Banja Luka
Erdei Zoltan, Technical University of Cluj Napoca
Gvero Petar, University of Banja Luka
Golubovic-Bugarski Valentina, University of Banja Luka
Guo Zhonghou, Military Economics Academy of Wuhan
Heput Teodor, University Politehnica Timisoara
Iagar Angela, University Politehnica Timisoara
Ilić Gradimir, University of Niš
Kaouka Allaedine, Amar Telidji University of Laghouat
Kosec Borut, University of Ljubljana
Volodymyr Kopei, Ivano-Frankivsk National Technical University of Oil and Gas
Labudzki Remigiusz, University of Poznan

Lesenciuc Adrian, "Henri Coanda" Air Force Academy of Brasov
Luculescu Doru, "Henri Coanda" Air Force Academy of Brasov
Lung Claudiu, Technical University of Cluj Napoca
Mandru Dan, Technical University of Cluj Napoca
Milašinović Aleksandar, University of Banja Luka
Milovanovic Zdravko, University of Banja Luka
Miron Liliana Ecaterina, "Henri Coanda" Air Force Academy of Brasov
Osaci Mihaela, University Politehnica Timisoara
Onysko Oleh, Ivano-Frankivsk National Technical University of Oil and Gas
Panchuk Vitalii, Ivano-Frankivsk National Technical University of Oil and Gas
Petkovic Snezana, University of Banja Luka
Pinca-Bretotean Camelia,University Politehnica Timisoara
Pop Sitar Petrica, Technical University of Cluj Napoca
Rusu-Anghel Stela, University Politehnica Timisoara
Serafimovski Marko, University of Skopje
Stoica Diana, University Politehnica Timisoara
Stoychitch Mihaylo, University of Banja Luka
Sziebig Gabor, UiT Arctic University of Norway
Tanasić Zorana, University of Banja Luka
Todic Mladen, University of Banja Luka
Vasiu Teodor, University Politehnica Timisoara
Vilceanu Lucia, University Politehnica Timisoara
Fras Zemljic Lidija, University of Maribor
Zhang Lijun, Military Economics Academy of Wuhan

We cordially thank the authors for their contributions and we express our gratitude to everyone who contributed to the success of our conference.

Ludovic Dan LEMLE
Editor of the Proceedings of ICAS2018

Peer review statement

All papers published in this volume of *IOP Conference Series: Materials Science and Engineering* have been peer reviewed through processes administered by the proceedings Editors. Reviews were conducted by expert referees to the professional and scientific standards expected of a proceedings journal published by IOP Publishing.

IOP Conf. Series: Materials Science and Engineering **477** (2019) 012001 doi:10.1088/1757-899X/477/1/012001

Structural research at high temperature on industrial semi-finished products made from thermostable stainless steels

I Ilca

Member of Technical Sciences Academy of Romania, Politehnica University of Timisoara, Faculty of Engineering Hunedoara, Revoluţiei 5, Hunedoara, Romania

E-mail: ilca@fih.upt.ro

Abstract. Austenitic stainless steels possess the best thermal shock resistance properties at low temperatures and the best resistance to elevated temperatures, with versatile demands in the economy.
For all stainless steel grades, the best hot workability is obtained when the structure of the steel at the hot-metal working temperature consists of a single phase. Small quantities of ferrite can be tolerated in the structure of the austenitic and martensitic steels as well as small quantities of austenite in the structure of ferritic steels, but these have to be maintained within certain limits, either through an appropriate change of the chemical composition of the steel, or through the control of the hot-metal working temperature.
Therefore, the research observes the evolution of the structure during the heating of austenitic stainless steels, stating the soaking zone favorable for reducing the existing content of ferrite down to values that do not worsen the plasticity.
Based on the conducted research, a new heating technology was designed, resulting in the reduction of the heating duration and a high efficiency in terms of the dissolution and spheroidization of the residual ferrite in the basic austenitic structure.

1. Introduction

For several years, stainless steels consisted of iron and chromium alloys or iron, chromium and nickel alloys. Incidental, small quantities of certain elements were added, with the purpose of increasing corrosion resistance and improving their mechanical properties or in order to immunize steels against the action of harmful impurities. Recently, new stainless steel grades were developed, in which the nickel content of the iron-chromium-nickel steels was entirely or partly replaced with combinations of nitrogen and manganese [1-3]. The slow reaction velocities characteristic to stainless steels have prevented the determination of accurate balance diagrams; however, the existing diagrams to date allow at least the drawing of certain qualitative conclusions regarding the structure of these steels.

In terms of the corrosion resistance, the chromium content appears to be the determinant variable and the effect of the chromium can be enhanced by adding molybdenum, nickel and other elements. The mechanical properties of stainless steels, much like those of other steels, are dependent upon the structure and composition of the material. Thus, the austenitic steels possess the best thermal shock resistance properties at low temperatures and the best resistance to elevated temperatures.

Consequently, in that they exist in a great variety of structures, austenitic stainless steels display a wide range of mechanical properties which, combined with an excellent corrosion resistance, make these steels multilaterally used in the construction sector [4], [5], transport equipment, sanitary equipment and cooking utensils, petroleum processing equipment [6], [7], chemical products processing equipment [8], paper and food products equipment.

2. Content

Research into the determination of a heating regime for 3.6 tonnes ingots that would allow the dissolution of the secondary phases in the structure and the definition of the range with optimal plasticity was conducted at the Arcelor Mittal Company and in the Materials Science Laboratories of the Hunedoara Faculty of Engineering.

At this company, the thermal condition for heating the austenitic stainless steel is determined depending on a series of factors, such as: the chemical composition, the phase composition, the heat distortion resistance, the type of furnace, the weight and sizes of ingots, the rolling diagram etc.

Given the fact that the structure of these ingots is not favorable from the point of view of deformability because of the extensive widening of the columnar crystal area and because of the strong segregation of the alloying elements which enable the appearance of constituents with a negative influence over plasticity, as well as a low thermal conductivity, it is recommended that the heating of the ingots up to approximately 800 °C be made at a low rate, while for higher temperatures it is possible to adopt higher heating rates. The optimal hot workability is obtained when the structure consists of a single phase.

Small quantities of ferrite can be tolerated in the structure of austenitic steels, but these must be maintained within certain limits, either by adjusting the processing temperature, or by the accelerated change of the chemical composition.

The segregated structure of the ingots is often a marked one, which entails the prolonged soaking at high temperatures in order to allow the homogenization of the structure and the diffusion of the segregation.

During heating, it is necessary to ensure an atmosphere low in sulphur, as sulphur can combine with nickel, resulting in nickel sulphide with a high tendency to break during working-up.

From the experience acquired to this date, we can mention that hot-metal working of austenitic stainless steels raises specific issues for each brand of steel and sometimes even for different batches of steel belonging to the same grade.

It is a well-known fact that the hot-deformability of a casted steel (in the form of ingots) is inferior to the deformability of the same steel plastically deformed beforehand. Not only that austenitic stainless steels make no exception from this rule, but the difference between the deformability of the two states is even more accentuated, as it can be seen in Figure 1 [9].

The chemical composition and an appropriate casting structure are prerequisites which influence the deformability of stainless steels, but these are not entirely controlled. Depending on the resulted chemical composition, as well as depending on other factors such as the sizes of the ingots, the casting and cooling conditions, the thermal heating conditions, the deformation process, emerges the presence of the second phase, namely ferrite, which leads to a sharp deterioration of plasticity.

A factor which worsens the steel's plasticity characteristics is the manner of conducting the melting process, more specifically the gas content and, in particular, the oxygen content. Thus, 0.01% O_2 leads to the deterioration and even to the rejection of the material.

For the manufacture of pipes from austenitic steels with molybdenum, highly stable, rich in nickel austenitic steels are recommended, such as 18-10; 18-12; 18-14 and not type 18-8 with Mo.

A 0.02 - 0.5% addition of cerium and lanthanum alloy is very conducive to these steels' deformability.

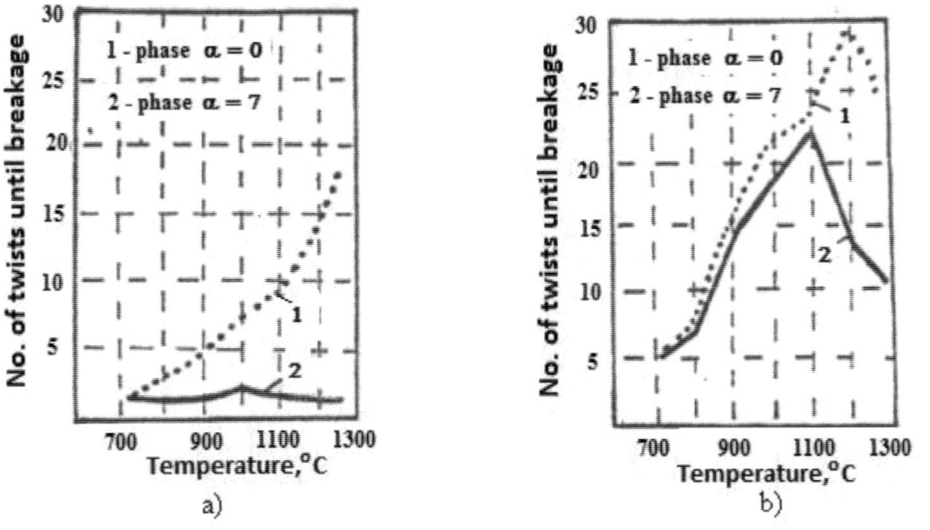

Figure 1. Deformability of austenitic stainless steels in casted (a) and deformed state (b)

Some post-deformation defects may occur as a result of certain inclusions, especially of titanium nitride which attaches to the edges of the granules and results in scale-shaped defects.

Lead content in quantities of 0.01% in the composition of steel can lead to defects in the shape of tears during deformation.

The research conducted on a number of 70 batches of austenitic stainless steel lead to the conclusion that, provided the contents of Cr, Ni and Mo are ensured within the limits established by standards, the manganese content had a considerable influence on the deformability. Thus, from a statistical processing represented in the diagram shown in Figure 2, it resulted that, with a Mn removal, content of under 1%, entire batches were rejected. As the percentage of Mn increased, so did the percentage of good product, progressing to a point where the maximum quantity is obtained at a manganese content between 1.2 and 1.6%. The influence of manganese probably consisted of the contributory action of this element on austenitic stabilization.

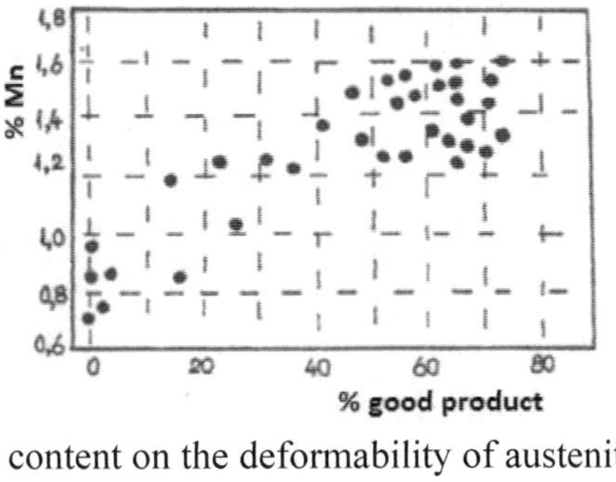

Figure 2. Influence of the Mn content on the deformability of austenitic stainless steels [9]

Even though Si is an alphagenic element, the results still showed that it needs to be present within certain limits, but no more than 0.50%, because its ferritizing effect can have a negative influence on plasticity.

Furthermore, the percentage of aluminium was limited to maximum 0.03%; taking into account the fact that it was not possible to ensure a content very low in non-metallic inclusions, especially oxides, it was necessary to limit the aluminium.

Once present in the structure, it should be possible to reduce the ferrite content by applying certain treatments for regenerating the austenitic structure, starting with prolonged soaking at temperatures of 1100 - 1200 °C, so that the ferrite matrix may dissolve or granulate.

The conducted research has shown that the experience acquired in this field in Romania matches the level of leading factories from abroad [10]. Thus, it has been observed that:

- Batches with a ferrite content of up to 5% have a good deformability. No important surface defects appear during their working-up;
- Batches with a ferrite content between 5 - 9% have a lower deformability. A more pronounced worsening of deformability can be seen at approx. 9% ferrite. Because of the formation of surface defects, in many cases the working-up of ingots must be interrupted at intermediate sections;
- Batches with a ferrite content over 9% have a very weak deformability. In most cases, their working-up must be interrupted because of marked surface defects. This leads to a low productivity, an increased energy, steel and manpower consumption.

At the same time, in order to achieve a good product percentage of approx. 70 - 72% finished products at these grades, only hot ingots are to be used (with a temperature of over 700 °C), taking into account the special precautionary measures that must be taken for heating cold ingots of these grades.

For this reason, the intended purpose of this research is to find a solution for dissolving the ferrite in the structural matrix of the steel and for improving the deformability, as well as for reducing the total heating-up period compared to the current technology.

An adequate study to reveal the influence of technological factors on the reduction of the ferrite content when working austenitic stainless steel would imply material costs and time, as well as the mobilization of productive capacities that cannot be made available to a research in operating conditions. This is because only the cross-sectional and longitudinal cutting of industrial semi-finished products could reveal the fluctuation of the structural secondary phases. Then, by applying various heating options, it would be possible to determine their efficiency in improving the structure [10].

For a more thorough knowledge of the phenomenon, the structural changes were investigated at elevated temperatures also using the "Vacutherm" microscope, using metallographic samples of the shape and sizes as the one in Figure 3, taken from the center of the experimental ingot.

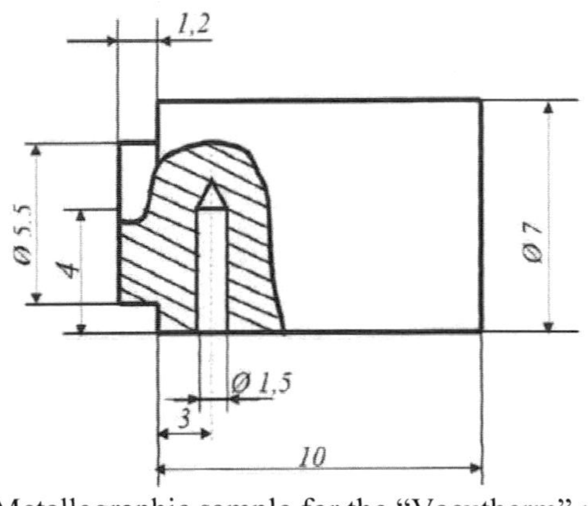

Figure 3. Metallographic sample for the "Vacutherm" microscope

It should be noted that vacuum heating facilitates the volatilization of certain steel components. This phenomenon is more pronounced at temperatures of over 950 °C. The volatile components settle down on the protection plate of the microscope's lens and prevent the examination of the structure in the temperature range of interest.

Since the settling down took place quite rapidly, even after changing the plate, not all the photographs of the microstructures could be taken clearly.

IOP Conf. Series: Materials Science and Engineering **477** (2019) 012001 doi:10.1088/1757-899X/477/1/012001

Samples with different initial content of secondary phases (ferrite) were studied and selected, for which a soaking at temperatures of 1150, 1180 and 1200 °C was applied. In addition to the structure, the reduction of the existing ferrite depending on the temperature and soaking period were also researched.

Figures 4 and 5 show clearer microstructures of some samples at different phases of the heating process. Thus, Figure 4a shows the microstructure of the sample with an initial ferrite content of 13.5%. By maintaining this sample at a temperature of 1150 °C for an hour, the content of phase α in the structure dropped to 5%. Figure 4b shows the structure of this sample after a two hour soaking at a temperature of 1150 °C. By increasing the temperature to 1180 °C, with a one hour soaking period at this temperature, Figure 4c, the ferrite content is maintained at 5% and we can observe a higher tendency of spheroidization of this phase in the base mass of the solid solution, a phenomenon favorable for the hot-metal working process. By continuing to raise the temperature to 1200 °C, with a one hour soaking period at this temperature, the ferrite content increased to 7.5%. The microstructure of this sample is shown in Figure 4d, resulting in austenite as a base constituent where the secondary phase, i.e. ferrite δ, is dispersed as elongated granules, an undesirable phenomenon, seeing that it is conducive to the creation of tensional fuses and a tendency for cracking, that is the appearance of cracks during the hot-metal working process.

Figure 5a shows the microstructure of the sample with an initial ferrite content of 9%. This microstructure shows several clusters of ferrite. After applying heat at 1150 °C, with a one hour soaking period, the microstructure shown in Figure 5b results, which shows a rather homogenous dispersion of the ferrite and the absence of this phase's separation on the edges of the austenite granules. By raising the temperature and by ensuring a one hour soaking period at 1180 °C, the ferrite content drops to 3%, being noticeable at the same time a more marked spheroidization tendency of this phase (Figure 5c). However, by increasing the temperature with just 20 °C and by maintaining the sample at 1200 °C for an hour (Figure 5d), the quantity of ferrite increases to 4.8%.

This increase is of no particular importance for plasticity, but in the microstructure it is possible to see that this separation occurs on the edges of the austenite granules, which worsens the deformability of this steel.

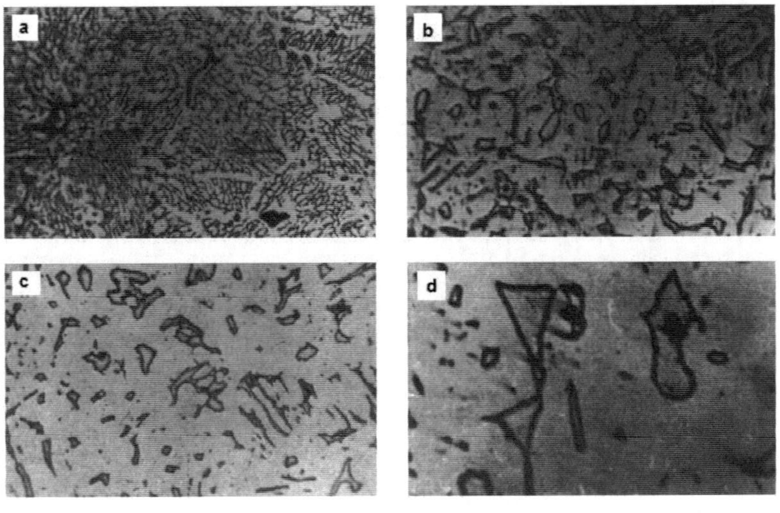

Figure 4. The evolution of the structure and of the ferrite content at elevated temperatures for austenitic stainless steel with 13.5% ferrite
a – after a 1 h soaking period at 1150 °C; b – after a 2 h soaking period at 1150 °C;
c – after a 1 h soaking period at 1180 °C; d – after a 1 h soaking period at 1200 °C.
(Atac Nital 2%, 200 x)

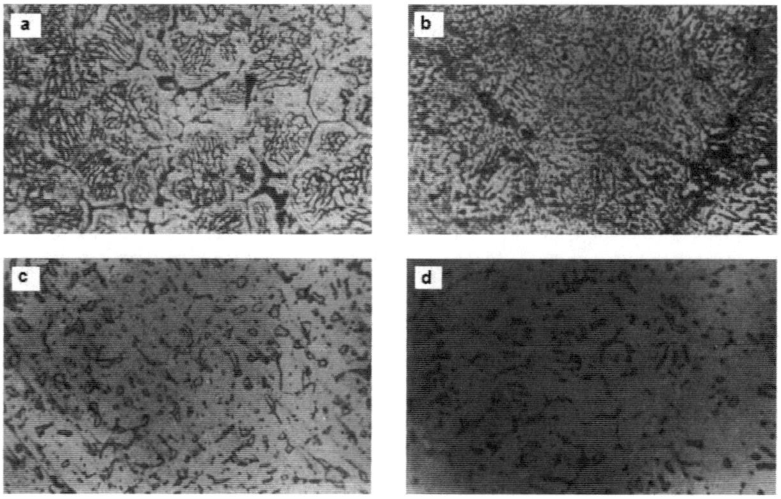

Figure 5. The evolution of the structure and of the ferrite content at elevated temperatures for austenitic stainless steel with 9% ferrite
a – after a 1 h soaking period at 1150 °C; b – after a 2 h soaking period at 1150 °C;
c – after a 1 h soaking period at 1180 °C; d – after a 1 h soaking period at 1200 °C.
(Atac Nital 2%, 200 x)

In order to define the best heating regime, deformability tests using the hot-twisting method in the 1000 - 1250 °C range were carried out.

The obtained technological characteristics, namely the plasticity expressed by the number of twists until breakage (N) and resistance to deformation - expressed by the value of the moment of breakage (M), are given in Figure 6.

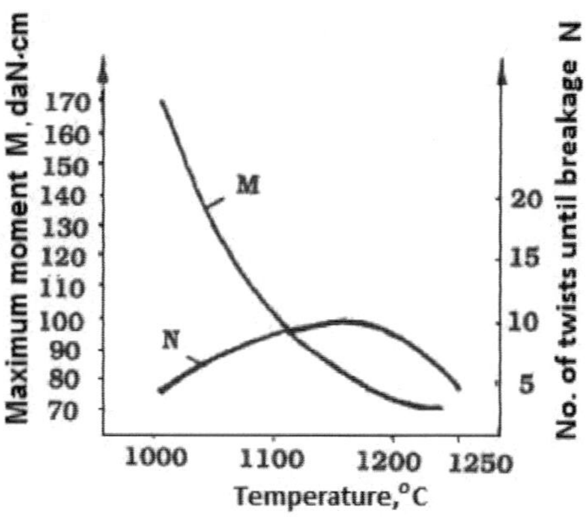

Figure 6. The technological characteristics resulted from the hot-twisting method

It can be observed that the plasticity values are not high, however the maximum level is around the temperature of 1150 - 1200 °C. At the same time, a growth tendency of the ferrite content could be observed in the structure of the samples twisted at 1200 °C.

Linking the results of the deformability tests with the structural effects obtained from the oscillating heating tests, it is visible that the best heating temperature for working-up (the maximum level) is around the temperature of 1180 °C.

With these values, a new technology for heating the steel of this grade was developed (indicated with a full line in Figure 7), compared to the current technology (indicated with a dotted line in the same figure), resulting in a reduction of the heating time and an increased efficiency in terms of ferrite dissolution and spheroidization in the basic structure.

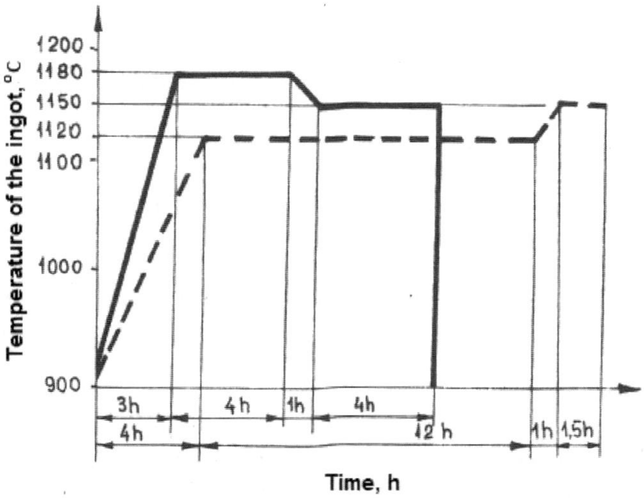

Figure 7. The optimized heating diagram for austenitic stainless steels for hot-metal working

The recommended temperature for the end phase of the working-up process is around 1000°C, a claim based on the high values of the steel's resistance to deformation and on the lower value of plasticity caused by the high degree of alloying which results in a decrease in the recrystallization speed during the hot-metal working process.

3. Conclusion

This paper is both a research and a practical tool for analysis and decision-making in the metal forming activity, for the purpose of ensuring the utilization rate of the austenitic stainless steel finished product.

For an optimal deformability through hot-metal working, the structure of these steels must be monophasic, which is austenitic.

Seeing that the non-uniformity of the ingots' structure caused by segregation also generates areas with notable ferrite occurrence which worsen the plasticity, the research also investigated the influence of the dynamics of the heating process and soaking at elevated temperatures in order to reduce and dissolve the secondary phases in the base structure, below levels that could interfere with subsequent working through hot plastic deformation.

In order to highlight the evolution of the structure during heating, it was examined with the Vacutherm microscope and the best soaking area for reducing the ferrite content was indicated.

In order to guarantee the utilization rate through a constant quality of semi-finished goods, metallographic research conducted at elevated temperatures, correlated with hot deformability tests, show the following conditions as advantageous and economic for the working up technology:

- the most suitable temperature for reducing the content of secondary phases in the structure is between 1150 - 1180 °C.
- if this range is exceeded, the result is the separation of elongated ferrite δ, unfavorable for plastic working, as they represent tensional fuses for triggering the formation of cracks and for reducing the good product percentage.

A new technology for heating austenitic stainless steels was developed, compared to the current technology (diagram shown in Figure 7), resulting in a reduction of the heating time and an increased efficiency in terms of ferrite dissolution and spheroidization in the basic austenitic structure.

References

[1] Lo K H et al. 2009 Recent developments in stainless steels, *Materials Science and Engineering* **65** 39-105
[2] Borgioli F et al. 2018 Corrosion behaviour of low temperature nitrided nickel-free, AISI 200 and AISI 300 series austenitic stainless steels in NaCl solution, *Corrosion Science* **136** 352-365
[3] Milititsky M et al. 2008 Characterization of the mechanical properties of low-nickel austenitic stainless steels, *Materials Science and Engineering* **498** 289-295
[4] Fan S et al. 2017 Experimental investigation of austenitic stainless steel material at elevated temperatures, *Construction and Building Materials* **155** 267-285
[5] Tao Z et al. 2017 Behaviour of three types of stainless steel after exposure to elevated temperatures, *J ConstrSteel Res*, https://doi.org/10.1016/j.jcsr.2018.02.020
[6] Mallick P et al. 2017 Effect of cryogenic deformation on microstructure and mechanical properties of 304 austenitic stainless steel, *Materials Characterization* **133** 77-86
[7] Huang Fu-xiang 2012 Microstructures of Austenitic Stainless Steel Produced by Twin-Roll Strip Caster, *Internationl Journal of Iron and Steel Research* **19**(2) 57-6
[8] Babu A et al. 2017 Hot deformation characteristics and processing map of a phosphorous modified super austenitic stainless steel, *Materials and Design* **115** 262-275
[9] Ilca I 2016 The theory and practice of heating quality steel ingots, *Journal of Engineering Science Innovation* **1**(1) 120-130
[10] Ilca I 2017 The determination of the gradient and thermal stresses for the heating of type 18-10 stainless steel ingots, *Journal of Engineering Science Innovation* **2**(2) 58-74

Evaluation of the brass CuZn39Pb3 resistance at vibratory cavitation erosion

I Lazar[1], I Bordeasu[1], M O Popoviciu[2], I Mitelea[1], C M Craciunescu[1], L D Pirvulescu[1], M Sava[1] and L M Micu[3]

[1]Politehnica University of Timisoara, Mechanical Machines, Equipment and Transportation, Mihai Viteazul str., no. 1, 300222 Timisoara, Romania
[2]Academy of Romanian Scientists, Bvd. Mihai Viteazul, Str., no. 1, Timisoara, Romania
[3]University of Agricultural Sciences and Veterinary Medicine of Banat King Mihai I from Romania, Calea Aradului str., no.119, 300645 Timisoara, Romania

E-mail: ilarica59@gmail.com

Abstract. The CuZnPb brasses are used on a large scale in manufacturing component details for hydraulic driving or regulation devices, as well as for small and huge vanes. Partially the material is used for manufacturing the bodies and partially for sealing seats. During normal operation those pieces are subjected to cavitation erosion. The present paper analyzes the cavitation erosion behavior and the resistance of the brass CuZn39Pb3 used for manufacturing the mentioned details. For the experimental researches was used the standard vibration device realized in the Timisoara Polytechnic University Cavitation Laboratory. The evaluations are made on the basis of the characteristic curves but also by analyzing the eroded areas with great performance microscopes. The behavior of the tested brasses is also compared with those of some etalon materials. The conclusions show that the researched brass can be used only for details working in condition of poor cavitation erosion. At normal or intense cavitation erosion it is necessary at least to use methods to increase the resistance of this materials.

1. Introduction

The brasses with lead content having biphasic structure (α of type CFC and β' of type CVC) are not used for details working in heavy corrosive environments or if can appear high mechanical stresses, but are used on a large scale for details subjected to friction wear [1], [2]. Although this material is used for casting valve bodies, which under certain working regimes are subjected to cavitation erosion, the literature give no data upon their behavior in those conditions. Being components of the systems transporting real liquids, some of having solid particles in suspension, such details can receive, in some parts, rapid detours as the result of the cumulative effect of the abrasive and cavitation erosions. Common, for details used in shipbuilding industry the brasses are under 0.5 % of lead [3]), even if their mechanical properties do not have greater values in comparison with the brasses with 2.5…3.5 % content of lead. Also, because of their good behavior during mechanical processing, the brasses with lead content are frequently used in manufacturing the seating for pressure valves or even for the whole valves of hydraulic control devices [4-7], details which at some working regimes are subjected to low intensity cavitation erosion. As a consequence we believe that the behavior to cavitation erosion must be well known, especially because the repair works by welding (common for cavitation eroded parts) is relatively difficult and is not recommended [8], [9].

2. Researched material. Laboratory devices

The researched material is brass with lead content, having the symbol CuZn39Pb3, in conformity with EN 10204:2004.
The specimens used in the research program were taken out from a cylindrical rod, with a diameter 20 mm and the active surfaces (subjected to vibratory cavitation erosion) were polished at a roughness of Ra = 0.02 μm.
In advance, there have been realized in the Timisoara Polytechnic University chemical and mechanical tests upon the material and have been obtained the following results:

- Chemical composition: 57.7 % Cu, 38.49 % Zn, 3.3 % Pb, 0.2 % Fe, 0.1 % Ni, 0.2 % Sn, 0.01 % Al.
- Mechanical and physical properties: fracture strength = 502 MPa, yield point $R_{p0.2}$ = 365 MPa, Brinell hardness =115 daN/mm^2, strength elongation A5 = 18 %, longitudinal elastic modulus E = 97 GPa, density ρ = 8.47 g/cm^3.
- Biphasic structure formed from the solid solutions α and β', Figure 1, [1].

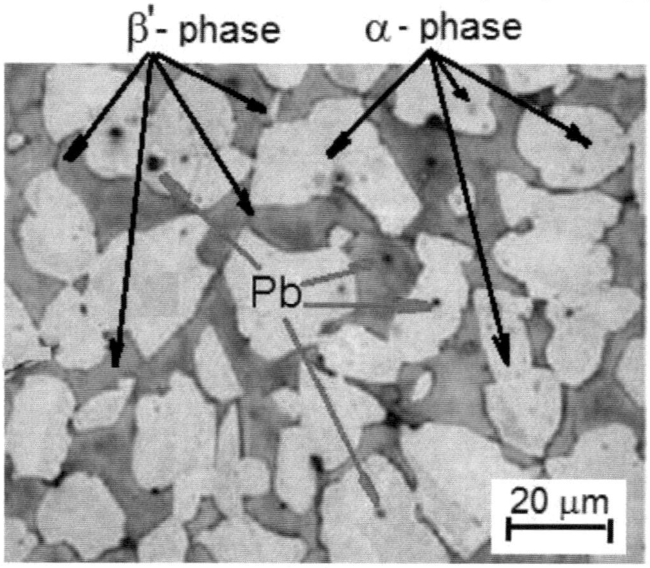

Figure 1. Structure of Brass CuZn39Pb3 (image taken from [1])

The cavitation erosion tests were carried out on the Standard Vibratory Device, with piezoelectric crystals, in the Timisoara Polytechnic University Laboratory of Cavitation [6], [7], [10]. The procedures regarding the specimen preparation before the beginning of the tests, the total exposure, and the intermediary measuring time, the registration of the mass losses and the maintenance of the specimens are in conformity with the Standard ASTM G32-2010 [11] and the laboratory customs [12-15].
The functional parameters of the device, having great influence upon the cavitation erosion intensity, were controlled by a special soft in order to maintain the prescribed values. These parameters are:

Double amplitude of the vibrations = 50 μm
Oscillation frequency = 20 ± 0.02 kHz
Power of the electronic ultrasonic generator = 500 W
Testing liquid: double distilled water
Liquid temperature = 22 ±1^0 C

In order to realize the cavitation erosion the used specimens, with cylindrical shape and a diameter of D = 15,8 mm, are fixed in the sonotrode of the vibratory system and immersed in double distilled water on a depth of 10 mm, Figure 2.

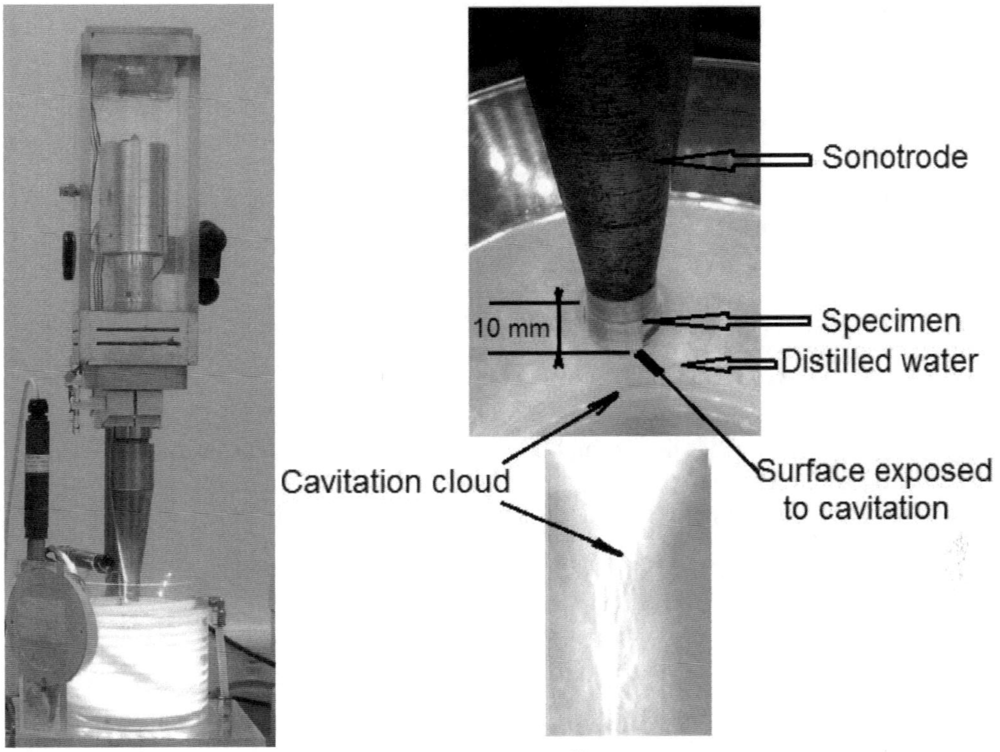

a) **b)**

Figure 2. Images taken during the exposure to vibrating cavitation
a) Ensemble image of the standard cavitation erosion device
b) Detail with the specimen fixed in sonotrode and immersed in water

The morphologic analyze of the eroded structure was realized using images obtained with the scanning electronic microscope Vega 3 Tescan.

3. Experimental results

The evaluation of the behaviour and resistance of the brass to vibratory cavitation respect the Standard ASTM G32 – 2010 and the experimental customs of the Timisoara Polytechnic University Cavitation Laboratory and is made using the specific curves for the mean depth erosion (MDE) and mean depth erosion rate (MDER). For eliminating aleatory differences there were tested three similar specimens and the curves are traced for the mean values of the results. Specification: *the total time exposure of a specimen is 165 minutes and there were made 12 mass loss measurements. The first two for smaller testing times (after 5 and 10 minutes) and the following 10 were made for 15 minutes of exposure. There were taken special measures to eliminate as much as possible the effect of corrosion.* The mass of the eroded material was measured by the use of an analytic balance type Zatklady with a precision of 10^{-5} g.

The parameters MDE and MDER were computed for each intermediate period with the relations:

- mean depth erosion penetration for a single interval

$$\Delta MDE_i = \sum_{i=1}^{12} \left(\frac{4 \cdot \Delta m_i}{\rho \cdot \pi \cdot d_p^2}\right) \tag{1}$$

- cumulative mean depth penetration

$$MDE_i = \sum_{i=1}^{12} \Delta MDE_i \tag{2}$$

- mean depth penetration rate for an interval:

$$MDER_i = \Delta MDE_i / \Delta t_i \tag{3}$$

The used symbols are:
i = 1…12 – testing period
Δm_i –is the erosion lost mass in the period "i", in grams,
ρ = 8.47 grams/mm^3- brass density,
Δt_i – cavitation exposure time for the period "i" (5 minutes, 10 minutes or 15 minutes)
d_p – diameter of the specimen (d_p= 15,8 mm),
To characteristic curves approximating the experimental results were obtained using the following analytical relations [16]:

- for mean depth erosion

$$MDE\ (t) = A \cdot t \cdot (1 - e^{-B \cdot t}) \tag{4}$$

- for mean depth erosion rate

$$MDER(t) = A \cdot (1 - e^{-B \cdot t}) + A \cdot B \cdot t \cdot e^{-B \cdot t} \tag{5}$$

The parameters A and B were established statistically, in conformity with previous researches [5], [16].

In Figures 3 and 4 are presented the experimental obtained values and the analytical approximation for MDE and MDER, which allow to examine the brass behavior during the exposure to vibratory cavitation. For the same purpose are given the images in Figure 5 which show the extension of the eroded surfaces.

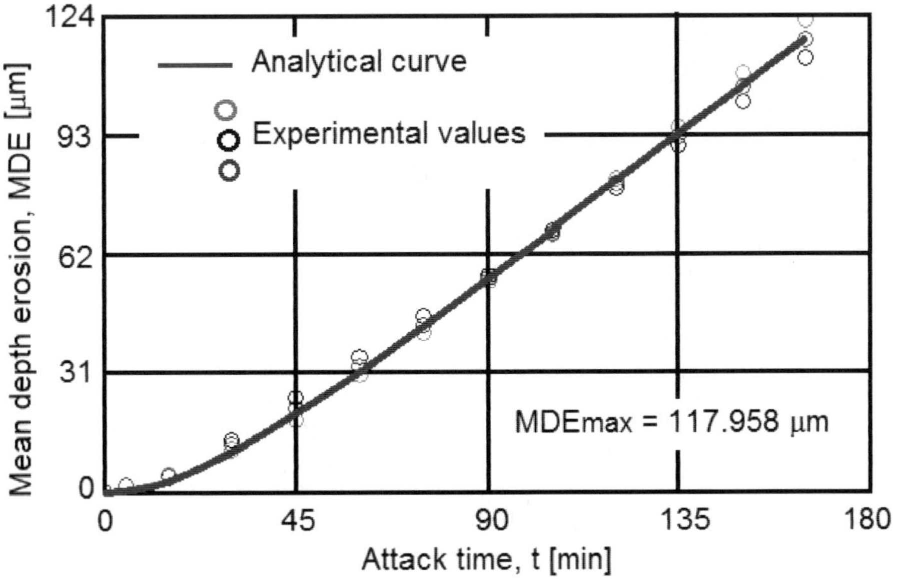

Figure 3. Evolution of mean depth erosion against cavitation exposure time

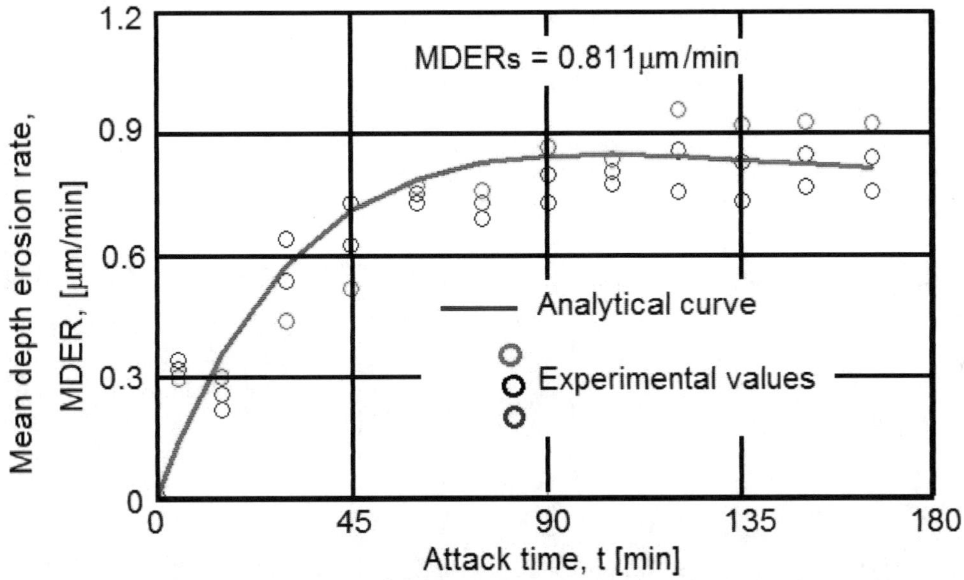

Figure 4. Evolution of mean depth erosion rate against cavitation exposure time

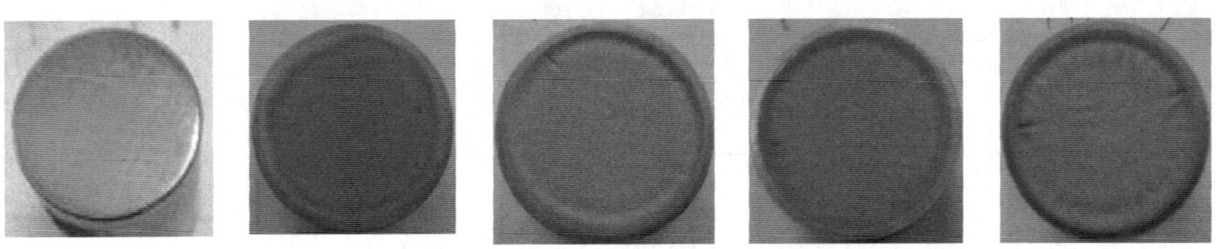

Min 0 Min 15 Min 90 Min 135 Min 165

Figure 5. Images of specimen surface at different cavitation exposure time

From Figure 3 it can be seen an approximatively symmetric scatter of the experimental points with regard to the analytic mediation curve while for Figure 4 the scatter is greater and the symmetry appear only after 120 minutes of tests (this difference appear for all tested materials). The scatter differences between the experimental points and the curves is caused by the fact that MDER(t) is the derivative of MDE(t). On the other hand, evidently the tested area presents a heterogeneous structural constitution formed by the α solid solution and the electronic compound β', which is fragile [8] and also differences between the grains dimensions and their border width, the grain border being the zone in which preferentially the first material losses occur (see Figure 6.). The variation of cavitation erosion resistances in different points of the surface is an assumption confirmed by the different values of the hardness on the same surface (the mean hardness is obtained by 8 measurements conducted in different points of the exposed surface) [5], [6], [10], [12], [13], [15], [17].
After the evolution of the approximation curve MDER(t), Figure 4, for which does not appear significate reduction after 90 minutes of exposure and taking into account the previous laboratory results, confirmed by industrial practice [1-16], it can be said that the tested brass can be classified through the materials which can be used for pieces subjected to reduced cavitation requirements (as for seats of high pressure valves). We also specify that the cavitation erosion resistance can be further improved by applying adequate technologies for surface engineering.
The images of Figure 5 show that in the first 15 minute the cavitation erosion begins with the formation of a peripheral ring. The explanation consists in the imploding mode of the cavitation cloud (see Figure 2) which determines caverns and erosions only in this ring and towards the inner zone of the exposed surface are only eliminated the roughness picks and generated cracks networks without visible caverns. For 90 minutes of cavitation, on the external ring are formed star like caverns and simultaneously on the whole exposed surface are visible pittings. This is clearly seen Figure 6 where are presented photographs obtained with a SEM microscope were it can be seen how the phase β' is expelled and how the cracks are propagated. All this facts determines the increase of the mean depth erosion rate which reached a maximum (see Figure 4).

a)-300X **b)-1000 x**

Figure 6. SEM images of the eroded microstructure, after 165 minutes of cavitation exposure (the images were taken in the inner zone of exposed area)

In the 90-135 minutes testing interval, the caverns with star shape increase their size (both in plane and depth), the small holes begin more evident, but in the same time the vapors and gases remaining in these small holes have a damping effect and determine a reduction of the erosion velocity. This phenomenon give the shape of the MDER(t) which receive a stabilization shape which is maintained till the end of the tests.

To evaluate the cavitation erosion resistance there are presented the diagrams of Figures 7 and 8, in which the mean values of the tested material are compared with the cavitation erosion behavior of the stainless steel OH12NDL (with a good cavitation erosion resistance [5], [6], [10], [15], used for manufacturing the blades of the hydraulic turbines for the Iron Gates I, for both power plants in Serbia and Romania) and of the bronze CuAl10.5Ni5Fe4.8Mn1.5 (with excellent cavitation erosion resistance [5], used for manufacturing the propellers for maritime ships); both are reference materials in our laboratory.

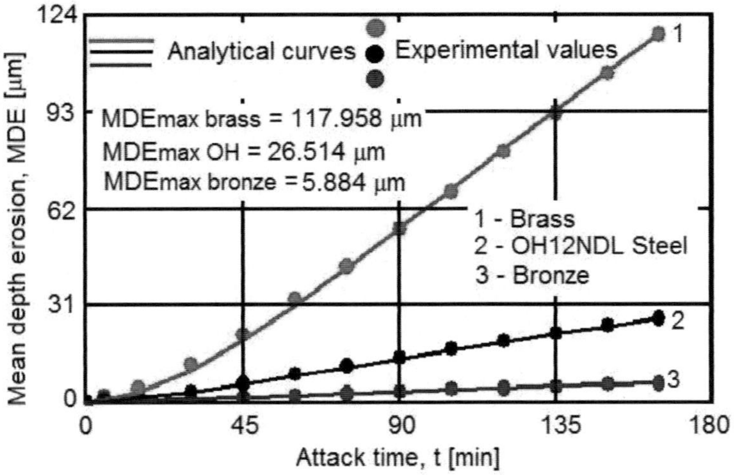

Figure 7. Comparisons of the cavitation erosion behavior taking into account the mean depth erosion

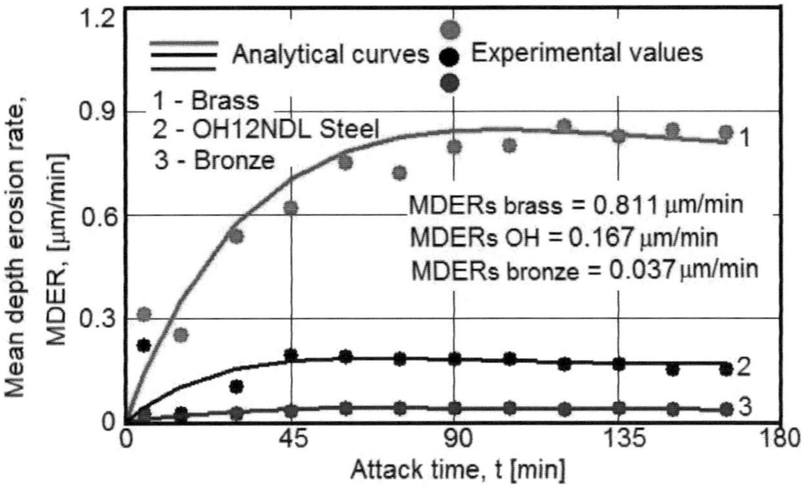

Figure 8. Comparisons of the cavitation erosion behavior taking into account the mean depth erosion rate

Comparing the curves of Figures 7 and 8 we can see that the brass CuZn39Pb3 has a much smaller resistance in comparison with the etalon materials.
This resistance is very well reflected quantitatively by the comparison of the material losses for the final exposure time (165 minutes):

- the maximum mean depth erosion MDE_{max} produced in the researched brass is approximately 4.5 times higher than that achieved in OH12NDL stainless steel and approximately 20 times greater than that achieved in CuAl10.5Ni5Fe4.8Mn1.5 bronze;
- the final mean depth erosion rate MDERs (frequently named the stabilization erosion rate [5], [11], [16]) is approximately 4.9 higher than that realized in the stainless steel OH12NDL and approximately 22 higher than that realized in the bronze CuAl10.5Ni5Fe4.8Mn1.5.

The similarity in the evolution of the curves MDE(t) and MDER(t) for the tested material and the reference ones, suggest that the cavitation erosion resistance of the brass can be improved. This can be done through the use of treatments for conferring to the surface subjected to cavitation, improved mechanical characteristics (especially for the yield point $R_{p0.2}$, the fracture strength R_m and especially for the hardness), homogenous distributed, as well as the reduction of the solid solution β' fragile component with weak resistance to the mechanical shocks produced by the bubble implosions.

4. Conclusions

The brass CuZn39Pb3 present a specific cavitation erosion behavior, for materials with uniform distributed structural constituents, but having a reduced mechanical resistance motive for which, the material researched can be used only for details working in hydrodynamic conditions with reduced cavitation intensity such as those generated in devices (pressure valves, pressure or flow controllers) working whit liquids having great viscosities (various kind of oils).
The erosion initiation and advancement take place at the boundary between the α solid solution and the electronic compound β', with the rapid destruction of the β' phase.
To can use the brass CuZn39Pb3 for details working in hydrodynamic regimes with normal intensity cavitation, such as the blades for ship propellers or hydraulic machinery runners, it is necessary to previously apply treatments for the increase of the mechanical characteristics (Rm, Rp0.2) and especially the hardness HB.

References

[1] Anton I 1985 *Cavitatia,* Vol. II, Editura Academiei RSR, Bucuresti, Romania
[2] Bordeasu I 2006 *Eroziunea cavitationala a materialelor*, Editura Politehnica, Timisoara, Romania
[3] *** Standard test method for cavitation erosion using vibratory apparatus *ASTM G32-2010*
[4] Mitelea I, Ghera C, Bordeasu I and Craciunescu C M 2015 Ultrasonic cavitation erosion of a duplex treated 16MnCr5 steel, *International Journal of Materials Research* **106** (4) 391–397
[5] Bordeasu I, Mitelea I, Micu L M, Lazar I and Karancsi O 2017 Cavitation Erosion Behaviour of Cooper Base Layers Deposited by HVOF Thermal Spraying, *Revista de Chimie* **68**(12) 2914-2917
[6] Franc J P, Kueny J L, Karimi A, Fruman D H, Fréchou D, Briançon-Marjollet L, Yves Billard J Y, Belahadji B, Avellan F and Michel J M 1995 *La cavitation. Mécanismes physiques et aspects industriels*, Press Universitaires de Grenoble, Grenoble, France
[7] Padurean I and Karabenciov A 2010 Influence of structural state on cavitational erosion of martensitic stainless steel GX4CrNi13-4, heat treated, *Metalurgia International* **XV** (10) 46-49
[8] Padurean I and Eugen G 2010 Experimental researches upon cavitation erosion resistance of the martensitic stainless GX4CrNi13-4, heat treated, *Metalurgia International* **XV** (1) 50-54
[9] Padurean I, Truşculescu M, Fay A, Nedelcu D and Padurean E 2008 About cavitation erosion resistance of the martensitic stainless steel from zones reconditioned by welding, *Metalurgia International* **XIII** (10) 60-66

[10] Micu L M, Bordeasu I and Popoviciu M O 2017 A New Model for the Equation Describing the Cavitation Mean Depth Erosion Rate Curve, *Revista de Chimie* **68**(4) 894-898
[11] Bordeasu I, Mitelea I, Salcianu L and Craciunescu C M 2016 Cavitation Erosion Mechanisms of Solution Treated X5CrNi18-10 Stainless Steels, *Journal of Tribology-Transactions of the ASME* **138**(3) 031102
[12] Mitelea I, Bordeasu I and Hadar A 2006 The effect of nickel from stainless steels with 13% chromium and 0.10% carbon on the resistance of erosion by cavitation, *Revista de Chimie* **56**(11) 1169-1174
[13] Mitelea I 2009 *Materiale inginereşti,* Editura Politehnica, Timişoara, Romania
[14] *** http://www.kleinmetals.ch/shop/Datenblatt/E/631.pdf
[15] Küçükömeroğlu T and Kara L 2014 The friction and wear properties of CuZn39Pb3 alloys under atmospheric and vacuum conditions, *Wear* **309** 21-28
[16] Pantazopoulos G A and Toulfatzis A I 2012 Fracture Modes and Mechanical Characteristics of Machinable Brass Rod, *Metallography, Microstructure and Analysis* **1**(2) 106-114
[17] *** 1995 *Registrul Naval Roman*, Reguli pentru clasificarea si constructia navelor romane, Bucuresti

IOP Conf. Series: Materials Science and Engineering **477** (2019) 012003 doi:10.1088/1757-899X/477/1/012003

Study on the quality of Boron micro–alloyed steels destined to applications in the automotive sector

I Ilca, I Kiss and D Miloştean

University Politehnica Timisoara, Faculty of Engineering Hunedoara, Department of Engineering and Management, Hunedoara, Romania

E–mail: imre.kiss@fih.upt.ro

Abstract. The paper presents a study of the possibilities of using 20MnB5q the Boron micro–alloyed steel in the assembly of the body components. It is investigated whether the steel in question can be deformed without being globulised but only chilled in flux, followed by a normalization treatment, in order to obtain a finer structure with a globulising tendency. The research has been carried out on samples of 20MnB5q indigenous steel, compared with samples from bars imported from the same French and Finnish origin. Finally there are specified the structural and qualitative differences for this micro–alloyed Boron steel, which in addition to the finishing effect of the structure, have contributed to the increase of the mechanical characteristics. This study examines the behaviour of 20MnB5q grade steel in the manufacturing process of assembly components. The effect of trace Boron in steels, especially the influence of Boron on microstructure and properties of these steels were summarized.

1. Introduction

Improving the safety of structures under specific service conditions has been the driving force for intensive research in the area of high strength steels. Therefore, in addition to producing steel with the required mechanical properties (strength, toughness), it is increasingly necessary to satisfy processing requirements like weldability, machinability, corrosion resistance, and more.[1], [2] To achieve improved mechanical properties and in the same time excellent weldability, it was necessary to produce an extra–low carbon steel with single phase of fine bainitic ferrite microstructure.[1], [3] This has been achieved by the addition of the elements for hardenability and thermo–mechanical controlled process and, therefore, addition of Boron plays an important role in increasing remarkably the hardenability of steel. Consequently, the effect of Boron content in steels and effect of heat treatment on mechanical and metallurgical properties of their has been investigated by various researchers.[1–20] Researchers have been paying attention to the Boron effect in steels for a long time due to its potential to increase the steel hardenability. However, the mechanism by which Boron increases the hardenability is not completely clear. At present, there are just a few studies strictly focused on the Boron effect on the hot ductility of steels, although the Boron segregation and its effect on austenitic grain boundary play an important role during high temperature deformation and dynamic recrystallization.[18] Mostly, the effect of Boron micro–alloying on the structural transformations, hardenability, and mechanical properties of building steel alloyed with Chromium, Manganese, Molybdenum, Vanadium, and Titanium are considered.[8], [9], [12], [16], [20] Also, the conditions of melting of Boron–bearing steel with most effective recovery of Boron are analysed.[9], [12], [19], [20]

The development of structural steels that can meet all requirements together with reasonably low costs of production is an eternal challenge.[1–8] Advanced structural steels used nowadays are high strength, micro–alloyed steels, obtained by means of a suitable combination of chemical composition and thermo–mechanical treatment parameters in order to have a correct balance between strength, toughness and weldability.[1], [2] Due to a steadily rising vehicle safety and crash requirements in the automotive industry, the use of high–strength steels in structural and safety components is rapidly increasing. The higher requirements for vehicle crash performance can be achieved only by using thick–gauge steel, which results in weight increase.[6–8] Therefore, the high–strength steels, like Boron alloy steels, which meet automotive safety and crash requirements, are being used increasingly in a number of end product applications in the automotive sector. In this sense, hot–forming of the quenchable Boron–alloy steels can produce several complex, crash–resistant structural parts such as front and rear bumper beams, door beams and pillars, with ultrahigh strength and reduced sheet thickness.[6–8]

Boron alloy steels are still under development and experience will show the applications for which they will have specific benefits.[5–10] Boron steels, ideal after heat treatment at high hardness values have reached, also the friction and wear very good resistance against which they often harsh working conditions are suitable for a variety of applications, as a wear material and as a high strength structural steel, and are of particular interest in the field of quality and the technical performances to be achieved. Extreme demands are imposed on the steel when machines and their components are exposed to high, dynamic stresses.[5–10] High–strength steels enable engineers to design lighter, higher–performance structures, offering good overall cost–effectiveness, lower fabrication costs and increased safety. The high–performance steels are recommendable whenever the steel is subject to extraordinary requirements, such as maximum purity, surface hardness and fatigue strength. The composition of these steels, therefore, was based on as low a carbon content, but exactly adjusting the entire chemical composition, must achieve optimum processing and service properties, made–to–measure for every application.[5–10]

Carbon–Manganese–Boron steels are generally specified as replacements for alloy steels for reasons of cost, these steels being far less expensive than alloy steels of equivalent hardenability.[1–8] Boron over 8 ppm the content is considered as an alloying element and it is dissolved in ppm quantities in solid state in steel.[2], [3], [9], [10] Recent research has highlighted the strong action on the feasibility of adding a few tenths of a thousandth of a percent of Boron.[1–20] It is well–known that a small amount of Boron dramatically increases the hardenability of steels, the basic effect of Boron in the steel being the enhancement of hardenability, which is evident already at a very small concentration, of the degree of 20ppm Boron.[8], [9], [11–14] This beneficial effect is attributed to grain boundary segregation previously mentioned, which retards the diffusional transformation of austenite to ferrite by lowering the interfacial energy at the austenite grain boundary.[15], [16] Therefore, Boron in steel in very low quantities (0.001–0.003) greatly increases hardness and strength, and also the hardenability of low carbon steels. With increasing addition of Boron to the extent of about 20ppm, both the tensile strength and hardness values increase, attributed to carbide forming tendency of the Boron.[17] Moreover, Boron have similar effect to those obtained with such common alloying elements as Manganese, Chromium, Nickel, and Molybdenum, but, unlike these elements, only a minute amount of Boron is required. Even in the small quantity of the degree of size up to 40 ppm, Boron gives the same effect of the hardenability enhancement as other more expensive elements which must be added in much bigger quantity.[9], [11]

The action of Boron is effective during quenching for low carbon steels, and when annealing is null. The Boron additive steels, easy to be processed and applied in very good mechanical properties after heat treatment are important features exhibitions. According to the same mechanical properties are cheaper than high alloy steels. In addition to the benefits of economy and alloy conservation, Boron steels offer significant advantages of better machinability compared with Boron free steels of equivalent hardness.[20] Moreover, steels containing Boron are also less susceptible to quench cracking and distortion during heat treatment. Consequently, Boron containing Carbon alloy steels are

widely used in automotive, constructional, and various other applications, including agricultural machinery and tools (discs, plough shares), public works machinery, mining, cutting equipment, safety beams in vehicles or concrete mixer drums, due to the several important advantages such:
— suitable for environmentally–friendly water quenching processes;
— excellent hardening performance, also possible with both oil and gas quenching;
— high wear and abrasion resistance – meaning more durability, less material weight loss, low maintenance level, longer lifetime and increased cost savings; and
— heat treatable Boron steel grades offer excellent resistance against abrasive wear.

Consequently, the Boron steel grades represent a breakthrough for durable steel – meaning they are a cost–effective solution to prolong the lifetime of your final products. Hardened via a Boron quenching process and hot rolling–controlled thermo–mechanical treatment, the Boron steels offers a remarkable degree of hardness, giving the final product:
— a uniform microstructure;
— outstanding mechanical loading; and
— excellent resistance to abrasive wear.

2. Boron micro–alloyed steels

Boron micro–alloyed steels are classified as low–alloy high–strength steels which are classified by their mechanical properties and especially their flow limit. These steels grades are: OLC15q, OLC35q, OLC45q and 20MnB5q. In addition, other steel grades are also used. To produce such a microstructure especially with a low–carbon micro–alloyed steel, the chemical composition had to be designed very carefully. The basic alloying system contains Nickel, Chromium, Molybdenum and the micro–alloying elements Vanadium, Niobium, and Titanium. Besides these the micro–alloying element Boron should also be effective.

Steel grade 26MnB5q, in accordance with [21], is one of the Boron–alloyed quenched and tempered steels. This grade is characterized in particular by their formability in the hot rolled state and their high strength after the heat treatment process, where the addition of Boron leads to the strain hardenability of this Carbon–Manganese–Chromium alloy. In fact, the strength characteristics after quenching and tempering are achieved in particular by the low Boron content, in addition to the Carbon and Manganese. This grade features homogeneous material properties that enable consistent and predictable performance of the final component. The material is very clean, meaning the levels of impurities are carefully controlled, which helps with consistency in formability and weldability. In Table 1, the recently established standard for 20MnB5q steel is presented.[21] Boron alloy steels are specified when the base composition meets mechanical property requirements (toughness, wear resistance, etc.), but hardenability is insufficient for the intended section size. Rather than call for a more highly alloyed and therefore more expensive steel, a user may simply specify the corresponding Boron grade, thereby ensuring suitable hardenability.

Table 1. Chemical composition [%] and the main mechanical properties of the structural carbon steel, grade 20MnB5 (1.5530): EN 10083–3–2006 [21]

The chemical composition [%]					
C	Si	Mn	P	S	B
0.17–0.23	max 0.4	1.1–1.4	max 0.025	max. 0.035	0.0008–0.005
The main mechanical properties					
Yield Rp0.2 (MPa)	Tensile Rm (MPa)	Elongation A (%)	Reduction in cross section on fracture Z (%)		Brinell hardness (HBW)
700 (≥)	900–1050	14	55		255–270

The transformation of austenite in the quality carbon steels is almost entirely determined by the Carbon and Manganese content. The Manganese content is chosen so as to obtain the desired microstructure and properties, the chosen Carbon content and the given cooling conditions.[1–5] The

IOP Conf. Series: Materials Science and Engineering **477** (2019) 012003 doi:10.1088/1757-899X/477/1/012003

effect of Phosphorus and sulfur is almost negligible, and the Silicon content is normally so low in order to obtain the desired mechanical strength.[1–5], [9–20] The synergistic effect of Boron with other elements can enhance the Boron effect.[19] The effect of Boron on hardenability also depends on the amount of carbon in the steel. The effect of Boron increases in inverse proportion to the percentage of carbon present. Boron may also become ineffective if its state is changed by incorrect heat treatment. It should also be noted that the properties of quality carbon steel can be influenced by the presence of gases, especially Oxygen, Nitrogen and Hydrogen, as well as their reaction products.[1–5] These secondary elements penetrate into the steel usually from the scrap iron used, from the deoxidants or from the furnace environment.[1–5] The gas content depends largely on the method used for melting, deoxidation and casting, so that the final properties largely depend on the method used to elaborate them. Since Boron has a strong affinity for oxygen and nitrogen, these elements either must be removed or controlled for Boron to have its full hardenability effect. Accordingly, it has been the general practice to add Boron to steel with titanium and zirconium present to protect the Boron from nitrogen, and aluminum to protect Boron from oxygen. In addition to effecting deoxidation and providing protection of Boron from oxygen, aluminum is an effective grain refiner in production of ingot cast fine–grained steel.

For steel grades OLC15q, OLC35q and OLC45q, the international standards have provided the chemical composition and the qualitative characteristics that can be found in Tables 2–4.[22]

Table 2. The chemical composition [%] and the main mechanical properties of the high quality structural carbon steel, grade OLC15q [22]

The chemical composition [%]							
C	Si	Mn	P	S	Cr	Ni	Al
0.12–0.18	0.17–0.37	0.35–0.65	0.045	0.045	max. 0.30	0.3	0.02
The main mechanical properties							
Yield Rp0.2 (MPa)	Tensile Rm (MPa)	Elongation A (%)	Reduction in cross section on fracture Z (%)	Brinell hardness (HBW)			
541 (≥)	982 (≥)	33	13	113			

Table 3. The chemical composition [%] and the main mechanical properties of the high quality structural carbon steel, grade OLC35q [22]

The chemical composition [%]							
C	Si	Mn	P	S	Cr	Ni	Cu
0.32–0.39	0.17–0.37	0.50–0.80	0.04	0.04	max. 0.30	0.30	0.3
Yield Rp0.2 (MPa)	Tensile Rm (MPa)	Elongation A (%)	Reduction in cross section on fracture Z (%)	Brinell hardness (HBW)			
873 (≥)	551 (≥)	34	22	134			

Table 4. The chemical composition [%] and the main mechanical properties of the high quality structural carbon steel, grade OLC45q [22]

The chemical composition [%]							
C	Si	Mn	P	S	Cr	Ni	Cu
0.42–0.50	0.17–0.37	0.50–0.80	0.04	0.04	max 0.30	0.3	0.3
The main mechanical properties							
Yield Rp0.2 (MPa)	Tensile Rm (MPa)	Elongation A (%)	Reduction in cross section on fracture Z (%)	Brinell hardness (HBW)			
528 (≥)	414 (≥)	11	32	321			

3. Research on Boron alloy steel 26MnB5

In 2010, a local steel company has produced the first Boron alloy steel 26MnB5 for the automobile industry, and soon this special mark of steel was also available for producers of machinery and equipment. In the meantime, were put up a few new brands of stainless steel with Boron, which are still in the process of being studied from the point of view of the practical utility: 27MnCrB5 and 33MnCrB5.

During the production of the 20MnB5q steel, there have appeared qualitative problems related to the realization of the calibre, the austenitic grain size, the microscopic purity, as well as the observance of the chemical composition and especially of the Carbon content occurred. The degree of chemical homogeneity in the 100–tones electric arc furnaces was scattered for segregating elements (especially Sulphur and Carbon), and Aluminum could not be dosed so as to be evenly distributed in the liquid steel. In order to avoid these qualitative deficiencies that have still been occurring in separate batches, it is important not to reduce the use of these high–grade steel, or to reduce the analysis gaps in the steelworks laboratories, to control the surfaces by the discharge test, or with non–destructive control equipment.

The most existing companies as well as the newest assemblies work only through the cold deformation process with the application of high reduction rates. In order to achieve these deformations in good conditions, without the appearance of cracks or other defects, the manufacturing conditions of round laminates without surface defects must be ensured. Therefore, it is necessary to observe strictly the technology of design, casting and lamination, in order to avoid surface defects in time. From the practice of manufacturing these grades of micro–alloyed Boron steels, it has been found that in many cases the defects of rolling in the finishing stands are defective with high weight in the case of lamination of Φ12mm profile. The manufacturing and control technology flow applied here provides very severe interfacial receptions to the finishing rolling mill with a sample on the bend, which leads to very good results from the quality point of view. For steel grades OLC35q and OLC45q there are frequent cases when it is not ensured an advanced globulating annealing in order to ensure cold deformation without the risk of cracks occurring. In the bolt factories there are executed in intermediate sections by which the material is emblazoned and introduced into this state in the manufacturing process of the assembly elements.

Following the sampling of the 20MnB5q grade steel with the chemical composition shown in Table 5, the batch characteristics shown in Table 6 were obtained.[21] The quenching was carried out in two mediums: water and oil. The French requirements for the same 20MnB5q grade steel provide the following features presented in Table 7 and Table 8.[23] Comparing our data with that obtained on the French steel samples, it is not consistent with the Re and Rm values. However, French indications do not prescribe the technological additions to be used.[23]

Table 5. Chemical composition [%] of the structural carbon steel with Boron, grade 20MnB5q, according to the Romanian requirements [21]

The chemical composition [%]											
C	Mn	Si	S	P	Cu	Ni	Cr	B	Al	As	Sn
0.20	1.7	0.25	0.02	0.023	0.07	0.08	1.15	0.0035	0.007	0.011	0.012

Table 6. The heat–treatment environment and temperature, and the main mechanical properties of the structural carbon steel with Boron, grade 20MnB5q, according to the Romanian requirements [21]

Quenching environment and temperature, [°C]	Tempering temperature, [°C]	Mechanical properties			
		Re [N/mm^2]	Rm [N/mm^2]	A [%]	Z [%]
Oil (900°C)	500	660	780	17	64
		640	760		64
		640	760		61

Water (900°C)	500	800	870	20	58
		870	860		
		820	870		

Table 7. Chemical composition [%] of the structural carbon steel with Boron, grade 20MnB5q, according to the French requirements [23]

The chemical composition [%]					
C	Mn	Si	S	P	B
0.16–0.22	1.10– 1.40	0.1–0.4	max.0.035	max.0.035	0,0008–0.005

Table 8. Main mechanical properties of the structural carbon steel with Boron, grade 20MnB5q, according to the French requirements [23]

Quenching environment and temperature, [°C]	Tempering temperature, [°C]	Mechanical properties		
		Re [N/mm^2]	Rm [N/mm^2]	A [%]
Oil (900°C)	500°C	min. 685	min. 830–1030	min.11

The tests on a piece of steel bar imported from Finland, grade 20MnB5q, led to the following results on chemical composition (Table 9).[24] In order to determine the feasibility of the imported product, heating was performed at 860°C followed by quenching in water. The resulting values are shown in Table 10. From these data it is observed that the hardness values after hardening are higher than those obtained on the native steel.

Table 9. Chemical composition [%] of the structural carbon steel with Boron, grade 20MnB5q, according to the Finnish requirements [24]

The chemical composition [%]										
C	Mn	Si	S	P	Cu	Ni	Cr	B	Ti	V
0.22	1.22	0.21	0.029	0.023	0.27	0.22	0.31	0.005	0.03	0.04

Table 10. The hardness values after hardening

Hardness measuring zones	Dimensions Φ mm	Hardness HRC (according colon 1)				
	22.5	43	44	47	45	43
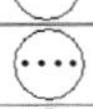	15.0	45	46	46	44	–
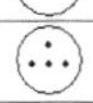	12.0	45	48	45	48	–
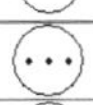	9.5	45	46	44	–	–
	8.0	47	47	48	–	–
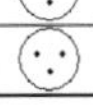	6.0	47	47	46	–	–

4. Results & Discussions

By making microstructures both for indigenous steel processed here, and for the foreign source in the state of delivery, it was found that the import steel has a slightly modified ferrite–perlite structure by an incomplete annealing treatment with tendencies of austenite formation in the globular state (Figure 1).

IOP Conf. Series: Materials Science and Engineering **477** (2019) 012003 doi:10.1088/1757-899X/477/1/012003

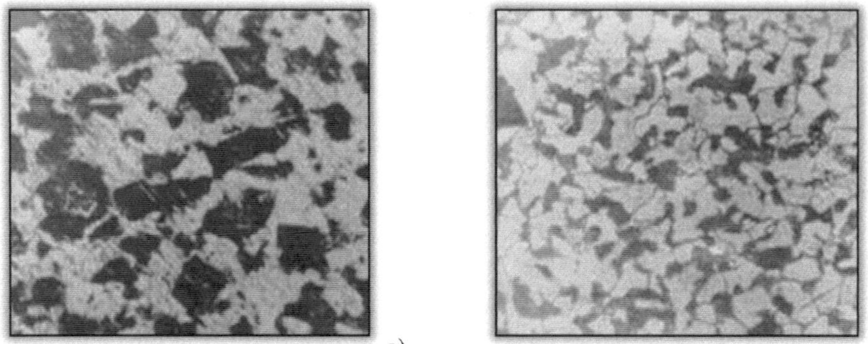

a) b)

Figure 1. Microstructure of Boron alloyed steel grade 20MnB5q (500x)
a) Finnish origin steel; b) indigenous steel

It was also found that austenitic grains are higher for imported steel than for indigenous steel, but in both cases they exhibit inhomogeneities with scores ranging from 1 to 5, a phenomenon evidenced by the microstructures of Figure 2.

The specialty literature suggests that it is very difficult to control the austenitic granulation in the case of Boron–micro–alloyed steels. After the 900ºC water hardening of the indigenous steel rolled bars, martensitic structures of the shape of Figure 3 were obtained.

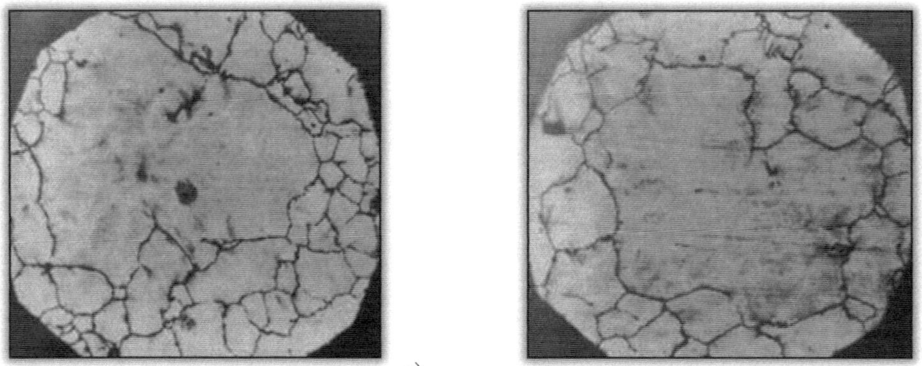

a) b)

Figure 2. Austenitic granulation of analysed Boron–micro–alloyed steels (500x)
a) Finnish origin steel; b) indigenous steel

Figure 3. Microstructure of indigenous steel samples after water quenching at 900ºC (500x)

Analysing the microstructures of Figure 3 it is observed that the resulting martensite is rough and comes from a high–grain steel. After normalization at 860°C, a normalized state of ferrite–perlite structure is obtained, which tends to globulate cementite (Figure 4), but the martensitic structure resulting from the quenching of the normalized samples is still rough (Figure 5).

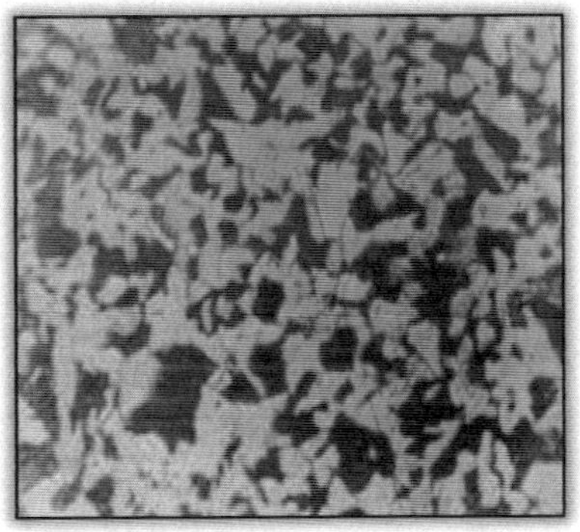

Figure 4. Microstructure of normalized indigenous steel (500x)

Figure 5. Microstructure of normalized indigenous steel after water quenching (500x)

Research on Finnish origin steel 20MnB5q revealed that after the quenching of the samples (with the dimensions Φ20,5; Φ12 and Φ6) there were very small structural differences, but compared to the indigenous steel of the same brand, it may be observed a higher degree of fineness of the structure (Figure 6). This is certified by the presence of technological additions of Vanadium and Titanium, which, besides the finishing effect of the structure, also contribute to the increase of the mechanical characteristics. For this reason, there was a difference between the resistance characteristics (Re and Rm) for indigenous steel as compared to the imported one – because, as it has been shown, the indigenous grade 20MnB5q steel produced in the first assimilation phases did not contain Vanadium and Titanium.

(a) (b) (c)

Figure 6. Microstructure of Finnish origin steel, after water quenching (500x):
(a) Φ20,5 mm samples; (b) Φ12 mm samples; (c) Φ6 mm samples

The analysis reveals an instability of the chemical composition for the 20MnB5q indigenous steel produced in the 100–tones basic electric furnace, resulting in the obligation to develop it in the 50–tones electric furnace, where the chemical composition is possible in a much narrower range.

4. Concluding remarks

The paper examines whether the studied steel can be cold–plastic processed without being globulised, but only cooled down stream, followed by a normalization treatment to obtain a finer structure with globular tendencies. It should be noted that for this low–carbon and low–carbon micro–alloyed steel this possibility exists, but the extent to which deformation degree can be used in the neglected condition must be established.

Research on low–carbon and low–carbon micro–alloyed steel made technological improvements in manufacturing, which consisted of:

— production of this type of steel in electric furnaces with limitation of Carbon and Manganese content to the maximum admissible limit;
— use of Ferro–Titanium and Ferro–Vanadium admixtures for the chemical analysis of steel 0.02–0.04 Titanium and 0.01–0.03 Vanadium;
— performing samples with a cooling in water or oil to establish a more appropriate cooling medium;
— performing in parallel steel samples: recoil, globulised – hardened, recovered and steel: rolled, tempered, rebound, to determine the weight of the globulisation influence.

Taking this into account, the research at both the Faculty of Engineering Hunedoara and the Hunedoara indigenous platform continues in order to improve the manufacturing technology of the 20MnB5q grade steel.

References

[1] ***2009 *Advanced Structural Steels: Part One* Key to Metals AG https://www.totalmateria.com
[2] Kuziak R, Kawalla R and Waengler S 2008 Advanced high strength steels for automotive industry, *Archives of civil and mechanical engineering* **8**(2) 103-117
[3] Merklein M, Lechler J and Geiger M 2006 Characterisation of the flow properties of the quenchenable ultra high strength steel 22MnB5, *CIRP Annals–Manufacturing Technology* **55**(1) 229–232
[4] Colan H, Tudoran P, Alinicăi G, Marcu M and Drugescu E 1983 *Metals Study,* Didactic and Pedagogic Publishing House, Bucuresti
[5] Marin B, Bell A, Idoyaga Z, Colla V and Fernandez L M 2007 *Optimisation of the influence of Boron on the properties of steel*, Technical Steel research, Directorate–General for Research and Innovation, European Commission
[6] Altan T 2006 Hot–stamping Boron–alloyed steels for automotive parts – Part I: Process methods and uses, *Stamping Journal, https://www.thefabricator.com*
[7] Altan T 2007 Hot–stamping Boron–alloyed steels for automotive parts – Part II: Microstructure, material strength changes during hot stamping, *Stamping Journal, https://www.thefabricator.com*
[8] Aspacher J 2006 New forms for modern autobody stamping: High–strength, low–weight requirements spur new developments, *Stamping Journal* http://www.thefabricator.com
[9] Gao Y L, Xue X X and Yang H 2015 Influence of Boron on initial austenite grain size and hot deformation behavior of Boron microalloyed steels, *Crystals* **5** 592-607
[10] Tsuji N, Matsubara Y, Sakai T and Saito Y 1997 Effect of Boron addition on the microstructure of hot–deformed Ti–added interstitial free steel, *ISIJ International* **37**(8) 797-806
[11] López–Chipres E, Mejía I, Maldonado C, Bedolla–Jacuinde A, El–Wahabi M and Cabrera JM 2008 Hot flow behavior of Boron microalloyed steels, *Materials Science and Engineering A* **480** 49-55
[12] Han F, Hwang B, Suh D, Wang Z, Lee D and Kim S J 2008 Effect of Molybdenum and Chromium on hardenability of low–carbon Boron–added steels, *Metals and Materials International* **14**(6) 667-672
[13] Titova T I, Shulgan N A and Malykhina I Y 2007 Effect of Boron microalloying on the structure and hardenability of building steel, *Metal Science and Heat Treatment* **49**(1) 39-44
[14] El–Shennawy M, Farahat A I, Masoud M I and Abdel–Aziz A I 2016 Heat treatment effect on micro–alloyed low carbon steel with different Boron content, *International Journal of Mechanical Engineering (IJME)* **5**(4) 9-20
[15] Taylor K A and Hansen S S 1990 The Boron hardenability effect in thermo–mechanically processed, direct–quenched 0.2 pct. Steels, *Metallurgical Transactions A* **21** 1697-1708
[16] Khare S, Lee K and Bhadeshia H K 2009 Relative effects of Molybdenum and Boron on ferrite and bainite kinetics in strong steels, *International Journal of Materials Research* **100** 1513-1520

IOP Conf. Series: Materials Science and Engineering **477** (2019) 012003 doi:10.1088/1757-899X/477/1/012003

[17] Düzcükoğlu H and Çetintürk S 2015 Effect of Boron addition on mechanical properties of 60SiCr7 stell, *International Journal of Materials, Mechanics and Manufacturing* **3**(2) 117-120
[18] López–Chipres E, Mejía I, Maldonado C, Bedolla–Jacuinde A and Cabrera J M 2007 Hot ductility behavior of Boron microalloyed steels, *Materials Science and Engineering: A* **460–461** 464-470
[19] Wang X M and He X L 2002 Effect of Boron addition on structure and properties of low carbon bainitic steels, *ISIJ International* **42**(Suppl) S38–S46
[20] Ghali S N, El–Faramawy H S and Eissa M M 2012 Influence of Boron additions on mechanical properties of carbon steel, *Journal of Minerals and Materials Characterization and Engineering* **11**(10) 995-999
[21] ***EN 10083–3: 2006 Steels for quenching and tempering – Part 3: Technical delivery conditions for alloy steel (Structural carbon steel with Boron)
[22] ***STAS 9382/4–89 & STAS 8949–82: OLC45q, OLC35q and OLC15q Standard Specification (Oţeluri de îmbunătăţire destinate fabricării organelor de asamblare prin deformare plastică la rece. Mărci şi condiţii tehnice de calitate)
[23] ***AFNOR NF A 35–557: Grade 20MB5q – Heat treatable steels for externally threaded, high performance fasteners for engineering purposes
[24] ***SFS 365: Grade G–20Mn5 – Low–carbon alloy cast steel (normalized, annealed) and HSLA steel with boron

IOP Conf. Series: Materials Science and Engineering **477** (2019) 012004 doi:10.1088/1757-899X/477/1/012004

Recycling of textile wastes into textile composites based on natural fibres: the valorisation potential

M P Todor [1], C Bulei[1], I Kiss[2] and V Alexa[2]

[1]University Politehnica Timisoara, Faculty of Engineering Hunedoara, Doctoral School, Timisoara / Hunedoara, Romania
[2]University Politehnica Timisoara, Faculty of Engineering Hunedoara, Department of Engineering & Management, Romania

E–mail: imre.kiss@fih.upt.ro

Abstract. Natural fibre reinforced composites are gaining interest because of its attractive properties such as low weight, higher stiffness and low cost. Fibre reinforced composites are lightweight, strong, and durable materials which are seeing increasing adoption in the transportation, construction and many other markets. Therefore, sustainability in their use phase is often a key driver for the selection of composites over traditional materials. In the meantime, the ever–growing amount of textile wastes leads to the necessity to utilize these waste materials and to develop further processing technologies for their beneficial application. Due to various application possibilities and its economic impact, the valorisation of textile wastes has attracted increasing attention. This review summarized the valorisation potential of a textile waste products, especially of flax, hemp and jute fabrics. In this context application of different textile wastes for development of some value added product has been thought of in this paper.

1. Introduction

Textile waste is one type of municipal solid waste growing rapidly in recent years. Abundant textile waste which includes the waste generated from streams of fibre, textile and clothing manufacturing process, commercial service and consumption has raised increasing concerns worldwide in developing novel circular textiles approach. In fact, disposal of textile waste and their management have risen increasing global concerns in the recent years.[1–6]

In many cases, major recycling options for textile wastes include only second–hand oversea trading and energy recovery by incineration. Moreover, the short lifecycles of apparel product due to rapid fashion cycles and increased buying power of consumers in urban areas is resulting in significant amounts of postconsumer textile waste in the form of used clothing or even second–hand clothing. Post–consumer textile waste mainly originates from the household sources and consists of textiles which the owner no longer needs as it was.[1–4] Currently, for textiles which are ripped or stained and are no longer wearable, disposal commonly consists of landfill or incineration as there are no other valorisation routes available. This waste is often made up of resources, which could be recycled and used.[5–10]

In recent years the world has been involved with environmental issues related to the continuous use of natural resources, including the textile wastes.[9–14] Therefore, several proposed solutions involve the use of natural, renewable and recyclable materials. In this respect, the use of natural fibres and their wastes is a successful example being investigated and industrially applied since the past decade. [1–6], [15–24]. In particular, the lingo–cellulosic fibres obtained from plants are increasingly being considered as reinforcement of composites for engineering applications, especially in automobile components. Therefore, applications of natural–reinforced composites (with mixture of particle, fibre and textile structures) are expected to have a significant positive environmental impact.[5], [6], [15–24] The availability of the technologies for producing of natural–reinforced composites and its high socio–economic value are solid arguments in favour of the future development of this product and can find a wide application as a substitute for non–biodegradable or non–ecological products.[5], [6], [22–24]

Bast fibres (flax, hemp and jute), for their versatility, rightfully deserves to be branded as the "fibres for the future", being another natural option for a cleaner environment.[9–15] Among all the natural fibres, bast fibres such as jute, flax, hemp, sisal and ramie appears to be a promising fibres and constitutes large area of investigation due to its good mechanical properties compared with other natural fibres.[16–20] Recently, cellulose fibres began to be used as a fibre–reinforcement material, especially in conjunction with polymers in fibre–reinforced composites, due to their similar properties to engineered fibres.[18–20]

A huge amount of these fibres is wasted and is gone to landfill every year, either in the form of textiles resulted from manufacturing of fabrics or in the form of used cloths after the end–of–life of the bags. In fact, bast fibres and their wastes can be used not only in its traditional materials, but also for the production of other value–added products such as geotextiles or composites.[15–24] Concern for the environment, both in terms of limiting the use of finite resources and the need to manage waste disposal, has led to increasing pressure to recycle materials at the end of their useful life. As well known, the bags at the end of their longevity will be as waste materials. Moreover the waste fabrics are not yet reused effciently.[5], [6], [14], [15] Therefore, several attempts were carried out in our previous research [5], [6], [23] to use the above–mentioned valuable properties of bast fibres, to reuse the wasted textiles, to recycle the end–of–life fibre bags for fabricating valuable reinforced textile/polymer composites or textile sandwich composites.

Figure 1. Textile packaging material as solid wastes

The textile industry needs to find imaginative solutions to produce environmental improvements.[4–9] One of the main environmental concern in the textile industry is about the solid wastes, coming from the packaging of raw textile material of different nature (in form of fibres, yarns or fabrics) or paper packaging wastes (boxes, bags, cardboard), considered as non–dangerous wastes.[5], [6], [10], [11], [15]

Recycling the waste for environmental protection has been an important challenge for the mankind. The fibrous waste in textile industry accounts for approximately 15% of the amount of fibre products used.[7], [8] Nearly 95% of used clothing and textiles can be reused and recycled. Any clothing, household textile or commercial linen textile can be reused and recycled. Only items that are wet or have been used with a solvent–type liquid cannot be recycled.[7], [8] Nearly 100% of all used clothing and household textiles can be sort and grade the used clothing based on quality, condition, and type, re–used or recycled in one of the following manners:[7], [8]

— 45 percent is re–used as apparel that are then sold to the second–hand clothing industry or are exported to emerging market nations where demand for top quality second–hand clothing is particularly high.
— 30 percent of the recovered textiles are cut and converted into wiping rags or polishing cloths that are then used in commercial and industrial settings.
— 20 percent is reprocessed into its basic fiber content which are then remanufactured to create furniture stuffing, home insulation, automobile sound–proofing, building materials and various other new products.
— 5 percent is unusable. If the textiles are wet or dirty, broken or contaminated with solvents they are not fit for any recycling process and are finally discarded.

Therefore, more than 95% of all textiles can be recycled or reused in some way, only 5% is unusable due to mildew or other contamination.[7], [8] In general, applications of re–used or recycled textile waste belong to the following three broad categories: apparel, home furnishing, and industrial. Most of the fibre products are for short term (e.g. disposables) to medium term (e.g. apparel, carpet, automotive interior) use, lasting up to a few years in their service life.

The current study focused on the valorisation potential of use of packaging textile waste as reinforcement for producing a fibre–reinforced polymer composites. This paper discusses about the recycling possibilities of the textile wastes as reinforcements in the polymer composites and the applications of these in different areas. At current study, recycled bags was used to fabricate ecologically friendly composites.

2. Secondary textiles: from source to treatment

However, large textile type material lost in landfilling or incineration are unavoidable in the currently wasteful, linear system which creates negative impacts on the environment. It is important to find innovative solutions to valorise this waste to ensure that it does not add to the already significant environmental impact created by textiles.[1], [6], [9–13]

Ensuring that there are markets for the secondary textiles to be processed reduces the amount of textiles which end up deposited in landfill or simply incinerated, hence improving the "circularity" of these valuable items. As the quality of secondary textiles decreases, the options available become of less value, as highlighted in Figure 2. The strategy to textile waste management must give priority in the first place the waste prevention, in the second place recycling, reusing and revalorization of different textiles and finally their dump deposition.[10], [12]

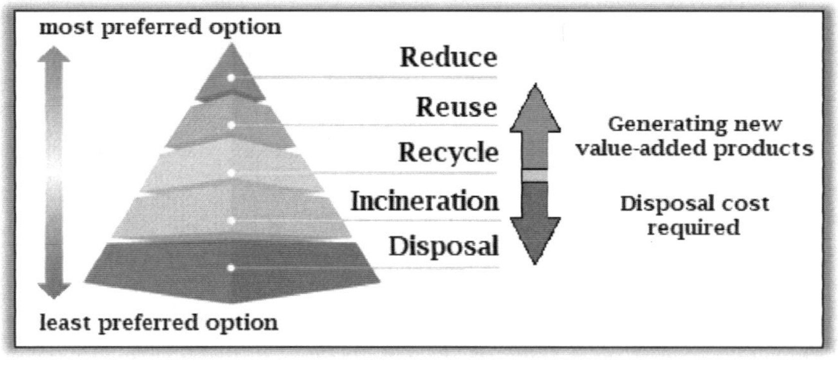

Figure 2. The textile waste hierarchy options

Different clothing and household textiles are a complex input materials for the textile wastes recycling process due to the variation in their composition, their quality at the point of disposal in landfills and the existence of alternative markets for textiles which have been recovered.[1], [3], [4], [11–14]

The available and most preferred options for waste textiles collected separately and sorted, considered as high–grade scope for material recycling, are highlighted in Figure 3. The least available options for waste textiles, separated or not at source, treated as municipal solid waste, considered as low–grade scope for material recycling which need disposal costs, are presented in Figure 4.

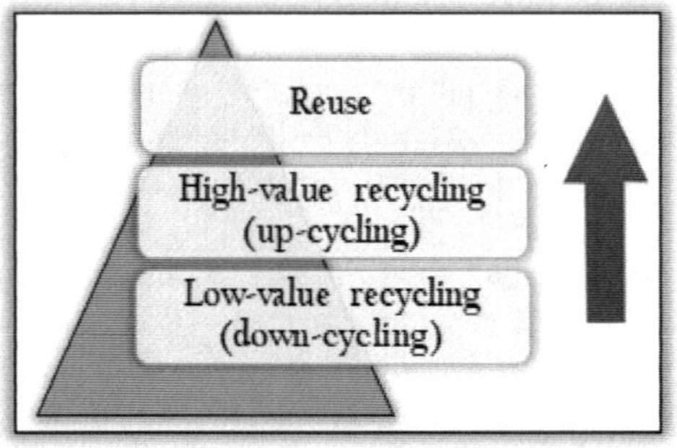

Figure 3. Flow diagram of waste textiles which can generate new value–added products

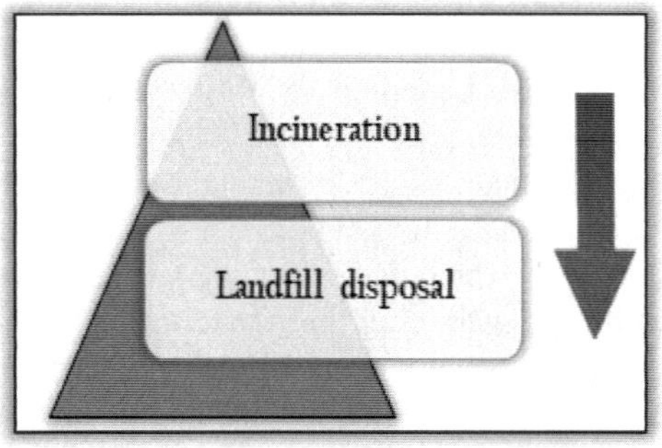

Figure 4. Flow diagram of waste textiles which disposal cost required

Also, the textile industry generates a large amount of waste and the textile manufacturing is worldly reported as the second most polluting sector and represents a complex, problematic waste stream. The growth of textile markets not only depends on population growth but also depends on economic and fashion cycles.[4], [9], [12–14] The fast fashion cycle in the textile industry has led to a high level of consumption and waste generation. This can cause a negative environmental impact since the textile manufacturing and clothing industry is one of the most polluting industries, being a chemical–intensive process. Wastewater and fibre or fabric wastes are the major wastes generated during the textile production process.[4], [10] On the other hand, the fibre and fabric waste was mainly created from unwanted clothes in the textile supply chain. This waste includes natural and synthetic fibre or fabric, and hybrid natural/synthetic blends.[6], [9], [12]

3. The recycling in practice

Concurrently, the valorisation of waste for useful materials production is a current practice, with particular focus on environmental indicators and sustainability goals and it is part of the larger endeavour of the circular economy (Figure 5). Outside of well–established reuse markets and mechanical recycling processes (where capacity limits exist), there remains a large quantity of low–value materials for which there is no market pull to utilize.

The move to a circular economy is becoming more widely discussed and adopted, largely thanks to its benefits. A circular economy approach to textiles involves moving away from our traditional "take, make and dispose" model and working to achieve a model whereby fibres or/and fabrics are kept in use and at their highest value. The ability to recycle used textiles – by use waste textiles to back into make new products – is of particular interest.

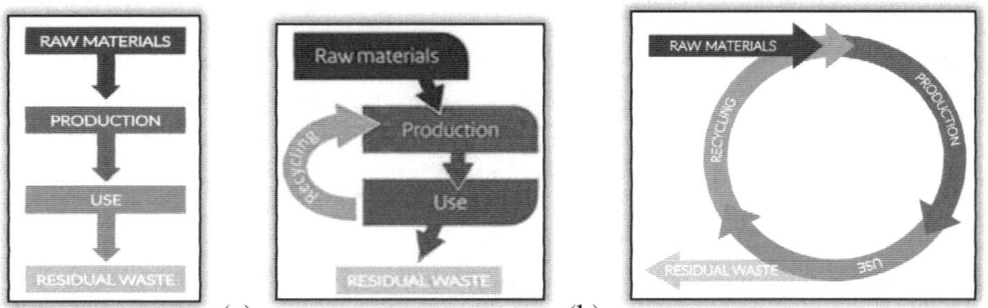

(a) (b) (c)

Figure 5. From linear to a circular economy (from waste to resource) (a) linear economy; b) reuse economy; (c) circular economy

Characterizing the linear economy, the economy with feedback loops, and the circular economy summarize the potential, and in some cases already proven, advantages of the circular economy approach in terms of three major categories.

— the production processes in this setup require significantly less newly produced or mined raw materials. Consequently, these processes become less sensitive to the growing need of many raw materials;
— the circular economy has the potential to generate innovations and new employment opportunities in the so called eco–industry, based on the development and application of eco–technology;
— the reduction of environmental damage due to less extraction of raw materials and significantly smaller waste disposal problems.

Products which have lost their functional value generally tend to form part of material left for recycling. Considering the impact of different measures, it appears that the greatest potential is to be found in the development of new technology. The challenge here is to develop value adding features. We all know what the basis of recycling is a practice that takes an item and targets it for reuse, returning it back to the cycle of daily contribution to society rather than discarding it to trash, meaning treat or process used or waste materials so as to make suitable for reuse, alter or adapt for new use without changing the essential form or use again in the original form or with minimal alteration.

Therefore, any recycling process is a first step in reaching a more sustainable solution of the waste management that can eventually limit the amount of new and virgin materials that need to be produced. In fact, recycling is undoubtedly useful, and is a conscious means for extending the useful lifetime of used materials. Therefore, any recycling process has simply prolonged the inevitable by stretching out waste stream and made the lifecycle costs of the material a bit less.

The circular economy has seen a significant increase in interest over the past few years, having several key elements as strategies, as follow:

— can prioritise the regenerative natural resources, and therefore the renewable and reusable resources are used as raw materials in an efficient way;
— can preserve and extend what is already manufactured, and, while resources are in–use, repair and upgrade or remanufactured them to maximise their lifetime and give them a second life through take back strategies when applicable;
— use the existent waste as a new resource, using waste streams as a source of secondary resources and trying to recover waste for reuse and recycling.

(a) (b) (c)

Figure 6. The recycling in practice: (a) recycling; (b) upcycling; (c) downcycling

In this way, materials destined for landfills or incineration can be recycled in scope to produce something completely different. The process where waste products are converted into a new materials of better quality and a higher environmental value is the upcycling i.e. remaking of textile material wastes to something new of greater value, like reinforcements in a composite materials. Upcycling involve creative ways of reuse old materials by using different pre–consumer or post–consumer waste or a combination of the two. In fact, the upcycling becomes dually important:

— first, the practice reduces the amount of waste that we produce and ultimately goes into the ground for longer than any of us will be around.

— secondly, it also reduces the need for virgin material to be harvested as feedstock for new generations of product.

Upcycling represents a truly cyclical, balanced process that all industries and companies should be aiming towards. All of products could be drastically changed if the beginning of their design started with the goal of not having them end up in a landfill, as is been thinking in the circular economy. A number of ways could be utilities to train the economy into an inherent practice of reuse. Upcycling is a process that can be repeated in perpetuity of returning materials back to a pliable, usable form without degradation to their latent value–moving resources back up the supply chain.

Upcycling is described by some as reusing a material without degrading the quality and composition of the material for its next use. As a result, these usually become products that will eventually also become trash. In fact, upcycling is equivalent in material terms to the "reuse" step in the classic waste hierarchy (reduce–reuse–recycle), though rather than being direct reuse of the same product, its reuse of the materials in an equally useful way. Therefore, upcycling is a process where waste or useless products are converted into new materials or products of equal or better quality or a higher environmental value. Moreover, by making use of already existing materials the consumption of new raw materials for new products is reduced. Hence, upcycling is an even greener way of recycling, better environmentally and cost efficiently.

4. Innovative recycling technologies for textiles – A potential solution

A potential solution is the development of innovative recycling technologies for textile materials. In this sense, technical textiles are a field envisaged by the European strategy, and consist of textile materials with technical applications, in fields like industry or agriculture. To improve and reach a significant valorisation, integrated processes for proper separation, detection and classification of textile waste from industrial and commercial packaging waste are promoted. In this sense, several research and industrial approach suggest the proposed technical solution, as a strategy for textile waste recovery from the disposal in landfills, with a great potential in treating textile waste and facilitating the recovery of natural textile wastes as value–added products into natural fibre reinforced composites.[6], [9], [15–24]

Textile wastes has gained a tremendous interest during the ten last years and will be a subject for significant research progress in the future, working on a recycling technology to enable low–value blended textiles to be collected, separated and processed into new raw materials as reinforcement into composites. The emerging applications of these materials (fibre and textile recycling, focusing on the case of natural and synthetic textile fabrics, which account for a large part of textile waste) as reinforcements in the manufacturing of natural fibre reinforced composites is one of the valorisation approach. The natural fibre is mostly comprised of cellulosic material, which can be used as a resource for producing fibre–based products like composites. The main challenge for utilization of various textile waste is finding the method that is able to recover it as reinforcements in composites. Another interesting contribution is proposed on geo–polymers used in agriculture. This category of new high–added value textile products comprises the properties of textile materials (light–weightiness, flexibility, resistance) and offers adequate solutions for the specific technical purpose.

Therefore, developing markets for the lowest grade textiles can have wider benefits, including supporting the manufacturing of natural fibre reinforced composites. Moreover, technologies of composite materials could be a feasible solution for the textile waste valorisation and the textile wastes consisting of natural and synthetic textile fibres or fabrics in various ratios can be used as low–cost raw materials in high–added value products.

In recent years, there have been collaborations between brands and innovator companies, with the overall aim of developing recycling technologies which deal with the textile waste problem. With the increase in clothing consumption, greater amounts of textiles are being discarded, some of which having hardly been worn. Knowledge of the markets considered as inputs (textiles wastes) and outputs (natural fibre reinforced composites) for the recycling process has been quantified to identify the overall opportunity, alongside technology development.

5. Concluding remarks

Due to environment and sustainability issues, this century has witnessed remarkable achievements in the composite technologies through the development of natural fibre composites. The development of high–performance materials made from natural resources is increasing worldwide. The greatest challenge in working with natural fibre reinforced polymer composites is their large variation in properties and characteristics. However, with appropriate attention to fibre and resin design and structural geometry, natural fibre composites may prove a viable alternative to traditional materials in the future. Industrial ecology, eco–efficiency, and green chemistry are guiding the development of the next generation of materials, products, and processes.

The current interest for natural fibres from textile wastes as an environmentally correct composite reinforcement has motivated the investigation of new possibilities. For instance, the textile fibres from the textile wastes were recently found to have adequate mechanical properties to reinforce polymer composites. Most natural fibre composites, including the textile waste inserted composites, however, are fabricated with traditional non–degradable polymer matrix but still presenting a recycling advantage over the common glass fibre reinforced polymer composites. Therefore, textile waste inserted composites stand out as a relevant class of engineering materials.

References

[1] Hawley J M 2000 Textile recycling as a system: A micro/macro analysis, *Journal of Family and Consumer Sciences* **92**(4) 40–43

[2] Bartl A, Hackl A, Mihalyi B, Wistuba M and Marini I 2005 Recycling of fibre materials, *Process Safety and Environmental Protection* **83**(4) 351–358

[3] Bhatia D, Sharma A and Malhotra U 2014 Recycled fibers: An overview, *International Journal of Fiber and Textile Research* **4**(4) 77–82

[4] Domina T and Koch K 1999 Consumer reuse and recycling of post-consumer textile waste, *Journal of Fashion Marketing and Management: An International Journal* **3**(4) 346–359

[5] Todor M P, Bulei C, Heput T and Kiss I 2018 Researches on the development of new composite materials complete / partially biodegradable using natural textile fibers of new vegetable origin and those recovered from textile waste, *IOP Conf. Ser.: Mater. Sci. Eng.* **294** 012021

[6] Todor M P, Bulei C and Kiss I 2018 Composite materials manufacturing using textile inserts with natural origins fibres, *IOP Conf. Ser.: Mater. Sci. Eng.* **393** 012088

[7] ***SMART Reports 2018 *Secondary Materials and Recycled Textiles Association Calls for Recycling Renaissance,* Secondary Materials and Recycled Textiles Association (https://www.smartasn.org)

[8] ***SMART Reports 2017 *Debunks textile waste dilemma,* Secondary Materials and Recycled Textiles Association (https://www.smartasn.org)

[9] Pensupa N, Leu S Y, Hu Y, Du C, Liu H, Jing H, Huaimin W and Lin C 2017 Recent trends in sustainable textile waste recycling methods: Current situation and future prospects, *Topics in Current Chemistry* **375**(5) 76

[10] Sule A D and Bardhan M K 2001 Recycling of textile waste for environment protection – An overview of some practical cases in the textile industry, *Indian Journal of Fibre & Textile Research (IJFTR)* **26**(1–2) 223–232

[11] Farrant L, Olsen SI and Wangel 2010 A environmental benefits from reusing clothes, *International Journal of Life Cycle Assessment* **15**(7) 726–736

[12] Sandin G and Peters M P 2018 Environmental impact of textile reuse and recycling – A review, *Journal of Cleaner Production* **184** 353–365

[13] Cuc S and Vidovic M 2011 Environmental sustainability through clothing recycling, *Operations And Supply Chain Management* **4**(2/3) 108–115

[14] Cuc S and Tripa S 2014 Fast fashion and second hand clothes between ecological concerns and global business, *Annals of the University of Oradea* **XV**(1) 163–167

[15] Saxena M, Pappu A, Sharma A, Haque R and Wankhede S 2011 *Composite materials from natural resources: Recent trends and future potentials, advances in composite materials – Analysis of natural and man–made materials*, InTech, http://www.intechopen.com/
[16] Faruk O, Bledzki A, Fink H and Sain M 2012 Biocomposites reinforced with natural fibers, 2000 – 2010 *Progress in Polymeric Science* **37** 1552–1596
[17] Faruk O, Bledzki A, Fink H and Sain M 2014 Progress report on natural fiber reinforced composites, *Macromolecular Materials and Engineering* **299**(1) 9–26
[18] Bledzki A K and Gassan J 1999 Composites reinforced with cellulose based fibres, *Progress in Polymer Science* **24** 221–274
[19] Summerscales J, Dissanayake N P J, Virk A S and Hall W 2010 A review of bast fibers and their composites. Part 1 – Fibers as Reinforcements, *Composites Part A* **41** 1329–1335
[20] Summerscales J, Dissanayake N P J, Virk A S and Hall W 2010 A review of bast fibers and their composites. Part 2 – Composites, *Composites Part A* **41** 1336–1344
[21] Pickering K L, Efendy M G A and Le T M 2016 A review of recent developments in natural fiber composites and their mechanical performance, *Composites Part A* **83** 98–112
[22] Sen T and Reddy H N 2011 Various industrial applications of hemp, kenaf, flax and ramie natural fibers, *International Journal of Innovation, Management and Technology* **2** 192–198
[23] Todor M P, Bulei C and Kiss I 2017 Inducing the biodegradability of polymeric composite materials using Bioranforts, *Applied Engineering Letters* **2** 84–90
[24] Nickel J and Riedel U 2003 Activities in biocomposites, *Materials todays* **6**(4) 44–48

Using MODELICA for calculating the power requirements of a mid-sized city water supply pumping system

O Nedelcu[1], C I Salisteanu[1], C V Oprescu[2] and P B Salisteanu[2]

[1]Valahia University of Targoviste, Department of Electronics, Telecommunications and Energy Engineering, Aleea Sinaia str., no 13, Targoviste, Romania
[2]Valahia University of Targoviste, PhD Student, Aleea Sinaia str., no 13, Targoviste, Romania

E-mail: otilia.nedelcu@valahia.ro

Abstract. The issue of water supply is one of the fundamental issues of the society, it is one of the pillars without which society would not work. The performance of the Targoviste water supply network has fallen over time, so solutions are being sought to optimize this network. A short description of the current situation has been made in the paper and solutions have been proposed for a certain area of capture and pumping of water to consumers. For the choice of pumps, water supply systems, calculations were made for the selection of system pipes, pumping height, pump motor power, and loss coefficients. With the help of the MODELICA calculation program, simulation of the pumping station was carried out using two separate calculation models by means of boundary conditions established with the help of water sources. The simulation of the water supply system was performed to verify that the pump power is sufficient to ensure the flow of water required by the consumer and to determine the operating mode of the drive system (continuous or intermittent).

1. Introduction

The water supply assures the basis for the operation of all social gears, and the water quality issues are a major priority. The control of the flow rate or the pressure of supply systems of water distribution networks is done manually or automated by the used of valves. The maximum efficiency is obtained by operating the pump motors at rated speed and the maximum allowable flow rate.

Although, the use of static converters and variable frequency electric drives is beneficial to pumping systems, more important than variable speed control of pump motors, it is the correct choice of the pumping systems [1].

The continuous increase in the number of connexion, by adding the neighbouring areas of Targoviste to the city network, and increasing the active population, basically doubling the population served in the last 40 years [2], led to the pressure reduction to values below 1 bar in certain areas of the network supply.

2. The performances of the pumping station. The current situation.

The pumping station construction should meet the need of water supply to the new policy of sustainable development at the same time [3].

The City Drinking Water Transport and Distribution Network of Targoviste has a total length of 140 km, of which 39.6 km were recently rehabilitated: 19.2 km from network for supplementing networks that no longer meet the consumption increase and 20.4 km from network for replacement of very old networks.

Targoviste water supply system was designed and executed in an annular system (Figure 1), having diameters between 100 - 600 mm (Table 1).

Table 1. The dimensions of the water network pipe-lines depending on the diameter

Diameter (mm)	Length (km)
50	16.5
80	14
100	38.5
150	29
200	14.5
225	5
250	3.5
300	1.5
400	9.5
600	8

To make the pumping system more efficient, the water networks have been connected and pressurized to two pressure levels, namely:

- *low pressure network*, which is realized annularly, supplies 35 pressure raising stations that fuel the buildings in Targoviste Municipality, private houses and the buildings of companies, in permanent regime, and 8 pressure raising stations for the neighbouring communes.
- high pressure network, not-interconnected, which supplies groups of housing blocks, public institutions, economic agents.

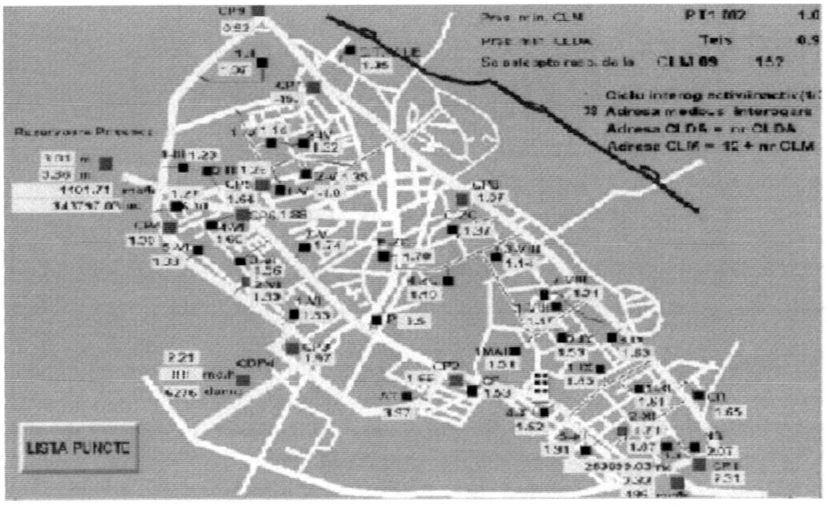

Figure 1. The network of distribution of Targoviste City, the points of pressure and the water flow rate at the network entrance, according to the measured data at the control panel of central dispatching of Water Company Targoviste, Dambovita.

The stations of pressure raising have 4 pumps each, one operated at variable speed, variable revolution speed, controlled by a speed variation, to maintain the constant pressure in the system under the conditions of variable consumption. The groups of pumping from hydrophore stations are HMLV 50.2x5 with $Q=80\ m^3/h$, HMLV 65x4 with $Q=100\ m^3/h$, and respectively HMLV 80x4 with $Q=180\ m^3/h$.

2.1. Dragomiresti North Water Capture Front

Dragomiresti North water capture front is monitored and is located to approximate 10 km from Targoviste, it contains 22 wells with drilling depths between 12 ... 60 m and they are equipped with Grundfos submersible electropumps of types SP46-4 7.5 kW, SP16-5 3 kW and steel pipes $D_n = 100$ *mm*. The distance between drillings is 120 m. The captured water is transported towards Dragomiresti station tank of 1250 m^3 through 2 telescopic metallic pipes $D_n = 100\ mm$ respectively $D_n = 300\ mm$, these pipes are connected to a metallic pipe $D_n = 500\ mm$, with the length of 500 m. The total length of the water capture front is 3600 m.

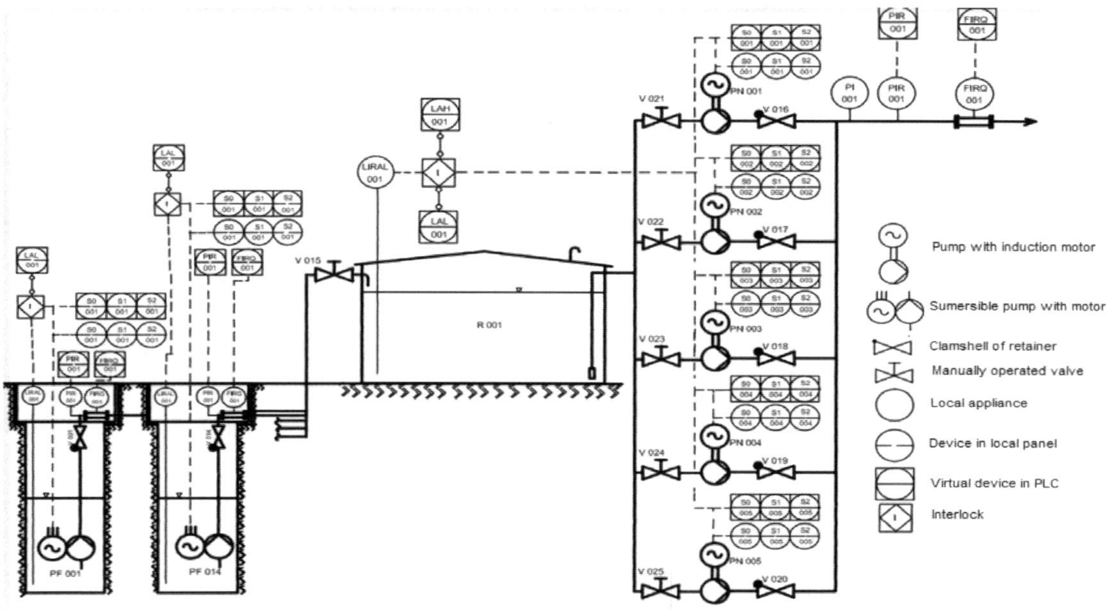

Figure 2. Dragomiresti North pumping station with the drilling wells, the storage tanks of the water and the horizontal submersible centrifugal pumps. Only two drilling wells are in the Figure.

Dragomiresti North pumping station still has a water capture front in the perimeter formed by 6 drillings, of which only 4 are functioning now. These drillings are equipped with prefabricated tubes of Dn = 100 mm. The drillings have depths of 10-12 m, except the drilling 5 which was been reactivated again to depth 64 m. This drilling is equipped with a Grundfos submersible electropump of type SP46-4 7.5kW and a steel pipe D_n =100 mm and the length 30-35 m. The drilling 2, 4 and 6 are equipped with the Grundfos submersible electropump of type SP27-3, 3kW. The distance among drillings is between 150…200 m. The total length for this is 1000 m.

The operating flow rate of Dragomiresti North water capture front is 170 l/s. The water accumulated into the tank 1250 m^3 is pumped towards the storage tanks from Priseaca by a, steel and PREMO pipe D_n = 600 mm, by means of a pumping station equipped with 5 horizontal electropumps of centrifugal type CM250x150x400 (Figure 2).

3. The choice of pumps, the water supply systems

The preliminary choice of the centrifugal pump and the water pipes will be made for a capture well of the pumping station of Dragomiresti North. As hypothesis of calculus is considered that the pumping system used a horizontal centrifugal pump, the water level into tank is kept constant to a level of 6 m, the capture tank (the case considered) has a depth of 60 m. The water level into the capture tank is imposed to 30 m and the imposed water flow rate at the exit is 45 m3/h. The length of the intake pipe is 100 m, and the length of discharge pipe is 1000 m.

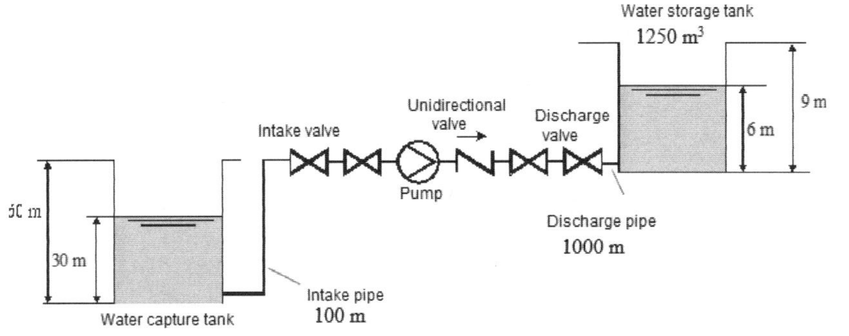

Figure 3. The water pumping system dimensioned for a well of Dragomiresti North station – when a horizontal centrifugal pump is used

3.1. Choosing the diameter of the intake and discharge pipes

The diameter of the intake and discharge pipes is determined by imposing the speed of flows of water into pipes: v = 1…2 m/s by the intake pipe and respectively v = 2…3 m/s by discharge pipe. For the flow rate 45 m^3/h is chosen a diameter of intake pipe d_{ab}= 101.16 mm and a diameter of discharge pipe d_{desc} = 76.2 mm.

3.2. The calculus of the pumping height

The total pumping height is calculated following the relation:

$$H_{tot} = H_s + (K_{fit} + K_{cond}) \cdot \frac{v^2}{2g} = H_s + \left(K_{fit} \cdot \frac{v^2}{2g} + h_f \right) = 36m + 92m = 128m \qquad (1)$$

where, the static height is H_S = 36 m, according to Figure 3. The difference of height, because the neglect of differences of pressure among the tanks and the difference due to the pressures of speed at pump, is $\Delta H = 0$. The calculus of the dynamical height depends of losses, meaning by the speed of the flow into pipe and the coefficients of losses K_{fit} and K_{cond}.

Table 2. The coefficient of losses K_{fit}

The way of flowing	*Numar componente* $\times K_{fit}$			
	Valve	Elbow	Tank	Directional valve
Intake pipe	2×0.16	-	0.5	-
Discharge pipe	2×0.25	3×0.35	-	1×2

Table 3. The height of losses by friction in pipes

The way of flowing	d in	h_f ft/100 ft	$v^2/(2g)$ ft
Intake pipe	3	2.27	0.395
Discharge pipe	4	8.9	1.17

3.3. The power of the pump's motor

The preliminary calculation imposes a motor of pump with the power:

$$P = \frac{Q \cdot H \cdot g \cdot \rho}{\eta} = \frac{0.0125 \cdot 128 \cdot 9.81 \cdot 1000}{0.7} = 22.5\,kW \tag{2}$$

where, for the efficiency of pump and motor, we choose $\eta = 70\%$. The optimization of pumping system is made by the calculus of the pump's power for several values of the diameter of pipes. So, the reducing pump power and the pumping cost can be done by the reducing power losses in the discharge pipe. If considering a discharge pipe with the diameter 4 inch, and we repeat the calculation process, we obtain total height of pumping 63 m and a pump with the power 10.7 kW, for an impose efficiency at 70%.

3.4. The coefficient of losses

The coefficient of losses is calculating K_{cond}, power pump's motor is calculating, the values obtained by calculation are shown in Table 4. The roughness coefficient of the pipe was chosen for a 0.025 mm steel pipe [4]. The height of pressure losses was calculated quite accurately to 61 m, instead of 63 m. The motor power is estimated at about 10.7 kW in both cases.

Table 4. Results of the pumping height calculation

Sizes	Symbol	Values	Measurement unit
Diameter of intake pipe	d_{abs}	0.1016	m
Section intake pipe	A_{abs}	0.00810732	m^2
Speed into intake pipe	v_{abs}	1.541816595	m/s
The roughness coefficient	-	2.50E-05	m
Dynamical viscosity of water	υ	1.31E-06	kg/s·m
Reynold Number	Re	119579.058	-
The coefficient of friction	f	0.018708461	-
The coefficient of losses due to fittings	K_{fit}	1.26	-
The coefficient of losses due to pipe	K_{cond}	202.55	-
Dynamically height	H_D	24.69	m
Statically height	H_S	36	m
Total height of pumping	$\boldsymbol{H}$	60.69	**m**
Power of motor	$\boldsymbol{P}$	10632.3	**W**

3.5. Choosing the pump using the manufacturer's calculation tools

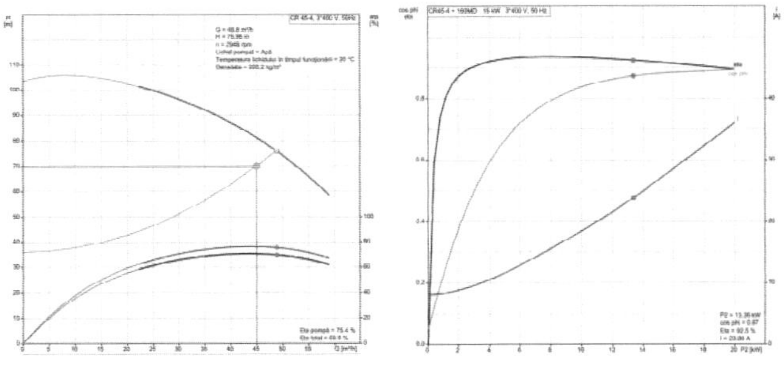

Figure 4. The characteristics $H = f(Q)$ and efficiency of pump, and the characteristics pump's motor CR45-4 15 kW

Water pumps allows for transfer of water to a higher height or a determined length [5].

Pump choice can be made directly from the catalogue using computational tools manufacturer. Example, according to Grundfos [6], imposing a minimum flow 45 m^3/h, it results that the pump CR45–4, 160MD 15 kW with the characteristics from Figure 4 can be used for the previous pumping system. In this case, the water flow into pipe will be 48.5 m^3/h. The power of motor will be 13.4 kW. Checks on this choice are still needed.

3.5.1. The cavitation and NPSH. The NSPH available from the system should be design than the NSPH required by the pump. The consideration of the NSPH value during the design stage will enhance the reduction for cavitation to occur [5].
The relation of calculation is:

$$NPSH = p \pm H_s - H_f - H_{vp} \tag{3}$$

where p is absolute pressure at surface of fluid in the intake tank, H_s is distance between the surface of fluid and the centre of impeller, H_f is the height of losses by friction in the intake line, H_{vp} is the vapor pressure of the liquid at the pumping temperature, expressed in m.
For the example shown above calculation (the pump CR45-4 15 kW operating at a flow rate of 45 m^3/h) the allowable value $NPSH_{adm}$ = 2.88 m is extracted from the manufacturer's diagram. The net positive absorption height is:

$$\text{NPSH} = 10.2\text{ m} - 30\text{ m} - 2.36\text{ m} - 0.23\text{ m} \ll \text{NPSH}_{adm} = 2.88\text{m} \tag{4}$$

The chosen pump CR45-4 cannot be used due to the cavitation phenomenon. For the previous system, the hypothesis of using a horizontal centrifugal pump at 30 m above the surface of the liquid was considered.

Table 5. Necessary data for calculation of the value NPSH for prediction of the cavitation of the pump

Absolute pressure	$1atm \Leftrightarrow 10.2m$
Absorption height	30 m
Liquid	apă: 20^0 C
Height of friction on absorption pipe	2,36 m
Vaporization pressure	$0.023atm \Leftrightarrow 0.23m$

A centrifugal pump can be used, if it is positioned below the ground level, and for safety reasons in service, in this case it must be installed below the water surface level in the capture tank to minimum 10 m deep. In many cases, the technical solution, does not allow this, because a room for pumps in the basement pumping stations does not exist [7].

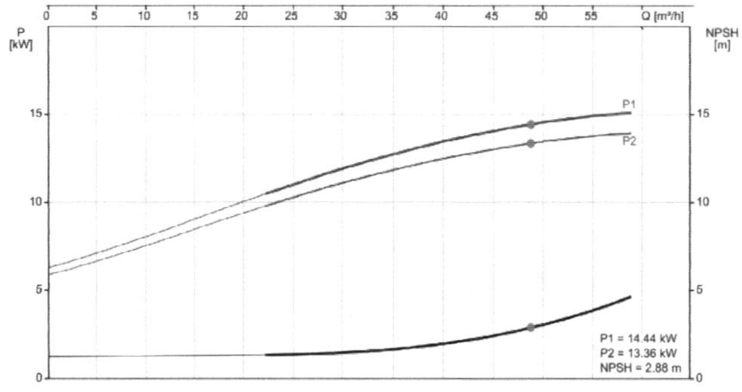

Figure 5. Power characteristics and NSPH characteristic of the pump CR45-4 15 kW

The diagram of pumping from Figure 3 is modified by used a submersible pump SP46-8. The characteristic H=f(Q) of the pump are shown in Figure 6. If eliminated the valves from the way of absorption, the losses by friction are reduce. The losses on the way of discharge increase due to increase the length of the pipe of discharge. The net positive height is recalculating:

$$NPSH = 10.2\ m + 30\ m - 2.36m - 0.23m >> NPSH_{adm} = 3.49\ m \quad (5)$$

The pump SP46-8 can be used. A pump of about 15 kW can be rise water in a well of capture with deep 60 m and it can be transporting water with flow rate of 45 m^3/h to 1000 m of the capture tank.

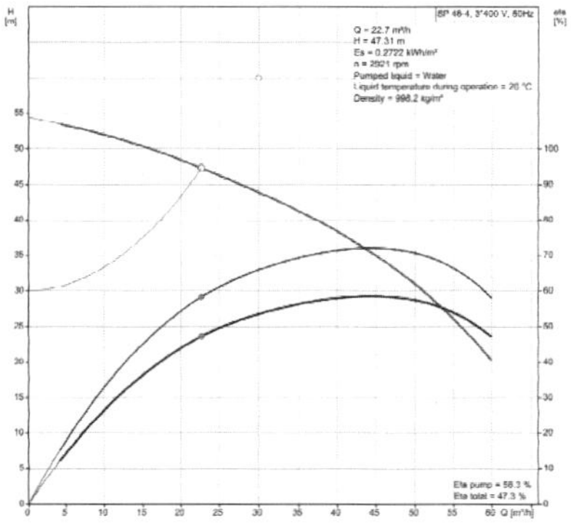

Figure 6. The characteristics of the pump SP46-8 7.5 kW, used to rise water in Dragomiresti North pumping station

To reduce the loss, a pipe with a minimum diameter of 4 inches should be used on the discharge path. The use of 7.5 kW pumps involves halving the flow rate or increasing the diameter of the discharge pipe, long enough to reduce the losses by more than 75% in pipe.

The water pumps SP46-8 used at Dragomiresti North work at about 50% the maximum efficiency. A possibility to increase pump efficiency would be to choose 15kW pumps that work alternately depending on the needs of the consumer.

IOP Conf. Series: Materials Science and Engineering **477** (2019) 012005 doi:10.1088/1757-899X/477/1/012005

The pumps of type SP17-5 (3kW) were installed in wells with deep of maximum 10 m, what makes for a total height of pumping of approximately 45 m it to operate a flow rate of approximately 14 m3/h = 3.9 l/s with the maximum efficiency (70%). However, the 3 kW pumps cover only a requirement of 13x3.9 l / s = 50.7 l / s of the total 170 l / s, i.e. 30% of the total. Nearly 70% of water requirements is covered by the other pumping systems, which require improvements, either by the change pumps' power, either by building some semi-buried tanks near the pumps of 7.5 kW. From the calculations, it results that the 7.5 kW submersible pumps have been placed at a depth of no more than 10 m from the water level, which makes their use possible only when the capture wells are filled with water, the variation margin of the water level not exceeding 6 m.

4. The simplified model of Dragomiresti North water pumping station used MODELICA

The Modelica.Fluid library includes the interface and basic components for object-oriented modelling of the convective flow of a fluid into a network containing reservoirs, pipes, pumps, fittings and valves. All components are implemented in such a way that they can be used for different environments [8]. The calculation hypotheses can be entered into an external object called a system. All assumptions can be redefined locally for each component [9], [10].

The simulation of the pumping station is carried out using two separate calculation models by means of boundary conditions established with the help of the water sources (Dragomiresti North capture tanks and Priseaca storage tank) and respectively the discharge valves on the consumers supply way. The separation of the system in two models makes possible, first the determination of necessary mechanical power of the pump to provide water with the imposed flow rate and to calculate the connexion times of pumps for to ensure the necessary water flow rate, and secondly, ensuring the stability of the simulation programs.

"Dragomiresti North 1" simulation program (Figure 7) includes the model of the water pumps, pipes connections between pumps and tank, and the automatic controllers by maintaining the pressure within the limits imposed of the height of the water level from capture tanks.

"Dragomiresti North 2" simulation program (Figure 8) includes the model with 5 water pumps, which are connected between the water tank from Dragomiresti (the first water tank of the network) and the storage tank of water from Priseaca (the second water tank of the network), the model with the water pipes and valve output to consumer (the city water network - entrance from Priseaca).

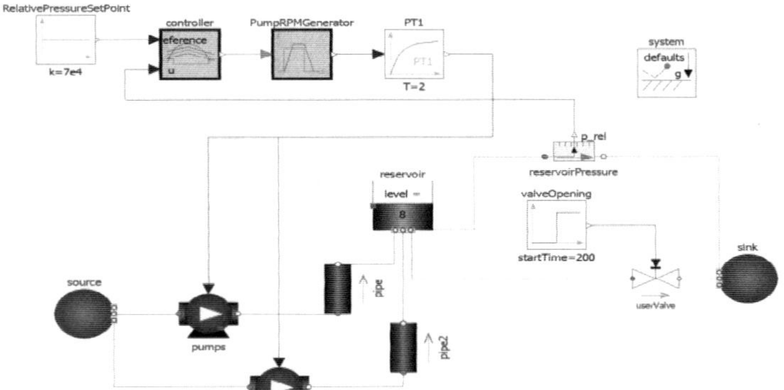

Figure 7. "Dragomiresti Nord 1" simulation model for Dragomiresti Nord pumping station (capture).

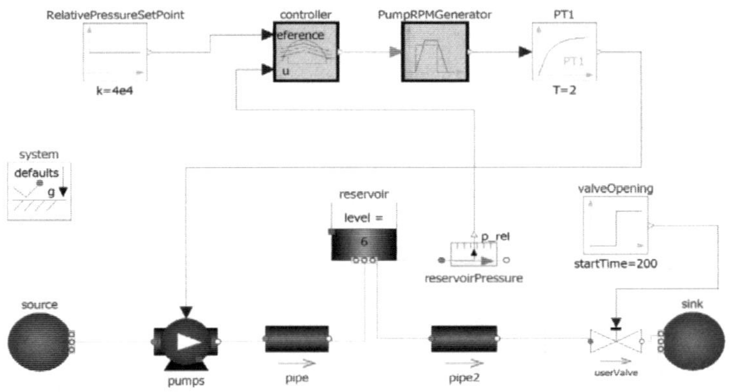

Figure 8. "Dragomiresti Nord 1" simulation model for Dragomiresti Nord pumping station (storage)

In Figure 7, "*Pumps*" is the model with 13 pumps of 3 kW which are connected in parallel, and "*Pumps2*" is the model with 14 pumps of 7.5 kW, which are connected in parallel, which are simultaneously controlled: on/off. The consumer is simulated using the valve "*userValve*".

In Figure 8 "*Pumps*" is 5 horizontal centrifugal pumps of 55 kW, which are connected in parallel, they pump the water from the Dragomiresti storage tank to Priseaca storage tank. The water flows from Priseaca tank towards the city network is modelled by imposing an output pressure of 1.23 atm at the border (*sink*).

4.1. Pumping station functioning description

The water flows to the consumer at an imposed flow rate through an output valve (*userValve*, Figure 7 & Figure 8). The pressure drop on the valve is set in the program to consider the losses of pressure in the passive circuits (fittings, elbows, junctions). The water storage tank (*source*) supplies necessary water to the consumer. When the valve is open, the water level begins to drop in tank. Because the water flow rate at consumer needs to be maintained within certain limits, there is realised an automatic control of the water level from the storage tank by means of pressure controller (*controller*), a pressure sensor is connected to the output of the water tank. Pressure control is performed by comparing the pressure measured by the sensor with a reference pressure, and when the difference between the required size and the measured value is detected, a control signal is transmitted to the pumps.

A PI-type regulator is used to linearize the control signal and to avoid possible control problems in the control and control circuit, it delays the signal with a time constant of a few seconds (2 seconds in this case). The pumps are commanded to start with a 6 second imposed speed ramp to reach speed from 0 at rated speed (3000 rpm for low power pumps and 1500 rpm for high power pumps). To stop, the same 6-second ramp is required.

4.2. The simulation results

The simulation of the water supply system was performed to verify that the pumps power is sufficient to ensure the flow rate of water demanded by the consumer and to determine the type of operation of the drive system (continuous or intermittent). Figure 9 shows the calculated power characteristics of the pumps. For the calculation of the power absorbed by the motor the efficiency is considered $\eta = 0{,}7$, resulting in the pump motor power of P_1= 2.6 kW / 0.7 = 3.7 kW and P_2= 5.2 kW / 0.7 = 7.4 kW. In this case, the service of operating is intermittent and permissible operating power is generally higher (in this case 10%, i.e $\sqrt{\frac{800}{650}} = 1.1$) than the rated power. It is noted that the installed power of

7.5 kW (pump SP46-4) and 3 kW respectively (pump SP17-5) covers the required power requested by the pumps to ensure a flow rate of 200 l/s. The imposed water flow rate is 170 l/s.

The improvement of the pumping system includes the successive start of the pumps using one or more inverters. After starting a pump and reaching the rated speed, the inverter is disconnected from the pump motor and connected to the motor of another pump, which is switched off and ready for starting. Once the pump runs at rated speed with the motor powered at rated voltage, it can be connected directly to the power supply.

The water level in the Dragomiresti Nord water storage basin is shown, Figure 11 shows the variation of the water flow rate to the consumer, Figure 12 shows the relative control pressure of the automatic regulation system.

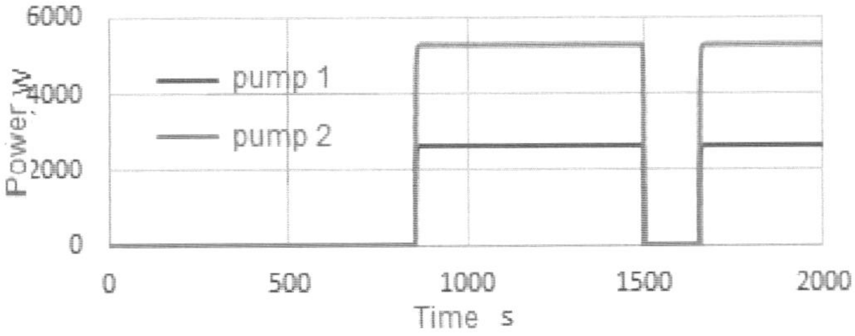

Figure 9. Calculated power of pumps SP46 / 4 and SP 17/5

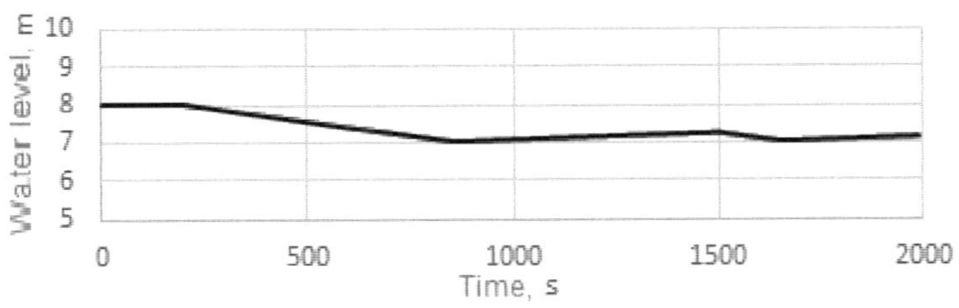

Figure 10. Water level in Dragomiresti Nord water storage tank

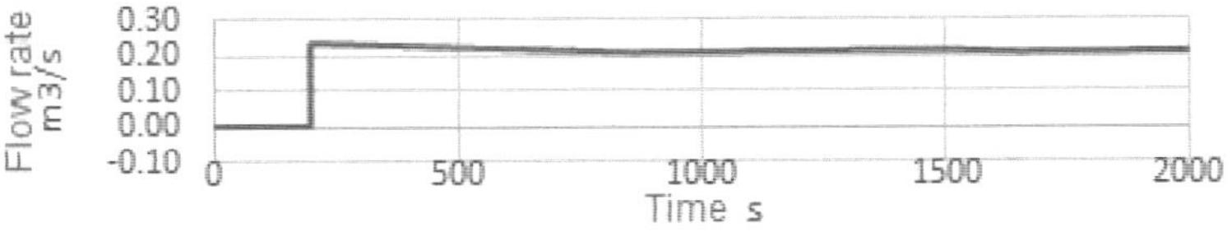

Figure 11. The water flow rate to consumer (water tank from Priseaca)

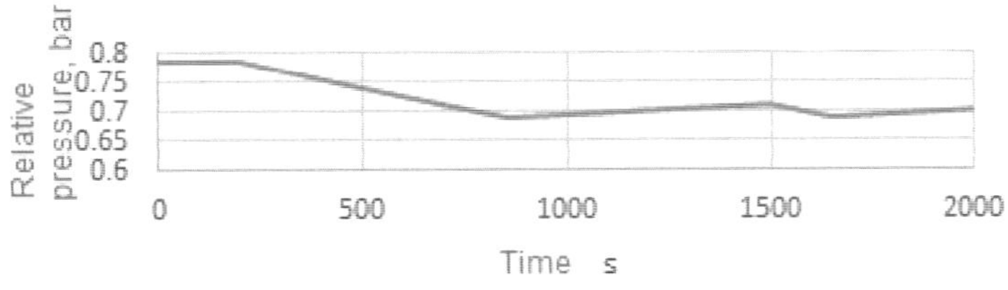

Figure 12. The relative pressure measured in the water tank (Dragomiresti Nord)

IOP Conf. Series: Materials Science and Engineering **477** (2019) 012005 doi:10.1088/1757-899X/477/1/012005

The simulation program no. 2 (Figure 8) is similar to the model shown in the simulation program no. 1.

The water flows from the Priseaca basin to Targoviste distribution network under the influence of gravitational force at a constant flow rate (Figure 13). The water level in the basin is maintained at a relatively constant value (Figure 14) by turning on and off the 5 horizontal centrifugal pumps. The adjustment of the starting and stopping of the pumps is done by imposing a relative pressure reference value of 0.4 bar with a control band of 0.02 bar (Figure 15). In the simulation program, all 5 pumps are switched on and off simultaneously. The maximum power absorbed by the single-pump motor is, according to the simulation results, of $P_1 \approx 50$ kW/0.7 = 71.4 kW. The intermittent power is about 80% higher than the rated power in continuous operation PN = 55 kW, assuming the effective power [11] valid rule, i.e. $\sqrt{\frac{450}{135}} = 1.8$.

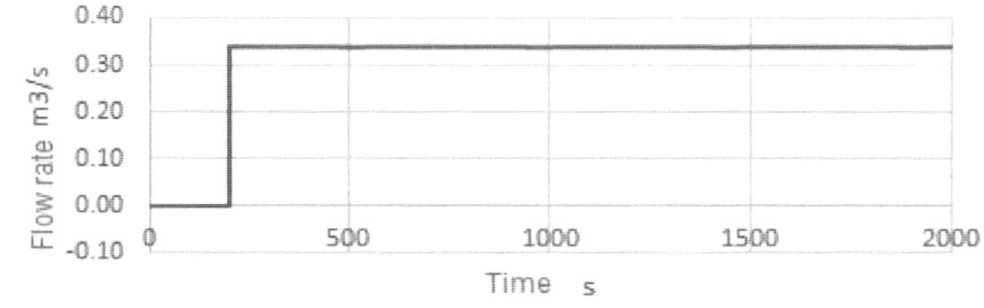

Figure 13. The calculated water flow rate on the city's water supply pipe from the Priseaca storage basin by imposing a pressure 1.23 atm to entry in network

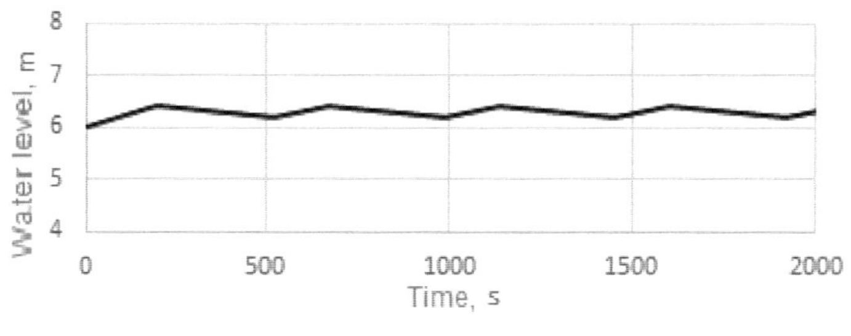

Figure 14. The calculated water level in Priseaca storage basin

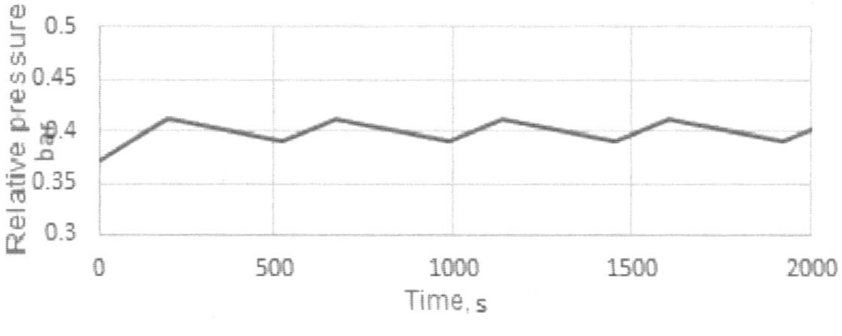

Figure 15. The relative pressure in the Priseaca storage basin

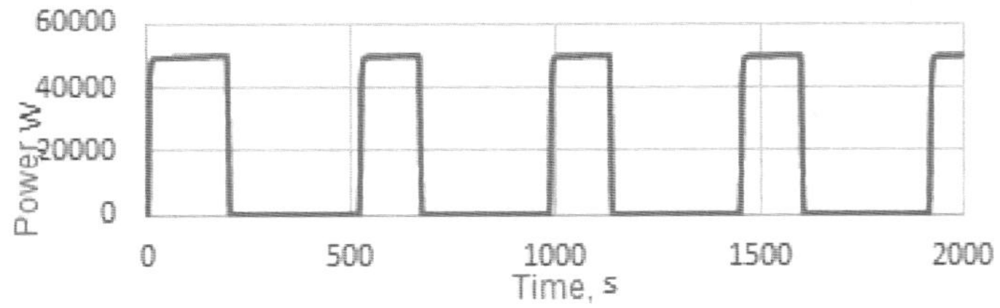

Figure 16. The power of a pump used for watering the Priseaca storage basin

5. Conclusions

The total installed power of the water supply pumps of the Priseaca storage basin is sufficient if the flow rate of water at the entrance to the city network does not exceed 0.32 m^3 / s equivalent to 1152 m^3 / h. Figure 1 shows the measured value at the entrance from Priseaca to the city's water supply network of 1100 m^3/ h. This value is not enough to ensure a constant pressure of more than 1.2 bar at all points of the city network, with values below 1 bar being observed at certain points of the network. Therefore, to improve the water supply of the city, as the Priseaca entry, the installed power needs to increase simultaneously with the efficiency of the operation of the Dragomiresti Nord submersible pumps, which feed the storage tanks from Dragomiresti Nord and Priseaca. A disadvantage of the pumping system is the large distance between the capture front and Dragomiresti Nord water storage tank. Possible water improvement may not only be an investment in increasing the pump power but also the possible construction of small storage tanks near the wells to allow submersible power pumps, such as 7.5 kW pumps greater depths.

The correct choice of pumps is more important than the potential control of pumps to reduce energy consumption. Because the northern area of the city necessitates an urgent improvement of the power system, the case of the Dragomiresti Nord pumping station and the operation of the water supply system in Priseaca water tanks has been treated as an example. Based on the available data, it was found that the 3 kW water pumps used in the water wells in the Dragomiresti Nord capture front were correctly selected.

References

[1] Halilovic H 2014 *Experimental transient behaviour characterization of induction motor fed by variable frequency drives for pump applications*, Master Thesis, Royal Institute of Technology, Stockholm

[2] ***http://www.catd.ro/index.php/companie/istoric

[3] Miao Qun, Wang Jiayuan and Liu Chao 2017 Energy-saving Optimization of water Supply Pumping Station Life Cycle based on BIM technology, *IOP Conf. Series: Earth and Environmental Science* **100** 012201

[4] Idelchik 2004 *Handbook of Hydraulic Resistance*

[5] Thiruchelvam V and Hong L M 2017 Fluids vertical transfer used VDF based centrifugal pumps, Journal of Physics: Conf. Series **822** 012062

[6] ***https://product-selection.grundfos.com

[7] Volk M 2005 *Pump Characteristics and Applications, Second Edition* CRC Press, Series: Mechanical Engineering

[8] Tiller M 2004 *Introduction to physical modelling with Modelica*, Kluwer

[9] Franke R, Casella F, Sielemann M, Proelss R, Otter M and Wetter M 2009 *Standardization of Thermo-Fluid Modeling in Modelica.Fluid*, Proceedings of 7th Modelica Conference, Como, Italy, September 20-22, pp 122-131

[10] Cellier F and Greifeneder J 2008 *ThermoBondLib – A New Modelica Library for Modeling Convective Flows*, Proceedings of 6th International Modelica Conference, Bielefield,

Germany, March 3-4, pp 1-10
[11] Tunsoiu Gh, Seracin E and Saal C 1982 *Acţionări Electrice*, Ed. Didactică şi Pedagogică, Bucureşti

IOP Conf. Series: Materials Science and Engineering **477** (2019) 012006 doi:10.1088/1757-899X/477/1/012006

Experimental studies regarding the influence of the intake manifold material on the thermodynamics processes in the internal combustion engines

C Birtok-Băneasă, S Rațiu, V Puțan and L Zgripcea

University Politehnica Timișoara, Faculty of Engineering Hunedoara, Department of Engineering and Management, 5 Revoluției Street, 331128 Hunedoara, Romania

E-mail: corneliugrup@gmail.com

Abstract. The paper presents an experimental stand designed to study the influence of the intake manifold material on the thermal transfer process during intake of air into the engine. For this purpose, the stand allows temperature measurements at different characteristic points for two geometrically identical intake manifolds, but made of different materials. It was taken into consideration the Aluminum alloy intake manifold of the Ford Puma 1.7i and the polyamide intake manifold of the Ford Puma 1.4i. The experimental results were presented in a comparative study where conclusions were drawn regarding the influence of the intake manifold material on the internal combustion engine processes.

1. Introduction

The tightening of global emissions standards and mandated minimum fuel economy requirements for automobile manufacturers have become the primary impetus behind internal combustion engine (ICE) research and development. Gasoline direct injection (GDI), multiple injection, gasoline-electric and plug-in hybrids, and common-rail (CR) diesel engines with sophisticated emissions-reducing after-treatment technology have all recently found their way onto public roads. These environmental constraints, along with the rapid advance of embedded computing power and advanced sensors, are pushing engine research into advanced combustion modes. Various low temperature combustion concepts, including homogeneous charge compression ignition (HCCI) and its variants, are being researched as viable future options for efficient internal combustion engines. The High-Efficiency Dilute Gasoline Engine (HEDGE) research currently focuses on knock-resistant combustion chambers, high-energy ignition approaches, and increases in high brake mean effective pressure (BMEP).[1]

Proper maintenance can help vehicles perform as designed, positively affecting fuel economy, emissions, and the overall driving. This effort investigates the effect of one maintenance factor, intake air filter replacement, with primary focus on vehicle fuel economy, but also examining emissions and performance.[2]

The engine behavior in engine performance, fuel consumption and exhaust emissions aspects are influenced by the magnitude of air intake pressure. It needs to form leaner air fuel mixture to extract more energy from the combustion process. The large amount of air increases the potentiality of fuel chemical elements to be burned with oxygen. As a result, the engine performance and fuel economy are increased while the unburned exhaust emissions components are reduced.[3]

The intake manifold of the Ford Puma, (Figure 1) is located behind the cooling radiator, which leads to a substation heating process resulting in raising the intake air temperature in direct connection with the thermodynamic losses.

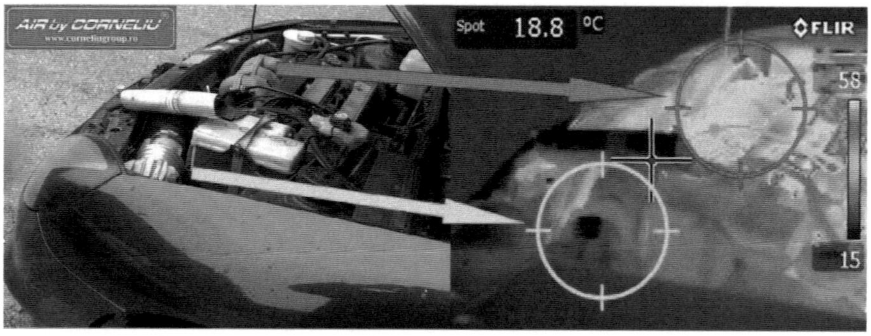

Figure 1. The thermal field in the engine compartment captured with a thermal imaging chamber

2. Experimental setup. Equipment and procedures

Resource intake manifolds stand (**R.I.M.S.**) (Figure 2) designed for the thermodynamic study of the influence of the intake manifold material on the heat transfer and gas-dynamic process during intake of air into the engine.

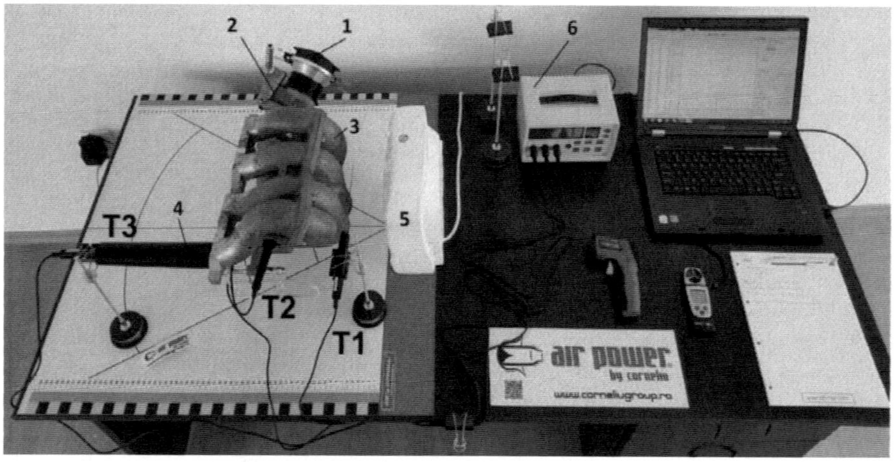

Figure 2. General view of experimental setup

For this purpose, the stand allows for temperature and pressure measurements at different characteristic points for identical or geometrically different intake manifolds made of various materials (aluminum alloy, polyamide, fiberglass, carbon fiber, etc.).

The R.I.M.S. (Figure 2) is composed of the following parts:
1 – air blower;
2 - throttle valve;
3 - intake manifold;
4 - straight pipe region;
5 - heating source;
6 - digital temperature indicator;
T1, T2, T3 - NiCr-Ni thermocouple sensors (T1 - outside heating air temperature, T2 – inside intake wall temperature, T3 – outlet air temperature).[4]

a.

b.

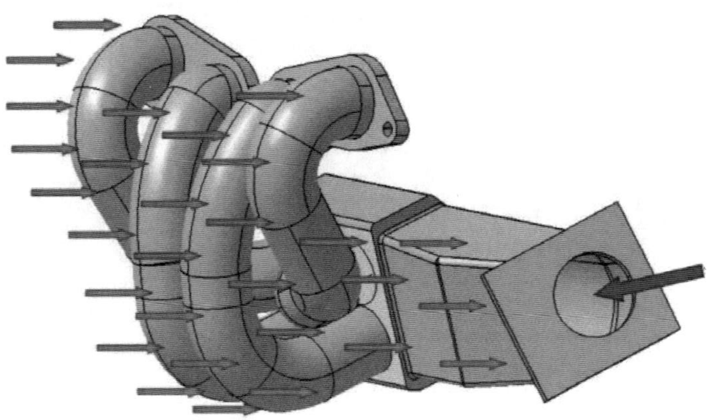

c.

Figure 3. Intake manifold configuration: a - Ford Puma 1.7i, b - Ford Puma 1.4i, c – virtual model Catia V5

An important factor is that the intake manifold subject to the comparative experimental study respects the mounting positions of the propulsion group (Figure 3).

It was taken into consideration the aluminium alloy intake manifold of the Ford Puma 1.7i (Figure 4.a) and the polyamide intake manifold of Ford Puma 1.4i (Figure 4.b).[4]

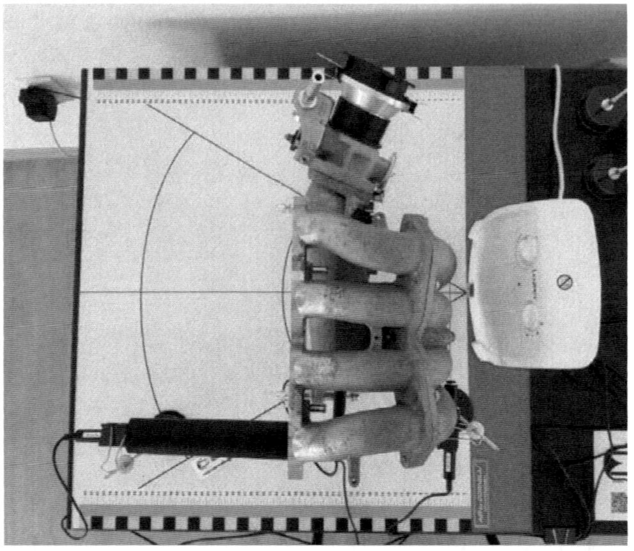

a.

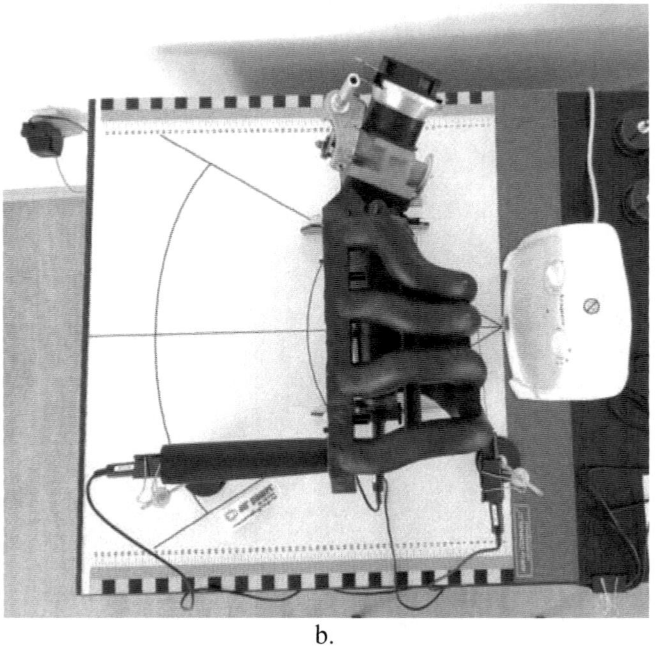

b.

Figure 4. Intake manifolds: a – aluminium alloy intake manifold Ford Puma 1.7i, b – polyamide (PA) intake manifold Ford Puma 1.4i

The program used to acquire the data is Cassy Lab2 software (Figure 5), the temperature was recorded for 40 minutes, second by second.[5]

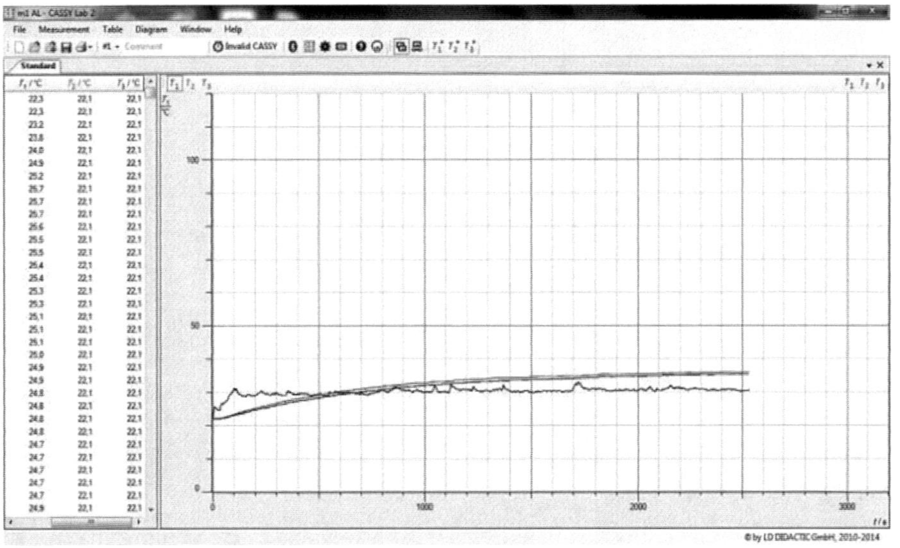

Figure 5. CASSY LAB 2 software

3. Results and conclusions

In the case of a gallery made of aluminum due to the relatively high heat transfer coefficient, the difference in temperature outside (T1) and the inside (T2) of the wall is negligible, unlike the polyamide in which case the difference is much higher (Figure 7), in the conditions where the thermal source generates the same value of temperature for both cases.

Figure 8 shows that the temperature values indicated by T1 at the outside of the intake manifolds are much higher for the polyamide (PA) than for the aluminium alloy (Al). The explanation: due to the low thermal transfer coefficient of the polyamide, it acts as a thermal barrier resulting in local temperature increase.

From Figure 9 it can be seen that the air temperature inside the aluminium intake manifold is much higher than the air inside the intake manifold of the polyamide with negative consequences on the cylinder filling efficiency.

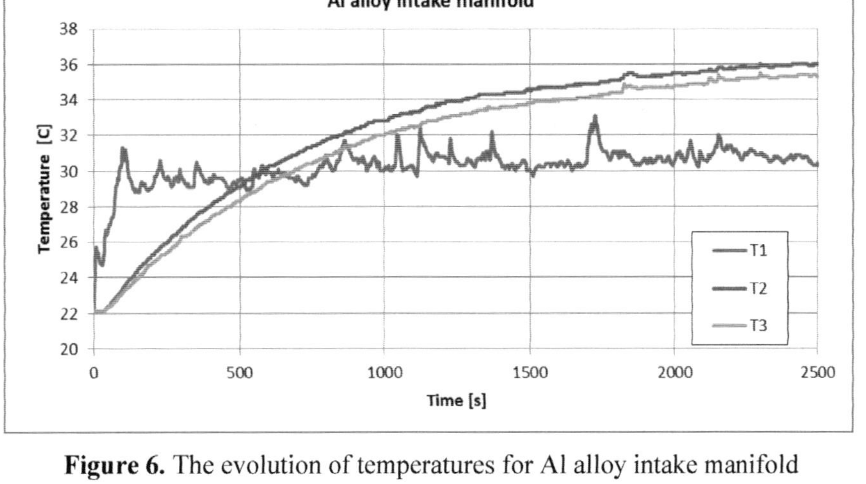

Figure 6. The evolution of temperatures for Al alloy intake manifold

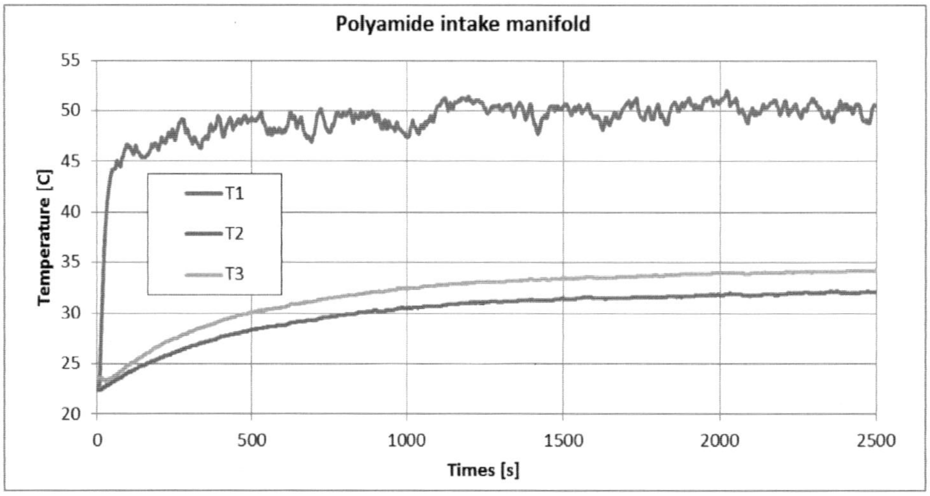

Figure 7. The evolution of temperatures for polyamide (PA) intake manifold

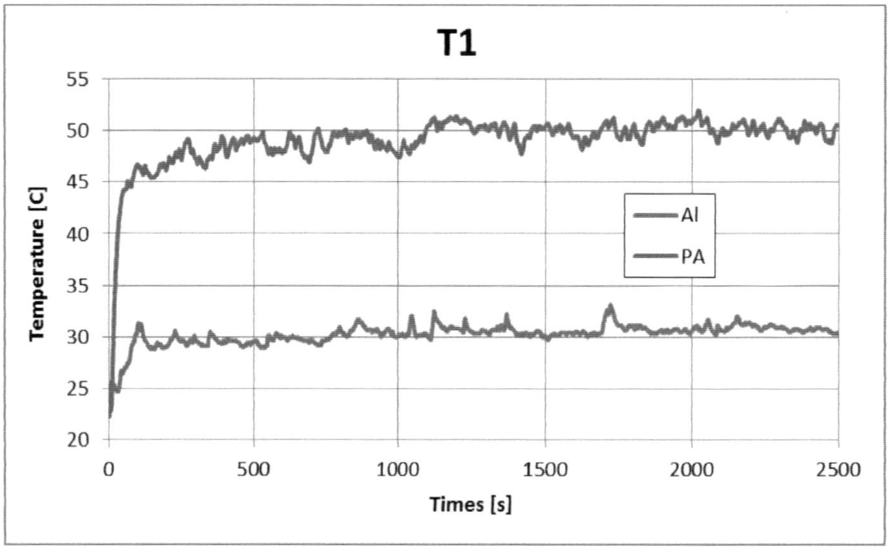

Figure 8. Comparative values for outside the intake manifold heated region (T1)

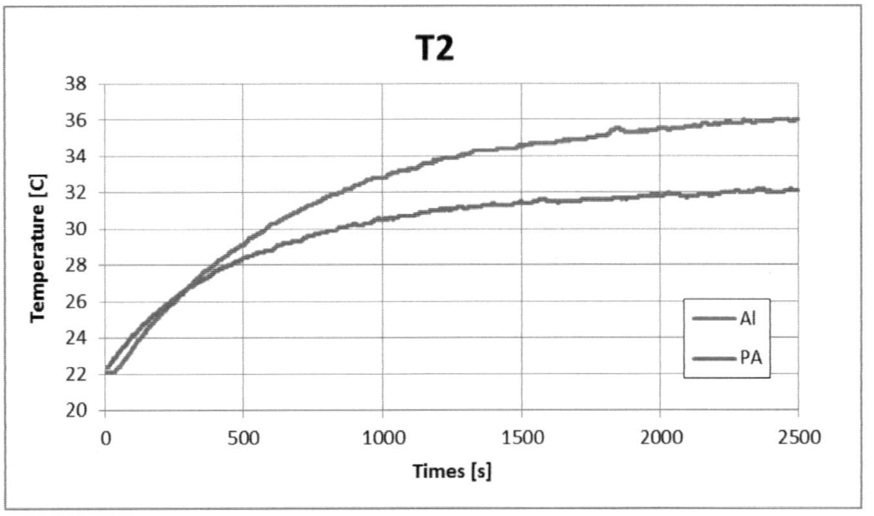

Figure 9. Comparative values for inside of the intake manifold (T2)

On the other hand, also in Figure 9, we notice that the air intake heating rate is higher for the aluminium alloy, which is an advantage when starting the cold engine.

Figure 10 shows the evolution of air temperature when leaving the intake manifolds. It is noticeable that the temperature of the inlet air to the engine cylinders is higher in the case of the intake manifold made by aluminium alloy.

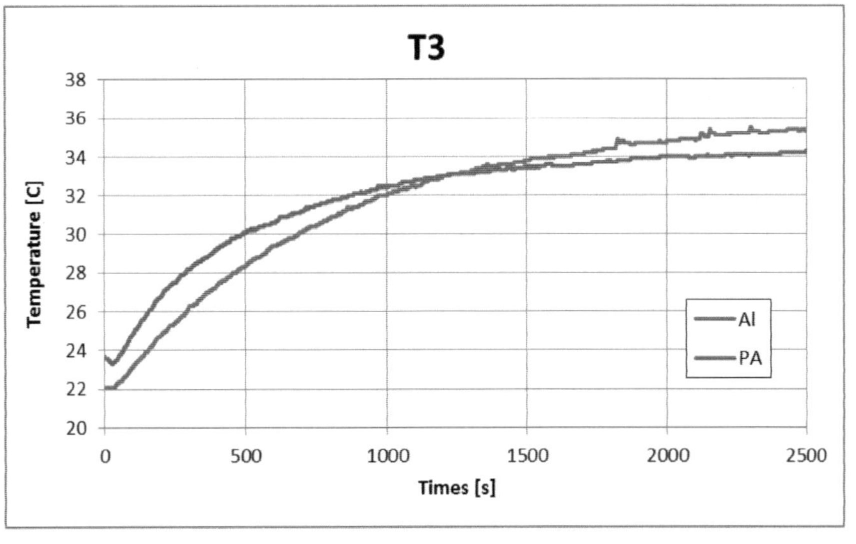

Figure 10. Comparative values for inside the straight pipe region (T3)

The analysis of the recorded data shows that the temperature variation in the case of the polyamide intake manifold is between (52-33) ^{0}C. In terms of thermal losses, the polyamide is a material that has the subunit heat transfer coefficient, which it recommends for the manufacture of thermally required components.

IOP Conf. Series: Materials Science and Engineering **477** (2019) 012006 doi:10.1088/1757-899X/477/1/012006

References

[1] ****Subsystems Required to Control Low Temperature Combustion Engines*, 2017, http://www.ni.com/white-paper/13516

[2] Thomas J, West B, Huff S, and Norman K 2012 *Effect of Intake Air Filter Condition on Light-Duty Gasoline Vehicles*, SAE Technical Paper 2012-01-1717.

[3] Abdullah N R, Shahruddin N S, Mohd A, Mamat I, Kasolang S, Zulkifli A, Mamat R 2013 Effects of Air Intake Pressure to the Fuel Economy and Exhaust Emissions on a Small SI Engine, *Procedia Engineering* **68** 278-284

[4] Birtok-Băneasă C, Rațiu S, Cioată V, Budiul-Berghian A 2018 *Thermodynamic Study of Airflow Through the Spark Ignition*, Engine Intake Manifold, Chapter 76, Springer Nature America, Inc

[5] ***https://www.ld-didactic.de/en/products-solutions/data-logging-and-measuring-echnology/cassy-software.html

IOP Conf. Series: Materials Science and Engineering **477** (2019) 012007 doi:10.1088/1757-899X/477/1/012007

Pulverous ferrous waste processing by agglomeration

O Lupu, L Zgripcea, A Socalici, E Popa and C Hărău

Politehnica University of Timisoara, Engineering and Management Department, 5 Revolution Street, Hunedoara, 331128, Romania

E-mail: oana.lupu@yahoo.com

Abstract. The current paper presents the results obtained after the laboratory testing of the of agglomeration process of pulverous ferous waste. The obtained products are destined for use in the steel industry, as raw material for the production of steel. When choosing the proceedure and technologies of capitalization we took into consideration the features of the waste, the destination of the obtained product and the existing processing facilities. The superior valorification of iron and steel waste represents an important matter because their transformation into by-products leads to a rational use of raw material and energy resources, thus ensuring both the satisfaction of human society and the protection of the environment.

1. Introduction

The worldwide strategies of metallurgical developments consist in developing performant technologies in order to reduce emissions while also grow the capability of by-product recovery and recycling. [1]

The ecological concept applied to metallurgical engineering implies the development of those technological fluxes in closed coil in which none of the resources are eliminated, all the materials are continuously reused, no dangerous waste or other product is evacuated in the environment [2], [3]. This field constantly tries to identify and implement the most efficient methods that retain all potential sources of iron inside the cycle of production – use- recycle with the purpose of protecting natural resources, cost reduction and to minimize the impact of wastes on the environment [4], [5].

The preoccupation about respecting the legislative requirements and the need to harmonize processes during economic growth with a rational segmentation of material and energetical resources, must lead to the capitalization of waste through technologies that offer the optimal solution both economically and ecologically [6] . It's necessary to promote technologies that ensure [7-10]:

- the rigorous segmentation of wastes;
- the controlled storage of all waste categories;
- the reduction of the quantities and harmfulness of the obtained waste.

2. Experimental research in the laboratory phase

The paper presents the capitalization methods of the ferrous sludge (10.02.12 – waste obtained after the clearance of cooling liquids) the came from a local plant that produces metallic powder.

The choice of the capitalization procedure must be made having in view the features of the waste, the destination of the obtained product and the existing processing equipment in the waste area. We chose processing the waste through agglomeration in order to valorificate it as agglomerate. The resulting product can be used as raw material in the steel processing ovens [11].

The chemical (Table 1) and granulometric (Figure 1) composition of the used sludge was determined.

Table 1. Chemical composition of ferrous sludge

Chemical composition, [%] ASTM E 1479/2016												
Mo	Ca	Cu	Cr	Mg	Mn	Ni	P	Fe	S	As	Zn	Alte
0.07	0.06	0.05	0.04	0.05	0.18	0.03	0.11	77.09	0.02	0.005	0.03	22.26

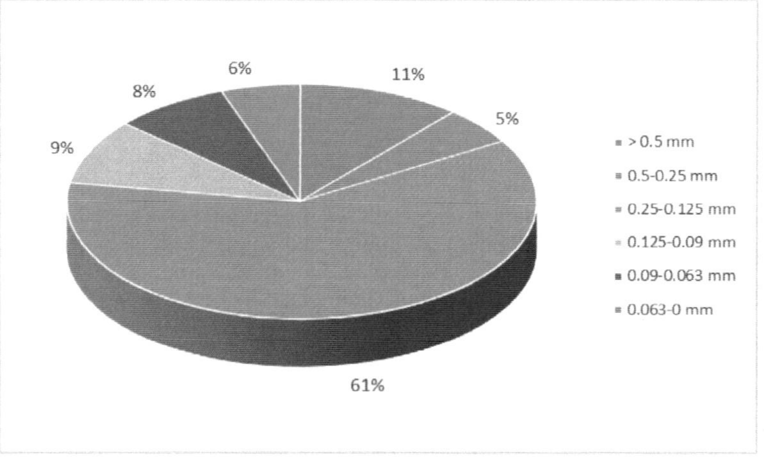

Figure 1. Ferrous sludge granulometric classes

By analysing the chemical and granulometric composition of the used waste that has to be put under the test must be pelletized. The ferrous sludge was put up for the pelletizing process described in Figure 2. These aspects from the time of experiments (pellets/micropellets) are shown in Figure 3.

In Figure 4 the obtained micro pellets are shown.

The obtained micropellets were used to obtain the agglomerate. The agglomeration recipe was made up from: 90% micro pellets and 10% charred coal.

The scheme of the agglomeration appliance that was used during the experiments is shown in Figure 5, and the technological flux of the agglomerate production is shown Figure 6.

The height of the agglomerate layer is decided based on its permeability, the aspiration capacity, the mineral composition of the mixture and the vertical speed of agglomeration.

The agglomeration process starts when the fuel is lit in the batch and the agglomeration happens in the superficial layer. Through the superficial layer of the agglomerate air is vacuumed from the atmosphere and the burn process is continued.

After all the fuel in the appliance is lit we can differentiate the zones shown in Figure 7.

The burning temperature in the superficial layer in which the fuel's ignition is energized can reach up to 1000-1200grC, after as the ignition zone lowers, the temperature slowly rises up to 1300-1400grC, in the inferior layers [12], [13].

The physical-chemical process that come along with agglomeration are [14], [15]:

- Processes that take place while the matter is gas and solid form during the temperature rise in the agglomeration mix from environmental temperature to the apparition if liquid phases (burning fuel, eliminating humidity, dissociating carbonates, dissociating and oxidating sulphides, decomposing hydroxide and the complex interaction between the components of the agglomeration mix and the resulted product).

- Processes that take place during the liquid-solid-gas phases from the moment the liquid phase appears and until the maximum temperature is reached (the soaking of the material, the formation of the liquid phase and the complex interaction between the three phases).
- Processes that take place during the cooling of the agglomerate between the inspired air and the solid phases, such as the interior of the solid phases from maximum temperature to environmental temperature (the formation of the mineralogical structure of the agglomerate).

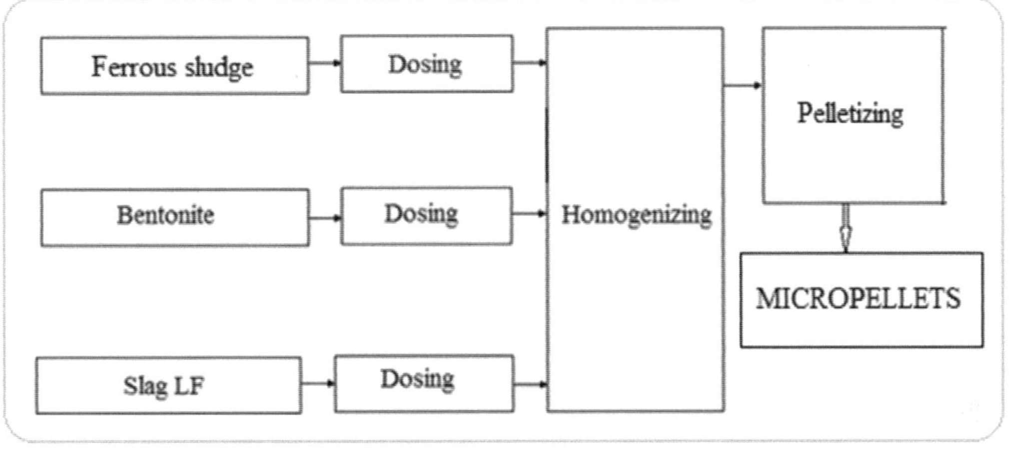

Figure 2. Technological flux of obtaining micropelets

Figure 3. Aspects during laboratory experiments (micropellets)

Figure 4. Micropelets

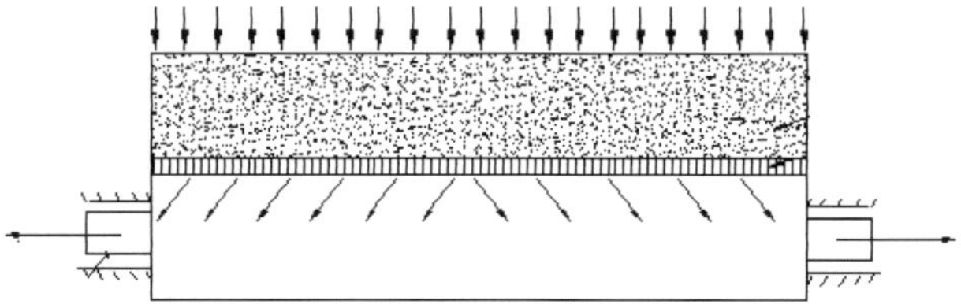

Figure 5. Schematic representation of the agglomeration installation

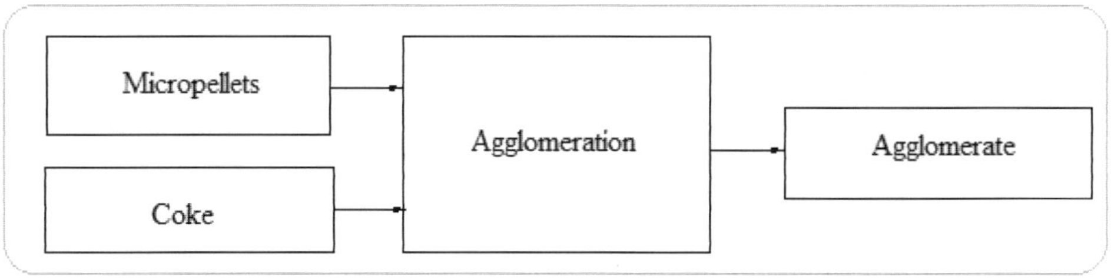

Figure 6. The technologic flux of obtaining agglomerate

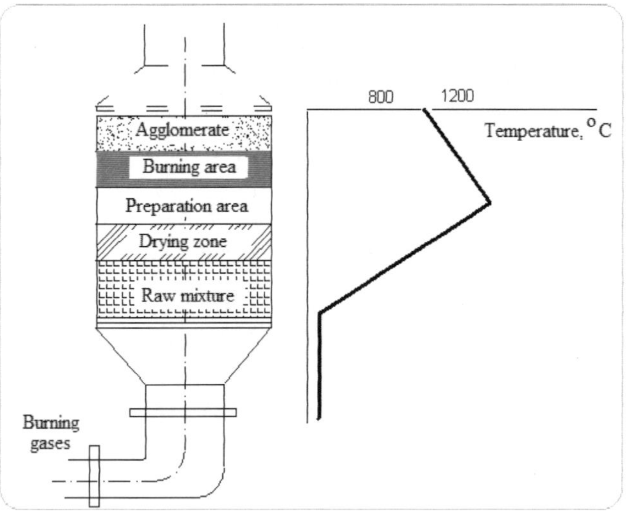

Figure 7. Areas specific to the agglomeration process

These aspects from the time of experiments (agglomeration) are shown in Figure 8 and SEM image are shown in Figures 9-10.

Figure 8. Aspects during laboratory experiments

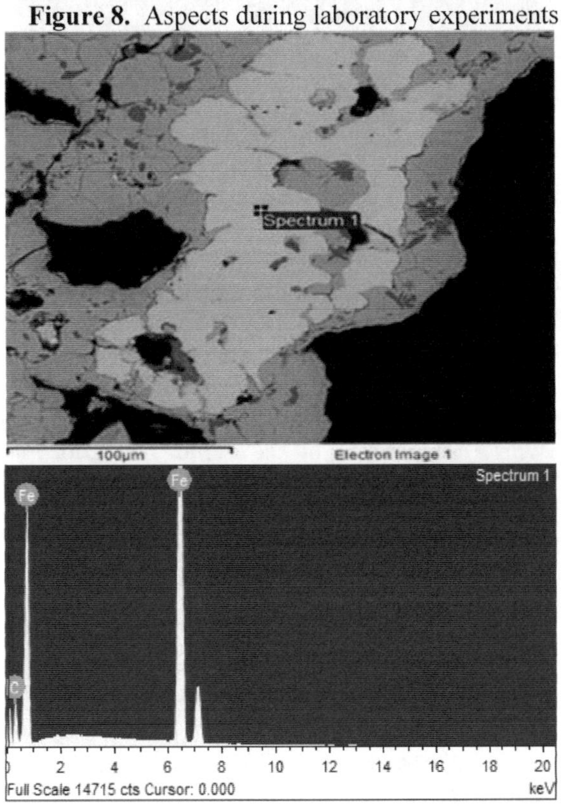

Figure 9. SEM Image (spectrum 1)

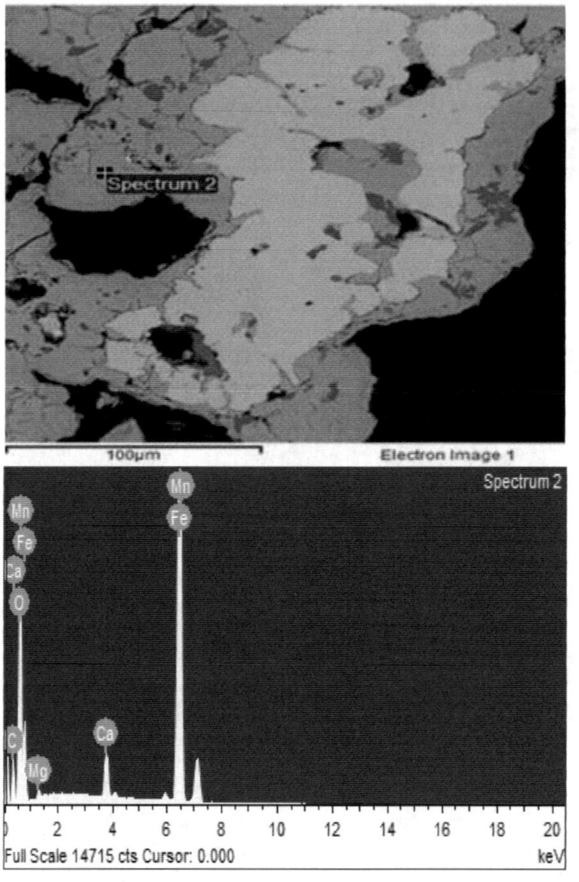

Figure 10. SEM Image (spectrum 2)

The lightly fusible materials that have been formed during the agglomeration process will melt firstly, dissolving with them the remaining compound of the layer. The schemes of the processes that forego the melting of the layer to agglomeration are typical for the agglomerate layer and the type of material that is being agglomerated. Through the agglomeration process a total and partial elimination of the harmful substances from the raw materials is obtained: sulphur, zinc and arsenic are eliminated.

The process of agglomeration, which is generally characterized by the agglomeration speed and the quality of the end product, is influenced by the following factors [11-16]:

- the quality of the prepared raw material that ensures a high level of permeability, which is also influenced by the granulometric composition of the waste, the shape of the granules and their equability, the level of moisture in the batch;
- the quality of the ignition in the batch that is influenced by the nature of the used fuel and adjacent mixture, the quality if the ignition air and the type of combustion appliances used;
- the use of certain influences in the agglomeration mixture;
- the mineral composition of the used waste.

3. Conclusions

Analysing the chemical and granulometric components of ferrous waste from metallic powders plant results show that this waste can be capitalized through recycling.

The procedures through which pulverous ferrous waste can be recycled in the iron industry are: agglomeration, briquetting and pelleting. The first two are recommended for wastes with a varied granulometric componence material.

This makes the experimental procedure easy and cost effective by companies which process and recycle these types of waste.

An important interest is shared in non-conventional recycling of waste in order to obtain a product with high metal iron composite. The experimental agglomeration obtained have a high amount of iron 60%Fe. It is necessary to intensify the capitalization process of these waste for both economical (they represent a source of iron), technological and ecological reasons

Wastes represent, on one hand, a problem for the environment and on the other hand a business opportunity. In order to efficiently solve these problems and also develop a successful business it's obvious that the market for waste reuse and recycle is necessary. This market is created through a simulation of all actors on the waste market and a precise regulation of the charges that fall unto each of them: waste generators, the ones involved in the capitalization and special companies that collect, by-treat, transport and deliver the waste in controlled conditions.

The ecological advantages are: cutting down the amount of powdery wastes by their continuous recycling, the diminishing of soil pollution with metallurgical waste by reducing the dumping areas, the valorization by recycling of these wastes, without any negative impact upon the environment.

References

[1] Costoiu M, Ioana A, Semenescu A, Constantin N, Florea B, Rucai V, Dobrescu C, Polifroni M and Păunescu L 2016 Environmental performance indicators for decision making and stakeholder interests, *Environmental Engineering and Management Journal* **15**(10) 2279-2284

[2] Constantin N, Stanasila O and Stanasila C 2009 Alternative iron making technologies, *Metalurgia International* **14**(7) 5-7

[3] Buzduga R, Constantin N and Ioana A 2018 Solutions to reduce the environmental pollution by the producers of refractories, *University Politehnica of Bucharest Scientific Bulletin, Series B-Chemistry and Materials Science* **80**(1) 231-244

[4] Rucai V, Constantin N and Dobrescu C 2015 Experimental research program regarding the influence of thermo-time treatment of multicomponent ni-base melting on their properties in solid phase, *University Politehnica of Bucharest Scientific Bulletin, Series B-Chemistry and Materials Science* **77**(4) 359-364

[5] ***Project no. 31-098/2007 *Prevention and fighting pollution in the steel making, energetic and mining industrial areas through the recycling of small-size and powdering wastes,* Program PN2, Consortium CO, Responsable Heput T, Beneficiary CNMP Romania

[6] Buzduga R V, Constantin N and Lazar E A 2015 Research on uses plant ashes in processing powders for classical moulding of steel, *IOP Conf. Ser.: Mater. Sci. Eng.* **85** 012008

[7] Andrei V, Hritac M and Constantin N 2017 Experimental research on the behavior of the pneumatic transport of fine-grained iron, *IOP Conf. Ser.: Mater. Sci. Eng.* **163** 012011

[8] Dragomirescu A, Constantin N and Stefan A 2017 Advanced study of thermal behaviour of CSZ comparing with the classic YSZ coating, *IOP Conf. Ser.: Mater. Sci. Eng.* **163** 012040

[9] Ioana A, Constantin N and Dragna E C 2017 EAF optimal managing elements, *IOP Conf. Ser.: Mater. Sci. Eng.* **163** 012004

[10] Andrei V and Constantin N 2015 Experimental research on quality features of metallurgical coke, *IOP Conf. Ser.: Mater. Sci. Eng.* **85** 012004

[11] ***Project no. BC 13/2018 Responsable Ardelean M, Beneficiary Johnson Solution SRL Deva

[12] Branescu E, Blajan A and Constantin N 2015 Experimental research on the characteristics of softening and melting of iron ores as significant factor of influence on gas permeability of blast furnace charge, *IOP Conf. Ser.: Mater. Sci. Eng.* **85** 012006

[13] Butnariu I, Constantin N and Dobrescu C 2018 Research on the Recycling of Pulverulent Waste from the Ferous and Non-Ferrous Industry in Order to Reduced the Pollution, *Revista de*

chimie **69**(5) 1066-1070
[14] Putan V, Putan A, Josan A and Vilceanu L 2016 Researches concerning influence of magnesium, aluminum and titanium lime on steel desulfurization, *IOP Conf. Ser.: Mater. Sci. Eng.* **106** 012033
[15] Birtok-Baneasă C, Ratiu S, Putan V and A Josan 2018 Study of materials used for the thermal protection of the intake system for internal combustion engines, *IOP Conf. Ser.: Mater. Sci. Eng.* **294** 012006
[16] Carlan B A, Constantinescu D and Constantin N 2016 Study regarding the thermal properties of the iron oxides and their application in the process of the heat transfer from the furnace hearth to the ingot, *University Politehnica of Bucharest Scientific Bulletin, Series B-Chemistry and Materials Science* **78**(4) 169-180

Researches for improving mechanical characteristics of high strength weld

L Zgripcea, O Lupu, V Putan and C Birtok Baneasa

Politehnica University of Timisoara, Engineering and Management Department, 5 Revolution Street, Hunedoara, 331128, Romania

E-mail: laurentiu.zgripcea@student.upt.ro

Abstract. The current paper presents application domains of high strength weld, fabrication technologies and tests. High strength weld is used for assembly of high strength low alloyed steel (HSLA) where the same or greater mechanical properties must be obtained in weld line. The method presented in this paper is based on micro alloying inside the root of the weld, with minimum supplementary cost and keeping the same productivity for welding process. The applications of this method are realizations of storage pressure vessels, shields, crane arms and generally, all applications where high strength at a low weight are required.

1. Introduction

Alloy steels known as high-strength low-alloy (HSLA) can assure better strength per weight ratio compared with conventional low-carbon steels with an acceptable, additional price of 24% in average. Because HSLA steels are stronger, they can be used in thinner sections, making them particularly attractive for buildings, bridges and automotives, where weight reduction is important. HSLA steels are available in all standard rolled forms as plates, bars, structural profiles, and special profiles.

Regarding carbon content steel classification, HSLA steels are low-carbon steels with carbon content between 0.05–0.25 % (Table 1). Second important alloy element is mangan up to 1.5 % , strengthened by small additions of elements, such as niobium, vanadium or titanium.

Beside of chemical composition special rolling and cooling techniques are required. Controlled chemical composition improve the microstructure of HSLA steel by reducing the size of ferrite / perlite crystals. This fine precipitate acts as crystallization points which refines grains size and increases the material's strength. Usually, decrease the grains size by 50% double the yield strength. Precipitation strength also plays a minor role [1].

Another point to assure the mechanical properties of HSLA is reduction of phosphor and sulphur inclusion, by addition of calcium during vacuum degassing treatment

Table 1. Chemical composition of HSLA grades [2]

	percentage %									
grade	C	Mn	Nb	Mo	Ti	N	P	S	Si	Al
X80	0.06	1.65	0.03	0.24	0.01	0.01	0.00	0.00	0.00	0.00
HSLA 65	0.06	1.24	0.06	0.01	0.00	0.01	0.01	0.00	0.05	0.04
HSLA 90	0.05	1.65	0.07	0.20	0.02	0.00	0.01	0.00	0.03	0.03

From chemical composition table results the correlation between Niobium (which is the most effective microalloying element) and yield strength. Usually, every 0.01% Niobium addition bring 35 to 40MPa in yield strength.

There are two ways such steel of high strength and high toughness steels are produced:

- By micro-alloying: Adding small amount of alloys which form carbide and nitride crystallization points
- By thermo-mechanical rolling: very accurate control of the rolling temperature and cooling time called thermo-mechanically controlled process (TMCP)

The highest strength is achieved by a combination of the two methods. Final objective is to produce smaller again size as possible, finer grains assure the best yield and impact properties.

The additional benefit is the increased weldability of the material. Fast cooling of steel after welding can induce fragility if hardenability of the steel is high. To minimize this effect, usually carbon content is around limited around 0.05% .

Main effect of increasing the yield strength of low alloyed steel is reduction of steel thickness used in various assembled structures. Although economically is a very positive point, the drawback is always reduction of weld area used in these structures.

Reduction of weld area makes the weld line the weak point in whole structure, weld cannot follow the strength of HSLA steel and from this reason is necessary to redesign the structure, make it again ticker and canceling partially the advantage of HSLA steel.

Several best practices are key to achieving success when welding HSLA steels. Key points are clean material surface, proper preheating and interpass temperatures [3]:

- **Clean material surface**: clean surface is important before welding and between weld layers. Grease, water, dirt or other contaminants can be decomposed by arc temperature and form hydrogen. Steel has a high hydrogen affinity which once entrapped in crystal matrix can induce cracks due to pressure at low temperature. Hydrogen reaction with carbon can amplify the effect because methane resulted has a bigger molecule and can induce higher pressure. It is recommended to remove all contaminants and oxides buy grinding, using abrasive paper. grinding wheel or sand blast gunning machine
- **Proper preheating:** Hardenability effect can be decreased by control of cooling rate. This must be increased to the maximum allowed by welding process. Slower cooling rate has also reduce the content of diffusible hydrogen, allows it to dissipate from weld line and reduces risk of hydrogen cracking.
  Preheating reduces shrinkage stress and forms more favourable microstructures that increase mechanical properties.
  Preheating temperature depends by thickness and carbon content (type) of HSLA steel used. Best procedure is to use equivalent carbon formula which take into account effect of carbon together with other alloy elements. Another important point is to preheat base material on the both side of the weld line in order to assure a uniform thermal pattern inside of weld line and heat affected zone.
- **Maintain interpass temperatures:** This is an important parameter and must be correlated with preheat temperature when HSLA steel is welded.
  Low interpass time can bring too much energy heat which together with preheating energy can alter the grains structure, make them larger. Mechanical properties, especially Charpy impact is reduced.
  Long interpass time together with a poor preheating can bring fast cooling which will increase fragility of the weld.

2. New researches and production trials

On top of best practices regarding welding of HSLA steel, the paper present new methods and technologies implemented experimentally in crane arms production.

Main idea was to improve mechanical characteristics of the fusion zone by microalloying during welding.

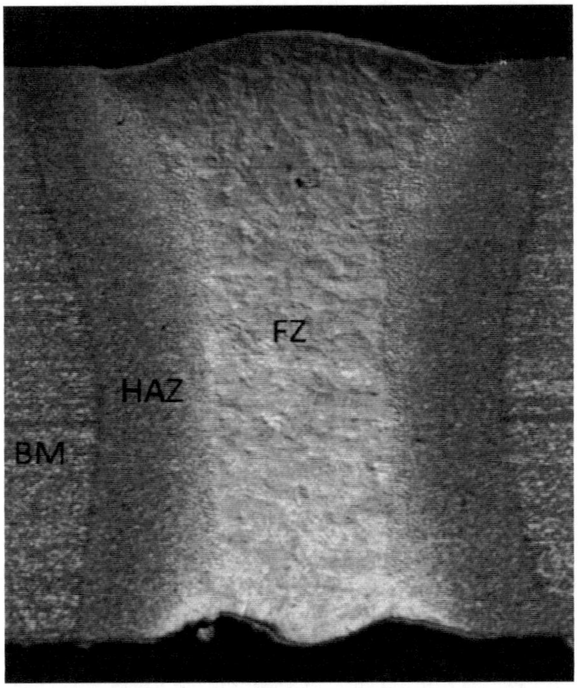

Figure 1. Weld transversal section

In the weld transversal section enhanced by acid attack are visible the 3 main zones (Figure 1):

- Bas material (BM) with fine granulation specific of HSLA steel
- Heat affected zone (HAZ)
- Fusion zone (FZ)

Trials were made using MAG welding technology for plates thickness between 3 and 10mm using an additional wire laid down in the weld root as in Figure 2.

In this case, during plates tack welding (red points) high alloyed wire was inserted (yellow color). Using this procedure only little supplementary maneuver is necessary and most important, does not change assembly technology in production line.

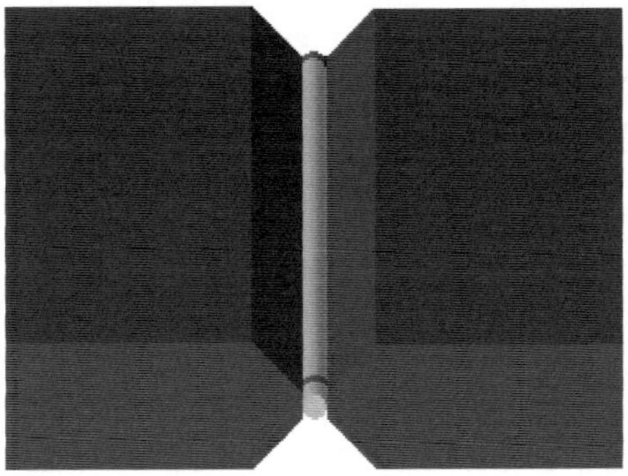

Figure 2. Alloyed wire position and fixation

Dilution of the welding wire is shown in Figure 3.

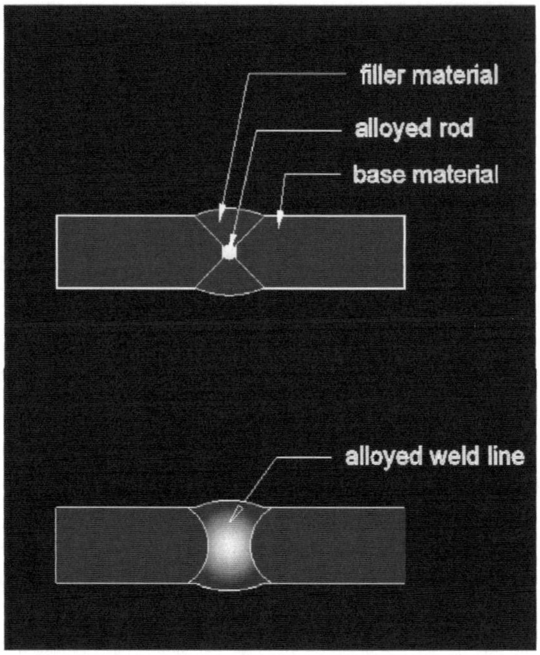

Figure 3. Dilution of alloyed element in fusion zone

Another important point was calculation of alloy element in fusion zone. For this purpose, was developed a calculation methodology and an excel spreadsheet calculation.

As input data, we consider plate thickness (s), alloy wire rod diameter (d) and concentration of alloy element in this wire.

Taking into account effect of various elements in steel we consider for first trial nickel element. Normally, around 1% nickel, provides a range of 60-ksi to 70-ksi tensile strength and has excellent toughness. Further increase of nickel content between 1.1% and 2.75% assure also high impact strength at low temperatures.

Joint was prepared in double V groove an welded on both sides as in Figure 4.

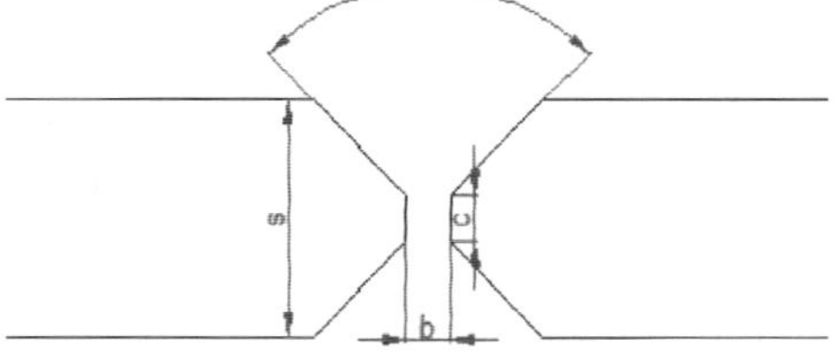

Figure 4. Joint bevel preparation

For calculation of alloying degree, first parameter is the geometrical area (A_g) of the bevel:

$$A_g = b \cdot s + \frac{1}{2}(c-s)^2 \cdot tg\frac{\alpha}{2} \quad (1)$$

For 10mm plate thickness case, Ag =75.4 mm^2

Because of 15% dilution rate (participation of base metal, melted by electrical arc) we define weld area (A_w):

$$A_w = 1.15 \cdot A_g \tag{2}$$

In this case, A_w=86.7mm^2.

In order to change the nickel content in fusion zone, we chose to weld plates with the same thickness (s=10mm) and use a nickel alloyed wire with different diameters, from 0.8mm to 3.2mm.

Regarding welding technology, aim is to keep as much as possible the layout of production line and not to be obliged to invest in expensive welding devices.

Main wire rod used for welding was classical 3GSi1 grade [4], and alloyed rod was Fe-Ni with 51% Nickel.

Welding procedure is metal activ gas (MAG) and sample was weld symmetrically on both sides with parameters [5] presented in Table 2.

Table 2. Welding parameters

Pass from-to	Welding position ISO6947	Pass type	Process	Filler diameter (mm)	Welding current (A)	Welding voltage (V)	Polarity	Wire speed (m/min)	Travel speed (cm/min)	Heat input (kJ/cm)
1	PA	side A	135	1.2	170	20	DC+	3.2	20	8.16
2	PA	side B	135	1.2	170	20	DC+	3.3	20	8.16

As a remark, versus classical weld additional energy is necessary to melt alloyed wire. In this case, if classical weld requires around 6kJ/cm, with alloyed wire the set point is 8.16 kJ/cm so 33% more heat.

Consequent trials with different alloyed wire diameter are presented in Table 3.

Table 3. Alloyed element percentage

wire diameter (mm)	0.8	1.0	1.2	1.6	2.0	2.5	3.2
wire section (mm^2)	0.5	0.8	1.1	2.0	3.1	4.9	8.0
weld area (mm^2)	86.7	86.7	86.7	86.7	86.7	86.7	86.7
Nickel content in alloyed wire (%)	51	51	51	51	51	51	51
Nickel content in weld (%)	0.3	0.5	0.7	1.2	1.8	2.9	4.7

The process is feasible for automation and. In case of high series of production and usage of welding robots, alloyed wire can be added automatically as second wire in the welding torch [4]. In this case, beside of high quality weld a higher productivity can be obtained.

Figure 5. New torch design for automatically alloying

Figure 5 shows a modification of welding torch, by adding second wire.
Regarding electrical regime of alloyed wire is possible to adjust it for different situations:

- For thick plated alloyed wire is equal in diameter with main wire and makes the own electrical arc. This situation is called "tandem" and brings higher productivity.
- For medium thickness plates, alloyed wire is submerged in melted pool and its electrical arc is very short or even does not exist. A controlled current is heating the alloyed wire to compensate partially the energy needed for melting.
- For ultra-thin plate the best results are obtained with "cold" weld. In this case alloyed wire is smaller in diameter than main wire and is only immersed in melted pool.

3. Conclusions

Trials were made with same rod and different diameter.

- From yield point of view, nickel micro alloying improved weld fusion line. All samples during traction test did not breake in weld zone.
- Resilience of weld was improved almost in all cases. Best results were obtain with 1.6mm rod when fusion line has 1.2% Nickel and resilience improved by 18%. Increasing nickel content to 4.7% decreased drastically weld resilience and shows a fragile rupture (Figure 6).

Ni 1.2% Ni 4.7%

Figure 6. Resilience test

- Trials were made at 20 Celsius, but as an observation for negative temperature higher nickel content gives better resilience versus low nickel content.
- Nickel microalloy does not affect the weldability of HSLA steel [3], [7]. The measure for weldability is given by carbon equivalent formula and is satisfactory up to values of 0.35% [8], [9]:

$$CE\ (IIW) = C + Mn/6 + (Cr + Mo + V)/5 + (Ni + Cu)/15$$

As we observe in CE formula, Ni factor is divided by 15 and is minimal as influence for total CE.

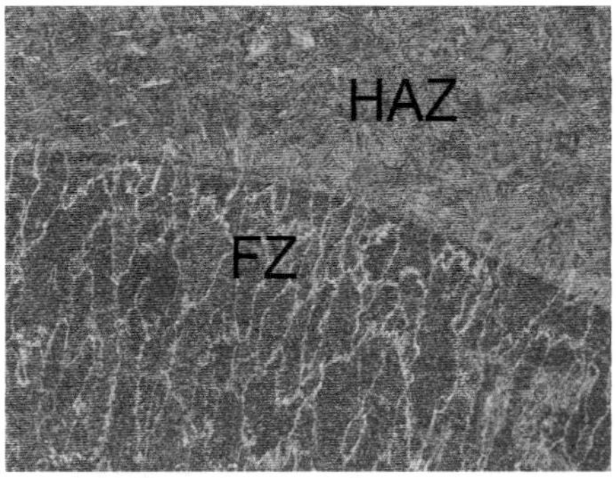

Figure 7. Microstructure, x100

- On microscopic structural analysis, fusion zone was carefully assessed. Structure is ferrite - perlite with acicular grains, fine enough to achieve better mechanical properties [1], (Figure 7).
- Weld micro alloying is suitable for industrial production. Alloyed wire is welded in points which usually are done during parts assembling, so no additional time is required
- Welding procedure modification is minimal, and alloy calculation can be done easy using excel files
- Method permit dosage of alloys in a large range, which is not possible for example by changing of filler material
- Dosage of alloys can target independently mechanic characteristics as resilience, toughness, behavior at low or high temperatures, or weldability
- Added cost with alloyed wire is minimal
- Method does not required changing of welding technology or procedures, so investments are minimal

References

[1] Lambert A, Drillet J, Gourgues AF, Stuel T and Pineau A 2000 Microstructure of martensite and austenite constituents in heat affected zones of high strength low alloy steel welds in relation totoughness properties, *Sci Technol Weld Join* **5**(3) 168e73
[2] ***ASTM A568 Standard specification for Steel, Sheet, Carbon, Structural and High-Strength General Requirements
[3] Omajene J, Martikainen J and Kah P 2014 Weldability of Thermo Mechanically Rolled Steels Used in Oil and Gas Offshore Structures, *The International Journal of Engineering and Science* 63-69
[4] Peters D J 1989 *Submerged arc welding consumables for HSLA-100 steel*, Massachusetts Institute of Technology
[5] Dunder M, Ivandic Z and Samardzic I 2008 Selection of arc welding parameters of micro alloyed steel, Metalurgia **47**(4) 325-330
[6] Buzduga R V, Constantin N and Lazar E A 2015 Research on uses plant ashes in processing powders for classical moulding of steel, *IOP Conf. Ser.: Mater. Sci. Eng.* **85** 012008
[7] Ragu Nathan S, Balasubramanian V S, Malarvizhi S and Rao A G 2015 Effect of welding processes on mechanical characteristics of high strength low alloy naval grade steel joints, *Defence Technology* **11** 308-317
[8] ***Equivalent carbon calculation, Atlas Steels, 2013
[9] Rucai V, Constantin N and Dobrescu C 2015 Experimental research program regarding the influence of thermo-time treatment of multicomponent ni-base melting on their properties in solid phase, *University Politehnica of Bucharest Scientific Bulletin, Series B-Chemistry and Materials Science* **77**(4) 359-364

Friction and wear characteristic of organic brake pads material

C Pinca-Bretotean, A L Craciun, A Josan and M Ardelean

Politehnica University of Timisoara, Engineering and Management Department, 5 Revolutiei Street, 331128 Hunedoara, Romania

E-mail: camelia.bretotean@fih.upt.ro

Abstract. Many parts of motor vehicles contribute actively or passively to increasing operation performance. The safety is closely related to the effectiveness of the braking system. Due to the fact that the automotive industry is a branch of economics that uses products obtained in almost all other industries, the current concerns are directed towards the application of the latest research in the field of modern materials and technologies in all fields involved in the manufacture of motor vehicles. The proposed and carried of this paper aimed to presented, obtained, characterize and test some organic composite materials, designed to make brake pads for the braking systems of small cars with medium performances. The experimental research strategy included the determination of the functional characteristics of organic composite materials developed. In this respect, experimental laboratory determinations were carried out on the friction and wear of composite materials produced, using standardized test methods. The knowledge of these elements provides the possibility of pertinent assessments on the durability in operation of the composite material used for the manufacture of brake pads.

1. Introduction

Brake friction materials are needed for aircrafts, helicopters and passagers cars [1-3]. They can be classified into semi-metalic, metallic, organic and ceramic types [3].

Over the time, the most used brake pads were those based on asbestos fibres [4]. Studies shown that asbestos fibres have harmful effects both on the environment and on human health [4], [5]. For this reason, it has been decided to ban this reinforcement fibre in the friction materials intended to make the brake pads. This created a scientific and technical problem that has caused by replacement of asbestos with other eco-friendly materials [6-9]. Paper [10] lists several materials that can replace asbestos, but none has the same properties as this one, but it offers some features with similar performance. The current trend in scientific research in the field of motor vehicles is to use composite materials with natural fibres because they offer numerous economic and environmental benefits [7].

The organic friction material for brake pads is a heterogeneous material and is composed of several elements. Each of these contributes to improving the friction properties at high temperatures, to increasing strength and stiffness, to prolong life, reduced porosity and noise. Ideally, the friction material for making the brake pads must have stable friction coefficient under different operating conditions such as: applied loads, temperature, speeds, braking mode in dry or wet conditions, so as to maintain the braking characteristics of the vehicles [4]. Also, besides the previously mentioned properties, the friction material must have: low wear rate, high thermal stability, low noise and not damaging the brake disc. From a practical point of view it is impossible for all these properties to be met. Therefore, some requirements need to be mitigated to meet others. In general, each recipe of friction material has its own unique friction behaviour and wear-resistant features [4], [11].

At present, different types of fibers are used in making brake pads, such as: calcium silicate, mineral fiber, ceramic, carbon, cellulose, fiberglass, vermiculite (aluminum silicate hydrated calcium), aluminum silicate, basalt fiber, steel or copper fibers polyacrylonitrile polyester fibers, etc., [4], [7], [10], [11]. Each of them has advantages and disadvantages, such as: braking capacity and dissipation of heat, braking behavior, noise and vibration, but the final properties of each type of brake pad depend on the abrasive used in their composition [6], [11].

A good number of studies has been done on the organic composites materials, [2-11]. Organic fibers used to reinforce composite materials can be obtained from a wide variety of plants: strains of cereals (cane, barley, rice), textile plants (flax, hemp, raffia, cotton), other plants with industrial uses (sugar cane, reed, sisal, etc. [6]. A big disadvantage of organic fibres is the differences in anatomical and chemical structure, which determines physico-chemical properties with large variations from one species to another, sometimes even within the same species due to vegetation and microclimate conditions [6]. On the other side, technical literature showed that there is no simple correlation between the frictional physical-mechanical properties of the friction material [9], [12]. Therefore, each newly developed composite material recipe developed must be subjected to a series of tests to assess the friction and wear properties [9], [12].

The proposed and carried of this paper aimed to presented, obtained, characterize and test some organic composite materials, designed to make brake pads for small cars with medium performances. The experimental research strategy included the determination of the functional characteristics of organic material developed. In this respect, experimental laboratory determinations were carried out on the friction and wear of composite materials produced, using standardized test methods. The knowledge of these elements provides the possibility of pertinent assessments on the durability in operation of the composite materials used for the manufacture of brake pads.

2. Developed and production in laboratory stage of composites materials

In order to establish the recipes was taken a mixture of raw materials, comprising eight components. Their choice was made in according to the destination of the brake pads (for small vehicles and average performance) and to the scientific literature. In order to establish prescriptions was started from a mixture of raw materials, including: aluminium, graphite, zirconium oxide, silicon carbide, titanium oxide, phenolic resin and coconut fibre. The recipes for composites materials produced in laboratory stage are shown in Table 1.

Table 1. Chemical composition of composite materials

Sample	Aluminium (%)	Graphite (%)	Zirconia oxide (%)	Silicon carbide (%)	Titan oxide (%)	Phenolic resine (%)	Hexametyl tetramine (%)	Coconut fibre (%)
C1	20	5	2	11	11	40	6	5
C2	15	5	2	11	11	40	6	10

The production of organic composite materials for brake pads is based on powder metallurgy [3], [5], [7], [11]. The main reason for using powder metallurgy is to obtain an increased homogeneity of composite materials [4]. The scheme for the production, characterization and testing of composite materials produced in the laboratory is shown in Figure 1.

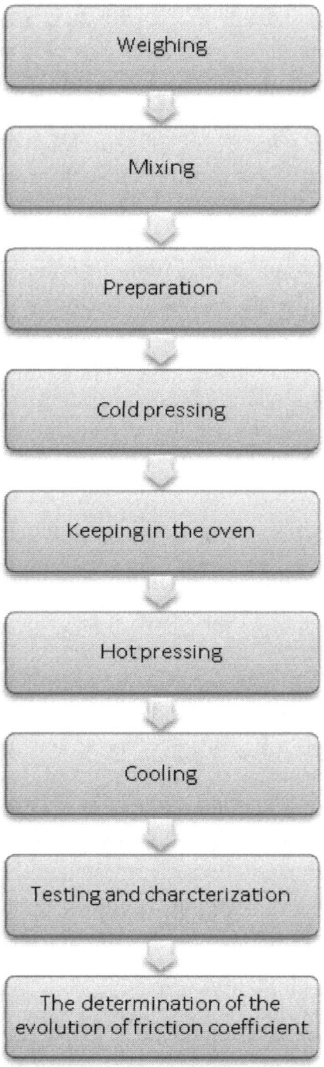

Figure 1. Scheme for laboratory production of composite materials

Parameters of sintering technology have been established as a result of several experimental tests, which allowed for critical analysis and successive improvements. Parameters of the sintering process are presented in Table 2.

Table 2. Parameters of sintering process

Heating temperature (^{0}C)	Time in oven (min)	Pressing force (KN)	Cooling			
			In oven		In air	
			Temperature (^{0}C)	Time (min)	Temperature (^{0}C)	Time (min)
200	60	20	100	480	25	2880

In order to characterize and test composite materials produced in laboratory stage, were made samples to determine physical-mechanical and functional characteristics of these friction materials.

3. Discusion and results

For composite materials developed and produced in laboratory stage were determined density, oil and water porosity, hardness HRS, Young modulus and compressive strength. The results are presented in Table 3.

Table 3. Physico-mechanical characteristics of organic friction material

Samples	Density (g cm -3)	Porosity(%)		Hardness HRS	Breaking force (N)	Young modulus (N mm^{-2})	Compresive strength (N mm^{-2})
		In oil	in water				
C1	1.57	0.29	0.75	59	12887	262999.142	-27.6141
C2	1.35	0.16	0.55	61	15766	32537.454	-38.7788

Table 3 shown that density values for composite materials are relatively low compared to metal densities. On the other hand, the density of the samples decreased with the increase in the amount of coconut fibre. In the literature there have also been produced friction materials with coconut fibre [7], [13], and the conclusions obtained are similar.

As for the water and oil absorption capacity, they decrease with the increase in the amount of coconut fibre. The reduced water and oil absorption capacity can be attributed to the increase in the bond between the binder and the filler, due to proper homogenization of the constituents. The variable content of coconut and aluminium fibres determines different hardness values of sample. Thus, it is observed that the highest value of the hardness was obtained for C2.This sample being made with 10% coconut fibre [13]. Table 3 shows that the compressive strength of composite C1 is lower than C2 sample.

The sample with the largest amount of coconut fibre (10%) has the highest compressive strength, respectively the highest value of the request effort. This is explained by the fact that the coconut fibre particles were evenly distributed over the aluminium matrix, making a corresponding interference between them [8], [9]. The compressive strength and the Young modulus of the composite materials increase with an increase in the amount of coconut fibre. Similar conclusions being reached also in paper [3].

In order to test composite materials, were performed experimental determinations which had the objective to study the influence of some parameters of the working regime (pushing force, slip speed, test time, friction mode) on the functional characteristics of composite materials and study the evolution of friction coefficient according to the pressing force. The knowledge of these elements provides the possibility of pertinent assessments on the durability in operation of the composite materials used for the manufacture of brake pads.

The study of the influence of some parameters of the working regime will be carried out on an experimental plant. The principle of this consists in pressing a pin made of composite materials (C1, C2) on a rotary disc made of cast iron, in order to determine wear characteristics [5], [13]. The applied load simulates the contact pressure between the friction couplers.

For the experiments, were made cylindrical samples (pins) from composite materials (C1-C2) and a cast iron disc made in laboratory assimilated to the G 2500, according to ASTM A 159 norms.

The disk and pins used in the experiments are shown in Figure 2.

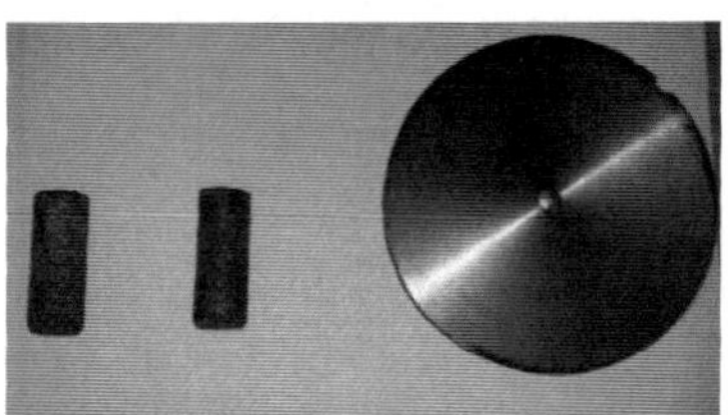

Figure 2. Iron disc and pins using in experimental determination

The main parameters used in these experiments are presented in Table 4.

Table 4. Parameters of experimental determination

Sliding speed ($m\ s^{-1}$)	Test range (mm)	Length of test (m)	Test time (min)
3.92	25	2000	8.5

The mass and dimensional characteristics of cast iron discs and composite pins at the beginning and to the end of experiments are presented in Table 5.

Table 5. Dimensional characteristics of samples

No.	Samples	Initial weight of samples (g)	Final weight of samples (g)	Initial height of samples (mm)	Final height of samples (mm)
1	Cast iron S1	50.8934	50.6214	-	-
2	C1	7.8657	6.3243	27.5	18.4
3	C2	7.2317	6.7532	26.4	17.8

The wear parameters for cast iron samples and composite samples are shown in Table 6.

Table 6. Wear parameters for cast iron and composite pins

No.	Samples	Wear mass (g)	Wear intensity ($g\ m^{-1}$)	Relative durability to wear
1	Cast iron S1	0,272	0.000136	-
2	C1	1.5414	0.0007707	0.1764
3	C2	0.4785	0.0002394	0.1839

Table 6, shown that the highest wear intensity was obtained for the composite C1, and the smallest for C2 composite. Also, the highest relative durability of wear was obtained for the composite containing 10% coconut fibre (C2).

During experiments, sample C2 was timed and weighed at intervals and lengths the results obtained are in Table 7, [13].

Table 7. Experimental results on variation in mass and height of sample C2

Time (min)	Length route (m)	Mass (g)	Absolute weight reduction (g)	Relative weight reduction (%)	Heigh (mm)	Absolute heigh reduction	Absolute heigh reduction
0	0	7.2317	0	0	26.4	0	0
2.12	500	6.7323	0.4994	6.9057	23.1	3.30	12.5
4.25	1000	6.0819	1.1498	15.8994	21.0	5.40	20.45
6.37	1500	5.9122	1.3195	18.2460	19.4	7.04	26.66
8.5	2000	5.7532	1.4785	20.4447	17.8	8.60	32.57

In order to establish simple correlation equations, the results presented in Table 7 were processed in Excel program and the graphs being presented in Figures 3,4 and 5.

IOP Conf. Series: Materials Science and Engineering **477** (2019) 012009 doi:10.1088/1757-899X/477/1/012009

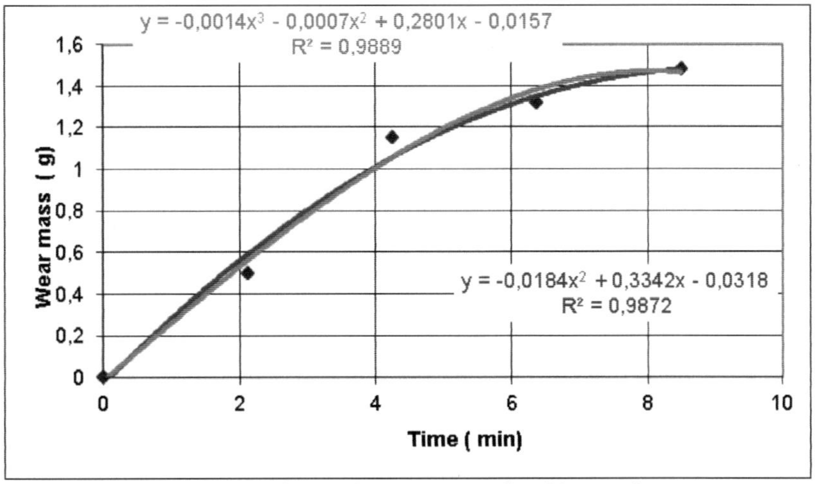

Figure 3. Wear mass for C2 sample

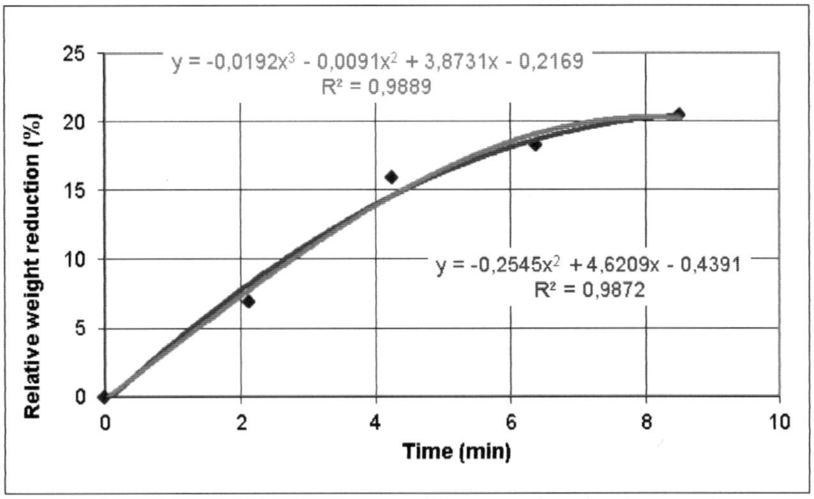

Figure 4. Relative weight reduction for C2 sample

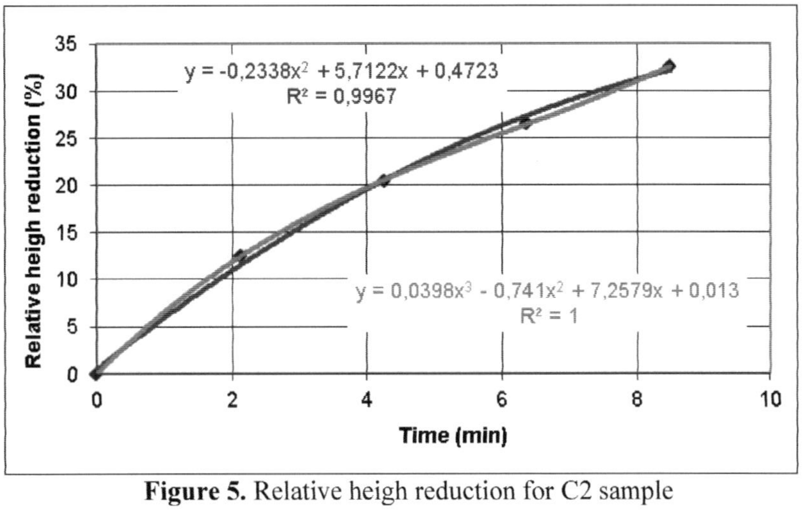

Figure 5. Relative heigh reduction for C2 sample

Analyzing the graphs presented in Figures 3, 4 and 5 it is observed that significant results were obtained from the correlation of coefficients [13].

The determination of the evolution of friction coefficient was performed on TR-20 test equipment, whose principle of operation is based on the "pin-on-disc" method. The pin equipment is a steel ball with 6 mm diameter and the experimentation regime is dry friction. For experiments the friction material was made of parallelepiped shape with dimensions 45x45x50mm. Each test was performed at a radius of 25 mm from the symmetry axis of the equipment. They were subjected to alcohol purging, then dried and weighed with a digital balance accurate to ± 0,01 mg. In the test, measurements were made for three different pressing forces (5N, 10N, and 15N) at each test weighing the samples before and after the test to determine the wear mass. Test parameters use in tests are presented in Table 8.

Table 8. Parameters for determining the friction coefficient

Wear trace diameter mm	Speed rot min^{-1}	Test time h	Test distance m
25	150	5	2200

The experimental data for the three values of the test results obtained in the text file in the tribome computer software are presented in Table 9.

Table 9. The wear and friction coefficient obtained in experiments

Samples	Friction coefficient			Wear (g)		
	F_1= 5N	F_2 = 10N	F_3 =15N	F_1= 5N	F_2 = 10N	F_3 =15N
C1	0.43	0.33	0.30	0.19	0.15	0.13
C2	0.50	0.45	0.39	0.07	0.05	0.03

At low pressing loads, friction coefficients have high values and as the workload increases, friction coefficient values decrease. Their values during the experiments are kept between 0.3 and 0.5. Technical literature states values of friction coefficients for friction materials from 0.15 to 0.6. Practically, for most vehicles the nominal values of friction coefficients range between 0.3-0.6 [4]. The values obtained for the coefficients of friction in the experiments are confirmed by specialized literature.

In Figure 6 was studied the effect of the pressing force on the friction coefficient and Figure 7 illustrates the influence of the pressing force on sample wear.

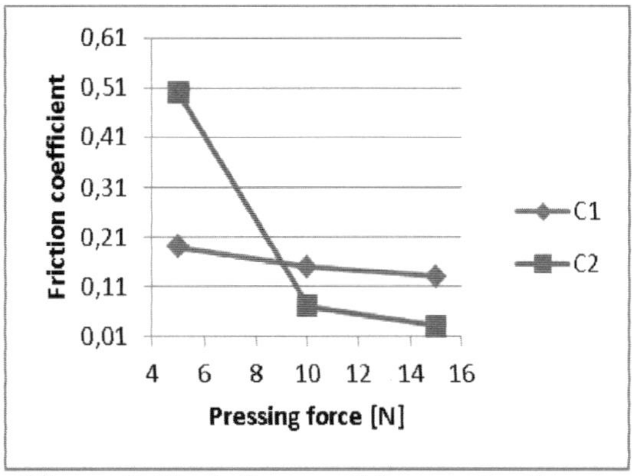

Figure 6. The influence of the pressing force on the friction coefficient

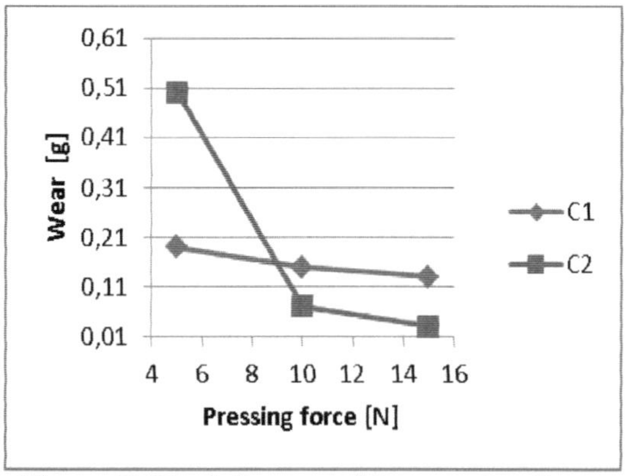

Figure 7. The influence of the pressing force on sample wear

Figure 6 shows that the friction coefficients decrease with the increase of the pressing force and Figure 7 shows that wear decreases with increased the pressing force. Also for sample C2 is observed the most pronounced decrease of the friction coefficient with the increase of the applied.

4. Conclusions

From the results and discussions it can be observed:

- organic composite materials are heterogeneous materials, composed of several components with different properties, therefore choosing the constituents and their percentage in recipes will significantly affect the behaviour in operation;
- physical-mechanical properties of composite materials with natural fibres depend on: behaviour, concentration, distribution and orientation of natural fibres;
- density, porosity, hardness, compressive strength, and Young modulus increase with the increase in coconut fibre;
- structural integrity of composite materials offers superior physical-mechanical characteristics of frictional materials;
- the highest wear resistance is obtained with 10% coconut fiber;
- cast iron disc wear is much less than the wear of composite materials, which shows that the wear resistance does not depend much on the amount of metal [13], this is very important for the practical situation because the brake pads are easier to replace than the disc;
- the best functional characteristics has C2 sample made with 10% organic fiber.

References

[1] Gieras J F, Podratzky A and Bansemir H 2007 Design and experimental characterization of modern helicopters rotor brakes, *Aero Sci Technol* **11** 360–369

[2] Atzeni E and Iuliano L 2008 Experimental study on grinding of asintered friction material, *J Mater Proc Technol* **196** 184–189

[3] Suruko M, Marwoto and Rustad S 2010 The mechanical properties of carbonized coconut char powdet-based friction materials, *Carbon* **48** 3616-3620

[4] Blau P J 2001 *Compositions, functions and testing of friction brake materials and their additives*, Contract DE-AC05-00OR22725, Oak Ridge, Tennesse, USA

[5] Crăciun A L, Pinca-Bretotean C, Uțu D and Josan A 2017 Tribological properties of nonasbestos brake pad material by using coconut fiber, *IOP Conf. Ser.:Mater. Sci. Eng.* **163** 012014

[6] Terciu O, Curtu I, Cerbu I and Stanciu M D 2012 Traction testing of lignocellulosic composites with applications in the automotive industry, *Buletinul AGIR* **1**(1) 9-16

[7] Maleque M A, Atiqah A, Talib R J and Zahurin H 2012 New natural fibre reinforced aluminium composite for automotive brake pad, *International Journal of Mechanical and Materials Engineering* **7**(2) 166-170

[8] Leman Z, Sapuan S M, Saifol S M, Maleque A M and Ahmad MM 2008 Mosture absorption behavior of sugar palm fibre reinforced apoxy composites, *International Journal of Materials and Design* **29**(8) 1666-1670

[9] Park J H, Chung J O and Kim H R 2010 Friction characteristics of brake pads with aramid fiber and acrylic fiber, *Industrial Lubrication and Tribology* **62**(2) 91–98

[10] Nicholson 1995 *Facts About Friction*, P&W Price Enterprises, Inc., Croydon, PA

[11] Vijay R, Jees J M, Saibalaji M A and Thiyagarajan V 2013 Optimization of Tribological Properties of Nonasbestos Brake Pad Material by Using Steel Wool, *Advances in Tribology* **2013** 165859

[12] Todorovic J 1987 Modelling of the tribological properties of friction materials used in motor vehicles brakes, *Proc. Instn. Mech. Engrs* **226** (C) 911-916

[13] Crăciun A L 2018 *Research on the use of composite materials in vehicle braking systems,* University Politehnica Timișoara, Romania, Doctoral Thesis

IOP Conf. Series: Materials Science and Engineering **477** (2019) 012010 doi:10.1088/1757-899X/477/1/012010

Researches on the influence of slag addition and secondary treatment duration on steel quality

E Ardelean, M Ardelean, A Socalici, V Puţan and A Puţan

Politehnica University of Timisoara, Engineering and Management Department, 5 Revolution Street, Hunedoara, 331128, Romania

E-mail: erika.ardelean@fih.upt.ro

Abstract. Steel physical, mechanical and usage characteristics are influenced by its elaboration, secondary treatment, casting and further treatment. But if elaboration and secondary treatment do not benefit from the appropriate qualitative technological steps, it will not be possible to correct the resulting fails during the casting and processing stages.
This paper introduces some laboratory researches related to the way steel quality can be influenced (in terms of sulphur and oxygen content) by the variation of the secondary treatment duration (inside the ladle) and that of the quantity of the reducing mixture in use (depending on the targeted metallurgical effect). The result consists in recommendable optimal variation intervals for the two parameters; these results can be easily implemented in industrial practice and therefore, with minimum costs, one can obtain maximal technological and economical effects.

1. Introduction

In order to obtain a high quality steel one should pay particular attention to the elaboration process, starting from the furnace load composition and going on through the economically appropriate means of melting, reducing to a minimum the alloy elements loss [1]. After melting and phosphorous oxide slag removal, the refining process makes for the increase of steel purity, by reducing the oxides resulted from the oxidizing of the accompanying element. Deoxidizing and desulphurization, operations that precede the refining stage, are meant to remove two of the elements that can damage steel quality: oxygen and sulphur [2].

Oxygen negatively affects quality by:

- causing hot steel brittleness, which makes it likely to overheat and leads to the appearance of cracks during hot plastic deformation, as a result of the oxide and oxide-sulphide deposits at the limits of the structural grains;

- the formation of blowholes during the solidification of the continuous cast ingots or semi-finished parts [2].

As a result, the content of oxygen dissolved in the steel may not exceed the limits of 40-80ppm (for calming steels).

Sulphur can cause red brittleness under 988°C (respectively 960°C) – when the products become brittle, or hot embrittlement, over 988°C (respectively 960°C), when sulphur rich products are melting and suppressing metallic continuity [3].

The contents of sulphur and oxygen can be cut down inside the elaboration aggregate, but not up to the prescribed standard limits. That is why modern technologies envisage advanced processing outside the elaboration aggregate [4], by means of synthetic slag [5], inert gas bubbling [6], [7], electromagnetic stirring and/or vacuum treatment [8], [9].

The paper intends to go deeper into the influence of synthetic slag addition [10] and duration of secondary treatment upon the rate of oxygen and sulphur removal from the steel meant for pipe production.

2. Laboratory experiments

The experiments have been done on steel elaborated in a 10kg electric induction furnace, during 16 charges. The charges consisted in pipe steel waste, more exactly samples from the steel load determination related to pipe steel elaboration in EBT-type electric arc furnaces [1].

After melting determination, steel was deoxidized by precipitation with FeMn (25g), FeSi (30g), SiMn (50g) and carburizing graphite (10g) after which, the reducing mixture was added. This was introduced into a metal plate cartridge, attached to a wooden pole and immersed down to the bottom of the furnace [1].

It was considered that after the deoxidizing stage, the furnace worked in a ladle-furnace regime (similar to LF). Steel casting was done into 2kg ingots, the casting scheme being shown in Table 1.

Table 1. Experimental ingot casting, depending on the duration of the secondary treatment

Charge	Duration of the secondary treatment with reducing mixture/min										
	4	6	8	10	11	12	13	14	15	16	17
1			X	X		X		X			
2			X		X			X			X
3	X		X			X				X	
4		X	X				X		X		
Total ingots	1	1	4	1	1	2	1	2	1	1	1

At the end of the melting process, we sampled the steel in order to determine its chemical composition (the content of sulphur and oxygen is mandatory); another sampling was done before casting, which allowed the determination of the sulphur, respectively oxygen removal extent. On treatment completion, slag was also sampled in order to determine its chemical composition, particularly the content of CaO.

During tapping, respectively during ingot casting, ferrous alloys were added to the jet, as follows: FeMn (8g), SiMn (8g), FeSi (8g), aluminum (6g), SiCa (8g), respectively for one ingot: FeMn (2g), SiMn (2g), FeSi (2g), aluminum (1.5g), SiCa (2g).

This experimental method was chosen in order to allow the practical implementation of the results for pipe steel elaboration in EBT-type electric furnaces; the steel was treated in an LF-type installation and was continuously cast.

3. Results, discussions

The resulting experimental data have been processed in EXCEL, in order to obtain the influence of the duration of treatment with synthetic slag and the specific reducing mixture added, upon the desulphurization and deoxidizing rate, for various contents of CaO into the slag (with various chemical compositions). Further on, the results of data processing meant to give the simple correlations are given in analytical and graphical form (Figures 1-4).

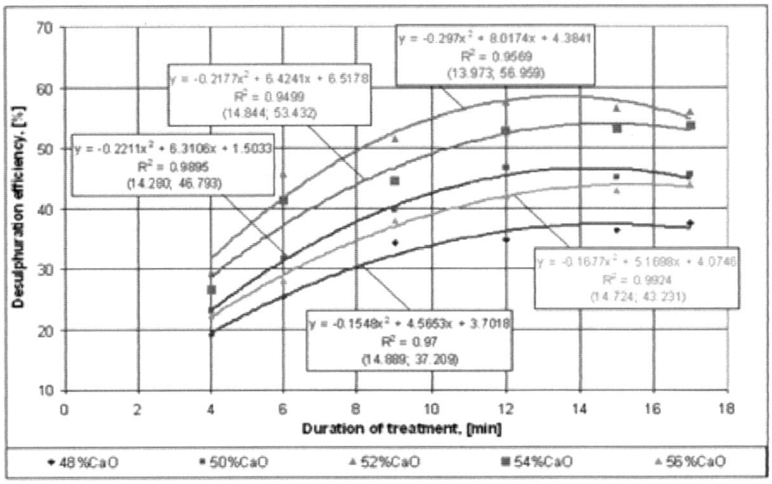

Figure 1. The influence of treatment duration upon the desulphurization rate

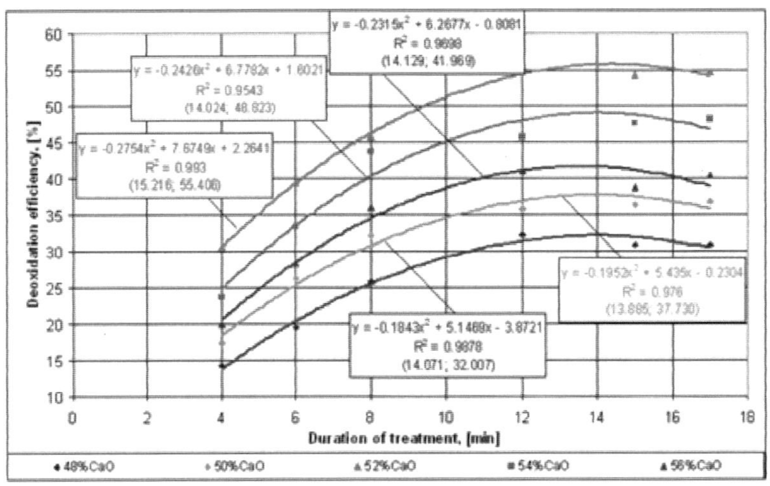

Figure 2. The influence of treatment duration upon the deoxidizing rate

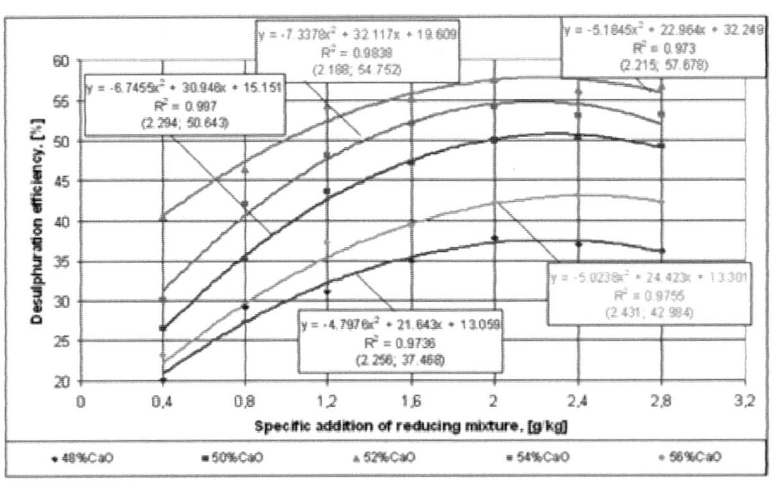

Figure 3. The influence of quantity of reducing mixture upon the desulphurization rate

IOP Conf. Series: Materials Science and Engineering **477** (2019) 012010 doi:10.1088/1757-899X/477/1/012010

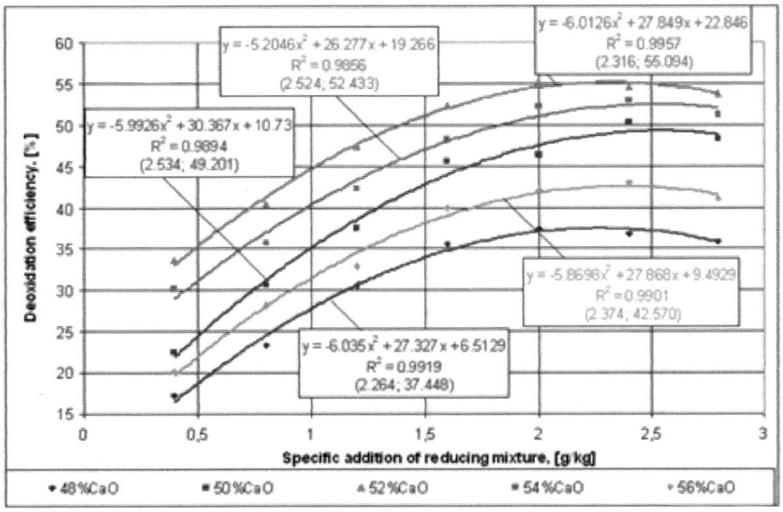

Figure 4. The influence of quantity of reducing mixture upon the deoxidizing rate

A global analysis of the resulting correlations shown above leads to the following results:

- for the cases under analysis, the 2nd degree polynomial correlations were studied, as it was considered that they are the most significant, since they have a maximum point, their coordinates being given for each equation describing the curve;
- both the desulphurization (η_S) and deoxidizing (η_O) output are directly proportional to the synthetic slag treatment duration (t) and also to the specific addition of reducing mixture (q) meant to form the synthetic slag;
- in all cases, a maximum point was obtained for each correlation; its coordinates range within the technological variation limits of the parameters under analysis;
- as to the duration of the treatment, it was found that the best values of the desulphurization, respectively deoxidizing output are obtained if the duration of this technological operation is ranging between 13.75-15.25 min (interval in which the output maximal values are reached: (13.973; 56.959) for η_S and (15.216; 55.406) for η_O);
- for the same duration of the treatment, the desulphurization and deoxidizing output values are also influenced by the content of CaO in the slag, as follows: the best values are obtained for 52%CaO, followed by 54%CaO and the lowest values, for 48%CaO;
- as to the specific addition of reducing mixture (g/kg of steel) it was found that the best values for the desulphurization and deoxidizing output are obtained if it ranges within the limits 2.0–2.5 (interval in which the maximal output value is reached);
- for the same reducing addition, the value of the desulphurization and deoxidizing output is also influenced by the content of CaO in the slag, as follows: the best values are obtained for 52%CaO, followed by 54%CaO and lower values for 48%CaO.

The conditions shown above can be observed technologically and η_S, respectively η_O can be granted values within the interval 30-58%.

The data processed in Excel were also processed in Matlab in view of obtaining the double, 1st, 2nd, 3rd and 4th degree correlations, the results being shown in Figures 5-20. After having processed the results, they will be analysed technologically.

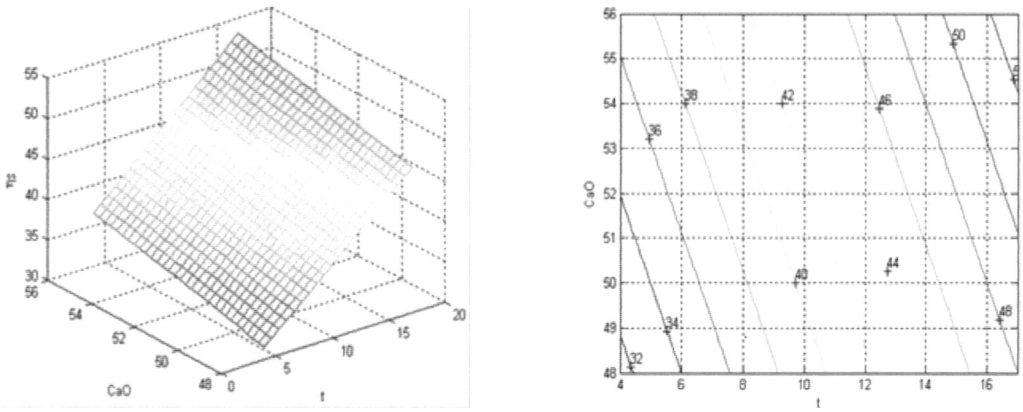

Figure 5. Correlation $\eta_S = f(t; \%CaO)$ – 1st degree

The correlation equation: $\eta_S = 1.271 \cdot t + 0.641 \cdot (\%CaO) - 4.397$

The correlation coefficient: $R_1 = 0.592$; Deviation: $S_1 = 7.984$

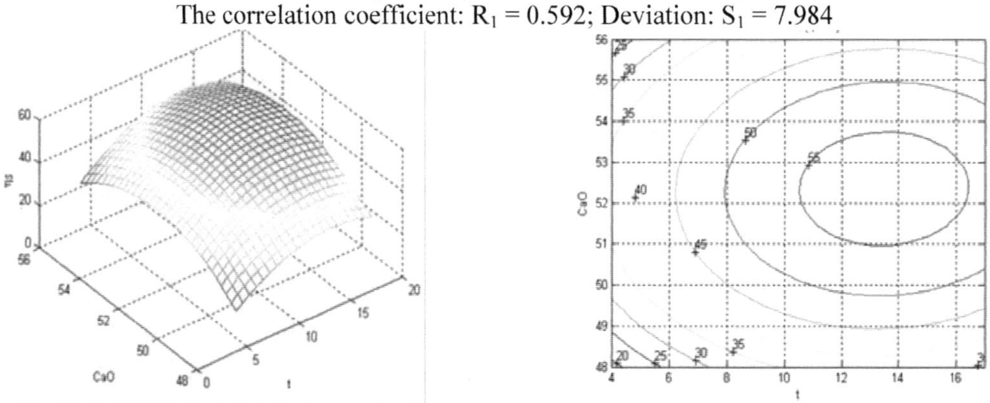

Figure 6. Correlation $\eta_S = f(t; \%CaO)$ – 2nd degree

The correlation equation: $\eta_S = -0.227 \cdot t^2 + 0.031 \cdot t \cdot (\%CaO) - 1.026 \cdot (\%CaO)^2 + 4.468 \cdot t + 106.998 \cdot (\%CaO) - 2773.921$

The correlation coefficient: $R_2 = 0.983$; Deviation $S_2 = 1.814$

Maximum point deviation: (%CaO) = 52.351%; t = 13.620 min; $\eta_S = 57.231\%$

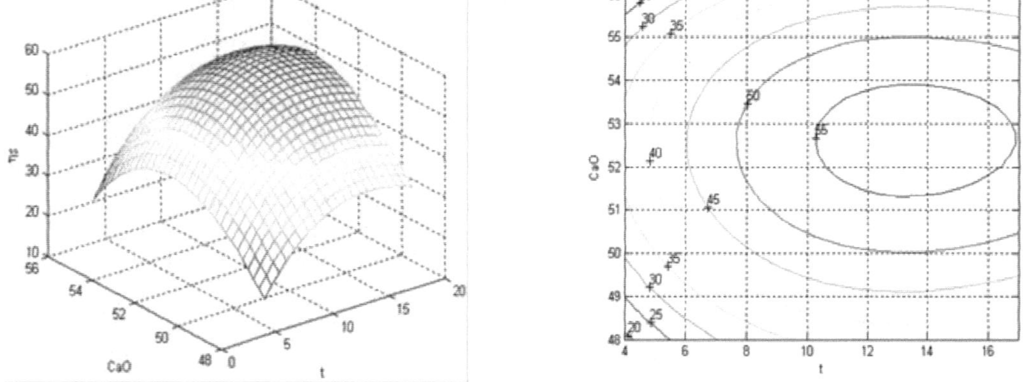

Figure 7. Correlation $\eta_S = f(t; \%CaO)$ – 3rd degree

The correlation equation: $\eta_S = 0{,}006 \cdot t^3 - 0{,}002 \cdot t^2 \cdot (\%CaO) - 0{,}0004 \cdot t \cdot (\%CaO)^2 - 0{,}047 \cdot (\%CaO)^3 - 0.322 \cdot t^2 + 0.116 \cdot t \cdot (\%CaO) + 6.402 \cdot (\%CaO)^2 + 3.058 \cdot t - 279.051 \cdot (\%CaO) + 3900.427$

The correlation coefficient $R_3 = 0.985$; Deviation $S_3 = 1.658$

Maximum point deviation: (%CaO) = 52.391%; t = 13.710 min; $\eta_S = 57.102\%$

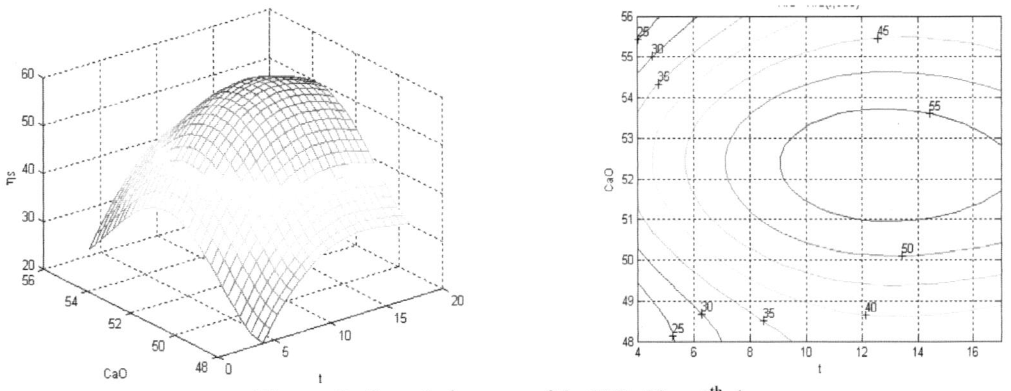

Figure 8. Correlation $\eta_S = f(t; \%CaO)$ – 4th degree

The correlation equation: $\eta_S = 0.0005 \cdot t^4 + 0.0002 \cdot t^3 \cdot (\%CaO) + 0.003 \cdot t^2 \cdot (\%CaO)^2 + 0.006 \cdot t \cdot (\%CaO)^3 + 0.044 \cdot (\%CaO)^4 - 0.004 \cdot t^3 - 0.338 \cdot t^2 \cdot (\%CaO)^2 - 1.079 \cdot t \cdot (\%CaO)^2 - 9.387 \cdot (\%CaO)^3 + 8.557 t^2 + 59.722 \cdot t \cdot (\%CaO) + 739.948 \cdot (\%CaO)^2 - 1089.643 \cdot t - 25864.411 \cdot (\%CaO) + 338292.863$

The correlation coefficient: $R_4 = 0.991$; Deviation $S_4 = 1.279$

Maximum point deviation: (%CaO) = 52.702%; t = 13.805min; $\eta_S = 57.872\%$

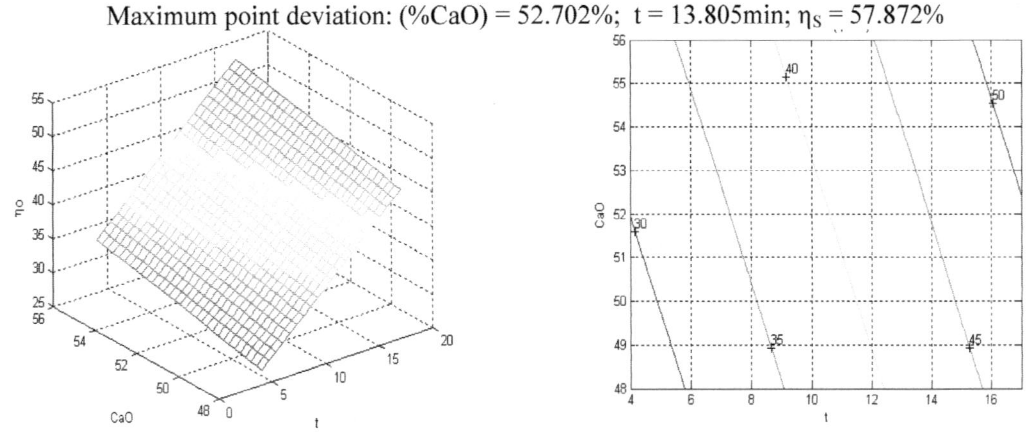

Figure 9. Correlation $\eta_O = f(t; \%CaO)$ – 1st degree

The correlation equation: $\eta_O = 1.514 \cdot t + 0.682 \cdot (\%CaO) - 11.526$

The correlation coefficient: $R_1 = 0.687$; Deviation $S_1 = 7.337$

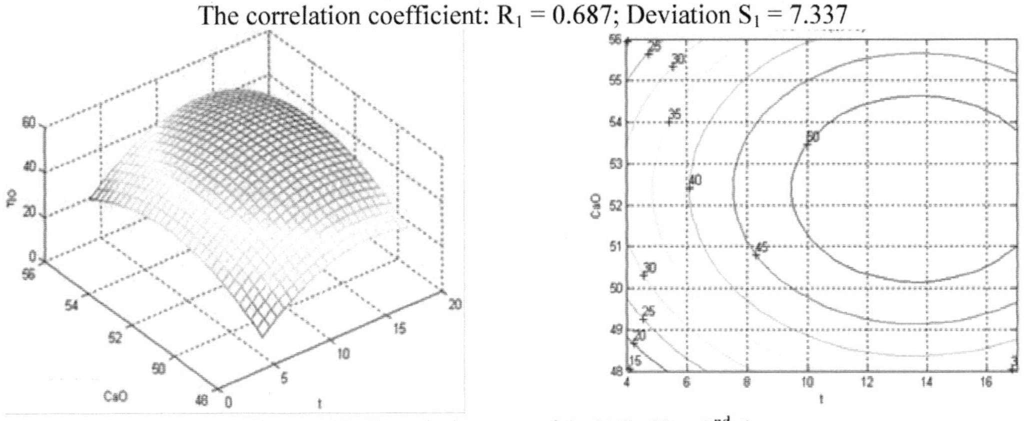

Figure 10. Correlation $\eta_O = f(t; \%CaO)$ – 2nd degree

The correlation equation: $\eta_O = -0.250 \cdot t^2 + 0.005 \cdot t \cdot (\%CaO) - 0.886 \cdot (\%CaO)^2 + 6.564 \cdot t + 92.791 \cdot (\%CaO) - 2421.275$

The correlation coefficient: $R_2 = 0.986$; Deviation $S_2 = 1.655$

Maximum point deviation: (%CaO)= 52.335%; t = 13,862 min; $\eta_O = 56.241\%$

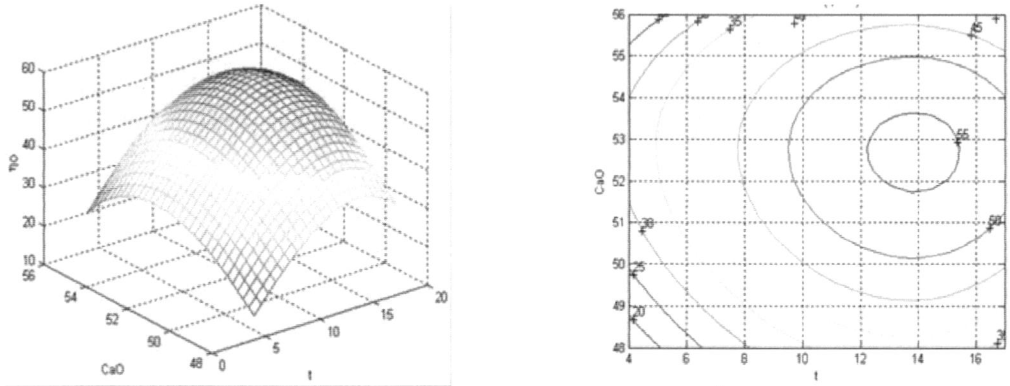

Figure 11. Correlation $\eta_O = f(t; \%CaO)$ – 3rd degree

The correlation equation: $\eta_O = -0.010 \cdot t^3 - 0.0002 \cdot t^2 \cdot (\%CaO) - 0.013 \cdot t \cdot (\%CaO)^2 - 0.051 \cdot (\%CaO)^3 + 0.098 \cdot t^2 + 1.417 \cdot t \cdot (\%CaO) + 7.239 \cdot (\%CaO)^2 - 33.368 \cdot t - 336.744 \cdot (\%CaO) + 5140.889$

The correlation coefficient: $R_3 = 0.991$; Deviation $S_3 = 1.328$

Maximum point deviation: $(\%CaO) = 52.681\%$; $t = 13.751$min; $\eta_O = 56.284\%$

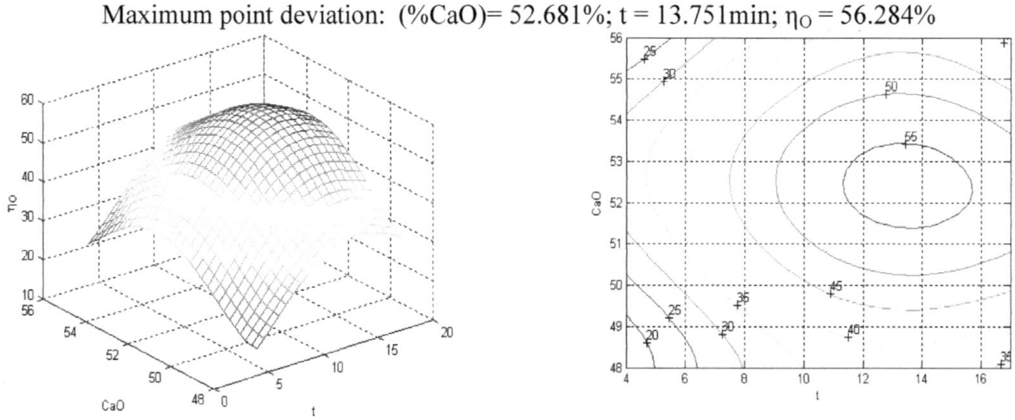

Figure 12. Correlation $\eta_O = f(t; \%CaO)$ – 4th degree

The correlation equation: $\eta_O = 0.002 \cdot t^4 - 0.001 \cdot t^3 \cdot (\%CaO) + 0.0004 \cdot t^2 \cdot (\%CaO)^2 + 0.0074 \cdot t \cdot (\%CaO)^3 + 0.033 \cdot (\%CaO)^4 - 0.011 \cdot t^3 + 0.007 \cdot t^2 \cdot (\%CaO)^2 - 1.191 \cdot t \cdot (\%CaO)^2 - 6.998 \cdot (\%CaO)^3 - 0.187 \cdot t^2 + 62.507 \cdot t \cdot (\%CaO) + 554.933 \cdot (\%CaO)^2 + -1087.221 \cdot t - 19503.508 \cdot (\%CaO) + 256355.498$

The correlation coefficient: $R_4 = 0.995$; Deviation $S_4 = 0.992$

Maximum point deviation: $(\%CaO) = 52.743\%$; $t = 13.710$min; $\eta_O = 56.621\%$

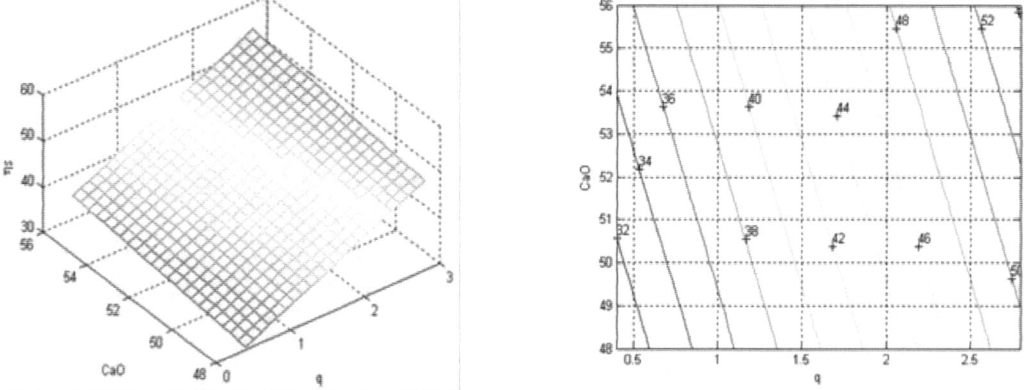

Figure 13. Correlation $\eta_S = f(q; \%CaO)$ – 1st degree

The correlation equation: $\eta_S = 7.904 \cdot q + 0.596 \cdot (\%CaO) - 1.352$

The correlation coefficient: $R_1 = 0.639$; Deviation $S_1 = 7.814$

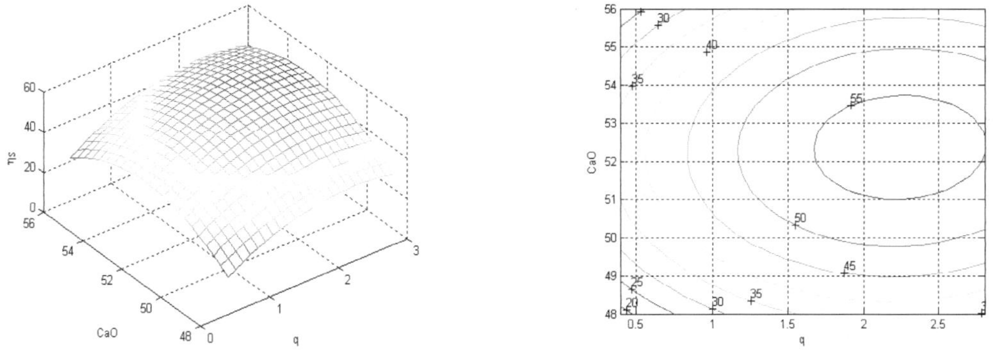

Figure 14. Correlation $\eta_S = f(q; \%CaO)$ - 2nd degree

The correlation equation: $\eta_S = -6.031 \cdot q^2 + 0.206 \cdot q \cdot (\%CaO) - 1.036 \cdot (\%CaO)^2 + 16.211 \cdot q + 108.106 \cdot (\%CaO) - 2792.035$

The correlation coefficient: $R_2 = 0.822$; Deviation $S_2 = 1.911$

Maximum point deviation: (%CaO)= 52.444%; q= 2.251g/kg; $\eta_S = 57.368\%$

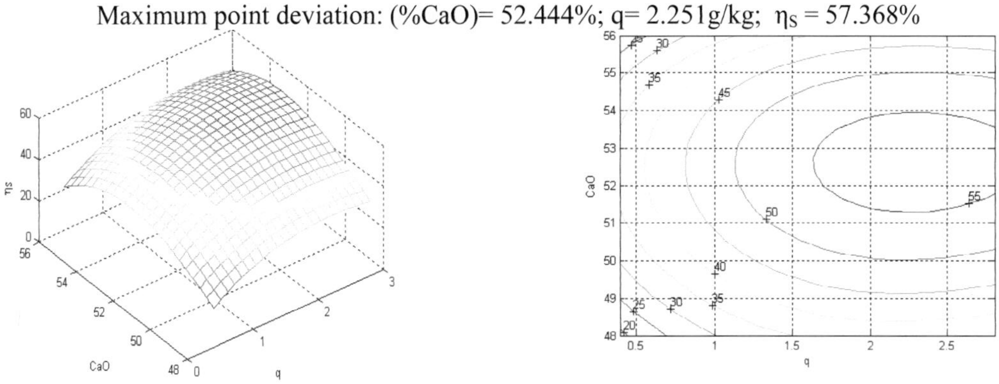

Figure 15. Correlation $\eta_S = f(q; \%CaO)$ – 3rd degree

The correlation equation: $\eta_S = 0.751 \cdot q^3 - 0.025 \cdot q^2 \cdot (\%CaO) - 0.019 \cdot q \cdot (\%CaO)^2 - 0.051 \cdot (\%CaO)^3 - 8.281 \cdot q^2 + 2.273 \cdot q \cdot (\%CaO) + 6.852 \cdot (\%CaO)^2 - 34.319 \cdot q - 303.169 \cdot (\%CaO) + 4341.912$

The correlation coefficient: $R_3 = 0.984$; Deviation $S_3 = 1.764$

Maximum point deviation: (%CaO)= 52.671%; q= 2.303g/kg; $\eta_S = 57.629\%$

Figure 16. Correlation $\eta_S = f(q; \%CaO)$ – 4th degree

The correlation equation: $\eta_S = 0.819 \cdot q^4 - 0.093 \cdot q^3 \cdot (\%CaO) + 0.115 \cdot q^2 \cdot (\%CaO)^2 + 0.034 \cdot q \cdot (\%CaO)^3 + 0.051 \cdot (\%CaO)^4 + 0.516 \cdot q^3 - 11.631 \cdot q^2 \cdot (\%CaO)^2 - 5.791 \cdot q \cdot (\%CaO)^2 - 0.687 \cdot (\%CaO)^3 + 292.968 \cdot q^2 + 320.039 \cdot q \cdot (\%CaO) + 840.278 \cdot (\%CaO)^2 - 5837.466 \cdot q - 29301.079 \cdot (\%CaO) + 382355.983$

The correlation coefficient: $R_4 = 0.991$; Deviation $S_4 = 1.305$

Maximum point deviation: (%CaO) = 52.384 %; q= 2.431g/kg; $\eta_S = 57.924\%$

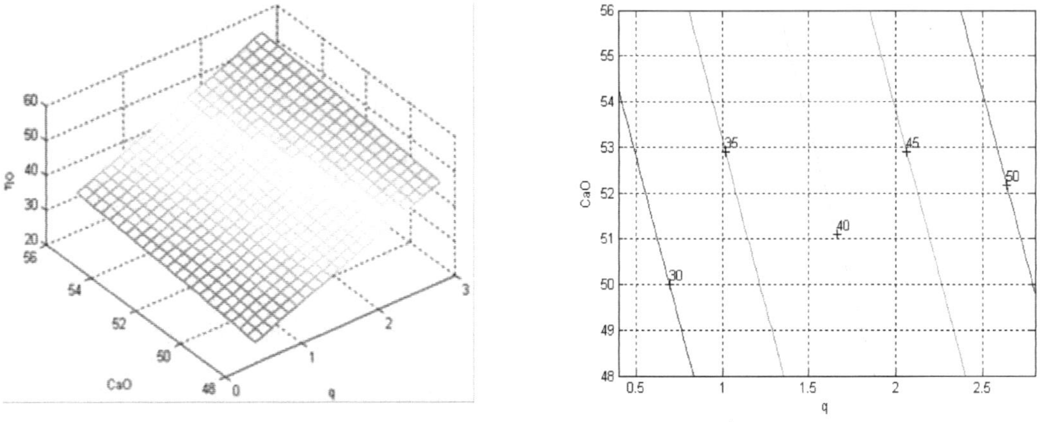

Figure 17. Correlation $\eta_O = f\,(q;\ \%CaO)$ – 1st degree

The correlation equation: $\eta_O = 9.559\cdot q + 0.657\cdot(\%CaO) - 9.492$

The correlation coefficient: $R_1 = 0.751$; Deviation $S_1 = 6.861$

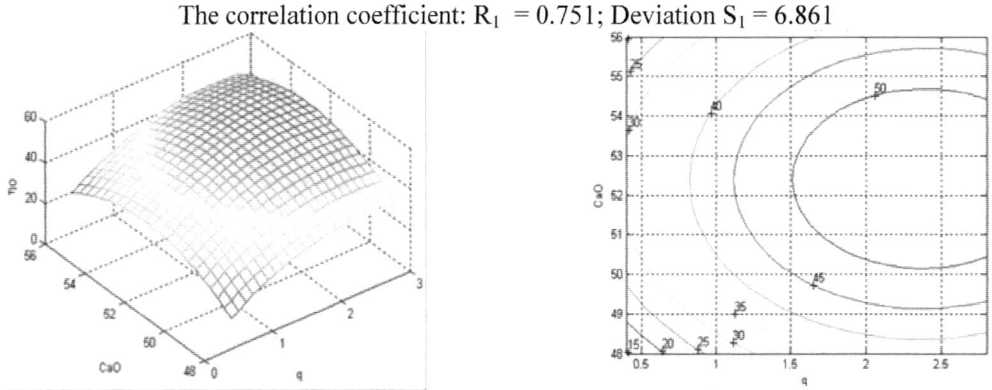

Figure 18. Correlation $\eta_O = f\,(q;\ \%CaO)$ – 2nd degree

The correlation equation: $\eta_O = -5.967\cdot q^2 + 0.058\cdot q\cdot(\%CaO) - 0.881\cdot(\%CaO)^2 + 25.398\cdot q + 92.224\cdot(\%CaO) - 2392.957$

The correlation coefficient: $R_2 = 0.985$; Deviation $S_2 = 1.747$

Maximum point deviation: (%CaO)= 52.513%; q= 2.463g/kg; η_O = 56.473%

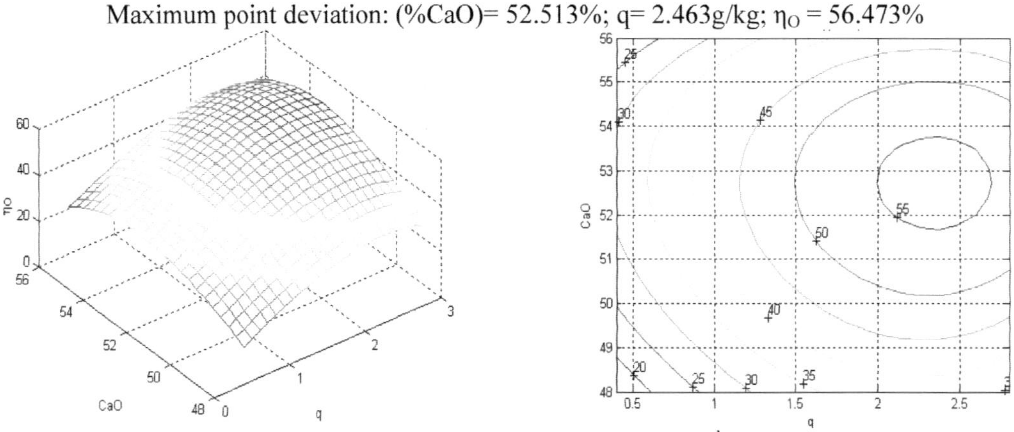

Figure 19. Correlation $\eta_O = f\,(q;\ \%CaO)$ – 3rd degree

The correlation equation: $\eta_O = -1.682\cdot q^3 + 0.084\cdot q^2\cdot(\%CaO) - 0.091\cdot q\cdot(\%CaO)^2 - 0.054\cdot(\%CaO)^3 - 2.412q^2 + 9.354\cdot q\cdot(\%CaO) + 7.736\cdot(\%CaO)^2 - 220.183\cdot q - 362.308\cdot(\%CaO) + 5584.811$

The correlation coefficient: $R_3 = 0.991$; Deviation $S_3 = 1.398$

Maximum point deviation: (%CaO) = 52.613%; q= 2.471g/kg; η_O = 56.937%

IOP Conf. Series: Materials Science and Engineering **477** (2019) 012010 doi:10.1088/1757-899X/477/1/012010

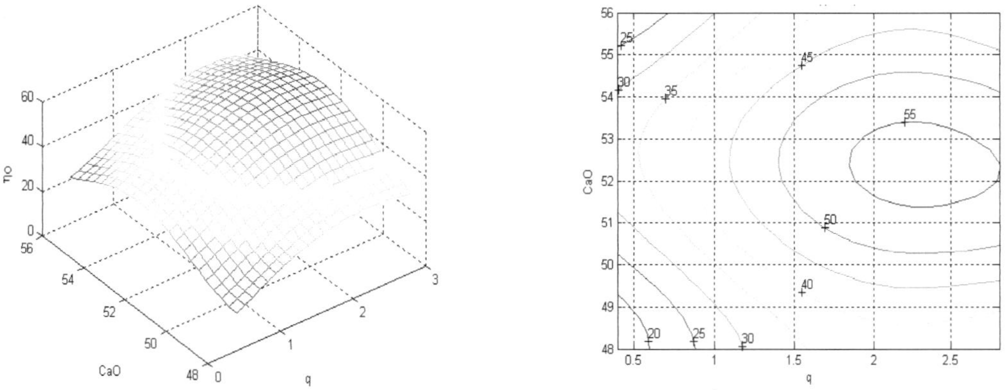

Figure 20. Correlation $\eta_O = f(q; \%CaO)$ – 4th degree

The correlation equation: $\eta_O = 1.464 \cdot q^4 - 0.194 \cdot q^3 \cdot (\%CaO) - 0.009 \cdot q^2 \cdot (\%CaO)^2 + 0.041 \cdot q \cdot (\%CaO)^3 + 0.036 \cdot (\%CaO)^4 - 0.902 \cdot q^3 + 1.917 \cdot q^2 \cdot (\%CaO)^2 - 6.431 \cdot q \cdot (\%CaO)^2 - 7.791 \cdot (\%CaO)^3 - 53.311 \cdot q^2 + 335.231 \cdot q \cdot (\%CaO) + 615.683 \cdot (\%CaO)^2 - 5790.541 \cdot q - 21570.321 \cdot (\%CaO) + 282676.383$

The correlation coefficient: $R_4 = 0.995$; Deviation $S_4 = 1.026$

Maximum point deviation: $(\%CaO) = 52.702\%$; $q = 2.476 g/kg$; $\eta_O = 56.937\%$

The analysis of the double correlation equations expressed analytically and graphically in Figures 5–20, leads to a series of observations and conclusions having both a general and a specific character, namely:

1. The variation of the independent parameters within the technological limits determines a variation of the dependent parameter also within the technological limits, it being located on a regression surface or in its vicinity, considering the standard dispersion, deviation and error.

2. The graphical representations of the analytical correlations expressed by 1st degree polynomial functions are plane surfaces and those expressed by 2nd, 3rd and 4th degree polynomial functions are non-plane surfaces.

3. The intersection of the correlation surfaces with the level planes (parallel to the horizontal) resulted in level curves (level lines for the 1st degree polynomial functions), which allowed setting the variation limits of the independent parameters.

4. The correlations expressed by a 1st degree polynomial function and shown in their graphical form in Figures 5, 9, 13 and 17, allow the choice of parameter values: t, q, (%CaO), so that for the desulphurization (η_S), respectively deoxidizing (η_O) output values obtained should be above the average values resulted for the charges under analysis.

5. The graphical representations of the 2nd, 3rd and 4th degree correlations show a maximum point, the values of all parameters (both independent and dependent) ranging within the technological limits. In order to obtain high values for the dependent parameter, it is desirable that these values be situated primarily within the sub-domain where the maximum point is located, and if this area is small (which means narrower limits for the independent parameters), in the nearby zones.

6. The analysis of the correlations $\eta_S = f(t, CaO)$ shown in their analytical and graphical form in Figures 6, 7 and 8 leads to the conclusion that for all cases, there is a maximum point within the technological variation interval of the technologically independent parameters. For (%CaO) within the limits 49–56% and for t =8–17min. can be obtained for η_S, values above the average obtained during the experiments (42.278%), and for (%CaO) =51–54% and for t =1–17 min. can be obtained for η_S values above 55%.

7. As to correlations $\eta_O = f(t, CaO)$ shown analytically and graphically in Figures 10, 11 and 12 it was observed that in all cases, there is a maximum point within the interval of technological variation of the technologically independent parameters. For (%CaO) and t within the same variation limits as for η_O values above the average, respectively for η_O above 55% (up to the maximum point 56-57%) for (%CaO) =51.5-53.5% and t=1.5-15.5min.

8. The study of correlations η_S = f (q, CaO) shown analytically and graphically in Figures 14, 15 and 16, proves they all have a maximum point. For the same variation limits of (%CaO) as in the previous case and for the specific synthetic slag consumption q = 1-1.15g/kg the values of η_S were above the average, while for %CaO= 51.0-53.5% and q =1.5–2.8g/kg the values of η_S were superior to 55% up to a maximum point.

9. As to correlations η_O = f(q, CaO) shown analytically and graphically in Figures 18, 19 and 20, the situation and the technological analysis are similar to the previous ones, except for the fact that in order to obtain for η_O values above 55% the variation domain is narrowed as follows: (%CaO)=51.5 - 53.5% and q=2.0-2.8g synthetic slag/kg steel.

4. Conclusions

The experiments, the data processing and the technological analysis of the results allow the following conclusions:

- lime, alumina and titanium oxide-based mechanical mixtures, respectively the slag resulting from their melting, grant favorable conditions for steel refining inside the ladle, through the so-called method of "synthetic slag treatment";
- obtaining a mechanical mixture having oxide values (respectively oxide summary) within the limits in the vicinity of the maximum point, leads to high values of η_S, respectively η_O;
- in the case of oxide ratio variation in the vicinity of the maximum point, it is possible to obtain for η_S, respectively for η_O values close to the upper limit;
- the results lead to the idea of continuing the experiments using for the mechanical mixture the slag resulted from ferro-titanium production and, according to some data in the reference literature, of the reducing mixture containing vanadium oxides and a higher content of magnesium oxide;
- the results obtained are applicable in steel elaboration research and practice;
- in order to obtain an advanced rate of desulphurization and deoxidizing, it is desirable to treat steel with slag containing CaO = 50–56%, the treatment time t = 10-17 min and the specific addition of synthetic slag of 1.5-2.8 g/kg;
- the values of the independent parameters (CaO), t and q can be established graphically, by use of any of the graphical representations (2nd, 3rd or 4th degree), minding that the results should be checked both graphically and analytically (after the 2nd degree double correlation, desirably).

References

[1] Tetileanu A 2013 *Researches on refining the steel elaborated along the flux arc furnace – ladle – continuous casting*, Politehnica Publishing House, Timişoara, Romania

[2] Hermanus Schrama F N, Beunder E M, Van den Berg B, Yang Y and Boom R 2017 Sulphur removal in ironmaking and oxygen steelmaking, *Ironmaking & Steelmaking , Processes, Products and Applications* **44**(5), 333-343

[3] Putan V, Putan A, Josan A and Vilceanu L 2016 Researches concerning influence of magnesium, aluminum and titanium lime on steel desulfurization, *IOP Conf. Ser.: Mater. Sci. Eng.* **106** 012033

[4] Dumitru M G, Ioana A, Constantin N, Ciobanu F and Pollifroni M 2018 Steel refining possibilities in LF, *IOP Conf. Ser.: Mater. Sci. Eng.* **294** 012018

[5] Dippenaar R 2005 Industrial uses of slag. The use and re-use of iron and steelmaking slags, *Ironmaking & Steelmaking; Processes, Products and Applications* **32**(1) 35-46

[6] Gizatulin R A, Valuev D V, Trifonov V A, Valueva A V and Serikbol A 2015 Oxidation Processes in Blowing Steel With Inert Gas into the Ladle, *IOP Conf. Ser.: Mater. Sci. Eng.* **91** 012037

[7] Gizatulin R A, Valuev D V, Valueva A V, Serikbol A and Borovikov I F 2015 Optimizing Technological Parameters of the Reduction Processes in Treating Steels in a Ladle Furnace, *IOP Conf. Ser.: Mater. Sci. Eng.* **91** 012039

[8] Ardelean E, Socalici A, Ardelean M, Şerban S and Vătăşescu M 2017 The analysis of the steel

deoxidation process in a vacuum installation, *IOP Conf. Ser.: Mater. Sci. Eng.* **163** 012025

[9] Steneholm K, Andersson M, Tilliander A and Jönsson P G 2013, Removal of hydrogen, nitrogen and sulphur from tool steel during vacuum degassing, *Ironmaking & Steelmaking , Processes, Products and Applications* **40**(3) 199-205

[10] Dragoi F, Socalici A, Ardelean E, Popa E and Hepuț T 2011 Researches on the influence of the slags formed in the installations on the hydrogen removal efficiency, *Revista de Metalurgia* **47**(6) 477-484

IOP Conf. Series: Materials Science and Engineering **477** (2019) 012011 doi:10.1088/1757-899X/477/1/012011

Comparative analysis of heat recovery methods from internal combustion engines

M E Filip, G Csaba and D Lelea

University Politehnica Timisoara, Department of Thermal Machines and Transportation, Timisoara, Romania

E-mail: fillip mihai@yahoo.com

Abstract. This paper discusses different ways of recovering wasted energy from the internal combustion's engine exhaust gas. It is well known that the mechanical work produced by an internal combustion engine is only a small amount related to the fossil fuel it uses. By recovering some of the wasted energy, the internal combustion engine's efficiency can be increased especially in the cold season. The amount of the toxic compounds and greenhouse effect gases that are released in the air can also be decreased by heating up the engine very quickly, lower the engine's load and reduce fuel consumption. The energy is recovered as heat using different types of heat exchangers but can be converted to mechanical energy, electrical energy or even air conditioning power. Because we are referring at vehicle engines, special care must be taken regarding safety in normal operating conditions and also in case of a crash. Some of the methods that combine recovering and storing heat are suitable for hybrid vehicles also. The conclusion of this paper is that the internal combustion engine can take advantage of some heat recovery systems.

1. Internal combustion engines efficiency

The biggest problem with the internal combustion engine is that most of the energy generated by combustion is wasted as heat rather than converted propulsion for the vehicle. The first Lenoir engine that was built in 1860 had 2 horsepower and only 4% efficiency. Four years later, in 1864 the first Otto engine had already 12% efficiency. Today's car engines reach about 40%-45% efficiency and the best efficiency achieved is 54,4% by the low speed diesel marine engine, MAN S80ME-C7. The turbine, the intercooler, the use of high-pressure stratified fuel injection, the recirculation gas valve and the downsizing of the engines all contribute to a better fuel efficiency. Also, the new fuels have better performances than the old ones.

2. Pollution, the main focus

2.1 Fossil-fuel based engines

Every engine that burns fuel will release exhaust gas that is toxic for humans, animals and the environment. If the engine is more efficient it will burn less fuel for the same amount of mechanical work and release less exhaust gas. A warmer engine will also have better efficiency and burn less fuel than a cold one. Exhaust gases produced by an internal combustion engine from diesel, bio-diesel, petrol and other alternate fuels consist in particulate matter, un-burnt hydrocarbons, oxides of nitrogen, carbon monoxide and carbon dioxide. Euro emission norms establish how much an engine can pollute to meet a specific euro standard. The newer cars, if maintained well, produce a smaller quantity of air pollutants but when we get into this equation a significant number of cars we still have a big air quality problem. The only visible air pollutant is the particulate matter which can also be observed only in certain conditions, especially in case of older diesel engines. Un-burnt hydrocarbons, oxides of nitrogen, carbon monoxide and carbon dioxide are our stealth enemies.

Other small particles can be also produced by tyres, brake discs, brake pads or brake floating shoe. If an engine is not correctly maintained and consumes oil it will release blue smoke which is very harmful for our health. In other cases, like clogged admission or very rich fuel mixture following a cold start even gasoline engines can produce visible smoke. A clean air filter is very important for the car to have enough intake air and burn all the injected fuel. The quantity of air pollutants is also influenced by other factors like traffic, driving style, ambient temperature, air humidity, fuel and oil quality. EU emissions standards are only a starting point for all new passenger cars. [1] Hydrocarbons, N_2O, CO_2, NOx, NH_3, CO and particulate matter have higher concentrations at cold start, until the engine reaches its optimal operating temperature. Since the Euro 3 regulations in 2000, performance has been measured using the New European Driving Cycle test (NEDC; also known as MVEG-B), with a "cold start" procedure that eliminates the use of a 40-second engine warm-up period found in the ECE+EUDC test cycle (also known as MVEG-A). The COMMISSION REGULATION (EC) No 692 / 18 July 2008, implementing and amending Regulation (EC) No 715/2007 of the European Parliament and of the Council on type-approval of motor vehicles with respect to emissions from light passenger and commercial vehicles (Euro 5 and Euro 6) [2] describes the COLD START procedure even better and also there are new specific measurements that car manufactures have to make before releasing a new vehicle on the market.

2.2 The copper heat exchanger

The goal is to heat up the engine as soon as possible by using energy that normally is dismissed in the air but we also need to be very careful where do we take that heat from. If we want to recover as much heat as possible we should mount a heat exchanger very close to the engine because the exhaust gas has over 500°C but at the end of the pipe it has less than 190°C.

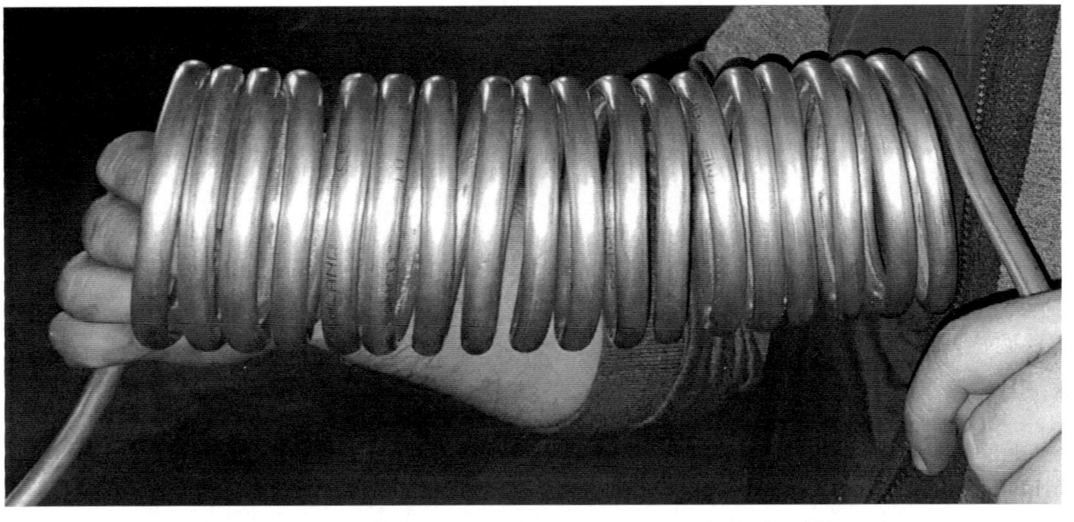

Figure 1. The copper heat exchanger used to recover heat from the exhaust gas

Figure 2. The copper heat exchanger mounted just in front of the exhaust pipe

2.3 The catalytic converter

The catalytic converter requires a temperature above 400 °C to efficiently convert harmful exhaust gases into inert gases, such as CO2 and water vapor. This is why, the first catalytic converters were placed close to the engine to ensure fast heating but this kind of placing caused several problems, such as vapor lock. It occurred a few minutes after the engine is shut off, because the heat from the catalytic converter influences fuel in the fuel lines, which makes it first boil and then aerate. This condition of fuel results in engine failing to start until the engine and fuel in the lines cool down. To resolve this, manufactures moved the catalytic converters to a third of the way back from the engine, and were then placed underneath the vehicle. In the desire to recover as much heat as possible and because it was enough room, we first placed the heat exchanger in front of the catalytic converter which was a very big mistake. We recovered a lot of heat but the catalytic converter failed to heat enough and after four tests the CHECK ENGINE lamp lighted up.

The next step was to re-mount the heat exchanger after the last sensor on the exhaust pipe to ensure the catalytic converter will heat fast enough and no other errors occur, as shown in Figure 3.

Figure 3. The copper heat exchanger correctly mounted after the catalytic converter

IOP Conf. Series: Materials Science and Engineering **477** (2019) 012011 doi:10.1088/1757-899X/477/1/012011

2.4 Heat energy recovered, experimental setup

The heat exchanger was filled with water (0,6l at 15°C).
The quantity of heat retrieved by the water is $Q = c \cdot m \cdot (\theta_2 - \theta_1)$ where
$c = 4{,}2\ kJ \cdot kg^{-1} \cdot K^{-1}$ is the water's specific heat;
$m = 0{,}60$ kg is the water weight;
$\theta_1 = 15°C$ is the initial water temperature;
$\theta_2 = 100°C$ is the final water temperature;
With this data the result is Q = 210 kJ;
Thermal power is thermal energy ÷ time;
In the first experiment the engine was cold, - 9°C.
The water needed 188 s to reach the boil temperature (100°C).

$Q = 210$ kJ, $\Delta t_1 = 188$ s, $P_1 = \frac{210}{188} = 1{,}1$ kW.

In the second experiment the engine was operating at optimal temperature (88°C)
The water needed 56 s to reach the boil temperature (100°C).

$Q = 210$ kJ, $\Delta t_1 = 56$ s, $P_1 = \frac{210}{56} = 3{,}8$ kW.

All tests were performed at 2500Rpm with the car on an elevator (Dacia Solenza 1.4MPI).

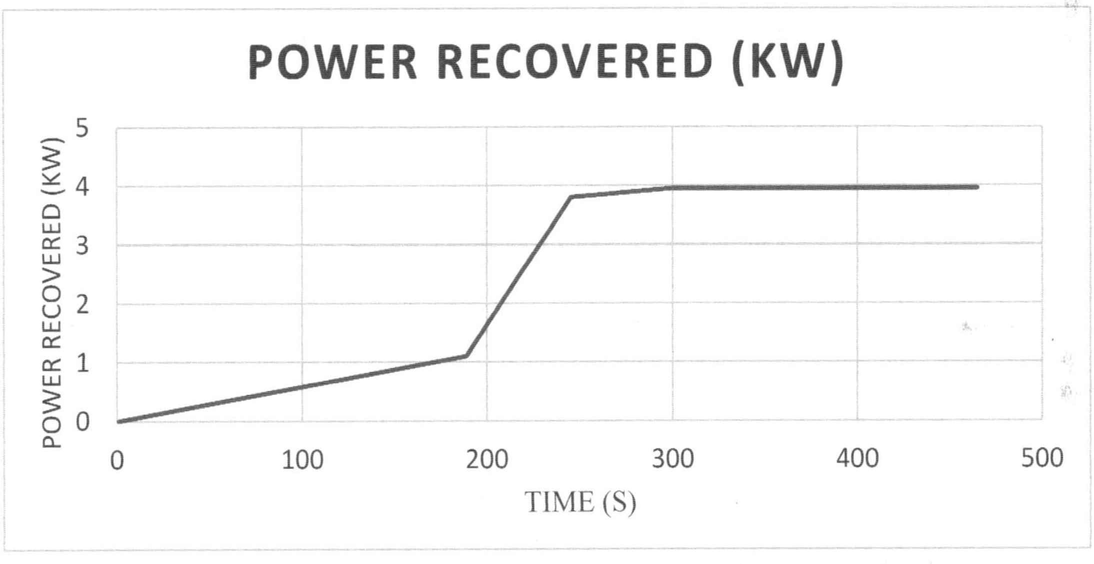

Figure 4. The amount of power recovered using the heat exchanger

2.5 Electric energy recovered, experimental setup

We used 16 pieces of SP1848 TEG, serial wired. The manufacturer gives the following specifications: at temperature difference of 100 degrees: open circuit voltage 4.8V, power current: 669MA.
The hot side was placed directly at the end of the exhaust gas pipe where temperature does not exceed 100°C because we wanted to protect the modules. The cold side was cooled with water blocks using water from the network at 15°C.

When electrical measurements were made, the temperature at the hot side was 92°C and at the cold side 17°C. The difference was 75°C.

As load, we used a H4 bulb, 110W@12v which didn't light up but we obtained the following results: closed circuit voltage 8.74v and closed circuit current 0.16A.

The energy recovered this way is only 1,4W

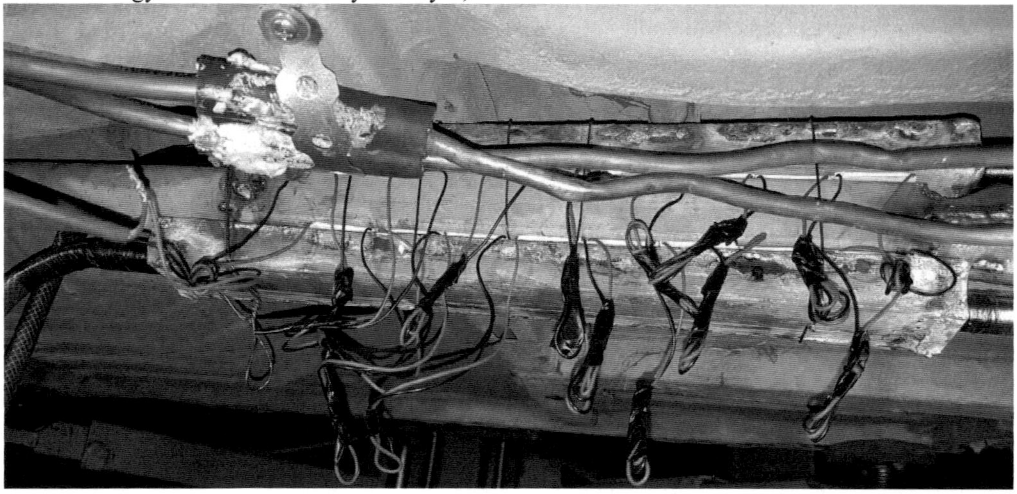

Figure 5. SP1848 Thermoelectric generator setup

Figure 6. Closed circuit obtained voltage (V)

Figure 7. Closed circuit obtained current (A)

3. Other means of recovering wasted energy from different producers

3.1 Alphabet EnergyPowerModule-γ™ [3]

Alphabet Energy Power Module-γ™ claims to produce up to 850W of electricity from exhaust gas, which is the best performance of any similar product found on the market. There are no independend tests to confirm this. We tried to contact them several times for more information using their website contact form, e-mail and the phone numbers provided on their website but nobody ever answered us.

Figure 8. Alphabet EnergyPowerModule

IOP Conf. Series: Materials Science and Engineering **477** (2019) 012011 doi:10.1088/1757-899X/477/1/012011

3.2 BMW turbosteamer [4]

This concept uses the principle of a thermal power plant, using a steam turbine. The steam circuit claims to produce 14 hp (10 kW) and 20 Nm of torque at peak (for a 1.8 Straight-4 engine), yielding an estimated 15% gain in fuel efficiency. BMW has been the pioneer of this concept as early as 2000 under the direction of Dr. Raymond Freymann, and while they were designing this system to fit to most current BMW models, the technology didn't reach production.

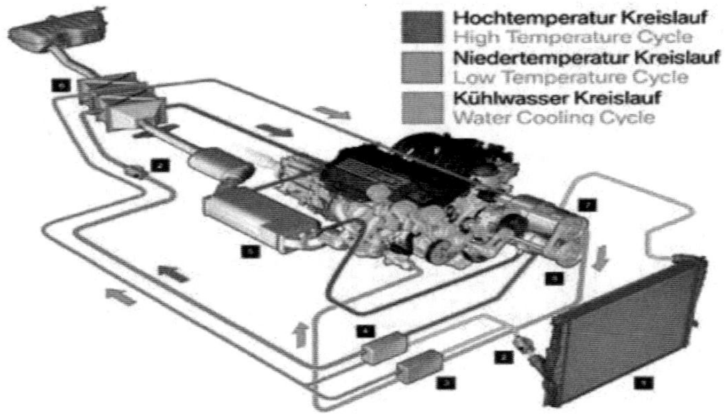

Figure 9. BMW Turbosteamer concept

4. Emission tests setup and results

4.1. Test setup

On the modified Dacia Solenza we have ran a series of tests at different engine temperatures, starting with the engine cold at -9°C. The exhaust gas analysis varies a lot when the engine is cold or when the engine is warm. It even smells very different from a cold engine to a warm one. At this moment the only thing we do with the recovered heat is to heat up the engine faster by warming up the coolant, the oil and the intake air. This reduces the time of heating up the engine to 90°C to 412 seconds. Without the heat recovery system, the engine needed 785 seconds to reach 90°C and only reached 57°C in 412 seconds. The engine also needed 278 seconds to reach 57°C if the heat recovery system was working.

It can also provide heat for the cabin and windows defrost quicker but during the tests all the ventilators and cabin heat were turned off. The car was parked outside in cold weather, -12°C. The ambient temperature in the garage where the tests were performed was +14°C. The engine was in idle mode except when performing the test because some tests required the engine to be accelerated to about 2500Rpm.

Time (s)	0	278	412
Engine Temp.°C	-9	57	93
CO %	1,76	0,11	0,01
CO_2 %	11,3	12,4	13,4
HC ppm	295	159	28
O_2 %	3,08	2,92	0,48
Lambda	1,091	1,15	1,023
COcorr %	2,02	0,13	0,01
AFR	16,13	17,02	15,1

Figure 10a. Measured values, engine COLD START, with the heat recovery system

Time (s)	0	412	785
Engine Temp	-9	57	93
CO %	1,76	0,11	0,01
CO_2 %	11,3	12,4	13,4
HC ppm	295	159	28
O_2 %	3,08	2,92	0,48
Lambda	1,091	1,15	1,023
COcorr %	2,02	0,13	0,01
AFR	16,13	17,02	15,1

Figure 11a. Measured values, engine COLD START, without the heat recovery system

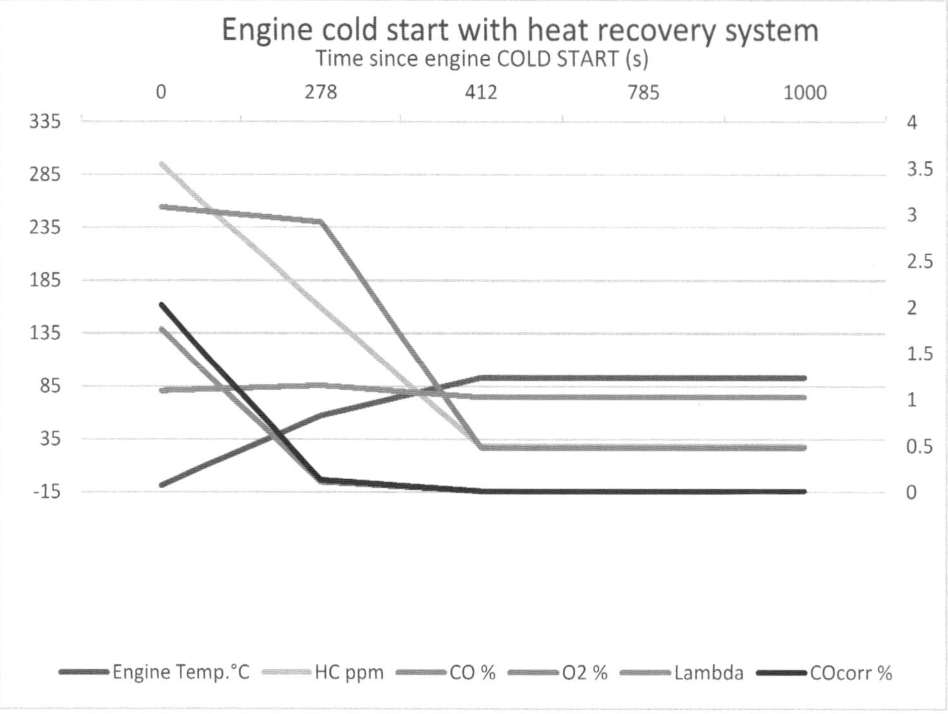

Figure 10b. Measured values, engine COLD START, with the heat recovery system

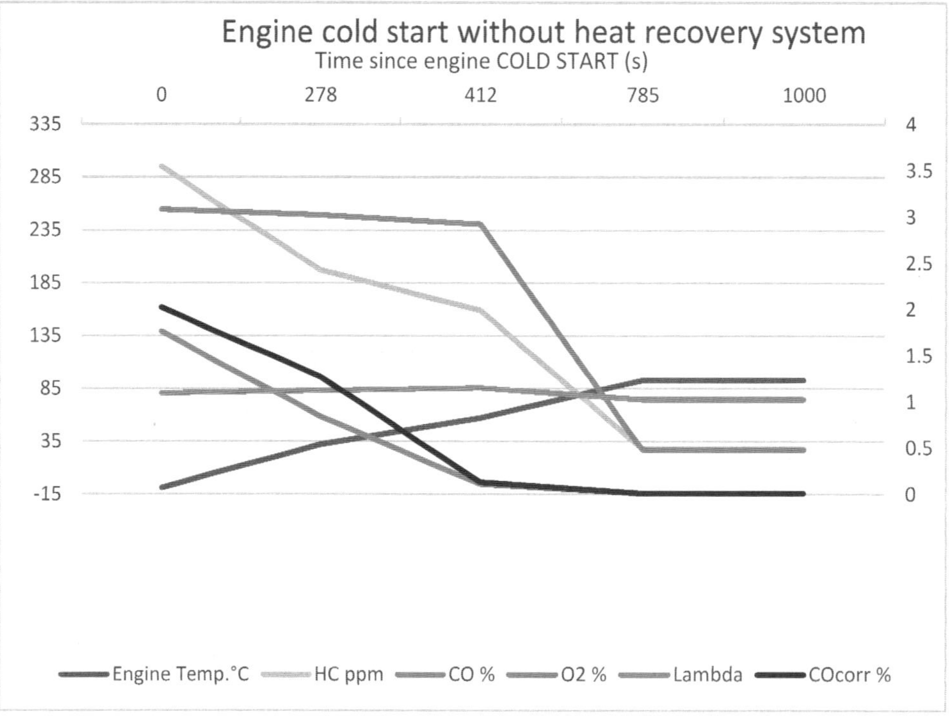

Figure 11b. Measured values, engine COLD START, without the heat recovery system

4.2 Test results
As shown in the results above, using the heat recovery system can almost halve the time the engine reaches the optimal operating temperature.

4.2.1 Hydrocarbon Emissions. The test vehicle's Hydrocarbon Emissions are almost linear and go from 295 ppm when the engine is very cold down to 28ppm which is less than one tenth when the engine reaches its optimal operating temperature. Hydrocarbons contribute to ground-level ozone formation leading to risk of damage to the human respiratory system. Some kinds of hydrocarbons, in addition, are both carcinogenic and indirect greenhouse gases.

4.2.2 Carbon monoxide. Carbon monoxide reduces the blood's oxygen-carrying capacity which can reduce the availability of oxygen to key organs. Extreme levels of exposure, such as might occur due to blocked flues in domestic boilers, can be fatal. At lower concentrations CO may pose a health risk, particularly to those suffering from heart disease. The test reveals that CO is very high when the engine is cold and decreases more than 170 times when the engine reaches its optimal operating temperature. The heat recovery system can help to obtain optimal values in half of the time required by normal operation.

5. Conclusions and future experiments
Internal combustion engines waste a lot of power as heat they give to the environment. If we could convert in usable energy only 1/3 of the wasted energy this could mean more than 30% increase in the overall engine efficiency. The large automotive companies didn't invest enough in research when it comes to fuel efficiency but invested a lot more in car electronics. The built recovery system helps to decrease fuel consumption and exhaust gas and is very effective on internal combustion engines and hybrid systems. The fuel exchanger has only 2kg and does not affect the weight of the car or the air flow in the exhaust system.

In the next experiment we will use a heat storage solution to heat up the engine even faster. We found a very good insulator, aerogel, that is commercially available but can be purchased only in large quantities.

References
[1] ***http://www.dft.gov.uk/vca/fcb/cars-and-air-pollution.asp
[2] ***Commission Regulation (EC) No 692/2008 of 18 July 2008 https://eur-lex.europa.eu/legal-content/EN/ALL/?uri=celex:32008R0692
[3] ***https://www.alphabetenergy.com/product/powermodule/
[4] ***https://www.press.bmwgroup.com/global/article/detail/T0119738EN/looking-for-the-next-gram?language=en

Test bench for studying the effects of water injection inside an internal combustion engine

A T Borborean, V Stoica and D Lelea

Politehnica University of Timisoara, Mechanical Engineering Faculty, Mihai Viteazu Blv, 1, Timisoara, Timis County, Romania

E-mail: andrei.borborean@student.upt.ro

Abstract. Under the increasingly drastic conditions imposed on motor vehicles, in terms of pollution, manufacturers and researchers are looking for solutions to reduce pollutant emissions, especially emissions of nitrogen oxides (NO_x) dissipated in the atmosphere. The problem of polluting emissions in the field of transport is becoming increasingly stringent, the pollution rules are becoming stricter, so there is a general interest in finding solutions to reduce pollutant emissions and fuel consumption. A possible solution to this problem can be achieved by direct or indirect water injection (through the valve port) inside the cylinder of an internal combustion engine. In this paperwork we present the test bench and we analyzed the effect of water injection inside a cylinder of an internal combustion engine. The water injection is made upstream of the throttle valve by single injector mounted on the intake. The injection timing and amount of water injected inside the engine have been controlled by CompactRIO hardware device and in-house algorithm implemented in LabView. The engine used for measurements is a 1.6 liter engine, manufactured by Daewoo Motor Company and for the moment it's running on wasted spark configuration. So in order to have water particles in all the cylinders, the water is injected two times at each rotation of the crankshaft. The main goal of this research was decreasing of nitrogen oxides (NO_x) by reducing the combustion chamber temperature.

1. Introduction

The increase of thermal efficiencies obtained by an internal combustion engine is mainly the reason of increasing the compression ratios. However, the use of high compression ratios leads to higher combustion temperatures and creates favorable conditions for the formation of nitrogen oxide (NO_x) emissions.[1] Additionally, the higher compression ratios used in internal combustion engines leads to higher temperature in the later stages of the compression stroke and a higher temperature in the earlier stage of the combustion and expansion stroke, resulting in a rapid NOx formation reaction rate. Minimizing NO_x is a well-known problem and it has been studied by many researchers.[2]

The method of implementing a water injection system would be one of the perfect solutions to inhibit NO_x formation. The heat is absorbed during the combustion stroke by the vaporization of water at high temperature which will form hydroxide and hydrogen. Besides the absorption of the heat during water vaporization, water provides oxygen for the burning process. The injected water also reduces the local temperature of the combustion flame and thus leads to lower NO_x emissions.[3], [4]

Fundamentally, water injection in the spark-ignition engine helps in controlling the temperature and pressure of the combustion process, thus giving an improvement in the volumetric efficiency of an engine and its power output.

Wu et al. [5] presented a new concept for combining the water injection process with an oxy internal combustion engine cycle for improving the thermal efficiency. The water injected into the cylinder after it was heated by passing it through the engine coolant and exhaust systems. Results showed that the thermal efficiency reached values from 53-67% when the water injected into the cylinder was at 120-200°C, moreover, the indicated thermal efficiency increased from 32.1% to 41.5%.

Boretti [6] used water injection in combination with a turbocharger, having ethanol as a fuel to explore the possibility of reducing the tendency to knock, increasing the charge efficiency and controlling the temperature of the air flowing into the turbine.

Cesur et al. [7] investigated the possibility of the engine using higher compression ratios and boost pressures and involved investigations of an original engine in combination with water injection.

Mingrui et al. [8] used water injection in combination with direct gasoline injection. Various tests with different water mass ratios were injected directly into the cylinder at the later part of the compression stroke. The optimum water ratio was determined at 15% for a certain cyclic dose, in terms of engine performance and emissions, the NOx emissions were reduced by 34.6%. These conclusions were drawn from the comparison of the results obtained from the measurements of a water injection engine with those obtained from the measurements of a standard gasoline engine.

Kohketsu et al. [9] studied a new water injection system applied to a diesel engine, and by using an optimal combination of water and fuel concluded that total NOx and particulate matter (PM) were reduced by 50% and respectively 25%, without an increase of fuel consumption.

The main purpose of direct water injection is the reduction of inlet temperature as a direct consequence of water vaporization, which results in a large decrease of temperature inside the combustion chamber at the latter stages of the compression stroke.

The present study presents preliminary experimental determinations using a four cylinder internal combustion engine, manufactured by Daewoo Motor Company connected to a water brake. The test bench and methodology is described in the next section.

2. Methodology and discussions

The test rig used for testing is presented below.

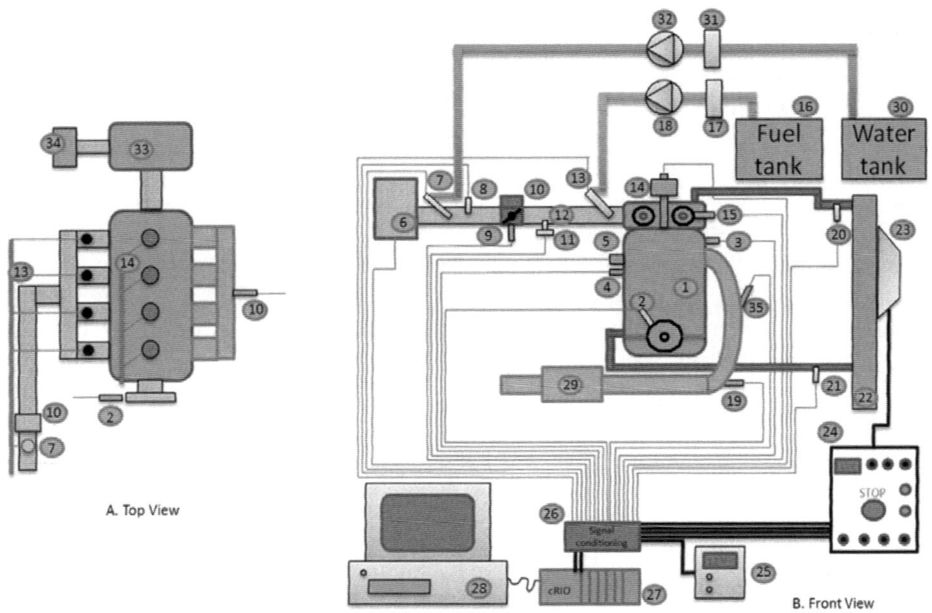

Figure 1. Schematic display of the experimental test bench

IOP Conf. Series: Materials Science and Engineering **477** (2019) 012012 doi:10.1088/1757-899X/477/1/012012

The system provides the possibility of using a standard fuelling and with the use of a single water injector mounted upstream of the throttle body as a function of the desired approach.

The experimental test engine has four cylinders and four valves per cylinder, and was manufactured by Daewoo Motor Company, with a power of 77kW and with a displacement of 1600cc connected to a water brake. The water brake 33 being set in motion by the water cooled internal combustion engine 1 which is started electrically by the startup system. The air enters in the internal combustion engine 1 through the intake manifold 12 after it passes the mass air flow sensor 6 and the throttle valve 10, and the exhaust gases are evacuated trough the exhaust manifold 29. For controlling the air-fuel mixture, a plurality of sensors are used. The temperature of the air inside the intake manifold 12 is measured with the air temperature sensor 8, the pressure drop after it passes the throttle valve 10 is measured with a pressure sensor 11. The water temperature is measured in three locations, one at the exit of the cylinder head with the coolant temperature sensor 3, and two more temperature sensors mounted on the inlet 20 and outlet 21 of the heat exchanger 22. The air circulation through the heat exchanger is assured by the fan 23. The temperature of the exhaust gases is measured with the exhaust temperature sensor 19, the amount of oxygen in the exhaust gases is measured with a binary lambda sensor 35.

Engine operation is ensured by the fuel injectors 13 and ignition coils 14. Fuel is provided to the fuel injectors 13 from the fuel tank 16 by the aid of the fuel pump 18 through the fuel filter 17. The water injector 7, is placed upstream of the throttle valve which is provided with water from the water tank 30 by the aid of a water pump 32 through a water filter 31. The position of the crankshaft and camshaft is measured with two inductive sensors 2 and 15 mounted on two independent trigger wheels.

To prevent engine damage two important sensors are used, one for monitoring the oil pressure 4, and one for detecting knock 5.

For emergency situations the control panel 24 can be used to power on and off various subsystems like ignition, injection, fuel and water pumps, cooling fans, etc. All electric signals from the sensor and to the actuators are passing through the signal conditioning box 26. The controller 27 is a Compact RIO system made by National Instruments consisting of a real time power PC embedded controller for Compact RIO of 533MHz, with 2Gb storage and 256 Mb DRAM, a TTL digital module NI 9401 with 8-Channels, with a response time of 100 ns, and a 16-Bit analogic input module NI 9205 with 32-Channels having the input range of ± 10 V, and an acquiring frequency of 250 kS/s.

The internal combustion engine runs on an in-house algorithm implemented in LabView. The computer 28 is used for programming the cRIO system and for modifying and visualizing the working parameters. For powering the fuel pump 16, water pump 30 and the actuators (injectors and ignition coil) an external 12 V power supply 25 is used.[10]

Different loads are ensured by the water brake 33 and their power are quantified by the scale 34. For the case presented in the current study there, were made determinations using RON 95 gasoline as standalone fuel for the testing part.

Additionally we measure the pollutant emissions using a Capelec 3201 gas analyzer. This gas analyzer can measure the concentrations of hydrocarbons (HC), carbon monoxide (CO), and carbon dioxide (CO_2), Oxygen (O_2), nitrogen oxide (NO_x), RPM and oil temperature.

Figure 2. Experimental test bench

3. Experimental results

The main purpose of the experimental part was to observe the obtained power and the obtained emissions through the modification of the advance to spark ignition and fuel injection time. By modifying the injection time, we modify the fuel quantity introduced inside the cylinder, thus modifying the air-fuel ratio.

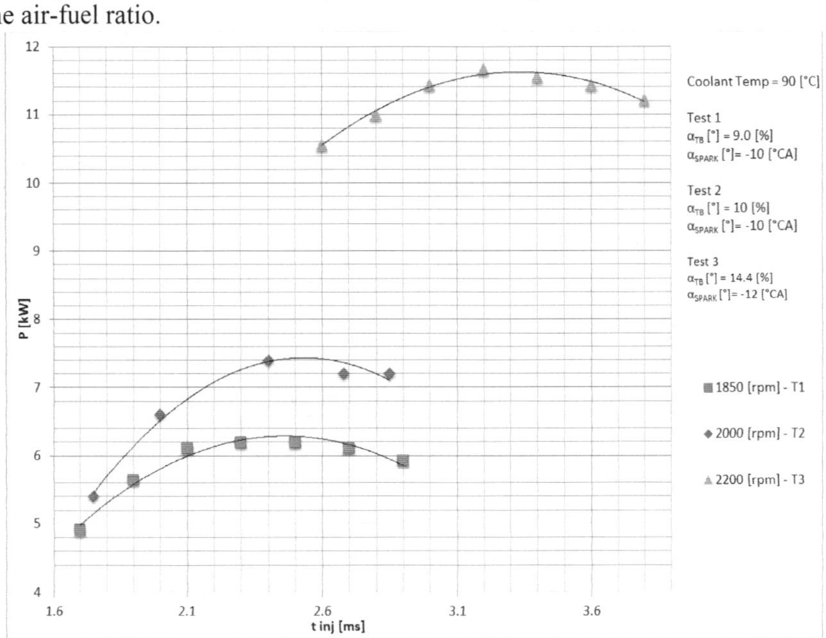

Figure 3. Power as a function of injection time

From Figure 3 we can observe that the engine power increases directly proportional with the increase of injection time, resulting in increased fuel consumption. In the literature [11] the maximum power of an internal combustion engine is achieved with an air-fuel ratio with lambda values of 0.95, this can be observed in Figure 4 that is achieved on the test bench at lambda values of ~0.95 [-].

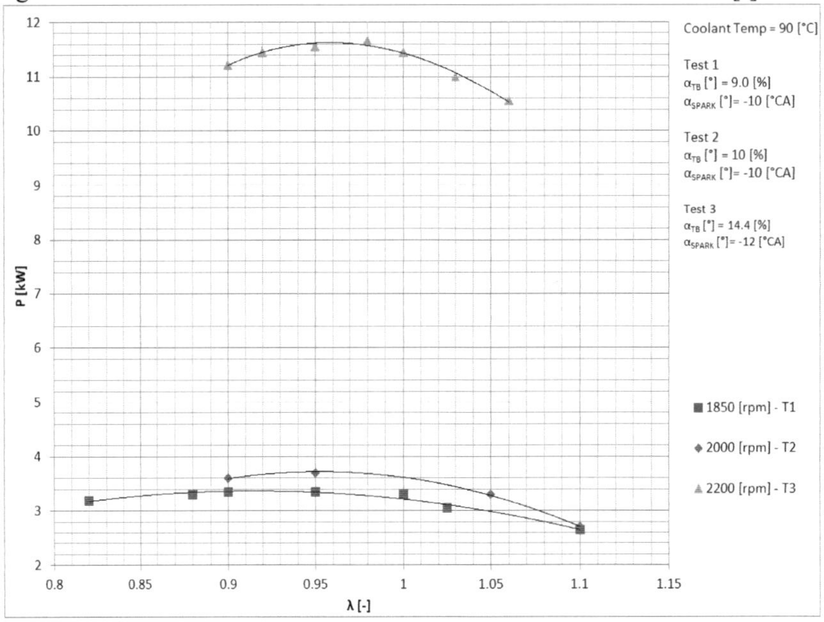

Figure 4. Power as a function of air fuel ratio

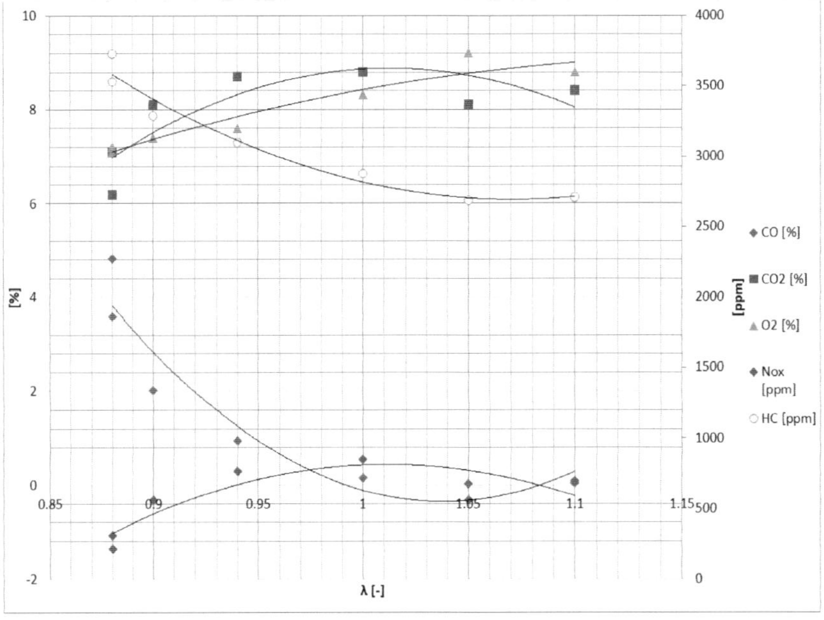

Figure 5. CO, CO_2, O_2, NO_x and HC emissions function of the air-fuel ratio

From Figure 5 the overall conclusion is that the CO emissions have a general decreasing tendency in relation with the increasing of the air-fuel ratio, while the CO_2 emissions have a general increasing tendency, while O_2 increases from a point where all the oxygen in the air-fuel mixture burns out to a point where is an excess of O_2.

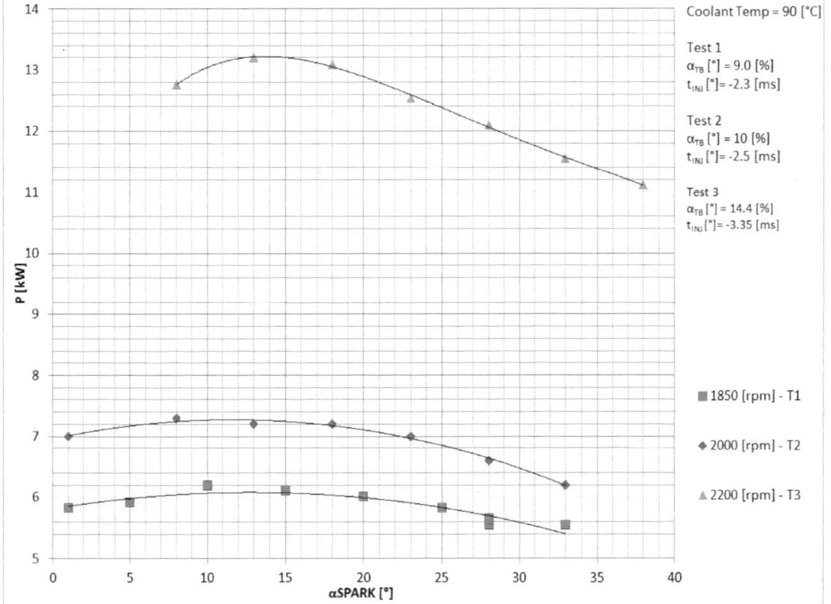

Figure 6. Power function of advance to spark ignition

From Figure 6 one can observe that the power has an increasing tendency proportional with the increase of the advance to spark ignition until a value of -14 °CA. After this value the power starts to decrease, resulting the maximum power for which the advance to spark ignition is optimum.

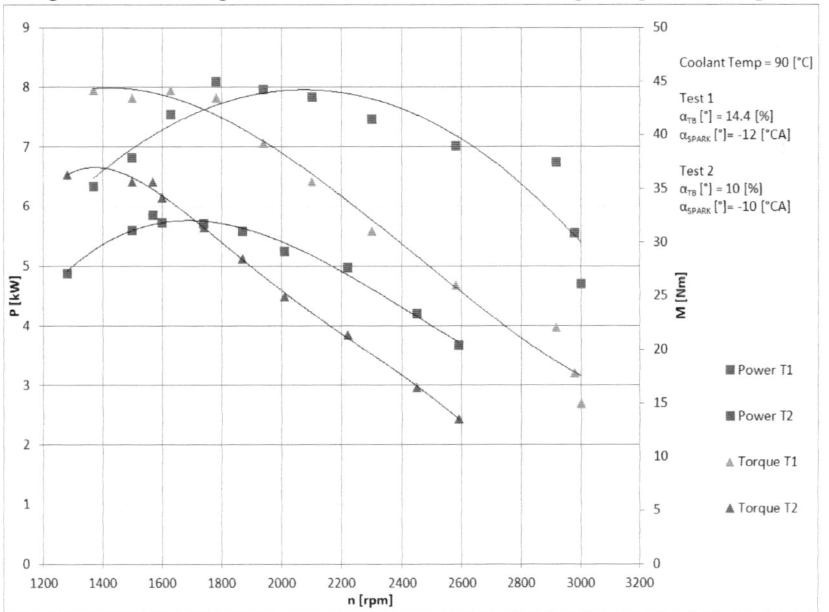

Figure 7. Partial load characteristics

From Figure 7 we can observe that both power and engine torque have a normal tendency as described in the literature.

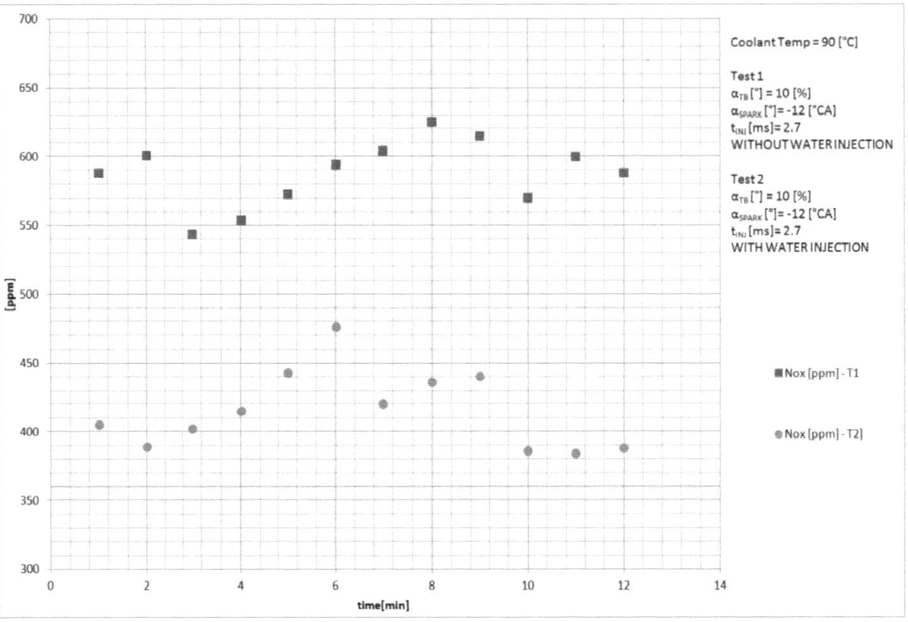

Figure 8. NO_x emissions with and without water injection

A simple test run was made under the same conditions with and without the water injection and in Figure 8 we can observe a decrease in NO_x emission values with an average of 30%. We also see that we have some high and some low peaks, they are due to the change of the water injection timing inside the combustion chamber, this means that we also have to find the correct timing of the water injection, not only the quantity.

4. Conclusions

After the tests it was determined that the spark ignition advance and injection time have a normal tendency as described in the literature [11], so we can observe that the test bench can provide us with constant valid data with little error deviations. The test bench will be improved by adding some extra sensors to record more parameters electronically.

According to Mingrui [8] he achieved a reduction of NO emissions up to 34.6% on average, for injecting 15% of water for a given mass of fuel. This thing can be observed that we reduced the NO emissions with an average of 30%.

A general conclusion is that water injection is a good field of study and without many published results.

Acknowledgment

This work was partial supported by research grants PCD-TC-2017 no.42 financed by the Politehnica University of Timisoara.

References

[1] Alkidas A C 2007 Combustion advancements in gasoline engines, *Energy Convers. Manag.* **48**(11) 2751-2761

[2] Fennell D, Herreros J and Tsolakis A 2014 Improving gasoline direct injection (GDI) engine efficiency and emissions with hydrogen from exhaust gas fuel reforming, *Int. J. Hydrogen Energy* **39**(10) 5153-5162

[3] Nande A M, Wallner T and Naber J 2008 Influence of Water Injection on Performance and Emissions of a Direct-injection Hydrogen Research Engine, *SAE International* 2008-01-

2377

[4] Subramanian V, Mallikarjuna J M, and Ramesh A 2007 Effect of water injection and spark timing on the nitric oxide emission and combustion parameters of a hydrogen fuelled spark ignition engine, *Int. J. Hydrogen Energy* **32**(9) 1159-1173

[5] Wu Z J, Yu X, Fu L Z, Deng J, Hu Z J and Li L G 2014 A high efficiency oxyfuel internal combustion engine cycle with water direct injection for wasted heat recovery, *Energy* **70** 110-120

[6] Boretti A 2013 Water injection in directly injected turbocharged spark ignition engines, *Appl. Therm. Eng.* **52**(1) 62-68

[7] Cesur I, Parlak A, Ayhan V, Boru B and Gonca G 2013 The effects of electronic controlled steam injection on spark ignition engine, *Appl. Therm. Eng.* **55**(1) 61-68

[8] Mingrui W, Sa N T, Turkson R F, Liu J P and Guo G L 2017 Water injection for higher engine performance and lower emissions, *Journal of the Energy Institute* **90** 285-299

[9] Kohketsu S, Mori K, Sakai K and Nakagawa H 1996 Reduction of Exhaust Emission with New Water Injection System in a Diesel Engine, *SAE Transactions* **105** 74-81

[10] Stoica V, Borborean A, Ciocan A and Manciu C 2017 Minimal algorithm for running an internal combustion engine, *IOP Conf. Ser.: Mater. Sci. Eng.* **294** 012036

[11] Reif K 2015 *Gasoline Engine Management*, Springer Vieweg, Friedrichshafen, Germany

NIDES as an IS Audit Tool

D P Drljača[1], D Starčević[2] and B Latinović[2]

[1]Independent University Banja Luka, Veljka Mlađenovića 12E, 78000 Banja Luka, Republic of Srpska, Bosnia and Hezergovina
[2]Paneuropean University APEIRON Banja Luka, Pere Krece 13, 78000 Banja Luka, Republic of Srpska, Bosnia and Hezergovina

E-mail: drljacad@gmail.com

Abstract. The aim of this paper is to present the role of the IDS system in the process of information system audit. The paper introduces the IDS systems, their organization and architecture, and the accent is placed on IDS systems of a network type that have in their architecture an associated expert system with the task of predicting malicious activities in the system in order to take protective or protective measures in a timely manner. The paper provides basic definitions, architecture and partially describes the functioning of these systems, the usefulness and need of consulting the IDS system in the process of information system audit. When auditing the information systems, a large part refers to the security or safety aspects of the functioning of the system, and therefore IDS systems are the ideal partner in the audit process, which can provide the auditor with very important data. These data can range from recognizing the established ways of using the system, specific actions in a system that can be considered potentially harmful to the data that definitely confirm malicious actions in the system.

1. Introduction

The basic problem of information systems is the security and protection of data integrity that circulates and communicates through the observed information system, as well as the components of that system as a whole. So far, a large number of papers have been published on the subject of security and security of information systems, there are a number of research problems related to technical and technological solutions to security and safety issues. However, surely the effort of each system administrator is to prevent or to anticipate possible disturbance of the security and integrity of the system it manages.

Although protection systems have been significantly improved over the past few years, there have been frequent infiltrations into systems and a deterioration in the integrity of their data, which often result either in loss of data or in high material costs. As the most effective tools currently available to system administrators, in practice, for detecting intrusion are *Intrusion Detection Systems* (IDS). These systems are rapidly evolving in order for system administrators to have the necessary level of insight into the security of their own system and related data.

The significance of this paper is that it helps to understand the functioning of such a system for the needs of the information system audit process, especially those systems where valuable data are stored, such as banking systems, military systems or public administration systems.

2. Goals of application of expert systems in audit

The auditor should, during and after the audit process, form a reasoned opinion on the integrity, reliability, usefulness and safety of the system being analysed, as well as audit techniques that are assisted by the computer. Therefore, it is very important to take this into account when preparing the audit. The auditor must anticipate the ways of collecting, processing and testing data and supporting documentation for support. It is important that the auditor is aware of the capabilities and potential of the expert system used in the audit process in order to make sure that the decision lines are adequate and best suited to the given environment and the system that is subject to the audit.

The practice of a financial audit is that the auditor starts the audit planning with one general audit objective. The auditor should obtain a reasonable assurance that the client's financial statements have been presented in a valid and correct manner in accordance with all material aspects and in accordance with the principles of international auditing standards. In order to achieve its objective, the auditor must obtain all the necessary and sufficient records of evidence that will serve as the basis for obtaining an audit opinion.

Similar to this definition, we can also define the objectives of the application of expert systems in the audit:

The auditor must be able ***to obtain audit evidence*** *using the expert system, which is* ***recorded in both electronic and written form*** *in the accompanying support documents. In addition, the expert system must draw* ***attention to potential inconsistencies*** *that can be identified using artificial intelligence techniques (such as neural networks).*

Section404 of so-called Sarbanes-Oxley Act of 2002 (SOX) and two directives of the European Parliament and of the Council of Europe (Directives 2006/43 / EC and 2008/30 / EC of the European Parliament and of the European Council - 'EuroSOX') commit to the establishment, documentation and management of the internal control system and their subsequent audit as part of the audit risk assessment [1], [2]. This means that it is necessary to take all necessary steps to ensure that the internal control system is reliable and provide adequate and credible findings. Also, such a system must satisfy the requirements of the said legal documents, but also to operate in accordance with best practices and standards in the field of information system audit. Therefore, the selected audit objectives, which are implemented with the help of expert systems, must be harmonized with these legal requirements and solutions.

Experience has shown that accounting and systems are most sensitive to data downturns and modifications, whether human error (intentional or accidental) or abuse. It is the purpose of the audit to identify potential forms of security breach and integrity of the system, and very often the revision of security aspects is concentrated in four basic stages:

- The **occurrence** of events,
- Event **detection**,
- Event **tracking**, and
- **Archiving** events.

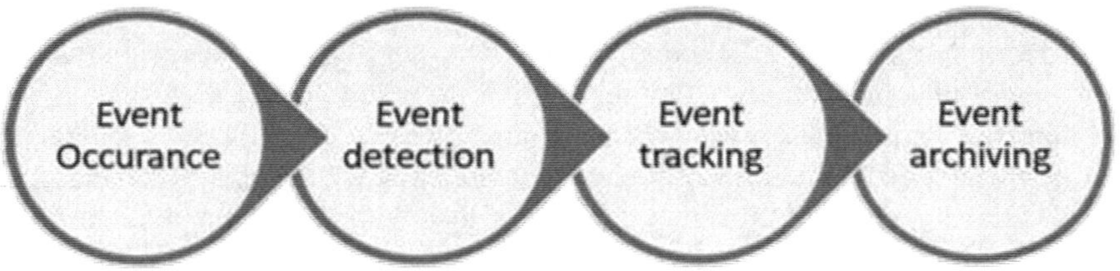

Figure 1. Audit of security incidents – flow diagram (author)

The **occurrence** of events often passes unnoticed in the system, since the challenger can be a part of the system (for example, a worker). However, it is very important that the system makes the event **detection**, or "to notice" that something is wrong, that is, that there is a latent (hidden) danger of changing, losing or misusing data or systems. The difference between occurrence and event detection, in time, can be significant, that is, relatively fair time from occurrence to detection of a particular event or incident. What does "relatively long time" mean? Such a statement is placed because very often this time frame is very small. That is, it is a very short time since the occurrence of the incident to the manifestation of the negative effects and the impact of that incident. In other words, depending on the severity and complexity of the incident, the negative effects can be manifested in a very short period of time, when the attacker is in direct contact with the system, or in a longer period of time when the attacker is waiting for a favourable moment to interact with the system and activate the unwanted an event (for example, an attacker is waiting for his working hours to access the system in order to avoid direct control). It is precisely because of this different approach that the detection of the phenomenon is very important, immediately after its emergence, if this can be achieved using an intrusion detection system.

Then the suspicious event is **monitored** to see its functioning (what is happening) in order to plan the way to eliminate the negative effects of an event or incident. Very often, this is the shortest phase because it should not be allowed the event to advance. The IDS contains certain predefined models of attacks that compares with the established and performs corrective measures already exist. However, it also happens that attacks or events are not recorded in the IDS, so it is necessary to react sensitively to the given incident. In general, it is avoided to allow further development of the incident, so the system reactions are mostly prompt and they occur immediately after identifying a negative phenomenon or intrusion and recorded in the relevant log files.

The system chooses an adequate response and neutralizes or prevents the further negative impact of event by various methods at its disposal and determines whether the incident is neutralized. If this is the case, the system **archives** an incident including archiving an event with the corresponding corrective measures, as well as adding an event to a list of records that will allow the system to prevent the re-occurrence of the same or a similar event.

In this regard, expert systems have the possibility of "*learning on the example*" and thus the possibilities of improving the technique of protection or detection of undesired changes in the system. These systems must also be designed to provide the possibility of preventive, online detection of intrusion into the system (or intrusions) in order to ensure timely protection of the system and data.

Dening proposes that such systems should be resident in the computerized accounting system, and they aim to monitor the behaviour of system users [3]. This should not be some *ad hoc* schemes that they provide a periodic check service, but the systems that constantly listen to communication and try to detect anomalies and irregularities. Dening has hypothesized that system exploitation also contains elements of abnormal use, so a security breach can be detected timely.

Examples of security breaches Deming states are:

- **Attempted decline** - a deliberate decline in the system that would allow the generation of more illegal credentials;
- **A salient decline** - represents a deliberate decline in order to make changes in the system, and the consequence of some lack of system security;
- **Penetration by a legitimate / registered user** - a common security activity where a registered user can deliberately or unintentionally endanger the integrity of the system and data, or may initiate a sequence of routines that are not in accordance with regular procedures;
- **"Leakage" of data by a legitimate / registered user** - an activity that also happens often, e.g. transmission of data by means of a transmission medium without prior authorization for such a move, printing on a network printer when it is contrary to the protection rules or "borrowing" its credentials to access the system to persons not authorized to access the data and system;

- **Virus** - also a very common phenomenon, which can manifest in several ways and in multiple manifestations;
- **"Phishing"** - the acquisition of a user's identity by hackers or other malicious persons who are not authorized to access data and system through interception and other methods.

Checking whether the system was in danger is very often tested using the so-called zero behaviour, or a specially created profile in the system for each user and for each object in the system. Then the current behaviour of the system is compared with the zero and determines whether the behaviour of the system is within the limits of normal or not. However, the question that arises is whether the artificial intelligence systems (that need to determine an intrusion) are sufficient considering the number of variables they handle or is still a human the type of detector that is the most reliable. The solution is, as always, a hybrid character, or a combination of these two modes.

3. Functioning and architecture of expert audit systems

Audit environment is a unique and very complex environment for decision making. There are many sources of errors and inconsistencies that are unique to the audit environment. Personal computers and other changes in technology have had an impact and will continue to affect the audit environment. In addition, the audit environment is focused on the process rather than on the results.

In the discussion of the audit environment complexity, Hansen and Mesir [4] argue that the audit problems are of a non-deterministic nature and as such have a large number of solutions. It is therefore very difficult (in some cases, and almost impossible) to examine all the solutions and to choose the best, but such problems are often solved by using heuristics to find a good or satisfactory solution that is not always the best and most optimal. Expert systems in this case have a key role precisely because of the involvement of heuristics in decision making, which will offer an alternative and feasible solution.

Hogart's study [5] showed that computer programs such as expert systems are very useful for improving the consistency of human thinking and reducing errors. Dillard and Macler [6] also noted that expert systems can contribute to consistency, thoroughness and credibility in the decision-making process on auditor's opinion. Expert systems are designed to provide continuous, online detection of downtime and system protection. Such systems are usually resident in computer systems and serve to monitor the behaviour of users of the given system [7].

The description of the architecture of expert systems for detection of network intrusions is mostly taken from the original document that started the development of these systems, which was developed by the group of scientists of the SRI International Computer Science Laboratory for the needs of the US Army. Although this architecture was made in 1995, it is still current and represents the essence of other systems that perform the same or similar functions.

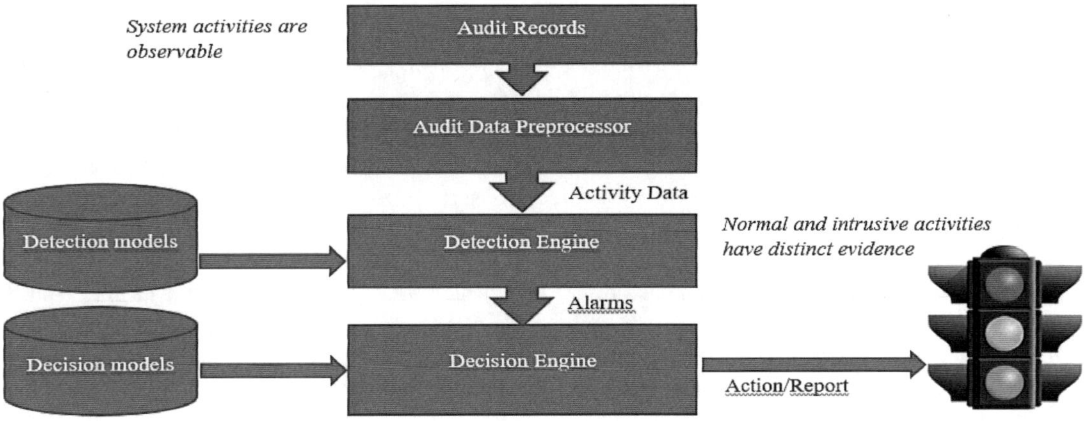

Figure 2. IDS Components (taken from [8])

The essence of NIDES is to connect several components that can interact with a remote call procedure (RPCs) or call the Library Calls. Anderson and others [9] define the following key components of NIDES:

- Persistent storage component,
- The program called *Agend*,
- *Agen* program,
- The program called *Arpool*,
- Component for statistical analysis,
- Analysis-based component analysis,
- The locking component (*Resolver*),
- Component for archiving results (*Archiver*),
- Batch analysis component
- User interface.

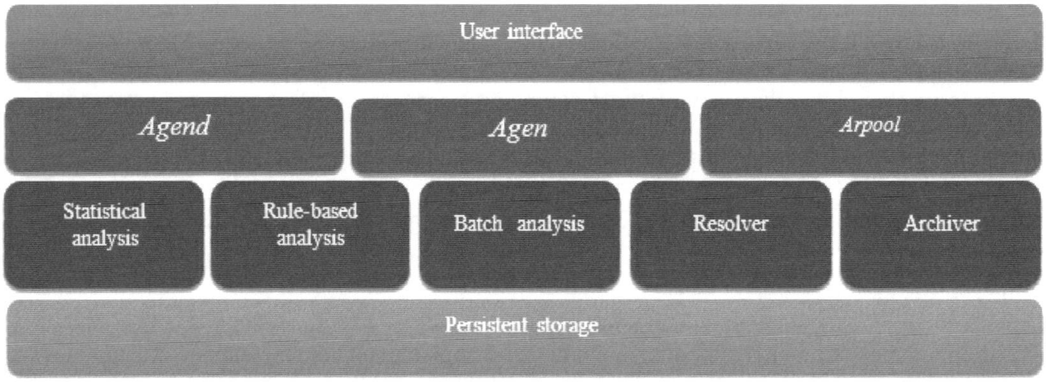

Figure 3. Four-layer NIDES architecture (adapted from [8])

The **Persistent storage component** includes a set of functions based on libraries that essentially perform the data management function for the needs of the NIDES process. The data contained in the permanent storage component are the archive data of the audit records, the archive of results, the user's statistical profile, and the analysis of the configuration information. This data set is created and stored for all instances of NIDES. NIDES instances use the concept of storing different versions of the same type of information. For example, each test that is made gets a unique name and represents a NIDES test instance that contains all the results, user profiles, and configurations analysis. Each test made has its own instance with which it is connected, and it is possible to re-use later, e.g. for multiple system tests. A special instance of NIDES is "Real Time" reserved for storing NIDES information in real time. This instance also contains all the elements - results, profiles, configurations, etc.

The **program named *Agend*** represents the background process (*daemon*) that is compulsory running on all targeted NIDES host systems. This program starts at every system boot-up for all hosts controlled by NIDES. *Agend* program works as the RPC server process and listens well-known ports regarding the request to start and stop the program for converting native auditing records (*Agen*) - that starts or stops through the user interface.

The ***Agen* program** is performed on each target host system that actively provides data for NIDES analysis. The work of this program is fairly easy to explain, although it is very complex in nature. The *Agen* program reads the original audit data, and then converts them into audit records in NIDES. After a successful conversion, it sends these records to the next program that processes them, which is the *Arpool* program. During the publication of the work (1995), NIDES functioned on the UNIX operating system that owned *Agen* programme too and supported three native audit data types: Sun OS BSM version 1, Sun OS C2, and standard UNIX accounting. The extension functionality has also been developed using the *Perl* script that has enabled users to develop their *Agen* scripts for other data sources. The task of this program is to process all available data types after the user prompts the user through the user interface and *Agend.*

The ***Arpool* program** processes the collected audit data from different target hosts and sends them to components for the analysis and detection of anomalies.

The **statistical analysis component** maintains historical statistical profiles for each user. If the system notices the activity of an individual that deviates from the established use, this component is triggering an alarm. In order for the component to function properly, it is necessary to regularly update historical profiles to help NIDES adaptively learn the behaviour of each user. The statistical analysis component can thus detect the incidents that are masked as legitimate users. On the other hand, statistical analysis can also detect attacks that use previously unknown vulnerabilities of the system, which cannot be otherwise detected. Statistical detection of anomalies can cause interesting and unusual events that can lead to discovery related to system security. Statistical analysis is flexible because the parameters and sense threshold can be easily changed from their default values, while specific crack detection measures (the behavioural aspects for which statistics are maintained) can be switched on or off. A series of papers has elaborated these statistical algorithms in detail. [10], [11]

A **rule-based analysis component** uses rules that characterize and describe known types of intrusions to enable the alarm to be triggered if the activity observed corresponds to any of the encoded rules. This type of analysis aims to uncover attempts to exploit known security vulnerabilities of supervised systems and intruders that show specific patterns of behaviour known to be essential or to violate the security policy of the system. Thus, any observed activity that corresponds to any of these predefined behaviours signifies. The rule database could be configured using the *rb_config* file, which supports the user configuration of the existing NIDES rules. The ability to customize the database of rules has been increased so that new rules can be defined and collected in an existing system, and existing rules can be turned on or off. Although the NIDES basic database is created for Sun UNIX operating systems, users should adjust the rule base for their environment and continually update the rules by adapting changes to the vulnerability of new system release and the newly discovered vulnerabilities of current releases.

The **resolver component** checks the alarms generated by the intangibles for analysis before registering them. Typically, one user action can generate a significant number of audit records. Therefore, an unusual action can lead to several tenths of hundreds of alarms reported by statistical analysis in a fast sequence. In order to avoid flooding the security administrator with redundant alarms, this component filters alarms to remove such redundancy. Alerts can be reported to NIDES consoles or emails lists, but some configuration filters can also be provided. For example, the administrator can shut out warning reporting for certain users if they know that they will do something unusual and that would otherwise generate a lot of false alerts. Although the filtered alerts have not been reported, they are have been recorded in the log.

The component for archiving the results controls and stores the audit records, the results of analyses and warnings. Archiving is possible through the user interface.

The **batch analysis component** allows the security administrator to experiment with new settings of statistical parameters or rules database settings before executing them as real-time operations in NIDES, and test results are archived for future comparison.

The **user interface** serves NIDES users to use all of the functionality. The user interface has a series of elements: drop-down menus, "*mark-and-click*" (enlarged *Point-and-Click*) selections and occasional text inputs. The interface also has a comprehensive multilayer context-sensitive help system, as well as system monitoring functionality that displays information about the supervised system, the status of the archive of audit data, an overview of the flow of the system every hour, and an overview of the generated alerts every hour.

Following is a brief explanation of the process itself, based on single event, while there is also possibility for batch event processing. The *Arpool* process retrieves data from multiple sources and combines them into a single stream of audit records by assigning a unique number of sequences to each record during this process. The NIDES generic format of the audit record can be easily adapted to new types of systems by writing simple routines for mapping audit data or using Perl scripts. Analysis components (statistical and rule-based analysis) receive audit data from the *Arpool* process that delivers audit records to all its users (those processes that have made the connection to obtain auditing data), and then discard them. Once the components have been analysed by individual audit records, these results are supplied to the locking components. The locking component analyses the results of both analytical components to determine whether a warning should be sent. If there is a need, this warning is sent to the user interface provided that the user has included at least one of the alert warning mechanisms (usually a popup notification window). Finally, all warnings issued are archived in the NIDES archive.

The basic problem of expert systems in the field of accounting data is the conditionality of their application and the domain character of expert systems. The use of the knowledge base is conditioned and based on a certain kind of symbolic logic that this knowledge can be presented. Thus, for example, logical programming is based on deduction, which also determines the goal of the program to make the problem applicable to deductive processes. The problem with the bases of knowledge of accounting and auditing information and data is not so pronounced when it comes to classical subsystems (auditing and accounting standards, etc.). But in the case of materiality or audit risk assessment, there is a significant problem due to their logical structures. In other words, such a form of knowledge may be more or less probable, and such knowledge is difficult to classify. It is questioned how the program will perform the classification of objects, regardless of which program classification rules will apply to the process of matching knowledge within the knowledge base.

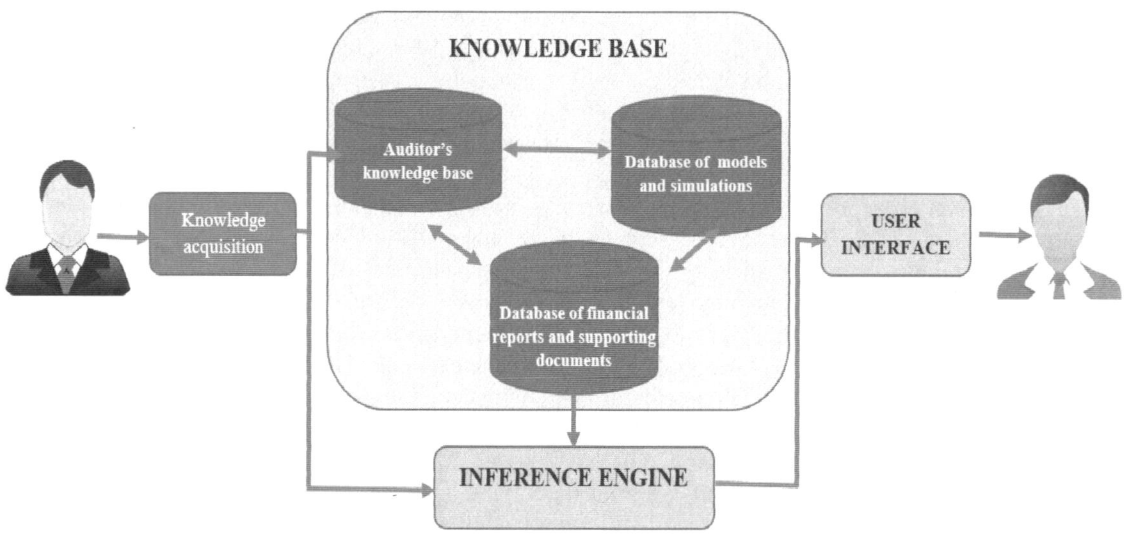

Figure 4. Generic model of expert systems' architecture for financial audit (author)

In the literature, three basic approaches to the development of this type of expert systems can be found:

- **Access to the Blackboard system approach [12-14],**
- **Frame-based approach [15],**
- **Object-oriented approach [16]**

The development of the system through **the "black board"** emphasizes the development of a set of independent domain specific modules called sources of knowledge, which interact with a common - shared data structure - "black board". In spite of the fact that it provides modularity, this approach does not say that a specific part of knowledge is being developed by other sources of knowledge - the "black board" does not "know" anything about the nature of the communication semantics. Just setting the hypothesis is a very limited form of communication when a comprehensive purpose is to mimic the processes of human thinking. Also, decisions on the further execution of the hypothesis are taken by an independent "planner", regardless of the knowledge of the hypothesis, which can reduce the coherence of the entire system [17].

A **frame-based approach** provides a more structured presentation of knowledge in the form of a framework. The box describes an object, consisting of slots containing the default values, pointers to other frames, rule sets, or procedures. Frames are related to ensuring inheritance and communication by transmitting messages. However, the modularity of knowledge presented in the frames can not be clearly defined, and the presentation lack flexibility [18]. Also, frame-based systems do not provide a way of defining fixed slots [19].

An **object-oriented approach** is an extension of a framework-based approach that allows the development of autonomous objects that communicate by transmitting messages to each other during troubleshooting. The mechanism for the transmission of messages in an object-oriented approach does not contain a provision on how the recipient must receive the message from the background and perspective of the object. Since human communication contains a context and intention other than content, only sending the content of a message becomes a very limited form of communication. In addition, the inclusion of new objects must be in line with the global scheme of the entire system, which can prove to be unfathomable in large and evolutionary systems.

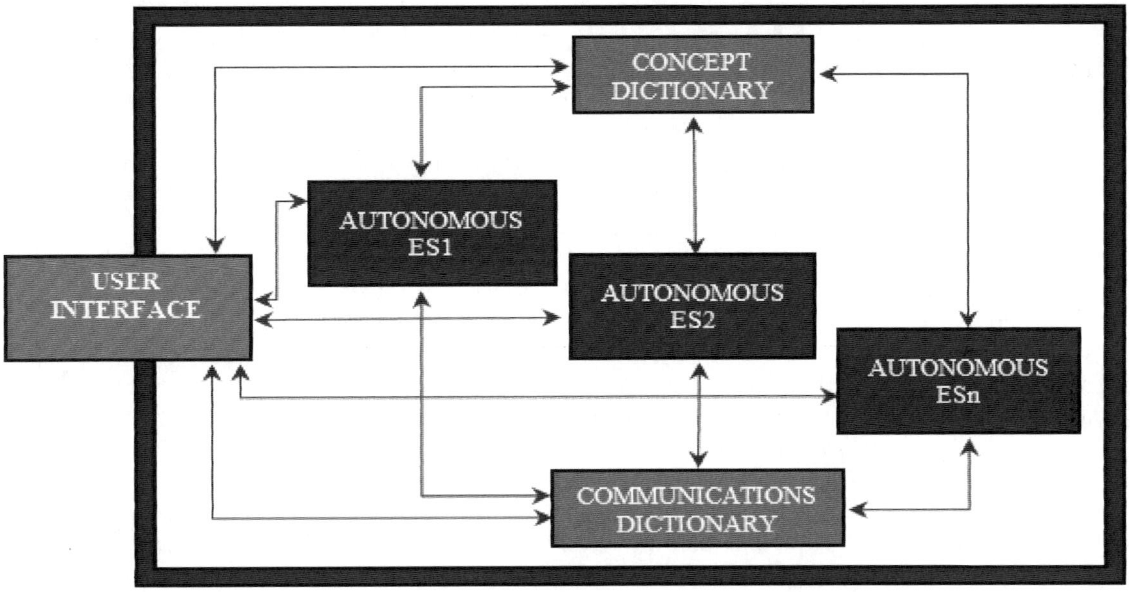

Figure 5. A diagrammatic scheme of the open architecture (adapted from [20])

In their work Kaula and Lander gave their vision of an alternative approach that would facilitate the development of large-scale expert systems to overcome the constraints of the aforementioned development approaches [20]. Their approach is based on the development of the system according to so-called ***Open-system approach*** which implies the development and existence of a system consisting of a large number of independently developed smaller subsystems that communicate and collaborate by forwarding messages (these messages can be knowledge, data, information) in the process of solving the problem [21], [22].

An open-based expert system architecture (OES) architecture is an expert system consisting of several independently developed smaller **autonomous expert subsystems (AESs)** that communicate during problem resolution. Each participant subsystem owns its own knowledge, i.e. assumptions,

IOP Conf. Series: Materials Science and Engineering **477** (2019) 012013 doi:10.1088/1757-899X/477/1/012013

limitations, and rules, and each is responsible for maintaining its own internal consistency. No subsystem can directly control the knowledge of the other, making communication and negotiation essential for solving the problem. Also, modularity is ensured, since adding or omitting the subsystem minimizes the impact on the work environment. Also, communication between the subsystems is facilitated by the use of a communication dictionary that contains procedures for implementing communication protocols called „the message acts“.

Such system architecture is very suitable for the construction of an expert systems in the field of auditing and accounting. Open architecture provides scalability and a modular approach to construction that allows for caste or system customization to audit or accounting needs. Namely, such an open architecture is very suitable for large economic systems consisting of several profit centres that need to be audited or controlled. The shell system can be modified in this way, depending on the complexity and size of the business entity or institution.

Such an expert system architecture is suitable even for the construction of an expert systems that would function in public administration systems, where each separate expert subsystem could represent a specific ministry, agency or unit of local self-government.

4. Conclusion

The beginnings of artificial intelligence date from the ancient past. All the time, one wondered if machines (computers) could dress and think instead of man and make decisions? With the advent of computers in the middle of the XX century and the understanding of the possibilities they open, the question has changed in - will the machines replace the man and his intelligence? Today, even the most famous scientists such as Stephen Hawking warn of the dangers of accelerated development of the artificial intelligence system. But the general impression is that human imagination and knowledge do not have a decent rival in the world of computers for now because it is created by the same man.

However, one of the founders of the artificial intelligence of Minsk, in his works, claims that the machines are not aware of what they are doing, but they are not aware of human beings either. Expert systems, as part of the artificial intelligence family, were made to make certain decisions instead of man. As such, they could not do this without the influence of a man - an expert, who was supposed to bring his knowledge into a system in the form of an understandable computer, which based on such a "learning" could then "draw" conclusions and make decisions based on this knowledge.

Expert systems have, first of all, found their application, except in the military industry, from where they originated, in the domain of economics, more precisely accounting. Namely, almost at the same time as they were created, they were immediately applied to solve economic problems, especially in the domain of accounting and auditing. Manipulation of a large number of numerical data required the auditor a long time to find irregularities or abuses.

This paper explains the basic concepts of artificial intelligence and expert systems, the revision and revision of information systems and attempted to make parallels between them in a historical context. Unfortunately, the author was not able to access the most recent research in this field because the results of these surveys are charged, so he had to satisfy the sources available on the Internet. Therefore, it is very possible that the presented models of expert system architecture are not very difficult, but they certainly reflect the basic concept of organization and implementation of expert systems in the field of accounting and auditing.

It is certain that further development of this area is expected, since the field of control and audit is an important area for each organization or institution, along with the overall security of the information system within which the financial module is located. Therefore, this paper put emphasis on the presentation of expert systems that perform system monitoring and the prevention of intrusion into the system, as well as expert systems that help the auditor to find potential problems and abuses in a large amount of data. There are still many other issues open in this field, but certainly the use of neural networks and genetic algorithms is something that will continue to be studied, and it can be expected that these systems will inevitably be part of future plans and training programs in the newly emerging area, which is called science Data Science.

IOP Conf. Series: Materials Science and Engineering **477** (2019) 012013 doi:10.1088/1757-899X/477/1/012013

References

[1] Ramos M 2004 Section 404 Compliance in the Annual Report, *Journal of Accountancy* **10** 43-48

[2] Dunn C L, Cherrington J O and Hollander A S 2005 *Enterprise Information Systems: A Pattern-Based Approach*, 3. ed., McGraw-Hill Irwin, Boston, MA

[3] Denning D 1987 An Intrusion Detection Model, *IEEE Transactions on Software Engineering* **13**(2) 222-232

[4] Hansen J V and Messier W F 1982 Expert Systems for Decision Support in EDP Auditing, International Journal of Computer and Information Sciences **11** 357-379

[5] Hogarth R 1985 *Judgment and Choice*, Wiley, Chichester, USA

[6] Dillard J and Mutchler J 1987 *Expertise in Assessing Solvency Problems, Expert Systems*, 170-178

[7] O'Leary D E and Watkins P R 1989 *Review of Expert Systems*, in Auditing, Expert Systems Review, University of Southern California, USA

[8] Stolfo S 2017 Teaching presentation slides, Columbia University New York, Columbia Engeeniring The Fu Foundation School of Enginnering and Applied Science, Computer Science Department, https://www.cs.columbia.edu/~smb/classes/f06/l19.pdf

[9] Anderson D, Frivold T and Vlades A 1995 Next-generation Intrusion Detection Expert System (NIDES) - A Summary, SRI International, Computer Science Laboratory SRI-CSL-95-07

[10] Harold J S and Valdes A 1994 The NIDES statistical component description and justification, Annual report, SRI International, Menlo Park, CA

[11] Valdes A and Anderson D 1994 Statistical methods for computer usage anomaly detection using NIDES, In Conference on Rough Sets and Soft Computing

[12] Balzer R, Erman L D, London P E and Williams C 1980 Hearsay-III: a domain-independent framework for expert systems, Proceedings AAAI

[13] Erman L D, Hayes-Roth F, Lesser D R and Reddy V R 1980 The Hearsay-II speech-understanding systems: integrating knowledge to resolve uncertainty, *Computing Surveys* **12** 213-253

[14] Nii H P 1986 Blackboard systems: the blackboard model of problem solving and the evolution of blackboard architectures, *The AI Magazine*

[15] Minsky M 1975 *A framework for representing knowledge*, in Winston P (Ed), The Psychology in Computer Vision, McGraw-Hill, New York, pp 211-217

[16] Yoon Y and Guimaraes T 1992 Developing knowledge- based systems: an object-oriented organizational approach, *Information Resources Management Journal* 15-32

[17] Kaula R and Ngwenyama O 1990 An approach to open intelligent information systems, *Information Systems: An International Journal* **15** 489-496

[18] Leung K S and Wong M H 1990 An expert-system shell using structured knowledge, *Computer* 38-47

[19] Giarratano J and Riley G 1989 *Expert Systems: Principles and Programming*, PWS-Kent Publishing Co, Boston, MA

[20] Kaula R and Lander L C 1995 A module-based conceptual framework for largescale expert systems, *Industrial Management & Data Systems* **95**(2) 15-23

[21] Hewitt C and de Jong P 1984 *Open systems*, in Brodie M L, Mylopoulos J and Schmidt J W (Eds), On Conceptual Modelling: Perspectives from Artificial Intelligence, Databases and Programming Languages, Springer-Verlag, Berlin, pp 147-164

[22] Kaula R 1990 *An open intelligent information systems architecture*, PhD Thesis, State University of New York, Binghamton

IOP Conf. Series: Materials Science and Engineering **477** (2019) 012014 doi:10.1088/1757-899X/477/1/012014

Hybrid power generator based on wind, hydro and solar for use as an alternative solution for power supply

V V Mihut, I Baciu, C Panoiu and M Panoiu

Politehnica University of Timisoara, Department of Electrical Engineering and Industrial Informatics, 5 Revolution Street, Hunedoara, 331128, Romania

E-mail: manuela.panoiu@fih.upt.ro

Abstract. The paper proposes the design of a hybrid generator based on wind, solar and/or hydro power. The proposed generator is intended to be used in areas where there is no power supply. Such a situation may be a house located in an isolated geographic area. Another application can be a portable hybrid generator. The existence of many geographic areas where it is not possible to expand the electric grids or isolated places in terms of relief, road access especially during the winter period leads to the necessity to use alternative electric sources. As an example, one can take into account the mountain area where the only access road is non-auto. Conventional alternative electrical sources require refuelling or are sources whose components are fixed and fitted permanently or are difficult to remove-mounted elsewhere. Often, component parts are considerable or require technical skill to be properly designed.

1. Introduction

Solar thermal is one of the main sources of renewable energy because it is the largest available globally. If only 1% of the desert area of the earth were to be used for the construction of plants to produce electricity, the energy needed for the year 2017 would have been assured. Also, solar radiation is more homogeneously spread in the tropical-equatorial area of the globe than wind or biomass, so more locations for plants are available than for the other two sources [1], [2]. Solar energy production technologies are the most economical in terms of costs, becoming competitive with conventional fossil-fueled power plants.

1.1. Why solar energy?

The amount of energy received from the sun corresponds annually to 1.5 billion million MWh, which represents about 23 000 times the current energy consumption, or 5-10 times the sum of all known fossil fuel reserves, including the ore uranium. The availability of this energy depends on the day-night cycle, the latitude of the place where it is captured, the seasons and the cloudy [3], [4]. Photovoltaic solar energy is based on the direct production of electricity through silicon cells [5]. When it shines and when the climate conditions are favorable, the sun provides a power of 1 kW / sqm [6]. Photovoltaic panels allow the direct conversion of electricity to 10 to 15% of this power [6]. For example, a photovoltaic roof of 5x4 meters has a power of 3kW and produces 2-6 MWh / year [7]. The photoelectric effect, namely the conversion of solar energy ("photon") into electrical energy ("volt"), was discovered in 1839 by physicist A. Becquerel [8]. This effect is based on three simultaneous physical phenomena, closely linked to each other: absorption of light by materials, transfer of energy from photons to electrical loads, collection of loads.

2. The proposed solution

The proposed solution takes into account the following mandatory features for such units:

- Easy, portable equipment that can be moved and mounted by a single person.
- All equipment should not exceed 20-25 Kg.
- Ability to capture and store as much electric power as possible in terms of overall dimensions and weight.
- Ability to capture any type of electricity, hydro, solar and wind.
- Operating life of at least 20 years for the system and 5 years for batteries (the batteries can be replaced).
- Easy to mount, exploit and weatherproof.
- Low cost.
- mini-turbine with vertical spindle 24V - 200W
- photovoltaic polycrystalline panel 30Wp
- 12V / 24V / 15W hybrid wind-solar controller
- Inverter 12V c.c. - 220V c.a., 500W
- Non-maintenance gel battery 12V - 24Ah
- 5V power supply c.c. (1 USB port), 12V cc (1 port), 220V c.a. (1 port).
- Portable aluminum frame, copper connectors and cables, fuse 10A (auto), on-off switch.

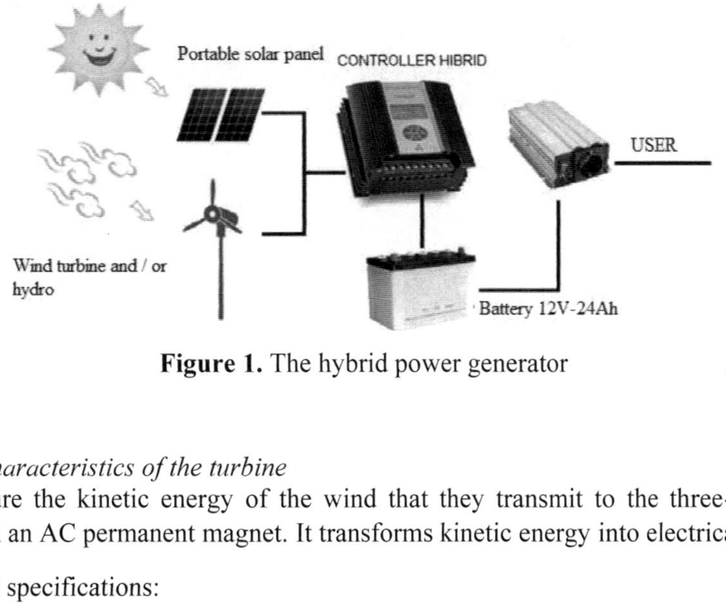

Figure 1. The hybrid power generator

2.1. Technical characteristics of the turbine

The blades capture the kinetic energy of the wind that they transmit to the three-phase generator, synchronous with an AC permanent magnet. It transforms kinetic energy into electrical energy.

Turbine technical specifications:

- Nominal power: 200 W
- Rated voltage: 24 V
- Starting Wind Speed: [m / s]: 2
- Nominal wind speed: 10 m / s
- Wind speed of survival: [m / s]: 45
- Minimum working temperature: [° C]: -40
- Maximum working temperature: [° C]: 80
- Weight: 5 Kg
- Generator type: three-phase, synchronous, permanent magnets

Figure 2. The wind turbine

2.2. Design of the hybrid generator

These turbines are used together with a hybrid controller that can be used in parallel on photovoltaic panels systems. The wind / hydro-solar controller is a device that simultaneously controls the efficient charging of the wind turbine-hydro and the solar photovoltaic panels. To design such a generator, calculate the average charge value for each alternative power source.

Technical characteristics: The load regulator has the ability to automatically detect the 12V / 24V voltage. Possibility of setting the type of solar batteries used: VRLA, AGM, GEL. The charging controller protects the system from overloading or discharging (minimum limit is 11.1V), overheat protection, short-circuit protection.

The hybrid generator nominal power is 350 W, and have connectors for: solar panel + wind turbine, accumulators and consumers.

Without using a controller between a solar panel and a battery, the panel would overload the battery by generating too much energy, causing serious damage to the battery.

When the battery falls below a certain voltage level, the controller disconnects the battery charge to prevent the battery from draining. A fully discharge battery will lose part of its total capacity over time. Low battery voltage may damage the battery if the load is connected. The controller has the function of stopping any current flowing back into the solar panel at night. This prevents damage to the solar charging kit.

It is proposed to use a PWM controller with a new technology to increase battery life, LED indicators to indicate battery charging status, and temperature compensation by automatically changing charging parameters, overload protection, discharge and short circuit and reverse polarity.

The USB port ensures the loading of mobile devices with a direct USB charger. The USB port can connect 2 consumers simultaneously with an absorbed current of max 1.2A. You can set the type of battery that can be charged to avoid damaging it.

The controller used will be a PWM and will be optimized to get the maximum battery charge current, depending on their model (VRLA, AGM, GEL), decoupling other functions that do not require use for this purpose (e.g. twilight sensor, various time-outs), functions that consume extra electricity from the two photovoltaic, wind / hydro units.

Depending on the load algorithm, the solar charge regulators are of two types: PWM (Pulse Width Modulation or Impulse Modulation) or MPPT (Maximum Power Point Tracking or Maximum Power Point Tracking).

PWM Charge Controllers constantly check the battery condition to determine the impulse transmission speed, but also their intensity. In the case of a charged battery, the charging regulator will transmit short pulses, and the impulse will be almost continuous when pulled out. Also, this type of controller can check the charge status of the battery between pulses and adjust them as needed. Basically, the PWM charging controller uses a simple operation system, being an on-off switch. Controllers - PWM load regulators are used in small photovoltaic systems and under high temperature photovoltaic panels.

MPPT Charge Controllers have the ability to manage voltage much better than PWM controllers and are typically used in photovoltaic systems whose output voltage across panels is higher than the battery bank voltage and systems operating under conditions of low photovoltaic panel temperature.

2.3. 30W solar panel

A 30W solar panel normally produces 4A/hr in bright sunshine. For the 12V-24Ah chosen battery, it will take about 24/4 = 6 hours of sunshine to theoretically charge from 0% to 100% of the battery. Knowing that the storage capacity of the battery will never fall below 20% due to the limitation of the controller we will calculate the average charging time under ideal conditions. Considering the energy losses on the photovoltaic panel-hybrid-battery controller it is considered that the full load time in bright sunshine conditions (ideal case) is about 6 hours. Under average sunshine conditions, the maximum charge time is 7 hours.

2.4. Hydro-wind turbine 24V - 200W

Considering that the electric circuit is equipped with 12V batteries and the controller detects this voltage, it results that this turbine will generate double the power, ie 400W at nominal speed. According to the formula $P = U * I$, taking into account these aspects result in nominal power:
200 W = 12 V * I resulting approximately I = 16.5 A.

Mathematically speaking, charging the 24Ah battery at a nominal rotation (ideal case) of the generator will take place in less than one hour. As in the case above, the battery will never fall below 20% of the power load and by adding the power losses on the hybrid-battery generator-circuit controller it is considered that the full charging time under nominal rotation conditions (ideal) is about one hour. Under average and fluctuating rotation times of the generator, the maximum loading time is considered 2 hours.

We propose 2 variants

Variant 1 = parallel operation of the photovoltaic panel and of a single hydro or wind generator, we can approximate the average charging time of the whole battery system in maximum one hour.

Variant 2 = parallel operation of the photovoltaic panel and 2 hydro + wind generators, we can approximate the average charging time of the entire battery system in 30 minutes.

Designing the whole assembly will be done taking into account the maximum weight a man can carry for a longer time, 20-25 Kg.

The formula of calculation takes into account the weight of the person wearing the equipment, knowing that a person can carry in average a quarter of his own weight. The average calculation was made for a person of 80-100 Kg.

Considering that the power discharged by two alternative power sources (variant 1) is sufficient to keep the battery loaded and considering that the entire assembly does not exceed the average weight, it will be used for a vertical shaft generator to which the propeller it can be used both in hydro-generator system and in wind system (according to generator image).

In Figure 3 the scheme of the proposed system is shown.

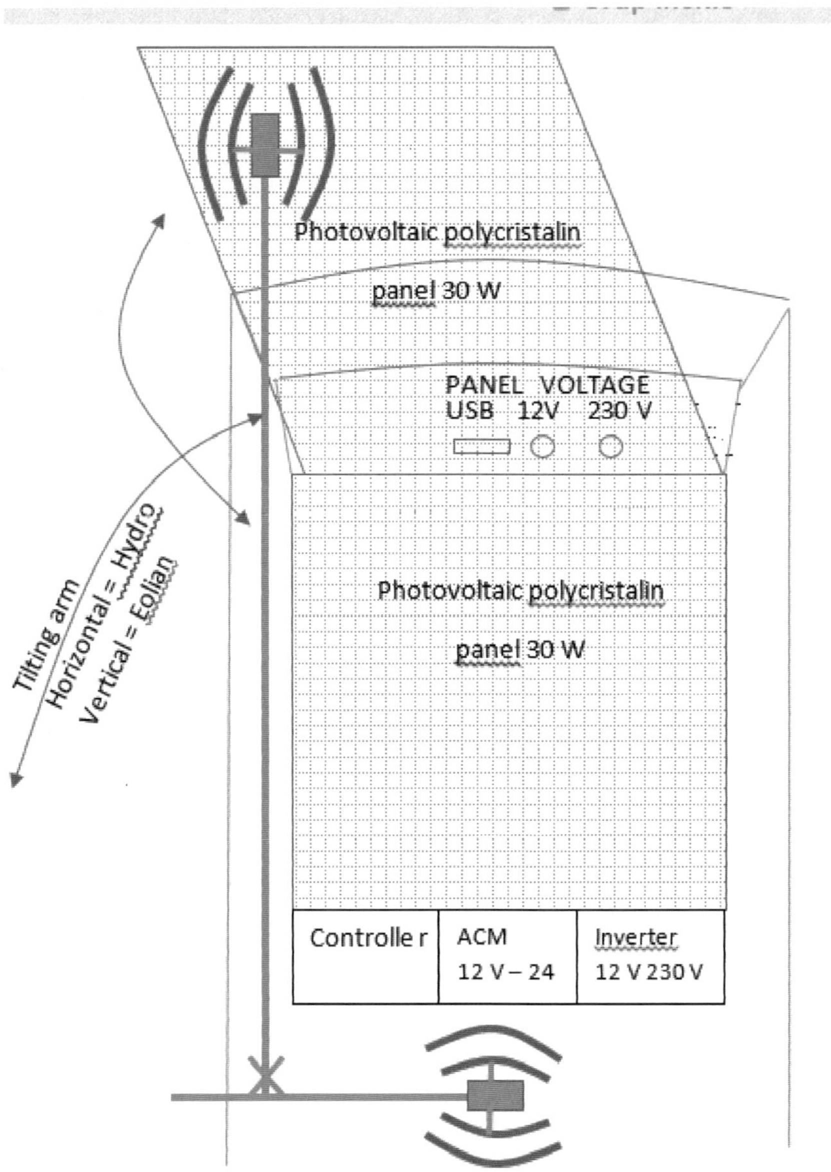

Figure 3. The structure of the propose portable hybrid generator

3. Conclusions

The need to use alternative energy sources is increasing. Transforming solar and hydraulic power especially small hydropower into electricity is not polluting [9], [10]. The paper proposes such a system to capture solar, hydro or wind energy as needed. The proposed solution can be used as a portable mini-generator for areas where there is no power supply. This way, you can supply some equipment that is absolutely necessary. With some modifications the proposed solution can also be used as a generator for a building located in an isolated geographic area where either there is no electricity or it happens frequently that the power supply is interrupted.

References

[1] Guo S, Liu Q, Sun J and Jin H 2018 A review on the utilization of hybrid renewable energy,

Renewable and Sustainable Energy Reviews **91** 1121–1147

[2] Mahmoudi H, Spahis N, Goosen M F, Sablani S, Abdul-Wahab Sabah A, Ghaffour N and Drouiche N 2009 Assessment of wind energy to power solar brackish water greenhouse desalination units: A case study from Algeria, *Renewable and Sustainable Energy Reviews* **13** 2149–2155

[3] Rogers J G, McManus M C and Cooper S J G 2013 Potential for reliance on solar water heating throughout the summerin northern cloudy climates, *Energy and Buildings* **66** 128–135

[4] ***https://www.worldenergy.org/wp-content/uploads/2016/10/World-Energy-Resources_FullReport_2016.pdf

[5] Yuan J, Farnham C, Emura K and Lu S 2016 A method to estimate the potential of rooftop photovoltaic power generation for a region, *Urban Climate* **17** 1–19

[6] Vartanyan Yu L, Grigoryan A K, and Shahinyan H A 2015 Maximum Mass of Strange Stars and Pulsars with the Most Accurately Measured Masses, *Astrophysics* (2015)

[7] Desideri U, Campana P E 2014 Analysis and comparison between a concentrating solar and a photovoltaic power plant, *Applied Energy* **113** 422–433

[8] ***http://solenergy.com.ph/solar-panel-philippines-edmond-becquerel/

[9] Deaconu S I, Babău R, Popa G N and Gherman P L 2018 Hydroelectric power plant with variable flow on drinking water adduction, *IOP Conf. Ser.: Mater. Sci. Eng.* **294** 012023

[10] Dinis C M, Popa G N and Iagar A 2017 Analysis of synchronous and induction generators used at hydroelectric power plant, *IOP Conf. Ser.: Mater. Sci. Eng.* **163** 012033

Connection time study for 400 kV circuit breakers with SF6

D Ciulica

Politehnica University of Timisoara, Department of Electrical Engineering and Industrial Informatics, 5 Revolution Street, Hunedoara, 331128, Romania

E-mail: dumitruciulica@yahoo.com

Abstract. This paper presents the calculation of the time for the SF6 circuit-breaker operation by studying the voltage characteristic before connecting. Simulated on the basis of the switch characteristics and the factors that interfere with the time, i.e. the temperature and supply voltage of the breaker coil. Connecting the circuit breakers has a very important role in eliminating the vibrating and electrodynamic loading regimes in high voltage electrical networks.

1. Introduction

The SF6 switches are among the used switches that are reliable. The choice of when to connect is very important in the switching process, so in our case it is considered that the optimal moment is at the maximum voltage of the bar. For our case, it is considered a switch model in a three-phase network and one phase is studied.

The switching time for the circuit breakers is calculated by a program installed in the Bay Control Unit (BCU) of the control unit located in each cell of the power station and receives the command from the user to close or open the switch at the point on the voltage or current taken as a reference. The method is also called synchronous switching or controlled switching [1-3]. Switching control provides an efficient solution taking into account the current values, the control unit optimizes the switching operations of the switch using the instantaneous voltage curve or the instantaneous current curve [4].

The operational principle of the OPEN operation is the same as for CLOSE with the difference that it can be used as a reference signal for a zero-pass current and that the recharged currents [5], [6] are monitored. The main advantage of this method is the reduction of the transient switching effects, with implicit reduction of voltages in the system and its equipment [7].

Closing: The time that passes between the electrical control emitted by the shut-off switch and the firm closing of the contacts at all switches of the switch is the closing time. The process of extinguishing the electric arc in the circuit breaker can cause a current flow that can rearm the arc during the arc-off period. The time elapsed between the command emitted by the circuit-breaker and the occurrence of the electric arc (starting current) at the first pole of the switch is considered the zero point (time origin). The switching process for the three poles of the circuit breaker must be viewed separately for switching control. To study the switching process, we define the following times:

- Closing time: The time elapsed between the electric switch at the pole of the breaker in question and the closing of the contacts (galvanic) in the circuit of the switch.
- Working time: The time elapsed between the electrical command at the pole of the circuit breaker (starting current) and the electric arc pole of the circuit breaker.

- Time pre-arcing: The time at which the arc burns at the switch's pole during closure.

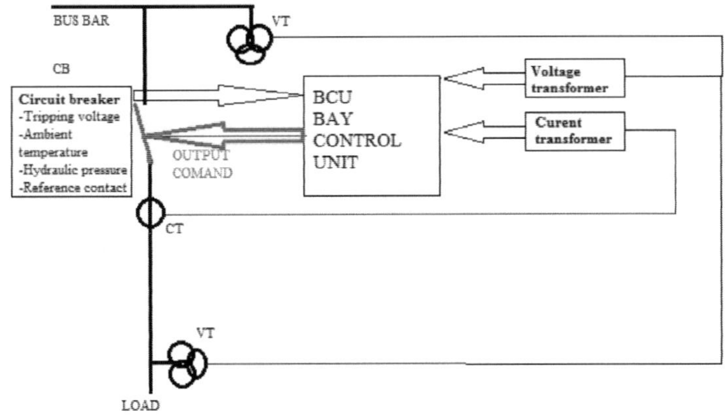

Figure 1. Switching Block Diagram [8]

2. The proposed solution

2.1. Mathematical model

For the study, one-phase system was used U=8,6 V , ϕU=-0.34 rad, f=50Hz, I=3A, ϕI=1.24 rad

$$u(t) = \sqrt{2}Usin(2\pi ft + \phi U) \quad (1)$$

$$i(t) = \sqrt{2}Isin(2\pi ft + \phi I) \quad (2)$$

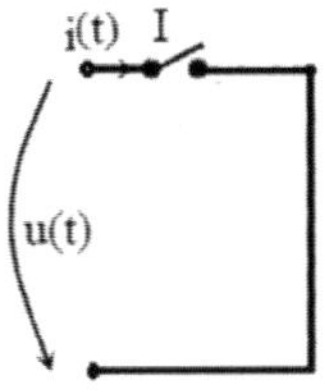

Figure 2. The monopolar scheme for the studied circuit breaker

2.2. Time model

The time model is shown in Figure 3.

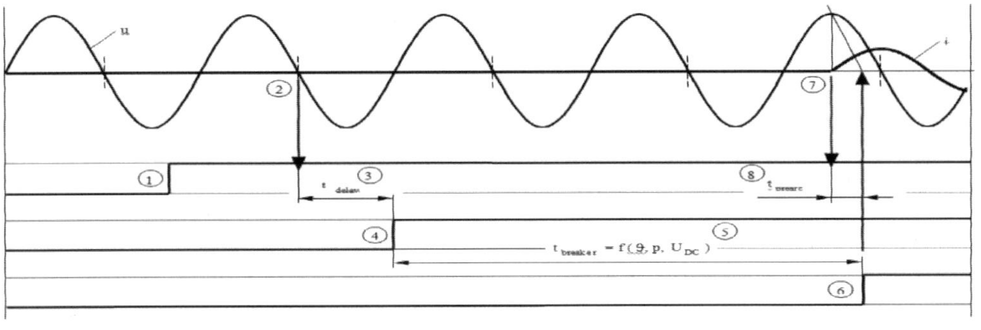

Figure 3. Closing operation of circuit breaker with a control system [9]
1 - Closing command
2 - Identification of the zero point of the stresses
3 - Delay time 4 - Circuit breaker control 5 - Closing time
6 - Touch contact 7 - Starting current 8 - Pre-arcing time

2.3. Calculation of time coefficients for switching the studied switch

Calculation of time coefficients dependent on:

tTemp=f(T) depending on ambient temperature

In Table 1 is presented the temperature dependent coefficient.

Table 1. Time-dependent temperature-dependent coefficient [10]

T[°C]	t(T)(ms)
40	71.5
20	72
0	72.5
-20	73
-30	74
-40	81

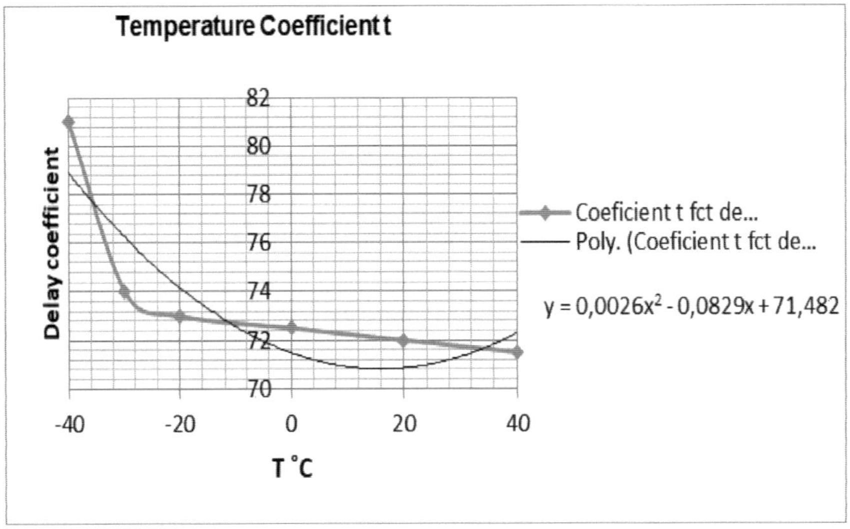

Figure 4. Evolution of the time factor due to temperature when switching on the circuit breaker

Since the switch operates under normal conditions, ie the temperature within the range of 0°C + 40°C, we can approximate this coefficient to 72 milliseconds for the closing operation.

tUDC =f(UDC) depending on the DC voltage (d.c.) Unom = 220V (d.c.) is the nominal voltage.

Table 2. Coefficient of time for U-auxiliary power supply coil (d.c.)

U (measured)[V]	Coefficient
U control (V)	*Coef (msec)*
187	6.6
192	5.6
203	3.4
209	2.2
221	-0.2
232	-2.4

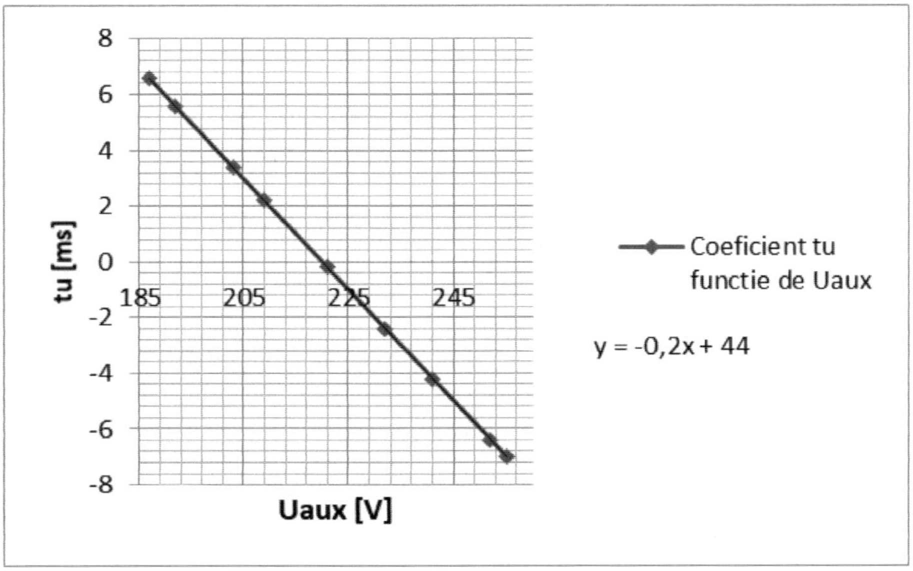

Figure 5. The evolution of the time factor that depends on the auxiliary voltage of d.c.

These values are measured by varying the supply voltage of the circuit breaker and measuring the switch-off time of the circuit-breaker. As can be seen from this graph, the coefficient of time dependent on the supply voltage supply voltage of the supply coil has a linear characteristic and has a null value of Unom = 220 V and decreases with increasing the voltage value.

This voltage must be in the range Umin = 187 V and Umax = 255 V if it exceeds this range and it is signaled by the monitoring relays in the internal d.c. services.

Time factor that depends on the pressure of the hydraulic agent (if the circuit breaker is operated with hydraulic oil actuators).

tP=f(P) depending on pressure Pnom=330 bar

In Table 3, the values of the coefficient according to the pressure of the hydraulic agent are shown.

Table 3. Values of the coefficient according to the pressure of the hydraulic agent.

P hydraulic (bar)	Coef (msec)
310	6
316	4.2
321	2.7
326	1.2
333	-0.9
337	-2.1
342	-3.6
346	-4.8
350	-6

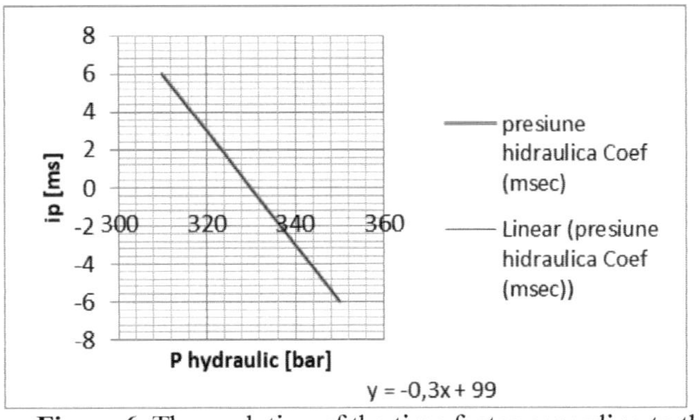

Figure 6. The evolution of the time factor according to the Hydraulic Pressure

The graph shows a linear evolution y = -03x + 99 with downward slope to increase the pressure of the hydraulic agent. These values of the closing time of the breaker are taken from the circuit breaker's technical sheet [10]

3. Graphics simulation and calculations

The logical diagram for the simulation of switch circuit breaker is shown in Figure 7. Graphics and simulation were done using Geogebra 4.4 and Excel.

4. Conclusions

The simulation is very useful in studying the operation of a high power circuit breaker [10-12]. Based on the results obtained, the following conclusions were identified: simulating the closing time for SF_6 circuit-breakers depends largely on the switch type; for each type of switch, its characteristics must be raised depending on the temperature, the supply voltage with d.c. the actuating coils, and the hydraulic agent pressure; since the switch operates under normal conditions, ie the temperature within the range of 0 ° C + 40 ° C, we can approximate this coefficient to 72 milliseconds for the closing operation; to check these times, it is necessary to use auxiliary contacts that copy the position of the circuit breaker contacts very fast and accurate; the calculation algorithm for the connection time is simple. It is all about calculating the seven time points and applying the command to the switch's coil as soon as possible. Measurement with high precision by means of existing transducers in the temperature and voltage. The simulation was done for a single phase, in case the circuit breaker control is monopolar, then the phase shift command must be transmitted in phase to the studied phase with a certain calculated delay.

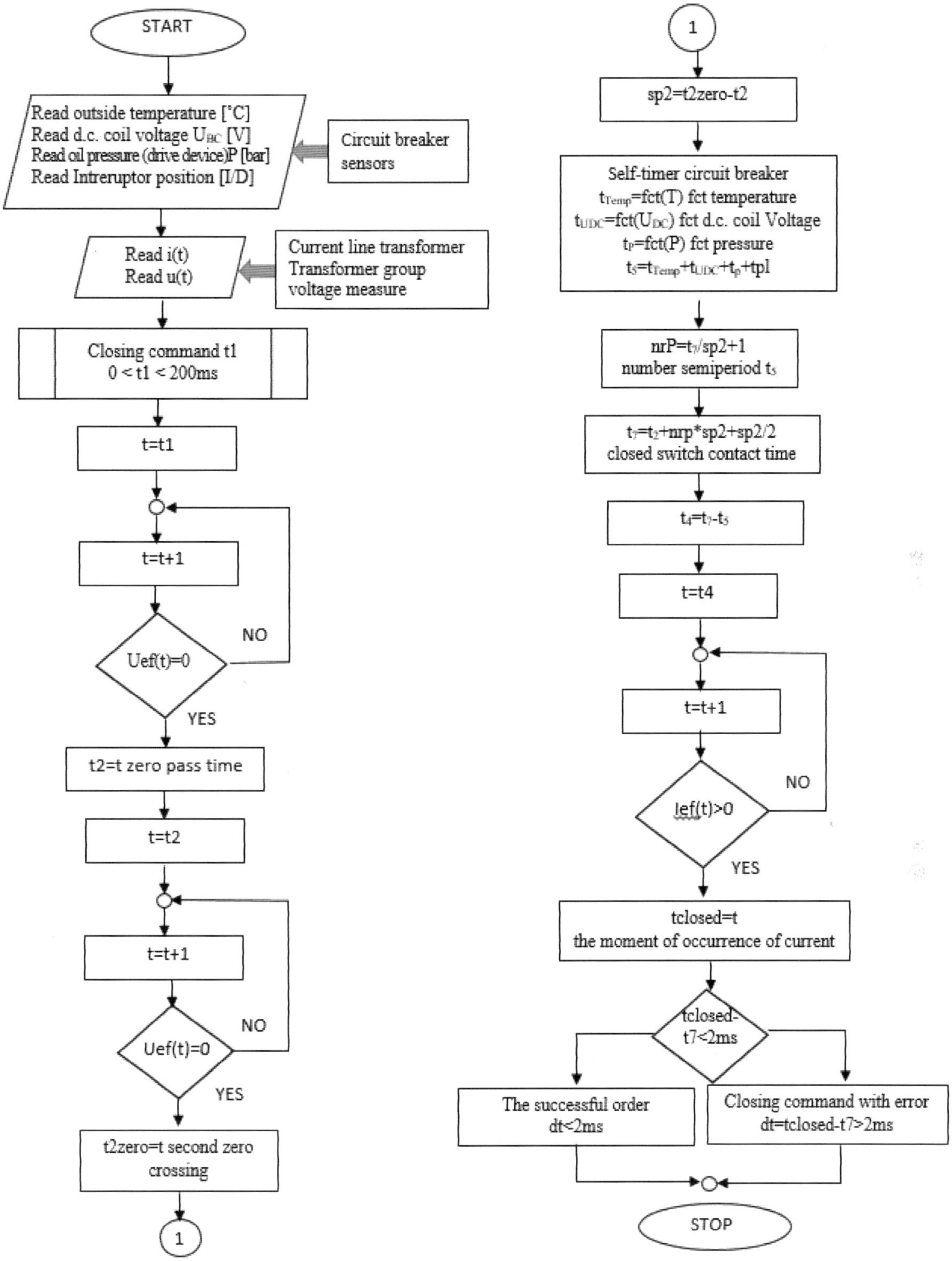

Figure 7. Logic circuit diagram of the circuit breaker

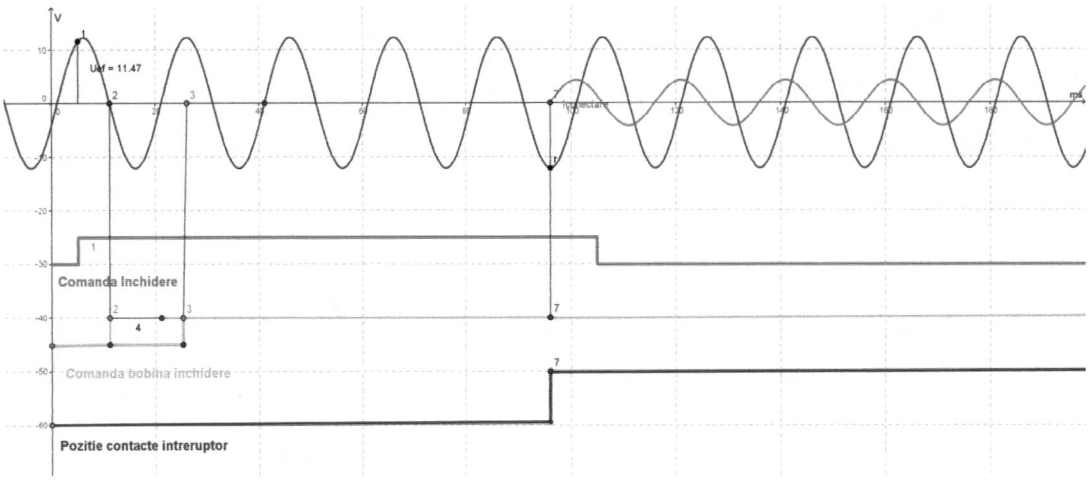

Figure 8. The single-phase connection feature for t1 = 5ms

Table 4. Calculation of time coefficients for connection characteristic t1 = 5ms

U	**8.6**	**V**		
f	**50**	**Hz**		
ϕU	**-0.34**	**rad**		
Temperature	**18**	tT	71	ms
U cc	**238**	tucc	-4	ms
P bar	**325**	tp	2	ms
		t5	69	ms

t1	5		ms
t2	6	12.158176	ms
	7	11.66022	ms
	8	10.02088	ms
	9	7.4006272	ms
	10	4.0559489	ms
	11	0.3142462	ms
	12	-3.4582172	ms
	13	-6.8921662	ms
	14	-9.6514619	ms
	15	-11.466005	ms
u(t)=0	t2=	11	ms
t=f(T,U,P)	t5=	69	ms
SP2=1000/f/3	sp2=	10	ms
nrp= t5/SP2 +2	nrp=	8	ms
t7=t2+nrp*SP2+Sp2/3	t7=	96	ms
t4=t7-t5	t4=	27	ms
t3=t4-t3	t3=	38	ms

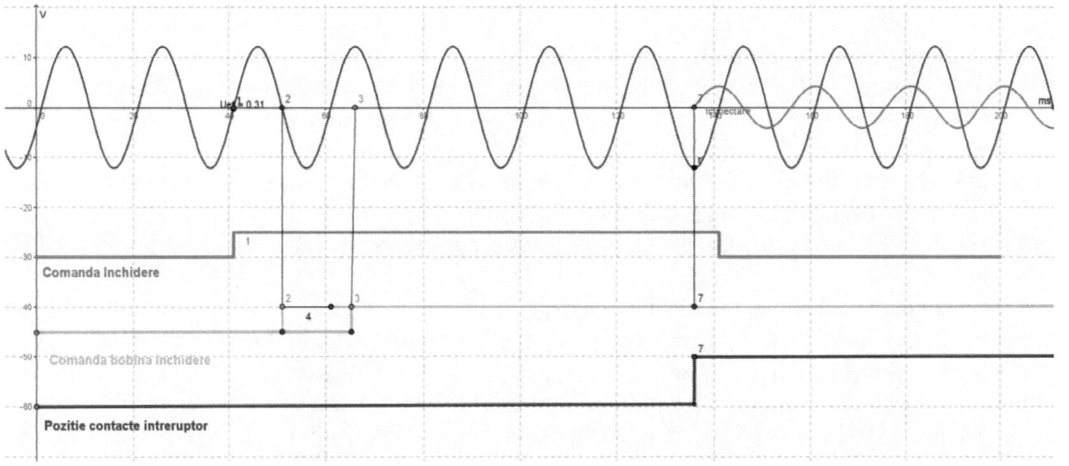

Figure 9. One-phase connection characteristic for t1 = 41ms

Table 5. Calculation of time coefficients for connection characteristic t1 = 41ms

U	**8.6**	**V**		
f	**50**	**Hz**		
ϕU	**-0.34**	**rad**		
Temperature	**23**	tT	71	ms
U cc	**242**	tucc	-4	ms
P bar	**330**	tp	0	ms
		t5	67	ms
	t1	**41**		ms
	t2	42	3.4582172	ms
		43	6.8921662	ms
		44	9.6514619	ms
		45	11.466005	ms
		46	12.158176	ms
		47	11.66022	ms
		48	10.02088	ms
		49	7.4006272	ms
		50	4.0559489	ms
		51	**0.3142462**	ms
	u(t)=0	t2=	51	ms
	t=f(T,U,P)	t5=	67	ms
	SP2=1000/f/3	sp2=	10	ms
	nrp= t5/SP2 +2	nrp=	8	ms
	t7=t2+nrp*SP2+Sp2/3	t7=	136	ms
	t4=t7-t6	t4=	69	ms
	t3=t4-t3	t3=	120	ms

References

[1] Liu J, Huang G M, Ma Z and Geng Y 2011 A Novel Smart High-Voltage Circuit Breaker for Smart Grid Applications, *IEEE Transactions on Smart Grid* **2**(2) 254-264

[2] Namjoshi Y 2013 *Auto-Calibration of a Sf6 Circuit Breaker for Point-on-wave Switching an Application of Signal processing to Switchgear*, International Conference on Electrical and Electronics Engineering, Goa, 14-18

[3] Sharma S and Bhardwaj H 2012 How to maintain SF6 circuit breaker, *International Journal of Scientific Research Engineering &Technology* **1** 030-035

[4] Tsudata H 2002 Controlled Switching System for Capacitor Bank and Transformer Switching, *Electrical Engineering* **5** 2125-2130
[5] ***ABB, Controlled Switching with Switchsync 2005, Presentation manual Edition 2005.
[6] ***Siemens 2001 *Operating instructions PSD01 controller Order No:927 00414 174 C*, Presentation manual Edition
[7] Ito H 2002 Controlled Switching Technologies, State-of-the-Art, Transmission and Distribution Conference and Exhibition, *Asia Pacific IEEE/PES* **2** 1455-1460
[8] Akesson U 1992 *Condition monitoring of SF6 circuit breakers*, CIGRE 1992 Session, Paper 13-104
[9] ***Breaker technical manual 3APFI 400 kV
[10] Panoiu M, Panoiu C and Sora I 2009 Modeling the Electromagnetic Polution of the Electric Arc Furnace, *Rev. Roum. Sci. Techn. – Électrotechn. et Énerg.* **54**(2) 165-174
[11] Panoiu C, Panoiu M, Muscalagiu I and Iordan A 2010 Visual Interactive Environment for Study the Power Electronics Using PSCAD-EMTDC Simulation Program, *Computer Aplications in Engineering Education* **18**(3) 469-475
[12] Panoiu M, Panoiu C and Ghiormez L 2013 Modeling of the electric arc behavior of the electric arc furnace, *Advances in Intelligent Systems and Computing* **195** 261-271

Proportional regulator-based approach for a three-phase electric arc furnace

L Ghiormez, M Panoiu, C Panoiu and I Muscalagiu

Politehnica University of Timisoara, Department of Electrical Engineering and Industrial Informatics, 5 Revolution Street, Hunedoara, 331128, Romania

E-mail: manuela.panoiu@fih.upt.ro

Abstract. The paper presents a proportional regulator-based control strategy for a three phase electric arc furnace. This regulator is used in order to obtain the maximum or a desired active power of the electric arc by modifying the speed and the direction of the electrodes from an electric arc furnace. The simulations are carried out in the Matlab / Simulink environment and are illustrated the responses of the system for a stationary regime or a dynamic one. In the case of the dynamic regime are illustrated the responses of the system for the variation of the reference parameter in a specific domain or for the case that disturbances are occurring in the process. Also, it is presented how was tuned the proportional gain of the regulator. The process is modeled using a model that is based on the representation of the voltage-current characteristic of the electric arc. The system is tested in a closed-loop and the responses of this system illustrate the performance of the system for the cases taken into consideration.

1. Introduction

Steel is the most important iron alloy, the main component being carbon in a concentration of maximum 2.11% by weight. Using the alloying elements can be changed the main properties of steels: hardness, elasticity, strength, flexibility, corrosion resistance, flow resistance, acid resistance. If there is more than 2.11% of carbon, steel becomes harder, but even more rigid than iron (becomes cast iron).

One of the equipment that is used in order to produce steel by melting scrap using an electrical supply as the main energy input is the electric arc furnace (EAF). Most of the EAFs currently in use are fed to alternating voltage and are of high and very high power [1]. These types of furnaces are nonlinear, unbalanced, time-varying and which generally cause power quality problems [2]. These include: the occurrence of harmonic currents, asymmetry on the three phases, reactive power of significant value, voltage gaps (flicker effect) [3]. The increased concern about the power quality has led to the need to take measures to reduce these effects. In these conditions, it is important to study the behavior of the electric arc because, being a nonlinear circuit element [4], this is the main cause of the disturbances produced by the EAF in the power systems.

Over time, technological improvements have been made, in particular regarding the elaboration time of steel, but still exists problems in influencing the power quality. Current concerns are related in obtaining reduced energy consumption, but also in rejecting disturbances that may affect other consumers which are connected to the same power network [2], [5].

EAF taken into consideration in this paper is a three-phase one with direct action, so, the electric arcs appear between each of the three electrodes and the metal that will be melted. These electrodes are supported by the electrode arms which are mounted on a hydraulic system that allows obtaining different arc lengths by changing the position of the electrodes.

In order to reduce the energy consumption and to melt the metals loaded in the furnace tank it is necessary to optimize the delivered power to the furnace. Arc power can be influenced in two ways: by modifying the position of the electrodes (during a melting stage) or by modifying the power delivered by the transformer to the EAF (changing the melting stages) [6].

2. Modeling of the Electric Arc Furnace

In order to analyze the operation of the EAF it is it is necessary to model the electric arc so that it is reflected as faithfully as possible the real arc behavior [7]. To analyze the behavior of the electric arc in the scientific literature are considered many types of models. The most used method for static and dynamic modeling of the electric arc behavior from the three-phase EAF is the method based on the voltage- current characteristic of the electric arc [8], [9] because this method is simple and direct and provides enough information regarding the operating of the technological installation taken into consideration. The method based on the voltage-current characteristic of the electric arc uses a numerical analysis procedure, which has the role of solving the differential equation which describes the behavior of the electric arc within the furnace [10].

The operating mode of the EAF can be defined by: electric arc current, electric arc voltage, voltage in secondary side of the furnace transformer, electrical resistance and circuit reactance through which is fed the EAF [11].

In this paper is used an exponential arc model illustrated in equations (1), model which was proposed and validated by the authors of this paper in [6].

In Figure 1 is presented the voltage-current characteristic of the EAF that will be obtained with the electric arc model. In this model, the voltage-current characteristic of the electric arc is divided in five sections in order to obtain a characteristic similar with the one from the real installation. In Figure 1 v_{ig} represents the ignition voltage, v_{ex} is the extinction voltage, v_m is the average between the ignition and the extinction voltage of the arc, v_{arc} is the voltage of the electric arc, i_1 is the current of the electric arc corresponding for the v_{ig}, i_2 is the current of the electric arc corresponding for the v_m, i_3 is the current of the electric arc corresponding for the v_{ex} and i_{arc} is the current of the electric arc [6]. R_1, R_2 and R_3 represents lines slope for the linear section of the voltage-current characteristic.

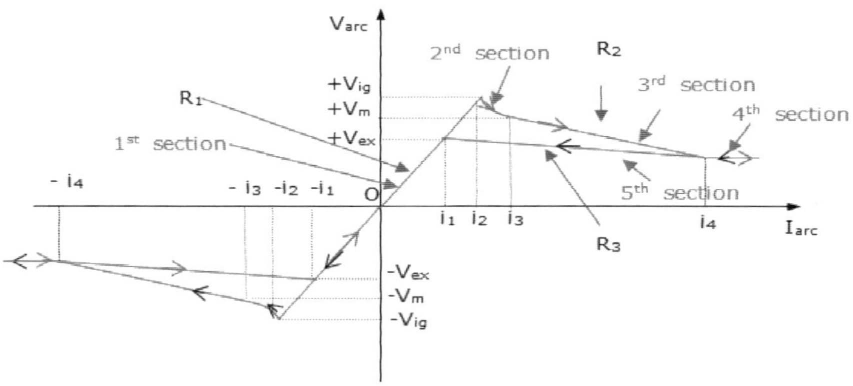

Figure 1. Voltage-current characteristic divided by sections

In the five sections of the voltage-current characteristic can be mentioned the followings:

- *1st section*: the voltage magnitude increases from extinction voltage to ignition voltage (EAF acts as a resistor);
- *2nd section*: the beginning of the arc melting process (exponential voltage drop across the electrodes);
- *3rd section*: the normal arc melting process;

- 4th *section*: stabilizing the voltage to a specific value;
- 5th *section*: changing the polarity of the supply voltage.

In the model from (1) were chosen values for the model parameters in order to accurately represent the voltage-current characteristic of the electric arc, $R_1 = 0.05$, $R_2 = -0.0007$ and $R_3 = -0.0003$. Limit i_1 is computed as presented in (2), limit i_2 is computed as presented in (3), limit i_3 is computed as presented in (4), limit i_4 is computed as presented in (5) and voltage v_m is computed as presented in (6).

$$v = \begin{cases} -v_{ex} + (i_1 - i_4)\cdot R_3, & i < -i_4 \text{ and } di/dt \geq 0 \\ & \text{or } i < -i_4 \text{ and } di/dt < 0 \\ -v_{ex} + (i + i_1)\cdot R_3, & i \in [-i_4, -i_1) \text{ and } di/dt \geq 0 \\ R_1 \cdot i, & i \in [-i_1, i_2) \text{ and } di/dt \geq 0 \\ & \text{or } i \in [-i_3, i_1) \text{ and } di/dt < 0 \\ v_{ex} + (v_{ig} - v_{ex})\cdot e^{(i_2 - i)/i_3} & i \in [i_2, i_3) \text{ and } di/dt \geq 0 \\ v_m + (i - i_3)\cdot R_2 & i \in [i_3, i_4] \text{ and } di/dt \geq 0 \\ v_{ex} + (i_4 - i_1)\cdot R_3 & i > i_4 \text{ and } di/dt \geq 0 \\ & \text{or } i > i_4 \text{ and } di/dt < 0 \\ v_{ex} + (i - i_1)\cdot R_3 & i \in [i_1, i_4] \text{ and } di/dt < 0 \\ -v_{ex} + (v_{ex} - v_{ig})\cdot e^{(i_2 + i)/i_3} & i \in [-i_3, -i_2] \text{ and } di/dt < 0 \\ -v_m + (i + i_3)\cdot R_2 & i \in [-i_4, -i_3] \text{ and } di/dt < 0 \end{cases} \quad (1)$$

where:

$$i_1 = v_{ex} / R_1 \quad (2)$$

$$i_2 = v_{ig} / R_1 \quad (3)$$

$$i_3 = 2 \cdot i_2 \quad (4)$$

$$i_4 = (v_m - v_{ex} + i_1 \cdot R_3 - i_3 \cdot R_2)/(R_3 - R_2) \quad (5)$$

$$v_m = (v_{ig} + v_{ex})/2 \quad (6)$$

3. Control Strategy of the Electric Arc Furnace

In this paper is proposed, designed and implemented a proportional regulator-based control strategy for a three-phase EAF. This regulator is used in order to obtain the maximum or a desired active power of the electric arc by modifying the speed and the direction of the electrodes from an electric arc furnace. The simulations are carried out in the Matlab / Simulink environment and are illustrated the responses of the system for a stationary regime or a dynamic one. In the case of the dynamic regime are illustrated the responses of the system for the variation of the reference parameter in a specific domain or for the case that disturbances are occurring in the process.

In Figure 2 is presented the block diagram of the proposed control closed-loop. In this block diagram I_{ref} is the reference arc current, e_I is the error computed as the difference between the reference value and obtained value of the arc current, c is the command for the hydraulic actuator in order to move the electrode up or down with a corresponding speed, *x* is a value that will be added in the position of the arc in this way obtaining the value of the reference current, because arc length influence the electric arc current. *I* is the current of the electric arc obtained by the model of the electric arc. *Controller* block stands for the proportional controller.

Electric arc block is the process in the control loop and the *hydraulic actuator* is the execution element.

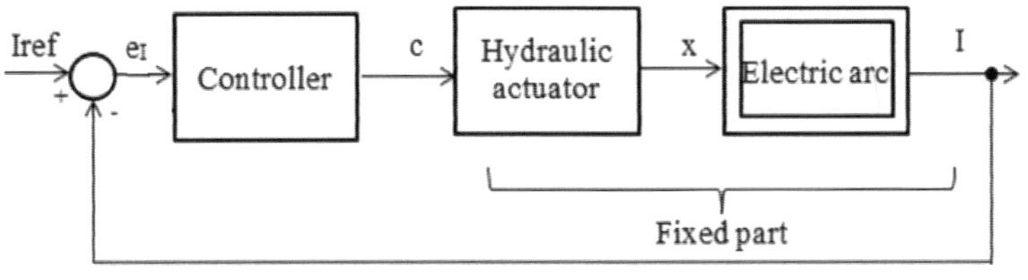

Figure 2. Block diagram of the proposed control closed-loop

4. The Control System in the Stationary Regime

In this case the input and output quantities are constant over time. The reference arc current is set to *56500 A* and the response of the system should reach the setpoint value and remain in the stable area.

The process is represented by the operating mode of the EAF which can be defined by: electric arc current, electric arc voltage, voltage in secondary side of the furnace transformer, electrical resistance and circuit reactance through which is fed the EAF (Figure 3).

In Figure 3 *ARC* block stands for the model of the electric arc that was presented in (1). *Rt* is the source resistance and *Rfc* is the flexible cables resistance. *Lt* is the inductivity of the source and *Lfc* is the inductivity of the flexible cables. These flexible cables are used to connect the transformer to the electric arc furnace. The *RMS block* will return the effective value of the response of the control system which is an electric arc current. This block is necessary in order to make comparison between the setpoint and the response of the system. *EE* is the execution element which is the hydraulic actuator in real plant and it is represented in the block diagram as a *PT1* system having time constant of 0.75 seconds. EE executes the commands received from the regulator which represents values that modify the arc length. The proportional regulator is represented by using of a gain block from Simulink.

Taking into account the block diagram presented in Figure 3 was obtained the response of the system in the stationary regime. It can be observed that the response of the control system reaches the value of the setpoint and after 0.2 seconds the response is stable.

4.1. The Tuning of the Proportional Gain of the Regulator

In order to tune the proportional gain (*Kp*) of the regulator was used the block diagram illustrated in Figure 3. The value of the *Kp* factor was chosen according to numerous tests performed by the authors of this paper. The tests have been made for negative and positive values of *Kp*.

The increase of the *Kp* determines the decrease of the stationary error of the system, having an important role in ensuring adequate amortization, implicitly on the system stability.

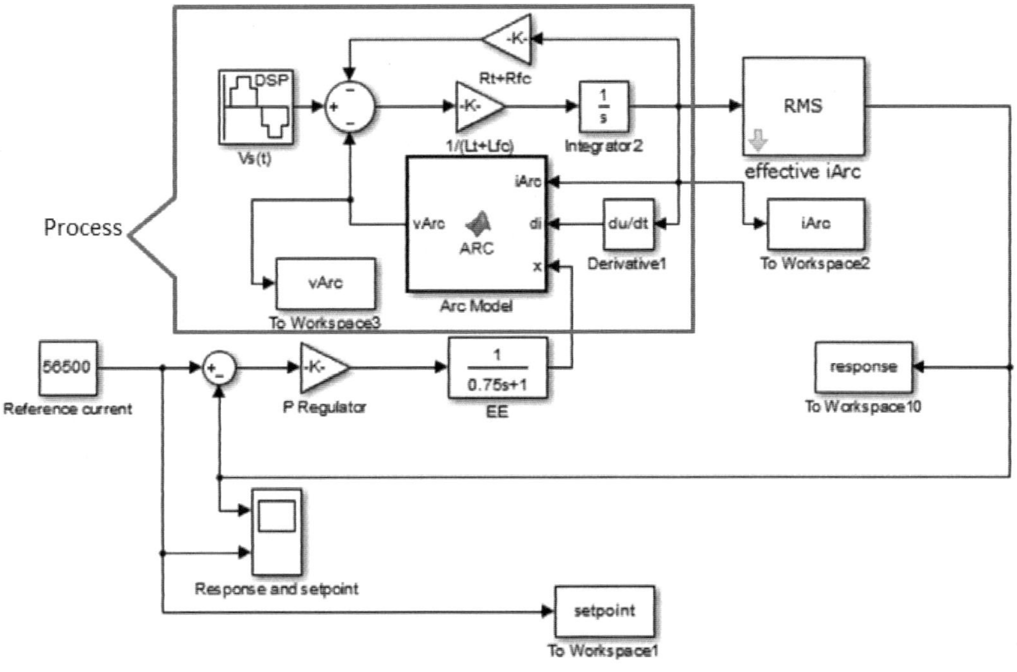

Figure 3. Simulink bloc diagram for the control system in the stationary regime

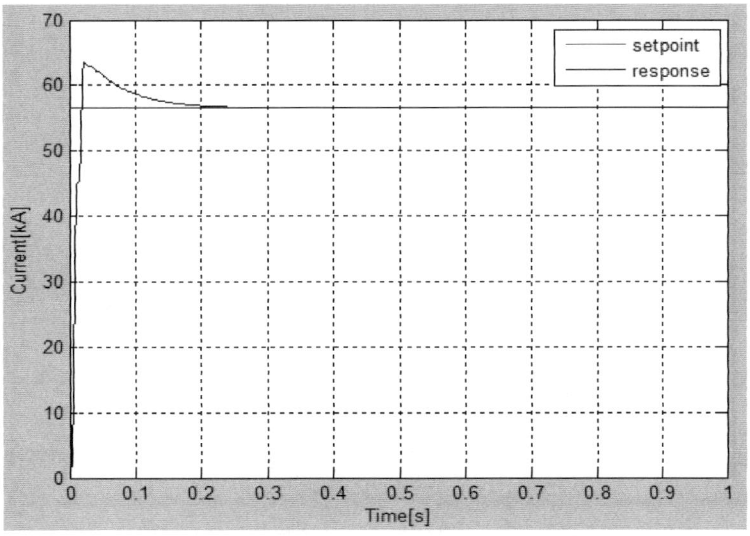

Figure 4. Reference and response of the system in the stationary regime

In Figure 5 are presented some of the tests results for negative values of the proportional gain.

If Kp = -5 it can be noticed that the response obtained at the output of the process has a large overshot and a residual oscillation around the reference.

Also the response does not stabilize for Kp = -5e-01. For Kp = -5e-02 the overshot remains high, but the response stabilizes after 0.4 seconds. For value Kp = -5e-03 the overshot becomes acceptable and the response stabilizes after about 0.25 seconds. For Kp = -5e-04 it can be observed that overshot is small, but the response stabilizes only after approximately 0.75 seconds.

Following the analysis of the results presented in Figure 5, another set of tests was carried out with values of *Kp* around the value *-5e-03*, the results being shown in Figure 6.

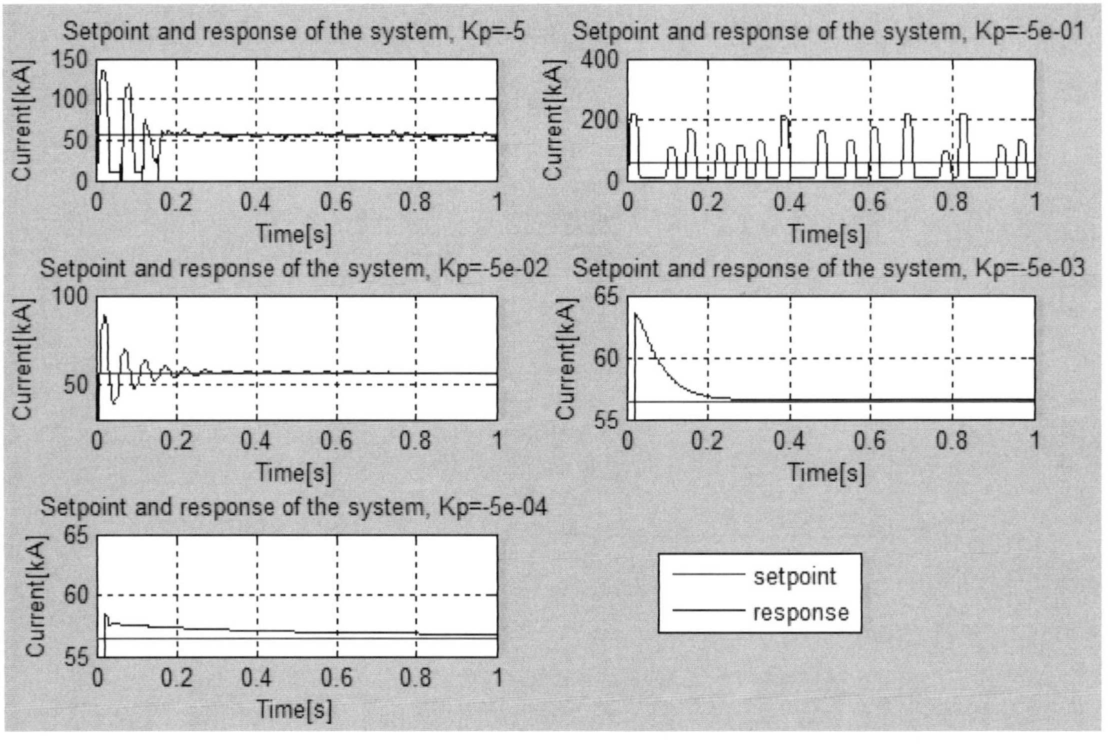

Figure 5. Obtained responses for using different negatives values of the proportional gain

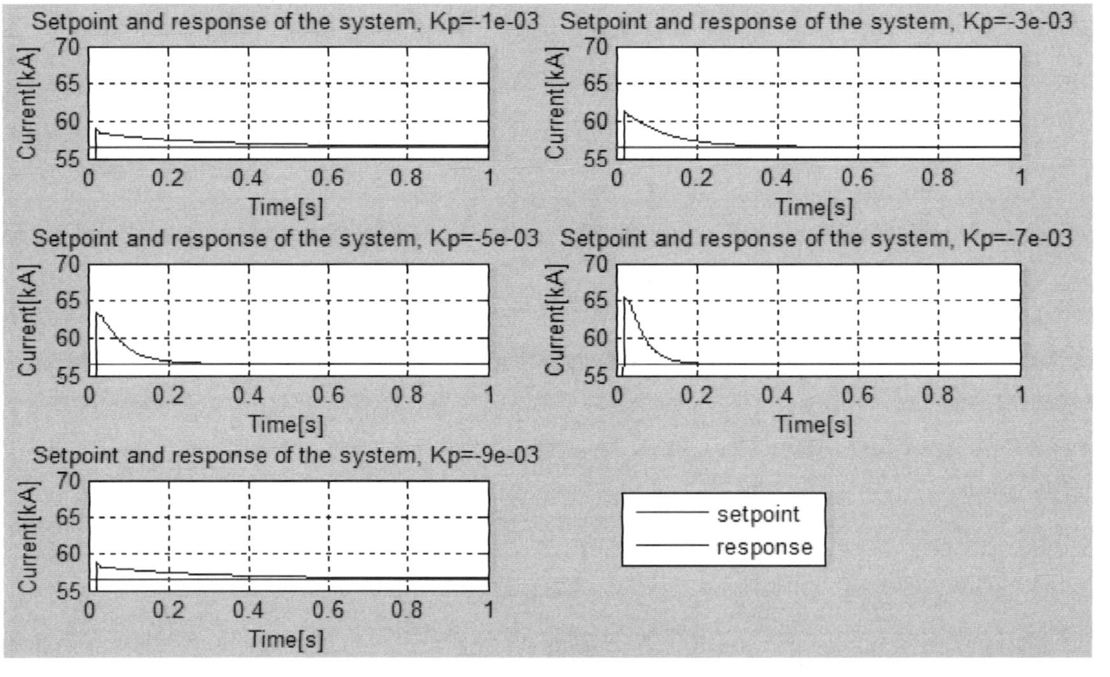

Figure 6. Obtained responses for using different negatives values of the proportional gain around the value -5e-03

IOP Conf. Series: Materials Science and Engineering **477** (2019) 012016 doi:10.1088/1757-899X/477/1/012016

The analysis of the results led to a choice of the proportional gain of *Kp* = -5e-03, taking into account the overshot and the response time.

Arc current and voltage waveforms for the chosen *Kp* are illustrated in Figure 7 and these corresponds to the ones from the scientific literature.

In order to choose a positive value for *Kp*, another set of tests was carried out for values of *Kp*. Some of the results obtained are shown in Figure 8. Analysis of all the results led to the conclusion that for *Kp* cannot be chosen positive values.

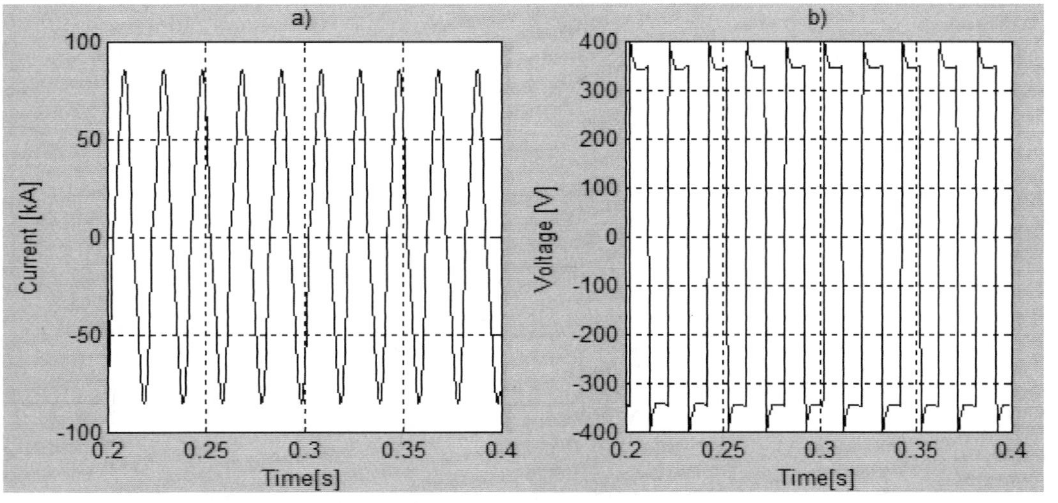

Figure 7. a) Arc current waveforms, b) Arc voltage waveforms for K_p = -5e-03

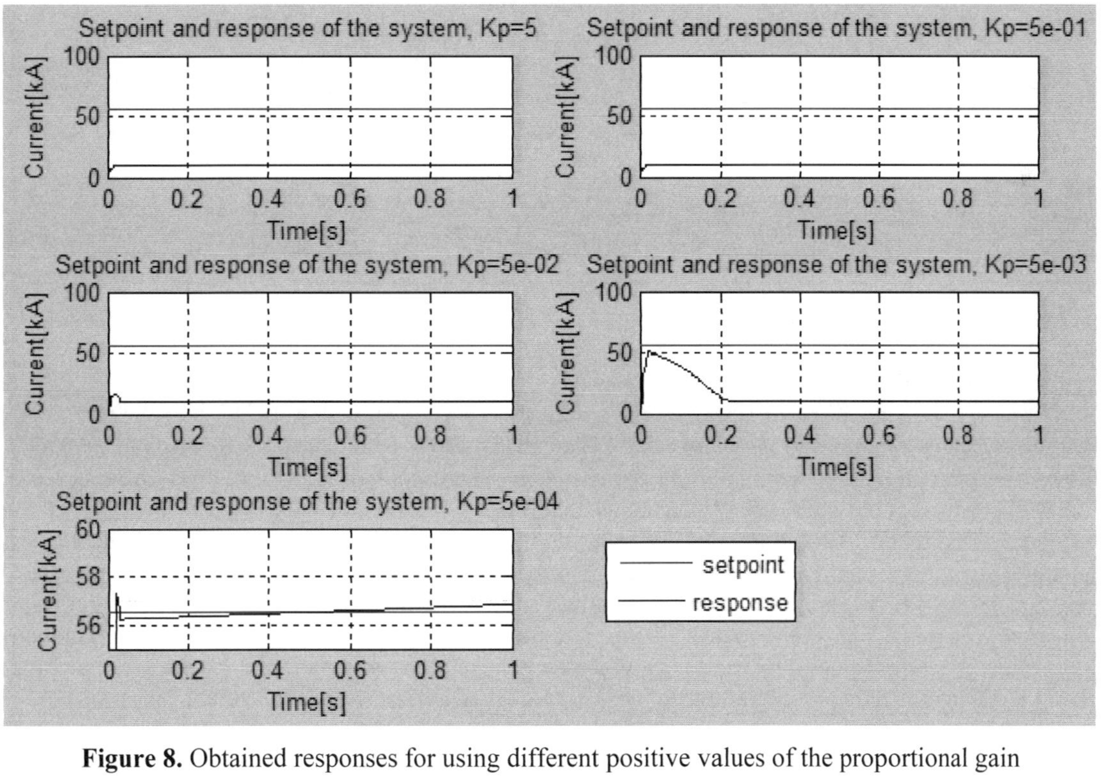

Figure 8. Obtained responses for using different positive values of the proportional gain

5. The Control System in the Dynamic Regime

In the case of the dynamic regime are illustrated the responses of the system for the variation of the reference parameter in a specific domain or for the case that disturbances are occurring in the process.

5.1. Step Variation of the Reference

In Figure 9 is presented the Simulink bloc diagram for the step variation of the reference in a specific range. A change of the reference value is obtained by using of a step type signal at $t = 0.5$ *sec*, leading to an increase of the reference current from 56.5 kA to 59.5 kA.

In Figure 10 a) is presented the setpoint and response of the system and in Figure 10 b) is illustrated the corresponding arc length variation. It can be noticed that after about *0.2 sec* after applying the step signal the response of the system manages to follow the reference value.

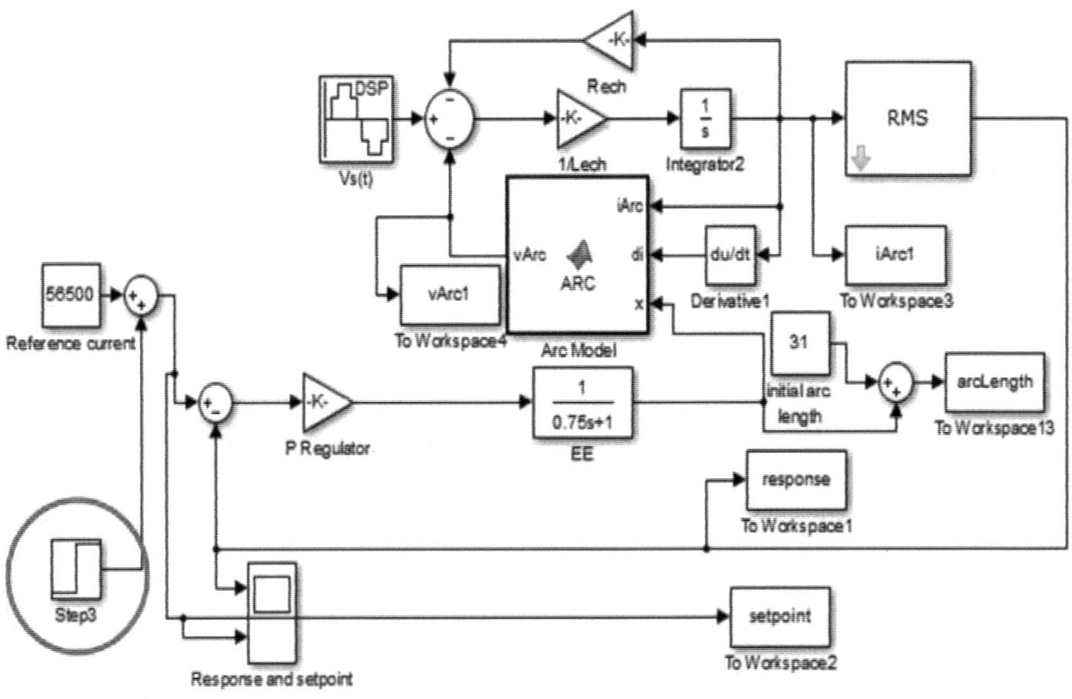

Figure 9. Simulink bloc diagram for the step variation of the reference

5.2. A Sequence of Step Variation of the Reference

In Figure 11 is illustrated the Simulink bloc diagram for a sequence of step variation of the reference. Multiple changes of the reference value are obtained by using of a *Slider Gain* component which allows changing the reference value in the form of step signals applied at random times, leading to an increase or a decrease of the reference current. The initial reference value is *56500 A*.

Figure 12 a) shows the variation of the setpoint and the response of the system and Figure 12 b) illustrates the corresponding variation in the length of the arc. Two variations of the reference value were considered which are modified in a step form.

It can be noticed that the response of the system follows the value of the reference parameter. When decreasing the value of the reference it can be observed an increase in the length of the arc and when increasing the value of the reference it can be observed a decrease in the length of the arc.

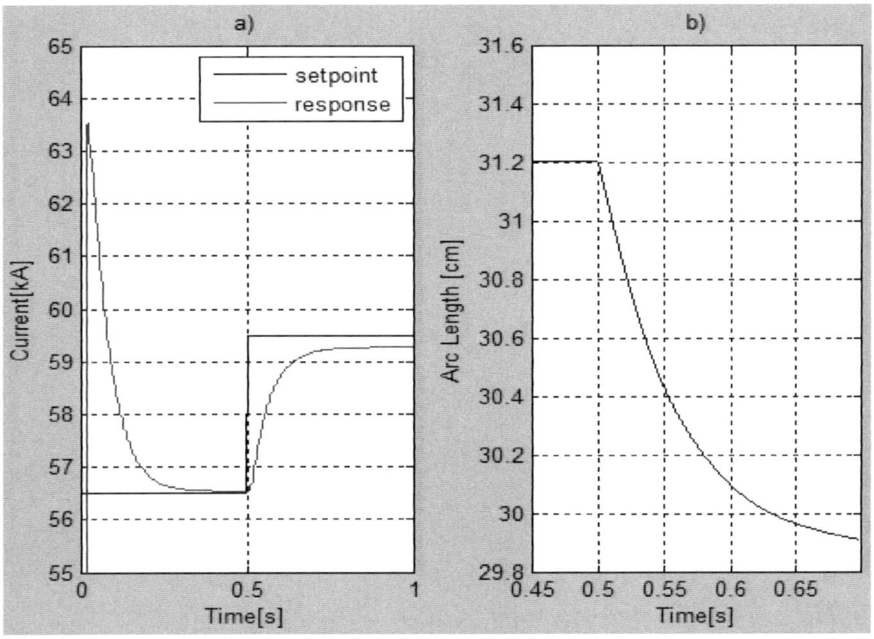

Figure 10. a) Setpoint and response of the system, b) Arc length variation

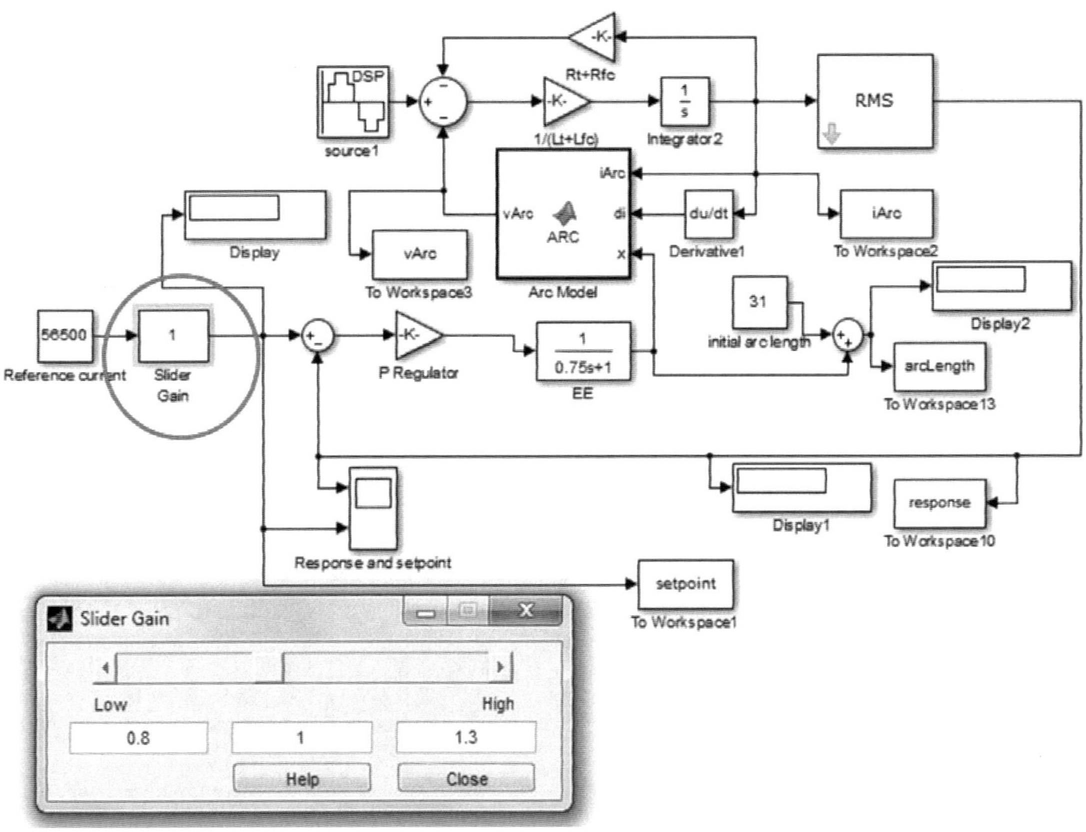

Figure 11. Simulink bloc diagram for a sequence of step variation of the reference

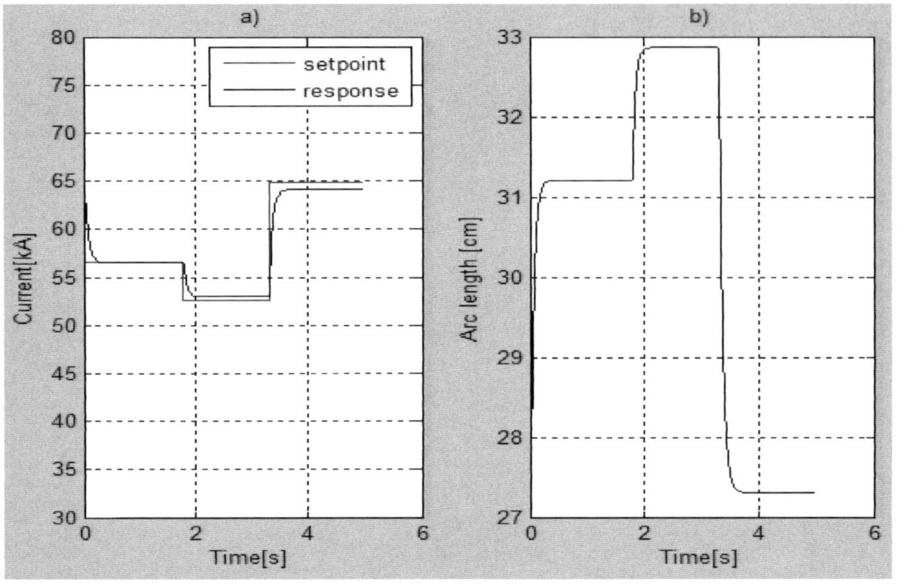

Figure 12. a) Setpoint and response of the system, b) Arc length variation

5.3. Step Disturbance in the Process

In Figure 13 is presented the Simulink bloc diagram for the step disturbance in the process. A disturbance is injected in the process by using of a step type signal at *t = 0.5 seconds* while the setpoint is kept constant at *56500 A*, leading to a decrease of the arc length which corresponds in real plant to a changing in position of the metals from the furnace tank.

In Figure 14 a) is presented the corresponding arc length variation and in Figure 14 b) is illustrated the setpoint and response of the system. It can be noticed that after about *0.25 seconds* after injecting the disturbance the arc length is approximately restored to the initial value.

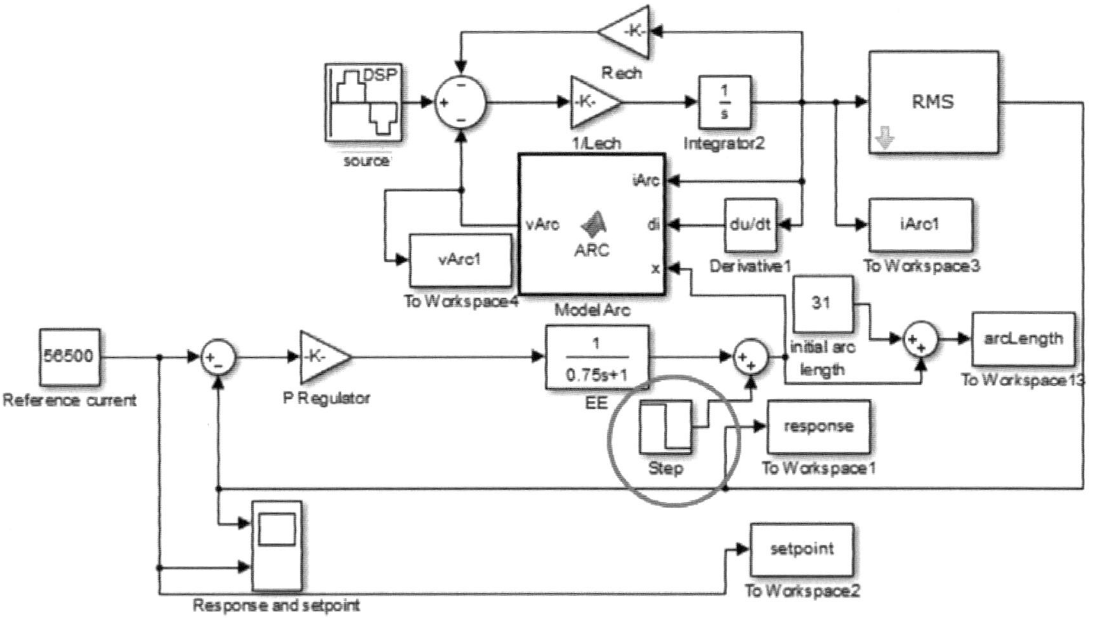

Figure 13. Simulink bloc diagram for a step disturbance in the process

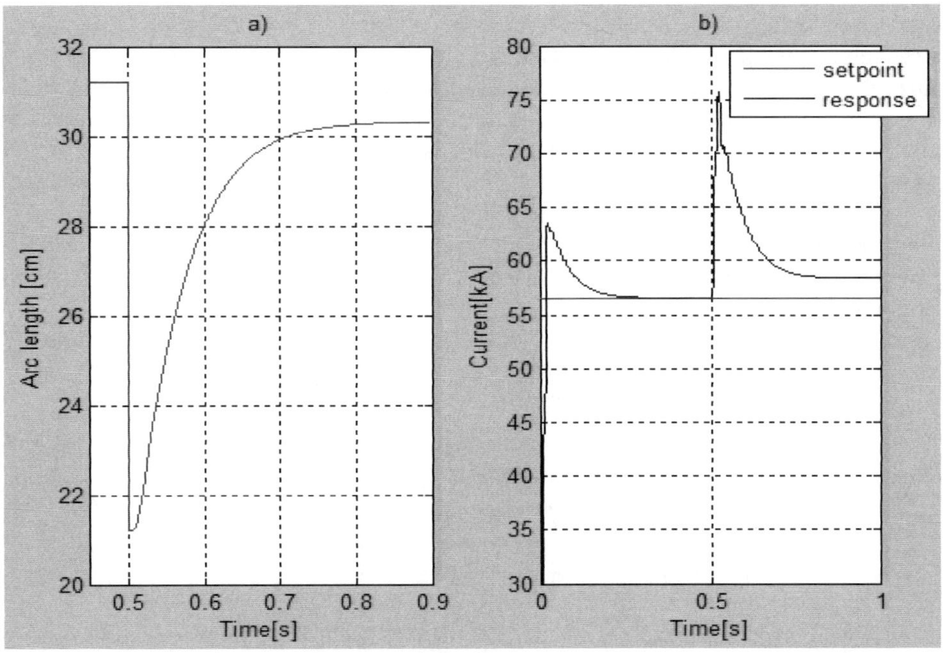

Figure 14. a) Arc length variation, b) Setpoint and response of the system

In Figure 15 a) is represented the arc current waveforms and in Figure 15 b) is illustrated the arc voltage waveforms for the presented case in Figure 14.

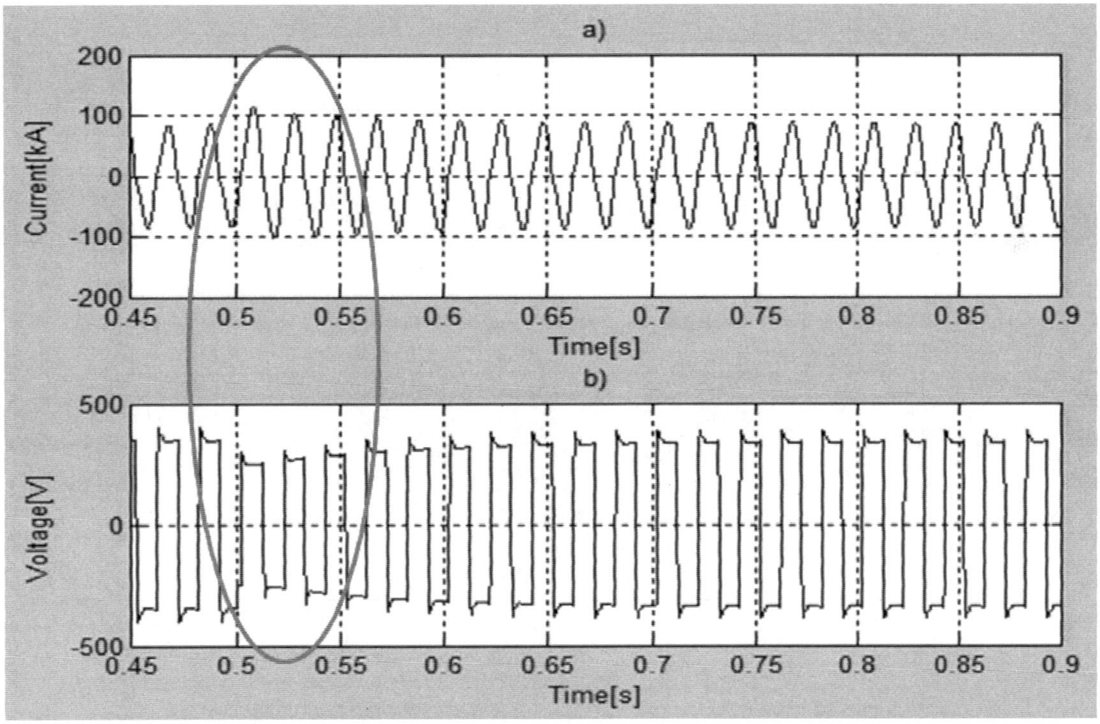

Figure 15. a) Arc current waveforms, b) Arc voltage waveforms

5.4. A Sequence of Step Disturbance in the Process

In Figure 16 is illustrated the Simulink bloc diagram for a sequence of step disturbance in the process. Multiple disturbances are injected in the process by using of a Slider Gain component which allows modifying the arc length in the form of step signals applied at random times, leading to an increase or a decrease of the arc length. The initial reference value is 31 cm.

In Figure 17 a) are illustrated the variation of the arc length when are injected disturbances in the process in the form of successive steps at random times.

It can be noticed that the control system compensates the effect of these disturbances, the system returning each time to the initial arc length of 31 cm (Figure17 a), respectively at the reference current value of 56.5 kA (Figure 17 b).

It can be observed that the response of the system is stable and does not have oscillations.

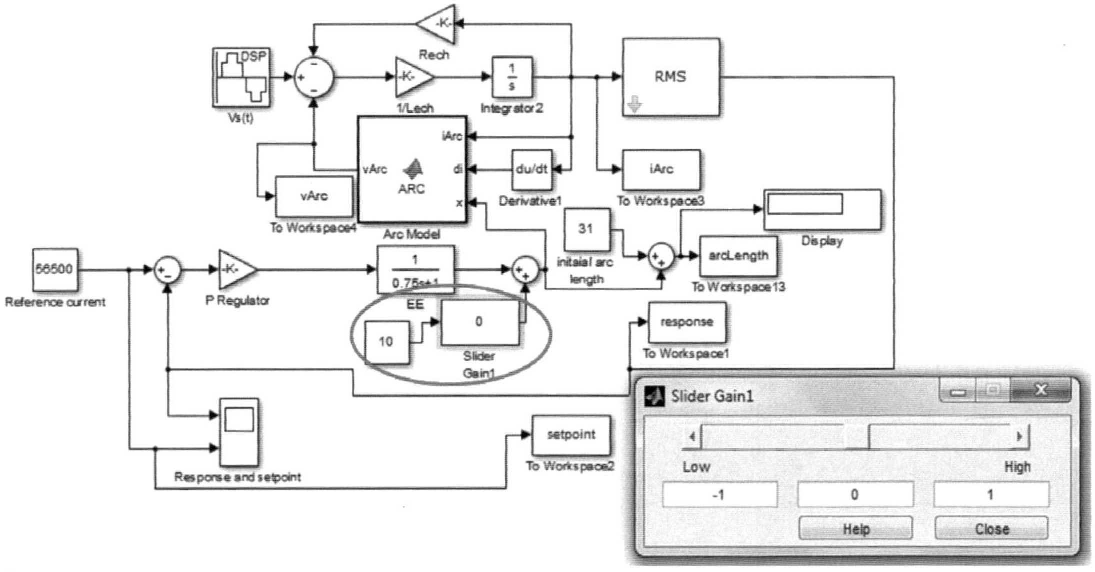

Figure 16. Simulink bloc diagram for a sequence of step disturbance in the process

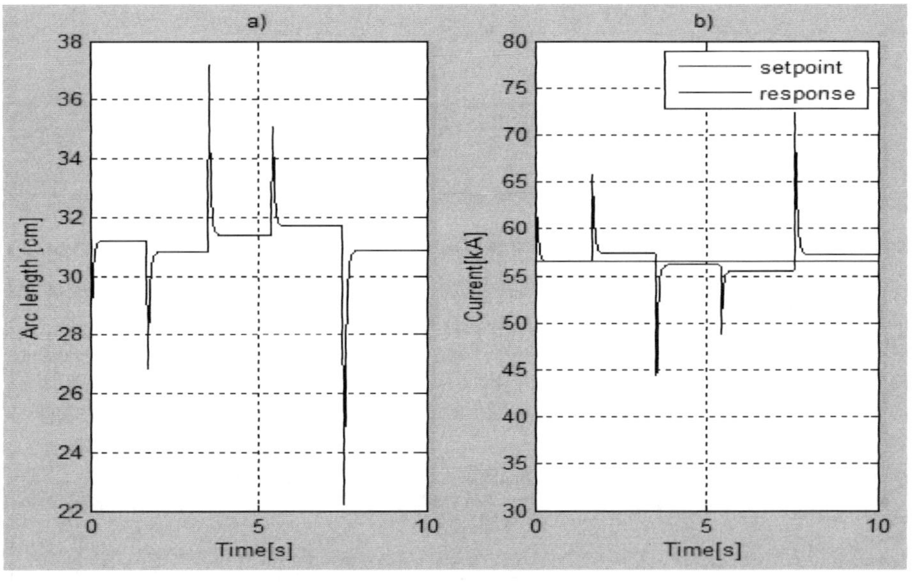

Figure 17. a) Arc length variation, b) Setpoint and response of the system

6. Conclusions

In this paper was used an exponential model of the electric arc for an electric arc furnace. The model is based on the voltage-current characteristic of the electric arc. This model is used in the implementation of the control strategy used in the current control of the electric arc.

It was proposed a proportional regulator-based control strategy for a three phase electric arc furnace. This regulator was used in order to obtain the maximum or a desired active power of the electric arc by modifying the speed and the direction of the electrodes from an electric arc furnace.

It was presented how was tuned the proportional gain of the regulator. The simulations were carried out in the Matlab / Simulink environment and were illustrated the responses of the system for a stationary regime or a dynamic one.

In the case of the dynamic regime were illustrated the responses of the system for the variation of the reference parameter in a specific domain or for the case that disturbances are occurring in the process.

By using the electrode system regulation it is reduced the energy consumption, it is avoided the appearance of damages caused by the breakage of the electrodes and it is compensated the effect of disturbances which influence the process.

Starting from the presented control strategy others control strategies can be implemented.

References

[1] Panoiu M and Panoiu C 2009 Modeling the Electromagnetic Polution of the Electric Arc Furnace, *Rev. Roum. Sci. Techn. – Électrotechn. et Énerg.* **54**(2) 165-174

[2] Hooshmand R A and Esfahani M T 2009 Optimal Design of TCR/FC in Electric Arc Furnaces for Power Quality Improvement in Power Systems, *Leonardo Electronic Journal of Practices and Technologies* **8**(15) 31-45

[3] Collantes-Bellido R and Gomez T 1997 Identification and Modelling of a Three Phase Arc Furnace for Voltage Disturbance Simulation, *IEEE Transactions on Power Delivery* **12**(4) 1812-1817

[4] Sousa J, Correia de Barros M T, Covasand M and Simoes A 1999 *Harmonics and Flicker Analysis in Arc Furnace Power Systems*, Proceedings of International Conference on Power Systems Transients IPST '99, Budapest, Hungary, June 20-24, pp 626-630

[5] Panoiu M and Panoiu C 2008 *Nonlinear processes modeling and simulation in electrothermics* Publishing House Mirton, Timişoara, Romania

[6] Ghiormez L, Prostean O, Panoiu M and Panoiu C 2016 *Fuzzy Logic and PD Control Strategies of a Three-Phase Electric Arc Furnace,* International Conference on Computational Collective Intelligence ICCCI 2016, Sithonia, Halkidiki, Grecia, September 28-30, pp 509-517

[7] Golkar M A, Tavakoli Bina M and Meschi S 2007 *A Novel Method of Electrical Arc Furnace Modeling for Flicker Study*, International Conference on Renewable Energies and Power Quality ICREPQ'07, Sevilla, Spain, March 28-30, pp 1-8

[8] Chang G W, Chen C I, Liu Y J 2010 A Neural-Network-Based Method of Modeling Electric Arc Furnace Load for Power Engineering Study, *IEEE Transactions on Power Systems* **25**(1) 138-146

[9] Chang G W, Shih M F, Chen Y Y and Liang Y J 2014 A Hybrid Wavelet Transform and Neural-Network-Based Approach for Modelling Dynamic Voltage-Current Characteristics of Electric Arc Furnace, *IEEE Transactions on Power Delivery* **29**(2) 815-824

[10] Andrei H, Cepisca C and Grigorescu S 2011 Power quality and electrical arc furnaces, *In Power Quality* INTECH 77-100

[11] Chang G W, Liu Y J and Chen C I 2008 Modeling Voltage-Current Characteristics of an Electric Arc Furnace Based on Actual Recorded Data: A Comparison of Classic and Advanced Models, *Power and Energy Society General Meeting-Conversion and Delivery of Electrical Energy in the 21st Century*, IEEE, pp 1-6

Identification of significant metrics and indicators for smart mobility

M D Pop and O Proştean

Politehnica University of Timişoara, Automation and Applied Informatics Department, 2 Vasile Pârvan Blvd., 300223 Timişoara, România

E-mail: madalinpop20@gmail.com

Abstract. Building a better life for citizens became a motto for many cities around the world. In this context, smart city concept was defined. The aim of this concept is to make use of intelligent technologies to improve the citizens quality of life. Smart mobility is considered one out of six characteristics which compose the smart city model. The purpose of this paper is to show the significant metrics and indicators which describe the smart mobility concept. These metrics and indicators have an important role in gained performance analysis for a smart city, facilitating the comparison to other smart cities. In current paper, an analysis will be made based on these metrics and indicators to see which is their impact in smart city score setting and implicitly, in establishing the influence on the inhabitants' quality of life.

1. Introduction

In a world that is in continuing development, the transportation problem shall be a priority. Looking around, we can see how the road networks are daily overloaded. This problem, known as congestion, has a negative impact by introducing delays in traffic, which affects the time spent in traffic. In this manner people are losing hours of their life when they are going to work, shopping etc. As we can see from these daily situations, the people quality of life is affected and is required to find solutions to avoid, or reduce, the congestion problem.

The purpose of our paper is to identify the significant metrics and indicators for smart mobility concept. More than that, in our paper will be highlighted the application of standardization on identified metrics and indicators to create a generalized framework that can be easily used to compare different cities around the world from smart mobility point of view.

The structure of this paper allows us to see how important the mobility is nowadays, starting with an overview about current studies and showing the level of interest on this research domain. In the third section will be defined the traffic congestion problem and how it influences the citizens quality of life.

Section four will present the smart city concept together with its characteristics. Next section will describe the smart mobility concept, highlighting its areas, indicators and metrics. In this fifth section will be also shown the drawbacks of having a big number of metrics and indicators.

The solution for the drawbacks presented in previous section will be illustrated in section VI. Here will be shown how a standardization process applied on smart mobility metrics and indicators can lead to a general framework definition. This framework allows characterizing a city, from smart mobility point of view, using a reduced number of metrics and indicators.

In section seven will be presented an analysis of smart mobility metrics and indicators for some cities around the world. On this analysis, we will see some rankings of European smart cities and a worldwide comparison between smart city general ranking and the ranking based on smart mobility.

The final section will highlight once more that the purpose of this paper was to show which are the significant metrics and indicators for smart mobility. In the same section, the conclusions will be issued, showing the importance of having standards which permits the creation of a common framework including only the significant metrics and indicators for smart mobility.

2. Literature overview

This paper can be seen as a continuation of previous work. In [1] was made a comparison between smart city approaches, from smart mobility point of view, after smart city modelling possibilities were highlighted. There was seen how important is this smart city component in establishing the city score and were shown some potential actions that can be taken by local administration to improve the road transport management. More than that, there were presented actions taken by Romanian smart cities to maintain their position in the European smart cities ranking [1], [2].

One of the most important things in smart mobility analysis can be considered the case studies for different cities around the world [3-7]. From these studies, can be selected common factors and metrics which can be used as features for creating an informal framework for an eventual comparison between them. If we need a general framework with a reduced number of factors and metrics that should fit to each city, independently of its location, there should be defined a list of them that respects some standards and regulations [8], [9].

An interesting approach in smart city studies is presented in [10]. There is shown how important is to take into consideration the climate change problem. Giving a little attention to collected meteorological data, several proposals are issued by authors. Four aspects are considered essential to achieve a smart city planning according to necessity of giving help on global climate change fight:

- energy saving and CO_2 emissions reduction;
- predictability of possible weather disasters caused by city development or planning;
- usage of IoT (Internet of Things) technologies to create a capable system to inform citizens, in good time, in case of a natural disaster will be produced;
- create a complex system that uses sensors placed in the city to monitor weather changes in comparison with urban planning and can give advices for some adjustments in the city planning, to prevent natural disasters.

A new category of studies that shall be mentioned, are the studies concerning the concept of smart mobility from urban planner point of view. At this level, cities can be characterized in terms of sustainable mobility using the concept of PCA (Principal Component Analysis) [11]. The first step on this analysis was to select a sample that includes several cities. Previous step was followed by a selection of relevant data for the studied cities which will lead to obtaining PCA factors. Authors considered next PCA factors, in order of their importance:

- inverse index of smart mobility;
- index of eco-mobility;
- index of car-free mobility.

In the same category of studies, as previous one, can be included the study based on new urbanism and compact city concepts [12]. New urbanism concept is very interesting because it puts on first place the people mobility as pedestrians instead of mobility as car drivers. This vision advocates to good connections between neighborhoods, walkability and quality architecture. Another concept presented is related to city compactness which promotes the car dependability reducing in favor of walkability, bike and public transportation usage.

3. Traffic congestion – problem definition

Traffic congestion can be defined as a situation that we experience daily, in many cities around the world, which is characterized by increased travel times, slower speeds and long queues of vehicles at the intersection entering point. If we look to a long-term evolution of transportation, the occurrence of this problem in road traffic has a negative impact on quality of life. It influences the environment, daily activities, different provided services and moving of goods.

A lot of studies are made on traffic congestion reducing field. Solutions as using LSTM (Long Short-Term Memory) networks, part of RNN (Recurrent Neural Networks), for traffic flow prediction are provided by scientists [13]. A more complex traffic management system, involving vehicles detection sensors, data analysis tool and traffic congestion prevention algorithm is presented in [14].

4. Smart city concept

Smart city concept is a complex system that has as final scope the citizens quality of life improvement. To achieve this goal is necessary to have a great involvement of local administrations through city development decision-making process. More than that, there is the need of an active involvement of citizens, helping in the application of local administrations decisions, or easier, by giving feedback to them.

A city can be considered smart if it respects some rules and achieves some defined goals. To a better understanding of how this concept is applied to a city, several modelling approaches were proposed. The most popular way to model a smart city is using Giffingers' approach [1], [2]. In this case, smart city can be seen as a combination of six characteristics, as it is illustrated in Figure 1. For each characteristic are attached several factors which will give the scores, for evaluated cities, to add them to smart cities ranking. We can see that is necessary to have an equilibrium between these characteristics to ensure a high quality of life for citizens.

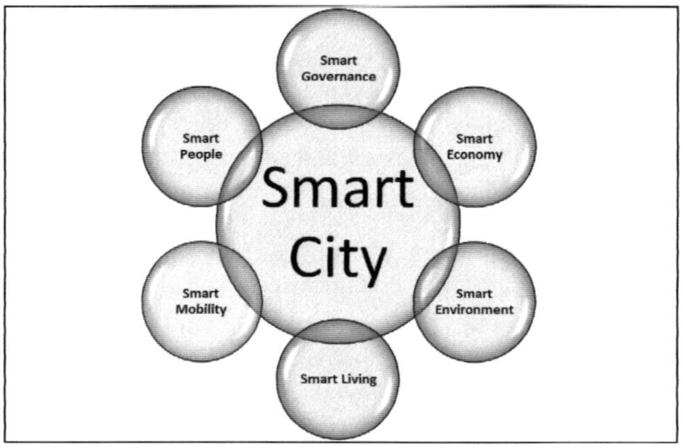

Figure 1. Smart city model based on Giffingers' approach

A good measure in quality of life improvement is to use the resources efficiently, to obtain the wastes and CO_2 emissions reduction. As Lattore-Biel mentioned in [5], we can observe the impact of traffic and transportation activities in the use of resources, leading to an increasing of noises, wastes and CO_2 emissions, which have a direct impact on our daily life.

Further, this study will be oriented to smart mobility concept definition and analysis of its components.

5. Smart mobility

Part of smart city concept, smart mobility has a big impact on cities ranking because it can influence other four characteristics: smart people, smart economy, smart environment and smart leaving. A smart mobility system includes the road infrastructure, the means of transport, the polices and regulations that are ensuring a good traffic flow in the city, with reduced travel times and CO_2 emissions.

After an analysis of relevant studies on smart mobility field, we propose the architecture from Figure 2 to describe how smart mobility concept can be applied to a city. In this case, using systems decomposition property, we can identify three levels of abstraction.

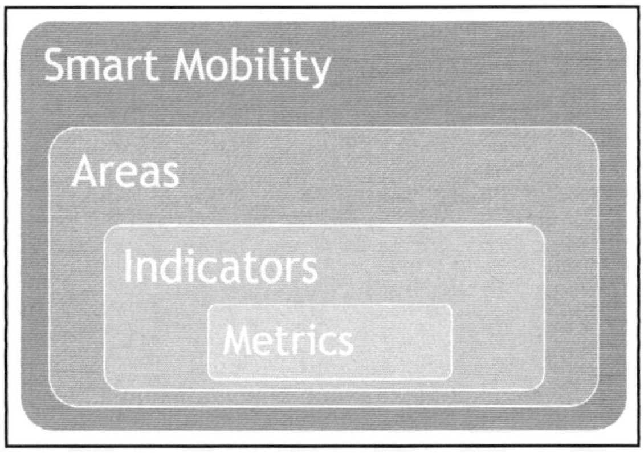

Figure 2. Smart mobility system architecture

First level of abstraction is represented by the general areas of interests that can give an overview of how is applied the smart mobility concept, such as: system management characteristics, application of intelligent technologies etc. Next level of abstraction is referring to some indicators that are specific for each area. These indicators are generally used to describe how smart is a city from mobility point of view and we can find them in many studies. The third level is represented by the metrics, which gives the potential ways of action to improve the current smart city score and implicitly the citizens quality of life.

Further we show an analysis of indicators of indicators and metrics for smart mobility, taking into consideration previous studies [1], [3-7], [12]. These indicators and metrics were tailored to the architecture proposed in Figure 2. We considered three areas as the most relevant to describe the smart mobility of a city. For each area, will be presented the indicators and metrics that were common in analysed studies.

One of the most relevant areas, which is illustrated in Figure 3, is the accessibility. Indicators as public transport, car sharing and rail networks define how easy people can reach their desired destination, saving time, money and protecting the environment. As important metrics shall be mentioned the ratio between demand and supply, the existence of special lanes for public transportation system and the covering routes.

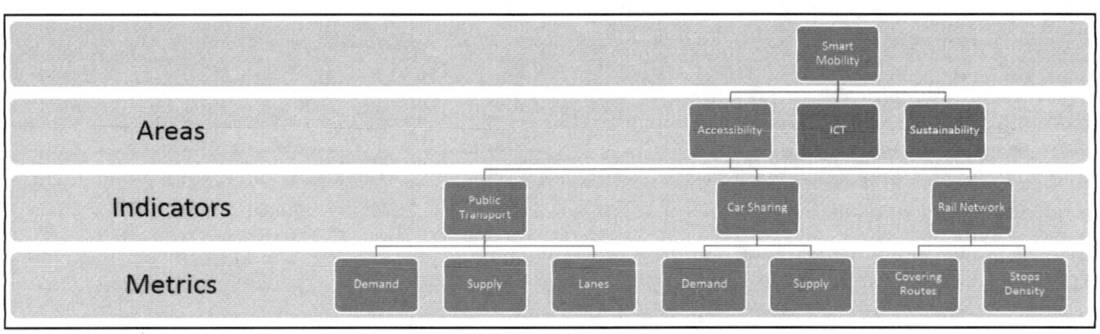

Figure 3. Accessibility area – indicators and metrics

Another area of interest is the ICT (Information and Communication Technology) and is shown in Figure 4. In this area are included as indicators the traffic coordination, SMS (Short Message Service) and electronic systems. A metric for traffic coordination is the traffic signals system, from intelligent green intervals scheduling point of view, case in which these timings are automatically updated based on real-time traffic data. Another metrics for traffic coordination are the variable message signs and routing applications, which are also related to real-time traffic status. SMS indicator includes metrics as traffic alerts and parking payment using this service. On electronic systems indicator category are included special devices which provide the opportunity of buying travel tickets for public transport or parking payment. Electronic systems can be also used as bus stop signs, general traffic signs or for showing messages as advices for alternative routes.

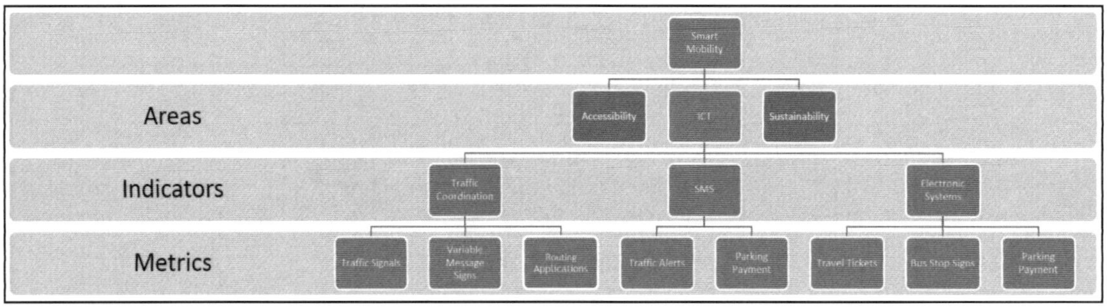

Figure 4. Information and Communication Technology (ICT) area – indicators and metrics

The last, but the most important area of smart mobility, is the sustainability (Figure 5). Here, were found as the most studied indicators the ecological fuels, bike sharing programs and the special zones that are existing in a city. As part of ecological fuels, we can mention as metrics the electric, natural-gas and LPG (Liquefied Petroleum Gas) buses provided by public transportation system. The sustainability indicator that is related to bike sharing, is ensured by a proper ratio between supply and density of bike sharing stations. Restricted zones indicator is referring to the transformation of some city zones to restricted for vehicles, known as pedestrian areas. In the same category of special zones is included the assurance of special lanes for bikes.

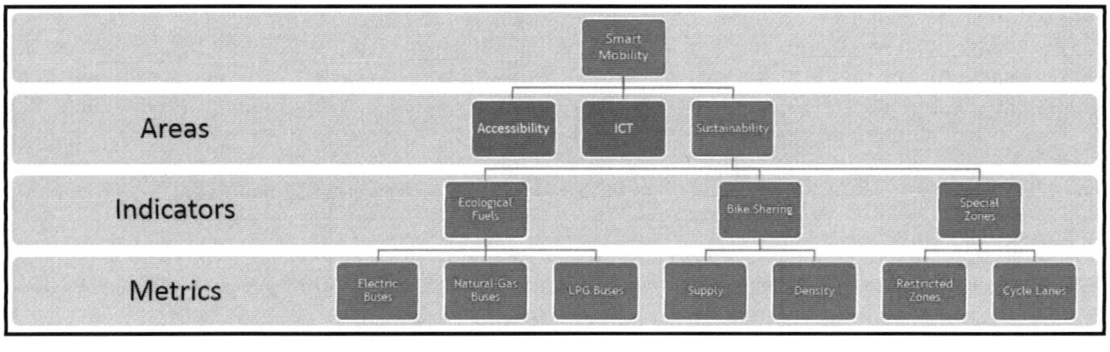

Figure 5. Sustainability area – indicators and metrics

6. Standardized metrics and indicators

As we seen before, there are many metrics and indicators to characterize a city from smart mobility point of view. It is needed to have a common framework which can make possible to select only some relevant metrics and indicators. This approach will simplify the comparison procedure between many cities around the world.

The solution for relevant metrics and indicators selection came through a standardization process. The ISO 37120:2014 standard defines indicators for a sustainable development of communities, with focus on quality of life improvement and quality of city services. ISO 37120:2014 approach is to define two categories of indicators: core indicators and supporting indicators, for each theme of interest, according to services and areas of application provided by the city. Core indicators are considered the relevant indicators which shall be followed to manage and evaluate the city services and citizens quality of life. The main difference between core indicators and supporting indicators is that the first type of indicators is considered as required to demonstrate the quality of services and quality of life provided by a city, while the second category of indicators is represented by some recommendations that should demonstrate the same features [9].

In Table 1 we can find the core and supporting indicators related to smart mobility according to [9]. As we mentioned before, we consider that smart mobility can be characterized as a mixture between four themes: environment, fire and emergency response, recreation and transportation.

Table 1. Significant indicators for transportation according to ISO 37120:2014

Theme	Core indicators	Supporting indicators
Environment	Fine Particulate Matter (PM2.5) concentration	Noise pollution
	Particulate Matter (PM10) concentration	O_3 (ozone) concentration
Fire and emergency response		Response time for emergency response services from initial call
		Response time for fire department from initial call
Recreation		Square metres of public outdoor recreation space per capita
Transportation	Kilometres of high capacity public transport system per 100 000 population	Percentage of commuters using a travel mode other than a personal vehicle
	Kilometres of light passenger public transport system per 100 000 population	Number of two-wheel motorized vehicles per capita
	Annual number of public transport trips per capita	Kilometres of bicycle paths and lanes per 100 000 population
	Number of personal automobiles per capita	Transportation fatalities per 100 000 population
		Commercial air connectivity (number of non-stop commercial air destinations)

A very good approach of a relevant set of indicators for the smart mobility areas of interests are proposed by Giffinger et al. [2]. The list of significant metrics and indicators for smart mobility, based on mentioned approach and tailored to our proposed method of smart mobility concept decomposability, can be found in Figure. 6.

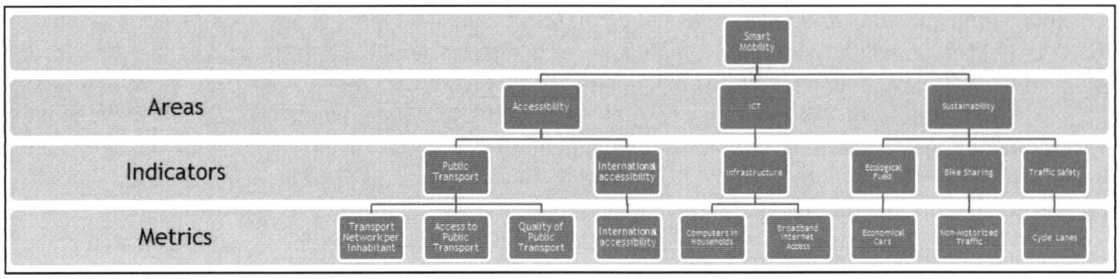

Figure 6. Common framework for smart mobility – indicators and metrics

7. Smart mobility worldwide analysis

Smart mobility metrics and indicators have a significant impact on smart cities ranking. In Table 2 we can see the top 5 European smart cities, from medium-sized cities perspective (100 000 to 500 000 inhabitants), according to [2]. In current analysis, we considered the scoring for each smart mobility area. The corresponding values for accessibility were obtained by summing up the indicators values for local accessibility, representing the public transport, and international accessibility.

Table 2. Top 5 European smart cities based on smart mobility – areas values

Rank	City	Country	Accessibility	ICT	Sustainability
1	Eindhoven	Netherlands	1.862	1.537	-0.276
2	Salzburg	Austria	2.382	0.354	0.077
3	Aarhus	Denmark	2.006	1.050	-0.293
4	Luxembourg-Ville	Luxembourg	2.672	0.622	-0.629
5	Leicester	United Kingdom	1.195	0.200	0.909

Looking from a worldwide perspective, we can see in Table 3 a top 10 worldwide smart cities. To see which is the impact of smart mobility on smart city scoring, for each city was attached its smart mobility ranking, according to [15], based on Cities in Motion Index (CIMI). In this case, were studied 165 cities. We can see some big differences in top 10 between smart city and smart mobility rankings. Singapore is on the 6th place as smart city and on 63rd place from smart mobility point of view. The biggest difference is on Hong Kong case, where the smart city ranking puts it on the 9th place and smart mobility on the 87th place.

The observed differences can be explained through better results of the metrics and indicators from the other smart city characteristics, that were presented in the on 2nd section of this paper.

Table 3. Top 10 worldwide smart cities based on CIMI scoring

Smart city ranking	City	Country	Smart mobility ranking
1	New York	United States of America	4
2	London	United Kingdom	2
3	Paris	France	1
4	Tokyo	Japan	22
5	Reykjavik	Iceland	7
6	Singapore	Singapore	63
7	Seoul	South Korea	3

8	Toronto	Canada	68
9	Hong Kong	China	87
10	Amsterdam	Netherlands	13

8. Conclusion

This paper has as starting point the presentation of traffic congestion problem that affects many cities around the world. This situation, that appears in daily road traffic has a negative impact on environment and on people quality of life by increasing travel times, vehicle queues on the road network and CO_2 emissions level. A solution for this worldwide problem can be considered the usage of smart city concept, especially the concept of smart mobility.

Smart mobility is a concept that starts to be used by many local administrations to increase the citizens quality of life by reducing traffic congestion and offering an intelligent framework to optimize the traffic flow in intersections. In this paper are identified the metrics and indicators that are characterizing the application of smart mobility concept. The most important thing that derives from the initial analysis, is the necessity of having a common framework that allows us to make a better comparison between different cities around the world, from smart mobility point of view.

The purpose of this paper was to identify the significant metrics and indicators for smart mobility and this was possible only by applying standardization, even if is not completely applied, being tailored to city necessities. ISO 37120:2014 defines these metrics and indicators and can be easily applied to each city and can offer an overview of the decisions that shall be taken by local administrations, or countries governments in some cases, to increase the smart mobility level.

References

[1] Pop M D and Proştean O 2018 *A Comparison Between Smart City Approaches in Road Traffic Management*, 14th International Symposium in Management 2017 (SIM), Timisoara, Romania, October 27-28, pp 29-36

[2] Giffinger R, Kramar H, Haindlmaier G and Strohmayer F 2015 *European Smart Cities*, Available at: http://smart-cities.eu/ [Last accessed on July 28, 2018]

[3] Battara R, Zucaro F and Tremiterra M R 2017 *Smart mobility: An Evaluation Method to Audit Italian Cities*, 5th IEEE International Conference on Models and Technologies for Intelligent Transportation Systems 2017 (MT-ITS), Naples, Italy, June 26-28, pp 421-426

[4] Hong D and Wong L W 2017 *A Study on Smart Mobility in Kuala Lumpur*, 2nd International Conference on Computing and Communications Technologies 2017 (ICCCT), Chennai, India, February 23-24, pp 27-32

[5] Latorre-Biel J I, Faulin J, Jiménez E and Juan A A 2017 *Simulation Model of Traffic in Smart Cities for Decision-Making Support: Case Study in Tudela (Navarre, Spain)*, 2nd International Conference Smart-CT 2017, Málaga, Spain, June 14-16, pp 144-153

[6] Nor N M and Wahap N A 2014 Workforce mobility: Contributing towards smart city, *IOP Conf. Ser.: Earth Environ. Sci.* **18** 012168

[7] Roda M, Giorgi D, Joime G P, Anniballi L, London M, Paschero M and Mascioli F M F 2017 *An Integrated Methodology Model for Smart Mobility System applied to Sustainable Tourism*, IEEE 3rd International Forum on Research and Technologies for Society and Industry 2017 (RTSI), Modena, Italy, September 11-13, pp 130-135

[8] ***European Commission 2012 *Smart Cities And Communities -European Innovation Partnership C(2012) 4701 final*, Brussels, Belgium, July 10, Available at: http://ec.europa.eu/transparenc y/regdoc/rep/3/2012/EN/3-2012-4701-EN-F1-1.PDF [Last accessed on July 28, 2018]

[9] ***International Organization for Standardization 2014 *ISO 37120:2014 - Sustainable development of communities -- Indicators for city services and quality of life*

[10] Deng D, Zhao Y and Zhou X 2017 Smart city planning under the climate change condition, *IOP*

Conf. Ser.: Earth Environ. Sci. **81** 012091

[11] Papa R, Gargiulo C and Russo L 2017 *The Evolution of Smart Mobility Strategies and Behaviors to Build the Smart City*, 5th IEEE International Conference on Models and Technologies for Intelligent Transportation Systems 2017 (MT-ITS), Naples, Italy, June 26-28, pp 409-414

[12] Varma G R 2017 *A Study on New Urbanism and Compact City and their Influence on Urban Mobility*, 2nd IEEE International Conference on Intelligent Transportation Engineering 2017 (ICITE), Singapore, Singapore, September 1-3, pp 250-253

[13] Zhong Y, Xie X, Guo J, Wang Q and Ge S 2018 A new method for short-term traffic congestion forecasting based on LSTM, *IOP Conf. Ser.: Mater. Sci. Eng.* **383** 012043

[14] Rath M 2018 Smart Traffic Management System for Traffic Control using Automated Mechanical and Electronic Devices, *IOP Conf. Ser.: Mater. Sci. Eng.* **377** 012201

[15] Berrone P and Ricart J E 2018 *IESE Business School - IESE Cities in Motion Index*, **ST-471-E**, pp 24-36

IOP Conf. Series: Materials Science and Engineering **477** (2019) 012018 doi:10.1088/1757-899X/477/1/012018

Loading three phased transformer following the quadergy

E Spunei[1], I Piroi[1] and F Piroi[2]

[1]Eftimie Murgu" University of Reşiţa, Department of Electrical Engineering and Informatics, Traian Vuia Street, str., no. 1-4, 320085 Reşiţa, Romania
[2]Technische Universität Wien, Institute of Information Systems Engineering, Viena, Austria

E-mail: e.spunei@uem.ro

Abstract. In this work, the authors propose a simple and fast method to establish the minimal real power for loading a three phased transformer of a specific power, such that power factor does not drop below a previously defined value. The method can be applied to transformers that are placed between a consumer's metering point and the electric power distribution network. In this situation, the billing energy and quadergy include the transformer energy and quadergy losses. This method establishes the moment where, due to low energy consumption, the transformer should be replaced with a lower power one, which still ensures the necessary energy for the consumers behind it. Our analysis has been done on a 100 kVA, an 80 KVA, and a 16 kVA transformer. Applying this method expedites the process of establishing the energy and the power factor where a transformer should be replaced by a lower power transformer, such that the power factor stays in acceptable ranges.

1. Introduction

Many economical units are connected to the electrical distribution network through 20 kV/0.4 kV transformers of various powers. As the electric load of these consumers varies irregularly, small energy loads appear, with negative technical effects (noise) [1] such that the associated quadergy causes a power factor below 0.65, having a negative impact on the consumed energy billing.

Most of the time, the metering point differs from the connection to the 20 kV energy network point, which is considered to be the billing point. In this situation, the specific legislation [2] establishes a methodology to compute the transformer energy and quadergy losses. Depending on the transformer's power, the bill will include energy and quadergy corrections, which, together with the metered energy, will make up the total sum for a given time period (usually a month).

When the metering point is not the same as the billing point [3], for a period of time, even though no energy consumption is registered, the energy bills include high energy and inductive quadergy correction values, as can be observed in Figure 1.

2. Theoretical Considerations on Correction Computations

According to Directive 24/2006 of the Romanian Energy Regulatory Authority (ANRE), when the metering point (PM) is different from the billing point (BP), and the meter can measure neither hourly energy consumption, nor the electric energy load curve, we have the following losses:

a. Constant energy losses, which can be:

- Electric energy ΔE_{ac}:

$$\Delta E_{ac} = P_0 \cdot T_f \quad \text{[kWh]} \tag{1}$$

- Electric quadergy ΔE_{rc}:

$$\Delta E_{rc} = \left(\frac{i_0\%}{100} \right) \cdot S_n \cdot T_f \quad \text{[kVArh]} \tag{2}$$

where P_0 [kW] stand for the transformer losses in the no-load operation state, T_f [h] is the time period for which the transformer is live, $i_0\%$ is the no-load operation current, and S_n [kVA] is the transformer's rated apparent output. For a consumer functioning without human staff, or for an automation equipment, we consider T_f 24 h (one day).

b. Variable energy losses, which can be:

- Electric energy ΔE_{av}:

$$\Delta E_{av} = P_{sc} \cdot \left(\frac{S_m}{S_n} \right)^2 \cdot \tau \quad \text{[kWh]} \tag{3}$$

- Electric quadergy ΔE_{rv}:

$$\Delta E_{rv} = \left(\frac{u_{sc}\%}{100} \right) \cdot \left(\frac{S_m}{S_n} \right)^2 \cdot S_n \cdot \tau \quad \text{[kVArh]} \tag{4}$$

where P_{sc} [kW] stand for the transformer losses in the short-circuit operation state, S_m [kVA] is the maximum apparent power, $u_{sc}\%$ is the short-circuit voltage, and τ [h] is the period on which the losses are computed, whose value is established by ANRE, depending on the number of functioning hours. During the short-circuit operation, even when very short, additional energy losses happen due to the high current values [4].

Specification	**MU**	**Const**	**Read index 31.03.2018**	**Read index 01.03.2018**	**Quantity**	**Corrections BP ≠ MP**	**Other corrections**	**Measured quantity**
Electric Energy	kWh	1	1685.8	1685.8	0	63	-	63
Inductive quadergy	kVArh	1	40.231	40.231	0	476	-	476
Capacitive quadergy	kVArh	1	57.493	55.314	2	-	-	2
			Inductive quadergy (kVArh)			**Capacitive quadergy (kVArh)**		
PF ≥ 0.90			0			0		
0.65 ≤ PF < 0.90			0			0		
PF < 0.65			445			0		

Figure 1. Energy corrections when the metered energies are zero

The consumer's power factor, φ_m, depends on the energy E_{am} and on the quadergy E_{rm} measured at the consumer location:

$$\cos\varphi_m = \frac{1}{\sqrt{1 + \left(\frac{E_{rm}}{E_{am}} \right)^2}} \tag{5}$$

To find the maximum apparent output, S_m, we must compute the maximum real output, P_m:

$$S_m = \frac{P_m}{\cos\varphi_m} = \frac{E_a}{\cos\varphi_m \cdot T_{sm}} \quad [\text{kVA}] \tag{6}$$

where T_{sm} is the duration use of the peak load whose value is established, depending on the number of operating hours, by the ANRE.

The energy and quadergy technical losses, ΔE_a and ΔE_a, show on the bills as corrections, are determined with the following equations:

$$\Delta E_a = \Delta E_{ac} + \Delta E_{av} \quad [\text{kWh}] \tag{7}$$

$$\Delta E_r = \Delta E_{rc} + \Delta E_{rv} \quad [\text{kVArh}] \tag{8}$$

All transformer losses can be determined already in the transformer design phase, during simulations of various regimes transformer operations [5-8].

The energy and quadergy which are used to compute the energy consumption at the billing point are computed using the following equations:

$$E_a = E_{am} + \Delta E_a \quad [\text{kWh}] \tag{9}$$

$$E_r = E_{rm} + \Delta E_r \quad [\text{kVArh}] \tag{10}$$

The lagging power factor at the delimitation point (including the transformer) is computed by:

$$\cos\varphi = \sqrt{\frac{1}{1+tg^2\varphi}} = \sqrt{\frac{1}{1+\left(\frac{E_r}{E_a}\right)^2}} \tag{11}$$

According the ANRE directives [2], [9] consumers pay the total capacitive quadergy, and for the inductive quadergy the consumers pay only the quantity that exceeds the power factor limit value at the delimiting point. The lagging power factor limit is 0.92 and the capacitive power factor limit is 1. When the consumer effects a power factor lower than 0.65, he will pay the rate difference between the recorded quadergy value and the value corresponding to the power factor limit value, multiplied by three. For lagging power factor values between 0.92÷0.65 the quadergy price rate is the one fixed by ANRE.

3. Case Study

In this work we analyse an industrial consumer where the metering point is different from the billing point, while at the consumer point an hourly time counter and a transformer with the following nominal values are installed:

- Nominal power S_n = 100 kVA;
- Primary rated voltage U_{1n} = 20 kV;
- Secondary rated voltage U_{2n} = 0.4 kV;
- Primary winding current flow I_{1n} = 20 A;
- Secondary winding current flow I_{2n} = 0.4 A.

According to ANRE directives [10] for this transformer type we have the following technical parameter values defined:

- No-load operation transformer losses P_0 = 0.6 kW;
- Short-circuit operation transformer losses P_{sc} = 2.76 kW;
- No-load operation current $i_0\%$ = 3.3;
- Short-circuit voltage $u_{sc}\%$ = 4;
- Loss computation time for a monthly continuous operation with 24 hours per day, τ = 203;

- Peak load use duration for a monthly continuous operation with 24 hours per day, $T_{sm} = 430$.

Table 1 presents the energy and quadergy values we measured, as well as their corrections according to the energy distribution provider bills.

Table 1. Power factor, energy and quadergy values, correction factors for 2008

Month	E_{am} [kWh]	ΔE_a [kWh]	E_{ar} [kVArh]	ΔE_r [kVArh]	Cos φ
Jan.2008	4268	451	301	2538	0.857
Feb.2008	2590	420	186	2370	0.762
Mar.2008	2385	448	263	2532	0.712
Apr.2008	2364	434	150	2451	0.732
May.2008	2576	449	406	2533	0.717
Jun.2008	1842	412	466	2450	0.612
Jul.2008	2504	448	913	2533	0.651
Aug.2008	2228	448	801	2532	0.626
Sep.2008	2086	433	872	2450	0.604
Oct.2008	2130	447	513	2532	0.646
Nov.2008	2060	434	507	2451	0.731
Dec.2008	1962	448	309	2531	0.647

Figure 2 shows the billed energy and quadergy variations and the power factor variations for 2008.

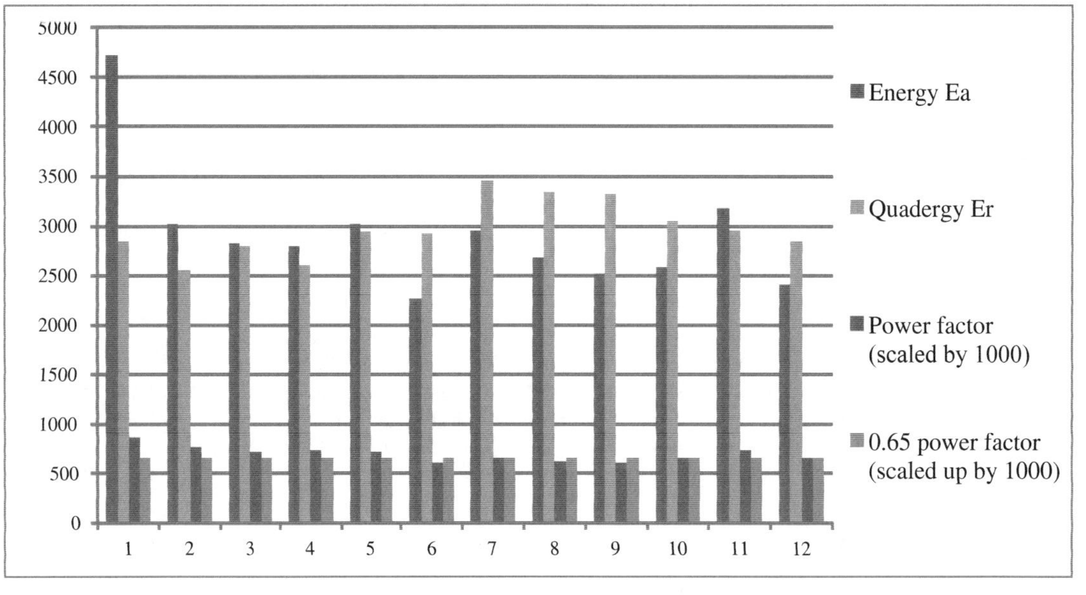

Figure 2. Billed energy and quadergy variations, power factors variations for year 2008

The power factor computed using equation (11) is scaled by a factor of 1,000 in order to represent it on the same coordinate system. On the same figure, scaled by a factor of 1,000 we show the 0.65 power factor where the quadergy is billed three times the regulated value.

We immediately notice that for June, August, September, October, and December the power factor value was below 0.65, which led to additional electricity costs.

To reduce these costs, the transformer was replaced with a 80 kVA apparent output transformer.

The measured energy and quadergy, as well as their corrections given by the electricity distribution provider are shown in Table 2.

Table 2. Power factor, energy and quadergy values, correction factors for 2012

Month	E_{am} [kWh]	ΔE_a [kWh]	E_{ar} [kVArh]	ΔE_r [kVArh]	Cos φ
Jan.2012	1859	0	69	0	0.999
Feb.2012	2092	261	63	1842	0.777
Mar.2012	1900	278	187	1968	0.711
Apr.2012	840	266	125	1902	0.479
May.2012	641	276	170	1965	0.395
Jun.2012	544	267	214	1901	0.358
Jul.2012	621	276	248	1965	0.376
Aug.2012	695	276	289	1965	0.396
Sep.2012	688	267	188	1901	0.416
Oct.2012	934	276	81	1965	0.509
Nov.2012	938	267	27	1902	0.530
Dec.2012	1350	276	25	1966	0.633

Figure 3 shows the energy and quadergy variations and the power factor variations for year 2012.

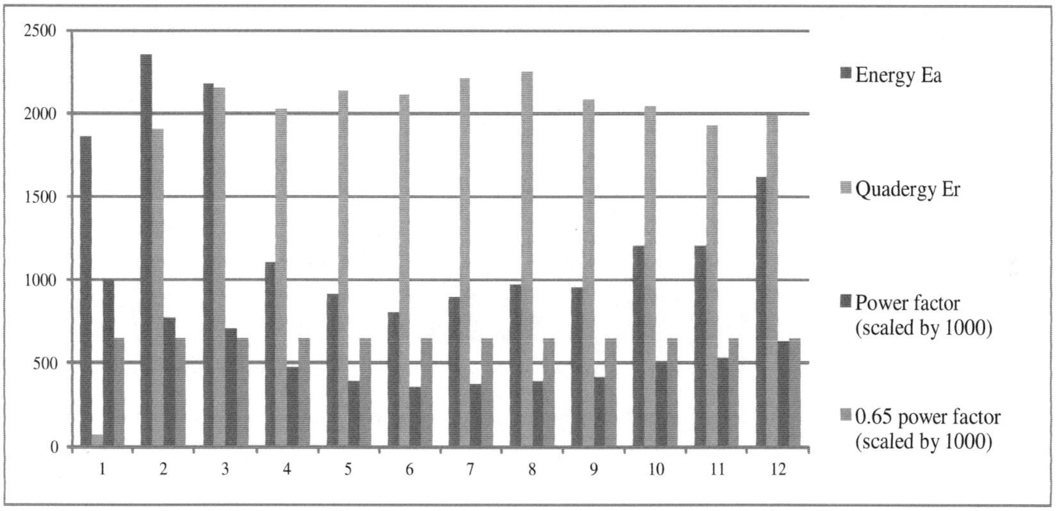

Figure 3. Billed energy and quadergy variations, power factors variations for year 2012

Using the data from Tables 1 and 2, and observing Figures 1 and 3, using equation (11), we obtain that for a 744 operation hours per month, to have a power factor higher than 0.65, the consumer's average power must not go below 3% of the 100 kVA transformer's nominal power, and not below 2.6% of the 80 kVA transformer's nominal power.

Beginning with 01.01.2017, with new ANRE directive [11] it was regulated that the 0.92 neutral power factor becomes the limit power factor, with a value of 0.90. This allowed the non-billing of inductive and capacitive reactive power up to a value of 0.90.

Analysing Figure 3 we note that in March 2012 the power factor is higher than 0.65, which followed a drastic decrease consumer energy use. By replacing the 80 kVA transformer by a 16 kVA transformer, at the same time when the energy consumer is suspended, the energy corrections are significant, considering the computed losses. For this type of transformer [10] we have the following technical parameters:

- No-load operation transformer losses $P_0 = 0.065$ kW;
- Short-circuit operation transformer losses $P_{sc} = 0.465$ kW;
- No-load operation current $i_0\% = 4$;
- Short-circuit voltage $u_{sc}\% = 4$.

Table 3 presents the measured energy and quadergy values, as well the corresponding corrections as they were calculated established by the energy distribution provider, as shown on the bill (Figure 1).

Table 3. The power factor, energy values, and corresponding corrections for 3 months in 2018

Month	E_{am} [kWh]	ΔE_a [kWh]	E_{ar} [kVArh]	ΔE_r [kVArh]	Cos φ
Feb.2018	0	57	2	430	0.131
Mar.2018	0	63	2	476	0.131
April 2018	0	61	2	461	0.131

Figure 4 shows the billed energy and quadergy variations and the power factor variations for February, March and April, 2018, when the consumer is using the 16 kVA transformer.

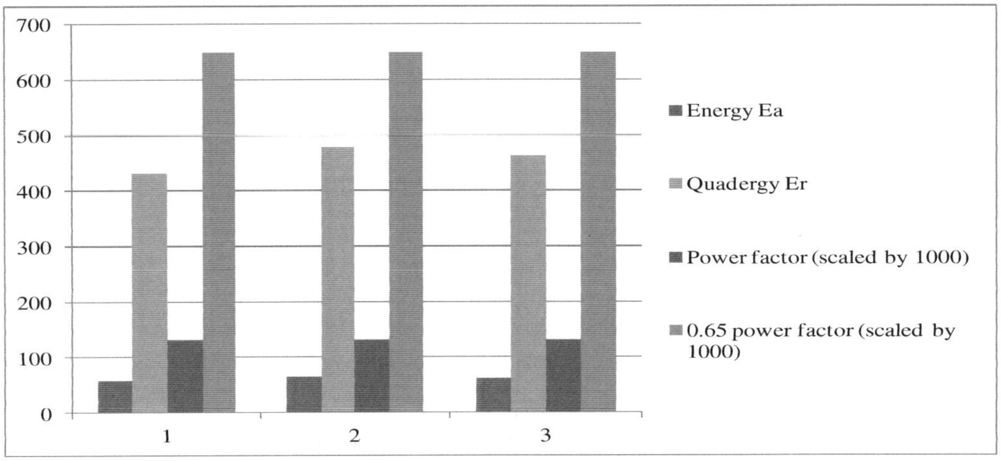

Figure 4. Billed energy and quadergy and the power factor for February, March and April, 2018

4. Conclusions

Analysing the data on the energy bills as well as the corresponding loss calculations we have determined the moment when, due to no energy consumption, a decision about a transformer replacement or about an energy contract change should be made.

The use of this method allows obtaining a quick, solid fact basis to show the energy distributor such that new, equally advantageous contracts between partners can be closed.

For a 100 kVA transformer, such that the power factor is higher than 0.65, the average consumer power must not drop below 3% of the transformer's nominal power. For the 80 kVA transformer, the consumer power must not drop below 2.6%.

We found that, although the original transformer has been replaced with a lower power one (16 kVA) for a no energy consumption the applied corrections are still rather high.

IOP Conf. Series: Materials Science and Engineering **477** (2019) 012018 doi:10.1088/1757-899X/477/1/012018

Looking at Figure 4 we see that for no energy nor quadergy consumption, in a three days interval, in two consecutive calendar months, the energy corrections are almost double. The cause of this must be spelled out and clarified by the energy distributor.

References

[1] Deaconu I D, Dragomir R C, Chirila A I, Deaconu A S, Saracin C G and Navrapescu V 2017 *Noise Analysis for No Load Operating Power Transformers*, 10th International Symposium on Advanced Topics in Electrical Engineering ATEE 2017, Bucharest, Romania, Mar 23-25, pp 128-131

[2] ***Ordinul 33/2014 *Metodologia privind stabilirea obligaţiilor de plată a energiei electrice reactive şi a preţului reglementat pentru energia electrică reactivă*, https://www.e-distributie.com/ro-RO/Documents/Ord%2033%2014.pdf, accessed at 13.04.2018

[3] ***Ordinul 24/2006 *Procedura privind corecţia energiei electrice în cazul în care punctul de măsurare diferă de punctul de decontare*, http://www.anre.ro/ro/legislatie/norme-tehnice/reglementari-tehnice 1387199450, accessed at 13.04.2018

[4] Deaconu I D, Ghita, C, Navrapescu V and Chirila A I 2010 *Radial and Axial Short-Circuit Electrodynamic Forces of Three Phase Power Transformers*, 12th International Conference on Optimization of Electrical and Electronic Equipment OPTIM 2010, Brasov, Romania, May 20-21, pp 185-190

[5] Munteanu A, Livadaru L, Simion A and Virlan B 2016 *An Efficient Approach for 3D Toroidal Transformers Simulation,*9th International Conference and Exposition on Electrical and Power Engineering EPE 2016, Iasi, Romania, Oct 20-22, pp 277-280

[6] Gong R H, Ruohan J J, Chen J Z, Quan Y, Wang J and Duan C H 2017 Analysis and Experiment of Hot-Spot Temperature Rise of 110 kV Three-Phase Three-Limb Transformer, *Energies* **10**(8) 1079

[7] Song Y D, Jia H D, Xu X F and Yu L 2017 *Simulation Analysis of Inrush Current of Three Phase Transformer Based on MATLAB*, 29th Chinese Control And Decision Conference CCDC 2017, Chongqing, Peoples R China, May 28-30, pp 3983-3985

[8] Quan F Y, Chen J, Liu Y, Wang Y, He W L and Wen X S 2017 *Calculation and Analysis of Transformer Transient Over-Voltage in no-Load Switching,* 2nd International Conference on Advances in Materials, Mechatronics and Civil Engineering ICAMMCE 2017, Guangzhou, Peoples R China, Jan 19-20, Book Series: AER-Advances in Engineering Research, Volume 121, pp 1-8

[9] ***Ordinul 35/2005 *Procedura privind asigurarea energiei electrice reactive şi modul de plată a acesteia*, http://www.electricafurnizare.ro/wp-content/uploads/2016/09/ordin-anre-35-2005-53.pdf, accessed at 13.04.2018

[10] ***Ordinul 75/2015 *Procedura privind corecţia datelor de măsurare în raport cu punctual de delimitare*, http://www.anre.ro/ro/legislatie/norme-tehnice/reglementari-tehnice1387199450, accessed at 13.04.2018

[11] ***Ordinul 76/2016 *Ordin pentru modificarea şi completarea Metodologiei privind stabilirea obligaţiilor de plată a energiei electrice reactive şi a preţului reglementat pentru energia electric reactivă*, http://www.anre.ro/ro/legislatie/metodologii-tarife/furnizare-energie-electrica, accessed at 13.04.2018

Latin square experiment design in R

D Hodzic, A Hodzic and E Bajramovic

University of Bihac, Faculty of Technical Engineering, Irfana Ljubijankica bb, 77000 Bihac, Bosnia and Herzegovina

E-mail: damir.hodzic@unbi.ba

Abstract. In this paper we will describe design of experiment by latin square method. Latin square is statistical test which is used in planning of experiment and is one of most accurate method. Programming language R is a tool which can be used for statistical tests and graphics. It has a wide range of applications both in research and practice. A practical example will be presented in this paper and its processing in the classical approach and in the programming language R.

1. Introduction

Design of Experiment is one of the new scientific disciplines whose object is the scientific-research experiment. The word experiment is inspired by the Latin word of the experiment, which translates into an opus or an excuse. There are several different definitions of experiments, but from the engineering point of view we can say that the experiment presented a scientifically designed opus that includes the operation system, algorithm and experimental technique, to test an object under well-defined regimes and conditions. The experiment is used in the final phase of the research when it is the key criterion for verifying the truth of theory and hypothesis. Experiment planning is a discipline associated with sample theory. Theory of samples evaluates population characteristics based on observation of individual parts of these populations without any change of population while planning of experiments affects one part of the population to assess the effect of these activities on the entire population. One experiment can have different meanings. Thus, in engineering analyzes, the experiment may be the preparation of pieces of a material and the measurement of the characteristics of that piece. It may also mean studying the moon by sending monthly footage or studying the Earth's Ozone Layer. Also, the experiment may also mean studying the effects of certain drugs in the treatment of certain diseases in humans. In all of these studies it is assumed that the visible part of the population or data is representative. Design of Experiment (DOE) is a powerful tool for achieving significant product quality improvements and process efficiency. For the design of the experiment it can also be said that it is the methodology of applying statistics in the process of performing the experiment. As such discipline is not just a engineering problem but is used in different social sciences [1].

A well-planned experiment allows us to get clear interpretations and avoid complicated analyzes. The poorly planned experiment gives us some wrong conclusions of a process. Design of Experiment is of particular importance in all quantitative studies, especially when multiple factors are studied at the same time to examine and compare their effects.

The basic terms we will use in this paper and which are the basis of each experiment planning are factors, treatments and blocks. Factors represent different types of influences that apply to experimental units. Treatment is a specific combination of levels of different factors and the blocks are homogeneous groups by which we compare the treatments.

2. Latin square method

If the experiment is performed so that another factor is presented beside the effect of the treatments and the blocks, and if there is no interaction between these three factors, then we can use Latin square method. The basic feature of the randomized block design, which is a method also used for design of experiment and is similar to Latin square method, is that at each block one treatment occurs at least once. In the Latin square method in each row and column every treatment occurs only once. In this way, it is often achieved that the variations of these basic groups are small, so it is a small experimental error.

For the statistical analysis of the Latin square, we will characterize the row with *i*, column with *j* and treatment with *k*. The results of measurement occurring under the i*th* row, j*th* column and k*th* treatment is denoted by y_{ijk} where $1 \leq i, j, k \leq p$.

In general The Latin square with dimensions *p* x *p* is a quadratic scheme in which Latin letters p means *p* raws and *p* columns, indicating treatments and each letter is repeated p times. Thus each letter occurs only once in each row and each column [2], [3].

Some of examples of Latin squares are:

A	B	C	D
B	A	D	C
C	D	B	A
D	C	A	B

A	D	B	E	C
D	A	C	B	E
C	B	E	D	A
B	E	A	C	D
E	C	D	A	B

A	D	C	E	B	F
B	A	E	C	F	D
C	E	D	F	A	B
D	C	F	B	E	A
F	B	A	D	C	E
E	F	B	A	C	D

Figure 1. Examples of Latin Square (*4x4, 5x5, 6x6*)

Latin squares are used in many fields of research work (agriculture, industry, medicine, sociology,...). The advantage of the Latin squares is that three factors can be studied in a small experiment. In addition to contributing to a reduction in the experimental error, the statistical analysis procedure in the Latin square is relatively simple. The lack of a Latin squares is that a greater number of treatments require a greater number of repetitions, making the experiment uneconomical and often less precise. Therefore, the squares are rarely applied with square dimensions greater than 10 x 10. Also, Latin squares with dimensions less than 4 x 4 are not being used because of the small number of degrees of freedom at the sum of squares for error.

Results can be displayed in several ways, some of which are:

8_A	10_C	12_B	9_D
7_B	8_D	10_A	11_C
9_D	10_A	8_C	9_B
12_C	11_B	9_D	13_A

A+8	C+10	B+12	D+9
B+7	D+8	A+10	C+11
D+9	A+10	C+8	B+9
C+12	B+11	D+9	A+13

8	10	12	9
7	8	10	11
9	10	8	9
12	11	9	13

Figure 2. Examples of displaying latin square results

In the statistical analysis of the Latin square, the Analysis of Variance (ANOVA) with three factors is applied, which will be presented in the next section.

3. Practical example of Latin square

We will assume that we have an example of four workers who have processed different workpieces in four periods. The work was graded from 1 to 15. The experiment was performed by random selection of the Latin square shown in Figure 3

C	B	D	A
D	C	A	B
A	D	B	C
B	A	C	D

Figure 3. Random selection of Latin square for example

and we got following results (Table 1).

Table 1. Results for the example

	Period			
Worker	t_1	t_2	t_3	t_4
u_1	12	9	8	7
u_2	6	7	6	9
u_3	5	10	6	9
u_4	11	8	7	8

It is necessary to examine whether there is a significant influence of the worker, the time at which the workpiece was processed or the processing method to the experimental results if $\alpha = 0{,}05$.
First, we will determine the sum of columns and rows and arithmetic means for the given data.

Table 2. Sum of columns and rows and arithmetic means for the example

	Period					
Worker	t_1	t_2	t_3	t_4	V_i	$\bar{y}_{i..}$
u_1	12	9	8	7	36	9
u_2	6	7	6	9	28	7
u_3	5	10	6	9	30	7,5
u_4	11	8	7	8	34	8,5
K_j	34	34	27	33	128	
$\bar{y}_{.j.}$	8,5	8,5	6,75	8,25		$\bar{y}_{.j.} = 8$

According to ANOVA calculation we have to calculate sum of squares

$$q = \sum_{i=1}^{p}\sum_{j=1}^{p}\sum_{k=1}^{p}\left(y_{ijk} - \bar{y}\right)^2 = (8-12)^2 + (8-9)^2 + \ldots + (8-9)^2 = 56 \quad (1)$$

$$q_V = p\sum_{i=1}^{p}(\bar{y}_{i..} - \bar{y})^2 = 4(1^2 + 1^2 + 0{,}5^2 + 0{,}5^2) = 10 \quad (2)$$

$$q_K = p\sum_{i=1}^{p}(\bar{y}_{.j.} - \bar{y})^2 = 4(0{,}5^2 + 0{,}5^2 + 1{,}25^2 + 0{,}25^2) = 8{,}5 \quad (3)$$

where:
q – Sum of squares total
q_V – Sum of squares of rows
q_K – Sum of squares of columns
Now we will show results according to treatments (Table 3).

Table 3. Results according to treatments

	Treatment A	B	C	D
	7	9	12	8
	6	9	7	6
	5	6	9	10
	8	11	7	8
Total	26	35	35	32
$\bar{y}_{..k}$	6,5	8,75	8,75	8

$$q_T = p\sum_{k}(\bar{y}_{..k} - \bar{y})^2 = 4(1{,}5^2 + 0{,}75^2 + 0{,}75^2 + 0) = 13{,}5 \quad (4)$$

$$q_R = q - q_V - q_K - q_T = 56 - 10 - 8{,}5 - 13{,}5 = 24 \quad (5)$$

where:
q_T – Sum of squares of treatments
q_R – Sum of squares of error
Now we calculate F values for rows, columns and treatments:

$$F_V = \frac{(p-1)(p-2)}{p-1}\frac{q_V}{q_R} = \frac{3\cdot 2}{3}\cdot\frac{10}{24} = 0{,}83 \quad (6)$$

$$F_K = \frac{(p-1)(p-2)}{p-1}\frac{q_K}{q_R} = \frac{3\cdot 2}{3}\cdot\frac{8{,}5}{24} = 0{,}71 \quad (7)$$

$$F_T = \frac{(p-1)(p-2)}{p-1}\frac{q_T}{q_R} = \frac{3\cdot 2}{3}\cdot\frac{13{,}5}{24} = 1{,}125 \quad (8)$$

where p is number of rows and columns.
Now we can make ANOVA table with all datas (Table 4)

Table 4. Analysis of Variance for Latin square example

Source of variation	Sum of squares	Degrees of freedom	Mean square	F value
Rows	10	3	3,33	0,83
Columns	8,5	3	2,83	0,71
Treatments	13,5	3	4,5	1,125
Error	24	6	4	
Total	56	15	3,73	

F values calculated in equations (6), (7) and (8) we compare with *F* from distribution table for α = 0,05 (Figure 4). We choose F according to degrees of freedom 3 and 6 from Table 4.

F - Distribution (α = 0.05 in the Right Tail)

df_2 \ df_1 (Denominator Degrees of Freedom \ Numerator Degrees of Freedom)	1	2	3	4	5	6	7	8	9
1	161.45	199.50	215.71	224.58	230.16	233.99	236.77	238.88	240.54
2	18.513	19.000	19.164	19.247	19.296	19.330	19.353	19.371	19.385
3	10.128	9.5521	9.2766	9.1172	9.0135	8.9406	8.8867	8.8452	8.8123
4	7.7086	9.9443	6.5914	6.3882	6.2561	6.1631	6.0942	6.0410	6.9988
5	6.6079	5.7861	5.4095	5.1922	5.0503	4.9503	4.8759	4.8183	4.7725
6	5.9874	5.1433	4.7571	4.5337	4.3874	4.2839	4.2067	4.1468	4.0990
7	5.5914	4.7374	4.3468	4.1203	3.9715	3.8660	3.7870	3.7257	3.6767
8	5.3177	4.4590	4.0662	3.8379	3.6875	3.5806	3.5005	3.4381	3.3881
9	5.1174	4.2565	3.8625	3.6331	3.4817	3.3738	3.2927	3.2296	3.1789
10	4.9646	4.1028	3.7083	3.4780	3.3258	3.2172	3.1355	3.0717	3.0204
11	4.8443	3.9823	3.5874	3.3567	3.2039	3.0946	3.0123	2.9480	2.8962
12	4.7472	3.8853	3.4903	3.2592	3.1059	2.9961	2.9134	2.8486	2.7964
13	4.6672	3.8056	3.4105	3.1791	3.0254	2.9153	2.8321	2.7669	2.7144
14	4.6001	3.7389	3.3439	3.1122	2.9582	2.8477	2.7642	2.6987	2.6458
15	4.5431	3.6823	3.2874	3.0556	2.9013	2.7905	2.7066	2.6408	2.5876
16	4.4940	3.6337	3.2389	3.0069	2.8524	2.7413	2.6572	2.5911	2.5377
17	4.4513	3.5915	3.1968	2.9647	2.8100	2.6987	2.6143	2.5480	2.4943
18	4.4139	3.5546	3.1599	2.9277	2.7729	2.6613	2.5767	2.5102	2.4563
19	4.3807	3.5219	3.1274	2.8951	2.7401	2.6283	2.5435	2.4768	2.4227
20	4.3512	3.4928	3.0984	2.8661	2.7109	2.5990	2.5140	2.4471	2.3928
21	4.3248	3.4668	3.0725	2.8401	2.6848	2.5727	2.4876	2.4205	2.3660
22	4.3009	3.4434	3.0491	2.8167	2.6613	2.5491	2.4638	2.3965	2.3419
23	4.2793	3.4221	3.0280	2.7955	2.6400	2.5277	2.4422	2.3748	2.3201
24	4.2597	3.4028	3.0088	2.7763	2.6207	2.5082	2.4226	2.3551	2.3002
25	4.2417	3.3852	2.9912	2.7587	2.6030	2.4904	2.4047	2.3371	2.2821
26	4.2252	3.3690	2.9752	2.7426	2.5868	2.4741	2.3883	2.3205	2.2655
27	4.2100	3.3541	2.9604	2.7278	2.5719	2.4591	2.3732	2.3053	2.2501
28	4.1960	3.3404	2.9467	2.7141	2.5581	2.4453	2.3593	2.2913	2.2360
29	4.1830	3.3277	2.9340	2.7014	2.5454	2.4324	2.3463	2.2783	2.2229
30	4.1709	3.3158	2.9223	2.6896	2.5336	2.4205	2.3343	2.2662	2.2107
40	4.0847	3.2317	2.8387	2.6060	2.4495	2.3359	2.2490	2.1802	2.1240
60	4.0012	3.1504	2.7581	2.5252	2.3683	2.2541	2.1665	2.0970	2.0401
120	3.9201	3.0718	2.6802	2.4472	2.2899	2.1750	2.0868	2.0164	1.9588
∞	3.8415	2.9957	2.6049	2.3719	2.2141	2.0986	2.0096	1.9384	1.8799

Figure 4. *F* distribution for α = 0,05

Tabular *F* value is 4,76 and we can see that this value is greater then all calculated *F* values ($F_K < F$; $F_V < F$; $F_T < F$). It means that the effects of columns, rows and treatments are not significant, there is no significant influence on the time of processing, the worker and the way of processing on the experimental results.

4. Practical example of Latin square in R

R is the programming language and environment for statistical calculations and visualization. It is free to use and provides a wide range of statistical methods for linear and nonlinear modeling, classical statistical tests and others. In this paper, it will be shown that the programming language R can easily solve an example of a previously performed classic experiment.First it is necessary to input values and names for rows, columns and treatments then to group the values and analyze the variance (Figure 5) [4-6].

```
# Latin square
# Damir Hodzic, Atif Hodzic, Esad Bajramovic -  Faculty of Technical Engineering Bihac
# Input values
worker <-c (rep("worker1",1), rep("worker2",1), rep("worker3",1), rep("worker4",1))
```

```
period <-c (rep("periodA",4), rep("periodB",4), rep("periodC",4), rep("periodD",4))
letter <-c ("C","D","A","B","B","C","D","A","D","A","B","C","A","B","C","D")
value <-c (12,6,5,11,9,7,10,8,8,6,6,7,7,9,9,8)

# Grouping the values
mydata <- data.frame(worker, period, letter, value)
mydata
matrix(mydata$letter, 4,4)
matrix(mydata$value, 4,4)
myfit <- lm(values ~ worker+period+letter, mydata)

#Analysis of variance
anova(myfit)
par(mfrow=c(2,2))
plot(value ~ worker+period+letter, mydata)
```

Figure 5. Program code in R for One way ANOVA

After *R* gets input values it will generate and arrange them into the matrix (Figure 6).

```
> mydata
     worker period  letter  value
  1 worker1 periodA    C      12
  2 worker2 periodA    D       6
  3 worker3 periodA    A       5
  4 worker4 periodA    B      11
  5 worker1 periodB    B       9
  6 worker2 periodB    C       7
  7 worker3 periodB    D      10
  8 worker4 periodB    A       8
  9 worker1 periodC    D       8
 10 worker2 periodC    A       6
 11 worker3 periodC    B       6
 12 worker4 periodC    C       7
 13 worker1 periodD    A       7
 14 worker2 periodD    B       9
 15 worker3 periodD    C       9
 16 worker4 periodD    D       8

> matrix(mydata$letter, 4,4)
     [,1] [,2] [,3] [,4]
[1,] "C"  "B"  "D"  "A"
[2,] "D"  "C"  "A"  "B"
[3,] "A"  "D"  "B"  "C"
[4,] "B"  "A"  "C"  "D"

> matrix(mydata$value, 4,4)
     [,1] [,2] [,3] [,4]
[1,]   12    9    8    7
[2,]    6    7    6    9
[3,]    5   10    6    9
[4,]   11    8    7    8
```

Figure 6. Generating the values in *R*

By executing the ANOVA command (Figure 7) we will get the results in the *R* which are identical to the calculated values in Table 4.

```
> anova(myfit)
Analysis of Variance Table
Response: value
Df Sum Sq Mean Sq F value Pr(>F)
worker     3  10.0  3.3333  0.8333 0.5226
period     3   8.5  2.8333  0.7083 0.5813
letter     3  13.5  4.5000  1.1250 0.4108
Residuals  6  24.0  4.0000
```

Figure 7. Results in *R*

Results can also been presented graphically.

5. Conclusion

Design of Experiment is an important scientific field in engineering disciplines. The Latin squares method is one of the methods that has a wide range of applications and is interesting for analysis in new software packages. Software program R is a free code package that is accessible to everyone and is therefore used for that reason. It is relatively easy to reach conclusions for a particular experiment and much faster than the classic way outlined in this paper.

References

[1] Montgomery D C 2015 *Design and Analysis of Experiments (8th Edition)*, Arizona State University, John Wiley and Sons
[2] Kovač P 2016 *Metode planiranja i obrade eksperimenta*, Univerzitet u Novom Sadu, Fakultet tehničkih nauka
[3] Petrović Lj 2013 *Teorija uzoraka i planiranje eksperimenta*, Univerzitet u Beogradu, Ekonomski fakultet
[4] Hodzic D 2017 *Planiranje eksperimenta metodom ANOVA u programskom jeziku R*, 11th International Scientific Conference on Production Engineering RIM 2017, Sarajevo, Bosnia and Herzegovina
[5] Kerns G J 2012 *Introduction to Probability and Statistics using R*, GNU Free Documentation
[6] Gerrard P, Johnson R M 2015 *Mastering Scientific Computing with R*, Packt Publishing, Birmingham-Mumbai

IOP Conf. Series: Materials Science and Engineering **477** (2019) 012020 doi:10.1088/1757-899X/477/1/012020

Economic evaluation of tourism infrastructure development in Ukraine

I P Kinash[1], L M Arkhypova[2], A S Polyanska[1], O G Dzoba[3], U Y Andrusiv[3] and Iu I Iuras[2]

[1]Ivano-Frankivsk National Technical University of Oil and Gas, Institute of Economics and Management, Department of Management and Administration, 15 Karpatska str., 76018, Ivano-Frankivsk, Ukraine
[2]Ivano-Frankivsk National Technical University of Oil and Gas, Institute of Tourism and Geosciences, Department of Tourism, 15 Karpatska str., 76018, Ivano-Frankivsk, Ukraine
[3]Ivano-Frankivsk National Technical University of Oil and Gas, Institute of Economics and Management, Department of Theory of Economics and Management, 15 Karpatska str., 76018, Ivano-Frankivsk, Ukraine

E-mail: iuliia.murava@ukr.net

Abstract. The article presents own evaluation methodology of the tourism infrastructure development at the state level. To obtain data on the state and trends of national tourism development the authors chose economic diagnosis based on an evaluation of industry infrastructure provision. The following objects were selected for study: sanatoriums and resorts, guest houses and lodges, hotels, child resort and health centers. Their state, features of their use and their material and technical base were evaluated. Using own evaluation methodology of level of tourism resource provision the indicator of objects density was calculated and the results of calculations were summarized. According to the results of determined partial indicators the integrated evaluation of tourism infrastructure provision was made and the dynamics of integral index of infrastructure provision according to the Ukrainian regions was presented.

1. Introduction

Tourism is one of the most dynamic sectors of the global economy. Despite the global crisis, tourism has shown almost uninterrupted growth. The number of international arrivals increased globally from 25 million in 1950, 278 million in 1980, 527 million in 1995 to 1.18 billion in 2015 which allowed tourism to enter the top five most profitable industries in the world [1].

Ukraine has one of the leading positions in Europe according to the availability of natural, historical and cultural resources. Resorts and recreational territories in our country occupy about 9.1 million hectares that is about 15% of the state territory. However, according to the World Economic Forum Ukraine currently uses less than a third of its existing tourist and recreational potential. According to the tourism competitiveness index ranking of 2016 Ukraine received 3.5 points out of seven possible and got 88th place in the ranking of 136 countries [2].

One of the problems that hinders the development of tourism in Ukraine, in our opinion, is the low infrastructure level of the industry [3]. That is why the need for information on trends in the development of national tourism industry is becoming more relevant. One of the instruments of getting such information is economic diagnosis which allows to evaluate the features and problems of the tourism industry and to identify the patterns and trends of its development.

The world scientific society is also interested in the studies of tourism development evaluation. There are some works that study structural models in general. For example, Hamish Low and Costas Meghir contrast the treatment effects approach with structural models, and present an example of how a structural model is specified and the particular choices that were made [4]. There are also some methods designed specifically for tourism development evaluation. Hong Zhang, Chao-lin Gu, Luwen Gu, Yan Zhang applied the Technique for Order Preference by Similarity to Ideal Solution to evaluate the tourism destination competitiveness of the Yangtze River Delta in China [5]. Paul Suman studied the tourism infrastructure as well as seasonal arrival of tourists in the Gangtok city and tried to develop the probabilistic travel model on the basis of tourist perception which will help the tourism department for the further economic development of the area [6].

The materials of authors' research indicate that scientists offered different methods of evaluating the tourism development, but they mostly focuse on the recreational attractiveness of the territories where there are tourist objects. The authors believe that more complete diagnosis of current development trends, in particular, on the basis of the infrastructure evaluation, is needed for better information support of the tourism sector management.

The purpose of the study is to develop a method of evaluation of tourism infrastructure provision, to apply such method for the tourism industry in Ukraine and to represent the evaluation results.

The main hypothesis of the study is the introduction of the indicator of objects density in the proposed methodical approach in the evaluation of the tourism infrastructure provision, which allows to determine the availability and accessibility of objects in one value, and the use of taxonomic method with the possibility to determine the sustainability of the development indicators trends and sustainability of series levels.

2. Method of infrastructure provision evaluation for tourism industry

A measurement system which takes into account simple indices was prepared in order to allow evaluation of the state and development of tourism facilities. Firstly, we used index method, coefficients method, summary and grouping method, indicators standardization, construction and analysis of time series, integral indices, comparative analysis and others. Secondly, we collected the analytical data, which indicates the trends of tourism infrastructure provision. We took the following indicators for this analysis: number of sanatoriums, number of resort, guest houses and lodges, the number of hotels and other accommodation facilities, and the number of beds in all these types of accommodation facilities.

In the work [7] the following indicator is offered for the social sphere objects density evaluation:

$$X_{ik} = \frac{x_{ik}}{\sqrt{\mathrm{N}_i \quad \mathrm{P}_i}}, \tag{1}$$

where x_{ik} – k-th indicator in the i-region;

P_i – the area of i-region;

N_i – population in i-region.

In the formula the density indicator is an average geometric number of the indicators of availability and provision, that is:

$$X_{ik} = \sqrt{d_{ik} Z_{ik}} = \sqrt{\frac{x_{ik}^2}{N_i P_i}} = \frac{x_{ik}}{\sqrt{N_i P_i}}. \tag{2}$$

The use of these indicators will enable us characterize more accurately the specific material and technical support of tourism industry in Ukraine, make conclusions about the saturation territory with these objects.

The procedure of calculating the level of tourism development in Ukraine is based on the taxonomic indexes calculation and is the following:

1. Formation of indicators that characterize tourism infrastructure availability in terms of the regions within the country.
2. Identification of the elements of X matrix observations. The matrix will look like:

$$X = \begin{bmatrix} X_{11}\ X_{12} & \dots & X_{1k} & X_{1k} \\ X_{21}\ X_{22} & \dots & X_{2k} & \dots X_{2n} \\ \dots\ \dots & & & \\ X_{i1}\ X_{i2} & & X_{ik} & \dots X_{in} \\ \dots & & \dots & \dots \\ X_{w1}\ X_{w2} & \dots & X_{wk} & \dots X_{wn} \end{bmatrix}, \tag{3}$$

where w is the number of regions; N is number of indicators that characterize the level of tourist objects development; Xik is an indicator k for the i-region.

3. Differentiating of the observations matrix features. The matrix features are divided into stimulators and non-stimulators. The feature that has a positive effect on the level of tourist infrastructural provision is added to stimulators. It should be noted that during calculations the authors took into account such non-stimulators of tourism development as depreciation of fixed assets in tourism; number of sanatorium and health centers that did not work; number of sanatoriums and health centers that need repair or are in dangerous condition.
4. Standardization of observations matrix. The authors suggest using the following formula:

$$Z_{ik} = \frac{\tilde{X}_{ik} - \overline{X}_k}{S_k}, \tag{4}$$

$$\overline{X}_{k_{kk}} = \frac{\sum_{i=1}^{w} \tilde{X}_{ik}}{w}, \tag{5}$$

$$S_k = \sqrt{\frac{\sum_{i=1}^{w} \left(\tilde{X}_{ik} - \overline{X}_k\right)^2}{w}}, \tag{6}$$

where: Z_{ik} is a standardized k-indicator for the i-region; - $\tilde{X}_{ik}$ is a k-indicator for the i-region; $\overline{X}_{ik}$ is a arithmetic average of k-indicator; S_k is a standardized deviation of k-indicator; w is a number of regions.

As a result of standardization of matrix observations features the matrix is the following:

$$Z = \begin{bmatrix} Z_{11}\ Z_{12}\ \ldots\ Z_{1k}\ \ldots Z_{1n} \\ Z_{21}\ Z_{22}\ \ldots Z_{2k}\ \ldots Z_{2n} \\ \ldots\ \ldots\ \ldots\ \ldots\ \ldots \\ Z_{i1}\ Z_{i2}\ \ldots\ Z_{ik}\ \ldots\ Z_{in} \\ \ldots\ \ldots\ \ldots\ \ldots\ \ldots \\ Z_{w1}\ Z_{w2}\ \ldots\ Z_{wk}\ \ldots\ Z_{wn} \end{bmatrix}. \quad (7)$$

The conducted standardization procedure erases differences in the importance of individual features, and all features are considered equivalent, that distorts reality. Mitigation of this adverse event can be achieved by the introduction of the hierarchy coefficients that will allow to divide the features according to their importance. These coefficients characterize the position of each feature, its role and importance in the study. The calculation of hierarchy coefficients is offered to be conducted basing on the so-called critical distances, the greatest distance ρ between neighboring features (α_i, α_j):

$$p(a_i a_j) = \sqrt{\sum_{k=1}^{27} (a_i^{(l)} - a_j^{(l)})}, \quad (8)$$

where α_i – is an i vector-indicator according to the regions in a given year;

α_j is a *j*-indicator according to the regions in the same year;

I is a number of the region;

K = max – min ρ (α_i α_j) is a critical distance.

The sequence of calculating the hierarchy coefficients is the following:

- determination of the distances that are not higher than the critical distances for the each feature of observations matrix according to the formula:

$$Q_i = \{(i, j) \mid p(a_i, a_j) \le k, j = 1,2,\ldots n\}, \quad (9)$$

where n is a number of the indicators

- the authors sum up the obtained distances for each element:

$$\omega_i = \sum_{(i,j)\in Q_i} p(a_i, a_j), \quad (10)$$

- choice of the feature, for which the calculated sum of the distances is the highest:

$$\omega_m = \max_i \omega_i, \quad (11)$$

- hierarchy coefficients calculation:

$$\lambda_i = \frac{\omega_i}{\omega_m}, \quad (12)$$

Then the indicator λ_k is multiplied at the corresponding indicator in the corresponding year $z_{i.}$. Adjusted features are used for taxonomic research.

5. Choice of the benchmark point for each of the indicators of tourism infrastructure provision. The benchmark point was found for each feature. The integrated level will be calculated according to the distance to the point. The highest stimulators and the lowest non-stimulators form the coordinates of the benchmark development of Z_{ok}:

$$Z_{ok} = \max_{t}, if, k \in I, \tag{13}$$

$$Z_{ok} = \min_{t} Z_{tk}, if, k \in I; (k = 1,..., n), \tag{14}$$

where **I** is a set of stimulators;

Z_{0k} is a standardized feature k in the period t.

Thus, we obtain the vector of benchmark numbers of the features which is a point P_0 with coordinates (number n): Z_{01}, Z_{02}, Z_{0n}.

The distance between the individual points-units (regions according to the studied feature) and point P_0 (benchmark one) is calculated according to the formula:

$$C_{i0} = \sqrt{\sum_{S=i}^{n} (Z_{ik} - Z_{0k})^2 \; (i = 1,..., w) \quad (s = 1,..., n)}. \tag{15}$$

Calculation of such indicator as a level of material and technical support development in the country's tourism industry is done.

On the basis of all previous transformations and calculations the integral indicator can be calculated according to the formulas:

$$\overline{C}_0 - \frac{\sum_{i=1}^{w} C_{i0}}{w}, \tag{16}$$

$$S_0 = \sqrt{\frac{\sum_{i=1}^{w} \left(C_{i0} - \overline{C}_0\right)^2}{w}}, \tag{17}$$

$$C_0 = \overline{C}_0 + 2S_0, \tag{18}$$

$$D_i = 1 - \frac{C_{i0}}{C_0}, \tag{19}$$

where D_i is an indicator of development;

C_{i0} is a distance between individual points-units and benchmark ones.

Development of the industry the authors understand as a change in the chosen indicators. The proposed method involves the evaluation of the sustainability development trends. Sustainability indicator will look like:

$$i_{\overline{T}} = \frac{\overline{T}auspicious}{\overline{T}adverse}, \tag{20}$$

where $\overline{T}auspicious$ is an average annual growth rate of the integrated indicator of the material and technical support of the social sphere during "auspicious" years when the growth rate exceeds the average of this rate for the entire period $T_{gen.}$;

$\overline{T}adverse$ is an average annual growth rate of the integrated indicator of the material and technical support of the social sphere during "adverse" years, when the growth rate is lower than the average annual rate T_{gen}

The closer the indicator of sustainable development $i_{\overline{T}}$ is to one, the less are fluctuations and, consequently, the higher is sustainability. The authors understand sustainability as increasing of the indicator's rate growth.

3. Main results of the study

The calculations were done according to the proposed methodology. It should be noted that the distinguishing feature of the proposed method is the calculation of such indicator, which, in author's opinion, allows determining the accessibility and provision by specific tourism industry objects in one value. The results of the analysis of infrastructure provision calculations for Ukrainian tourism industry are shown in Table 1.

Table 1. The accessibility of tourist facilities and their regional availability in Ukraine [a]

Region	Main indicators [b]					
	Density of sanatoriums 2016	Density of places in sanatoriums 2016	Density of resort, guest houses and lodges 2016	Density of places in resort, guest houses and lodges 2016	Density of hotels and other accommodation facilities 2016	Density of places in hotels and other accommodation facilities 2016
Ukraine	0,086	23,985	0,424	52,219	0,603	29,390
Crimea	n.d.	n.d.	n.d.	n.d.	n.d.	n.d.
Vinnytsia	0,101	23,527	0,019	1,523	0,221	9,217
Volyn	0,042	9,427	0,484	17,282	0,318	11,060
Dnipropetrovsk	0,052	6,238	0,403	40,467	0,433	25,273
Donetsk	n.d.	n.d.	n.d.	n.d.	n.d.	n.d.
Zhytomyr	0,031	7,275	0,046	2,967	0,246	10,699
Zakarpattia	0,158	33,328	0,261	13,483	1,257	54,816
Zaporizhia	0,072	20,529	0,811	116,258	0,525	27,591
Ivano-Frankivsk	0,079	12,318	0,051	5,437	2,491	62,325
Kyiv	0,059	10,195	0,277	36,958	0,255	14,371
Kirovohrad	0,019	2,165	0,217	11,711	0,204	8,266
Luhansk	n.d.	n.d.	n.d.	n.d.	n.d.	n.d.
Lviv	0,200	67,226	0,183	7,278	0,969	53,490
Mykolaiv	0,059	23,513	1,057	147,786	0,411	15,243
Odesa	0,113	45,407	1,351	135,827	0,769	33,709
Poltava	0,063	17,972	0,034	4,106	0,417	17,686
Rivne	0,053	8,036	0,046	4,635	0,269	12,986
Sumy	0,024	3,744	0,097	10,785	0,290	8,569
Ternopil	0,082	13,882	0,033	2,948	0,377	17,395
Kharkiv	0,051	10,336	0,392	26,503	0,351	16,317
Kherson	0,046	8,975	0,928	174,313	0,421	18,643
Khmelnytsk	0,085	10,472	0,006	0,218	0,364	19,046
Cherkasy	0,055	10,662	0,282	28,411	0,428	15,368
Chernivtsi	0,047	5,454	0,047	4,087	0,782	38,408
Chernihiv	0,038	7,111	0,150	15,160	0,236	9,547
Kyiv (city)	0,316	60,296	0,527	49,843	3,288	388,90
Sevastopol (city)	n.d.	n.d.	n.d.	n.d.	n.d.	n.d.

[a] authors' calculations based on [8]

[b] objects density (m^2 per 1 person/ ths. km^2)

According to the calculations the accessibility of tourist facilities and their regional availability in Ukraine is different. The most accessible tourist facilities for the population are resort, guest houses and lodges. The least accessible ones are sanatoriums. After determining the density of hotels and other places for temporary residence the authors found out that this indicator is the highest in the capital of Ukraine, Ivano-Frankivsk and Zakarpattia regions. The lowest density of hotels and other places for temporary residence is in Kirovohrad and Vinnytsia regions.

Changes of the integral index of tourism facilities development in Ukraine are shown in Table 2.

Table 2. Changes of the integral index of tourism facilities development in Ukraine in 2010-2016[a]

Region	2010	2011	2012	2013	2014	2015	2016
Ukraine	0,247	0,306	0,246	0,247	0,244	0,251	0,211
Crimea	0,481	0,586	0,437	0,372	n. d.	n. d.	n. d.
Vinnytsia	0,118	0,120	0,127	0,123	0,125	0,138	0,159
Volyn	0,212	0,310	0,230	0,231	0,223	0,221	0,183
Dnipropetrovsk	0,153	0,261	0,146	0,145	0,147	0,164	0,218
Donetsk	0,273	0,538	0,295	0,310	n. d.	n. d.	n. d.
Zhytomyr	0,110	0,140	0,108	0,104	0,103	0,110	0,135
Zakarpattia	0,139	0,193	0,121	0,159	0,168	0,197	0,190
Zaporizhia	0,344	0,412	0,330	0,340	0,346	0,350	0,385
Ivano-Frankivsk	0,129	0,141	0,126	0,122	0,123	0,137	0,155
Kyiv	0,223	0,363	0,202	0,202	0,198	0,201	0,222
Kirovohrad	0,156	0,162	0,149	0,148	0,144	0,150	0,151
Luhansk	0,157	0,227	0,144	0,144	n. d.	n. d.	n. d.
Lviv	0,180	0,183	0,194	0,200	0,207	0,226	0,209
Mykolaiiv	0,396	0,531	0,366	0,384	0,385	0,391	0,439
Odesa	0,405	0,461	0,406	0,411	0,407	0,434	0,431
Poltava	0,121	0,137	0,124	0,120	0,119	0,130	0,152
Rivne	0,131	0,172	0,121	0,118	0,117	0,124	0,146
Sumy	0,134	0,218	0,128	0,124	0,119	0,123	0,151
Ternopil	0,114	0,134	0,119	0,113	0,113	0,128	0,152
Kharkiv	0,183	0,227	0,177	0,175	0,173	0,175	0,199
Kherson	0,331	0,298	0,344	0,289	0,262	0,237	0,300
Khmelnytsk	0,100	0,115	0,107	0,112	0,114	0,120	0,144
Cherkasy	0,195	0,255	0,196	0,193	0,187	0,186	0,200
Chernivtsi	0,097	0,129	0,097	0,102	0,102	0,118	0,139
Chernihiv	0,144	0,202	0,143	0,135	0,133	0,141	0,162
Kyiv (city)	0,245	0,298	0,254	0,307	0,298	0,305	0,308
Sevastopol (city)	0,369	0,356	0,422	0,465	n. d.	n. d.	n. d.

[a] authors' calculations

Taking into account the specifics, problems, and interests of each region of the state in the development of tools aimed at the successful tourism management, the reasonable grouping of the regions in terms of their development is needed. The calculated data show that in general there is an unstable growth trend of the tourism industry development indicator in Ukraine. The authors conducted the grouping of the regions according to the sustainability of the development indicators trends (Figure 1).

Figure 1. Grouping of the regions according to the sustainability of the development indicators trends

4. Conclusions and prospects for further research

Nowadays Ukrainian tourism industry is developing and is in a state of quantitative and qualitative changes. The methodological approaches of Ukrainian and foreign scientists to the analysis and evaluation of tourism infrastructure provision were summarized. According to the advantages and disadvantages of existing methodological approaches the own method of tourism development evaluation was developed. The distinguishing feature of the proposed method is the calculation of the density indicator that allowed determining the accessibility and availability of material objects in the single value. The proposed hypothesis of the study on the feasibility of the introduction of the indicator of objects density confirmed.

The methodical approach to determining the integral indicator of tourism infrastructural provision of country was proposed. The use of the taxonomic method allowed determining the sustainability of the development indicators trends of the studied industry and sustainability of series levels. It revealed the disparities in the state of such provision in the chosen industry. The obtained results indicate that there is a slowdown in the growth rate of the integral indicator of tourism infrastructure provision in Ukraine. Identified patterns and tends of its development and results of the infrastructure provision evaluation require further research in the field of justification of mechanisms of the tourism industry development that will facilitate the implementation of the proposed public policy directions in the tourist resorts on the vector of tourism infrastructure development.

References

[1] ***UNWTO barometer of world tourism, available at: http://mkt.unwto.org/en/content/tourism-highlights (15.02.2017)

[2] ***The Travel & Tourism Competitiveness Report, available at: www3.weforum.org/docs/WEF_TTCR_2017_web_0401.pdf (29.04.2017)

[3] Mandryk O M, Arkhypova L M, Pobigun O V, Maniuk O R. 2016 Renewable energy sources for sustainable tourism in the Carpathian region, *IOP Conf. Ser.: Mater. Sci. Eng.* **144** 012007

[4] Low H and Meghir C 2017 The Use of Structural Models in Econometrics, *Journal of Economic Perspectives* **31**(2) 33-58

[5] Zhang H, Gu Ch, Gu L, Zhang Y 2011 The evaluation of tourism destination competitiveness by TOPSIS & information entropy – A case in the Yangtze River Delta of China, *Tourism Management* **32**(2) 443-451

[6] Suman P 2013 Analysis of Tourism Attractiveness Using Probabilistic Travel Model: A Study on Gangtok and Its Surroundings, *European Journal of Geography* **4**(2) 46-54

[7] Hladkyi I N and Chystobaiev A I 1998 *Basics of Regional Policy,* Izdatelstvo Myhailova V A

[8] ***The official website of the State Statistics Committee of Ukraine: *Statistical Yearbook of Ukraine for 2016*, available at: www.ukrstat.gov.ua (21.02.2017)

The Reverse Engineering technique performed on a Francis Runner Geometry through Photogrammetry

S L Bogdan[1], D Nedelcu[1] and I Pădurean[2]

[1]"Eftimie Murgu" University of Reşiţa, Romania, The Faculty of Engineering and Management, Department of Mechanical Engineering and Management, Traian Vuia square, no.1-4, 320085, Reşiţa.
[2]Politehnica University of Timişoara, Romania, Mechanical Faculty, Mechanical Machines, Equipments and Transportations Department, "Mihai Viteazu" street, 300222, Timişoara.

E-mail: d.nedelcu@uem.ro

Abstract. The Reverse Engineering is a reliable technique to convert a real geometry of a part into 3D computer file. One reason for scanning a part may be the lack of the drawings to manufacture a new one that will replace the old part. The paper illustrates the Reverse Engineering of a Francis runner geometry with a diameter of 1160 mm and 19 blades, which, after 50 years of operation, has pronounced wear marks, high porosity surfaces and cracked areas. The geometry will be scanned through Photogrammetry technique and processed with the following software packages: Agisoft Photoscan and Geomagic Design X (formerly Rapidform XOR); finally, the solid geometry and the drawing of the runner were generating using the SolidWorks software.

1. Introduction

Reverse engineering relies on the reconstruction of a part geometry using 3D scanning process, the usual areas of application being: architecture [1], [2], cultural heritage [3], medicine [4-6], soil erosion assessment [7], [8], paleontology [9]. The Photogrammetry technique is applied in the mechanical field: for deformation measurement [10-12], for quality control and inspection [13-15], for components reconstruction geometry [16-18]. There are some of reasons to use the reverse engineering process [19]: "*to compare products, in preparation for imitating a product, to get technical data that do not exist or the original supplier is no longer willing or able to provide, to shorten market entry times, to enhance existing data, to perform product verification, to aid in product design*". Common scanning technologies are Laser scanning and Photogrammetry technique, many of published papers including as a subject the comparison between them [20], [21]. Photogrammetry is the technique used to extract geometric information from two-dimensional images/videos, by taking multiple overlapping photos and obtaining measurements from them in order to create 3D models of objects or scenes.

One reason for scanning a part may be the lack of the drawings to manufacture a new one that will replace the old part. This is the case of the Francis runner, the subject of the present paper, which, after 50 years of operation, has pronounced wear marks, high porosity surfaces and cracked areas as shown in Figure1.

Figure 1. The existing runner with wear marks, high porosity surfaces and cracked areas

Using the reverse engineering technology, the geometry of a Francis runner with an 1160 mm diameter and 19 blades will be reconstructed by following the next steps:

- 3D scanning of the runner using Photogrammetry and the Agisoft Photoscan software;
- the runner reconstruction using the Geomagic Design X and SolidWorks software packages.

For 3D scanning of the runner geometry, the conditions to be ensured by the beneficiary are as follows:

- placing the rotor on a horizontal support;
- providing a working space with light around the rotor over a radius of at least 4 m;
- the possibility of manipulating the rotor with the crane.

2. The 3D scanning of the runner

The 3D scanning of the runner requires the following stages [22]:

- *Acquire photos* - with a high quality camera; the input data was captured from the platform where the runner was placed; a number of 113 photos were taken with a CANON EOS 600D camera from a ≈ 4 m mean distance; the mean ISO-value for each photo is 800 by shutter 1/50 ÷ 1/60 and focal length 28; some images of the runner are presented in Figure 2;
- *Import Photos* - loads all the raw images into the Agisoft Photoscan interface;
- *Photo's inspection* – the quality of the images must be a minimum of 0.5, but 0.7 is recommended; the *Quality* factor is calculated by the Agisoft Photoscan software;
- *Align Photos and build sparse cloud* - this processing step compares the pixels in the photos to find matches and build a 3D geometry from them; a sparse point cloud of 5,326 points is generated (Figure 3); the software tries to identify common points between pairs of images and estimate camera locations;
- *Build Dense Cloud* - once satisfied with the alignment, the sparse point cloud is processed into a dense cloud, with 12,891,212 points, where each matchable pixel will get its own X, Y, Z location in 3D space (Figure 4);
- *Build Mesh* - this step connects each set of three adjacent points into a triangular face, which combine seamlessly to produce a continuous mesh over the surface of the model; this step is optional, because the mesh was generated using Geomagic Design X software;
- *Build Texture* - in this step, the original images are combined into a texture map and wrapped around the mesh, resulting in a photorealistic model of the original object; this step is optional;
- *Scale the geometry* – the geometry was scaled based on 3 distances between three markers and verified for other multiple points (Figure 5);
- *Export the geometry* – export the dense point cloud to Geomagic Design X software.

IOP Conf. Series: Materials Science and Engineering **477** (2019) 012021 doi:10.1088/1757-899X/477/1/012021

Figure 2. Some captured images of the Francis runner

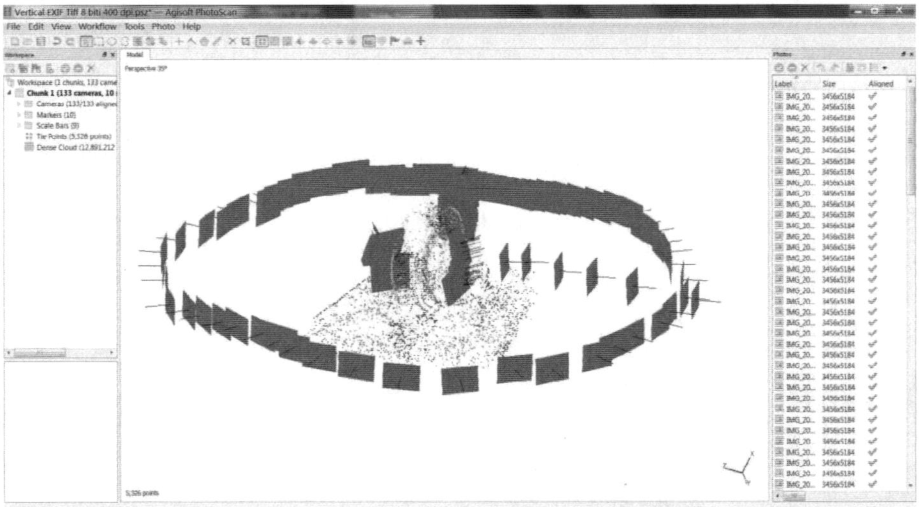

Figure 3. Images aligned and the sparse cloud with 5,326 points

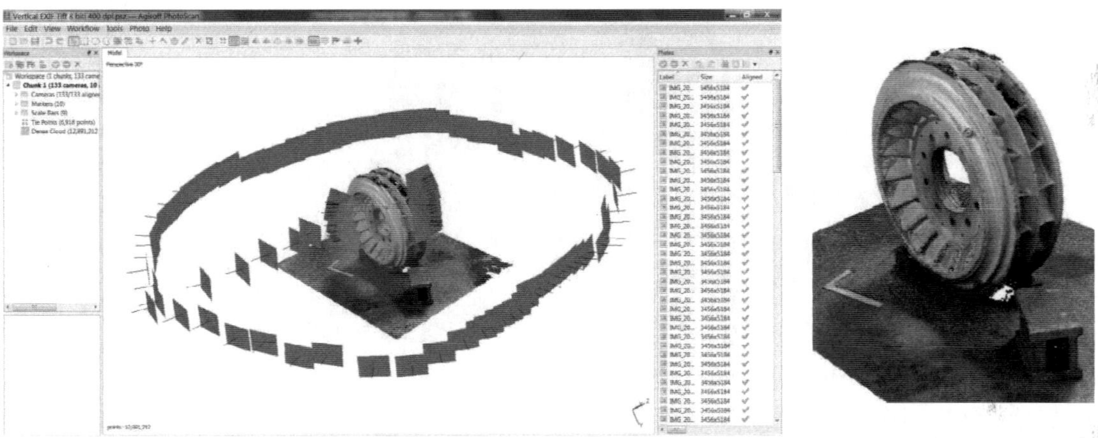

Figure 4. The dense cloud with 12,891,212 points

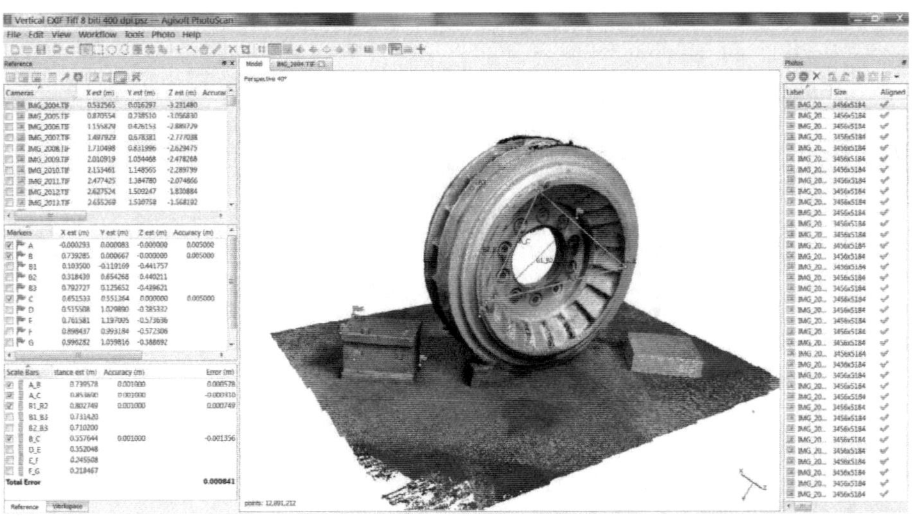

Figure 5. Scale the geometry using Agisoft Photoscan software

IOP Conf. Series: Materials Science and Engineering **477** (2019) 012021 doi:10.1088/1757-899X/477/1/012021

3. The Francis runner reconstruction

The aim of this stage was to generate the runner geometry as a solid format, by following the next steps:

- import the dense point cloud into Geomagic Design X software (Figure 6);
- generate the mesh based on a dense point cloud; the Geomagic Design X software creates a network with 6,491,472 triangles similar to the mesh from the finite element software;
- generate the median disc, the two outer disks and 1 blade using Geomagic Design X software (Figure 7);
- export the median disc, the two outer disks and 1 blade to the SolidWorks software;
- generate the geometry of the runner as a solid format in the SolidWorks software (Figure 8);
- generate the runner drawing using the SolidWorks software (Figure 9).

Figure 6. The dense point cloud imported into Geomagic Design X software

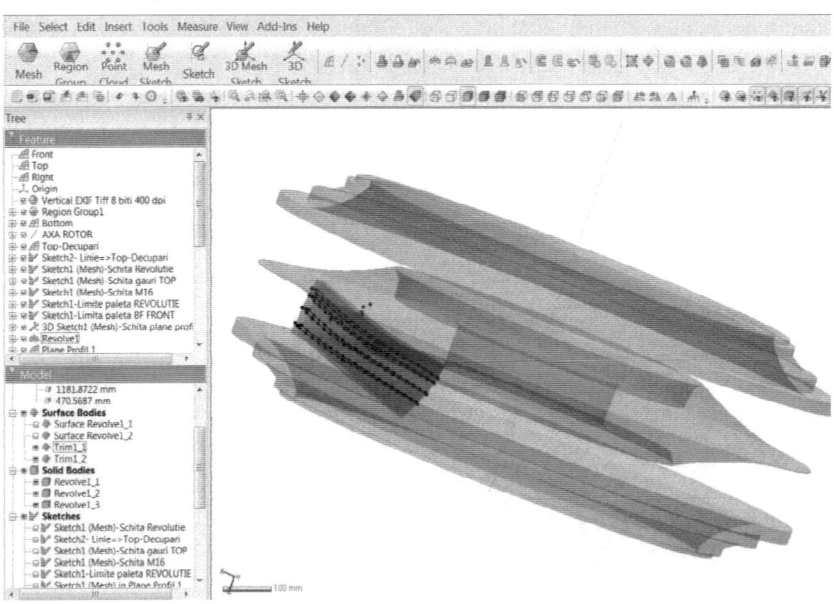

Figure 7. The median disc, the two outer disks and 1 blade in Geomagic Design X software

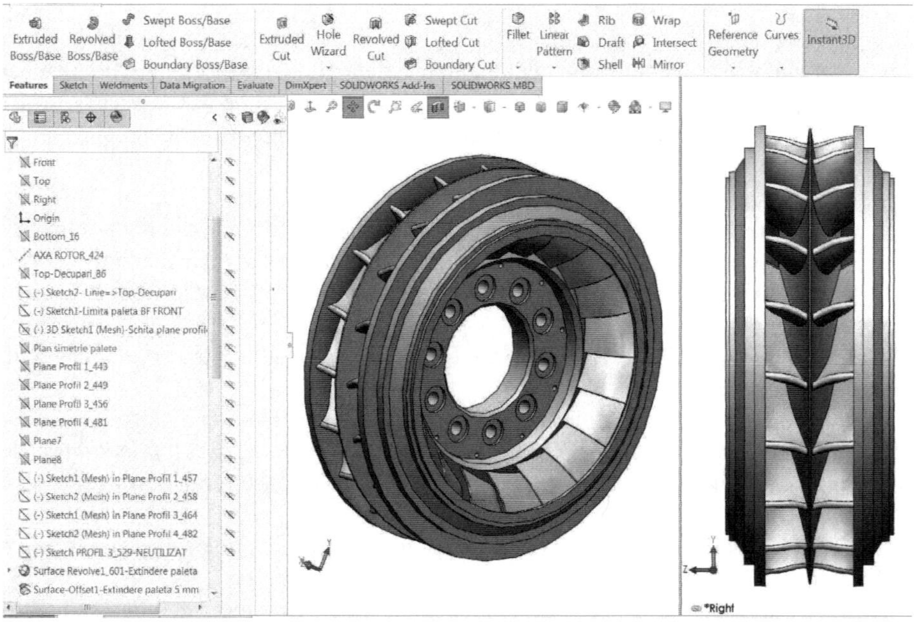

Figure 8. The solid format of the runner in SolidWorks software

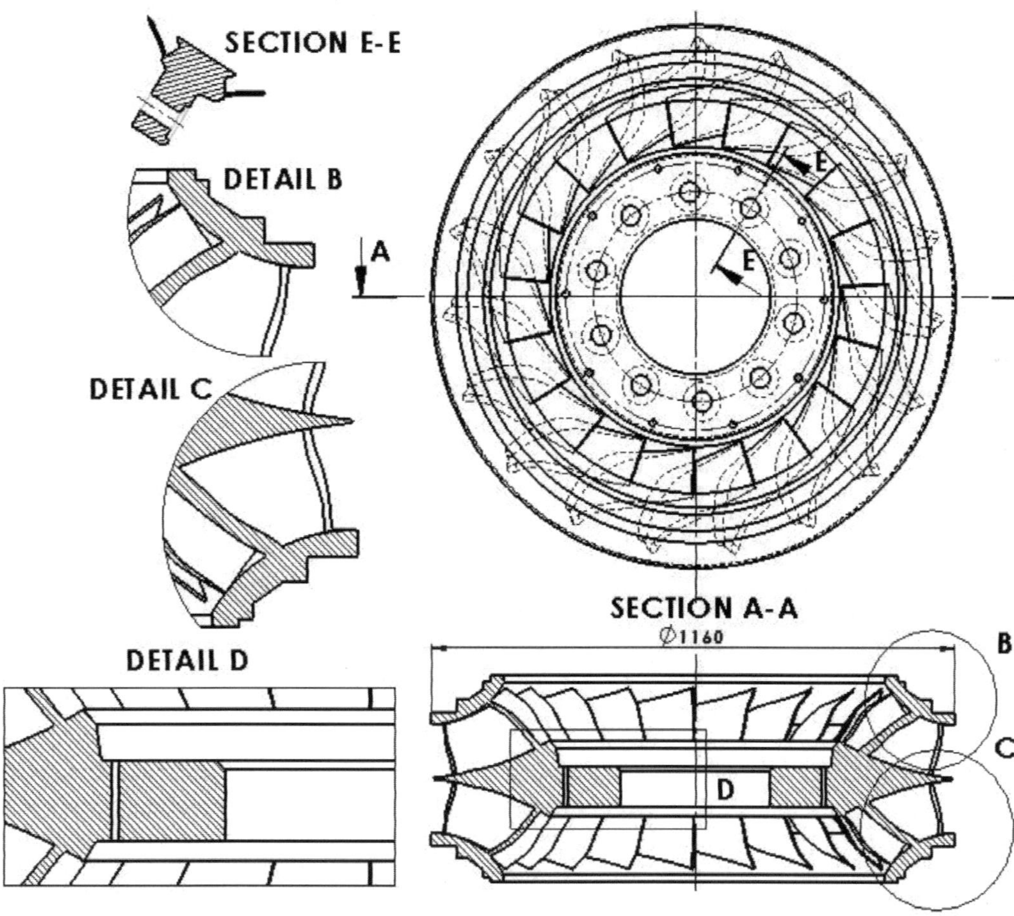

Figure 9. The runner drawing in SolidWorks software

4. Conclusions

In the case of a complex geometry like the Francis runner and in the absence of the drawings to replicate it, the reverse engineering process represents a more efficient alternative compared to contact measurement methods. The use of Photogrammetry technique enables to get the 3D geometry of the runner at a more reasonable cost than the Laser scanning technique, with comparable accuracy.

References

[1] Aparicio-Sánchez Luis Javier, Villarino Alberto, Gago-García Jesús and Aguilera-González Diego 2016 Photogrammetric, Geometrical, and Numerical Strategies to Evaluate Initial and Current Conditions in Historical Constructions: A Test Case in the Church of San Lorenzo (Zamora, Spain), *Remote Sens.* **8** 60, doi:10.3390/rs8010060

[2] Riveiro B, Caamaño J C, Arias P and Sanz E 2011 Case study: byzantine land walls-Istanbul, *Automation in Construction* **20** 380–388

[3] Kersten T P 2006 *Combination and Comparison of Digital Photogrammetry and Terrestrial Laser Scanning for the Generation of Virtual Models in Cultural Heritage Applications*, The 7th International Symposium on Virtual Reality, Archaeology and Cultural Heritage, VAST (2006), pp 207–214

[4] Chong Albert K 2009 New developments in medical photogrammetry, *Geoinformation Science Journal* **9**(1) 41-50

[5] Mitchell H L and Chadwick R G 2008 Challenges of photogrammetric intra-oral tooth measurement, *The International Archives of the Photogrammetry, Remote Sensing and Spatial Information Sciences* **XXXVII** Part B5, Beijing

[6] Rodrigo Munhoz, Cícero André da Costa Moraes, Harki Tanaka and Maria Elizete Kunke 2016 A digital approach for design and fabrication by rapid prototyping of orthosis for developmental dysplasia of the hip, *Research on Biomedical Engineering* **32**(1) http://dx.doi.org/10.1590/2446-4740.00316

[7] Dirk H Zapp-Rieke 2005 Digital close range photogrammetry for measurement of soil erosion, *The Photogrammetric Record* **20**(109) 69–87

[8] Nouwakpo Sayjro, Huang Chi-hua, Frankenberger Jim and Bethel James 2010 *A simplified close range photogrammetry method for soil erosion assessment*, 2nd Joint Federal Interagency Conference, Las Vegas, NV, June 27 - July 1

[9] Mallison Heinrich and Wings Oliver 2014 Photogrammetry in paleontology – a practical guide, *Journal of Paleontological Techniques* **12** 1-31

[10] Barazzetti L and Scaioni M *Photogrammetric tools for deformation measurements*, https://www.researchgate.net/publication/228828761

[11] Maas Hans-Gerd 1998 *Photogrammetric techniques for deformation measurements on reservoir walls*, IAG SC4 Symposium Geodesy for Geotechnical and Structural Engineering Eisenstadt, Austria, 20 - 22 April

[12] Chinn Tori 2013 *Strain Measurement Validation Using 3D Photogrammetry Techniques*, (2013), Master's Theses, 4263 http://scholarworks.sjsu.edu/etd_theses/4263

[13] Marcis M, Frastia M and Augustin T 2017 Measurement of flat slab deformations by the multi-image photogrammetry method, *Slovak Journal of Civil Engineering* **25**(4) 19-25

[14] Yao A W L 2005 Applications of 3D scanning and reverse engineering techniques for quality control of quick response products, *Int J Adv Manuf Technol* **26** 1284-1288

[15] Urbanic R J 2015 A design and inspection based methodology for form-function reverse engineering of mechanical components, *Int J Adv Manuf Technol* **81** 1539-1562

[16] She Chen-Hua and Chang Chun-Chi 2007 Study of applying reverse engineering to turbine blade manufacture, *Journal of Mechanical Science and Technology* **21** 1580-1584

[17] Mohaghegh K, Sadeghi M H and Abdullah A 2007 Reverse engineering of turbine blades based on design intent, *Int J Adv Manuf Technol* **32** 1009-1020

[18] Koppel P *Agisoft PhotoScan: Point Cloud accuracy in close range configuration*, www.koppel-

engineering.de/
[19] Curtis Shane K, Harston Stephen P and Mattson Christopher A 2011 The fundamentals of barriers to reverse engineering and their implementation into mechanical components, *Springer Res Eng Design* **22** 245-261
[20] Radosevic G 2010 *Laser Scanning Versus Photogrammetry Combined with Manual Post-modeling in Stecak Digitization*, Proceedings of CESCG 2010: The14th Central European Seminar on Computer Graphics
[21] Bayrama B, Nemli G, Özkanc T, Oflaz O E, Kankotan B and Çetin İ 2015 *Comparison of laser scanning and photogrammetry and their use for digital recording of cultural monument case study: byzantine land walls-Istanbul*, ISPRS Annals of the Photogrammetry, Remote Sensing and Spatial Information Sciences, Volume II-5/W3, 2015 25th International CIPA Symposium 2015, 31 August – 04 September, Taipei, Taiwan
[22] ***Agisoft LLC 2017 *PhotoScan*, *Fully automated professional photogrammetric kit*, http//:www.agisoft.com

Innovative natural rubber bearing for earthquake isolation

V Iancu and D Nedelcu

"Eftimie Murgu" University of Reşiţa, The Faculty of Engineering and Management, Department of Mechanical Engineering and Management, Traian Vuia Square, no.1-4, 320085, Reşiţa.

E-mail: d.nedelcu@uem.ro

Abstract. In the paper are presented two seismic isolation systems. One is classic, made out of two rubber plates, with a metallic plate vulcanized between the two. For fastening, the isolator has another two clamping plates vulcanized on it. The proposed model has same components as the classical one, having on top and under, attached to the rubber, two thin plates with bossages. For fastening the insulator there are two plates for clamping, which are not vulcanized to the isolator. The contact between the insulator and the clamping plates is made just by the bossages, the rest of the surface being free. As a result of the simulations done in the SolidWorks softwarewe concluded that, in the case of the proposed model, the rubber has no longer the tendency to detach from the clamping plates. As long as the bossage plates are moving together with the rubber, the whole isolator becomes more elastic and durable in time.

1. Introduction

Given the significant importance of the built environment, solutions have been sought permanently to keep them in good working order. For reinforcement and post-consolidation, various stiffening methods were used using reinforced concrete. Due to the consolidation work, buildings have become stiffer [1]. Accelerations and displacements during earthquakes have not diminished, and damage has increased (Figure 1).

Figure 1. Attempts to strengthen the Canyon Channel Bridge (source: Institute of Earthquake Research 1996)

IOP Conf. Series: Materials Science and Engineering **477** (2019) 012022 doi:10.1088/1757-899X/477/1/012022

The solution to these problems is the seismic isolation [2]. A good seismic isolation is achieved by inserting flexible bearings into the structure, between the substructure and the superstructure (Figure 2) [3], [4]. It shifts the vibration period of the system, permit relatively large displacements and provide good energy dissipation [5]. Natural rubber materials are characterized by their capacity to take large amounts of strain with relatively small stress [6]. The following properties of the natural rubber material were imposed from the material library of SolidWorks software: elastic modulus 0.01 N/mm^2, Poisson's ratio 0.45, mass density 960 kg/m^3, tensile strength 20 N/mm^2 [7].

Figure 2. Example of building isolated with natural rubber bearings (source: nova next tech rubber bearings seismic protection 2013 [4])

2. The geometry of the classic and innovative natural rubber bearing (NRB)

The geometry and dimensions of the classic and innovative NRB are presented in (Figure3) and (Figure 4). The innovative NRB has two steel plate with label 2, cylindrical bossage and 100 mm diameter. The central steel plate with label 4 has no bossage.

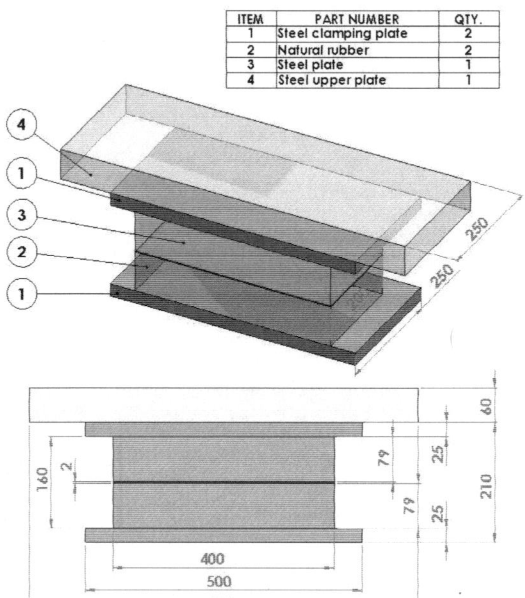

Figure 3. The geometry of the classic NRB

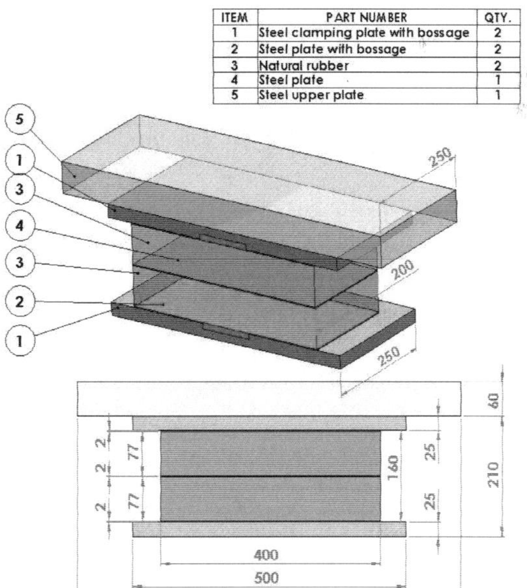

Figure 4. The geometry of the innovative NRB

3. The FEM simulation of the classic NRB

The FEM simulation of the classic NRB was made in the SolidWorks Simulation software. A nonlinear study was performed on the half of the classic NRB in order to reduce the simulation time.

The following boundary conditions were applied to the classic NRB (Figure 5):

- the **Fixed Geometry** restrain are applied on the bottom face of the base steel clamping plate;
- the **Symmetry** restraints are applied on all faces of the parts included in the symmetry plane;
- the **On flat faces 0 mm** restraint are applied in +X direction on the steel upper plate right face, to block the X movement of this part;
- the vertical **force F=30 N** are applied on -Y direction on the top face of the steel upper plate, which increases linearly from zero to the 30 N nominal value in the first second, after which it remains constant until the 2 second end of the analysis (Figure 6);
- the horizontal **force F=60 N** are applied on +X direction on the right face of the superior steel clamping plate, with null value in the first second, after which increases linearly from zero to the 60 N nominal value until the 2 second end of the analysis (Figure 7);
- the **Bonded** contact are imposed between all parts of the classic NRB, except between the top face of the superior steel clamping plate and the bottom face of the steel upper plate, where **No penetration contact** with friction are imposed, using the following values for the friction coefficient $\mathbf{C_f}$: 0, 0.25, 0.5, 0.75 and 1.

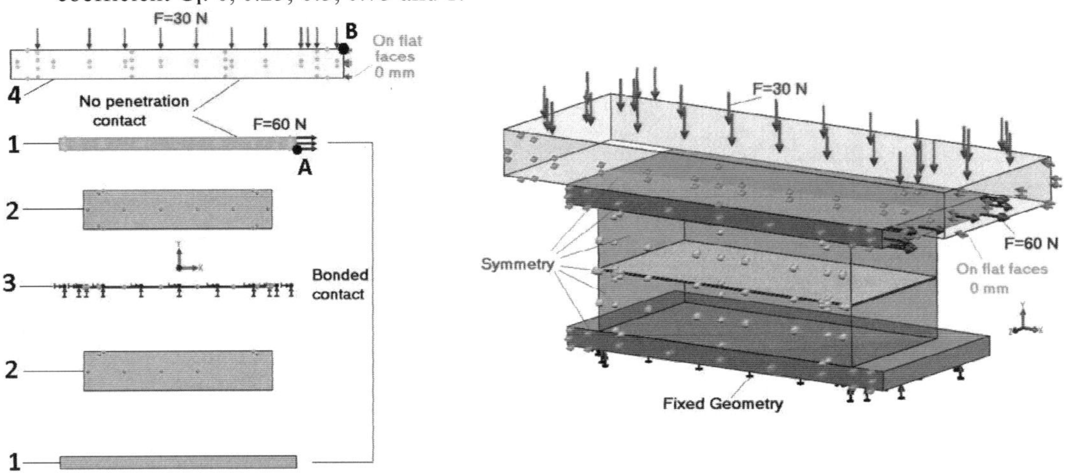

Figure 5. The boundary conditions applied to the classic NRB

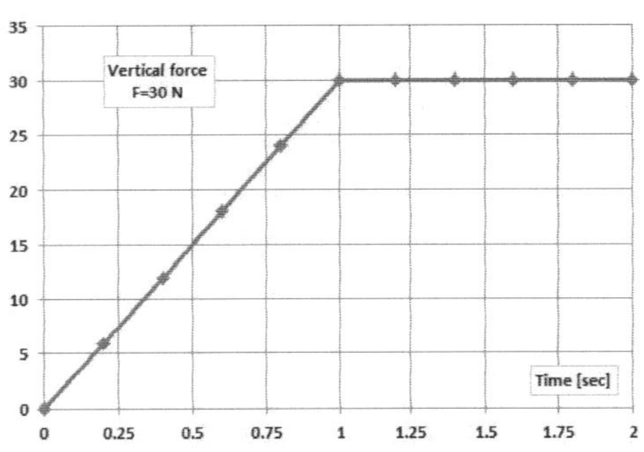

Figure 6. The time distribution of vertical force F=30 N

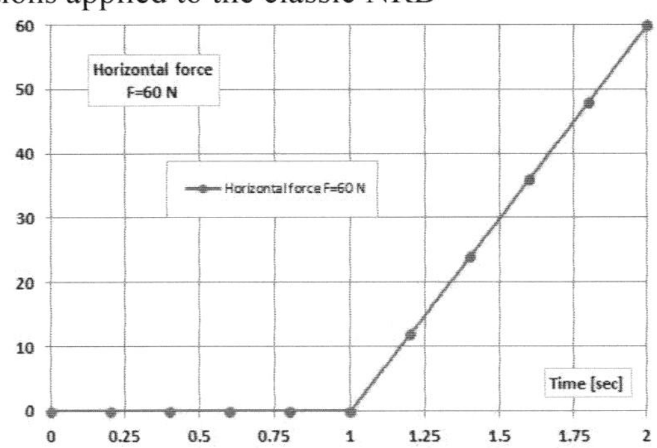

Figure 7. The time distribution of horizontal force F=60 N

(Figure 8) show the mesh of the classic NRB made from 28328 elements and 6132 nodes. The simulation results of the classic NRB (**von Mises** stress, the maximum resultant displacement $\boldsymbol{\delta_{rez\ max}}$

and the maximum displacements on X, Y and Z directions - $\delta_{X\ max}$, $\delta_{Y\ max}$, $\delta_{Z\ max}$) are presented in Table 1 at the end of the 2 seconds simulation time. The classic NRB distribution friction force from the top face of the superior steel clamping plate as a function of time for imposed values of the friction coefficient $\mathbf{C_f}$ are shown in (Figure 9). Until the 1 second time the friction force is not present. Between 1 second and the time corresponding to the intersection point the friction force increase linearly from zero to the maximum value, calculated by multiplying the friction coefficient $\mathbf{C_f}$ with the value of the vertical force **F=30 N**. Finally, until the end of the 2 seconds simulation time the friction force remains constant at maximum value.

Table 1. The simulation results of the classic NRB at the end of the 2 seconds simulation time

Parameter	UM	Friction coefficient C_f				
		0	**0.25**	**0.5**	**0.75**	**1**
von Mises	MPa	0.200955	0.176176	0.069451	0.126389	0.101
$\delta_{rez\ max}$	mm	36.73	31.933	26.021	22.617	18.074
$\delta_{X\ max}$	mm	36.711	31.932	26.005	22.596	18.015
$\delta_{Y\ max}$	mm	-3.136/4.415	-2.908/3.337	-2.832/2.438	-2.951/1.945	-2.863/1.375
$\delta_{Z\ max}$	mm	-2.152/1.038	-1.994/0.789	-1.612/0.698	-1.580/0.271	-1.361/0.057

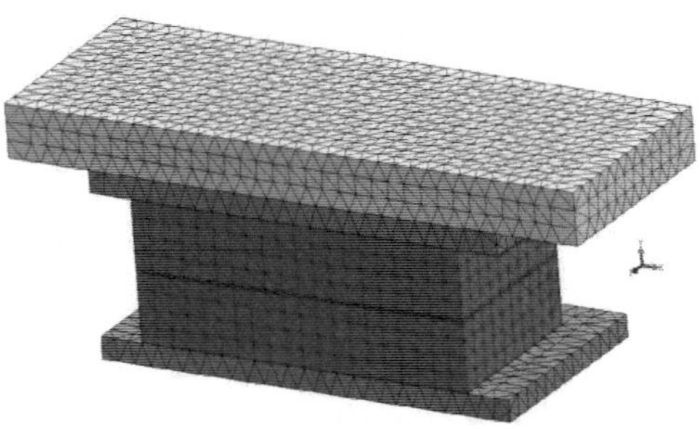

Figure 8. The mesh of the classic NRB

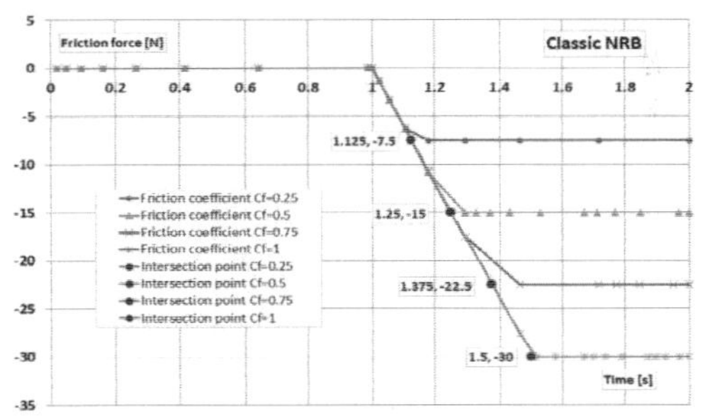

Figure 9. The classic NRB friction force

The time displacements distribution on X and Y directions - δ_X, δ_Y are presented in (Figure 10) and (Figure 11), for points **A** and **B** (Figure 5). In the absence of horizontal force, until the time corresponding to the intersection point marked in (Figure 9) there is no displacement on X direction, then the δ_X displacement increase until maximum value simultaneously with the horizontal force. The displacement on Y direction δ_Y decreases linearly simultaneously with the vertical force until 1 second, remain constant until the time corresponding to the same intersection point, then increase to a maximum value.

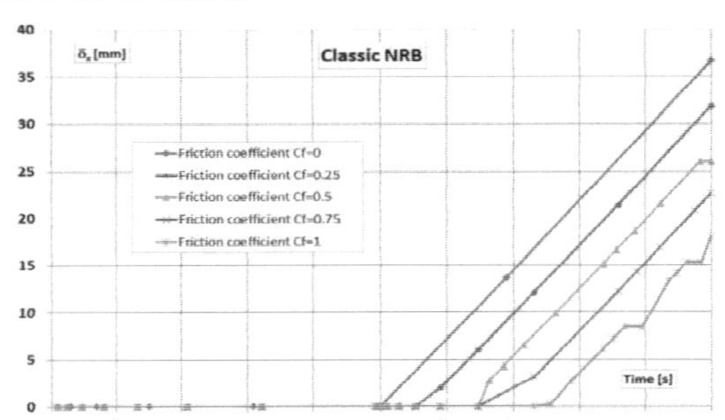

Figure 10. The time distribution of the classic NRB displacements on X direction

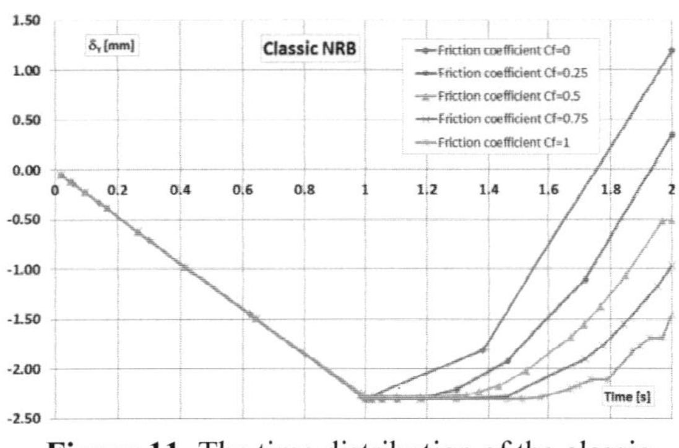

Figure 11. The time distribution of the classic NRB displacements on Y direction

4. The FEM simulation of the innovative NRB

The FEM simulation of the innovative NRB was made in the SolidWorks Simulation software. A nonlinear study was performed on the half of the innovative NRB in order to reduce the simulation time. The following boundary conditions were applied to the innovative NRB (Figure 12):

- the **Fixed Geometry** restrain are applied on the bottom face of the base steel clamping plate;
- the **Symmetry** restraints are applied on all faces of the parts included in the symmetry plane;
- the **On flat faces 0 mm** restraint are applied in +X direction on the steel upper plate right face, to block the X movement of this part;
- the same vertical **force F=30 N** are applied on -Y direction on the top face of the steel upper plate;
- the same horizontal **force F=60 N** are applied on +X direction on the right face of the superior steel clamping plate;
- the **Bonded** contact are imposed between 2, 3, 4 parts of the innovative NRB;
- three **No penetration contact** with friction contacts are imposed, using the following values for the friction coefficient $\mathbf{C_f}$: 0, 0.25, 0.5, 0.75 and 1:
  - between the top face of the base steel clamping plate with bossage with label 1 and the bottom face of the inferior steel plate with bossage with label 2, except the cylindrical face of the bossage where **Bonded** condition is applied;
  - between the bottom face of the superior steel clamping plate with bossage with label 1 and the top face of the superior steel plate with bossage with label 2, except the cylindrical face of the bossage where **Bonded** condition is applied;
  - between the top face of the superior steel clamping plate with label 1 and the bottom face of the steel upper plate with label 5.

(Figure 13) show the mesh of the innovative NRB made from 31258 elements and 7672 nodes. The simulation results of the innovative NRB (**von Mises** stress, the maximum resultant displacement $\delta_{rez\ max}$ and the maximum displacements on X, Y and Z directions - $\boldsymbol{\delta}_{X\ max}$, $\boldsymbol{\delta}_{Y\ max}$, $\boldsymbol{\delta}_{Z\ max}$) are presented in Table 2 at the end of the 2 seconds simulation time. The innovative NRB distribution friction force on the top face of the superior steel clamping plate as a function of time for imposed values of the friction coefficient $\mathbf{C_f}$ are identical with the curves of the classic NRB, presented in (Figure 9). Small friction forces with values smaller than 1.3 N appear on the other contact surfaces where **No penetration contacts** were applied. (Figure 14) exemplify the friction force for the $\mathbf{C_f}$=0.5 value of the friction coefficient on the exploded view of the innovative NRB.

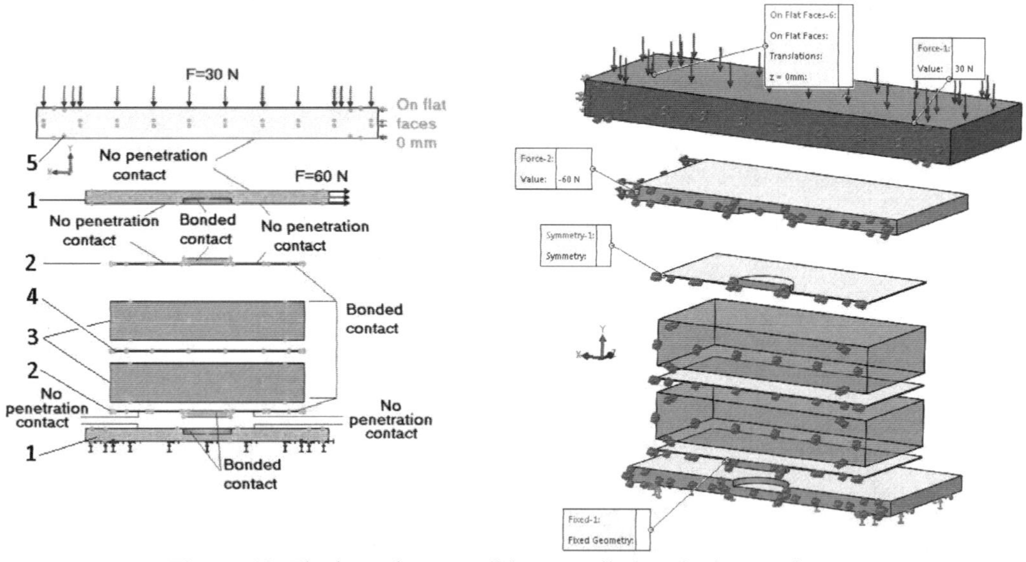

Figure 12. The boundary conditions applied to the innovative NRB

Table 2. The simulation results of the innovative NRB at the end of the 2 seconds simulation time

Parameter	UM	Friction coefficient C_f				
		0	**0.25**	**0.5**	**0.75**	**1**
von Mises	MPa	0.227	0.185	0.146	0.107	0.075
$\delta_{rez\ max}$	mm	35.754	31.082	26.505	22.009	17.586
$\delta_{X\ max}$	mm	35.734	31.08	26.503	21.99	17.531
$\delta_{Y\ max}$	mm	-3.004/4.555	-2.788/3.460	-2.788/2.531	-2.790/1.937	-2.716/1.377
$\delta_{Z\ max}$	mm	-1.673/1.801	-1.352/1.69	-0.040/1.583	-0.738/1.479	-0.446/1.38

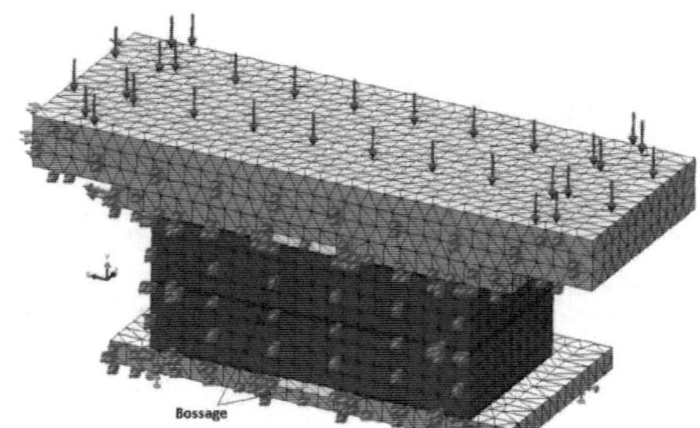

Figure 13. The mesh of the innovative NRB

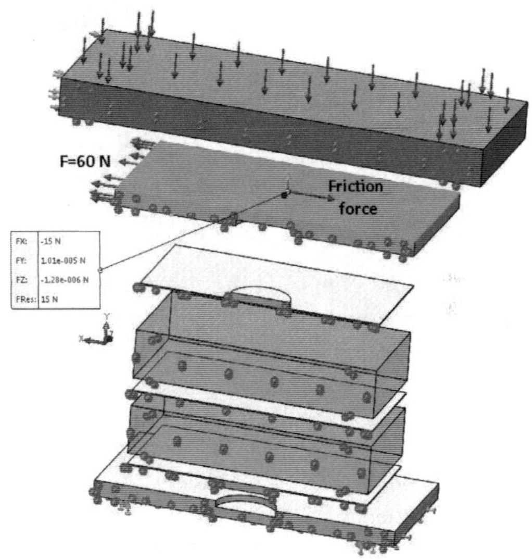

Figure 14. The friction force for C_f=0.5

The time displacements distribution on X and Y directions - δ_X, δ_Y for the innovative NRB are presented in (Figure 15) and (Figure 16). The curves present the same allure as those in (Figure 10) and (Figure 11) respectively with small differences of numerical values between X/Y displacements of the classic and innovative NRB.

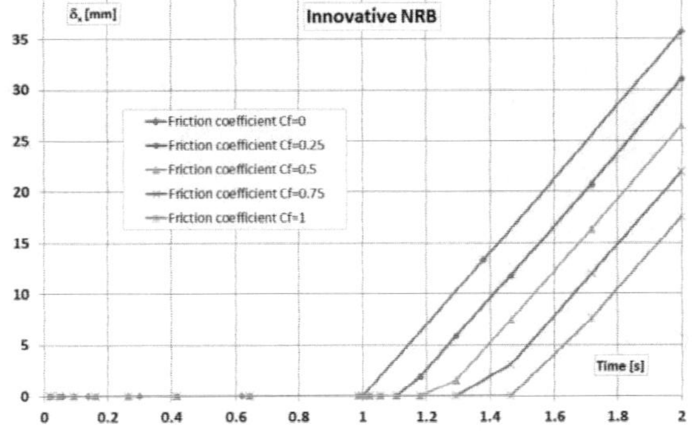

Figure 15. The time distribution of the innovative NRB displacements on X direction

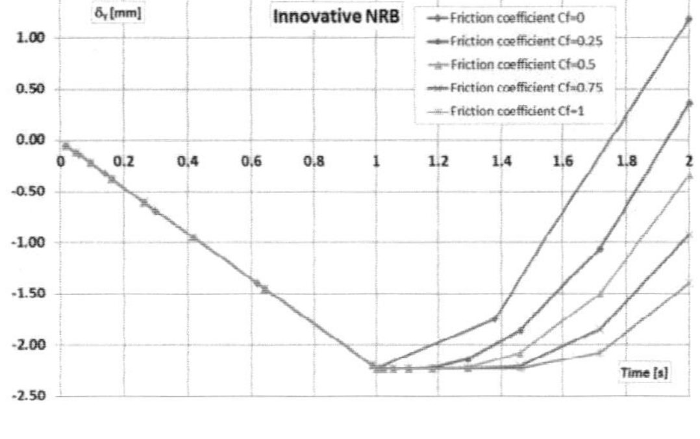

Figure 16. The time distribution of the innovative NRB displacements on Y direction

5. The innovative NRB behaviour

Although the stress, displacements and friction forces are relatively close for the two geometry variants, a remarkable and important difference compared to classic NRB, is the appearance of the two detachments of the bossage plate from the clamping plate shown in (Figure 17).

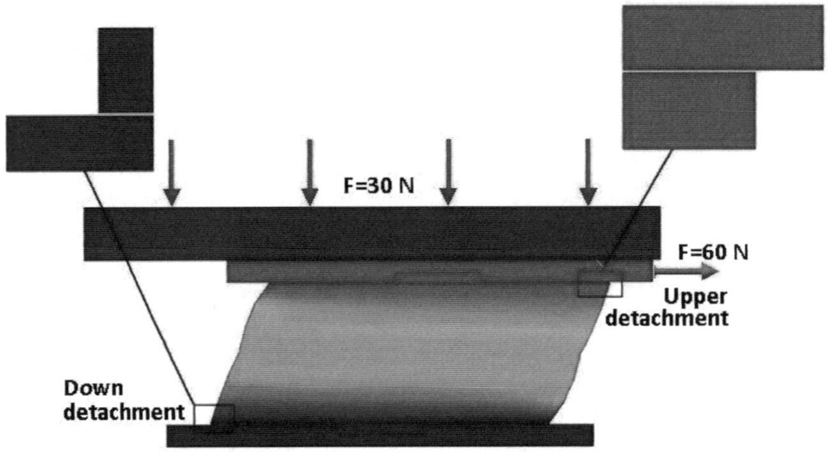

Figure 17. The upper and down detachments of the bossage plate from the clamping plate

(Figure 18) exemplify the detachments of the steel plate with bossage with label 2 from the base steel clamping plate with bossage and label 1 at the 2 second end of the analysis, for the value of the friction coefficient $\mathbf{C_f}$=1. The values of the upper and down detachments are presented numerically in Table 3 and graphically in (Figure 19) and (Figure 20). For both detachments the values increases as the friction coefficient decreases, while the down values are bigger then upper values.

6. Conclusions

The possibilities of adjusting the dynamic response of constructions that contain isolation systems are limited, due to the small number of control parameters. The innovative isolation system allows for the proper combination of component parts and optimal elastic characteristics for a broad band of exciting earthquake frequencies.

At the new innovative NRB, the phenomenon of detachment of the bossage plate from the clamping plate, offer to the seismic system an increased durability.

Table 3. The values of the upper and down detachments

Friction coefficient $\mathbf{C_f}$	Upper [mm]	Down [mm]
1	0.000002	0.000215
0.75	0.000003	0.000527
0.5	0.000005	0.000867
0.25	0.000007	0.001215
0	0.000008	0.001567

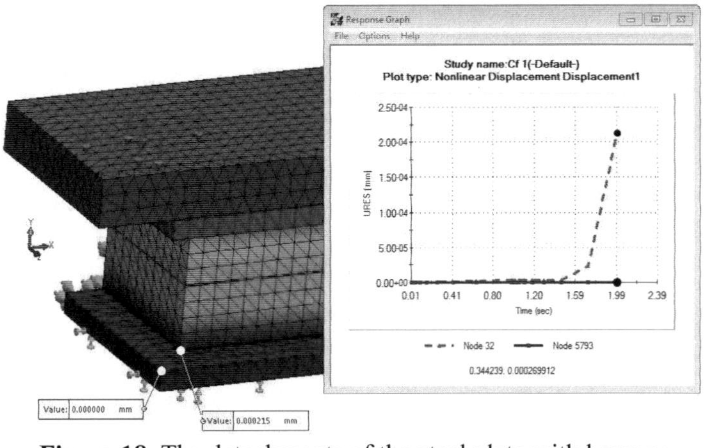

Figure 18. The detachments of the steel plate with bossage from the base steel clamping plate for friction coefficient $\mathbf{C_f}$=1

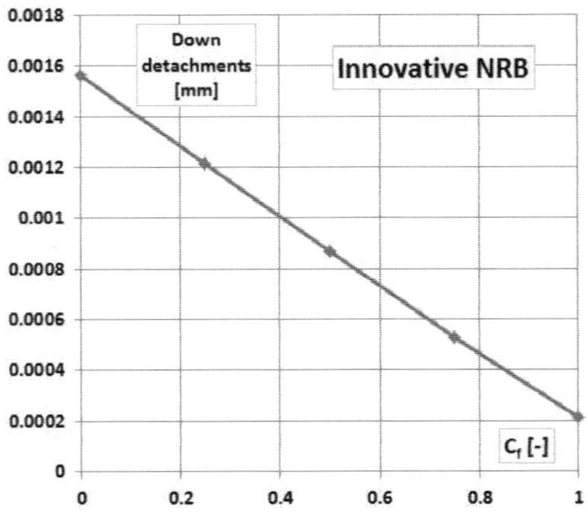

Figure 19. The down detachments

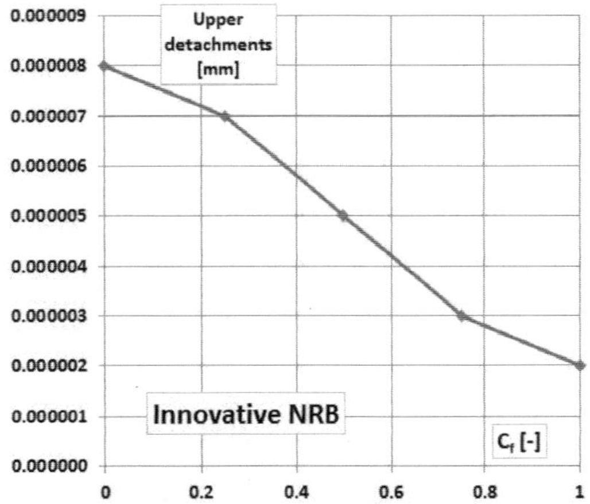

Figure 20. The upper detachments

Acknowledgments
This paper was supported by "EftimieMurgu" University of Reşiţa, "TraianVuia" Square, No. 1-4, 320085, Caraş-Severin county, Reşiţa, Romania.

References

[1] Papazoglou A J and Elnashai A S 1996 Analytical and field evidence of the damaging effect of vertical earthquake ground motion, *Earthquake Engineering and Structural Dynamics* **25**(10) 1109-1137

[2] Gillich G R, Amariei D, Iancu V and Jurcău C Ş 2009 *Aspects behavior of bridges which use different vibration isolating systems*, 10th WSEAS International Conference on Automation & Information (ICAI'09) Prague, Czech Republic, March 23-25, pp 140-145

[3] Gillich G R, Bratu P P, Răduca M, Amariei D and Iancu V 2009 *Behavior of composite bearings used in bridge isolation*, KORSD 2009 Vilnius, Lithuania, 30 September-03 October, pp 489-494

[4] Kelly J M 2013 *A Tested, Inexpensive Way to Protect Buildings from Earthquakes*, Nova Next

[5] Gillich G R, Samoilescu Gh, Berinde F and Chioncel C P 2007 Experimental determination of the rubber dynamic rigidity and elasticity module by time-frequency measurements, *Materiale Plastice* **44**(1) 18-2

[6] Gillich G R, Bratu P P, Frunzăverde D, Amariei D and Iancu V 2010 *Identifying mechanical characteristics of materials with non-linear behavior using statistical methods,* Proceedings of the 4th WSEAS international conference on Computer engineering and applications, Harvard, Cambridge, January 27-29, pp 99-103

[7] Iavornic C, Praisach Z I, Vasile O, Gillich G R and Iancu V 2011 Study *of Stress and Deformation in Elastomeric Isolation System Using the Finite Element Method,* The 11th WSEAS International Conference (ISTASC'11), Florence, Italy, August 23-25, pp 239-244

IOP Conf. Series: Materials Science and Engineering **477** (2019) 012023 doi:10.1088/1757-899X/477/1/012023

The calculation of the Kaplan Turbine Hill Chart using the HydroHillChart software

D Nedelcu[1], I Pădurean[2] and A Bostan[3]

[1]"Eftimie Murgu" University of Reşiţa, Romania, The Faculty of Engineering and Management, Department of Mechanical Engineering and Management, Traian Vuia square, no.1-4, 320085, Reşiţa.
[2]Politehnica University of Timişoara, Romania, Mechanical Faculty, Mechanical Machines, Equipments and Transportations Department, "Mihai Viteazu" street, 300222, Timişoara.
[3]Resita Reductoare & Regenerabile SA, Romania.

E-mail: d.nedelcu@uem.ro

Abstract. The HydroHillChart software is a complex one used to calculate the hill chart for hydraulic turbines. The application was developed in the Python programming language, while the mathematical tool used for interpolation is the cubic spline type function. The paper presents the HydroHillChart - Kaplan module application, used to calculate the hill chart of the Kaplan hydraulic turbine models, by processing the data measured on the test rig. The hill chart expresses graphically the functional dependence of the Kaplan turbine parameters (n_{11}, Q_{11}, a_o, φ, η) and is the instrument on which the industrial turbine is designed. This dependence arises from extensive experimental research carried out on Kaplan turbine models, leading to a library of optimized characteristics for the specific flows and heads of these turbines types.

1. Introduction

The HydroHillChart software is used to calculate and plot the hill chart for Pelton [1], [2] Francis [3], [4] and Kaplan turbines [5]. The Kaplan turbine operates with mobile blades of the wicked gate and the runner based on an optimal correlation between the opening of the wicked gate and runner blades. The paper describes the HydroHillChart application - Kaplan module, designed for the calculation of the hill chart for Kaplan hydraulic turbine models, by processing the measured data on the test rig. The Kaplan module interface and menu are also described, input data are presented graphically and helical characteristics are calculated for each runner blade "φ" position, ultimately getting the hill chart of the model. This diagram is the basis for calculating the optimal correlation between the opening of the wicked gate and runner blades. Because the entire turbine operating range cannot be explored through measurements, the measurements are punctually made at a constant parameter and from the parametric curve interpolation, the hill chart of the model arises. The HydroHillChart application - Kaplan module is a computer tool specialized in obtaining the hill chart diagram using cubic spline interpolation functions. Based on measurements made on several Kaplan turbine models, it can be a computerized library usable in the design of Kaplan industrial turbines. The application was developed in the Python programming language, while the mathematical tool used for interpolation is the cubic spline type function.

IOP Conf. Series: Materials Science and Engineering **477** (2019) 012023 doi:10.1088/1757-899X/477/1/012023

2. The Kaplan module

The Kaplan turbine option from the main menu displays a window with a specific Kaplan module interface, (Figure 1), composed of: toolbar, measured data table called **Measured points** (in which measured data for a model runner is loaded) and the table called **Intersection with efficiency constant values** (where the application stores values that result from the intersection of primary curves with constant efficiency).

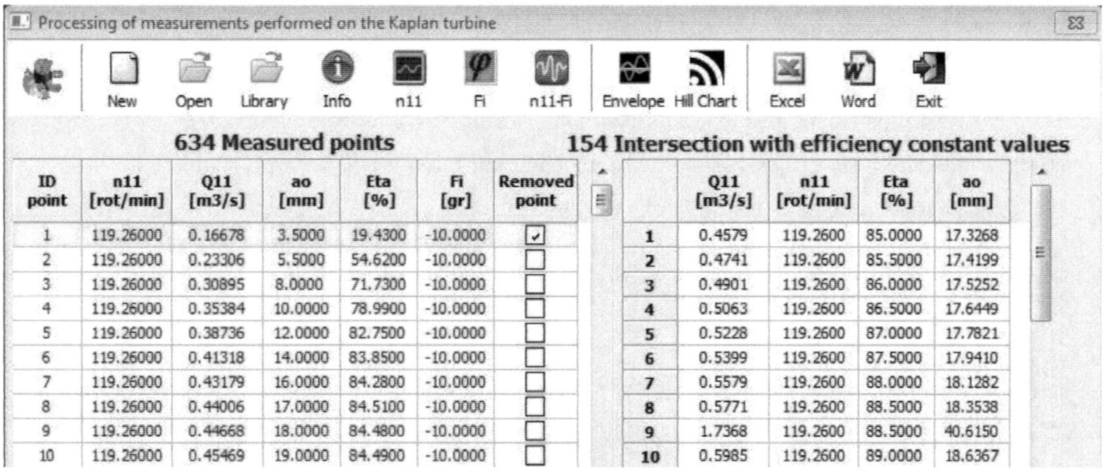

ID point	n11 [rot/min]	Q11 [m3/s]	ao [mm]	Eta [%]	Fi [gr]	Removed point
1	119.26000	0.16678	3.5000	19.4300	-10.0000	☑
2	119.26000	0.23306	5.5000	54.6200	-10.0000	☐
3	119.26000	0.30895	8.0000	71.7300	-10.0000	☐
4	119.26000	0.35384	10.0000	78.9900	-10.0000	☐
5	119.26000	0.38736	12.0000	82.7500	-10.0000	☐
6	119.26000	0.41318	14.0000	83.8500	-10.0000	☐
7	119.26000	0.43179	16.0000	84.2800	-10.0000	☐
8	119.26000	0.44006	17.0000	84.5100	-10.0000	☐
9	119.26000	0.44668	18.0000	84.4800	-10.0000	☐
10	119.26000	0.45469	19.0000	84.4900	-10.0000	☐

	Q11 [m3/s]	n11 [rot/min]	Eta [%]	ao [mm]
1	0.4579	119.2600	85.0000	17.3268
2	0.4741	119.2600	85.5000	17.4199
3	0.4901	119.2600	86.0000	17.5252
4	0.5063	119.2600	86.5000	17.6449
5	0.5228	119.2600	87.0000	17.7821
6	0.5399	119.2600	87.5000	17.9410
7	0.5579	119.2600	88.0000	18.1282
8	0.5771	119.2600	88.5000	18.3538
9	1.7368	119.2600	88.5000	40.6150
10	0.5985	119.2600	89.0000	18.6367

Figure 1. The interface of Kaplan module

The measured data is taken from Excel and placed in the table called **Measured points,** by completing the following fields:

- **ID point** - represents the current number for the measured point;
- **n_{11} [rot/min]** - represents the unitary speed;
- **Q_{11} [m^3/s]** - represents the unitary flow;
- **a_o [mm]** - represents the wicked gate opening;
- **Eta=η [%]** - represents the efficiency;
- **Fi [gr]** - represents the angular opening of the runner blades;
- **Removed point** – allows the removal of a measured point, by selecting a **Check Box.**

3. The Kaplan module toolbar

The Kaplan module toolbar is located at the top of the window and includes control buttons marked with specific icons, (Figure 1), which fulfill the following functions:

- informative icon for the Kaplan runner, without a related function;

New - create a new database for Kaplan runners; the following information need to be completed: turbine type name, database filename, the period when the measurements took place, the person responsible for the performed measurements, additional information; the measurement values are taken from an Excel file.

Open - opening and loading an existing database for Kaplan runners; after this operation, the **Measured points** table, (Figure 1), will be emptied and then rewritten with the values from the selected database.

Library - to open a Kaplan database from the saved library;

Info - provides information about the current database; turbine type name, database filename, the period when the measurements took place, the person responsible for the performed measurements, the file from which the measured data was taken, number of measured and eliminated points, the minimum and maximum values for unitary speed, unitary flow, efficiency, wicked gate and blade runner opening;

n11 - primary data visualization in graphic form for n_{11}=const: 3D curves $\eta = f(n_{11}, Q_{11})$ and $a_o = f(n_{11}, Q_{11})$, respectively 2D parametric curves $\eta = f(Q_{11})$ and $a_o = f(Q_{11})$, (Figure 2), (Figure 3), (Figure 4), (Figure 5);

Fi - primary data visualization in graphic form for φ=const: 3D curves $\eta = f(n_{11}, Q_{11})$ and $a_o = f(n_{11}, Q_{11})$, respectively 2D parametric curves $\eta = f(Q_{11})$ and $a_o = f(Q_{11})$, (Figure 6), (Figure 7), (Figure 8), (Figure 9);

n11-Fi - primary data visualization in graphic form for n_{11}=const & φ=const: 3D curves $\eta = f(n_{11}, Q_{11})$ and $a_o = f(n_{11}, Q_{11})$, respectively 2D parametric curves $\eta = f(Q_{11})$ and $a_o = f(Q_{11})$;

Envelope - calculating the envelope for n_{11}=const value;

Hill Chart - calculating and plotting the hill chart or the helicoidal hill chart for a number of specified efficiencies values;

Excel - export results in an Excel file: input data and the numerical and graphical processing carried out;

Word - graphics export in a Word file;

Exit - return to the main window of the **HydroHillChart** software.

4. The Graphics Toolbar

For each graph generated by the **HydroHillChart** software, at the bottom of the window, a toolbar with command buttons marked with specific icons can be found, which perform the following functions:

Home - Return to initial view;

Back - Back to previous view;

Forward - Forward to the next view;

Pan - Left click and hold to zoom, zoom in/out with the right mouse button pressed;

Zoom - Enlarge selected area;

Subplots - Chart configuration;

Save - Save chart format: EPS; JPG, PGF, PDF; PNG, PS; RAW, SVG, TIF.

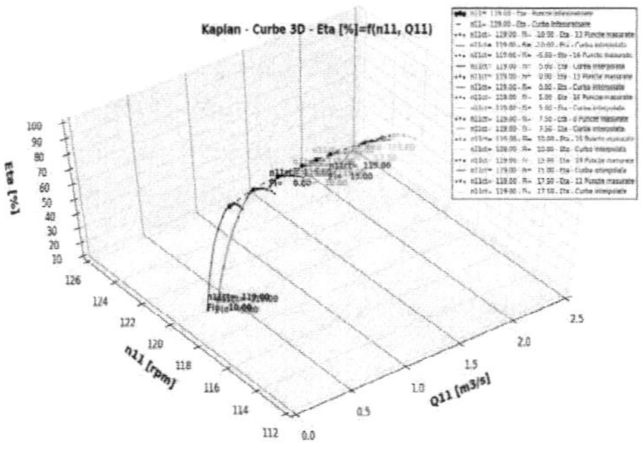

Figure 2. $\eta = f(n_{11}, Q_{11})$ at n_{11}=119 rpm

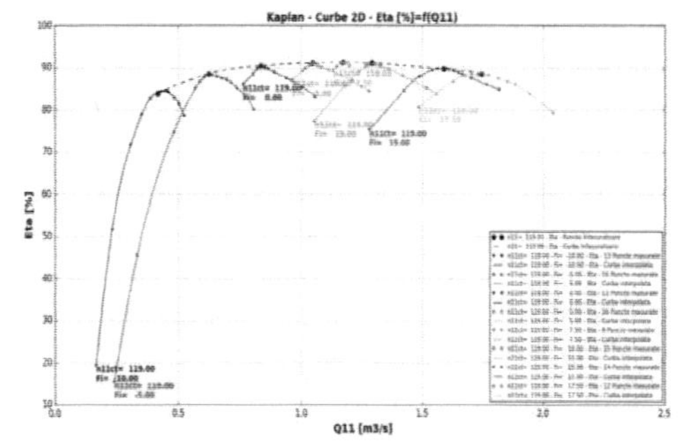

Figure 3. $\eta = f(Q_{11})$ at n_{11}=119 rpm

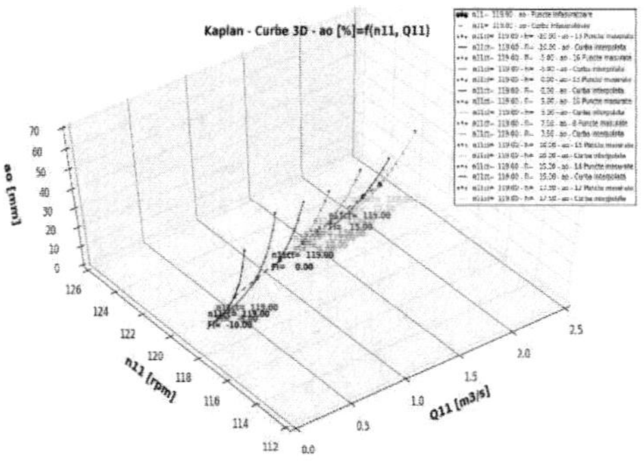

Figure 4. $a_o = f(n_{11}, Q_{11})$ at n_{11}=119 rpm

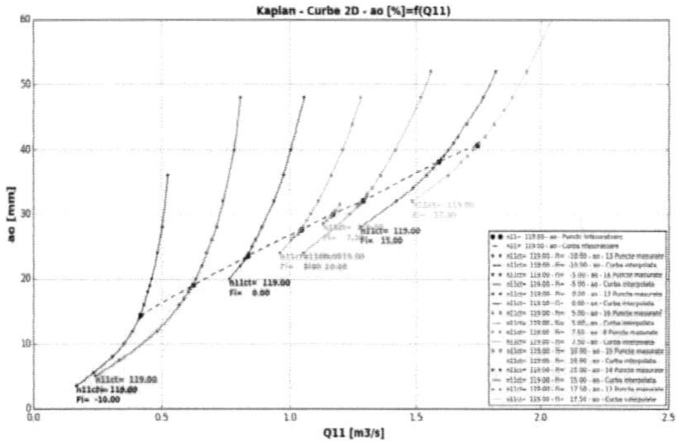

Figure 5. $a_o = f(Q_{11})$ at n_{11}=119 rpm

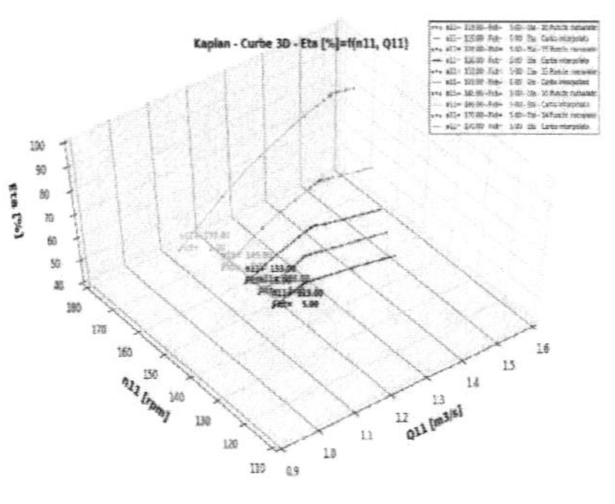

Figure 6. $\eta = f(n_{11}, Q_{11})$ at $\varphi = +5^o$

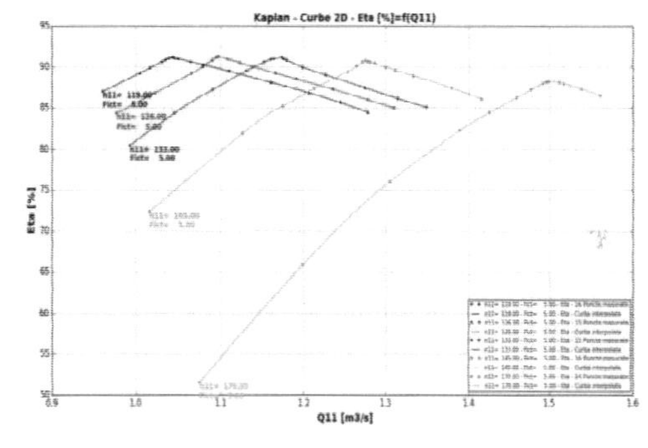

Figure 7. $\eta = f(Q_{11})$ at $\varphi = +5^o$

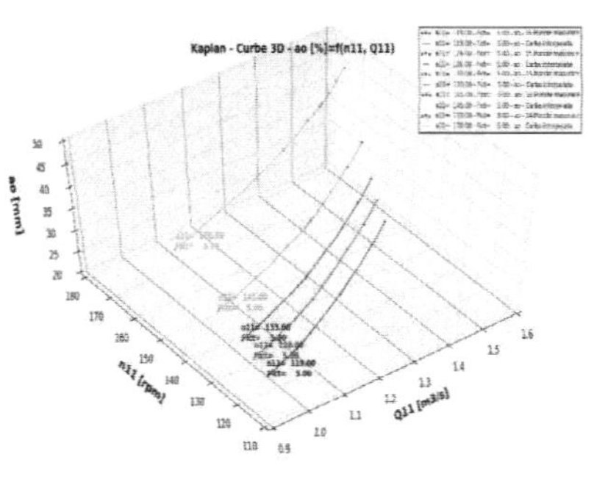

Figure 8. $a_o = f(n_{11}, Q_{11})$ at $\varphi = +5^o$

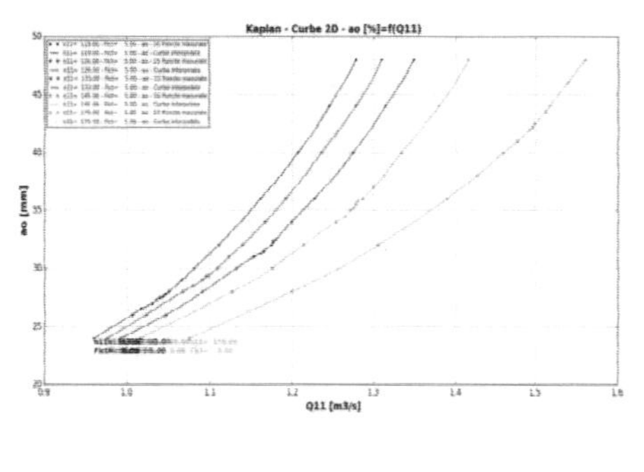

Figure 9. $a_o = f(Q_{11})$ at $\varphi = +5^o$

5. The Envelope option

The **Envelope** button is dedicated to generate the envelope curve for multiple runner angle positions φ at an imposed n11 unitary speed. For each position "φ" of the runner blades, the efficiency curves and wicked gate opening are plotted as a function of the flow rate. The further step is to trace the curve that follows the maximum points corresponding to the "φ" positions of the runner blades. Pressing the **Envelope** button opens a window whose interface is shown in (Figure 10).

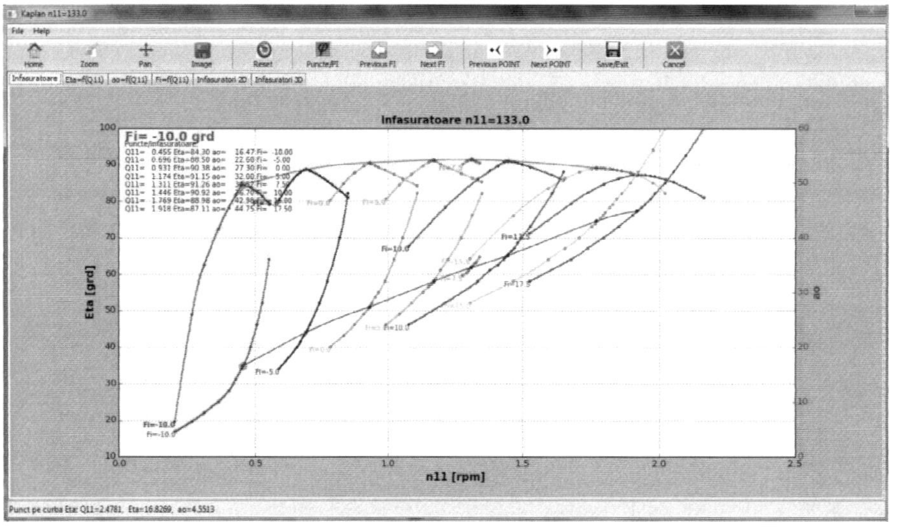

Figure 10. The **Envelope** interface

The window is provided at the top of the screen a toolbar with command buttons marked with specific icons, which performs the following functions:

Home - Return to initial view;

Zoom - Enlarge selected area;

Pan - Left click and hold to zoom, zoom in/out with the right mouse button pressed;

Save - Save chart format: EPS; JPG, PGF, PDF; PNG, PS; RAW, SVG, TIF.

Reset – To reset the envelope curve;

Fi points – To view the points of the blade runner angle φ;

Previous Fi – To return to the previous blade runner angle φ;

Next Fi - To acces to the next blade runner angle φ;

•(**Previous POINT** – To select the previous point on the curve as the point of the envelope for the current blade runner φ;

)• **Next POINT** - To select the next point on the curve as the point of the envelope for the current blade runner φ;

Save / Exit – To save the envelope curve to Kaplan runner database;

Cancel – Exit without save the envelope curve.

Within this window is shown the envelope at the selected unitary speed n_{11}, where the point of tangency between the curve at a certain position φ of the runner blades and the envelope is initially calculated as the maximum point on the efficiency curve. This point can be changed by positioning the cursor on the efficiency curve in the location where the intersection point between the envelope and the efficiency curve will be moved. Defining the new point is done either by pressing the **Previous POINT** or **Next POINT** icon or by pressing the middle button of the mouse. The hill chart calculation for a Kaplan runner involves determining the envelope for each individual unitary speed n_{11}. The envelope generated for the unitary speed n_{11}=119 rpm is shown in (Figure 11). Additionally, this window shows the following curves: $\eta = f(Q_{11})$ (Figure 12), $a_o = f(Q_{11})$ (Figure 13), $\varphi = f(Q_{11})$ (Figure 14) and also the 2D / 3D envelopes for all unitary speed values, (Figure 15), (Figure 16).

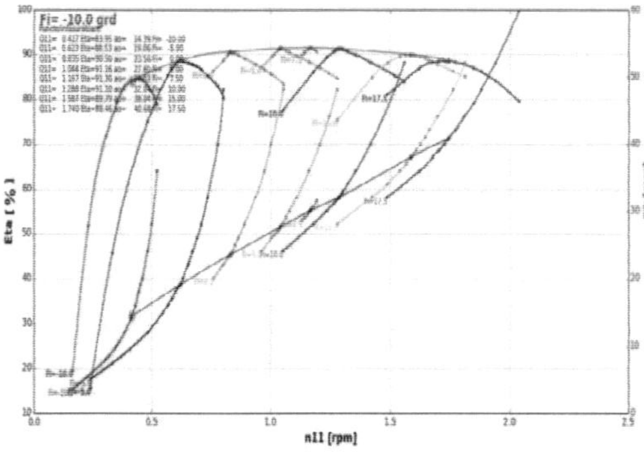

Figure 11. The envelope at n_{11}=119 rpm

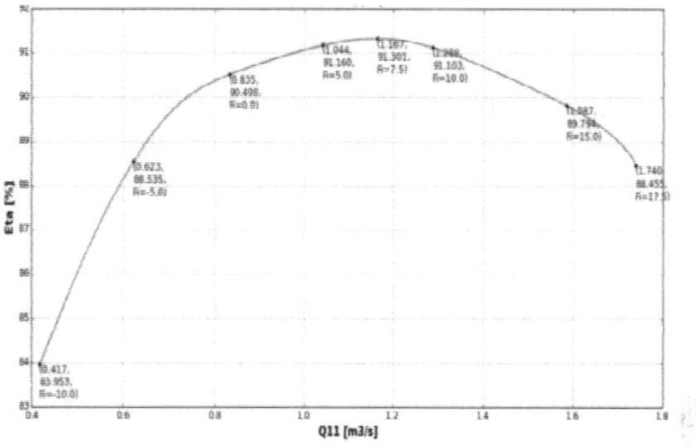

Figure 12. $\eta = f(Q_{11})$ at n_{11}=119 rpm

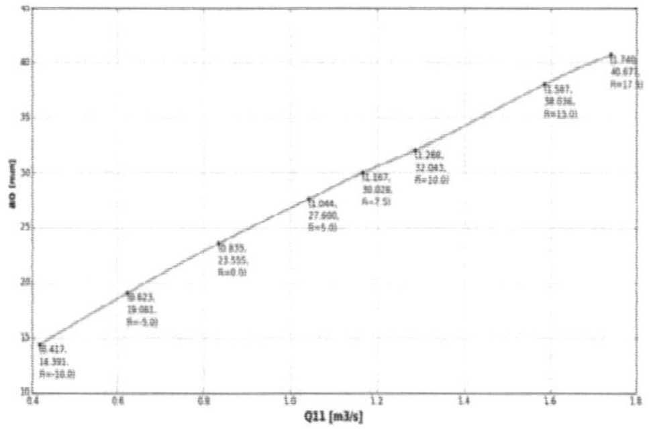

Figure 13. $a_o = f(Q_{11})$ at n_{11}=119 rpm

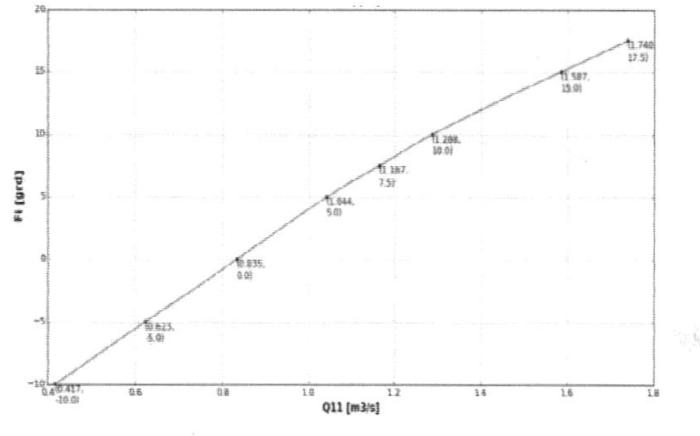

Figure 14. $\varphi = f(Q_{11})$ at n_{11}=119 rpm

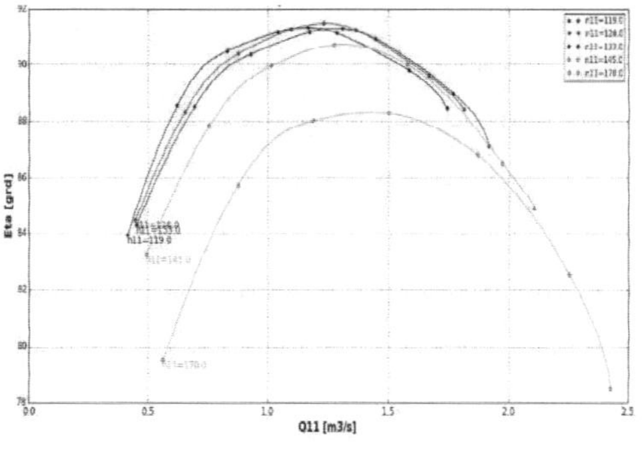

Figure 15. The 2D envelopes for Kaplan runner

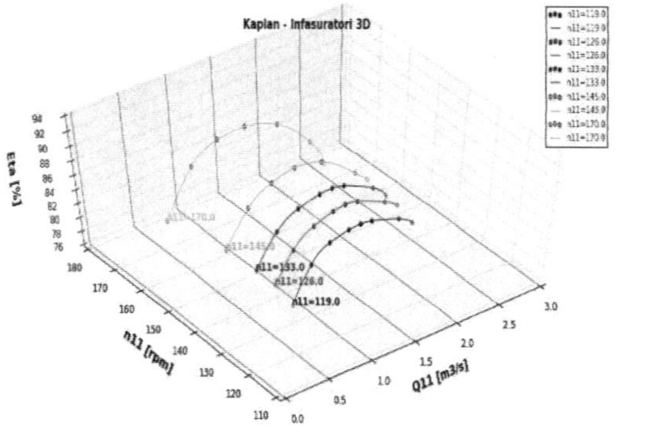

Figure 16. The 3D envelopes for Kaplan runner

6. The Hill Chart option

The **Hill Chart** button is dedicated to generate the helical or the hill chart. Accessing this button will open a window that provides information about the minimum and maximum efficiency in the current database and allows to impose the minimum, the maximum and the step efficiency, for which the helical or the hill chart will be calculated and plotted. Also in this window, particular values of intersection efficiency values can be specified and the display color scheme can be selected. The 3D surface $\eta = f(n_{11}, Q_{11})$ and the helical chart generated for blade runner angle φ=17.5° is shown in Figure 17 and Figure 18. The hill chart is shown in the Figure 19 ÷ Figure 21 as surface, 3D and 2D curves.

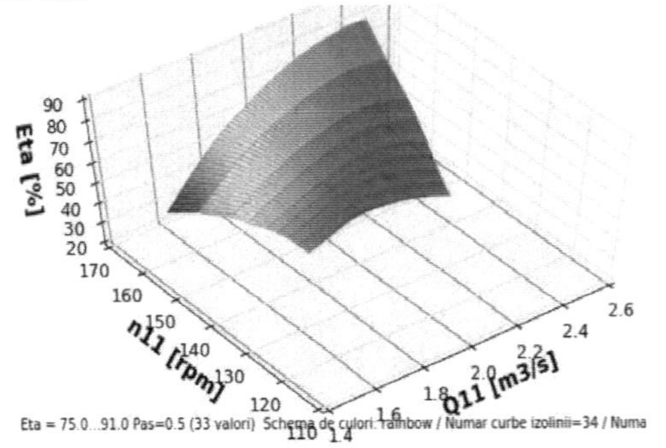

Figure 17. The 3D surface $\eta = f(n_{11}, Q_{11})$ at φ=17.5°

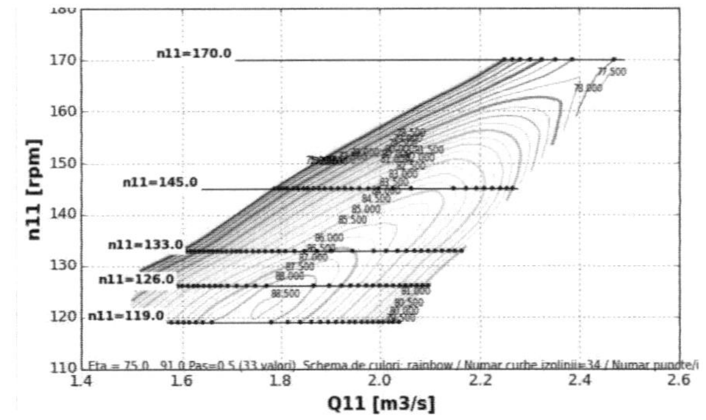

Figure 18. The helical chart $\eta = f(n_{11}, Q_{11})$ at φ=17.5°

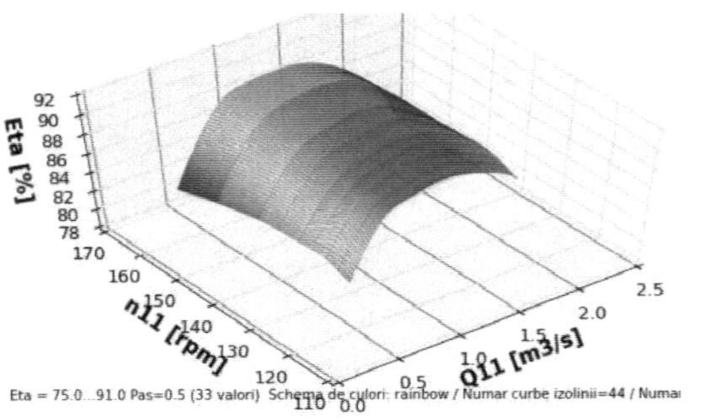

Figure 19. The 3D hill chart surface $\eta = f(n_{11}, Q_{11})$

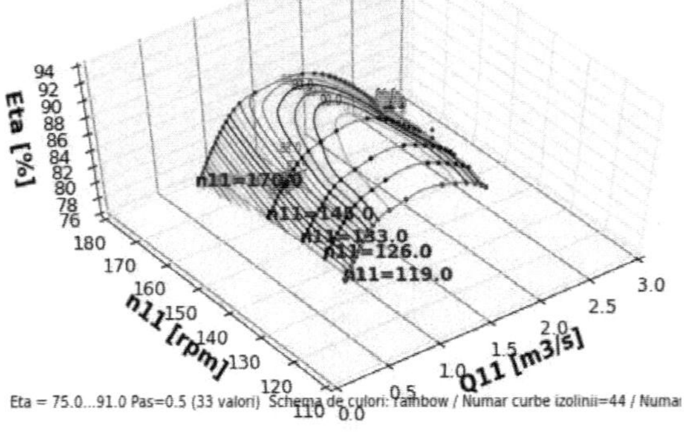

Figure 20. The 3D hill chart curves $\eta = f(n_{11}, Q_{11})$

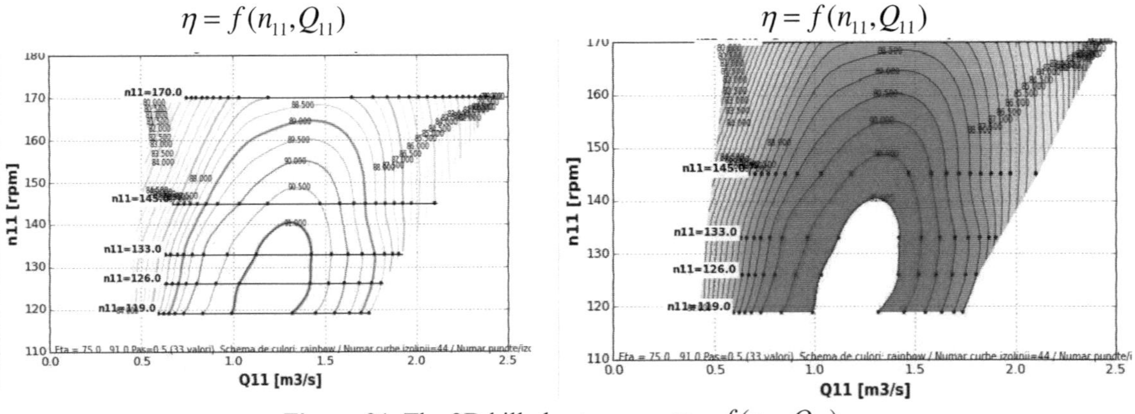

Figure 21. The 2D hill chart curves $\eta = f(n_{11}, Q_{11})$

IOP Conf. Series: Materials Science and Engineering **477** (2019) 012023 doi:10.1088/1757-899X/477/1/012023

7. Conclusions

The paper presents HydroHillChart application [6] - Kaplan module that allows the calculation of the hill chart for Kaplan hydraulic turbine models. For turbines with double settings, the tests are based on the stepwise change of the runner blade angle or the opening a_o of the wicked gate. The efficiency hill chart can be obtained through graphical packages, like general graphic processing and by computer-aided design programs, or through specialized programs like [7]. The HydroHillChart software was created using **Python** – a high-level object-oriented language and related modules: **wxPython** - a graphical user interface toolkit for the Python language [8], **matplotlib** - a python 2D plotting library which produces publication quality figures [9], **SQLite** – a database engine, **SciPy** - a Python-based ecosystem of open-source software's for mathematics, science and engineering.

References

[1] Nedelcu D, Ghican (Bostan) A and Periş-Bendu F 2015 HydroHillChart - Pelton module. Software for calculating universal characteristic of Pelton hydraulic turbines, *"Eftimie Murgu" University Annals* **XXI**(1)

[2] Ghican (Bostan) A, Nedelcu D and Periş-Bendu F 2015 Calculation of universal characteristics for Pelton runner models using the HydroHillChart - Pelton module software, *"Eftimie Murgu" University Annals* **XXI**(1)

[3] Nedelcu D, Ghican (Bostan) A and Periş-Bendu F 2015 HydroHillChart – Francis module. Software for calculating universal characteristic of Francis hydraulic turbines, *"Eftimie Murgu" University Annals* **XXI**(1)

[4] Ghican (Bostan) A, Nedelcu D and Periş-Bendu F 2015 Calculation of universal characteristics for Francis runner models using the HydroHillChart - Francis module software, *"Eftimie Murgu" University Annals* **XXI**(1)

[5] Ghican (Bostan) A 2015 *Contributions regarding the improving of experimental measurement methods and processing the results on the test rig for hydraulic turbine models*, PhD Thesis, "Eftimie Murgu" University of Reşiţa, Romania

[6] ***https://www.youtube.com/results?search_query=hydrohillchart

[7] Aggidis A and Zidonis A 2014 Hydro turbine prototype testing and generation of performance curves: Fully automated approach, *Renewable Energy* **71** 433-441

[8] Rappin N and Dunn R 2006 *wxPython in Action*, Manning Publications Co

[9] Hunter J D 2007 Matplotlib. A 2D graphics environment, *Computing In Science & Engineering* **9**(3) 90-954

IOP Conf. Series: Materials Science and Engineering **477** (2019) 012024 doi:10.1088/1757-899X/477/1/012024

Analysis of two-dimensional (2D) Fruit Drying Process through Heat and Mass Transfer Model

N Shahari , H A Hasnan , A Y Hanan and N A H Noor Ishak

Faculty of Computer and Science Mathematics, Universiti Teknologi MARA, Seremban, Negeri Sembilan, Malaysia

E-mail: norazni@ns.uitm.edu.my

Abstract. The production of dried fruit occurs when there is removal of water content, either naturally, through sun drying, or through the use of the specialized dryers. However, the amount of water left and the temperature inside the fruit cannot be accurately known and only a rough idea is obtained by observation during the drying process. In this paper, two-dimensional fruit drying was analysed through Fick's Law of Diffusion and Fourier's Law of Heat Conduction to allow the effect on the front and rear faces of the fruit to be shown during drying. These equations were used and solved using the Finite-Difference Method (FDM). The result shows that, as time increases, the moisture content decreases and this occurs at each location of the fruit. The reduction rate of moisture in the region near the surface is higher than the interior of the fruit because of the high initial moisture gradient in this region, which drives diffusion from the inside to the surface. The model allows us to predict temperature and moisture distribution as time increases. A Graphical User Interface (GUI) was developed, which can present the results graphically and can aid the development of a food product. The model will be presented in such a way that it can be used by those less familiar with the process of drying, and so that it can be adapted for different food products.

1. Introduction

One of the benefits of the drying process is to prevent the growth of bacteria, yeast and mould. In general, drying is the removal of water content and involves two phases: the initial phase is called the constant-rate drying period while the second phase is called the falling-rate drying period. The drying rate is constant during the initial phase when the water inside the fruit will be transferred to the surface. Once the fruit has dried sufficiently, the falling-rate drying period begins and the drying rate continuously decreases. Thus, drying can be described as the reduction of fruit moisture to the necessary dryness values as a specific process [1]. Necessary dryness value is important since it is one of the aspects of product quality. Furthermore, drying can also be considered as the process whereby heat and mass transfer occurs simultaneously [2].

During drying, heat is gained by convection when the outer surface is warmed up by the surrounding air. From this, the heat transfer coefficients exist at the surface of the fruit. Then, the inner part of the fruit will warm up towards the region that is in contact with the outer surface as the heat is transferred from a hotter region to a colder region by conduction. Generally, diffusion is the movement of water from a higher concentration to a lower concentration. In the drying process, the water diffuses out from the inner part of the fruit to the surface. Then, heat changes the water to water vapour at the surface. It causes the moisture inside the fruit to reduce. Thus, a mass transfer coefficient exists at the surface.

The most common modelling in food involves a one-dimensional model (1D), in which food can be assumed to be of infinite length with moisture and temperature given across the thickness of the food. This method has been used to analyse drying behaviour [3-6]. For the drying of slices of food with the cross-sectional area of a rectangle, a two-dimensional model (2D) is used to allow the effect of geometry to be included. Some researchers have investigated the two-dimensional model [2], [7-9]. The various models proposed for convective drying of a food product using heat and mass transfer (1D, 2D or 3D) can be found in [6]. Various theoretical models proposed in the literature for the drying of foods linked the heat equation with the mass equation by using the latent heat evaporation boundary condition at the surface in the heat equation. However, the driving gradient from the surface to the air that involves partial pressure and temperature at the surface of the mass equation was not considered. In this research, the analysis focused on two dimensional fruit with the application of coupled boundary conditions, both for mass and heat equations ([10]).

2. Methodology

2.1. Mathematical model

The diffusion and heat equation is applied to describe the two-dimensional movement of moisture (*M*) and temperature (*T*) in fruits. In the Cartesian coordinates with mass diffusivity of fruit, D (m^2/s) and thermal diffusivity of heat, α (m^2/s), the general equation for moisture and temperature at any time, t (s) is given by:

$$\frac{\partial M(x,y,t)}{\partial t} = D\left(\frac{\partial^2 M}{\partial x^2} + \frac{\partial^2 M}{\partial y^2}\right), \quad 0 < x < L_x, \quad 0 < y < L_y, \tag{1}$$

$$\frac{\partial T(x,y,t)}{\partial t} = \alpha\left(\frac{\partial^2 T}{\partial x^2} + \frac{\partial^2 T}{\partial y^2}\right), \quad 0 < x < L_x, \quad 0 < y < L_y, \tag{2}$$

where x and y represent the thickness and length of the fruit (mm).

The boundary conditions are given below:

$$x = 0, \quad \frac{\partial M(0,y,t)}{\partial x} = 0, \quad \frac{\partial T(0,y,t)}{\partial x} = 0, \tag{3}$$

$$y = 0, \quad \frac{\partial M(x,0,t)}{\partial y} = 0, \quad \frac{\partial T(0,y,t)}{\partial x} = 0 \tag{4}$$

$$x = L_x, \quad -D\frac{\partial M(L_x,y,t)}{\partial x} = h_m(C_{sur} - C_{air}), \tag{5}$$

$$-h(T_{sur} - T_{air}) = k\frac{\partial T(L_x,y,t)}{\partial x} - \lambda D\rho_s \frac{\partial M(L_x,y,t)}{\partial x},$$

$$y = L_y, \quad -D\frac{\partial M(x,L_y,t)}{\partial y} = h_m(C_{sur} - C_{air}), \tag{6}$$

$$-h(T_{sur} - T_{air}) = k\frac{\partial T}{\partial y}(x,L_y,t) - \lambda D\rho_s \frac{\partial M(L_x,y,t)}{\partial y}.$$

The initial condition is as follows:

$$t = 0, \quad M(x,y,0) = M_0, T(x,y,0) = T_0. \tag{7}$$

2.2. Non-Dimension

In order to solve and produce an equation with scale $0<\bar{x}<1$, and scale of $0<\bar{y}<1$, the equation is then dimensionalized in a non-dimensional formulation. The non-dimensional scaled variables used are representative of diffusion timescale, τ ; scaled moisture content, $\overline{M}$; and scaled temperature, $\overline{T}$.

$$\tau=\frac{Dt}{L_x^{\;2}},\quad \overline{M}=\frac{M}{M_0},\quad \overline{T}=\frac{T-T_0}{T_{air}-T_0},\quad \bar{x}=\frac{x}{L_x},\quad \bar{y}=\frac{y}{L_y}.$$

By applying the dimensional scaled variables, the non-dimensional variables of moisture and temperature equations, ((1)-(7)) are simplified and arranged as follows:

$$\frac{\partial\overline{M}(\bar{x},\bar{y},\tau)}{\partial\tau}=\frac{\partial^2\overline{M}}{\partial\bar{x}^2}+\beta^2\frac{\partial^2\overline{M}}{\partial\bar{y}^2},\quad \text{where } \beta=\frac{L_x}{L_y}, \tag{8}$$

$$\frac{\partial\overline{T}(\bar{x},\bar{y},\tau)}{\partial\tau}=Le\left(\frac{\partial^2\overline{T}}{\partial\bar{x}}+\beta^2\frac{\partial^2\overline{T}}{\partial\bar{y}}\right). \tag{9}$$

With the boundary equations,

$$\frac{\partial\overline{M}(0,\bar{y},\tau)}{\partial\bar{x}}=0,\quad \frac{\partial\overline{T}(0,\bar{y},\tau)}{\partial\bar{x}}=0,\quad \frac{\partial\overline{M}(\bar{x},0,\tau)}{\partial\bar{y}}=0,\quad \frac{\partial\overline{T}(\bar{x},0,\tau)}{\partial\bar{y}}=0, \tag{10}$$

$$\frac{\partial\overline{M}(1,\bar{y},\tau)}{\partial\bar{x}}=-Bi_m\frac{C_{air}}{M_0}(\overline{C}_{sur}-1), \tag{11}$$

$$\frac{\partial\overline{T}(1,\bar{y},\tau)}{\partial\bar{x}}=Bi\left(1-\overline{T}_{sur}\right)+\bar{\lambda}\frac{\partial\overline{M}(1,\bar{y},\tau)}{\partial\bar{x}}, \tag{12}$$

$$\frac{\partial\overline{M}(\bar{x},1,\tau)}{\partial\bar{y}}=-\frac{1}{\beta}Bi_m\frac{C_{air}}{M_0}(\overline{C}_{sur}-1), \tag{13}$$

$$\frac{\partial\overline{T}(\bar{x},1,\tau)}{\partial\bar{y}}=\frac{1}{\beta}Bi\left(1-\overline{T}_{sur}\right)+\bar{\lambda}\frac{\partial\overline{M}(\bar{x},1,\tau)}{\partial\bar{y}}. \tag{14}$$

And the initial conditions are

$$\overline{M}=1,\quad \overline{T}=0. \tag{15}$$

where $Bi_m=\frac{h_mL_x}{D}$, $Bi=\frac{hL_x}{k}$, $Le=\frac{\alpha}{D}$, $\bar{\lambda}=\frac{\lambda D\rho_s}{k}\frac{M_0}{(T_{air}-T_0)}$, $\overline{C}_{sur}=\frac{C_{sur}}{C_{air}}$.

2.3. Numerical Method

Discretization equations (8) and (9) with a central difference approximation become (the over bar notation is dropped as clarity)

$$T_{i,j}^{n+1}=T_{i,j}^n+sx(T_{i,j+1}^n-2T_{i,j}^n+T_{i,j-1}^n)+\beta^2sz\left(T_{i+1,j}^n-2T_{i,j}^n+T_{i-1,j}^n\right), \tag{16}$$

$$M_{i,j}^{n+1}=M_{i,j}^n+mx(M_{i,j+1}^n-2M_{i,j}^n+M_{i,j-1}^n)+\beta^2mz(M_{i+1,j}^n-2M_{i,j}^n+M_{i-1,j}^n), \tag{17}$$

at general index point *i* and *j*, for *i*=2,3,…*Nz*-1 and *j*=2,3,…*Nx-1*.

with $sx=Le*\frac{\Delta t}{\Delta x^2}$ and $sz=Le*\frac{\Delta t}{\Delta y^2}$, $mx=\frac{\Delta t}{\Delta x^2}$ and $mz=\frac{\Delta t}{\Delta y^2}$.

To apply the symmetry condition, a fictitious grid point is involved and the equation is eliminated using central difference approximation. This gives the condition,

$$T_{i,1}^{n+1} = T_{i,1}^{n} + sx(2T_{i,2}^{n} - 2T_{i,1}^{n}) + \beta^2 sz(T_{i+1,1,}^{n} - 2T_{i,1}^{n} + T_{i-1,1}^{n}), \quad i = 2,3,....Nz-1. \tag{18}$$

$$T_{1,j}^{n+1} = T_{1,j}^{n} + sx(T_{1,j-1}^{n} - 2T_{1,j}^{n} + T_{1,j+1}^{n}) + \beta^2 sz(2T_{2,j}^{n} - 2T_{1,j}^{n}), \quad j = 2,3,....Nx-1. \tag{19}$$

$$M_{i,1}^{n+1} = M_{i,1}^{n} + mx(2M_{i,2}^{n} - 2M_{i,1}^{n}) + \beta^2 mz(M_{i+1,1}^{n} - 2M_{i,1}^{n} + M_{i-1,1}^{n}), \quad i = 2,3,....Nz-1, \tag{20}$$

$$M_{1,j}^{n+1} = M_{1,j}^{n} + mx(M_{1,j-1}^{n} - 2M_{1,j}^{n} + M_{1,j+1}^{n}) + \beta^2 mz(2M_{2,j}^{n} - 2M_{1,j}^{n}), \; j = 2,3,....Nx-1. \tag{21}$$

Applying the surface condition equation by introducing a fictitious grid point at x=Nx and y=Nz and using central difference approximation gives

$$\begin{aligned} T_{i,Nx}^{n+1} &= T_{i,Nx}^{n} + sx(2T_{i,Nx-1}^{n} - 2T_{i,Nx}^{n} + 2\Delta x Bi(1 - T_{sur}) + \bar{\lambda} * 2\Delta x(3M_{i,Nx}^{n} - 4M_{i,Nx-1}^{n} \\ &\quad + M_{i,Nx-2}^{n})) + sz\beta^2(T_{i+1,Nx}^{n} - 2T_{i,Nx}^{n} + T_{i-1,Nx}^{n}), \quad i = 2,3,....Nz-1. \end{aligned} \tag{22}$$

$$\begin{aligned} T_{Nz,j}^{n+1} &= T_{Nz,j}^{n} + sx(T_{Nz,j-1}^{n} - 2T_{Nz,j}^{n} + T_{Nz.j+1}) + \beta^2 sz\{2T_{Nz-1,j}^{n} - 2T_{Nz,j}^{n} + \frac{1}{\beta}[2\Delta z Bi \\ &\quad (1 - T_{Nz,j})] + \bar{\lambda} * 2\Delta z(3M_{Nz,j} - 4M_{Nz-1,j} + M_{Nz-2,j})\}, \; j = 2,3,....Nx-1. \end{aligned} \tag{23}$$

$$\begin{aligned} M_{i,Nx}^{n+1} &= M_{i,Nx}^{n} + mx(2M_{i,Nx-1}^{n} - 2M_{i,Nx}^{n} - 2\Delta x Bi_m \frac{C_{air}}{M_0}(C_{sur} - 1)) \\ &\quad + mz\beta^2(M_{i+1,Nx}^{n} - 2M_{i,Nx}^{n} + M_{i-1,Nx}^{n}), \; i = 2,3,....Nz-1. \end{aligned} \tag{24}$$

$$\begin{aligned} M_{Nz,j}^{n+1} &= M_{Nz,j}^{n} + mx(M_{Nz,j-1}^{n} - 2M_{Nz,j}^{n} + M_{Nz.j+1}) \\ &\quad + mz\beta^2[2M_{Nz-1,j}^{n} - 2M_{Nz,j}^{n} - \frac{1}{\beta}2\Delta z Bi_m \frac{C_{air}}{M_0}(C_{sur} - 1)], \; j = 2,3,....Nx-1. \end{aligned} \tag{25}$$

3. Results and Discussions

Solution of the model is obtained by solving two coupled Partial Differential Equations (PDE) corresponding mass and heat equation (equation (8)-(9)), together with initial and boundary conditions (Equation (10)-(14)). Equations (8) and (9) are numerically discretized using the FDM (Equation (16)-(25)) while the derived algebraic equations are solved using MATLAB. For the first simulation, the thickness and the length are taken as aspect ratio (AR) $\frac{L_x}{L_y} = 1$. The simulation is then repeated with aspect ratio $\frac{L_x}{L_y} = 2$. The thickness of the food is maintained at constant value $L_x = 1$ and the length of the food is changed according to the corresponding aspect ratio (AR).

The moisture profile at different time intervals is shown in Figure 1. At the beginning, the moisture drops significantly at the surface of the fruit and there is still much water content inside the fruit. At this time, the fruit is saturated with water at the surface and moisture removal occurs at a faster rate. Once the surface moisture becomes less, moisture from the inside will be transferred by diffusion onto the surface. The process of diffusivity occurs from the inside to the surface of the fruit. As the drying proceeds, the moisture level starts to drop substantially compared to the beginning of the process. The experiment conducted by [11] shows a similar finding: the moisture contents are relatively low towards the surface facing the surrounding air, compared to the inner part of the object. Based on the Figure 1, the longer the time period, the lower the moisture content left inside the fruit.

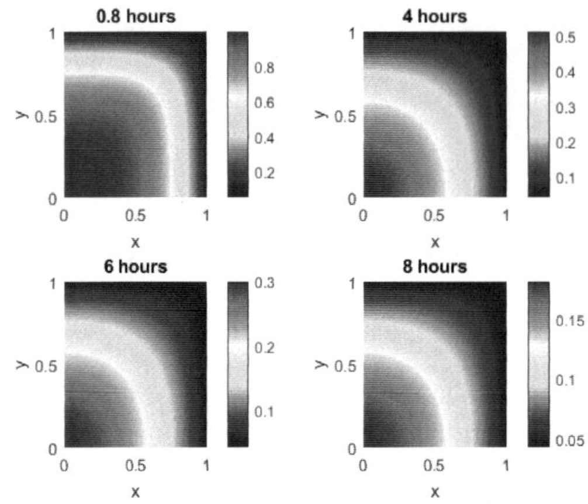

Figure 1. Predicted moisture distribution inside the moist fruit at different time intervals (a) $t = 0.8$ hours, (b) $t = 4$ hours, (c) $t = 6$ hour (d) $t = 8$ hours

Figure 2 shows the distribution of temperature inside the slice at different times. There is a difference between the temperature at the surface and at the core. Initially, the surface temperature starts to increase and the movement of the heat from the surface to the core rises, showing that the temperature inside the fruit has increased. The substantial increase shows that the surface is getting hotter and at the core there is a clear difference between the current temperature inside the fruit and the earlier temperature. This is consistent with findings by [11], which stated that the temperature distributions are higher at the surface of the moist object and these progressively decrease towards the interior portion of the moist object. From these results, the part of the fruit that is most exposed to the surrounding air tends to warm up faster. Thus, the longer the drying period, the higher the temperature at both the surface and the core of the fruit T.

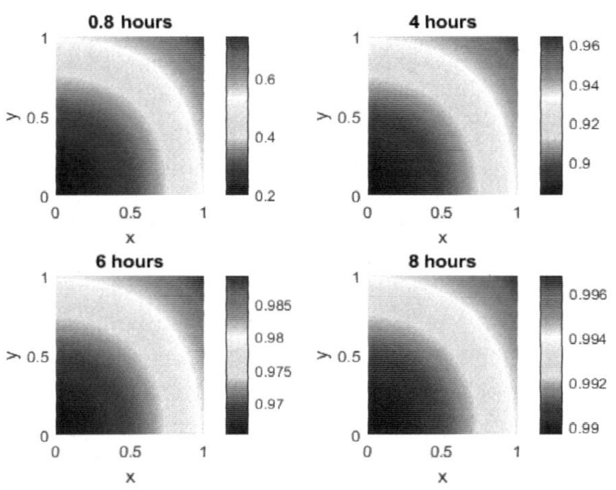

Figure 2. Predicted temperature distribution inside the moist fruit at different time intervals (a) $t = 0.8$ hour, (b) $t = 4$ hour, (c) $t = 6$ hour (d) t=8 hour

Figure 3 clearly shows that the processes of heat and mass transfer occur simultaneously. When the temperature at the surface of fruit rises, the moisture starts to drop sharply near the surface. However, the core of the fruit shows only a small increase in temperature, while the moisture content drops a little. The temperature and moisture stop when equilibrium is achieved. The thin surface region, having a steep moisture profile, causes a shell to develop on the outside [12]. Normally, heat enters the surface of the fruit by conduction, while water moves outwards from the inside of the fruit by diffusion. At this stage, heat is required to change water to water vapour at the surface of fruit. After that, the water vapour loss to the surrounding area causes the moisture at the surface to decrease. From the result above, the higher the temperature, the lower the moisture level inside the fruit. It indicates that the temperature increases, in contrast to the moisture content.

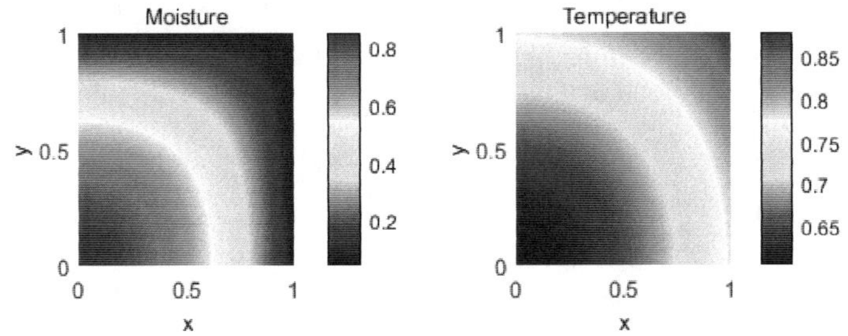

Figure 3. The temperature and moisture distribution after 2 hours when the parameters: Bi_m=20, Bi=1.5 and Le=5

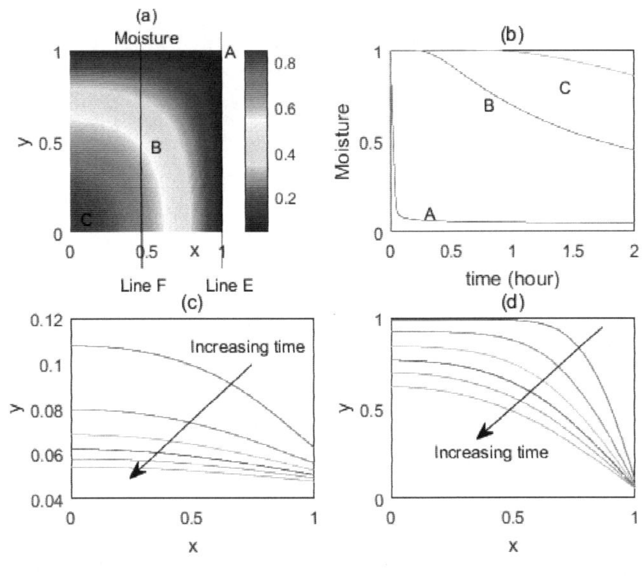

Figure 4. (a)Surface plot of moisture evolution after 2hours. b) Moisture decreasing at selected points A, B and C c) Moisture across a line passing through the surface (line E). d) Moisture across a line passing through the surface (line F)

Figure 4 and Figure 5 show three characteristic points inside the fruit that were selected: point A-located at the top surface; Point B-located in the middle of the fruit; point C-located in the symmetry point of the fruits. Figure 4(b) shows that the removal of water content at A is higher compared to B and C since point A represents the top surface of the fruit, which is more exposed to the surrounding air. Point C has the lowest water removal because it is located in the middle core of the fruit. Figure 4(c) shows the distribution of moisture at line E, which is located at the outer surface. Figure 4(d) shows the distribution of moisture at line F, which is located in the middle of the surface. From the graph observation, the rate of water removal at line E is higher as time increases, compared to line F.

This is supported by [7] who state that, at the surface, the drying rate is faster than the drying rate inside the fruit since the moisture gradients between the surrounding air and the surface are higher.

Figures 5(a) and 5(b) show that the increase in temperature at A is higher compared to B and C since point A represents the top surface of the fruit, which is more exposed to the surrounding air. Point C has the lowest increase in temperature because it is located in the middle core of the fruit. Figure 5(c) shows the temperature distribution at line E, which is located at the outer surface. Figure 5(d) shows the temperature distribution at line F, which is located in the middle of the surface. From the graph observation, the rate of temperature increase at line E is higher as time increases compared to line F. This finding from [11] states that the heat transfer coefficients, *h* are higher at the surface because of immediate contact with the hot air.

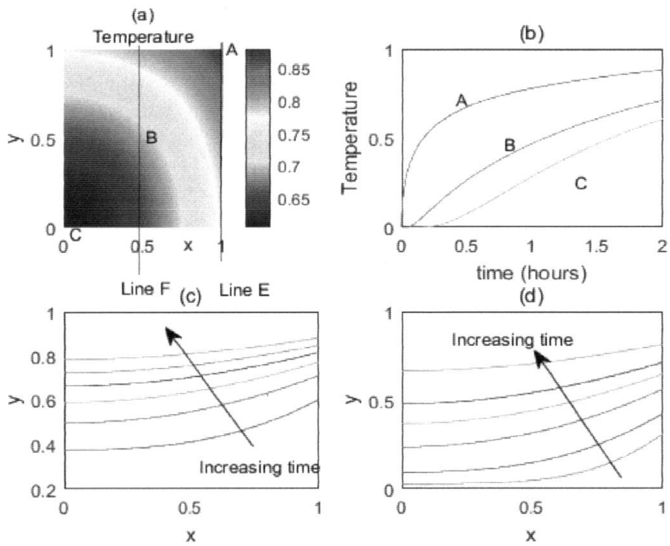

Figure 5. a) Surface plot of temperature evolution at 2.4 hours. b) Temperature is increasing at selected points A, B, and C c) Temperature across a line passing through the surface (line E). d) Temperature across a line passing through the surface (line F)

Variation of moisture inside the fruit with 2 different aspect ratios (AR) is shown in Figure 6. It was observed that moisture reduces more quickly for smaller AR, which concludes that smaller aspect ratios (AR) provide shorter drying times than higher AR. This is consistent with findings by [7], which showed that the drying rate decreases by 4.5% with a 1 mm increase in mango thickness.

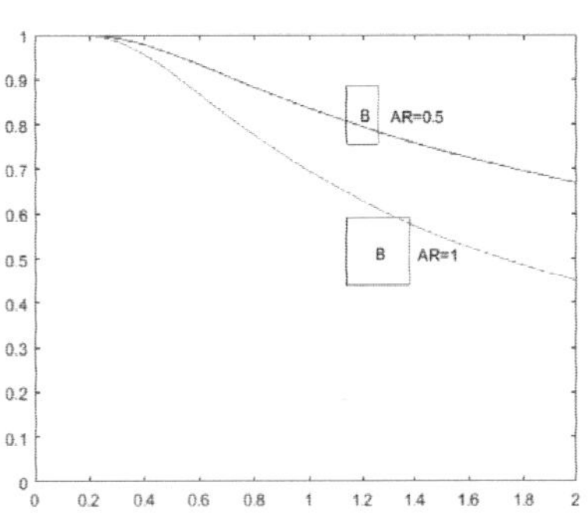

Figure 6. Effect of aspect ratio(AR) on the moisture at a selected points B

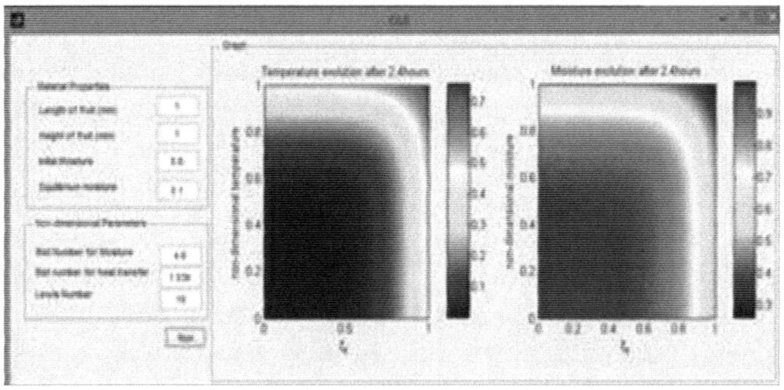

Figure 7. : Graphical User Interface for two-dimensional fruit drying

Figure 7 shows the Graphical User Interface (GUI) that was developed using MATLAB Graphical User Interface Development Environment (GUIDE). The interface was designed to have one major subpanel, material properties and non-dimensional parameters, which compute the material properties of fruit and its non-dimensional parameters. The Graph subpanel has the responsibility of obtaining the results from the GUI core and interpreting the simulation results in human friendly language. After receiving the simulation results, the Graph presents the results of the "Temperature evolution after 2 hours" axes and the "Moisture evolution after 2 hours" axes. The image will be used as a system for monitoring the moisture and temperature inside the fruit.

4. Conclusion

A 2D heat and mass transfer numerical model was developed to study the transient behaviour of temperature and moisture content during the drying of fruit. Using a square to represent the fruit, this model well described the mass and heat transfer during the drying process. The conduction process is considered to occur as a heat transfer inside the fruit, while the liquid diffusion is governed solely as mass transfer. The model included the constant diffusivity and thermal diffusivity as fixed properties. In this work, a mathematical model for heat and mass transfer during the drying process has been discretized using finite difference and implemented in MATLAB along with the initial and boundary conditions.

It was observed that the temperature is higher at the surface corner of the leading edge and lower at the symmetry leading edge. The depletion of water is more effective at the higher temperature of the surface, where the surface moisture is less than that at the surface corner. The fruit that is exposed to the surrounding air tends to warm up faster, causing the moisture content to decrease. The proposed model was able to provide information about moisture and temperature at all times, thus allowing detection of the region within the fruit, and a very important aspect in terms of food safety. The model allows us to predict temperature and moisture distribution as time increases.

Modelling of overall food behaviour is important and local and transient features can help to establish firmer knowledge for modelling applications. Simple processes allow researchers to base their development on a trial-and-error approach; more complex processes will quickly increase the required number of experiments. Moving from a trial-and-error approach to a quality by design approach requires the development of a model of the production process linking variables to critical quality attributes of the final product. Development of a user-friendly model for food drying can aid the development of a food product by shifting from a trial-and-error towards a quality-by-design approach.

References

[1] Ceylan I 2009 Energy analysis of PID controlled heat pump dryer, *Engineering* **1** 188-195

[2] Kaya A, Aydın O and Dincer I 2008 Experimental and numerical investigation of heat and mass transfer during drying of Hayward kiwi fruits (Actinidia Deliciosa Planch), *Journal of Food Engineering* **88** 323-330

[3] Barati E and Esfahani J 2011 A new solution approach for simultaneous heat and mass transfer during convective drying of mango, *Journal of Food Engineering* **102** 302-309

[4] Karim M A and Hawlader M 2005 Mathematical modelling and experimental investigation of tropical fruits drying, *International Journal of Heat and Mass Transfer* **48** 4914-4925

[5] Shahari N, Jamil N, and Rasmani K A 2016 Comparative Study of Shrinkage and Non-Shrinkage Model of Food Drying, *Journal of Physics: Conference Series* **738** 012087

[6] Tzempelikos D A, Mitrakos D, Vouros A P, Bardakas A V, Filios A E, and Margaris D P 2015 Numerical modeling of heat and mass transfer during convective drying of cylindrical quince slices, *Journal of Food Engineering* **156** 10-21

[7] Villa-Corrales L, Flores-Prieto J, Xamán-Villaseñor J and García-Hernández E 2010 Numerical and experimental analysis of heat and moisture transfer during drying of Ataulfo mango, *Journal of food engineering* **98** 198-206

[8] Oztop H F and Akpinar E K 2008 Numerical and experimental analysis of moisture transfer for convective drying of some products, *International Communications in Heat and Mass*

Transfer **35** 169-177

[9] Esfahani J, Majdi H and Barati E 2014 Analytical two-dimensional analysis of the transport phenomena occurring during convective drying: Apple slices, *Journal of Food Engineering* **123** 87-93

[10] Shahari N, Rasmani K, and Jamil N 2016 Comparison between analytical and numerical solution of mathematical drying model, *AIP Conference Proceedings* **1705** 020043

[11] Mohan V C and Talukdar P 2010 Three dimensional numerical modeling of simultaneous heat and moisture transfer in a moist object subjected to convective drying, *International Journal of Heat and Mass Transfer* **53** 4638-4650

[12] Gulati T and Datta A K 2015 Mechanistic understanding of case-hardening and texture development during drying of food materials, *Journal of Food Engineering* **166** 119-138

IOP Conf. Series: Materials Science and Engineering **477** (2019) 012025 doi:10.1088/1757-899X/477/1/012025

Modelling and simulation of the asymmetric rolling process – establishing the optimal technology parameters to asymmetric rolling

V Alexa, I Kiss, V G Cioată and S A Raţiu

University Politehnica Timisoara, Faculty of Engineering Hunedoara, Department of Engineering and Management, Hunedoara, Romania

E–mail: vasile.alexa@fih.upt.ro

Abstract. Asymmetric rolling is a technique used to control both the texture and the grain refinement of metallic materials. The aim of asymmetric rolling is to apply a large shear strain uniformly through the thickness of the plate, by maintaining a high degree of friction between the sheet and the rolls. It can be used to improve the formability of material. One of the advantages of asymmetrical rolling is that the rolling force and torque can be decreased. An economic and technological reason, in particular the longitudinally asymmetric rolling is very important to determine the optimum process parameters. The paper presents a method of simulating the asymmetric rolling process. This simulation is carried out to establish the main parameters of rolling particular asymmetric regime. The main goal is to establish a link between the peripheral speed of rolls, pressure and contact length.

1. Introduction

Any well managed operation of rolling needs a precise determination of the force factors. It is a complicated problem, to solve it means knowing the basic rules of process of plastically deformation of metallic materials, generally, and also to know the physical and mechanical characteristics of the mentioned materials. One knows that the efforts applied to a metallic material during its plastic deformation are conditioned not only by its properties, but also by stress status applied to it. The metal material deformed between the rollers of rolling mill suffer high compression stress due to the action of the rollers, and it also bears tangential surface tensions because of the friction between the rollers and the metal material [1].

For the achievement of the rolling process parameters with an electronic calculation technique, their mathematical description has to be precisely formulated. This can be obtained by the elaboration of proper algorithms. Most of the times, the algorithm is written under a graphic form of a block–scheme, or under a textual form, in words. By textual expression of the algorithms is used the mathematical consecrated symbols and expressions that are attended by explanatory text. But, such writing is hard to understand and too complicated. Therefore, the algorithm's representation under a block–scheme form is usually used. The essence of the algorithm elaboration consists of choosing and establishing of such specific usage order of the mathematical expressions that would mandatory lead to the problem solution through any combinations of the initial data [2].

The methods for measuring the rolling forces are classified according to the type of equipment and the possibility to determine them for a particular type of rolling mill, as follows [3]:

- direct measurement of the forces occurring during the rolling process, using resistive transducers placed in the point in which these forces are acting:
- measurement of deformations or strains that develop in a certain part of the working stand, with subsequent recalculation of these deformations based on the values of the rolling forces.

In addition to the forces, the rolling moments represent one of the most important parameters, whose knowledge is necessary for the design and operation of the rolling lines. The electric motor torque, developed during the metal material deformation, determines the loading level of the machine, as well as the resistance of some parts and assemblies. The assessment of motor workload by taking into account the power consumption for deformation is not complete, because this parameter does not characterise the electrical and mechanical capacity of the equipment.

2. Methods and materials

The aim of asymmetric rolling is to apply a large shear strain uniformly through the thickness of the plate, by maintaining a high degree of friction between the sheet and the rolls. It can be used to improve the formability of material. One of the advantages of asymmetrical rolling is that the rolling force and torque can be decreased. An economic and technological reason, in particular the longitudinally asymmetric rolling is very important to determine the optimum process parameters.

The paper presents a method of simulating the asymmetric rolling process. This simulation is carried out to establish the main parameters of rolling particular asymmetric regime. The main goal is to establish a link between the peripheral speed of rolls, pressure and contact length. Since the intended purpose was to study the qualitative aspect of phenomena related to the asymmetric longitudinal rolling, for eliminating the inevitable influence of the iron oxides (scale) on the process, the tests were carried out on samples of aluminium and copper, with the following dimensions [4]:

$$h_0 = 12; 6; 2 \text{ and } 1\,\text{mm} \quad b_0 = 40\,\text{mm} \quad l_0 = 150\,\text{mm}$$

The mechanical properties of the material used (copper) to modelling and simulation are given in Figure 1.

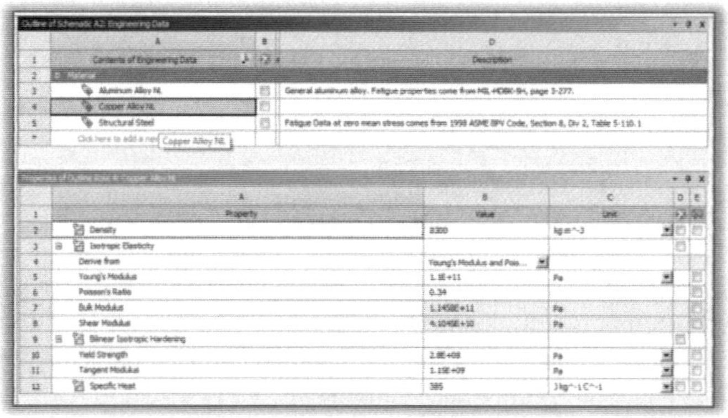

Figure 1. The mechanical properties of the material used to modelling and simulation

Therefore, experimental research on pressure distribution is of particular importance for both the theory and the practice of lamination processes. They can serve directly for the experimental confirmation of the fairness of the theoretical conclusions regarding the general law of pressure distribution on the contact spring, as well as the influence of the main pressure factors on the pressure, such as: the external friction coefficient, the initial thickness of the metal material, the degree of reduction and temperature.

The symmetrical and asymmetric rolling process was obtained by equipping work cylinders with segments executed under different radius that allowed the following ratios to be obtained between the diameters of the upper D_s and lower D_i cylinders:

$$\frac{D_s}{D_i} = \frac{170}{170}; \frac{140}{200} [\text{mm}]$$

The cylinders are mounted in such a way that the measurement of the rolling pressure from each cylinder is made in the same plane (Figure 2).

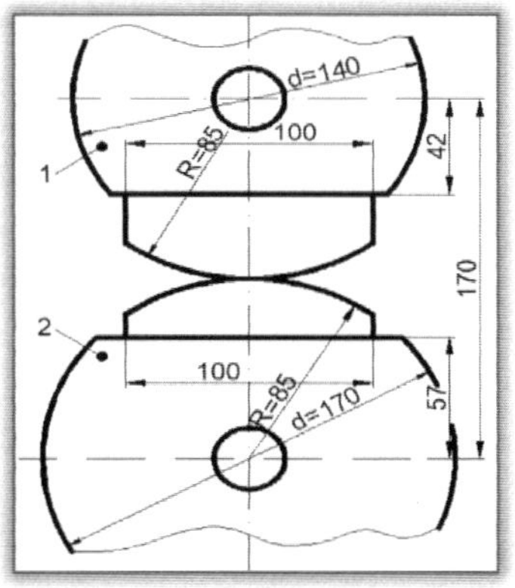

Figure 2. The cylinders work equipping with segments executed under different radius

The deformation of the metal by rolling is possible only in the presence of frictional forces on the contact surfaces of the cylinders. These forces exert influence on the entire volume of metallic material that deforms. As a result, any elementary particle of the respective material on the height and length of the deformation zone is found under volumetric tension conditions characterized by the differential equations of the plasticity theory.

3. Results and discussions

It is known that the efforts applied to a metallic material during its plastic deformation are conditioned not only by its properties but also by the state of tension it is subjected to. In the deformation of the metal material between the rolling mill cylinders it is subjected to high compressive stress due to the action of the cylinders and superficial tangential stresses due to the friction between the cylinders and the metal. Friction forces are also the cause of the reduction in metallic material between the cylinders.

Usually, the problems of the content of modern theory are given in a complicated mathematical form, which is sometimes difficult to understand, although their main ideas are quite simple [5], [6].

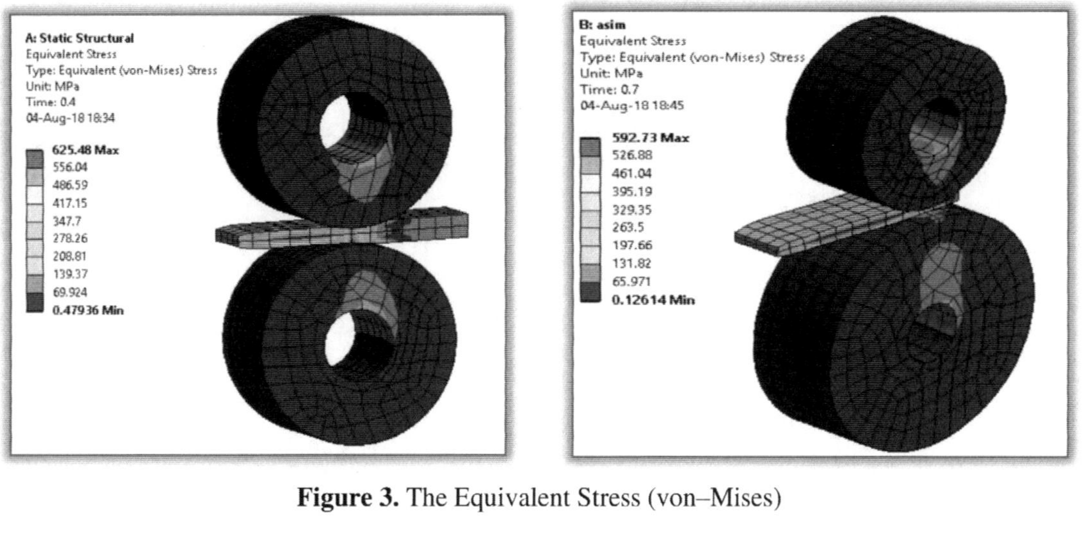

Figure 3. The Equivalent Stress (von–Mises)

IOP Conf. Series: Materials Science and Engineering **477** (2019) 012025 doi:10.1088/1757-899X/477/1/012025

For modelling and simulation we use ANSYS EXPLICIT DYNAMICS, it is a transient explicit Dynamics Workbench application that can perform a variety of engineering simulations, including the modelling of nonlinear dynamic behaviour of solids, fluids, gases and their interaction. A typical simulation consists of setting up the model, interactions and the applied loads, solving the model's nonlinear dynamic response over time for the loads and interactions, then examining the details of the response with a variety of available tools.

The relatively flat rolling of strips, in the general case, the geometric distortion is composed of two parts (Figure 4):

— the metal sliding portion on the cylinders, where the contact friction is subject to Coulomb's law; here are actually two portions, in the immediate vicinity of the geometric deformation zone, i.e., the inlet and outlet of the metallic material between the cylinders;
— the grip portion where the material slip on the surface of the cylinders is missing, i.e. the metal surface layer moves at a tangential velocity equal to the peripheral speed of the cylinders, or the particles of metallic material as if they stick to the surface of the cylinders.

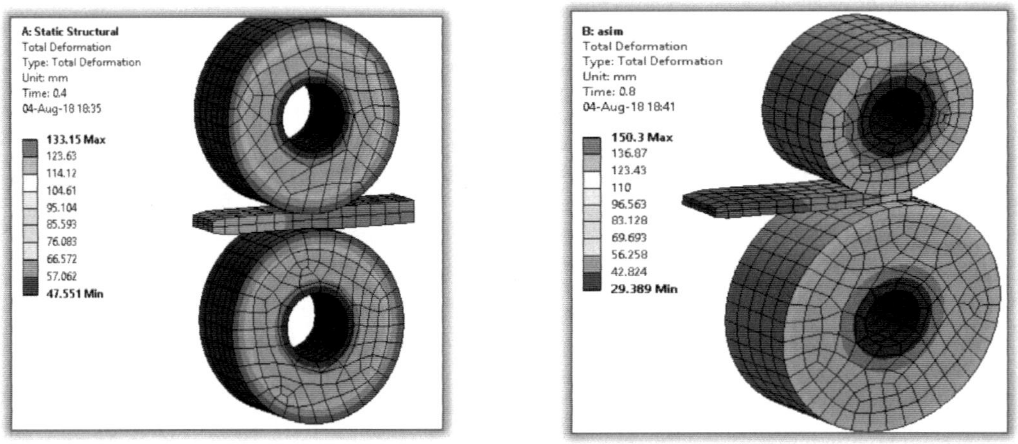

Figure 4. The total deformation

The typical diagram showing the unevenness of the deformation in the reduction of the outer and inner structures of the metallic material in the deformation area for the medium thickness laminates is given in Figure 5. As can be seen from the curve alloy shown in Figure 5, the metal layer on the roller contact surface (1–2) begins to thin, and the central layer (3) thickens until the geometric boundary of the plane of entry into the deformation zone.

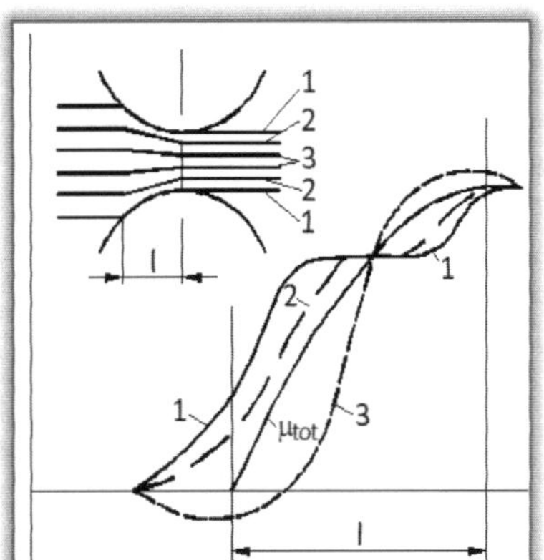

Figure 5. The graph of variation reductions elementary layers relative length of the deformation

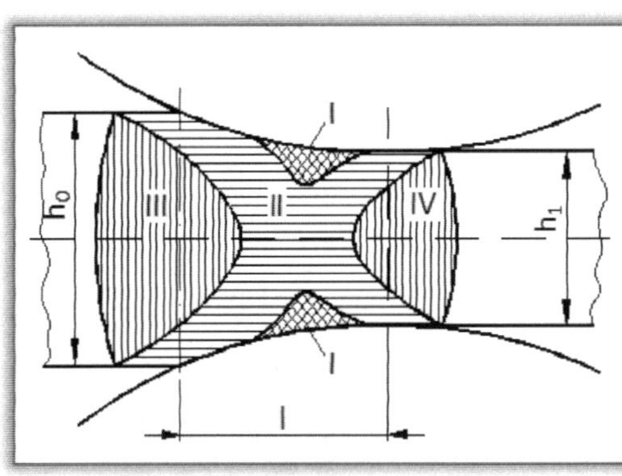

Figure 6. Diagram of distribution of plastic deformations in metallic material

The unevenness of the deformation as well as the length of the grip portion increases with the increase of the coefficient of friction and the thickness of the laminate (Figure 6):

- **I** – the area with difficult deformation;
- **II** – the range of plastic deformation by compression in height and stretching in the longitudinal direction;
- **III, IV** – the range of plastic deformation by longitudinal compression and height thickening.

Figures 7 and 8 show the elongation, respectively the width of the rolled strip according to the deformation direction:

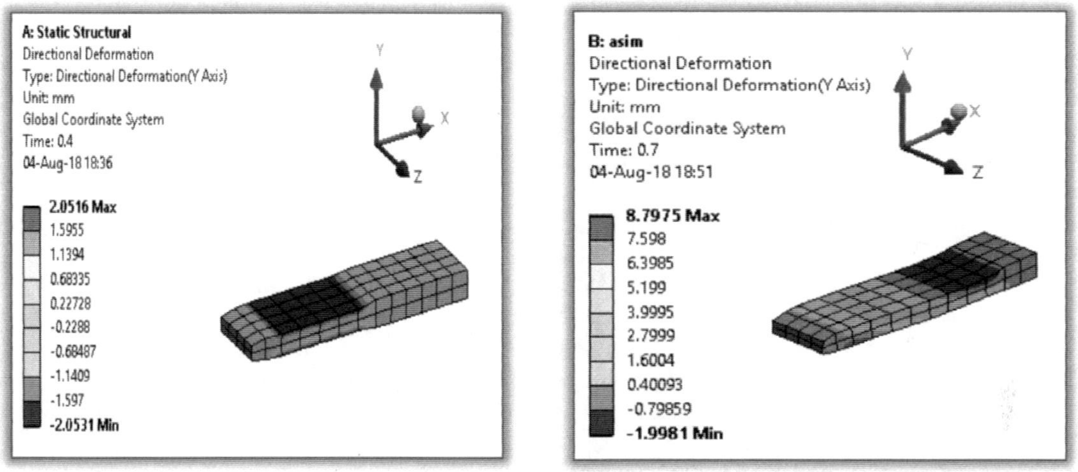

Figure 7. Directional deformation Y – axis

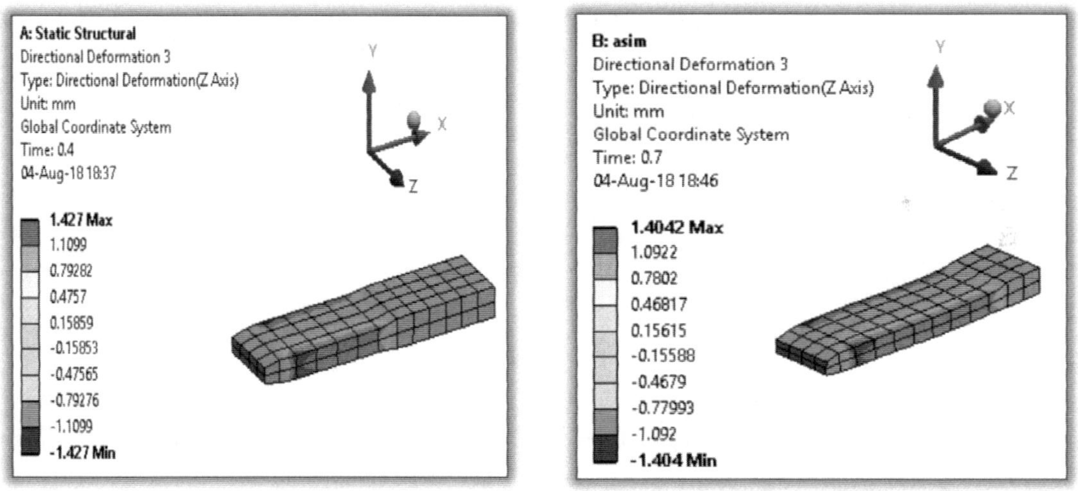

Figure 8. Directional deformation Z – axis

4. Conclusions

The stress state criteria depend only on the deformation method, the stress action character, and especially the friction conditions on the contact surfaces, i.e., for a given deformation method, dimensions and shapes of the deformation zone, stress state criteria do not depend on the fact that it deforms steel, copper or aluminium.

It should also be noted that some dependence of the stress state criteria on the nature of the material and its deformation conditions exists because the dependence of the external friction coefficient is known, depending on these factors, but this dependence is indirect and in the calculations thinks easy.

It has also been clarified by the present study the qualitative passage laws from the symmetrical to the asymmetric process, depending on the change in the lamination conditions, confirming the idea that the symmetric process is only a particular case of the asymmetric process that is widespread in the industry's rolling practice.

This interpretation of the functional dependence of the deformation resistance is based on the analysis of the physical content of the plasticity equation, allows for the division of the complicated dependence of the resistance to deformation towards the various factors in elementary parts, which considerably simplifies the study of the problem and allows for a more in– to obtain generalizations. In particular, it is possible to extend the results of research to any process to others, and vice versa, to generalize the data obtained by various methods of deformation.

As the linear deformation resistance and the stress state criteria influence the unitary pressure at the plastic deformation independently of each other, they can be studied separately. Therefore, the tendency to consider all the factors influencing both the state of tension of the respective material and its resistance to linear deformation (in particular, the extrusion in the rolling process) complicates without any justification the problem.

References

[1] Alexa V 2004 Analysis of the lateral efforts In asymmetrical longitudinal rolling, *Annals of the Faculty of Engineering–Hunedoara* **2** 45-50

[2] Alexa V, Ratiu S and Kiss I 2003 Calculation program for the determination of force parameters in rolling, *Annals of the Faculty of Engineering–Hunedoara* **3** 149-154

[3] Alexa V 2002 *Contributions to research asymmetric rolling process*, Politehnica University of Timişoara, Romania, Doctoral Thesis

[4] Alexa V, Ratiu S and Kiss I 2005 Theoretical and experimental analysis of the rolling strain in the deformation area, *Metalurgia International* **5** 20-28

[5] Kiefer T and Kugi A 2008 An Analytical Approach for Modelling Asymmetrical Hot Rolling of Heavy Plates, *Mathematical and Computer Modelling of Dynamical Systems* **14** 1-19

[6] Pesin A, Chukin M and Pustovoytov D 2015 Finite Element Analysis of Symmetric and Asymmetric Threeroll Rolling Process, *MATEX Web of Conference* **26** 03006

IOP Conf. Series: Materials Science and Engineering **477** (2019) 012026 doi:10.1088/1757-899X/477/1/012026

Fuzzy classification based on Combinative Algorithms

N A M Saffie[1], K A Rasmani[1] and N H Sulaiman[2]

[1]Faculty of Computer and Mathematical Sciences, Universiti Teknologi MARA Seremban Campus, 70300 Seremban, Negeri Sembilan, Malaysia
[2]Faculty of Computer and Mathematical Sciences Universiti Teknologi MARA, 40450 Shah Alam, Selangor, Malaysia

E-mail: khairulanwar@ns.uitm.edu.my

Abstract. Classification models are used in order to predict the class label of objects for which the class label is unknown. The performance of a single classification model can be determined on the basis of the classification accuracy. However, it is difficult to determine which single-model is the best classification technique in a specific application domain since a single learning algorithm may not uniformly outperform other algorithms over various datasets. Furthermore, most classification algorithms, using either fuzzy or non-fuzzy approaches, produce results in the form of crisp or categorical classification outcomes. Unfortunately, in certain applications, the classification outcomes that represent the class labels may involve categorisations which are fuzzy in nature. In this study, a new technique is developed to handle the classification task based on combinative algorithms by adopting multiple-model concepts. The aggregation process is based on a modified fuzzy Ordered Weighted Averaging (OWA) operator. The performance of the proposed technique has been analysed using the Domestic Water Consumption (DWC) dataset obtained from a previous study. The findings show that the proposed method consistently outperforms all selected single-models and performs better than a statistical multiple-model approach, known as the Pool Method. This result indicates that the proposed technique has the potential to assist decision-making, especially in the following situations: i) when it is difficult to decide which model is better for a specific classification problem and, ii) to decide selected unconfirmed cases.

1. Introduction

The classification model, also known as a classifier, is used in order to predict the class label of objects for which the class label is unknown. The performance of a single-model (classifier) can be determined on the basis of the classification accuracy. However, it is difficult to determine which single-model is the best classification technique in a specific application domain since single learning algorithms may not uniformly outperform other algorithms over various datasets.

The multiple-model concept that combines different single-models has been introduced by many researchers with the aim of improving prediction performance. In general, there are three mechanisms used to combine different single-models, which are: i) models created using different subsets of training data based on a single learning method, ii) models created using different training parameters based on a single training method, and iii) models created by combining different learning methods [1]. From the literature, it can be observed that little work has been focused on the classification technique based on the combination of different learning methods (termed as combinative algorithms in this study).

Moreover, most classification algorithms using either fuzzy or non-fuzzy approaches produce results in the form of crisp or categorical classification outcomes. Unfortunately, in certain application domains such as business administration and management science, the classification outcomes that represent class labels may involve categorisations which are fuzzy in nature. For example, the classification outcomes from the single algorithm can be in the form of several classes such as 'low', 'moderate' and 'high', which are obviously more suited to being represented in fuzzy sets. Presently, few techniques have been developed to combine the outcomes from several different algorithms which take into consideration the elements of fuzziness and uncertainties in the classification outcomes. Therefore, this study aimed to develop a new technique to handle the classification task based on combinative algorithms by adopting the multiple-model concepts and a well-known fuzzy aggregation for multiple criteria, namely fuzzy Ordered Weighted Averaging (OWA) operator [2].

The rest of the paper is planned as follows: Section 2 reviews the multiple-model concepts and fuzzy OWA operator. Section 3 describes the proposed fuzzy classification technique based on combinative algorithms. Section 4 presents the implementation of the proposed technique using the Domestic Water Consumption dataset and provides comparison of classification performance of the proposed technique with the performance of every participating single-model as well as with the performance of a multiple model based on the traditional statistical approach. Finally, the work is concluded in Section 5.

2. Background

This section describes the multiple-model concepts and fuzzy Ordered Weighted Averaging (OWA) operator.

2.1. Multiple-model concepts

Various studies show that multiple-models generally have similar purposes, which are to improve prediction over a single-model or to make the resulting ensemble more representative of the observed data, but they differ in their technique of combination. The simplest multiple-model is the classification model constructed on the basis of a statistical approach with the assumption that each participating method is equal in terms of weight. This approach is known as the Pool Method [3]. The multiple-model approach has been claimed to be better than the single-model in many aspects such as accuracy, consistency and reliability [3].

A simple probabilistic multiple-model, namely the Pool Method [3], [4], can be generated by simply pooling together the participating single-models, with all ensemble members having equal weight. The probabilistic multiple-model prediction under *k*th category, $y_k{}^{pool}$ is constructed by:

$$y_k{}^{pool} = \frac{\sum_{\eta=1} \tilde{m}_{\eta,k}}{\sum \tilde{M}_{\eta}} \tag{1}$$

where

$\tilde{m}_{\eta,k}$ = number of single - model under category *k*th,

$\tilde{M}_{\eta}$ = number of participating single - models.

2.2. Fuzzy Ordered Weighted Averaging (OWA) operator

Two extreme cases arise for aggregating criteria functions that have been considered important in order to perform an overall decision in many disciplines. The first case is when 'all the criteria' need to be satisfied and the other case is when 'at least one of the criteria' is satisfied. Fuzzy Ordered Weighted Averaging (OWA) operator [2] has been introduced as a technique to overcome the problem of multiple-criteria aggregation in decision making. The fuzzy OWA operators provide connections between the logical 'or' and 'and' for the aggregation process [5].

The fuzzy OWA operator of j dimension is defined as a mapping $\hat{F}: I^j \rightarrow I$ where $I = [0,1]$ that has associated weighting vector, $\hat{W}$ of dimension j

$$\hat{W}_J = [\hat{w}_1 \quad \hat{w}_2 \quad \cdots \quad \hat{w}_j]$$

with the following properties

i. $\hat{W}_J \in [0,1]$

ii. $\sum_{J=1}^{j} \hat{W}_J = 1$

such that

$$\hat{F}(a_1,...,a_j) = \sum_{J=1}^{j} \hat{W}_J R_J = \hat{w}_1 r_1 + ... + \hat{w}_j r_j \tag{2}$$

where R is an ordered argument vector consisting of the arguments of $\hat{F}$, where r_j is ranked accordingly to a general criterion $\hat{F}(a_J)$.

3. Proposed fuzzy classification based on combinative algorithms

This section describes the proposed fuzzy classification based on combinative algorithms. The two main concepts adopted to develop the proposed fuzzy classification technique are the multiple-model concepts [3] and fuzzy Ordered Weighted Averaging (OWA) operator [2]. The multiple-models concept is used to combine different learning methods, whereas the concept of fuzzy OWA operator is used to aggregate the fuzzy outcomes produced by each participating single-model. The proposed method will be used to perform classification tasks. The framework of the proposed fuzzy inference method for combinative algorithms is illustrated in Figure 1 and the process can be described as follows:

Step 1. Algorithm selection based on classification performance.

Let $\hat{A}_\delta = \{\hat{A}_1, \hat{A}_2, ..., \hat{A}_m\}$ be classification algorithms where m is the number of algorithms under study. The percentage of classification accuracy of each algorithm, $\Phi(\hat{A}_\delta)$ defined as

$$\Phi(\hat{A}_\delta) = \frac{\text{correctly classified instances}}{\text{total number of instances present in the dataset}} \times 100\% \tag{3}$$

Then, $\hat{A}_\lambda$ where $\lambda = 1,2,...,n$ such that n is the number of selected algorithms under study with $n \leq m$, representing the selected algorithm if its performance or its classification accuracy is greater than a predefined degree of confidence, ε (determined by expert of the study) such that $\Phi(\hat{A}_\delta) \geq \varepsilon$. Therefore, algorithms with $\Phi(\hat{A}_\delta) < \varepsilon$ will be omitted.

Step 2. Creation of fuzzy weight vector.

Fuzzy weight vector is based on the classification accuracy of the selected algorithms. The weight vector, W_λ of dimension λ has the same properties as the weighting vector in OWA operator [2] where

$$W_\lambda = [w_1 \quad w_2 \quad \cdots \quad w_\lambda \quad \cdots \quad w_n] \tag{4}$$

such that

$$W_\lambda \in (0,1) \text{ and } \sum_{\lambda=1}^{n} W_\lambda = 1.$$

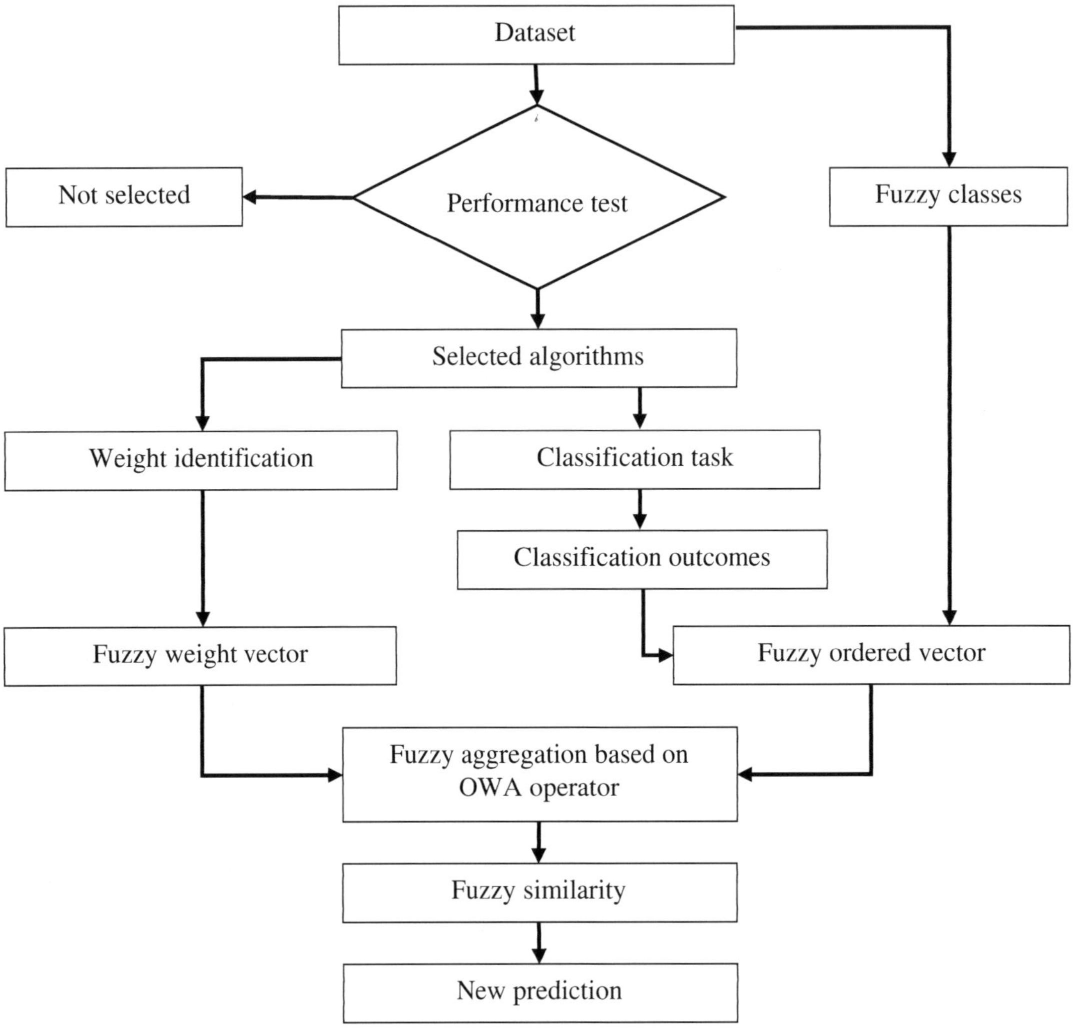

Figure 1. Methodological framework of fuzzy inference method for combinative algorithms

Step 3. Creation of fuzzy ordered vector.

A fuzzy ordered vector is created based on the relationship between dataset classes and classification outcomes of each selected algorithm, $\hat{A}_\lambda$. The fuzzy ordered vector, σ is defined as

$$\sigma = \begin{bmatrix} C(\hat{A}_1) \\ C(\hat{A}_2) \\ \vdots \\ C(\hat{A}_\lambda) \\ \vdots \\ C(\hat{A}_n) \end{bmatrix}, \tag{5}$$

where the fuzzy classification outcomes of each selected algorithm

$$C(\hat{A}_\lambda) \in \{C_t\},\ t = 1,2,\ldots,l,$$

are such that l is the number of dataset classes.

This process will transform the crisp categorical classes into fuzzy classes using linguistic terms in the form of normal Generalized Trapezoidal Fuzzy Number (GTFN). Each class can be represented as

$$C_t = \left(c_1^t, c_2^t, c_3^t, c_4^t\right), \tag{6}$$

where

$$C_1 \prec C_2 \prec \cdots \prec C_l,$$

with $\prec$ representing the ordering of fuzzy classes in ascending order.

Step 4. Aggregation of the fuzzy classification outcomes of selected algorithms.

The aggregated fuzzy classification outcomes, $F\left(\hat{A}_\lambda\right)$ are given as

$$F\left(\hat{A}_\lambda\right) = W_\lambda . \sigma = \left(f_1, f_2, f_3, f_4\right) \tag{7}$$

where W_λ is the fuzzy weight vector and σ is the fuzzy ordered vector.

Step 5. Determination of fuzzy class.

A fuzzy similarity measure will be used to determine the final classification outcomes. For the purpose of this research, the fuzzy similarity measure proposed in [6] will be employed. Other fuzzy similarity measures may be employed depending on the suitability of the technique with regards to the application domain. The similarity degrees will be calculated between the aggregated fuzzy classification outcomes, $F\left(\hat{A}_\lambda\right)$ and the fuzzy classes, C_t. The degree of similarity $S(F, C_t)$ is given by

$$S(F, C_t) = \left(1 - \frac{\sum_{i=1}^{4}\left|f_i - c_i^t\right|}{4}\right)\left(1 - \left|x_F - x_{C_t}\right|\right)^{B\left(S_F, S_{C_t}\right)} \left(1 - \frac{\left|h_F - h_{C_t}\right| + \frac{\left|y_F - y_{C_t}\right|}{\max(y_F, y_{C_t})} + \left|Ar(F) - Ar(C_t)\right| + \frac{\left|Pe(F) - Pe(C_t)\right|}{\max(Pe(F), Pe(C_t))}}{4}\right) \tag{8}$$

where

$$y_F = \begin{cases} \dfrac{w_F\left(\dfrac{f_3 - f_2}{f_4 - f_1} + 2\right)}{6} & if \quad f_1 \neq f_4 \\ \dfrac{w_F}{2} & if \quad f_1 = f_4 \end{cases}, \tag{9}$$

$$x_F = \begin{cases} \dfrac{y_F\left(f_3 + f_2\right) + \left(f_4 + f_1\right)\left(w_F - y_F\right)}{2w_F} & if \quad w_F \neq 0 \\ \dfrac{f_4 + f_1}{2} & if \quad w_F = 0 \end{cases}, \tag{10}$$

$$B\left(S_F, S_F\right) = \begin{cases} 1 & if \quad S_F + S_{C_t} > 0 \\ 0 & if \quad S_F + S_{C_t} = 0 \end{cases}, \tag{11}$$

such that $S_F = f_4 - f_1$ and $S_{C_t} = c_4{}^t - c_1{}^t$. The heights of normal GTFN is $h_F = 1$ and $h_{C_t} = 1$.

$$Ar(F) = \frac{1}{2} w_F (f_4 + f_3 - f_2 - f_1) \quad (12)$$

$$Pe(F) = \sqrt{(f_1 - f_2)^2 + w_F^2} + \sqrt{(f_3 - f_4)^2 + w_F^2} + (f_3 - f_2) + (f_4 - f_1) \quad (13)$$

4. Experiments and findings

A series of experiments was conducted to investigate the performance of the proposed method using the existing Domestic Water Consumption (DWC) dataset obtained from a previous study [7]. This dataset consists of 718 instances classified into two classes which are 'unlikely excessive' and 'likely excessive'. For the purpose of this research, the dataset has been reclassified into three classes which are 'unlikely excessive', 'likely excessive' and 'most likely excessive'.

Seven fuzzy classification algorithms were used for the experimentations. The performance of each single-model for the 3-class DWC dataset is presented in Table 1. The experiments were carried out with two different degrees of confidence: sets as $\varepsilon = 65\%$ and $\varepsilon = 70\%$, respectively for both Combination-1 and Combination-2.

Table 1. Performance of the fuzzy classification algorithms for 3-class DWC dataset

Algorithm	Performance (%)
$\hat{A}_1$	73.69
$\hat{A}_2$	75.63
$\hat{A}_3$	77.18
$\hat{A}_4$	79.27
$\hat{A}_5$	79.27
$\hat{A}_6$	81.22
$\hat{A}_7$	67.85

The 3-class dataset is labelled as C_1, C_2 and C_3, respectively to represent 'unlikely excessive', 'likely excessive' and 'most likely excessive'. Three sets of selected predefined fuzzy numbers for the 3-class dataset were used for the experimentation (Table 2).

Table 2. Selected predefined fuzzy numbers for the 3-class DWC dataset

Linguistic term for fuzzy class	Predefined fuzzy number		
	Set 1	Set 2	Set 3
Unlikely excessive	$C_1 = (0.000, 0.000, 0.045, 0.256)$	$C_1 = (0.000, 0.000, 0.210, 0.316)$	$C_1 = (0.000, 0.000, 0.045, 0.231)$
Likely excessive	$C_2 = (0.045, 0.210, 0.256, 0.843)$	$C_2 = (0.135, 0.210, 0.256, 0.316)$	$C_2 = (0.151, 0.231, 0.248, 0.271)$
Most likely excessive	$C_3 = (0.210, 0.843, 1.000, 1.000)$	$C_3 = (0.136, 0.248, 1.000, 1.000)$	$C_3 = (0.241, 0.323, 1.000, 1.000)$

The performance of single-models, the Pool Method and the proposed method using three sets of predefined fuzzy numbers were investigated under Combination-1 and Combination-2. The results of the experiments for Combination-1 and Combination-2 are presented in Table 3 and Table 4, respectively. Note that the results were obtained on the basis of 10-fold cross-validation methods.

Table 3. The performance of single-models, the Pool Method and the proposed method under Combination-1

Algorithm	Single-model	Pool Method	Proposed method Set 1	Set 2	Set 3
$\hat{A}_1$	0.7369				
$\hat{A}_2$	0.7563				
$\hat{A}_3$	0.7718				
$\hat{A}_4$	0.7927	0.8760	0.9053	0.8928	0.8886
$\hat{A}_5$	0.7927				
$\hat{A}_6$	0.8122				
$\hat{A}_7$	0.6785				

Table 4. The performance of single-models, the Pool Method and the proposed method under Combination-2

Algorithm	Single-model	Pool Method	Proposed method Set 1	Set 2	Set 3
$\hat{A}_1$	0.7369				
$\hat{A}_2$	0.7563				
$\hat{A}_3$	0.7718				
$\hat{A}_4$	0.7927	0.8649	0.8719	0.8747	0.8788
$\hat{A}_5$	0.7927				
$\hat{A}_6$	0.8122				
$\hat{A}_7$	-				

The first objective of this research is to compare the performance of the single-model with the performance of the proposed technique. Results from the analysis of classification performance show that the highest performance of a single-model under Combination-1 and Combination-2 is produced by $\hat{A}_6$ with classification performance of 0.8122. However, the performance of the combinative algorithms for both the proposed method and the Pool Method showed that the multiple-model concept, using either the fuzzy approach or the non-fuzzy approach, is better than the performance of each participating single-model. Hence, it can be concluded that the performance of the proposed technique is better than each participating single-model. The second objective is to compare the performance of the proposed technique with the probabilistic Pool Method. The results show that the Pool Method produced higher performance under Combination-1 with 0.8760 compared to Combination-2 with 0.8649. For Combination-1, the performance of the proposed method using Set 1, Set 2 and Set 3 is 0.9053, 0.8928 and 0.8886, respectively, whereas for Combination-2, the performance of the proposed method is 0.8719, 0.8747 and 0.8788, respectively. The findings clearly how that the performance of the proposed method for both Combination-1 and Combination-2 is higher than the performance of the Pool Method. These results indicate that the proposed method produced better performance compared with that of all selected single-models and the Pool Method.

5. Conclusion

This paper has presented a fuzzy classification for combinative algorithms by adopting the concepts of the multiple-model and the fuzzy Ordered Weighted Averaging (OWA) operator. It has been shown that the proposed combinative algorithms have been successfully developed and implemented. The findings of this research showed that the proposed method is better in terms of classification accuracy compared with the best single-model as well as the probabilistic Pool Method and hence can be used to improve prediction. Additionally, the finding also showed that the proposed method has the potential to be used for datasets with fuzzy classes such as the Domestic Water Consumption (DWC) dataset. These results indicate that the proposed technique has the potential to assist decision-making, for example, in the situation when it is difficult to decide which model is better for a specific classification problem or to decide selected unconfirmed cases decided by several single algorithms.

Acknowledgements
This research work is funded by the Ministry of Education, Malaysia under the Fundamental Research Grant Scheme (FRGS) with reference number 600-RMI/FRGS TD 5/3 (1/2015). The authors would like to thank the Institute of Research Management and Innovation (IRMI), Universiti Teknologi MARA, Malaysia.

References
[1] Kotsiantis S B, Zaharakis I and Pintelas P 2007 Supervised machine learning: A review of classification techniques, *Emerging Artificial Intelligence Applications in Computer Engineering* **160** 3-24
[2] Yager R R 1988 On ordered weighted averaging aggregation operators in multicriteria decisionmaking, *IEEE Transactions on Systems, Man, and Cybernetics* **18** 183-190
[3] Weigel A P, Liniger M A and Appenzeller C 2008 Can multi-model combination really enhance the prediction skill of probabilistic ensemble forecasts?, *Quarterly Journal of the Royal Meteorological Society* **134** 241-260
[4] Hagedorn R, Doblas-Reyes F J and Palmer T N 2005 The rationale behind the success of multi-model ensembles in seasonal forecasting – I. Basic concept, *Tellus* **57** 219–233
[5] Yager R R 1993 Families of OWA operators, *Fuzzy Sets and Systems* **59** 125-148
[6] Mat Saffie N A, Rasmani K A and Sulaiman N H 2017 Similarity measure for fuzzy number based on distances and geometric shape characteristics, presented at *The 3rd International Conference on Computing, Mathematics and Statistics 2017 (iCMS 2017)*, Langkawi Malaysia, 7-8 November
[7] Jalaluddin A S, Shahari N A and Rasmani K A 2016 Excessive water usage estimation based on domestic per capita water consumption and consumer data, *Advanced Science Letters* **22** 2959-2962

Using LabVIEW Linx for Creating 3d Objects

R Rob and C Panoiu

Politehnica University of Timisoara, Department of Electrical Engineering and Industrial Informatics, 5 Revolution Street, Hunedoara, 331128, Romania

E-mail: raluca.rob@fih.upt.ro

Abstract. LabVIEW Linx is a LabVIEW add-on for programming applications on external board devices like Arduino, Beagle Bone, Raspberry, etc. Present paper presents a hard-soft application which acquires in real time a 3-axis accelerometer using an Arduino Uno board. The 3d movement generates in real time a virtual object whose moving follows the accelerometer.

1. Introduction

Linx is an open source project by Digilent and is designed to make it easy to develop embedded applications using LabVIEW. LINX includes VIs for over 30 of the most common embedded sensors as well as hardware agnostic APIs for accessing peripherals like digital I/O, analog I/O, PWM, I2C, SPI, and UART.

Whether the user is remotely controlling a chipKIT or Arduino over USB/Serial, Ethernet or Wi-Fi, or deploying VIs to run on BeagleBone Black or Raspberry Pi 2/3, LINX and LabVIEW make it easy visualize the data to work with, debug the code, and create advanced embedded applications faster than ever before.

2. How Linx works

Linx is a hardware abstraction layer that allows user to have a single LabVIEW interface to a variety of different hardware devices. Depending on the device, it can be accessed in one of two ways: Remote or Local I/O.

2.1. Remote I/OA subsection. Figures

Remote I/O (Figure 1) works by running LabVIEW VIs on the development PC. In this configuration, the LINX VIs send commands to LINX firmware running on a device connected to the development computer by a USB, WiFi, or Ethernet connection. These commands communicate the firmware to perform pre-defined actions. Devices that use Remote I/O are Arduino, chipKITs, ESP8266s, and many others.

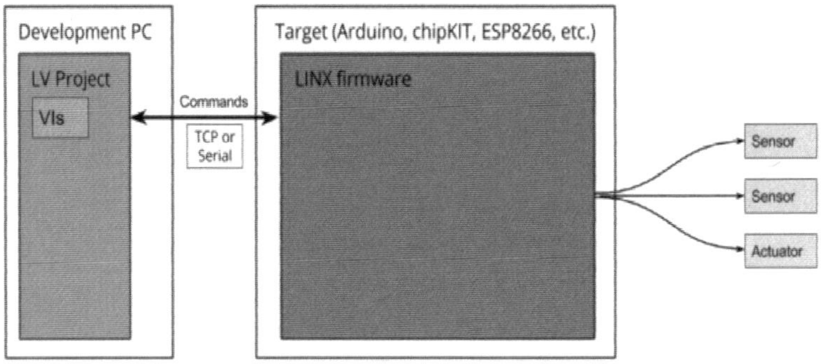

Figure 1. Remote I/O mode

2.2. *Local I/O*

Local I/O (Figure 2) is used on devices that can execute LabVIEW VIs directly without help from a development PC. Devices that use Local I/O are Linux-based devices like myRIO, Raspberry Pi 2/3, and BeagleBone Black.

In this configuration, VIs are written on the development PC, then deployed and run on the target via a TCP connection with the LabVIEW run-time engine running on the target. While the VIs are running on the target, debugging information and front panel values is transferred between the target and the development PC. When a VI is running on the target with live front panel data on the device PC, this is called Interactive Mode.

The LabVIEW run-time engine on the target runs inside a chroot, which is a Linux construct that is similar to a virtual machine. This allows the LV run-time to be installed and run safely on many different Linux operating systems. LabVIEW VIs running on the target utilize the LINX firmware, compiled as a Linux shared object, in order to access I/O on the target. The LINX shared object contains the logic necessary to access I/O on the specific target board.

2.3. *Local I/O with Startup Exacutable*

Local I/O (Figure 3) has a final mode it can be used in via a startup executable. As shown in the diagram above, a development PC is no longer needed in this configuration, since the startup executable is run automatically when the target boots.

The only difference in this mode is that the front panels are no longer visible since the development PC is not controlling the execution of the application's VIs.

This mode is "headless" in the sense that there is not a built-in user interface. Note that although startup executable are called "executable" they do not function like normal Linux programs and cannot be invoked from the command line.

Creation and deployment of the startup application are controlled from the development PC.

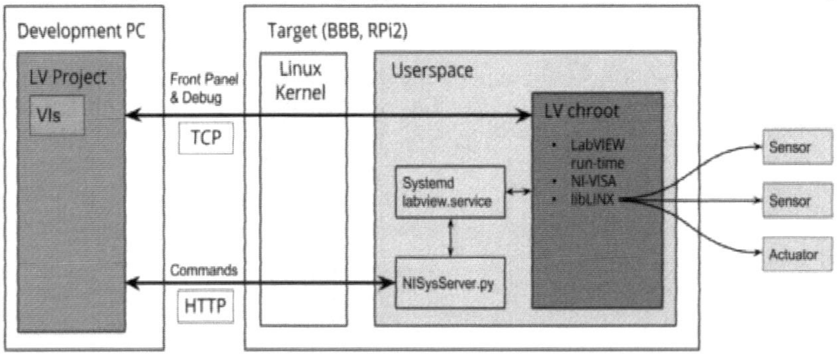

Figure 2. Local I/O mode

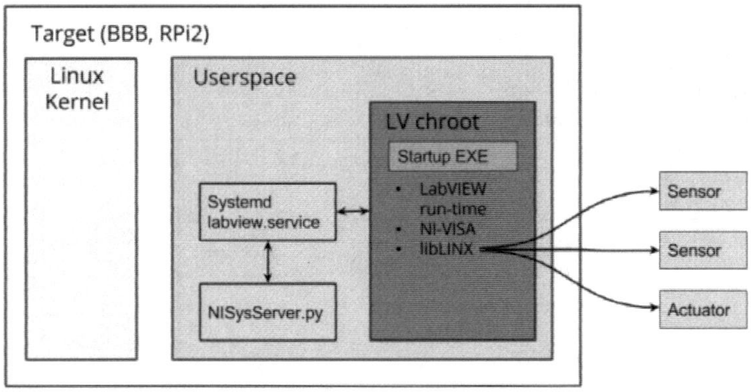

Figure 3. Local I/O with a Startup Executable

3. Using accelerometers in applications (ADXL 345)

Accelerometers are devices that measure acceleration, which is the variation of the velocity of an object. They measure in meters per second squared (m/s2) or in G-forces (g). A single G-force for us here on planet Earth is equivalent to 9.8 m/s2, but this does vary slightly with elevation. The accelerometer movements are presented in Figure 4.

The application which is presented in this paper uses 3-axis ADXL 345 accelerometer with high resolution (13-bit) measurement at up to ±16 g. Digital output data is formatted as 16-bit twos complement and is accessible through either a SPI (3- or 4-wire) or I2 C digital interface. The ADXL345 is well suited for mobile device applications. It measures the static acceleration of gravity in tilt-sensing applications, as well as dynamic acceleration resulting from motion or shock. Its high resolution (4 mg/LSB) enables measurement of inclination changes less than 1.0°. In Figures 5a and 5b the ADXL 345 accelerometer is presented.

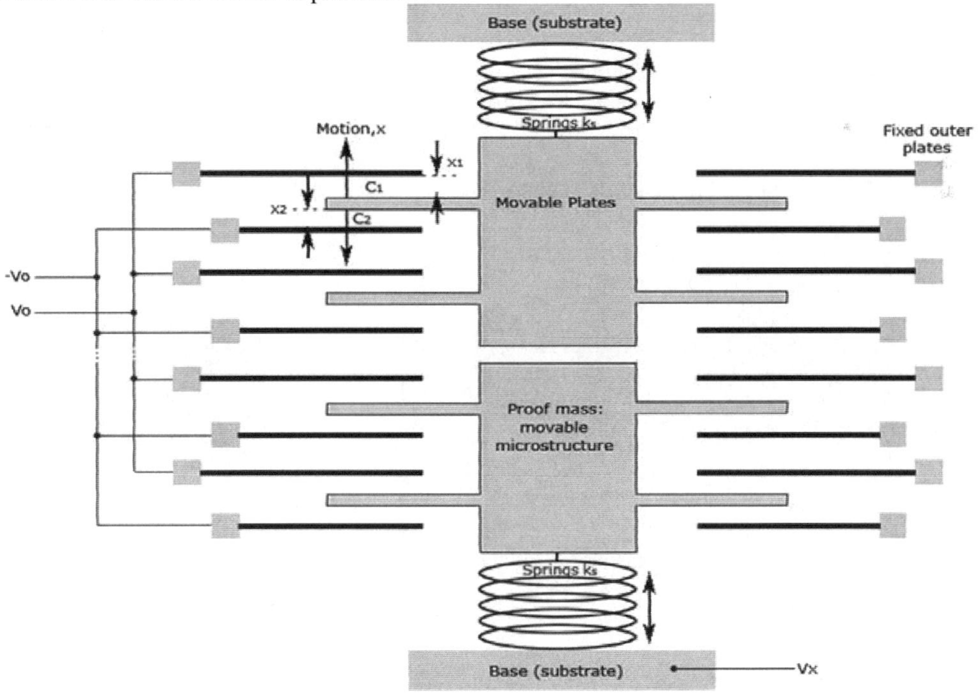

Figure 4. Accelerometer movements

Figure 5a. ADXL 345 accelerometer – Digilent device

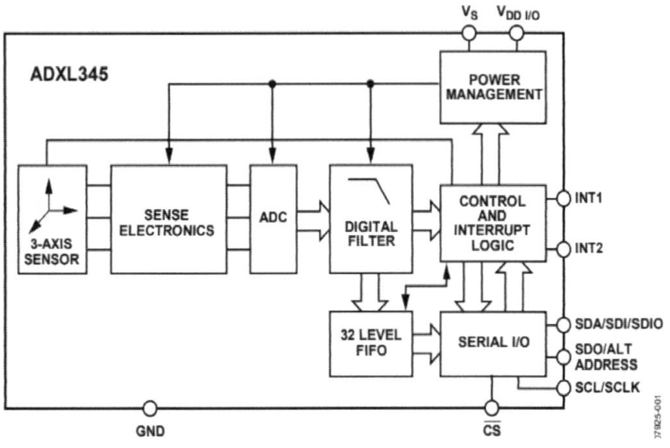

Figure 5b. ADXL 345 accelerometer – Block scheme

4. Application description

Arduino permits to connect the accelerometer to the I2C bus (Figure 6). Using the appropriate firmware, the code can be loaded into the Arduino microcontroller [1]. The code is programmed in LabVIEW Linx. The source code for acquiring accelerometer signals is presented in Figure 7. It contains specified Linx subroutines.

The subroutines which ensure the communication with the devices connected to the I2C bus are the followings:

-Linx_OpenSerial.vi ensures opening the session for serial port communication by specifying the port for loading the program into the microcontroller and also the real time communication with external devices,

-Linx_I2C_Open.vi is presented in Figure 8. It opens the communication session on I2C bus. The slave adress of the accelerometer is 1Dh. The I2C channel is 0 by default due to the fact that the Arduino Uno has one I2C channel [2].

Figure 6. Arduino connection

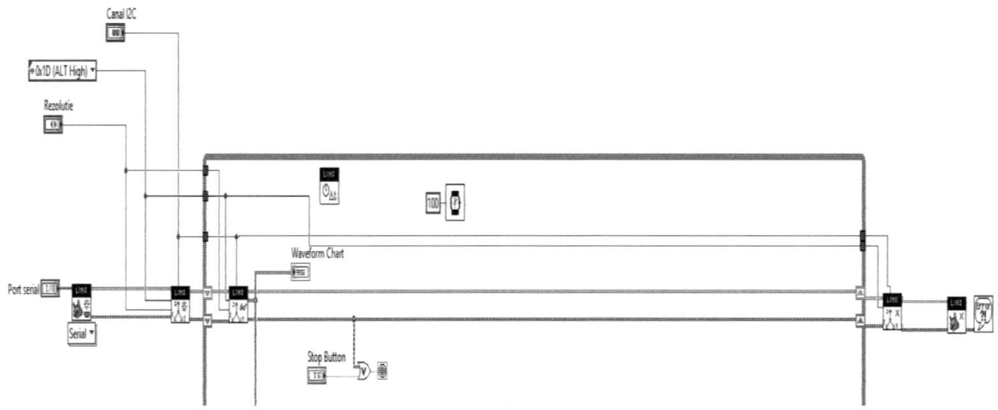

Figure 7. Source code for the accelerometer signals acquisition

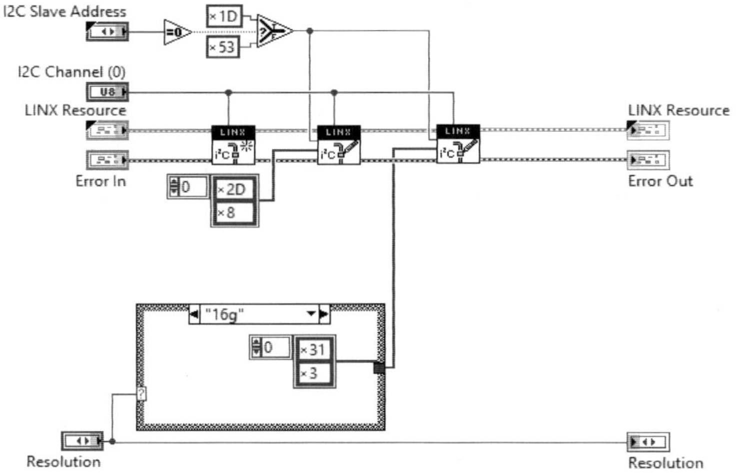

Figure 8. Linx_I2C_Open.vi

-Linx_I2C_Write.vi ensures data transmission at specified addresses [3]. Resolution selection implies using of a case structure with the specified addresses.

-Linx_I2C_Read.vi reads the information provided by accelerometer (Figure 9). This subroutine uses the accelerometer resolution and the used I2C bus.

-Linx_I2C_Close.vi closes I2C session.

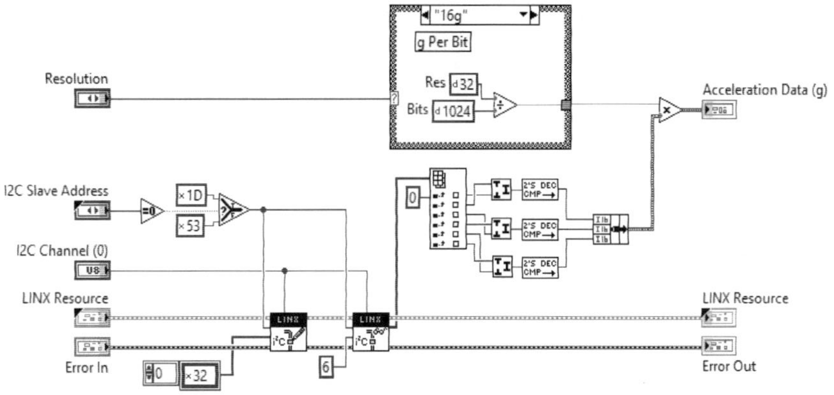

Figure 8. Linx_I2C_Read.vi

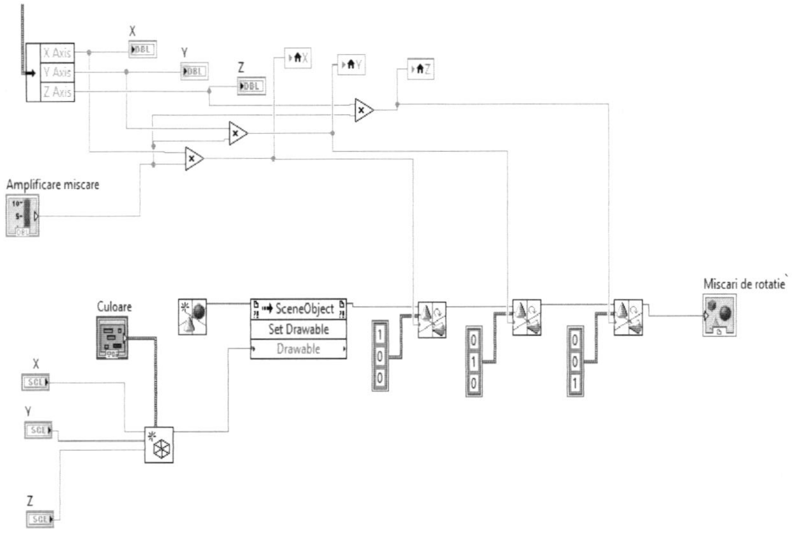

Figure 9. Source code for rotation movement generation

The application designed in this way is compiled and loaded into the microcontroller using Linx Firmware Wizard which offers to the used a wide range of embedded devices that can be connected to LabVIEW Linx.

The application generates 3d cuboid objects with different colour which follows in real time the accelerometer movements and transforms them into rotation and translation movements of the virtual objects. The source code for rotation movement generation is presented in Figure 9. The displacements on three axes of the accelerometer are acquired using three local variables named X, Y and Z. Using the specified subroutines Create_Object.vi and Create_Box.vi, the application generates movements around each axis also using Rotate_Object.vi subroutine. With Property_Node option, these properties are added in series. A case structure is used for generation of the translation movement. The first case generates a synchronous movement of the all cuboid objects (Figure 10a). The second case generates an asynchronous movement of the all cuboid objects (Figure 10b), each one is accomplishing a translation along one axis. Figure 11 presents the 3d object moving in real time with the accelerometer movements [4], [5].

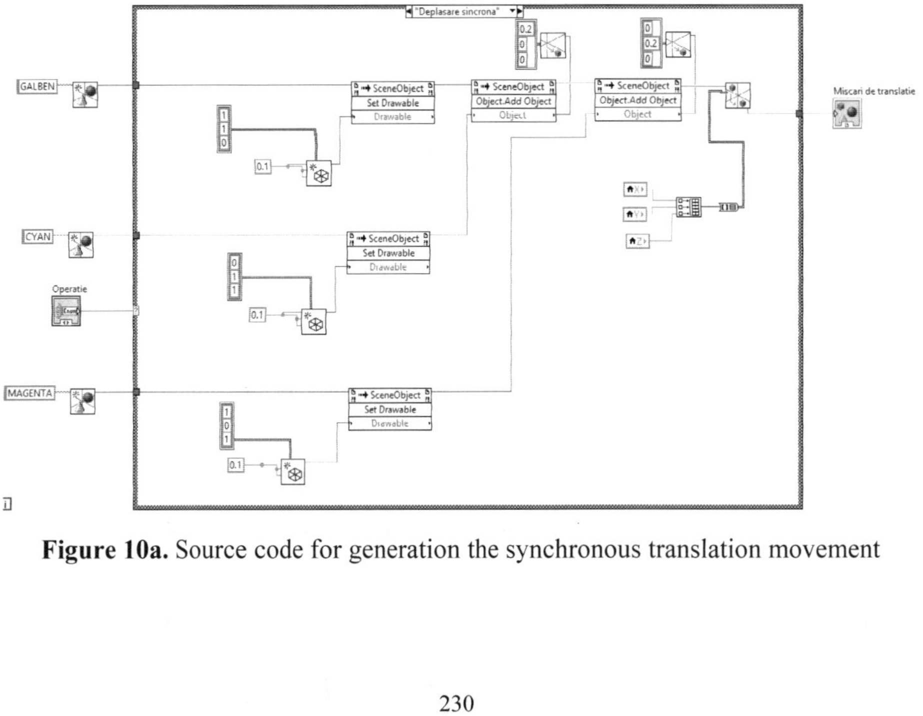

Figure 10a. Source code for generation the synchronous translation movement

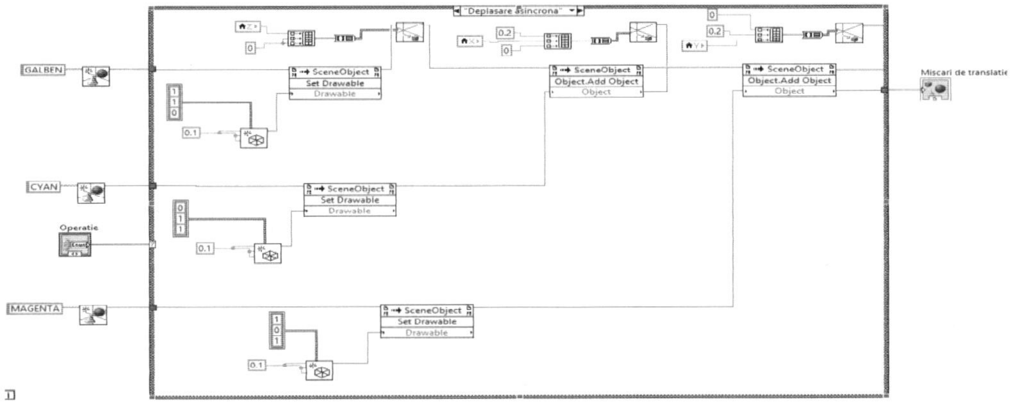

Figure 10b. Source code for generation the asynchronous translation movement

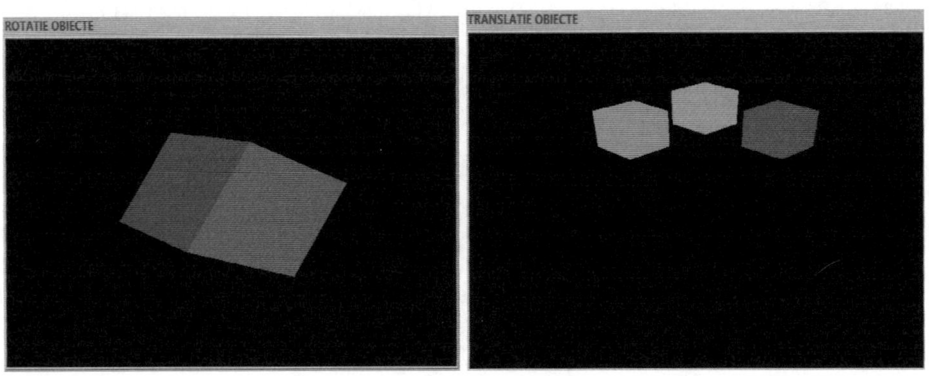

Figure 11. 3D object movement

5. Conclusions

The application presented in this paper presents a new method for controlling in real time external devices using embedded platform with LabVIEW. Arduino board becomes a simple extension board used for acquiring external signals.

References

[1] Changalasetty S B, Badawi A S and Ghribi W 2014 *Identification and feature extraction of moving vehicles in LabVIEW*, International Conference on Communication and Signal Processing, 3-5 April, Melmaruvathur, Indiapp 1885-1889

[2] Elya M N, Mohd Noor S B, Ribhan Zafira A R and Azrad S 2017 *Implementation of extendedhigh-gain observer in low-cost otitrack motion system for UAV control*, IEEE 15th Student Conference on Research and Development (SCOReD), 13-14 December, Putrajaya, Malaysia, pp 29-34

[3] Zhen Z, Yang Z, Yuchou L, Youjie Y and Xurui L 2017 *IP camera based LSPIV system for on-line monitoring of river flow* 13th IEEE International Conference on Electronic Measurement & Instruments (ICEMI), 20-22 October, Yangzhou, China, pp 357-363

[4] Dinis C, Popa G N and Iagar A 2017 *Automation heating and pumping for water using Zelio PLC* 10th International Symposium on Advanced Topics in Electrical Engineering, 23-25 March, Bucharest, Romania, pp 545-550

[5] Cunţan C D, Baciu I and Osaci M 2015 Operational Study of a Frequency Converter with a Control Sequence Using Xilinx Software, *Acta Polytechnica Hungarica* **12**(6) 201-212

The dependence of ellipsometric parameters Δ and Ψ on refractive index of superficial film

A D Berdie[1], A A Berdie[2] and S Jitian[1]

[1]Politehnica University of Timisoara, Department of Electrical Engineering and Industrial Informatics, 5 Revolutiei Str., Hunedoara, 331128, Romania
[2]"Lucian Blaga" University, Faculty of Letters and Arts, 5-7 Bd. Victoriei, Sibiu, 550024, România

E-mail: jitian s@yahoo.com

Abstract. The ellipsometrical analysis of external specular reflection of light on non-absorbing superficial films allows us to know factors which influence the ellipsometric measurement of the analyzed system. For optical non-absorbing superficial films the curves $\Delta = f(\Psi)$ are closed, the curves $\Delta = f(d_f)$ and $\Psi = f(d_f)$ are periodical, while the curves $\Delta = f(n_f)$ and $\Psi = f(n_f)$ are pseudo-periodic. Observations on the dependence of the ellipsometric parameters Δ and Ψ on n_f are presented.

1. Introduction

A change in the light polarization state occurs in the case of external reflection ($n_f > n_0$) of a monochromatic light radiation on a homogeneous and isotropic, optical non-absorbing, superficial film deposited on a solid reflector support.

The model of external reflection of a monochromatic radiation on a non-absorbing optical superficial film is schematically represented in Figure 1.

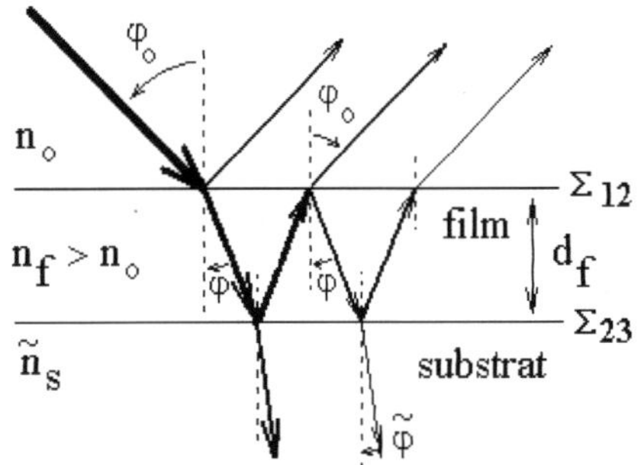

Figure 1. The model of the external reflection of a plan-polarized monochromatic radiation on the substrate-superficial film system

The change in the polarization state of the reflected radiation can be measured by ellipsometry [1]. The measurable ellipsometric parameters are Δ and Ψ. The first parameter Δ expresses the phase shift of the electrical component of the light radiation parallel to the plane of incidence relative to the perpendicular component on the incidence plane, after reflection and tanΨ represents the relative attenuation of the two components after reflection [2-6].

$$\tilde{\rho} = \tan\Psi \cdot \exp(i\Delta) = \frac{\tilde{E}_p^{(r)}}{\tilde{E}_p^{(i)}} \cdot \frac{\tilde{E}_s^{(i)}}{\tilde{E}_s^{(r)}} = \frac{\tan\psi^{(r)} \cdot \exp(i\delta)^{(r)}}{\tan\psi^{(i)} \cdot \exp(i\delta)^{(i)}} = \frac{E_p^{(r)}}{E_s^{(r)}} \cdot \frac{E_s^{(i)}}{E_p^{(i)}} \cdot \exp\left[i\left(\delta_p^r - \delta_s^r - \delta_p^i + \delta_s^r\right)\right] \quad (1)$$

The complex size $\tilde{\rho}$ can be calculated according to the ellipsometric parameters Δ and Ψ by the relation:

$$\tilde{\rho} = \tan\Psi \cdot \exp(i\Delta) = \tan\Psi\cos\Delta + i \cdot \tan\Psi\sin\Delta = \mathrm{Re}\,\tilde{\rho} + i \cdot \mathrm{Im}\,\tilde{\rho} = \frac{\tilde{R}_p}{\tilde{R}_s} \quad (2)$$

taking into account the reflections and therefore the attenuations and phase shifts thus produced on all interfaces.

The complex parameters Δ and Ψ are expressed according to the Fresnel complex reflection coefficients corresponding to the reflections on the two interfaces Σ_{12} and Σ_{23}: $\tilde{r}_p^{12}$, $\tilde{r}_p^{23}$, $\tilde{r}_s^{12}$, $\tilde{r}_s^{23}$ [7], [8].

$$\tilde{R}_p = \frac{\tilde{r}_p^{12} + \tilde{r}_p^{23}\exp\tilde{D}}{1 + \tilde{r}_p^{12} \cdot \tilde{r}_p^{23}\exp\tilde{D}} \text{ and } \tilde{R}_s = \frac{\tilde{r}_s^{12} + \tilde{r}_s^{23}\exp\tilde{D}}{1 + \tilde{r}_s^{12} \cdot \tilde{r}_s^{23}\exp\tilde{D}} \quad (3)$$

when:

$$\tilde{D} = -\frac{4\pi i d_f}{\lambda}\sqrt{n_f^2 - n_0^2\sin^2\varphi_0} \quad (4)$$

The expression $\exp\tilde{D}$ is a periodic function which depends on: the refractive index of the superficial film n_f, its thickness d_f, the incidence angle φ_0 and the wavelength λ of the incident radiation [3].

The complex parameters: $\tilde{R}_p$, $\tilde{R}_s$ and $\tilde{\rho}$ should be repeated after particular values of the film thickness d_f or the refractive index n_f for which $\tilde{D}$ is a multiple of $2\pi i$.

$$\tilde{D} = -\frac{4\pi i d_f}{\lambda}\sqrt{n_f^2 - n_0^2\sin^2\varphi_0} = m \cdot 2\pi i,\ \text{for } m = 1,2,3,\ldots \quad (5)$$

The value of film thickness for which the exponential function $\exp\tilde{D}$ is repeated is d_{min}.

$$d_f = m\frac{\lambda}{2\sqrt{n_f^2 - n_0^2\sin^2\varphi_0}} = m \cdot d_{min};\ d_{min} = \frac{\lambda}{2\sqrt{n_f^2 - n_0^2\sin^2\varphi_0}};\ m = 1,2,3,\ldots \quad (6)$$

For thicker films, the thickness shall be expressed by the relationship:

$$d_f = d_0 + m \cdot d_{min};\ m = 1,2,3,\ldots \quad (7)$$

where d_0 is the thickness, determined by ellipsometrical measurement (which is called standard solution) and m is an integer [9].

For example, if $\Delta = 186.37^0$ and $\Psi = 51.317^0$, the film thickness with $n_f = 1.4$ can be 120nm or $d_f = (120 + m \cdot 256.4)$ nm; m = 1, 2, 3,... [8].

2. The dependence of ellipsometric parameters Δ and Ψ on refractive index of superficial film

The ellipsometric parameters Δ and Ψ can be expressed with the relations:

$$\Delta = \mathrm{arctg}\frac{\mathrm{Im}\,\tilde{\rho}}{\mathrm{Re}\,\tilde{\rho}};\ \Psi = \mathrm{arctg}\left|\tilde{\rho}\right| \quad (8)$$

Both parameters are periodic functions that depend on the thickness of the surface film and its refractive index, as shown in Figure 2.

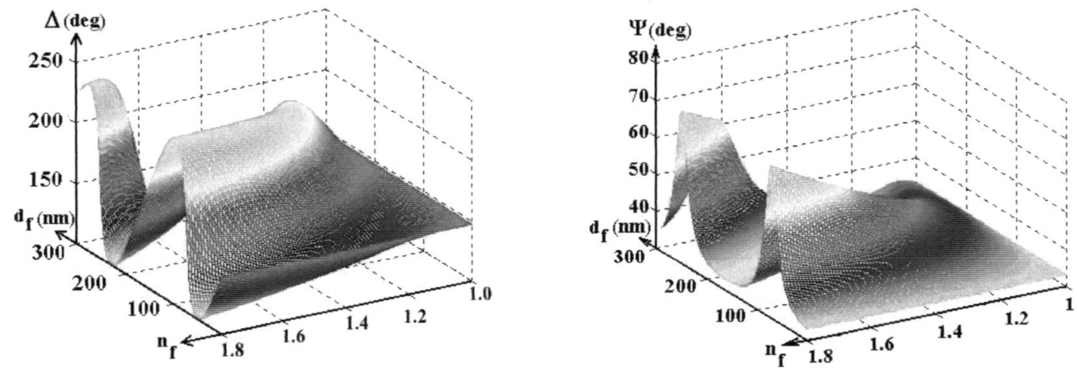

Figure 2. The dependence of Δ and Ψ on the refractive index and superficial film thickness ($\tilde{n}_s$=2.67(1-1.18i); φ_0=60^0; λ=562.5 nm)

The ellipsometric parameters Δ and Ψ are periodic functions toward the refractive index of the surface film, with the same period. Therefore, the curves Δ = f (Ψ) shown in Figure 3 are closed curves [10], [11].

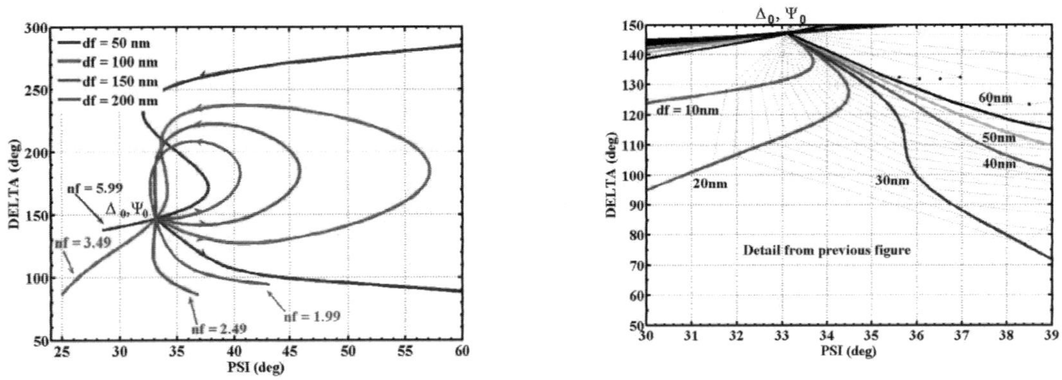

Figure 3. The plots Δ = f(Ψ) for superficial films with n_f = 1.2 ÷ 2.5, deposited on a metal substrate. $\tilde{n}_s$=2.67(1-1.18i); φ_0=60^0; λ=562.5 nm; Δ_0=147.118^0 and Ψ_0=33.116^0

Δ_0 and Ψ_0 are the ellipsometric parameters measured for the film-free substrate.

The curves are more scattered as the film is thinner, so that n_f can be determined with better precision.

The complex parameters: Δ and Ψ are repeated after a certain value of the refractive index n_f of the film for which $\tilde{D}$ is an integer multiple of 2πi, that is, when the refractive index of the superficial film fulfills the condition:

$$n_f = \sqrt{(n_0 \sin\varphi_0)^2 + \left(m\frac{\lambda}{2d_f}\right)^2}, \text{ for } m = 1,2,3,\ldots \quad (9)$$

The curves Δ=f(n_f) and Ψ=f(n_f) are pseudo-periodic for this reason, as can be seen from Figures 4 and 5 [9].

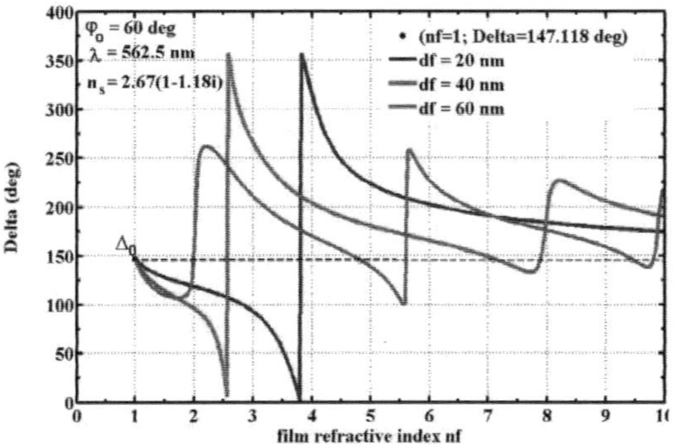

a) thin films

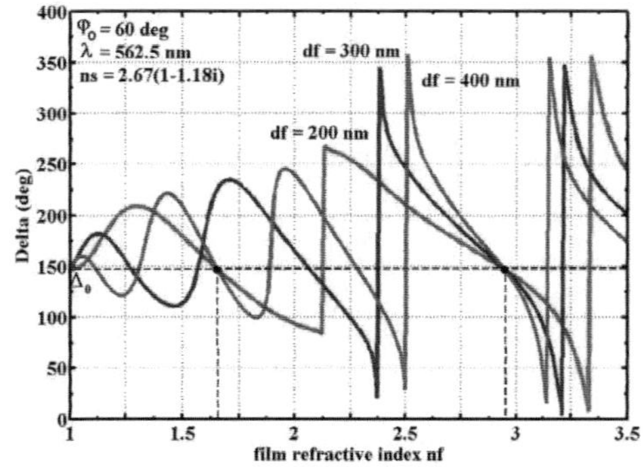

b) thick films

Figure 4. The dependence of Δ on the superficial film refractive index n_f

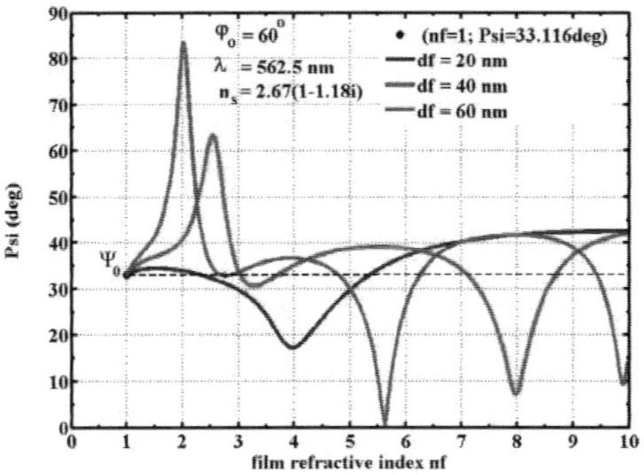

a) thin films

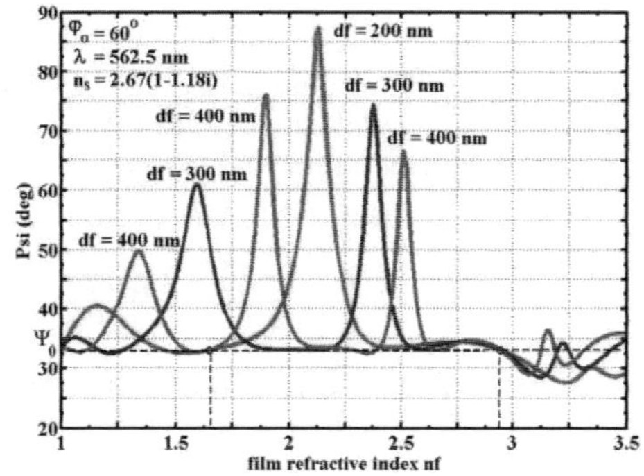

b) thick films

Figure 5. The dependence of Ψ on the superficial film refractive index n_f

All curves Δ=f(n_f) or Ψ=f(n_f) pass through the same point (Δ_0 = 147.12^0 ; Ψ_0 = 33.116^0 ; n_{f0} = 2.942) for the superficial film thickness an integer multiple of 100 nm (d_f = m·100 nm; where m = 1,2,3…. ; d_0 = 100nm) as seen in Figures 4b and 5b. This point corresponds to n_{f0} = 2.9425. Also, all Δ=f(n_f) or Ψ=f(n_f) curves pass through the same point (Δ_0 = 147.12^0 ; Ψ_0 = 33.116^0 ; n_{f0} = 1.65153) for the superficial film thickness an integer multiple of 200 nm (d_f = m·200 nm; where m = 1, 2, 3…. ; d_0 = 200nm). This point corresponds to n_{f0} = 1.6515. If in relation (9) the condition is that m = 1 then we obtain:

$$d_0 = \frac{\lambda}{2\sqrt{n_f^2 - n_0^2 \sin^2 \varphi_0}} \tag{10}$$

Values of n_{f0} for different values of d_0 are shown in Table 1.

Table 1. Values of n_{f0} for different values of d_0

φ	(deg)			60			
λ	(nm)			562.5			
d_0	(nm)	350	300	250	200	150	100
n_{f0}	-	1.18	1.28	1.48	1.65	2.06	2.94

If the thickness of the surface film is small compared to the wavelength of the light radiation (tens of nanometers), then equation (9) can be approximated as:

$$n_f \cong m\frac{\lambda}{2d_f} = m \cdot n_{f\,min}; \quad m = 1,2,3,.... \qquad (11)$$

and the curves Δ=f(n_f) and Ψ=f(n_f) are periodic. The value of $n_{f\,min}$ from equation (11) is all the greater as d_f is smaller, as shown in Figures 4a and 5a. For nano-sized film, the value of $n_{f\,min}$ is in the order of hundreds.

The curves Δ=f(n_f) and Ψ=f(n_f) are pseudo-periodic for films with thicknesses comparable to wavelengths.

Figures 4 and 5 show that there may be a great uncertainty in determining d_f. For example all films with multiple thicknesses of 200 nm and n_f = 1.65 have the same Δ and Ψ.

3. The choosing experimental conditions in an ellipsometric measurement in order to obtain minimal errors in determining the refractive indices of superficial films

The ellipsometric parameters Δ and Ψ may have greater or smaller variations with the change of the refractive index n_f of the superficial film depending on: the thickness d_f of the film, the angle of incidence φ_0 during the ellipsometric measurement or the type of the substrate on which it is located (expressed by the value of the refractive index ns of the substrate). The analysis of how Δ and Ψ vary when changing these parameters allows their optimal choice to obtain minimal errors in determining the refractive index of the superficial film [12].

This paper presents an evaluation of errors in the determination of the refractive index n_f of superficial films due to experimental errors of measurement of Δ and Ψ.

3.1. Films with thicknesses in the order of magnitude of the wavelengths

The precision in determining the refractive index n_f of the superficial film is even better as the thickness of the film is larger, as can be seen from Figures 4 and 5. The variation of Δ and Ψ versus the refractive index n_f of the superficial film is higher. Thus, for films with a thickness of 100 nm the period of variation of n_f is 2.8 and for films with a thickness of 1000 nm the period of variation of n_f is 0.26. For films with thicknesses of 300 nm or 500 nm the variations of Δ and Ψ versus n_f are quite large for superficial films with n_f = 1.55÷1.65 as shown in Figure 6. At 200 nm and 400 nm the curves Δ = f(n_f) show a minimum around n_f = 1.65. The variation of Ψ with n_f is very small so that errors in the measurement of Ψ cause great errors in the determination of n_f.

In order to see how much is influenced the determination of n_f by the experimental errors in the measurement of Δ and Ψ the derivatives of the refractive index n_f of the superficial film according to the ellipsometric parameters Δ and Ψ are analyzed. For large variations of Δ and Ψ versus the refractive index n_f of the superficial film, small errors are obtained in the determination of the refractive index.

The derivatives of refractive index n_f of the superficial film according to Δ and Ψ are shown in Figure 7.

In determining the refractive index n_f of the superficial film, for some film thicknesses and for certain refractive index values, large errors are obtained. If errors in determining the refractive index n_f of the superficial film are too high, some experimental conditions, such as incidence angle or incident wavelength, may be changed in order to obtain lower errors.

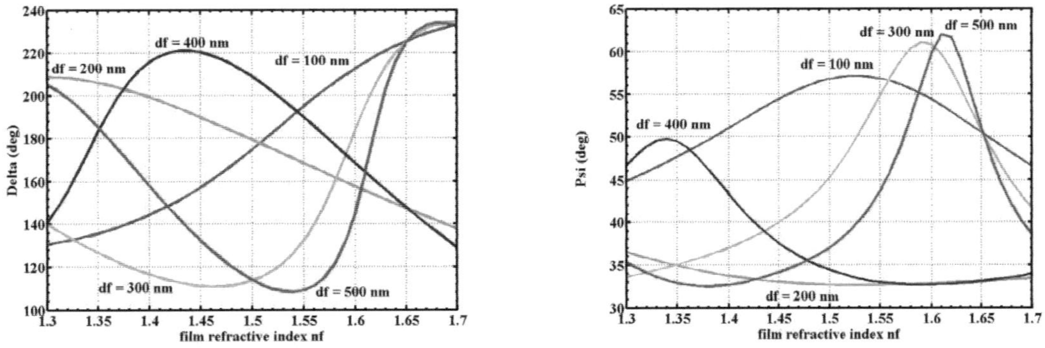

Figure 6. The variation of Δ and Ψ values for the refractive index n_f of the superficial film in the range $n_f = 1.3 \div 1.7$; ($\tilde{n}_s$ =2.67(1-1.18i); φ_0=60^0; λ=562.5 nm)

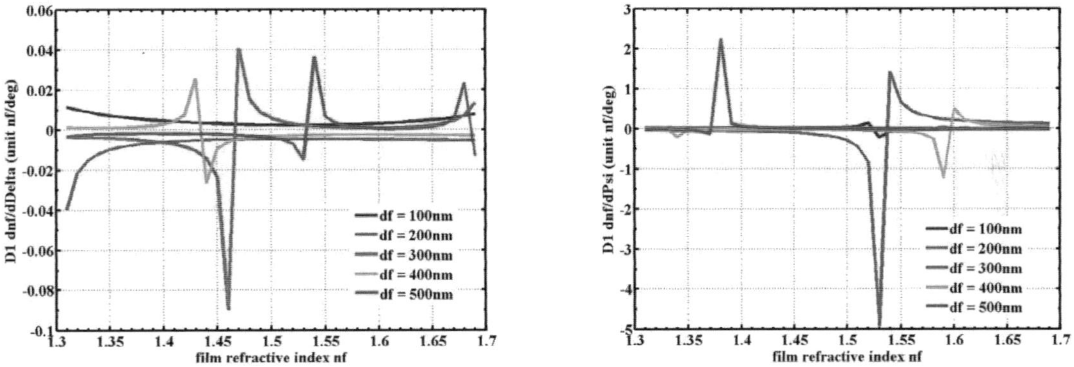

Figure 7. The first derivative of refractive index n_f of the superficial film according to Δ and Ψ; $\tilde{n}_s$ =2.67(1-1.18i); φ_0=60^0; λ=562.5 nm

3.2. Thin film thicknesses of the nanometer size order

In the range of the refractive index values of usual materials n_f = 1.3÷1.7, the variation of Δ and Ψ with the refractive index is small so that the experimental errors of the Δ and Ψ produce quite high errors in the appreciation of the n_f or d_f values.

Variations of Δ and Ψ versus the refractive index of the superficial film, for thin films are shown in Figure 8.

The first derivatives of the refractive index n_f of the superficial film versus Δ or Ψ for films with thickness of the order of magnitude of nanometers or tens of nanometers are shown in Figure 9.

For thin films with thicknesses in the order of nanometer tens, there is an error of 0.01 in the determination of n_f when the error in the measurement of Δ is 0.2 degrees. Large errors can occur due to the experimental measurement of the ellipsometric parameter Ψ. In this case, some experimental conditions such as incidence angle or wavelength of incidence radiation should be modified.

For nanometer-sized film thicknesses, the same experimental measurement errors of Δ and Ψ will produce errors in the determination of n_f approximately 10 times higher, since the variation of the ellipsometric parameters Δ and Ψ with the refractive index of the superficial film is smaller.

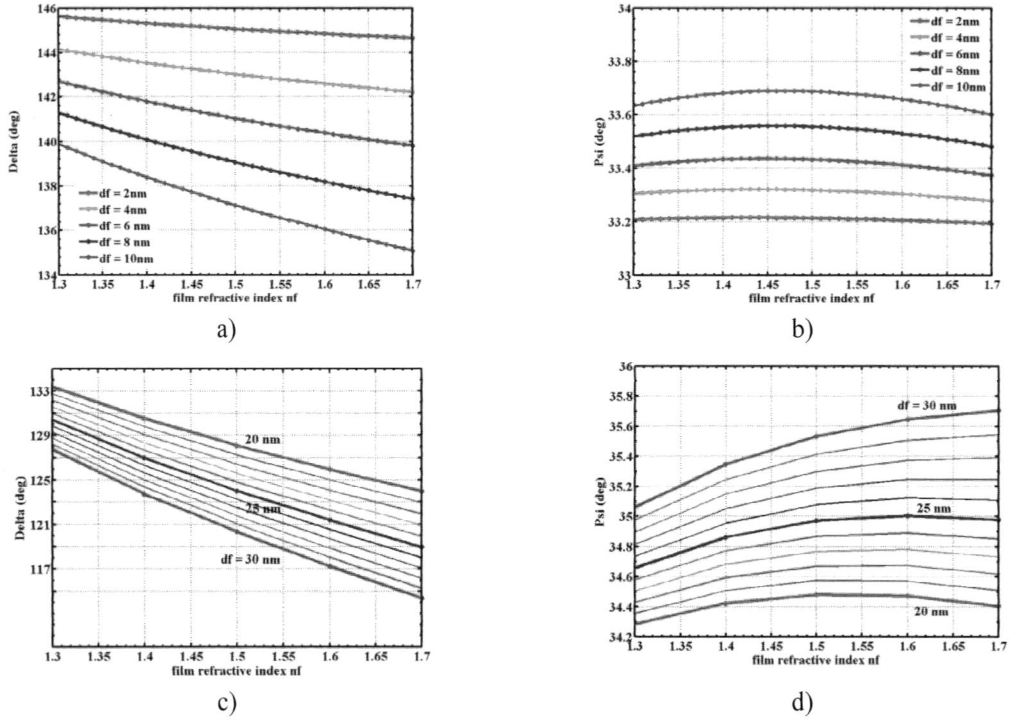

Figure 8. The variation of Δ and Ψ values for the refractive index n_f of the superficial film in the range $n_f = 1.3 \div 1.7$; ($\tilde{n}_s = 2.67(1\text{-}1.18i)$; $\varphi_0 = 60^0$; $\lambda = 562.5$ nm)

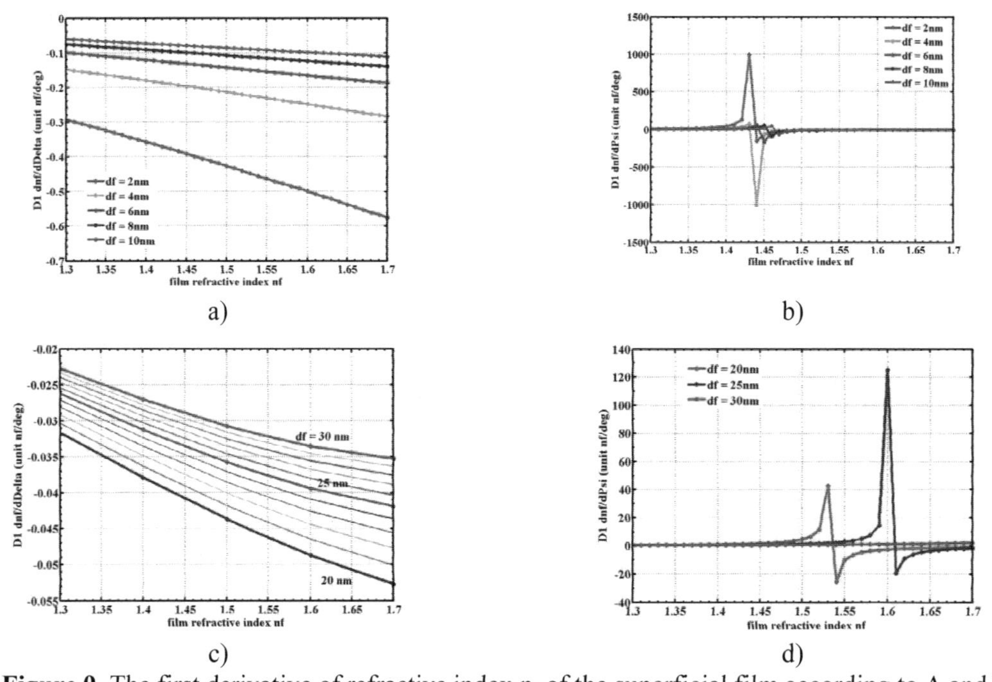

Figure 9. The first derivative of refractive index n_f of the superficial film according to Δ and Ψ; $\tilde{n}_s = 2.67(1\text{-}1.18i)$; $\varphi_0 = 60^0$; $\lambda = 562.5$ nm

3.3. The influence of the incidence angle on the shape of the curves $\Delta=f(n_f)$ and $\Psi=f(n_f)$

The value of the incidence angle in the ellipsometric measurement may be changed to avoid large errors in determining the refractive index n_f of the superficial film due to a small variation of the ellipsometric parameters Δ and Ψ versus the refractive index of the film. The change in the appearance of the curves $\Delta = f(n_f)$ and $\Psi = f(n_f)$ with the change of incidence angle is shown in figure 10. If for a particular superficial film the error in the determination of the refractive index n_f is too high, the slope of the curves $\Delta = f(n_f)$ and $\Psi = f(n_f)$ can be increased by changing the incidence angle. Thus, areas with large errors can be avoided.

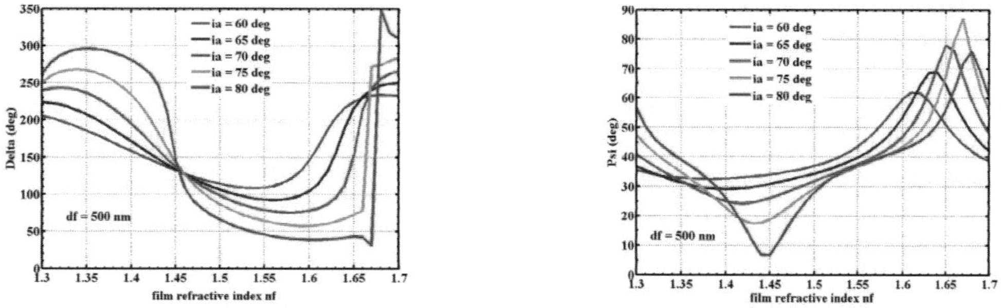

Figure 10. The variation of Δ and Ψ depending on the refractive index n_f of the superficial film for different incidence angle; $\tilde{n}_s$ =2.67(1-1.18i); d_f=500nm; λ=562.5 nm

3.4. The influence of the substrate refractive index $\tilde{n}_s$ on the shape of the curves $\Delta=f(n_f)$ and $\Psi=f(n_f)$

The change in the appearance of the curves $\Delta = f(n_f)$ and $\Psi = f(n_f)$ with the change in the nature of the substrate on which the superficial film lies for two different thicknesses of the superficial film is shown in Figure 11. Changing the nature of the substrate can change the range in which the error in determination of the refractive index n_f of the superficial film is high.

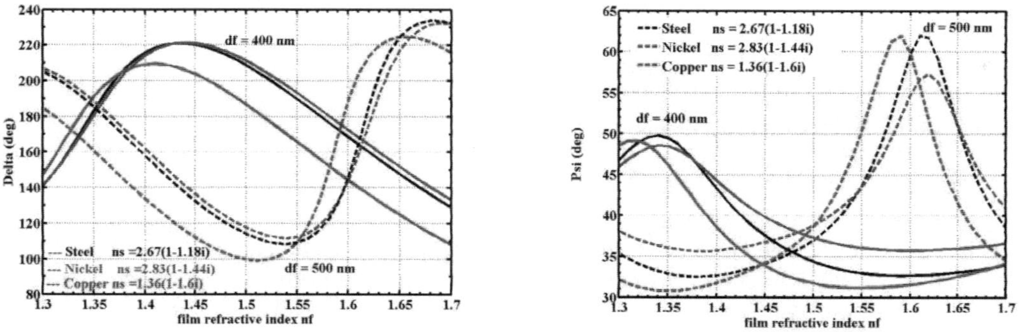

Figure 11. The variation of Δ and Ψ depending on the refractive index n_f of the superficial film for different substrates with different refractive indices (d_f = 400 nm and 500 nm; φ_0 =60^0; λ=562.5 nm)

All calculations were performed based on the Fortran's McCrackin program, modified accordingly and the graphics were processed in Matlab.

4. Conclusions

The ellipsometric analysis of the specular external reflection of light radiation on superficial films optical non-absorbing allows to know the experimental factors on which the analyzed system depends and which influence the error of determining the refractive index of superficial film. For these films the curves $\Delta = f(n_f)$ and $\Psi = f(n_f)$ are pseudo-periodic and curves $\Delta = f(\Psi)$ are closed.

The analysis of the periodicity of these curves allows accurate determination of the refractive index of superficial films. Conclusions on areas where experimental measurement errors of the ellipsometric parameters Δ and Ψ produce small errors in the determination of the refractive index n_f of the superficial film can be obtained from the curve shape.

The experimental parameters: incidence angle φ_0, wavelength λ, and refraction index of incidence medium n_0 can be properly modified to obtain minimal errors in determining the refractive index of superficial film.

It is important to perform an evaluation of errors in the determination of the refractive index n_f of superficial films due to experimental errors of measurement of Δ and Ψ.

Information from the shape of the curves in the areas where experimental errors in the measurement of ellipsometric parameters Δ and Ψ produce small errors in determining the refractive index n_f of a superficial film was obtained.

Changing experimental conditions in an ellipsometric measurement such as: the incidence angle φ_0 or the nature of the substrate on which the superficial film lies (expressed by the value of the refractive index $\tilde{n}_s$ of the substrate) can avoid large errors in determining the refractive index n_f of superficial film.

References

[1] Azzam R M A and Bashara N M 1977 *Ellipsometry and Polarized Light*, North-Holland publ. Comp., Amsterdam - New York - Oxford

[2] Jitian S 1994 Modification de l'etat de polarisation de le lumiere a la reflection sur les surfaces solides, *Bul. Şt. Univ. Politehnica Timişoara* **39**(53) 101-106

[3] Moisil G and Moisil D 1973 *Teoria si practica elipsometriei*, Ed. Tehnica, Bucuresti

[4] Jitian S and Chifu E 1986 The Ellipsometric Study of Polimer Films on Metals, *Studia Univ. Babeş-Bolyai. Chemia* **31**(2) 69-75

[5] Chen K H, Hsu C C and Su D C 2003 A method for measuring the complex refractive index and thickness of a thin metal film, *Appl. Phys. B – Lasers and Optics* **77** 839-842

[6] Azzam R M A and Khan M E R 1983 Complex reflection coefficients for the parallel and perpendicular polarizations of a film–substrate system, *Applied Optics* **22**(2) 253-264

[7] Jitian S 2013 The ellipsometrical study of adsorption-desorption of the corrosion inhibitors on metallic surfaces, *Romanian Reports in Physics* **65**(1) 204-212

[8] Jitian S 2015 The Ellipsometrical Analysis of External Reflection of Light on Superficial Films on Solid Substrates, *American Journal of Optics and Photonics* **3**(4) 48-53

[9] Jung T H J, Bork J, Holmgaard T and Kortbek N A 2004 *Ellipsometry. Detection of nanostructures*, Technical report, Aalborg University, Institute of Physics and Nanotechnology, Aalborg

[10] Pristinski D, Kozlovskaya V, and Sukhishvili S A 2006 Determination of film thickness and refractive index in one measurement of phase-modulated ellipsometry, *J. Opt. Soc. Am. A* **23**(10) 2639-2644
Manallah A and Bouafia M 2016 Determination of Optical Constants of Semiconductor Thin Films by Ellipsometry, *International Science Index, Materials and Metallurgical Engineering* **10**(7) 927-930

[11] Chandler-Horowitz D, Nguyen N V and Ehrstein R J 2003 *Assessment of Ultra-Thin SiO_2 Film Thickness, Measurement Precision by Ellipsometry, CP683, Characterization and Metrology for VLSI Technology*, 2003 International Conference, edited by D. G. Seiler, A. C. Diebold, T. J. Shaffner, R. McDonald, S. Zollner, R. P. Khosla, and E. M. Secula, American Institute of Physics 0-7354-0152-7/03, pp 326-330

[12] McCrackin F L 1969 *A Fortran Program for Analysis of Ellipsometer Measurement*, Natl. Bur. Std., Technical Note 479, Washington, USA

Characterization and corrosion resistance of boride layers on carbon steel

A Kaouka[1] and O Alaoui[2]

[1]Ecole Normale Supérieure de Laghouat, BP 4033, 03000 Laghouat, Algeria
[2]Laboratoire de Génie Des procédés, Laghouat University, 03000 Laghouat, Algeria

E-mail: a.kaouka@lagh-univ.dz

Abstract. Boriding treatment was carried out at 850, 900, 950, 1000 and 1050 °C for 2, 4 and 8 h on carbon steel. The effect of some treatment parameters such as the chemical composition of the mixture, boriding temperature and time were investigated. X-ray diffraction (XRD), scanning electron microscopy (SEM) and microhardness tests were used to describe the formed boride layers. Some properties of boride layers have been studied such as corrosion and wear resistance. The obtained results confirm that the corrosion resistance of boride carbon steel was improved compared to no-boriding steel. The dry abrasive wear resistance of boride alloy samples is about 5 times greater than that of non-boride ones.

1. Intoduction

Boriding or boronizing is well known as a thermochemical diffusion surface treatment in which boron atoms are diffused into the surface of samples to form borides on the surface of material. This treatment can be useful for several different types of steels as carbon steels, alloy steels, stainless steels and tool steels and some non-ferrous materials such as titanium alloy and niobium [1-5].

Temperatures process ranging between 850 and 1050 °C in either solid, liquid or gaseous, salt media [6]. Boriding treatment is a prominent optimal for a wide range of tribological applications. Boron atoms can enter smoothly into the samples forming a single or double phases FeB and Fe_2B.
The boride layers on carbon steels and low alloy steels have an accentuated acicular form oriented perpendicular to the surface being treated. This serrated formed ensures a good snap of the boride layers to the base metal. In the case of highly alloyed steels, this perforation is muted or non-existent steels. Scaling is done during the quenching oil, where the substrate should be hardened.

To assess the nature and the quality of a given boride layer, simply a metallographic test that allows checking the presence of one or two phases and evaluating the depth of the layers and the nature of the interface between the substrate and the boride layer and making a classification.

Depending on time and temperature process, chemical composition of the substrate, boron potential of medium, a single phase Fe_2B or double phases (Fe_2B, FeB) achieved by diffusing boron atoms into the surface of the substrate [7-10].

In this study, boriding treatment based on the information gathered from the literature taking into consideration the availability of chemicals and their cost. Searching of treatment must be made to ensure a good quality of the layers of borides formed on the carbon steel and a powder whose constituent chemical products may be available on the local market, a reasonable cost of the mixture depending on the proportion and nature of constituents used. Once the boriding treatment is realised, nature and quality of the borides layers formed on the carbon steel substrate were identified, the influence of parameters such as boriding temperature and time of on the nature, quality and microstructure of the obtained boride layers were examined. The microhardness of the layers of borides obtained on carbon steel and that obtained on the underlying layers and substrate were investigated. The corrosion resistance of the boride layers made by the technique of powders in various aggressive environments (acidic, basilicon carbide media and salt water) and the wear on abrasive and erosive of borides layers formed were studied.

Thus, the aim of present study is realised boriding treatment and then characterize the boride layers formed and measure the thickness and hardness of boride layer after different boriding treatments, and study some properties of boride layers such as wear and corrosion resistance.

2. Experimental procedure

2.1 Microstructure of the boride layer

The carbon steel essentially containing 0.36 wt. % C, 0.04 wt. % Cr, 0.28 wt. % Mn, and 0.13 wt. % Si was used as substrate. Substrate used had a rectangular shape and were 6×6×2.5 mm3. Boriding treatment was performed in a liquid medium based on borax at temperatures of 850 to 1050 °C include the time intervals for 2, 4 and 8 h.

The nature and the morphology of boride layers formed on the carbon steel substrate were established by optical microscopy using LEICA microscopy. The attendance of boride phases on the carbon steel substrate was determined by X-ray diffraction (XRD) analysis. X-ray diffraction patterns were taken immediately from the surface of samples using Cu Ka radiation with a wavelength of 1.7902 Å was used over a 2θ range of 20–90°. X-ray diffraction shapes were taken on each boride sample [11-13].

Microhardness measurements on metallographic cross-sections was carried out using SHIMATZU tester with the diamond pyramid (Vickers indenter). The microhardness of boride layers was measured six times for each point for the same distance between the surfaces and indentation points by means of Vickers indenter with applied load of 1 N (100 g/force).

Abrasive wear resistance tests were carried out on P180 Silicon carbide emery paper tape (main fraction grain size, 50 mm; maximum, 80 mm) using an AWRD device. The load is 50 N (5100 g/force). The wear resistance was determined by means of weighing the samples and measuring their height.

A digital measurement instrument devoted to the optical microscope and via a second method by ImageJ Launcher software to calculate the thickness of the boride layer. The thickness average of six different measurements.

3. Results and discussion

Figures 1 and 2 shown the cross-sectional micrographs of boride layers formed on carbon steel substrate at 900 °C for 2 h and 950 °C for 4 h, respectively. Boriding treatment of carbon steel mains to the formation of two borides, FeB and Fe_2B. FeB near to the surface and Fe_2B approximately close to the carbon steel substrate (Figure 1).

The boride layers were obtained on the entire surface of the treated samples, even in curved areas showing samples of hexagonal shape edges, it uniforms clear that the obtained boride layer covers the entire surface of the sample with an equivalent thickness across the entire surface. Depending on boriding temperature and time, the boride layers obtained on carbon steel are single-phase or double-phase in silicon carbide as shown in Figure 2.

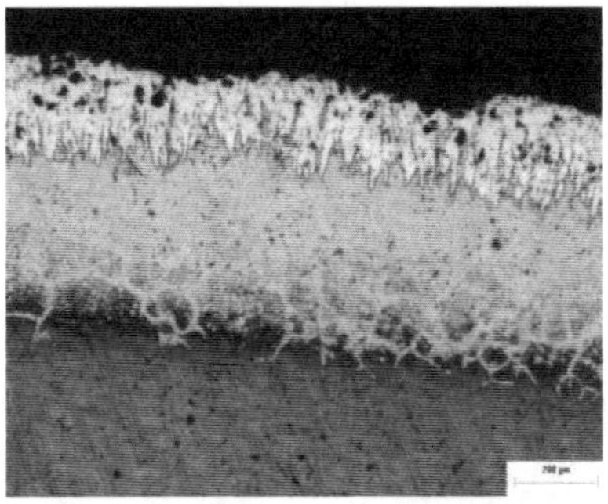

Figure 1. Cross-section micrographs of boride carbon steel at 950 °C for 4 h

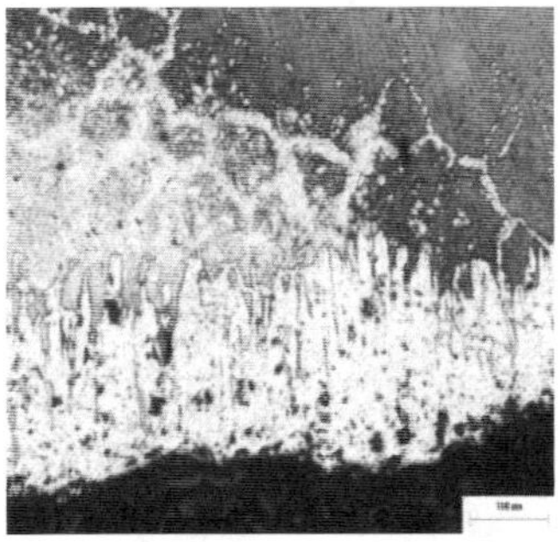

Figure 2. Cross-section micrographs of boride carbon steel at 900 °C for 2 h

When higher temperature and hold-time is prolonged, the proportion of boride FeB is more important. Therefore, we can say that the choice between a layer single-phase boride consisting of the unique boride Fe_2B or another or double-phase in silicon carbide consisting of two FeB borides and Fe_2B can occur depending on the operating conditions. It should be mentioned that the boride layer FeB at the top, close to the surface was darkly etched in comparison to the luminously etched inner Fe_2B and the highly etched substrate, as shown in Figure 3.

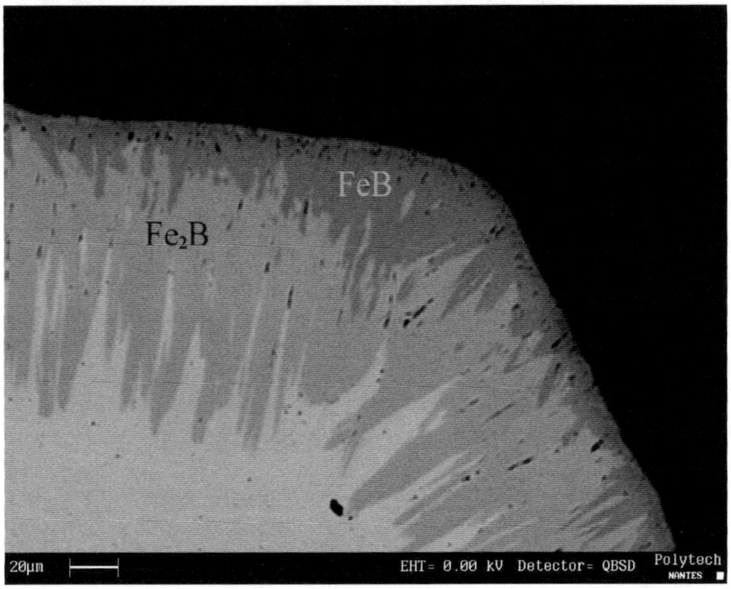

Figure 3. SEM image of microstructure of FeB and Fe_2B borided carbon steel at 900 °C for 2 h

At present study, the attendance of borides was identified and confirmed using XRD analysis. Optical microscopy examination of borides formed on the surface of carbon steel substrate shown a columnar and saw-tooth morphology at the interface of the substrate without any evidence of break. Depending on boriding temperature and time there is a semi parabolic relationship between boride layers thickness and boriding parameters (time and temperature). Increasing the boriding temperature not only reinforced to increase the boride layers thickness but also affected to roughening the microstructure in the substrate.

As estimated before, higher boron contents were detected at the boride layers than the substrate. The saw-tooth morphology of boride layers was clearly detected between Fe_2B and FeB layers during high magnification SEM examinations. The cross-sectional SEM micrographs of the samples boride are presented in Figure 4 along with the EDX line scan profile of boron as shown in Figure 5. There is no piks of boron due to the low porcetage of this element.

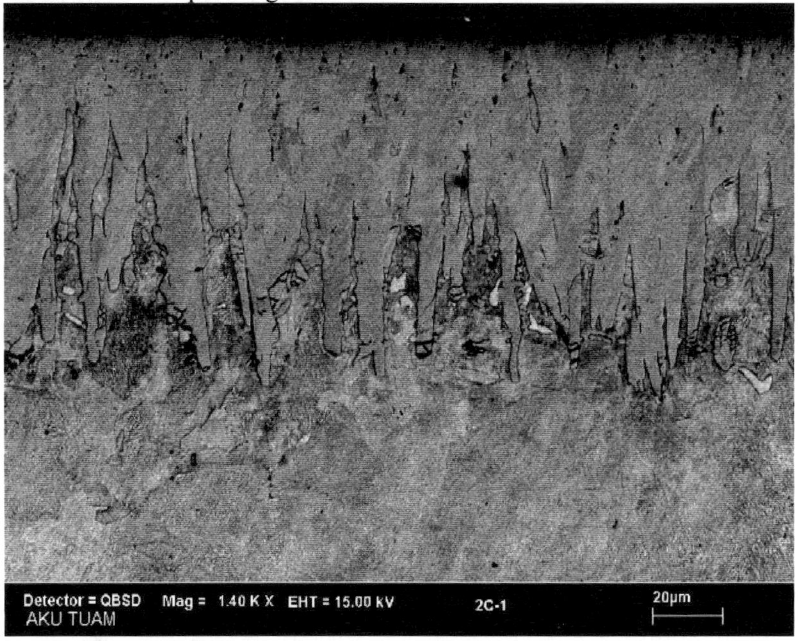

Figure 4. SEM image of microstructure boride carbon steel at 900 °C for 4 h

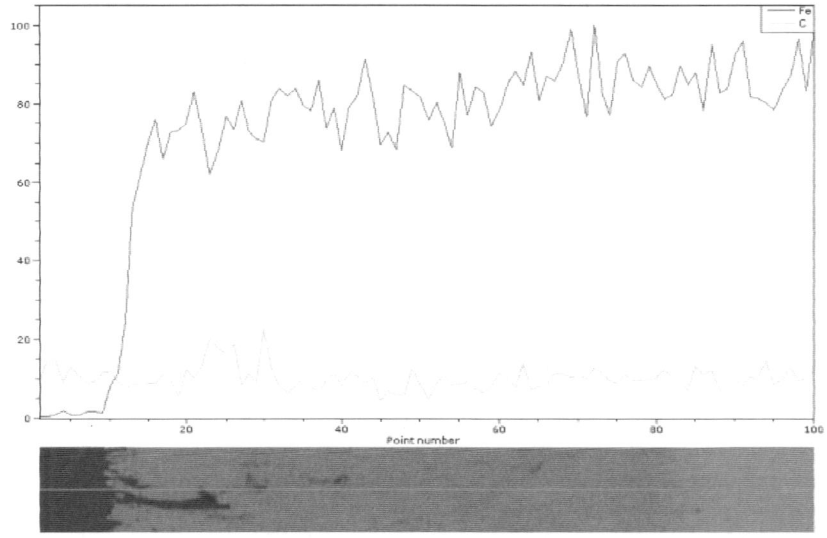

Figure 5. SEM image of microstructure of FeB and Fe_2B borided carbon steel at 900 °C for 2 h

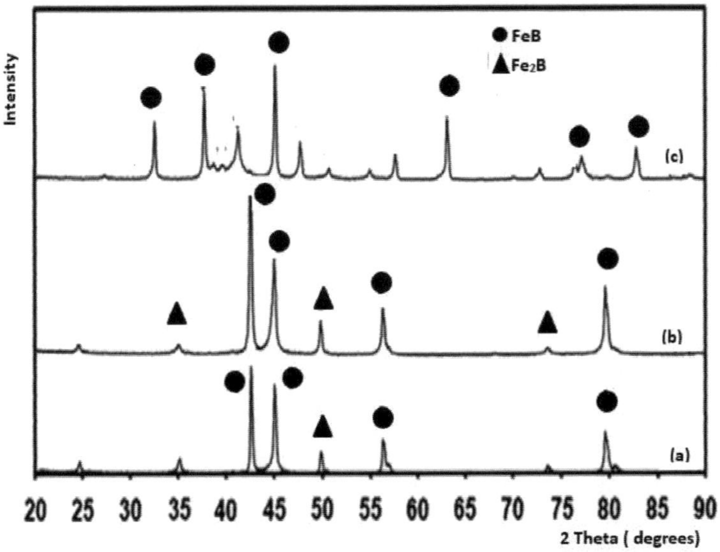

Figure 6. XRD patterns of carbon steel at 950°C for 4 h.
(a)borax–B_4C bath, (b)borax–Al bath,(c)borax–Silicon carbide bath

The is no remarkable difference was registered with regard to the form of the boride layers obtained on carbon steel substrate, the borides layers obtained have a wild form with needles that are perpendicular to the treated surface orientation. The wild form of the boride layers formed on steel presented a good way of hanging for these layers on the substrate. Decohesion at the coating/substrate interface problems often encountered in surface treatments are virtually absent in this case. FeB Boride appears darker than boride Fe_2B. The proportion of Fe_2B boride is greater than FeB boride. It is clairly to observe the difference between the boride and the rest of the material that appears with a color slightly lighter than FeB boride as confirmed previously. It should be noted also that FeB boride which is located towards the outside of the formed layer is characterized by its wild form.
Microstructural and properties of boride layers depended on the boriding temperature and the treatment time. Longer boriding time and higher boriding temperature resulted in a thicker boride layer and the boride layer thickness was ranged between 20 and 387 µm [14].

On the XRD patterns of the samples borided based on borax-Al or borax- B_4C, the peaks of the Fe_2B phase were detected as shown in Figure 6.a and 6.b. However, on the XRD patterns of the samples boride based on borax- Silicon carbide only FeB peaks were present (Figure 6.c). The XRD results exhibited that the dominant phases formed on the surface of carbon steel are FeB and Fe_2B phases, depending on the temperature, time of process and mixture of compounds. The average thickness of boride layers along with the thicknesses of Fe_2B and FeB phases is grouped in Table 1.

Table1. The average thickness of boride layers at 900 °C for different times

Boriding time (h)	Fe_2B Thickness (µm)	FeB Thickness (µm)	Total thickness (µm)
2	128	-	128
4	137	12	149
8	146	19	265

The higher boriding temperatures at 1000 °C encouraged the growth of boride layers. It is perceptible that in the case of double boride layers phases, the thickness of the Fe_2B was about approximately more than the double of FeB thickness.

The results of micro hardness measurements exhibited on the boride surfaces are design in Figure 7 with respect to the indentation load. Depending on the phase structure of the boride layer, hardness of the boride surfaces followed two different pathways; boride layers which the hardest zone, transition zone and matrix.

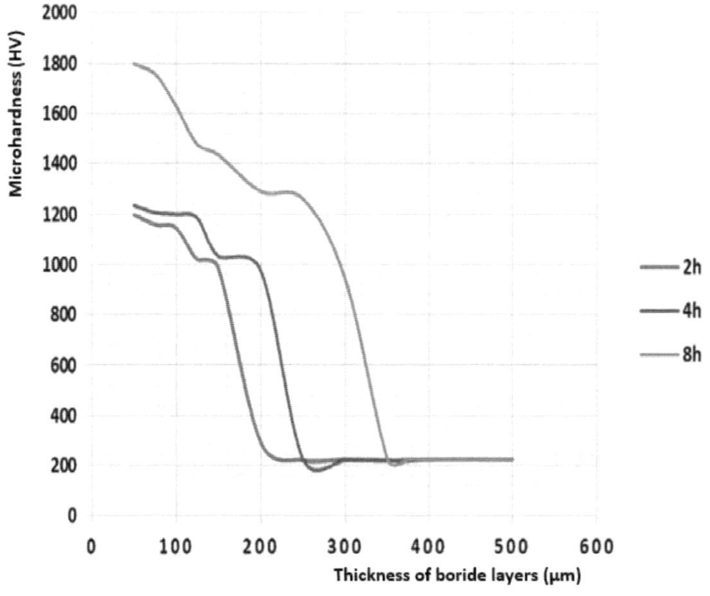

Figure 7. Hardness profiles of carbon steel

Higher boriding temperatures creating a double phase boride layer provided higher hardness than lower boriding temperatures inducing a single-phase boride layer [15-17]. At the minimum indentation load applied in the present study (100 g), hardness values obtained from the single (1600 HV) and double phase boride layer (1850 HV) coated surfaces were in the range reported in the literature for the Fe_2B and FeB phases, respectively. While the hardness of the double phase boride layer coated surface reduced about 1000 HV. Vickers indenter estimated the values of fracture toughness of boride layers of carbon steel. It depends on the boriding parameters (time and temperature), indentation loads applied and the distance from the surface. Their values were located between 3.42 and 4.57 MPa $m^{1/2}$ depending on time and temperature of treatment [14].

The effect of the chemical composition on boriding mixture was studied and confirmed that the mixture boriding provide boride layers where boron carbide is replaced by borax as a boron source. Increasing the content of activator in the boriding mixture where the boron source is boron carbide growths, the thickness of the boride layer formed. While it does not change the non-effectiveness of the powder where the boron source is borax. Addition of $NaBF_4$ and/ or NH_4Cl as an additional activator to the boriding mixture does not change the effectiveness of this mixture. The boron activation energy on carbon steel surface was found about 227.51 kJ mol^{-1} and the kinetic studies of boride layers showed a semi parabolic relationship between the boride layer thickness and the process time [18].

The boriding treatment of carbon steel greatly improves its corrosion resistance in acidic environments in general and particularly in sulfuric acid. This treatment clearly ameliorates the corrosion resistance of carbon steel in the aqueous solution of NaCl (30 g/l). The boriding has not the same effect on the corrosion resistance in basilicon carbide solutions used, it improves in the case of normal aqueous NaOH solution and slightly in the case of normal aqueous KOH solution. The

IOP Conf. Series: Materials Science and Engineering **477** (2019) 012029 doi:10.1088/1757-899X/477/1/012029

corrosion rate of boride steel in acidic solutions is much higher than basic aqueous solutions. The boride steel has a good resistance to corrosion in the aqueous solution of NaCl (30g / l).

Figure 8 represents the effect of corrosive environment on the surface of boride layers, it can be seen that the areas where the boride layer is relatively weak are completely corroded, while the areas where the borured layer is quite large remain intact. Boride layers of carbon steel for dry abrasive wear resistance tests was carried out at 950 °C for 4 h. An optical microscope observation of different samples after the wear tests revealed that the deterioration of boride surface by abrasive silicon carbide particles (SiC) is of the purely abrasive type as revealed in Figure 9. Figure 9 shows clearly the damage caused by the silicon carbide abrasive particles leaving visible scratches on the test surface. There is no evidence of adhesion was observed on the tested samples to consider that the wear tests performed are abrasive wear tests. A comparison of the wear resistance of boride carbon steel samples with that non boride samples investigated around 5 times greater than non-boride samples.

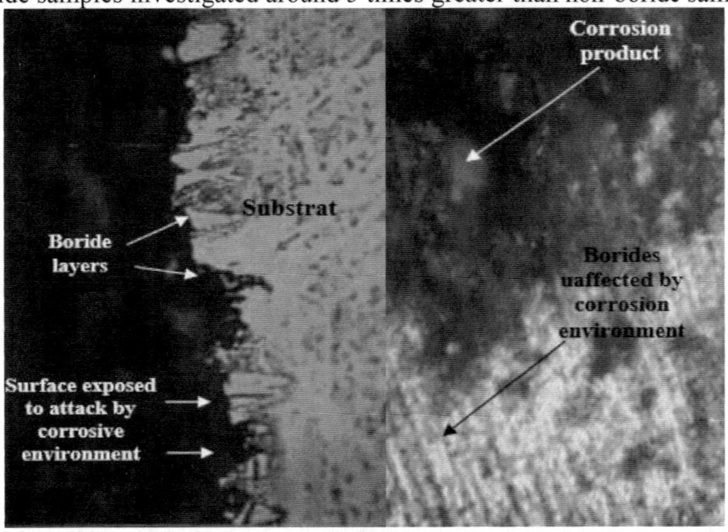

Figure 8. Optical microscopy of a boride surface attacked by corrosive environment

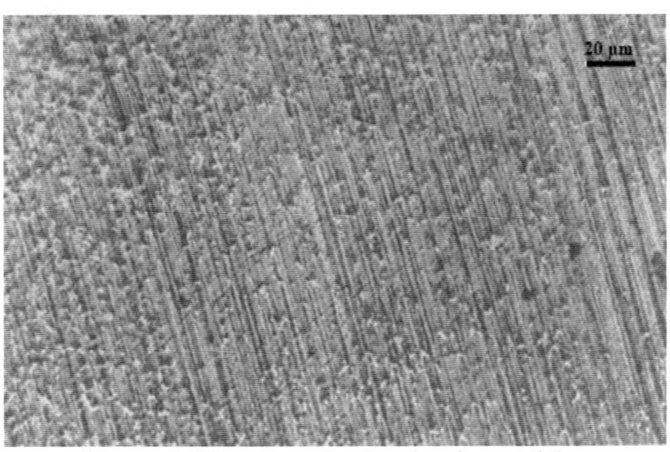

Figure 9. Wear traces by the abrasive particles of silicon carbide on the boride surface

4. Conclusion

The following conclusions can be derived as:

In the present study, carbon steel was borided at five different temperatures between 850, 900, 950, 1000 and 1050 °C for 2, 4 and 8 h.

The progress of boride layer thickness can be increase as function of boriding time, depending on the temperature.

Boriding treatments compounds based on borax and Silicon carbide formed a single-phase of boride layer (Fe_2B), while the compound based on boron carbide (B_4C) formed double phase (Fe_2B and FeB). Single and double phase boride layers shown almost similar tribological performance and superior wear resistance.

The micro hardness values of the boride layers formed on carbon steel were ranging from 1600–1850 HV, depending on parameters boriding treatment (time and temperature) and the componds of process, while the non-boride steel substrate is 250 HV. Micro hardness measurements conducted on the borided surfaces revealed higher micro hardness.

The wear resistance shown a tendency to decrease with increasing temperature. The double phase boride layer yielded better wear resistance than the single-phase boride layer at high temperatures and the wear resistance of boride carbon steel samples is greater than non-boride sample about 5 times.

Acknowledgement
The authors would like to acknowledge the financial support of Ecole Normale Supérieure de Laghouat- Algeria.

References

[1] Voroshnin L G and Lyakhovich L S 1978 *Boriding stell*, Metallurgiya, Moscow
[2] Sinha A K 1982 *Boriding (boronizing)*", in "Metals handbook" (ed A K Sinha) 844 Metals Park, OH, ASM International
[3] Yücel O, Alkan M and Turan A 2017 *Concise Encyclopedia of Self-Propagating High-Temperature Synthesis* 39-41
[4] Campos-Silva I and Rodríguez-Castro G A 2015 *Thermo. Surf. Engi. of Steels* 651-702
[5] Hihara L and Bakkar H 2016 Reference Module in *Mater. Sci. Mater. Eng.*
[6] Kunst H, Schroll H, Luetje R, Wittel K, Lugscheider E, Weber T, Eschnauer H R and Raub C 1991 A16 **427** Weinheim Verlag Chemie
[7] Campos-Silva I, Bravo-Bárcenas D, Meneses-Amador A, Ortiz-Dominguez M, Cimenoglu H, Figueroa-López and Andraca-Adame U 2013 *Surf. Coat. Technol.* **237** 402-414
[8] Kulka M, Makuch N and Popławski M 2014 *Surf. Coat. Technol.* **244** 78-86
[9] Campos I, Palomar M, Amador A Ganem R and Martinez J 2006 *Surf. Coat. Technol.* **201** 2438-2442
[10] Spies H J 2015 *Thermo.l Surf. Engi. of Steels* 267-309
[11] Campos I, Palomar-Pardavé M, Amador A, VillaVelázquez C and Hadad J 2007 *Appl. Surf. Sci.* **253** 9061-9066
[12] Keddam M, Kulka M, Makuch N and Pertek A 2014 *Appl. Surf. Sci.* **298** 155-163
[13] Muhammad W, Hussain K, Tauqir A, Haq A and Khan A 1999 *Mater. Trans. A* **30** 670-675
[14] Kaouka A, Allaoui O and M. Keddam **467** *Applied Mech. Mater.* 116-120
[15] Campos I, Rosas R, Figueroa U, VillaVelázquez C and Guevara A 2008 *Mater. Sci. Eng. A* **488** 562-568
[16] Genel K, Ozbek I and Bindal C 2003 *Mater. Sci. Eng. A* **347** 311-314
[17] Kwok C T, Cheng F T and Man H C 2000 *Mater. Sci. Eng. A* **290** 55-73
[18] Kaouka A, Allaoui O and Keddam M 2013 *Matériaux & Techniques* **101**(7) 705

IOP Conf. Series: Materials Science and Engineering **477** (2019) 012030 doi:10.1088/1757-899X/477/1/012030

Study on a maximum power point tracking controller for photovoltaic panels

G N Popa, C M Diniş, A Iagăr and S I Deaconu

Politehnica University of Timisoara, Department of Electrical Engineering and Industrial Informatics, 5 Revolution Street, Hunedoara, 331128, Romania

E-mail: gabriel.popa@fih.upt.ro

Abstract. The maximum power point tracker is a DC-DC electronic converter that optimizes the match between the solar array (photovoltaic panels) and the battery bank (stand-alone type) or utility grid (grid-tie type). They convert a higher and variable voltage DC output from photovoltaic panels down to the lower voltage needed to charge batteries or for on-grid inverters. The MPPT controller depends on the operating conditions of the photovoltaic panels, but also on the electrical characteristics of the electrical loads. The aim of the MPPT controller is to keep the operating point as close to the maximum power point. The paper presents a MPPT controller with 8-bit microcontroller. The microcontroller monitors the voltage and current of the photovoltaic panel as well as charging the battery. The controller is a buck converter, having a P channel MOSFET transistor. A microcontroller that also manages the charging of the batteries with the three charging stages, bulk, absorption, float and, also, the manual equalization option. Also, the battery bank is monitoring by measuring the temperature with a NTC thermistor.

1. Introduction

The global energy context leads to an intense concern in the field of unconventional energies. Of these, solar power occupies an important place.

Solar technologies can generally be [1], [2]:

- passive (orientation of a building to the sun, optimal design of spaces, etc.);
- active (with photovoltaic panels and thermal collectors).

The maximum power point tracker (MPPT) is a DC-DC electronic converter that optimizes the match between the solar array (photovoltaic panels) and the battery bank (stand-alone type) or utility grid (grid-tie type). They convert a higher and variable voltage DC output from photovoltaic panels down to the lower voltage needed to charge batteries [3-7].

The MPPT controller depends on the operating conditions of the photovoltaic panels, but also on the electrical characteristics of the electrical loads [8].

The aim of the MPPT controller is to keep the operating point as close to the maximum power point.

2. Photovoltaic panels and systems

Photovoltaic (PV) or solar cells are semiconductor devices that convert solar energy into DC energy. Photovoltaic cells directly convert solar radiation into electricity.

IOP Conf. Series: Materials Science and Engineering **477** (2019) 012030 doi:10.1088/1757-899X/477/1/012030

A photovoltaic cell is similar to a diode (Figure 1). The upper layer is type N, and the lower layer is type P [9], [10].

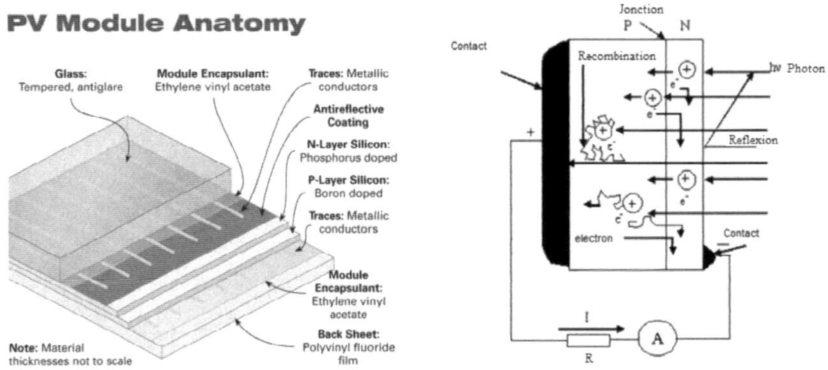

Figure 1. Structure of a photovoltaic junction P-N: design and operation [11]

When the photons hit the cell, the electrons are released. This causes a potential difference between the two material types (P and N). If a load is connected, an electric current will occur.

Basically, a PN junction is made up of a N-type silicon-tin pill, about 1 mm thick and 5 cm in length / width, placed on a metal base having a P-type thickness material just a few thousandths of a centimeter.

There have been different PVs called first generation, second generation and third generation (Table 1). PV efficiency has reached tens of percent efficiency (third generation).

In practice, different PV architectures are used (Figure 2): there are PV systems without battery bank and without connection to the grid - PV direct (Figure 2.a.), PV systems with battery bank and without grid connection - PV stand-alone (Figure 2.b), PV systems without battery bank and with grid connection with an on-grid inverter - Grid TIE without battery (Figure 2.c), and PV systems with battery bank and grid connection - Grid TIE with battery backup (Figure 2.d). The systems of Figure 2.a are the cheapest and easiest to use, and the systems in Figure 2.d are the most complex.

Table 1. Commercial PV Module Types [12]

Technology	First generation	Second generation	Third generation
Cell efficiency (%)	14-22	5.4-12.7	30-38
Module efficiency (%)	11-19.7	20-22	24-25
Area/kW	7-8	10-15	12

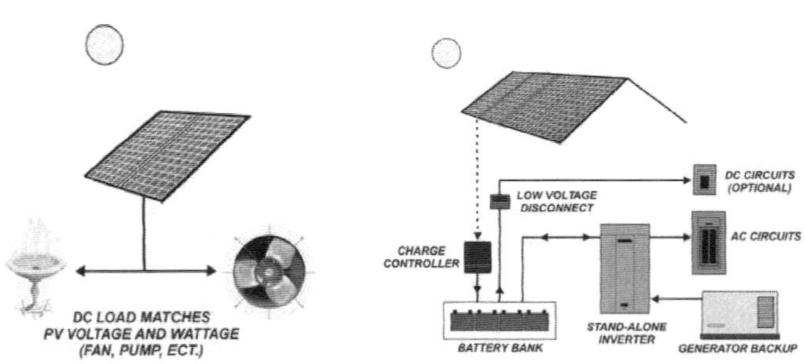

a. b.

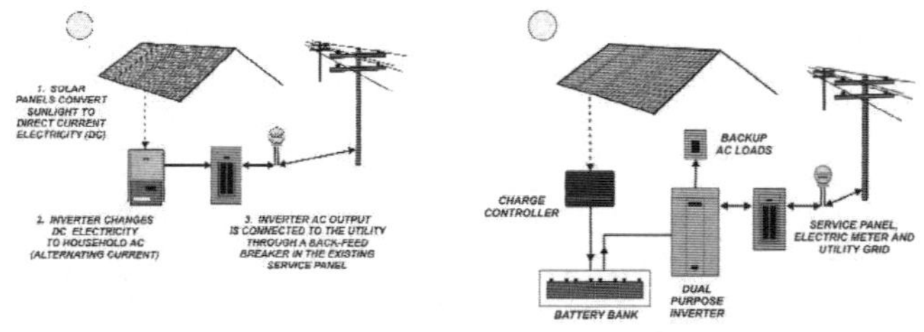

c. d.

Figure 2. Types of photovoltaic systems [1], [13], [14]
a. PV direct; b. PV stand-alone; c. Grid TIE without battery;
d. Grid TIE with battery backup

3. Charge controllers for photovoltaic systems

Charging regulator is used to charge the batteries in photovoltaic systems. The correct dimensioning of the capacity for an accumulator battery is made according to the rated power of the consumers and the consumption diagram on hourly intervals.

The battery life of a battery depends on how much it is discharged and the working temperature [15]. If it is not discharge more than 30%, the batteries resist several thousand charge/discharge cycles. If the batteries discharged regularly up to 80%, it takes about a thousand cycles. It is very important how the voltage of a battery varies during charging and discharging, because the charger, which regulates the current flow from the panel to the batteries, uses voltage as a "control signal" to protect batteries and extend their life [3-5].

The electrical power delivered by a photovoltaic cell is not sufficient for most domestic or industrial applications. For this reason, the photovoltaic cells are connected in series to increase the voltage value at the terminals and thus a panel (module) is made. Then the modules connect either in series to increase the voltage even more, or in parallel to increase the current through the circuit (series-parallel connections form the photovoltaic systems).

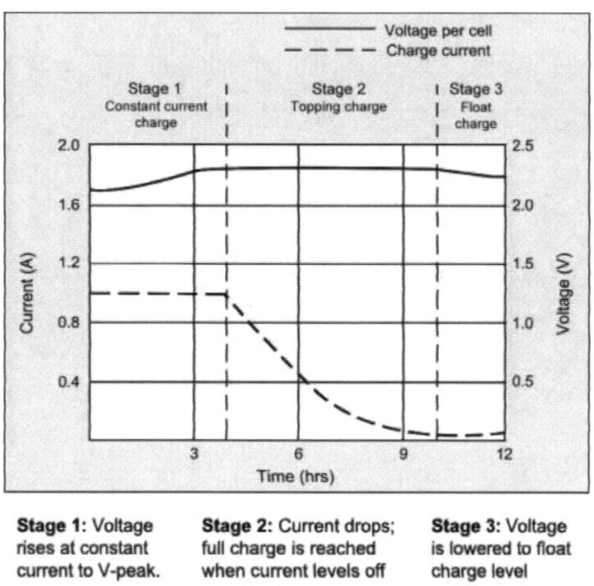

Figure 3. Typical charging of lead acid battery [16]

IOP Conf. Series: Materials Science and Engineering **477** (2019) 012030 doi:10.1088/1757-899X/477/1/012030

The voltage-current characteristic (U-I) of a photovoltaic panel depends mainly on the intensity of solar radiation and cell temperature [8].

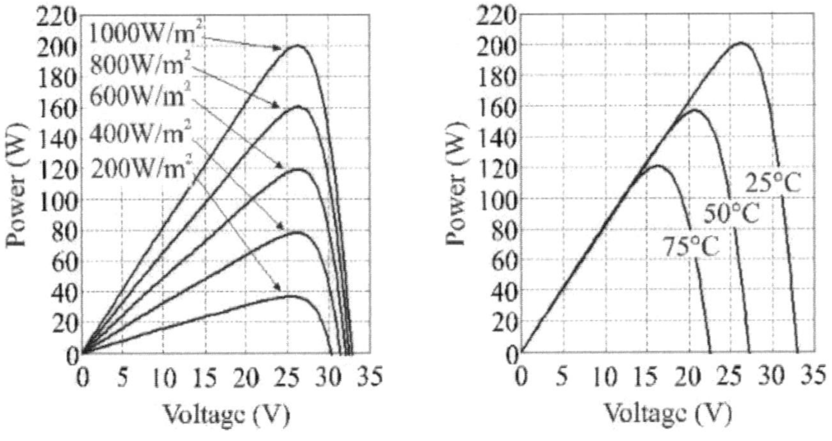

Figure 4. PV panels depends on solar incident radiation and temperature [17]

At the intersection of the PV U-I characteristic with the consumers characteristic of the photovoltaic panel terminals, there is the operating point (PF). This point generally differs from the maximum power point (MPP) at which the system can work when the optimum power transfer is achieved between the photovoltaic panel and the load.

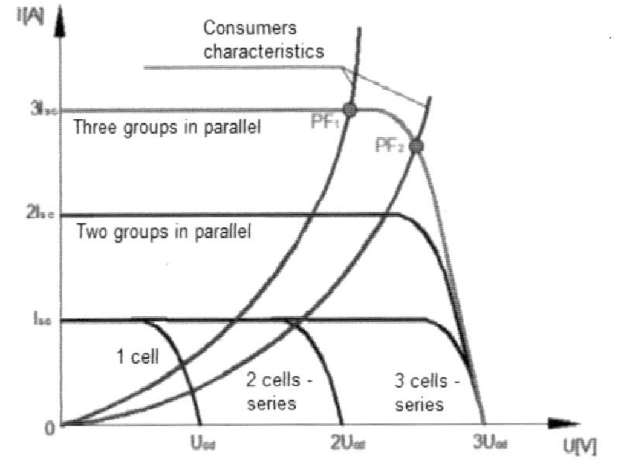

Figure 5. PV cells connected in series and parallel

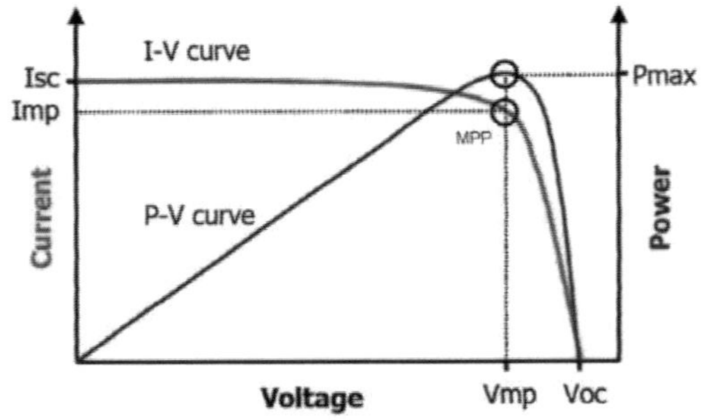

Figure 6. PV characteristics and MPP

As a result, MPP depends on the operating conditions of the photovoltaic panel, but also on the electrical characteristics of the terminal loads. The purpose of MPPT is to keep the operating point as close MPP.

4. MPPT for photovoltaic systems

Figure 7 shows the basic scheme of a MPPT solar charger. The key element is a DC-DC converter that allows the panel to operate at a voltage different from that of the battery. The MPPT controller quickly determines, in real-time, the maximum power point (MPP) of the solar panel system, depending on the power of the consumers and the battery charge status, at different times of the day, and in various stages of the weather [5-7], [18], [19].

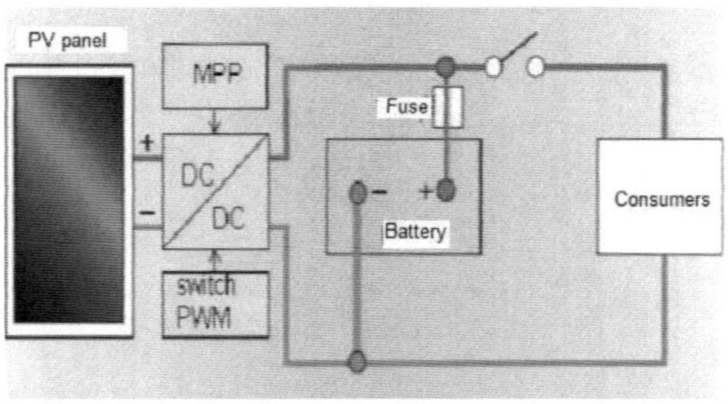

Figure 7. The schematic diagram of an MPPT controller

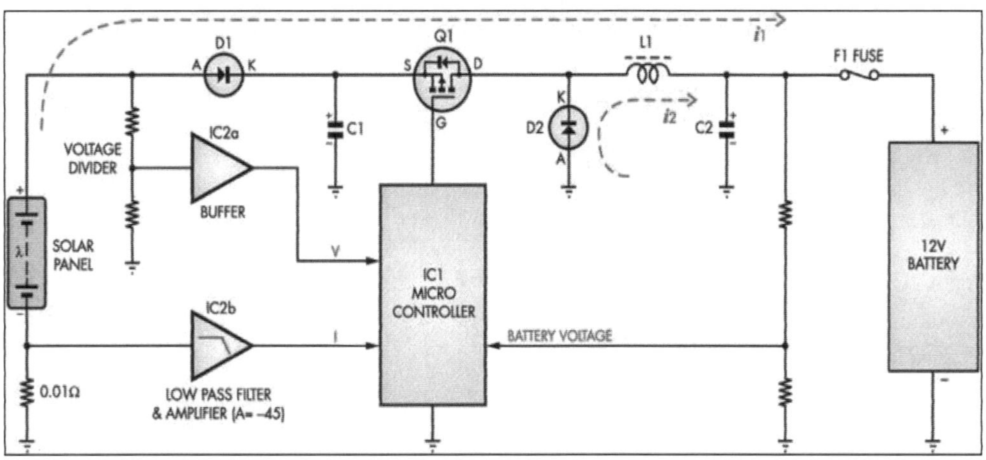

Figure 8. Block diagram of a MPPT controller with microcontroller

The microcontroller monitors the voltage and current of the photovoltaic panel as well as charging the battery.

When the MOSFET Q1 is on-state the current i1 passes through the coil L1 to the capacitor C2 and the battery. When Q1 is off-state, the energy stored in the L1 coil is transmitted via the D2 diode to the battery (current i2).

The rate of change of I_L can be calculated from:

$$V_L = L \cdot \frac{dI_L}{dt} \tag{1}$$

with equal to during the on-state to during the off-state.

Therefore, the increase in current during the on-state is given by:

$$\Delta I_{L_{on}} = \int_{0}^{t_{on}} \frac{V_L}{L} \cdot dt = \frac{(V_i - V_o)}{L} \cdot t_{on} \text{ , } t_{on} = DT \tag{2}$$

where D is a scalar called the duty cycle with a value between 0 and 1.

Conversely, the decrease in current during the off-state is given by:

$$\Delta I_{L_{off}} = \int_{0}^{T=t_{on}+t_{off}} \frac{V_L}{L} \cdot dt = -\frac{V_o}{L} \cdot t_{off} \text{ , } t_{off} = (1-D) \cdot T \tag{3}$$

The proposed MPPT controller is made with microcontroller (PIC 16F88) [17], [20]. The microcontroller monitors the voltage and current of the photovoltaic panel as well as charging the battery. The controller is a buck converter, having a P channel MOSFET transistor.

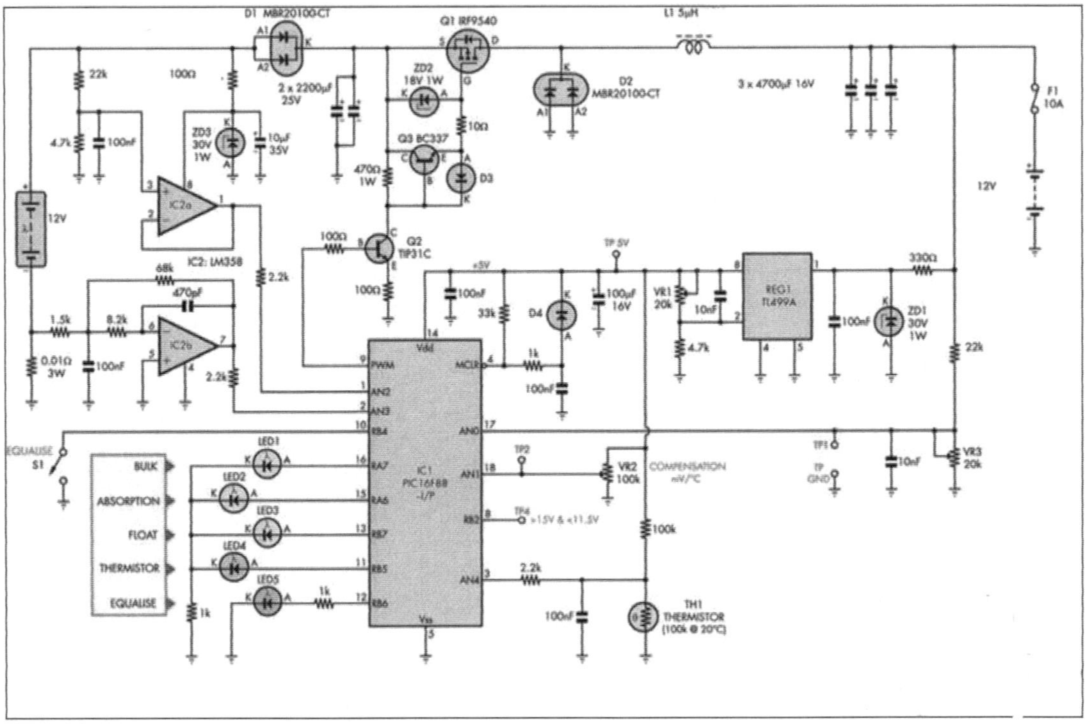

Figure 9. The electronic diagram of a MPPT controller made with microcontroller

A microcontroller that also manages the charging of the batteries with the three charging stages, bulk, absorption, float (Figure 10) and the manual equalization option (Figure 11). Also, the temperature of battery is monitoring with a NTC thermistor to prevent overheating.

The photovoltaic panel chosen is MWG-10, having the following features: maximum power: P_{max} = 10 W; tolerance = + 3%; voltage at maximum power: V_{mp} = 17.49 V; current at maximum power: I_{mp} = 0.57 A; open circuit voltage: V_{oc} = 21.67 V. The following tables (Tables 2 and 3) show the variation of voltage and charging current according to the vertical inclination angle of the photovoltaic panel, as compared to the orientation to the cardinal points for two accumulators with different A·h capacities.

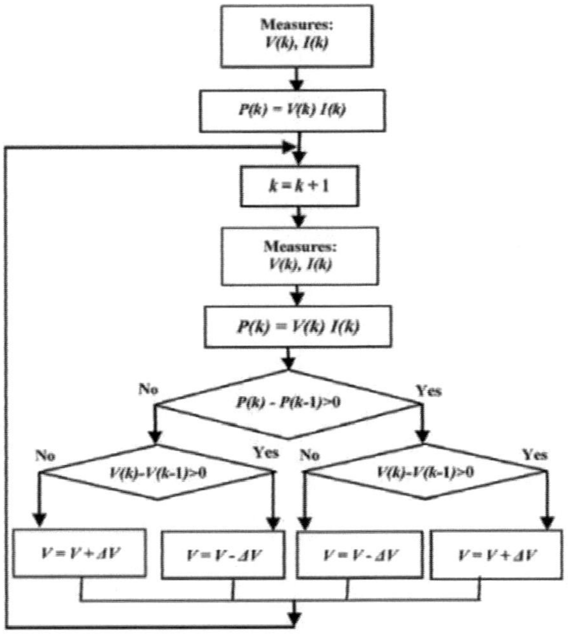

Figure 10. The main flowchart for program

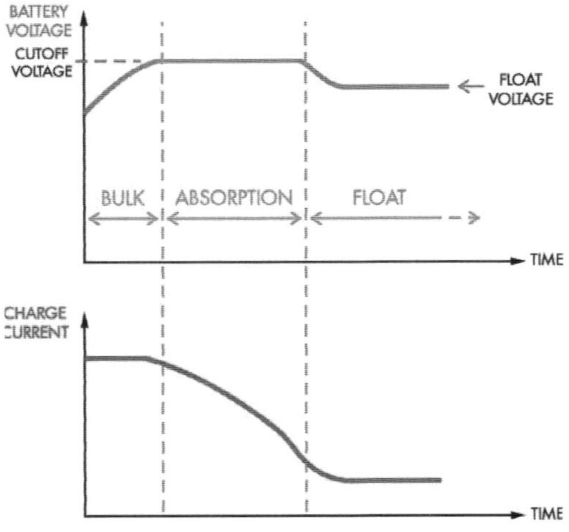

Figure 11. Standard charging

Table 2. Experiments with battery 7 Ah, U=13.14 V

Elevation angle (°)	N-E Orientation				E orientation			
	U_{PV} (V)	I_{PV} (mA)	U_{bat} (V)	I_{bat} (mA)	U_{PV} (V)	I_{PV} (mA)	U_{bat} (V)	I_{bat} (mA)
15	14.03	120	13.67	60	14.33	240	13.87	180
30	14.05	130	13.69	80	14.36	280	13.88	220
45	14.46	170	13.72	120	14.20	290	13.72	220
60	14.07	100	13.65	50	14.36	310	13.87	240
90	13.94	40	13.61	20	14.19	250	13.74	190

Elevation angle (°)	S-E Orientation				S Orientation			
	U_{PV} (V)	I_{PV} (mA)	U_{bat} (V)	I_{bat} (mA)	U_{PV} (V)	I_{PV} (mA)	U_{bat} (V)	I_{bat} (mA)
15	14.43	290	13.98	210	14.32	270	13.86	210
30	14.24	160	13.82	100	14.23	260	13.77	200
45	14.35	220	13.89	150	14.36	190	13.65	140
60	14.54	260	14.09	200	14.14	110	13.57	70
90	14.29	180	13.88	120	13.96	50	13.52	10

Table 3. Experiments with battery 9 Ah, U=11.12 V

Elevation angle (°)	N-E Orientation				E orientation			
	U_{PV} (V)	I_{PV} (mA)	U_{bat} (V)	I_{bat} (mA)	U_{PV} (V)	I_{PV} (mA)	U_{bat} (V)	I_{bat} (mA)
15	12.60	230	11	210	12.33	330	11.83	250
30	12.12	270	11.65	210	12.38	330	11.86	270
45	12.53	240	11.57	180	13.02	330	11.88	270
60	12.20	230	11.57	160	12.30	300	11.80	240
90	12.06	150	11.55	100	12.06	220	11.52	150
Elevation angle (°)	**S-E Orientation**				**S Orientation**			
	U_{PV} (V)	I_{PV} (mA)	U_{bat} (V)	I_{bat} (mA)	U_{PV} (V)	I_{PV} (mA)	U_{bat} (V)	I_{bat} (mA)
15	12.33	300	11.83	240	12.23	280	11.74	220
30	12.37	330	11.86	260	12.61	270	11.72	230
45	12.34	320	11.83	260	12.12	250	11.65	200
60	12.26	300	11.77	240	12.57	220	11.57	180
90	12.07	250	11.61	190	12.04	210	11.6	160

The measurements were made only with PV, MPPT and an lead acid battery (without any electrical load). The two batteries used in experiments were not preloaded. When the measurements were made, it was a sunny summer day, between 9^{00} h to 12^{00} h. The place where the measurements were made was Romania, at the latitude of 45^0, for which an ideal elevation angle is 30^0 for summer and 60^0 for winter for fixed PVs (without automatic solar tracking systems).

The current absorbed by the PV depends on the panel orientation, the elevation angle, and the lead acid battery charge state. It was found that the maximum charge (by maximum current) of lead acid batteries was obtained for orientation S-E, S and E (depending on the time interval when the measurements were made) at elevation angles between 15^0-45^0. The charging current is higher for the larger capacity battery.

5. Conclusion

Despite the price and dependence on external factors, solar panels for power generation are a solution for the future. The most significant advantage is that, at least theoretically, they have a long service life and a very low maintenance cost. Initial investment is the major drawback of all alternative energy systems and photovoltaic panels.

There are, of course, other drawbacks, among which we can list the low photovoltaic cell yield and the acute dependence on solar radiation (radius angle, temperature, intensity, etc.). The MPPT Charge Controller is essential under the following conditions:

- in winter and/or sky with rain, when it is always necessary to have extra power;
- cold weather - solar panels work better at low temperatures;

- when the battery is very low, the lower the voltage on the batteries, the MPPT controller introduces a higher current.

References

[1] Sankar R, Jery Althaf P K and Sreevas S 2012 A Study on Low Cost Electrification Using Solar Energy a Field Work, *Int. Journal of Computer Theory and Engineering* **4**(6) 1015-1019
[2] Zhao J, Wang A, Abbaspour-Sani E, Yun F and Green M A 1997 Improved Efficiency Silicon Solar Cell Module, *IEEE Electron Device Letters* **18**(2) 48-50
[3] Darla R B 2007 *Development of Maximum Power Point Tracker for PV Panels Using SEPIC Converter*, Telecommunications Energy Conference, INTELEC 2007, 29th International Conference, 30 September- 4 October, Rome, Italy, 6 pp
[4] Lopez-Lapena O, Penella M T and Gasulla M 2010 A New MPPT Method for Low-Power Solar Energy Harvesting, *IEEE Transactions on Industrial Electronics* **57**(9) 3129-3138
[5] Han L and Jia L 2012 *Photovoltaic Power Generation System MPPT*, IEEE 5th Int. Conference on Advanced Computational Intelligence, October 18-20, Nanjing, Jiangsu, China, 6 pp
[6] Raja B, Satheesh Kumar M R, Vikash S and Hariharan K 2016 *Maximum Power Point Tracking in Solar Panels under Partial Shading Condition using Equilibration Algorithm*, International Conference on Communication and Signal Processing, April 6-8, India, 5 pp
[7] Wadghule T and Aranke V R 2016 *Efficiency Improvement of Photovoltaic Panel by Tracking Method*, Int. Conference on Electrical, Electronics, and Optimisation Techiques, 6 pp
[8] Yandt M D, Hinzer K and Schriemer H 2018 Representative Atmospheric Parameters in Multijunction Solar Cell Design, *IEEE Journal of Photovoltaics* **8**(3) 895-902
[9] Lin J T, Lai C C, Lee C T, Hu Y Y, Ho K Y and Haga S W 2018 A High Efficiency HIT Solar Cell with Pillar Texturing, *IEEE Journal of Photovoltaics* **8**(3) 669-675
[10] Gao Q, Zhang Y, Yu Y, Meng F and Liu Z 2018 Effects of I–V Measurement Parameters on the Hysteresis Effect and Optimization in High-Capacitance PV Module Testing, *IEEE Journal of Photovoltaics* **8**(3) 710-718
[11] ***https://www.homepower.com/articles/solar-electricity/equipment-products/peek-inside-pv-cell
[12] ***http://www.deparsolar.com/en_sayfa1.asp?id=594&title=pv-technologies:cells-and-moduls
[13] ***http://www.aladdinsolar.com/pvsystems.html
[14] ***http://www.insolergy.com/blog/solar-basics/types-and-methods-of-solar-power-installation/
[15] Platon R, Martel J, Wooddruff N and Chau T 2015 Online Fault Detection in PV Systems, *IEEE Transactions on Sustainable Energy* **6**(4) 1200-1207
[16] ***http://batteryuniversity.com/learn/article/charging_the_lead_acid_battery
[17] http://e2e.ti.com/support/microcontrollers/c2000/f/171/p/455407/1641265
[18] Cunţan C D, Baciu I and Osaci M 2015 Operational Study of a Frequency Converter with a Control Sequence, Utilizing Xilinx Software, *Acta Polytechnica Hungarica* **12**(6) 201-212
[19] Baciu I and Cunţan C D 2012 *Operation Analysis of a Frequency Converter with Control Realized in LabView*, IEEE International Conference on Industrial Technology (ICIT), March 19, pp 432-437
[20] ***http://www.electronicslovers.com/2017/05/pic16f88-based-mppt-charge-controller.html

IOP Conf. Series: Materials Science and Engineering **477** (2019) 012031 doi:10.1088/1757-899X/477/1/012031

An analyse on analogue ESR-meter

C M Diniş, G N Popa, C D Cunţan and A Iagăr

Politehnica University of Timisoara, Department of Electrical Engineering and Industrial Informatics, 5 Revolution Street, Hunedoara, 331128, Romania

E-mail: corina.dinis@fih.upt.ro

Abstract. The paper presents an analogue ESR meter, made with a JFET integrated circuit, which can be used to identify defective electrolytic capacitors. It is powered by continuous voltage and consists of a differential voltage source, a 100 kHz oscillator, a Wheatstone bridge with resistors supplied with alternating voltage, in a connecting arm of bridge being connected the electrolytic capacitors to be checked, a voltage amplifier-comparator with optical warning, a high-pass filter and a voltage-current converter. At the output of voltage-current converter is a milliammeter pointing out the ESR value (up to 20 Ω). Measurements were made on a batch of electrolytic capacitors (new and used) and a comparative analysis was performed between the measured values with an analogue ESR-meter and a digital one.

1. Introduction

In present day, electrolytic capacitors are widely used in power electronics circuits and are the main sensitive components with the shortest life. Usually, the main functions in power sources of electrolytic capacitors are in power sources, bus voltage, stabilization, etc. [1-5]. Primary wear-out mechanism in electrolytic capacitors failures are deterioration of the electrolyte between electrodes and loss of electrolyte by vapours (dry-up).

The electrolytic capacitors have series internals resistances (ESR – equivalent series resistance) that increase with operating time, temperature and voltage [6-9]. Increasing of ESR conducts to heating of electrolytic capacitor and decreasing of capacitor value and increasing of tangent of loss angle (tan δ) [10]. Today, to fix the problems in electronic circuit must be used an oscilloscope, an analogue multimeter, a digital multimeter, and an ESR meter to verify electrolytic capacitors life.

2. Electrolytic capacitors. Design, using, and electrical models

Aluminum electrolytic capacitor is widely used in power electronics circuits. The main functions are: bus voltage, stabilization, etc. Most of electrolytic capacitors have maximum temperature ratings: 85 ^{0}C and 105 ^{0}C, and a few rated 125 ^{0}C [11], [12]. The aluminum electrolytic capacitor consists of (Figure 1): plastic insulation, aluminum, metal plate, dielectric, positive and negative charge connections.

There are different types of capacitors (Figure 2): ceramic capacitors, plastic capacitors, aluminum electrolytic capacitors, power capacitors, tantalum capacitors, double-layer capacitors and supercapacitors [13].

The electrolytic capacitors life is finite depending on electrolyte evaporation (dry-up of electrolytic capacitors). E.g.: at 35 ^{0}C, the electrolytic capacitors life is 16 years [10]. The electric conductivity depends of the type of capacitors (Figure 3).

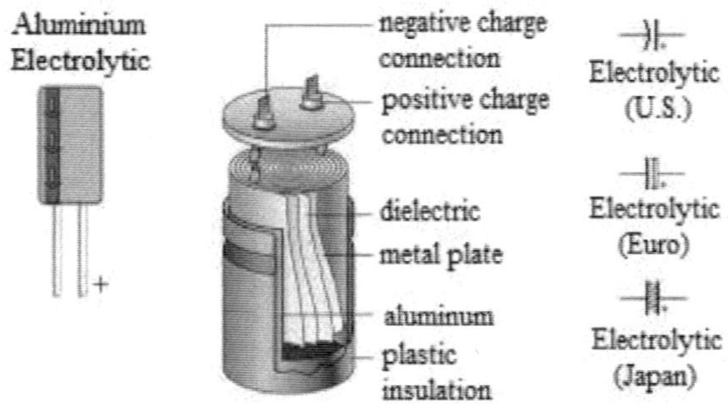

Figure 1. Design of electrolytic capacitor

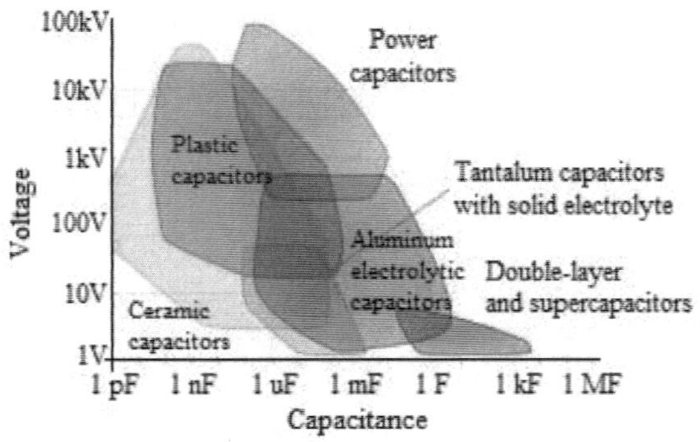

Figure 2. Typical capacitors used in practice

Capacitor life is defined as the time laps to the point when the capacitance value is decreased to 70% of the initial value. Usually, the capacitors users could not determine when should replace fault capacitors before they fail completely. Capacitor capacity decreases and tan δ increases after approximately 10000 hours of operation.

Electrolytic capacitors are sensible to: especially temperature and voltage, and humidity, temperature gradients, shocks and vibrations [11], [14], [15].

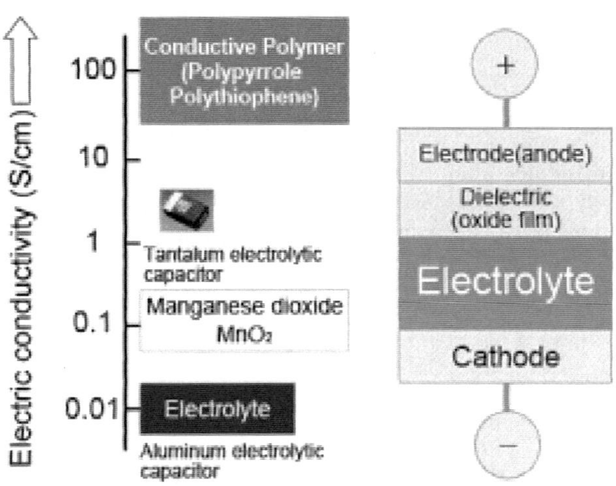

Figure 3. Electric conductivity depending on capacitors type

For switch mode power sources, the fault rate is [12]:

- 59% for electrolytic capacitors;
- 30% MOSFET;
- 5% coils;
- 2% diodes.
- 4 % others.

Capacitors have equivalent series internal resistance ESR that conduct to heating when ripple current flows [16]. The failure mechanisms and modes in aluminum electrolytic capacitor is presented in Figure 4 [17].

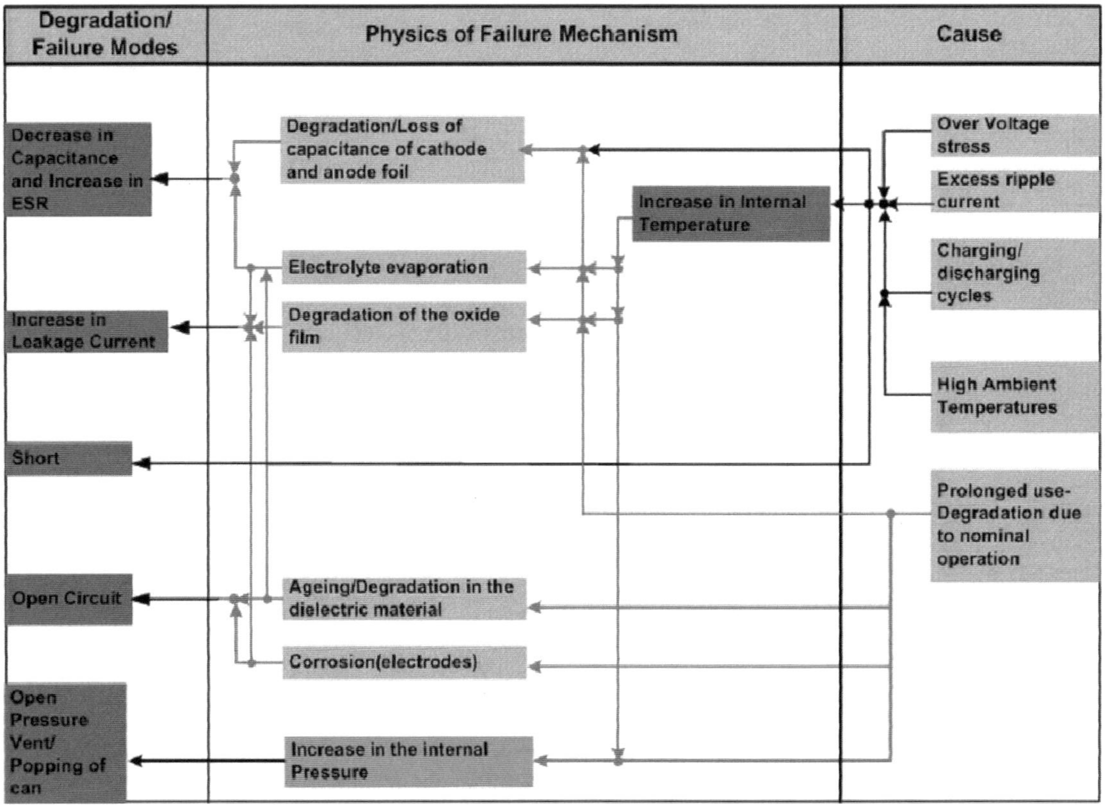

Figure 4. Failure mechanisms and modes in aluminum electrolytic capacitor [17]

Primary wear-out mechanisms in electrolytic capacitors is mainly due the loss of electrolyte by vapor diffusion and deterioration of the electrolyte. Typically, lowering the mass of the electrolyte leads to ESR growth (sometimes even 20 times), and capacitor capacity decreases by up to 40%. The detailed electrical model for electrolytic capacitor is presented in Figure 5.

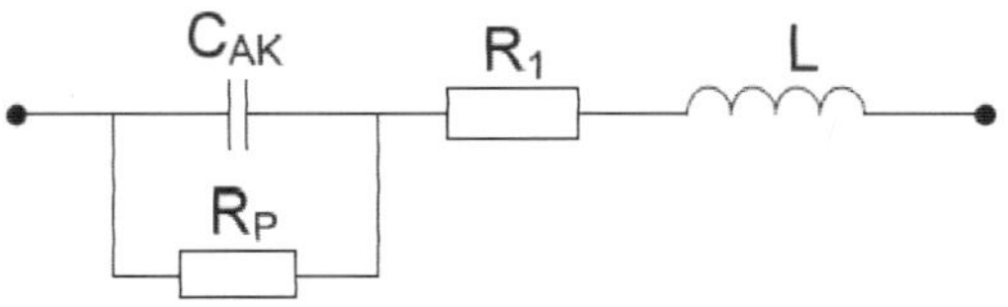

Figure 5. Detailed electrical model for electrolytic capacitor [4]

In Figure 5: C_{AK} is ideal anode-cathode capacitance (main element of the capacitor), R_P is the parallel resistance due to alumina layers, R_1 is series resistance of connections, frames and separator, and L is equivalent inductance series of connections and windings.

In practice, usually, is used simplified electrical model for electrolytic capacitor – Figure 6.

Figure 6. Simplified electrical model for electrolytic capacitor

$$Z_C = \sqrt{ESR^2 + \left(ESL \cdot \omega - \frac{1}{C \cdot \omega} \right)^2} \quad (1)$$

$$C = C_{AK} \cdot \left(1 + \frac{1}{R_p^2 \cdot C_{AK}^2 \cdot \omega^2} \right) \quad (2)$$

$$ESR = R_1 + \frac{R_p}{1 + R_p^2 \cdot C_{AK}^2 \cdot \omega^2} \quad (3)$$

$$ESL = L \quad (4)$$

3. ESR-meters

Troubleshooting power electronic circuits requires an investment in quality equipment. Today, to fix the problems in power electronic circuits must be used: an oscilloscope, a good analogue multimeter, a high quality digital multimeter, and an ESR-meter to verify electrolytic capacitors.

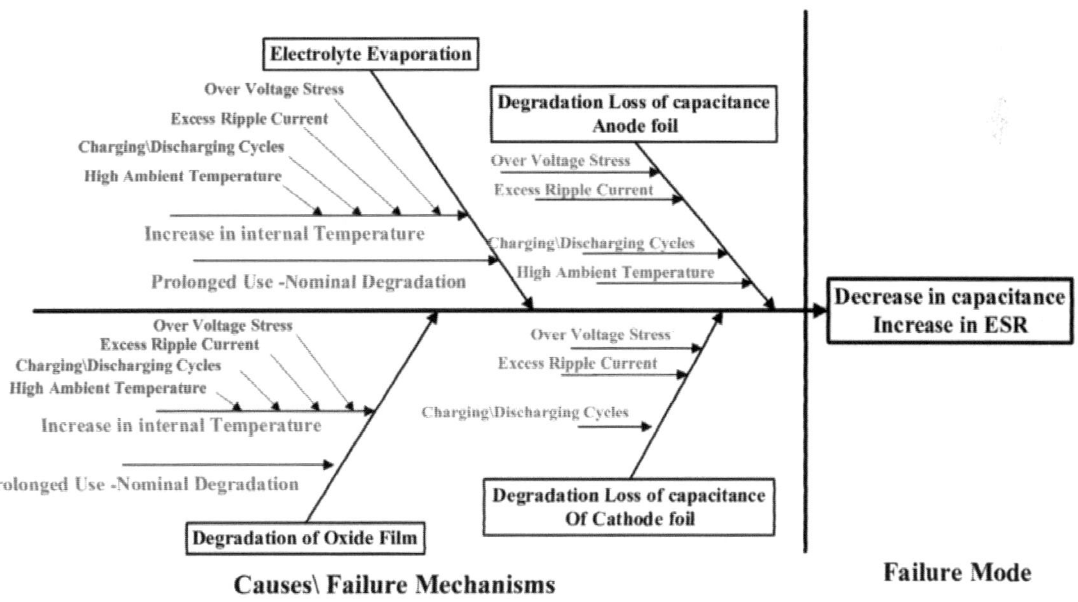

Figure 7. Fishbone diagram of failure mechanisms in aluminium electrolytic capacitor [18]

During along a short time or a long time operation of aluminum electrolytic capacitors there are a lot of failure mechanisms which switch off the operation of electronic circuits (Figure 7).

An ESR meter is an instrument that measure of equivalent resistances for electrolytic capacitors (ESR). It is an alternating current meter using pulse or high frequency signals (tens of kHz). It can be used with AC measurement bridges.

The capacitor is measured either directly or in a circuit free of voltage, with the capacitor discharged, because the measured resistance will be less than any component from the circuit.

An ideal capacitor has a value close to 0 Ω (ESR). A second verification can be done with an ohmmeter.

If the ESR value is high, over 10 Ω, the capacitor is defective, regardless of capacity of capacitor.

Table 1. Maximum value for ESR (Ω) depending on nominal capacity of electrolytic capacitor and nominal voltage [9], [16]

Capacity (μF)	10 (V)	16 (V)	25 (V)	35 (V)	50 (V)	100 (V)	160 (V)	250 (V)
1					5.0	7.0	10.0	14.0
2.2					4.0	6.0	8.0	10.0
4.7			3.0	3.0	3.0	4.0	4.0	3.5
10		2.0	2.0	2.0	2.0	1.2	1.5	2.8
22	1.3	1.3	1.3	1.3	1.3	0.66	1.1	1.2
47	1.3	1.3	1.3	0.6	0.6	0.32	0.46	0.60
100	1.3	0.6	0.6	0.33	0.33	0.16	0.24	0.30
220	0.6	0.33	0.33	0.25	0.19	0.09	0.14	0.27
470	0.33	0.25	0.19	0.14	0.09	0.06		
1000	0.19	0.14	0.09	0.07	0.06			
2200	0.09	0.07	0.06	0.05	0.04			
3300	0.07	0.06	0.05	0.04				
4700	0.06	0.05	0.04	0.03				
10000	0.04	0.03						

4. Making an analogue ESR-meter

The analogue ESR meter is made with a JFET integrated circuit, which can be used to identify defective electrolytic capacitors (Figure 8).

It is powered by continuous voltage and consists of a differential voltage source, a 100 kHz oscillator, a Wheatstone bridge with resistors supplied with alternating voltage, in a connecting arm of bridge being connected the electrolytic capacitors to be checked, a voltage amplifier-comparator with optical warning, a high-pass filter and a voltage-current converter [16].

At the output of voltage-current converter is a milliammeter pointing out the ESR value (up to 20 Ω). Measurements were made on a batch of electrolytic capacitors (new and used) and a comparative analysis was performed between the measured values with an analogue ESR-meter and a digital one.

Components list (in Figure 8):

R1=1K5	C1,C2,C5,C6=1μF	R19=680 Ω	
R2,R3,R4,R5=10K	C3=1nF, POLY	R20=100 Ω	
R6=68K	C4=100nF, POLY	R21=500 Ω	
R7=4K7	DS1,DS2=1N4007	R10,R12=5,6 Ω, 1%	R14,R15 =1K
R8=12K	DS3=1N4148	T1,T3=BC547	T2=BC557
R9,R11=1K, 1%	DL1=LED	R13,R16,R17=47K	IC1=TL084

In Table 2: C(μF) – nominal capacity; U (V) – nominal voltage; C_m (μF) – measured capacity; ε_V (%) – percentage voltage loss; ESR_{max} (Ω) – the maximum ESR; ESR_d (Ω) – measured ESR with digital ESR-meter, LCR-T5 Multifunction Tester; ESR_a (Ω) - measured ESR with analogue ESR-meter; ε_{rESR} (%) – relative error for measuring ESR between analogue and digital ESR-meters.

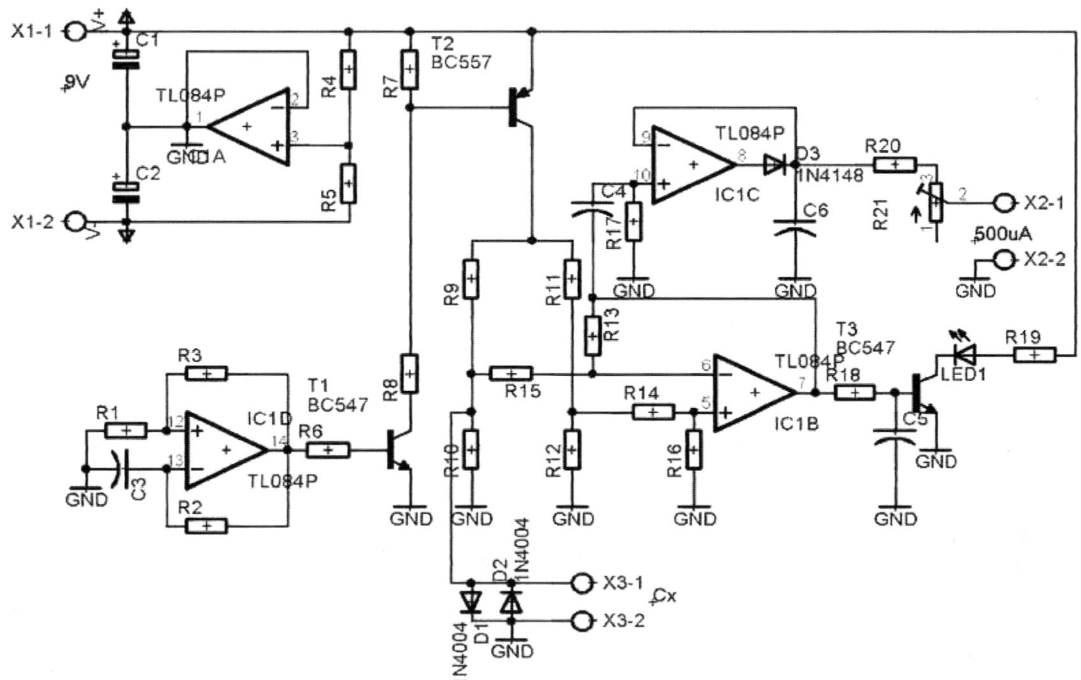

Figure 8. The electronic diagram for analogue ESR-meter [16]

Table 2. Measurements made on electrolytic capacitors (new and used) with digital and analogue ESR-meters

No.	C(μF)	U (V)	C_m (μF)	ε_V (%)	ESR_{max} (Ω)	ESR_d (Ω)	ESR_a (Ω)	ε_{rESR} (%)	Cap.
1	4.7	50	5.272	12.17	3	3.64	2.64	-27.47	√
2	10	25	11.81	18.10	2	3.94	3.44	-12.69	X
3	10	63	10.43	4.30	1.3	0.86	2.39	177.91	?
4	22	10	25.49	15.86	1.3	14	19	35.71	X
5	22	100	24.64	12.00	0.66	0.43	0.39	-9.30	√
6	47	10	48.73	3.68	1.3	13	> 20		X
7	47	10	57.03	21.34	1.3	17	> 20		X
8	47	10	61.57	31.00	1.3	1.14	1.29	13.16	√
9	47	16	46.1	-1.91	1.3	2.8	2.65	-5.36	X
10	100	10	102.8	2.80	1.3	1.24	0.99	-20.16	√
11	100	10	76	-24.00	1.3	11	18	63.64	X
12	100	10	109.3	9.30	1.3	2.6	3.2	23.08	X
13	100	10	97.91	-2.09	1.3	3.5	3.6	2.86	X
14	100	10	112.3	12.30	1.3	0.96	0.96	0.00	√
15	100	10	104.6	4.60	1.3	0.84	0.99	17.86	√
16	100	10	102.2	2.20	1.3	0.69	0.89	28.99	√
17	100	10	120.7	20.70	1.3	0.84	0.94	11.90	√

18	100	50	-		-	0.33	> 20		X
19	220	6.3	217.2	-1.27	1.3	0.46	0.49	6.52	√
20	220	10	164.6	-25.18	0.6	0.49	0.59	20.41	√
21	220	10	214.3	-2.59	0.6	0.15	0.49	226.67	√
22	220	10	214.5	-2.50	0.6	1.74	1.91	9.77	X
23	220	10	244.6	11.18	0.6	0.43	0.69	60.47	√
24	220	10	202.1	-8.14	0.6	0.32	0.59	84.38	√
25	470	10	506.4	7.74	0.33	0.23	0.24	4.35	√
26	470	25	128	-72.77	0.19	13	21	61.54	X
27	470	25	-		0.19	> 1000	> 20		X
28	470	25	-		0.19	> 1000	> 20		X
29	470	25	26x10-6		0.19	> 1000	> 20		X
30	1000	16	639.8	-36.02	0.14	0.21	0.54	157.14	?
31	1000	35	973.5	-2.65	0.07	0.12	0.04	-66.67	√
32	1000	50	705.1	-29.49	0.14	0.19	1.04	447.37	X
33	1000	50	932.3	-6.77	0.06	0.09	0.01	-88.89	?
34	1000	50	844.5	-15.55	0.06	0.1	0.09	-10.00	?
35	1000	50	932.3	-6.77	0.06	0.09	0.04	-55.56	?
36	2200	25	2219	0.86	0.06	0.08	0.04	-50.00	?
37	2200	25	2233	1.50	0.06	0.08	0.04	-50.00	?
38	2200	25	2219	0.86	0.06	0.08	0.04	-50.00	?
39	2200	25	2240	1.82	0.06	0.09	0.01	-88.89	?
40	3300	16	3447	4.45	0.06	0.08	0.01	-87.50	?
41	3300	16	3435	4.09	0.06	0.07	0.09	28.57	?
42	3300	16	3332	0.97	0.06	0.09	0.04	-55.56	?
43	3300	16	3493	5.85	0.06	0.09	0.04	-55.56	?
44	4700	25	5458	16.13	0.04	0.07	0.04	-42.86	√

Increasing the ESR over the maximum value is not related to a decrease in capacitor capacitance.

Some faulty capacitors have been identified with the analogue ESR-meter (the digital ESR-meter considers them good).

5. Conclusion

ESR-meter is an electronic device, relatively new introduced in power electronic troubleshooting. Using the ESR-meter to troubleshoot electronic circuits becomes an indispensable device because most of the power electronics failures are due to electrolytic capacitor failure. Sometimes measurements can be made without removing the capacitors in the circuit, so the repair time is greatly diminished.

ESR (Equivalent Series Resistance) is the sum of all internal resistors of a capacitor measured in ohms (reactance is not included). ESR is a dynamic size and must be measured at alternating voltage of tens of kHz.

Development of capacitors with long-life, high allowable ripple current and low impedance are effective to make high performance power supplies.

References

[1] Sankaran V A, Rees F L, A and C S 1997 *Electrolytic Capacitor Life Testing and Prediction*, IEEE Industry Applications Society Annual Meeting, October 5-9, New Orleans, Louisiana, USA, pp 1058-1065

[2] Zhao K, Ciufo P and Perera S 2010 *Lifetime Analysis of Aluminum Electrolytic Capacitor*

Subject to Voltage Fluctuations, Research Online, University of Wollongong, Australia, 7 pp
[3] Lahyani A, Venet P, Grellet G and Viverge P J 1998 Failure Prediction of Electrolytic Capacitors During Operation of a Switchmode Power Supply, *IEEE Transactions on Power Electronics* **13**(6) 1199-1207
[4] Pănoiu C, Rob R, Baciu I, Pănoiu M 2011 *IGBT Command in Active Filtering of the Harmonics Current*, WSEAS European Computing Conference Paris, France, April 28-30, pp 158-163
[5] Pănoiu C, Rob R, Baciu I and Pănoiu M 2011 Shunt Active Filter Command Designed in Labview, *International Journal of Circuits, Systems and Signal Processing* **5**(5) 513 – 520
[6] Harada K, Katsuki A and Fujiwara M 1993 Use of ESR for Deterioration Diagnosis of Electrolytic Capacitor, *IEEE Transactions on Power Electronics* **8**(4) October 355-361
[7] Hewitt D, Green J, Davidson J, et al. 2016 *Observation of Electrolytic Capacitor Ageing Behaviour for The Purpose Of Prognostics*, IECON 2016, 42nd Annual Conference of the IEEE Industrial Electronics Society, October 24-27, Florence, Italy, 6 pp
[8] Kulkarni C S, Biswas G, Celaya J R and Goebel K 2013 Physics Based Degradation Models for Electrolytic Capacitor Prognostics under Thermal Overstress Conditions, *International Journal of Prognostics and Health Management* **4** 1-17
[9] Albertsen A 2008 *Electrolytic Capacitor Lifetime Estimation*, Jianghai Europe, January
[10] Kiuchi K and Yanagibashi M 1983 *Operating Life of Aluminum Electrolytic Capacitor*, IEEE, Telecommunications Energy Conference, INTELEC 03, Fifth International Conference, October 18-21, Tokyo, Japan, pp 535-540
[11] Wang H et al. 2014 Transitioning to Physics of Failure as a Reliability Driver in Power Electronics, *IEEE Journal of Emerging and Selected Topics in Power Electronics* **2**(1) March 97-114
[12] Laadjal K, Sahraoui M, Marques Cardoso A J and Amaral A M R 2017 *On-Line Estimation of Aluminum Electrolytic-Capacitor Parameters Using a Modified Prony's Method*, Diagnostics for Electrical Machines, Power Electronics and Drives (SDEMPED), 2017 IEEE 11th International Symposium on, 29 August - 01 September, Tinos, Greece, pp 387-393
[13] Sood B 2013 *Root-Cause Failure Analysis of Electronics*, Philadelphia, University of Maryland, USA, March 14, 2013, pp 1-19
[14] Parler S G 1999 *Thermal Modeling of Aluminum Electrolytic Capacitors*, IEEE Industry Applications Society Conference, Thirty-Fourth IAS Annual Meeting, October, pp 2418-2429
[15] ***https://enphase.com/sites/default/files/Electrolytic_Capacitor_Expert_Report.pdf
[16] ***www.epsicom.com/kits.php
[17] Kulkarni C, Biswas G, Celaya J and Goebel K 2011 *Prognostic Techniques for Capacitor Degradation and Health Monitoring*, The Maintenance and Reliability Conference, Knoxville, TN, USA, September 13-15, 2011, pp 1-14
[18] Kulkarni C, Biswas G, Celaya J and Goebel K 2013 *Physics Based Degradation Models for Capacitor Prognostics under Thermal Overstress Conditions*, International Journal of Prognostics and Health Management **4**(005) 1-17

IOP Conf. Series: Materials Science and Engineering 477 (2019) 012032 doi:10.1088/1757-899X/477/1/012032

Computerized system based on FreeCAD for geometric simulation of the oil and gas equipment thread turning

V B Kopei, O R Onysko and V G Panchuk

Ivano-Frankivsk National Technical University of Oil and Gas, Department of Computerized Machine-Building Production, Karpatska str., no. 15, 76019 Ivano-Frankivsk, Ukraine

E-mail: vkopey@gmail.com

Abstract. This paper describes the principles of the computerized system for creating geometric models of threaded joints with various geometric deviations, which arise during the thread turning, and demonstrates the use of these models to determine the dependencies of reliability indicators on these deviations by the finite element method. Using the developed system, based on Python and FreeCAD, axisymmetric finite element models of drilling tool joints and sucker rod couplings with deviations of the profile angle and the thread pitch were created and analyzed.

1. Statement of the problem

Requirements for reliability of threaded connections of oil and gas equipment are constantly increasing. It is possible to determine the dependencies of reliability indicators on different geometric parameters by finite-element (FE) modeling of the stress-strain state [1-3]. But for conducting of such studies it is necessary to construct a parametric geometric model of a threaded connection.

One of the main parameters that affects the reliability of threaded connections is their geometric accuracy. Geometric accuracy depends on the machining errors of thread turning (machine errors, load induced errors, thermal growth errors, error caused by tool wear and other errors). The purpose of this work is the development of the computerized system, which is designed to build models of threaded joints with various geometric deviations arising during the thread turning, and the demonstration of the use of these models to determine the dependencies of reliability indicators on these deviations. This rather complicated geometric problem can be solved with the help of modern geometric modeling kernels, in particular Open CASCADE Technology, on which FreeCAD is based.

2. Description of the system

The computerized system being developed is based on the open-source software Python 2.7 and FreeCAD 0.17. Python is a high-level general-purpose programming language. FreeCAD is a parametric 3D CAD based on Open CASCADE Technology 7.2.0 geometry kernel with Python API. Open CASCADE Technology is a software development kit intended for development of applications dealing with 3D CAD data. Note that if the FreeCAD capabilities for creating a model are not enough, you can use the PythonOCC library [4], in which the Open CASCADE kernel is fully available.

The developed system has the ability to create the geometrical models of threads with different computational complexity and adequacy (Table 1). The most adequate models for the FE simulation of the stress-strain state are 3D models with helical threads. But they have the highest computational complexity for geometric and FE simulation. Planar geometric models are often used to create axisymmetric FE models [5].

Table 1. Computational complexity and adequacy of thread models

3D real thread (helical)	3D quasi-thread (circular grooves)	2D planar
high	medium	low

We describe the program using the drilling tool joint ZN-80 GOST 5286 model as an example. This is the joint with the Z-66 GOST 28487 thread (2 3/8 REG API Spec 7 equivalent). First import the necessary modules.

```
import FreeCAD as App
import FreeCADGui as Gui
import Part
import numpy as np
from dims import d
```

The functions in the `FreeCAD` module allow working with FreeCAD documents. `FreeCADGui` module contains everything related to the Graphical User Interface and the 3D views [6]. The `Part` module is the direct connection between FreeCAD and the Open CASCADE kernel. With this module, you can create BREP (boundary representation) topological shapes (Vertex, Edge, Wire, Face, Shell, Solid, Compsolid, Compound) [6]. The `numpy` module provides the routines for fast operations on multidimensional arrays. The `dims` module contains an object `d` which attributes are equal to different parameters of the joint. If you are not using an internal Python interpreter, you must specify the path to the FreeCAD modules.

```
import sys
sys.path.append("e:\\FreeCAD 0.17x64\\bin")
```

We can use parametric FreeCAD sketches to simplify the development of the models. The sketches of tools, workpieces of pin and box are shown on the Figure 1.

The function `rebuildSketch(dim, sk)` can be used to rebuild the parametric sketch. The `dim` parameter is a dictionary with pairs (sketch constraint ID, dimension value), and the `sk` parameter is the name of a sketch. The function calls `setDatum(k, v)` method for each pair in `dim` to set value `v` of a distance or angle constraint with ID `k`, rebuilds the document and returns the face.

```
def rebuildSketch(dim, sk):
    doc=App.getDocument("Sketches") #get the document object by name
    sketch=doc.__getattribute__(sk) #get the sketch object by name
    # for each pair (scketch constraint ID, dimension value)
    for k,v in dim.iteritems():
        sketch.setDatum(k,v) # set dimension value
    doc.recompute() # rebuild the document
    w=sketch.Shape.Wires[0] # get the first wire
    f=Part.Face(w) # create the face
    return f
```

The `revolve` function creates the solid shape by rotating the face `f` around the Y axis.

```
def revolve(f):
    return f.revolve(App.Vector(0,0,0), App.Vector(0,1,0))
```

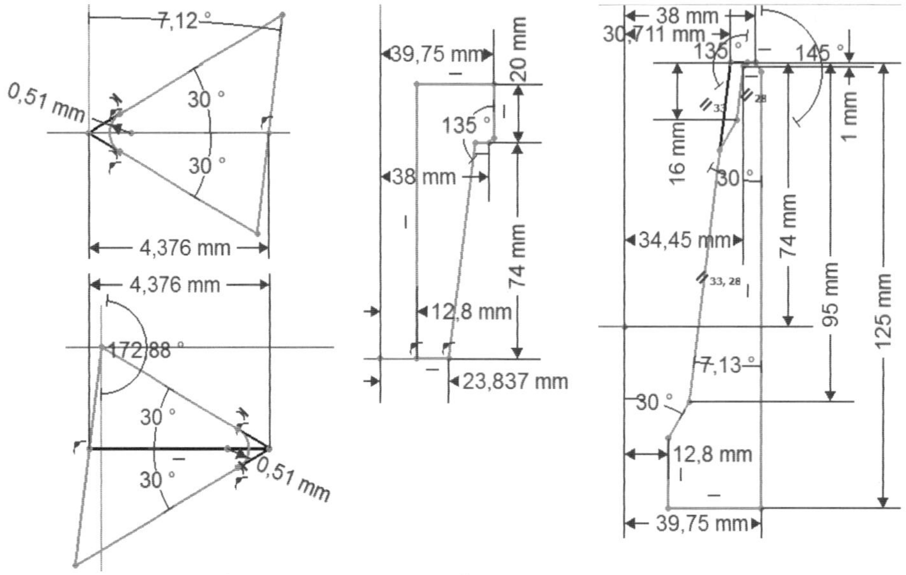

Figure 1. Parametric sketches of tools (a), pin workpiece (b) and box workpiece (b)

To create the ideal helix, use the `makeLongHelix` function from the `Part` module. To avoid errors when building long threads, do not use the `makeHelix` function. It is necessary to rotate the helix so that its axis coincides with the Y axis. These operations are executed by the function `helix(r,h,p,fi)`, where `r` is the radius, `h` is the height, `p` is the pitch, `fi` is the cone angle (rad).

```
def helix(r,h,p,fi):
    w=Part.makeLongHelix(p,h,r,np.degrees(fi))
    w.rotate(App.Vector(0,0,0),App.Vector(1,0,0),-90)
    return w
```

The thread on the pin (or box) is created by the `makeThread(f, h, s)` function, where `f` is a tool face, `h` is a helix wire, `s` is a workpiece solid.

```
def makeThread(f,h,s):
    s2=h.makePipeShell(f.Wires, True, True) # the helical solid
    s=s.cut(s2) # Boolean cut operation
    return s
```

With the help of these functions, it is possible to create nominal solids of the pin or box (Figure 2).

```
App.open(u"D:/3/Sketches.FCStd") # open the document with sketches
fA0=rebuildSketch(dim={6:d.fi, 9:d.H}, sk='Sketch') # tool face
fA0.translate(d._v2) # move the tool to the starting position
sA0=revolve(rebuildSketch(dim={20:d.fi, 16:d.d3/2}, sk='Sketch001'))
# workpiece solid
hA0=helix(r=d._r, h=d.l4-12, p=d.P, fi=d.fi) # helix wire
hA0.translate((0,d.l3-d.l4,0)) # move the wire
```

```
sA1=makeThread(fA0, hA0, sA0) # pin with thread
```

Figure 2. The 3D model of the drilling tool joint ZN-80 (GOST 5286)

Now simulate threads with different deviations. To simulate various machining errors in the program, it is possible to change the values of the geometric parameters of the workpiece and tool and the parameters of the tool path relative to the workpiece. The trajectory of the relative movement of the workpiece and the tool is a conical helical curve with parametric equations (1). In this equations $a = \tan(\varphi)p/(2\pi)$, $b = p/(2\pi)$, t - parameter (rad), r - start radius, φ - cone angle (rad), p - pitch.

$$x = (r+at)\cos(-t),$$
$$z = (r+at)\sin(-t), \quad (1)$$
$$y = bt.$$

The function `helixPoints(r, h, p, fi, n)` returns an array of points of the conical helix. Parameter `r` is the smaller radius, `h` - height, `p` - step, `fi` - cone angle, `n` - the number of points in one turn. Here `t`, `x`, `y`, `z` - numpy arrays.

```
def helixPoints(r, h, p, fi, n):
    a=np.tan(fi)*p/(2*pi)
    b=p/(2*pi)
    k=h/p # number of turns
    N=int(n*k) # total number of points
    t=np.linspace(0,h/b,N)
    x=(r+a*t)*np.cos(-t)
    z=(r+a*t)*np.sin(-t)
    y=b*t # helix axis
    return zip(t,x,y,z)
```

You can simulate various systematic or random machining errors by using the helix equation with deviations (2), where x_0, y_0, z_0 - coordinates of points of a nominal helix, A_x, A_y, A_z, Δx, Δy, Δz - constants or random variables. For nominal thread $A_x = A_y = A_z = 1$, $\Delta x = \Delta y = \Delta z = 0$. For example, the constant A_x can be used to simulate the spindle radial run-out, and the variable Δy - to simulate the pitch error. To simulate the random machining errors, the variables A_x, A_y, A_z, Δx, Δy, Δz must be random variables with some distributions.

$$x = A_x x_0 + \Delta x,$$
$$z = A_z z_0 + \Delta z, \quad (2)$$
$$y = A_y y_0 + \Delta y.$$

The function `perror(points)` returns the points coordinates of the helix with the deviations, where `Δx`, `Δy`, `Δz` are random variables with a triangular distribution. The `points` parameter is the points of the nominal helix. The user can modify this function to simulate other errors.

```
def perror(points):
    epoints=[] # points of the helix with the deviations
    d=0.2 # triangular distribution parameter
    for t,x,y,z in points:
        x=x+np.random.triangular(-d,0,d)
        y=y+np.random.triangular(-d,0,d)
        z=z+np.random.triangular(-d,0,d)
        epoints.append((t,x,y,z))
    return epoints
```

The thread creation algorithm is based on the approximation of the tool movement trajectory by splines or polylines. To approximate a helix with deviations by a polyline, use the `helix2(points)` function, where `points` are the points of the helix with deviations.

```
def helix2(points):
    pts=[(x,y,z) for t,x,y,z in points]
    h=Part.makePolygon(pts) # polyline wire
    return h
```

It is possible to create a thread with deviations in different ways. For example, the `makePipeShell` function can be used to create a helical body using a spline (or polyline) and a list of tool profiles at each point. But the most simple and universal way is to use shape operations "loft" and "cut" for each point `helix2(points)` (Figure 3). The function `makeThread2(f, h, s)` creates such thread by using `copy`, `rotate`, `translate`, `makeLoft` and `cut` functions. The parameter `f` is the face of the tool, `h` is the helix wire, `s` is the workpiece solid. Calculations can take a long time if there are many points.

```
def makeThread2(f, h, s):
    w=f.Wires[0] # tool wire
    points=[(v.X,v.Y,v.Z) for v in h.Vertexes] # helix points
    W=[] # profiles at different points of the helix
    for x,y,z in points: # for each point
        w_=w.copy() # tool wire copy
        # tool rotation angle relative to the workpiece
        t=np.arctan2(-z,x)
        # rotate the tool around an Y axis
        w_.rotate(App.Vector(0,0,0),App.Vector(0,1,0),np.degrees(t))
        w_.translate(App.Vector(x,y,z)) # move the tool to x,y,z
        W.append(w_.copy()) # append the tool profile to the list W
        if len(W)>1:
            # create a solid shape using two adjacent profiles
            st=Part.makeLoft([W[-2],W[-1]],True)
```

```
        s=s.cut(st) # Boolean cut for workpiece and shape st
    return s,W
```

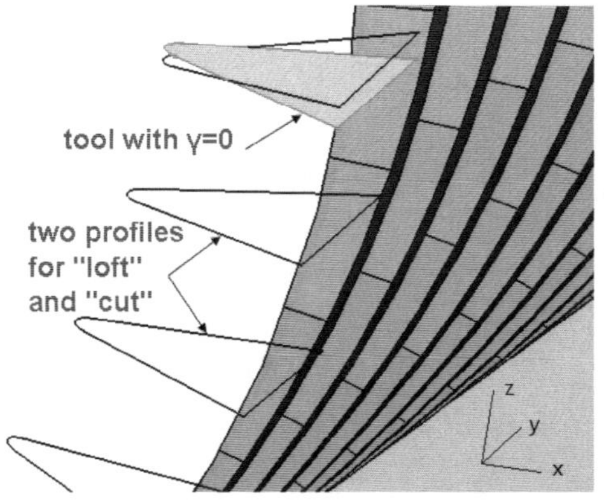

Figure 3. Creating threads with deviations

You can also simulate the slope of the rake face of the cutting tool and nonzero rake angle γ. To do this, rotate the face by an angle γ around the line, that parallel to the helix axis and passes through the middle of the profile (Figure 3). The following code creates a pin with the thread that has deviations.

```
fA3.rotate(App.Vector(0,0,0),App.Vector(0,1,0),-10) # rake angle
points=perror(points) # points of a helix with deviations
hA3=helix2(points) # helix with deviations
hA3.translate((0,d.l3-d.l4,0)) # move the helix
sA4,W=makeThread2(fA3, hA3, sA0) # pin with thread
```

It is possible to calculate the volume or area of the symmetrical difference (XOR or union without the intersection) for nominal and real threads. Below you can see the corresponding FreeCAD API code. The testing of the thread suitability with deviations can be realized by Boolean operations on the maximum permissible and real shapes.

```
# XY plane:
f=Part.makePlane(400,400,App.Vector(-200,-200,0),App.Vector(0,0,1))
f1=sA3.common(f).Faces[0] # axial section of a pin
f2=sA4.common(f).Faces[0] # axial section of a box
f3=f1.cut(f2).fuse(f2.cut(f1)) # Boolean XOR operation
print f3.Area # XOR area
```

Figure 4 shows the maximum deviations between nominal (γ=0°, A_x=0, Δy=0) and real profiles when there is combination of different errors. The nominal profile is indicated by an arrow, and the XOR area is between the profiles. Deviations are almost not visible in Figure 4(a). In this case, according to the data [7], the deviation of the profile angle is not more than 0.47°.

Figure 5 shows the 3D nominal thread (Δx=0, Δy=0, Δz=0) and the 3D thread with deviations by equation 2, where Δx, Δy, Δz - random variables with a triangular distribution (with parameters a=-0.2 mm, c=0 mm, b=0.2 mm). Calculated XOR volume is equal 133.79 mm^3.

3. Use of geometric models for finite element simulation

The models of threaded connections constructed with this program can be used to simulate the stress-strain state by the finite element method and justify the values of the thread tolerances. We used an open source finite element mesh generator Gmsh 2.7 [8] and a FEA software CalculiX 2.12 [9].

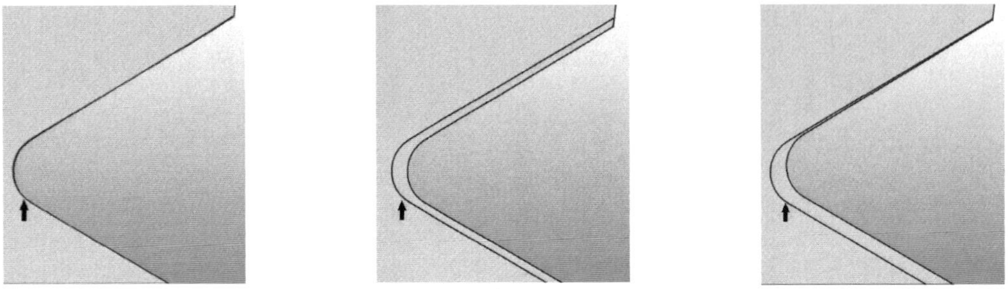

A b c

Figure 4. Deviations between nominal and real profiles when there is combination of different errors: γ=-10°, Δy =0, A_x =0, XOR area=0.0353 mm^2 (a); γ=-10°, Δy =0, A_x =1.01 mm, XOR area=0.8412 mm^2 (b); γ=-10°, Δy =0.1 mm, A_x =1.01 mm, XOR area=0.7744 mm^2 (c)

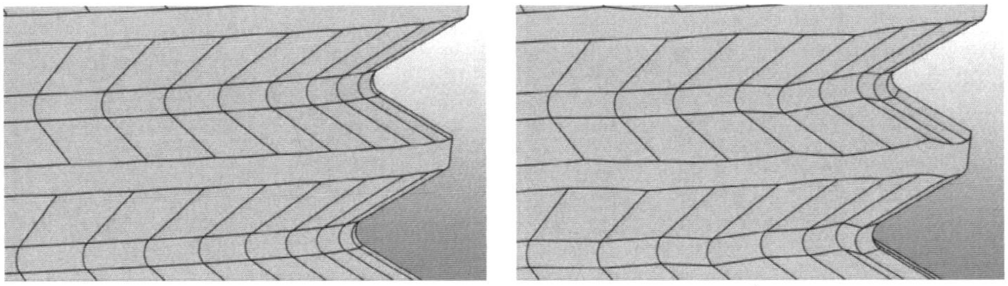

a b

Figure 5. The nominal thread (a) and the thread with random deviations (b)

Based on the planar geometric model of the drilling tool joint ZN-80, obtained with the help of the developed program, an axisymmetric FE model is constructed. The material of the parts is steel 40XH (GOST 4543) with the Young's modulus E=2.1e5 MPa, Poisson's ratio ν =0.28, yield strength σ_y =735 MPa, tensile strength σ_t =882 MPa. Material plasticity and friction are simulated. The size of the mesh elements is 0.2 mm. Two steps of the external axial tensile load were created: $F_{\min}$ =0 N and $F_{\max}$ =1 MN. The joint make-up was simulated by axial deformation Δ of the box shoulder. For each tool joint design, a value of Δ was set such that the contact pressure on joint shoulders was 274 MPa for $F_{\max}$. For the standard design Δ = 0.2 mm. For the approximate calculation of the fatigue safety factor *D*, the dependence of Sines [10], [11], with a material endurance limit of 207 MPa was used. Note that *D* values can only be used for the relative comparison of different designs, but not for calculating cyclic durability values [1]. A smaller value of *D* only indicates a lower cyclic durability.

The Figure 6 shows the values of equivalent von Mises stress σ_v and fatigue safety factor *D* as a function of the deviation of the thread profile angles of the pin from the nominal value. Deviations are determined by the formula $\Delta\varphi_{1,2} = \varphi_{1,2} - 30°$, where $\varphi_{1,2}$ is the angle between the pin profile side (index 1 means unloaded side, 2 - loaded side) and the line perpendicular to the axis, 30° is the nominal value. Much larger contact stresses arise on the loaded side 2 (the bottom side in Figure 7) if the joint is subjected to an axial tensile load $F_{\max}$. Deviations $\Delta\varphi_{1,2}$ can be caused by a nonzero rake angle γ of the tool and machining error, caused by tool wear. The deviation of the unloaded side $\Delta\varphi_1$ of the pin profile can be caused by the wear of the left cutting edge of the tool. As a rule, the wear of the left cutting edge near the corner is greatest, and this increases the angle of the unloaded side of the thread profile φ_1.

IOP Conf. Series: Materials Science and Engineering **477** (2019) 012032 doi:10.1088/1757-899X/477/1/012032

Negative values of the deviations are the cause of a sharp increase of σ_v values and decrease of D values (Figure 6). Areas with very small D values appear in the root of the box thread, and D values of the pin decreases insignificantly (Figure 7). Positive values of the deviations of both sides $\Delta\varphi_{1,2}$ are somewhat smaller decrease D values. The D value of the first root in the pin thread decreases most of all (Figure 7). Positive values of the deviations of the unloaded side $\Delta\varphi_1$ almost do not change the stress and fatigue strength of the joint (Figure 7). Thus, the tool wear along the left cutting edge has no influence on the fatigue strength of the joint.

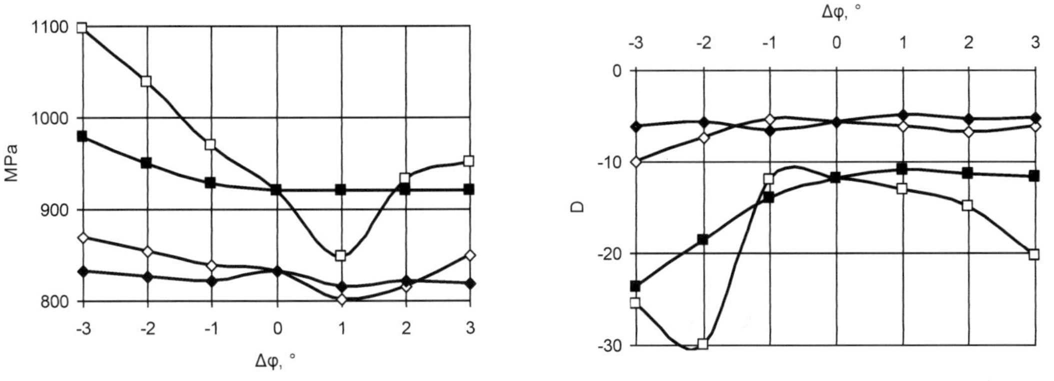

A b

Figure 6. Von Mises stress at F_{max} (a) and fatigue safety factor D (b) as a function of $\Delta\varphi_1$ (■, ♦) and $\Delta\varphi_{1,2}$ (□, ◊): □, ■ - maximum value in the joint; ◊, ♦ - value at the first loaded root of the pin thread

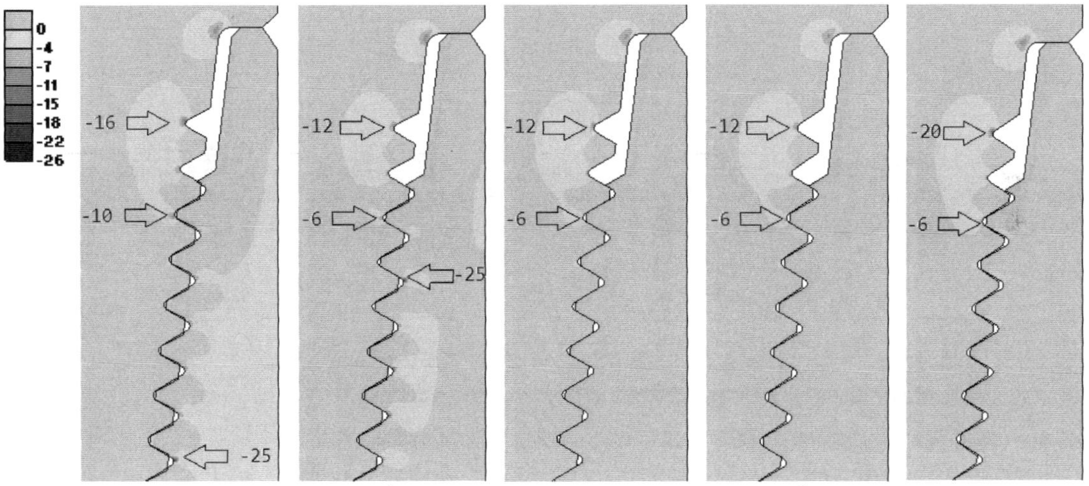

A b c d e

Figure 7. Distribution of the fatigue safety factor D for different $\Delta\varphi_{1,2}$ values: $\Delta\varphi_{1,2} = -3°$ (a); $\Delta\varphi_1 = -3°$ (b); $\Delta\varphi_{1,2} = 0°$ (c); $\Delta\varphi_1 = 3°$ (d); $\Delta\varphi_{1,2} = 3°$ (e)

Using the developed program axisymmetric FE models of 19 mm GOST 13877 (3/4 in. API Spec 11B equivalent) sucker rod couplings were also constructed. According to GOST 13877, the pin thread must be rolled, and the box thread can be rolled or manufactured by another method, for example, by a single-point cutting tool. The material of the parts is steel 40 (GOST 1050) with

E=2.1e5 MPa, ν=0.28, σ_y=314 MPa, σ_t=559 MPa. Material plasticity and friction are simulated. The size of the mesh elements is 0.2 mm. Two steps of the external axial tensile load were created, which form in the rod body the stress P_{min} = 0 MPa and P_{max} = 170 MPa. The joint make-up was simulated by axial deformation Δ of the box shoulder. For each coupling design, a value of Δ was set such that the contact pressure on joint shoulders was 312 MPa for P_{max}. For the standard design Δ = 0.12 mm.

In Figure 8 the von Mises stress and fatigue safety factor D as a function of angle deviation $\Delta\varphi = \Delta\varphi_1 = \Delta\varphi_2$ of box thread profile are shown. The index 1 denotes the unloaded profile side and 2 - the loaded one. It is noticeable that negative $\Delta\varphi$ values cause a slight decrease of stress values and an increase of D values in the first loaded root of the pin thread. But contact stresses grow in the zone of the minimum diameter of the box thread (Figure 9(a)). Positive $\Delta\varphi$ values sharply reduce the D values in the first root of the pin thread, but almost do not change the von Mises stress values in it for P_{max}.

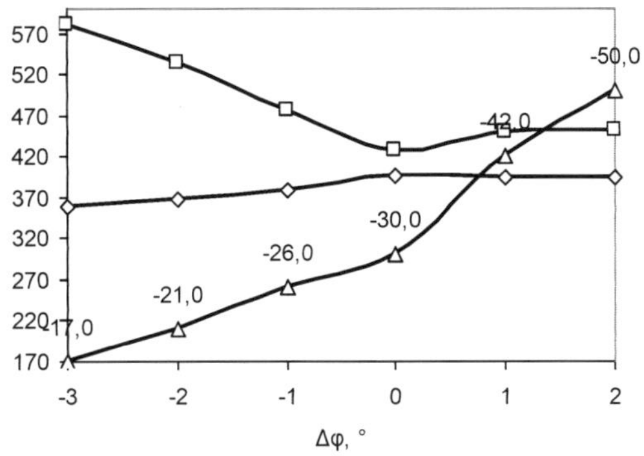

Figure 8. Von Mises stress at P_{max} (□, ◊) and fatigue safety factor D (Δ) as a function of $\Delta\varphi$: □ - maximum value in the joint; ◊, Δ - value at the first loaded root of the pin thread

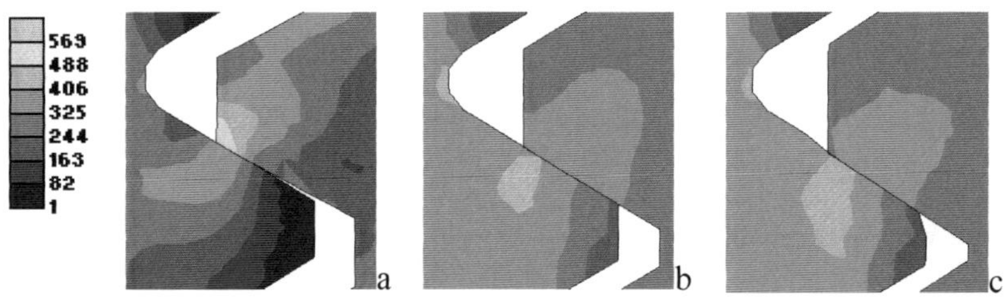

Figure 9. Von Mises stress distribution (MPa) in the first thread of sucker rod coupling at P_{max} for different $\Delta\varphi_{1,2}$ values: -3° (a); 0° (b); 2° (c)

The Figure 10 shows the von Mises stress and fatigue safety factor D as a function of Δp - the magnitude of decrease the box thread pitch on the each thread turn (starting from the first). Positive Δp values decrease the pitch of the box thread, and negative values increase the pitch. Positive Δp values are the cause of a sharp decrease of D values due to non uniform load to each thread (Figure 11, 12). Stresses, which are shown in the Figure 12(a), may occur after multiple make-up of the joint as a result of plastic deformation and wear of the first threads. Negative Δp values equalize loads in the thread, somewhat reduce the von Mises stresses in the first loaded root of the pin thread (Figures 10-12) and somewhat reduce the area with negative D values (Figure 13). Therefore, you should avoid

Δp >0 but try to provide Δp <0. For single-point threading on a CNC machine, this can be achieved by increasing the feed by the Δp value per one revolution of the spindle.

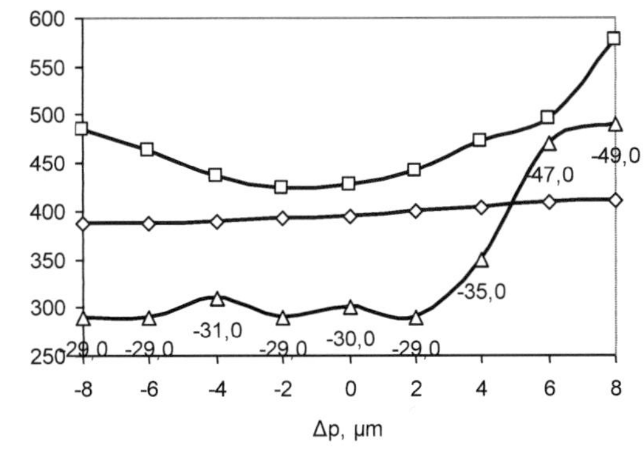

Figure 10. Von Mises stress at P_{max} (□, ◊) and fatigue safety factor D (Δ) as a function of Δp: □ - maximum value in the joint; ◊, Δ - value at the first loaded root of the pin thread

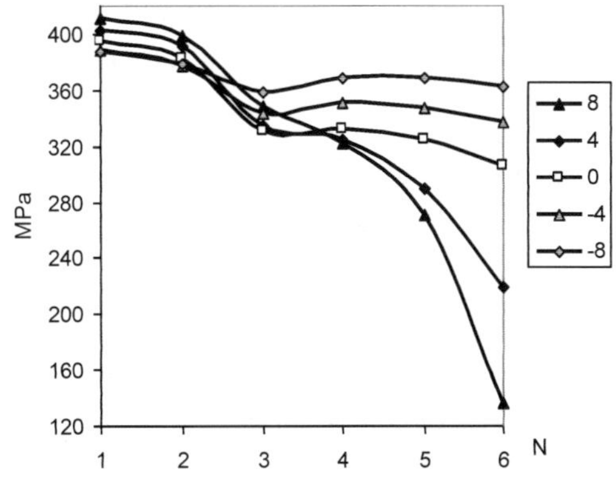

Figure 11. The values of the von Mises stress at P_{max} in the root N of the pin thread for different Δp values (μm)

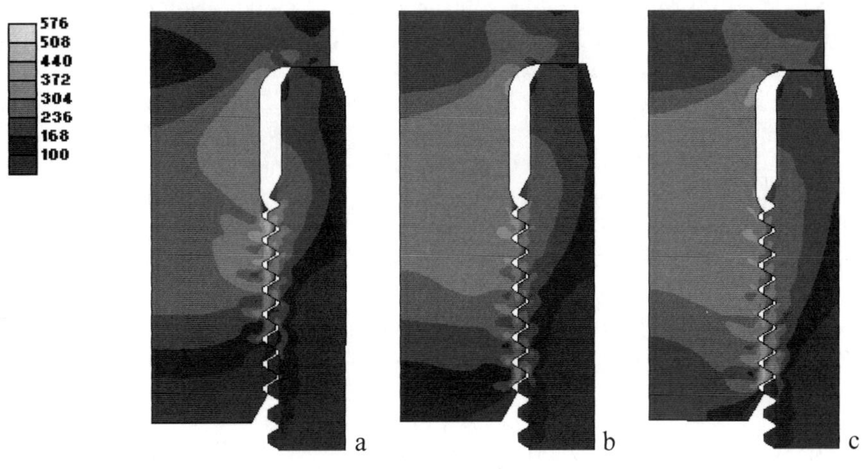

Figure 12. Von Mises stress distribution (MPa) in the sucker rod coupling at P_{max} for different Δp values: 8 μm (a); 0 μm (b); -8 μm (c)

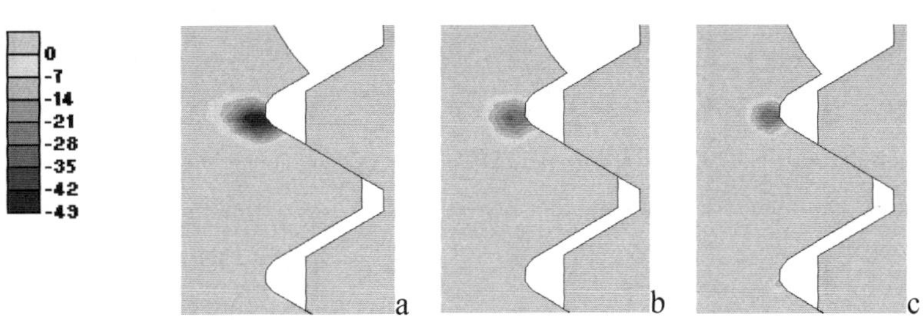

Figure 13. Distribution of the fatigue safety factor D in the first thread of sucker rod coupling for different Δp values: 8 μm (a); 0 μm (b); -8 μm (c)

4. Conclusions

The developed system can be used to construct 3D and 2D geometric models of threaded connections with deviations, justify their tolerances and optimize the geometric parameters with additional software for finite element analysis. You can download the source code by the following address: https://github.com/vkopey/Thread-turning-simulator

References

[1] Hoffman E L 1997 *Finite element analysis of sucker rod couplings with guidelines for improving fatigue life* Sandia report SAND97-1652 (Albuquerque, New Mexico: Sandia National Laboratories) p 66

[2] Mackerle J 2003 Finite element analysis of fastening and joining: A bibliography (1990–2002) *International Journal of Pressure Vessels and Piping* **80** 253–71

[3] Galle T, Van Wittenberghe J, Camelia Jula F, De Waele W and De Baets P 2013 Effect of load flank angle modifications on the structural integrity of buttress threaded connections *Sustainable Construction and Design* **4**(1)

[4] ****pythonOCC, 3D CAD/CAE/PLM development framework for the Python programming language* Available from: http://www.pythonocc.org

[5] Kopei V B 2010 Skinchenno-elementnyi analiz ta optymizatsiia rizbovykh ziednan *Visnyk Natsionalnoho tekhnichnoho universytetu Ukrainy «Kyivskyi politekhnichnyi instytut». Seriia mashynobuduvannia* **58** 42-7 in Ukrainian

[6] ****FreeCAD Documentation* Available from: http://www.freecadweb.org/wiki/Main_Page

[7] Onysko O R 2017 Pro funktsionalnu zalezhnist velychyny polovynnoho kuta profiliu zamkovoi narizi vid velychyn perednoho kuta, kuta nakhylu ta polovynnoho kuta profiliu rizalnoi kromky riztsia *Visnyk natsionalnoho universytetu «Lvivska politekhnika». Optymizatsiia vyrobnychykh protsesiv i tekhnichnyi kontrol u mashynobuduvanni ta pryladobuduvanni* **867** 19-28 (in Ukrainian)

[8] Geuzaine C and Remacle J F 2009 Gmsh: a three-dimensional finite element mesh generator with built-in pre- and post-processing facilities *International Journal for Numerical Methods in Engineering* **79**(11) 1309-31

[9] Dhondt G 2004 *The finite element method for three-dimensional thermomechanical applications* (Wiley) p 362

[10] Sines G 1955 *Failure of materials under combined repeated stresses with superimposed static stresses* Technical Note 3495 (Washington: National Advisory Committee for Aeronautics) p 70

[11] Sines G 1959 Behavior of metals under complex static and alternating stresses, in *Metal Fatigue* eds G Sines and J L Waisman (New York: McGraw-Hill) 145-69

Researches and studies regarding the casting of brake drums from cast iron

A Josan, C Pinca Bretotean and S Raţiu

Politehnica University of Timisoara, Department of Engineering and Management, 5 Revolution Street, Hunedoara, 331128, Romania

E-mail: ana.josan@fih.upt.ro

Abstract. Construction of machinery and technological equipment, but also the construction industry and household parts are the main branches using cast iron castings. This is due to the special properties of gray cast iron castings. Thanks to these properties gray cast iron is very good for casting the brake drum type. The paper presents the molding-casting technology for this type of piece, with the presentation of its advantages and disadvantages as well as the possibilities of improving the molding-casting technology.

1. Introduction

The Brake drum casting is parts of the category of railway equipment type which are cast at the Romanian foundry [1].

The piece "brake drums" is cast of unalloyed grey cast iron with lamellar graphite, EN-GJL-350, according to the standard SR EN 1561 [2].

This type cast iron is very well suited for pouring this type of casting due to its special grey cast iron properties, respectively high vibration damping capacity, excellent cutting machinability, high corrosion resistance, low fatigue strength etc. [3].

Establishing the moulding-casting technology is based on casting configuration; based on it, the pattern set is dimensioned (the pattern of casting, the cores, the core box etc.).

An important role in obtaining a quality casting is the establishing the separation plane of the mould and the casting, which is the starting point of the moulding-casting technology [4-8].

These separation surfaces may be flat, curved and concave (Figure 1), depending on the configuration of the finished piece. Thus, the separation plane for the analyzed part (Brake drum) is placed the upper part of the piece (Figure 2) [4-8].

In this case, the solution chosen allows obtaining the mould only in the lower mould part [4-8].

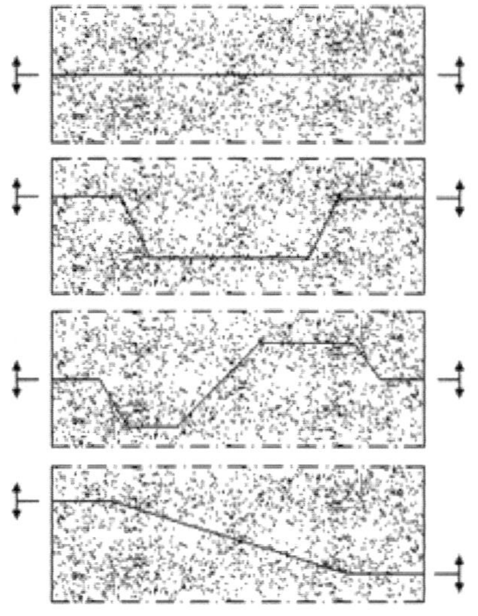

Figure 1. Choosing the separation plane

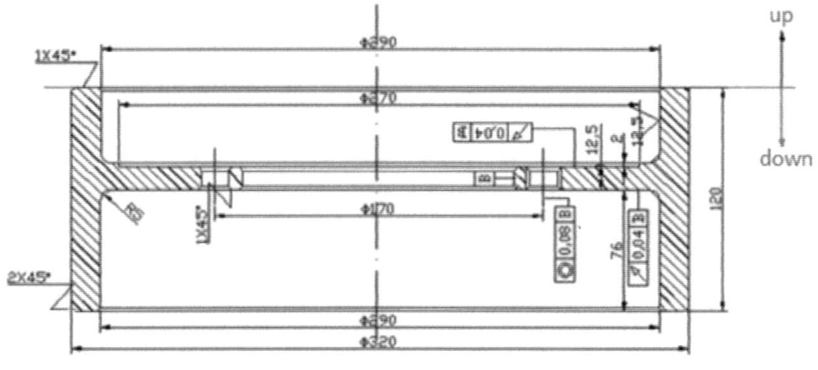

Figure 2. The separation plane of the casting

The following aspects to be taken account when establishing the moulding-casting technology are: establishing of the allowances (processing, technological and for contraction) as well as constructional inclinations and connections [4-7].

The dimensioning of the gating system in the present case is not necessary because it has been chosen to pour the alloy directly through the rising head (such as open rising head type) [4-7].

Industrial practice requires the use of cores as the optimal method of making of the mould. The role of the cores is to ensure the realization of the the internal cavity or external configuration of the casting, to very precise dimensions. For this, besides a dimensionally correct construction, the core must resist all stresses that appear during its realization, drying and transport as well as the casting of the liquid alloy [7].

The configuration of the core and the core box for the analyzed casting is presented in Figure 3. Also, their nominal sizes are presented.

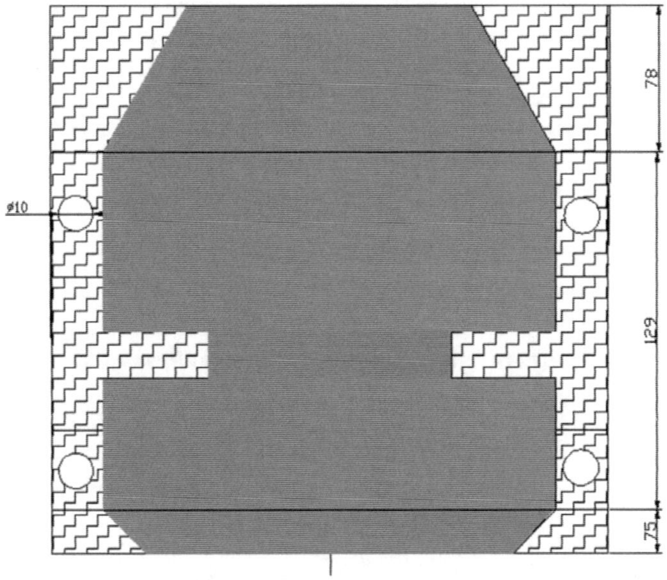

Figure 3. Configuration of core and core box

The dimensions of the mould jackets and the possibility of placement the counterweights are presented in Figure 4. The method chosen for mould consolidation depends on the method moulding and the size of the moulds [4-7].

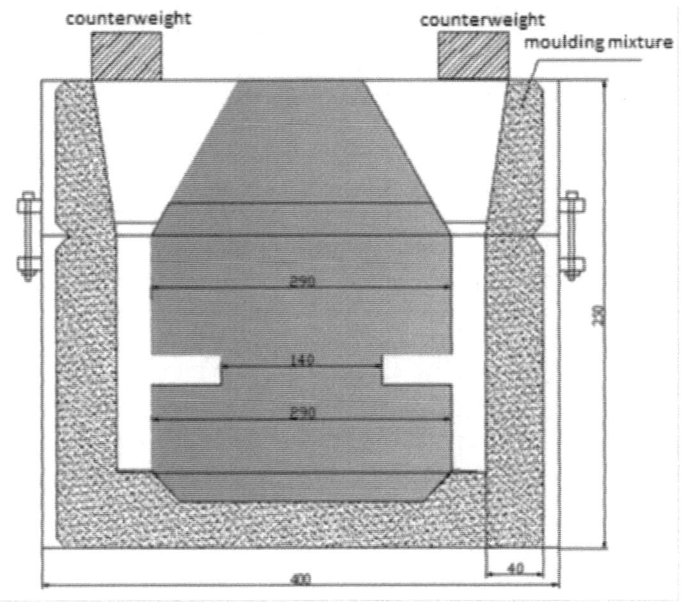

Figure 4. Dimensions of the mould jackets and the posibility of placement the counterweights

The role of counterweights is to consolidate the two mould parts, because the liquid alloy tends to lift the upper mould part (due to the metalostatic pressure that the liquid alloy exerts). Otherwise, the liquid alloy will pour through separation surface causing irreparable defects (reject of the casting, obtaining the casting with burrs, unnecessary increase of liquid alloy, accidents) [4-7].

2. Analysis of defects registered in industrial practice

Performing the critical analysis on a batch of castings of the same type, such as the *Brake drums*, for a certain period of time, we find that the following typologies of defects have been registered (Figure 5) [9-11]:

a) blisters, caused by mould;
b) blisters, caused by metallic material in the furnace charge;
c) shrinkage cavity and micro shrinkage cavity (microporosities).

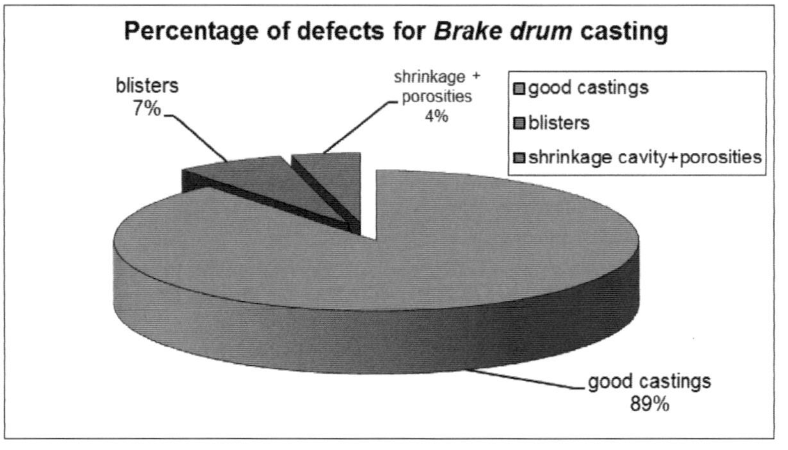

Figure 5. Categories of defects registered for the analyzed casting

The blisters are inclusions of gas incorporated into the body of the casting (exogenous blisters), due to the gases released from moulds and cores. The formation of the blisters inside the castings occurs when gas inclusions can not be eliminated durind cooling and solidification of the liquid alloy [9-11]. In case of castings there are two types of blisters namely blisters caused by mould and blisters caused by the mettalic material in the charge.

The blisters caused by mould are due both to large amounts of gas released from the mould and to the non-evacuated air of the mould during casting.

The blisters caused by the mettalic material in the charge usually appear in the entire mass of the casting and do not have oxidized surfeces. In the case of the raw casting these blisters are harder to observe (they ussualy occur in the cutting processing) [9-11].

The shrinkage cavity is a direct consequence of the shrinkage phenomenon and occurs as holes with harsh walls due to the formation of the dendrites.

Micro shrinkage cavity (or microporosities) are defects that occur when the alloy is solidified. These occur between crystalline dendrites as a result of the lack of feed of molten metal during the solidification process and are due to the large crystallization interval of some alloys. These micro-shrinkage cavity reduce the physical and mechanical characteristics of the casting and lead to loss of tightness [9-11].

Thus, out of 120 castings for a period of 3 months, 13 defect castings were registered which representing a 11% (Figure 6,7).

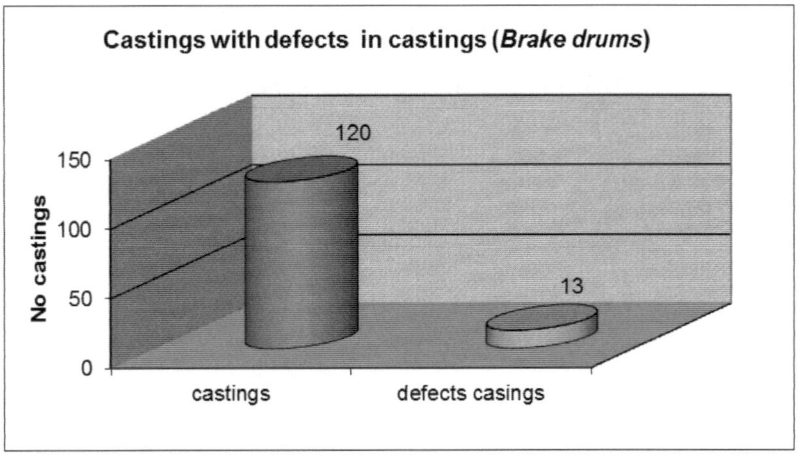

Figure 6. Castings with defects in castings

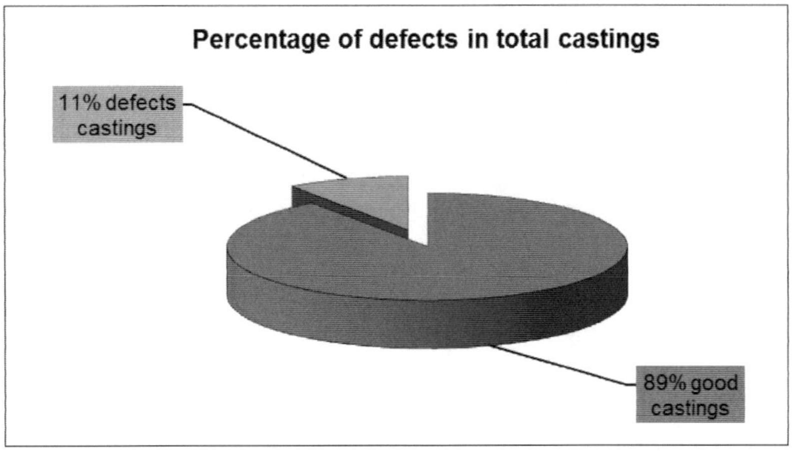

Figure 7. Percentage of defects in total castings

3. Applying technological measures to improve the moulding-casting technology of casting studied

The analysed pieces (*Brake drums*) cast from grey cast iron at Romanian Foundry is not of very high complexity however, due to the high demand for this type of casting, it is necessary to review the moulding-casting technology.

This is necessary because, besides this piece, the foundry also receives other commands for similar pieces, so that the reduction of defects in the moulding technology is fully justified and is directly related to the costs of the enterprise.

Thus, according to the data registered in the industrial practice, it can be said that the casting in green-sand moulds (although it is economical) can not be used for the casting of high-rise parts. The explanation of this is the following: after filling the mould, the liquid alloy exerts on the walls of the mould a metallostatic pressure. This pressure is even greater as:

- The casting height is higher and
- The specific weight of the cast alloy is higher.

Casting in green-sand moulds is also limited by the fact that immediately after casting there is an excessive increase of moisture of the moulding sand at a certain distance in the wall of mould. The consequence of this phenomenon is the decrease of the resistance of the moulding sand in this layer.

The current technology of obtaining these castings presents some inconveniences, namely:

- The use of a inappropriate moulding mixture lead to occurrence of the blisters
- The use of a inappropriate moulding mixture lead to damage cores and moulds
- Due to the fact that the mould are of the green-sand mould type, there is the possibility of the occurrence of the micro shrinkage cavities and of the pores (due to the moisture absorbed by the molds during their stationing up to the casting of the liquid alloy)
- Uncorrelation of casting parameters (speed, temperature, duration) lead to inappropriate solidification of the liquid alloy.

In order to remove these defects, it is necessary to apply technological measures to prevent their occurrence, respectively:

1. In order to improve the quality of castings and to reduce the defects due to the non-drying forms, it is necessary to use the anti-adherence paints with self-drying for moulds and cores [12], [13]

The main advantage of self-drying paints is their stability over time and, on longer storage, the accumulated substance at the bottom of the container can be easily homogenized by mixing. In term of costs, anti-adherence paints belong to the expensive materials category. The cost of purchasing these types of paints are amortized due to [12], [13]:

- Improving the commercial aspect of castings;
- Shortening cleaning and finishing operations;
- The decrease of the mechanical processing allowances (due to increased resistance to the surface of the cores and moulds to which they apply);
- Fuel economy;
- Elimination of the need to neutralize the gases that are released from the dryers.

Industrial practice indicates that in most foundries in industrially developed countries over 70% of foundries use self-drying paints in the manufacturing processes of castings [12], [13].

2. Modification of the recipes of the mixtures used and replacement with self-hardening mixtures (these are recommended to be used together with anti-adherence paints) [14]

The self-hardening mixtures are a very veried group of mixtures due to the different binders used [14].

In 1970 it began the assimilation of the cold-hardening mixtures based on synthetic resins for the manufacture of cores and moulds used for the casting of pieces (from cast iron and steel) [14].

These mixtures have a number of advantages compared to cold-hardening mixtures based on sodium silicate. These advantages relate to:

- The castings surface quality,
- Easily shaking-out of cores
- The possibility of regenerating the used moulding mixture.

However, it is still continuing to discuss about the dangers of using furanic/phenolic resins (and other resins) on humans and the environment (due to the noxious substances they release).

Besides the above, the process of using the cold-hardening mixtures can also be applied under cold weather conditions when the temperatures in many foundries are negative [14].

3. Covering the shrink head with the mixtures, in order to guide the solidification

Elimination of the shrinkage cavities is achieved through a judiciously designed technology and by strict compliance to it. For this purpose is performed the sizing of the rising head in accordance with the data registered in industrial practice. In addition, for the analyzed casting, the cover of open rising head (after casting) with exothermic mixture is used.

This is actually a method of heating the rising head and uses a special material, respectively a exothermic powder [14].

The exothermic mixture, upon contact with the liquid metal, ignites and the heat ceded increases the time of liquid storage of the rising head, guiding the solidification.

4. the correct correlation of the casting parameters (speed, temperature, duration);
5. the correct management of the technological process of molding-casting.

4. Conclusions

As a result of the decline in demand for castings in the last decade, the vast majority of surplus production capacities are focused on the production of unique and small series parts. In these conditions the preponderance of the castings belongs to the production department, equipped with equipments and universal machines which are based on technologies that easily adapt to the market requirements [14].

The criteria for choosing one or other noulding-casting technology include mainly quality and price, which in fact determines competitive oportunities of the internal and external markets. But, these criteria are necessary but not sufficient.

Thus, in the curentstage of the casting production the ecological criterion is extremely important because, without taking into account the volume of gases emanating from casting, cooling and shaking-out of the castings, no technology can be used [14].

Through a series of simple, practical and easy to implement measures applied permanently, spectacular reductions in consumption per ton of casting are achieved, with positive implications for costs.

These include:

- correct management of the elaboration process of the liquid alloy;
- maintenance of foundry installations and equipments (especially mixers);
- correct corelation of casting parameters (speed, temperature, duration);
- sand temperature control and sand parameters control
- the correctness of the moulding process.

Similarly, on a permanent basis, a particularly important role must be accorded to customers / beneficiaries' requirements regarding the provision of castings that correspond technically and quality standards.

These technological measures, which modify the molding-casting technology of the parts, can lead to a decrease in the quantity of rejected castings, so that the rejects registered in the industrial practice do not exceed 4.5% of the total good cast parts. These aspects lead implicitly to important business savings.

References

[1] ***http://www.cfi.ro/
[2] ***SR EN 1561:2012 Foundry – Grey cast irons
[3] Sofroni L 1981 *The elaboration and pouring of alloys*, Publishing House Didactic and Pedagogical, Bucharest
[4] Josan A 2012 *Casting and moulding technologies of metal parts – the project guide-book*, Politehnica Publishing, Timişoara
[5] Josan A 2002 *The technology of moulding and casting alloys*, Politehnica Publishing, Timisoara
[6] Ştefănescu Cl and Cosneanu C 1984 *Tehnologii de executare a pieselor prin turnare*, Ed. Tehnică, Bucureşti
[7] Stefanescu C and Cazacu I 1980 *Technologies of realization of the parts by casting*, Technical Publishing, Bucharest
[8] Josan A, Pinca Bretotean C and Ratiu S 2018 Critical analysis of the influnce of the posibilities of establishing the moulding technology on obtaining the castings, *IOP Conf. Ser.: Mater. Sci. Eng.* **294** 012038
[9] Josan A 2016 Impoving the quality of castings by optimization of the moulding-casting technlogy, *ACTA TECHNIA CORVINIENSIS – Buletin of Engineering* **IX**(4) 75-78
[10] ***1977 *International Atlas of casting defects*, Technical Publishing, Bucharest
[11] Josan A 2010 *The analysis of casting defects recorded in the metallurgical enterprises*, International Symposium on Advanced Engineering & Applied Management - 40th Anniversary in Higher Education (1970-2010), Hunedoara, Romania, 7pg, CD
[12] Davidov N I 2004 Refractory daies for moulds and cores, *Romanian Foundry Journal* **3,4** 17-

20 (translation C Cosneanu)

[13] Josan A and Pinca Bretotean C 2017 Using the antiadherence paints to manufacturing of the moulds intended for iron castings, *IOP Conf. Ser.: Mater. Sci. Eng.* **163** 012020

[14] Tkacenko S S 2005 The forecast of using cold setting phosphatic moulding, *Romanian Foundry Journal* **1,2** 14-16 (translation C Cosneanu, from Journal Liteiscik Rossiiv no 8, pp 23/2004)

IOP Conf. Series: Materials Science and Engineering **477** (2019) 012034 doi:10.1088/1757-899X/477/1/012034

Design of remote electro-pneumatic control system using microcontroller

M Y Stoychitch and B Z Knezevic

University of Banja Luka, Faculty of Mechanical Engineering, Vojvode Stepe Stepanovica 71, 78000 Banja Luka, Republic of Srpska, Bosnia and Herzegovina

E-mail: mihajlo.stojcic@mf.unibl.org

Abstract. In this paper we considered design and realization of remote control system that is used for control of movement of pneumatic cylinders. It is assumed that each cylinder is controlled by electro pneumatic monostable 5/2 (or 4/2) valve. Also, each cylinder has a limit switches in its two final positions. The control system is implemented as a web application, where the server is realized using the ESP8266 or ESP32 microcontroller and Arduino software, while as the client uses some Windows or Android device. The use of this control system is very simple, so (after client-server connection) it is only necessary on the client side (according to the already defined rules) to enter the desired operation sequence of the cylinders. It is possible, also, to control the times between the adjacent movements of the all cylinders. Software and necessary documentation for practical realization of described solution is given in this paper.

1. Introduction

Electro-pneumatic systems (EPS) are very common in industry, especially in industry of food and beverages as well as in industry of furniture. Industrial application of the EPS increases the production productivity and decreases costs. From technical point of view, EPS are integration of pneumatic and electric technologies. So, in EPS actuators are pneumatic elements (cylinders) while sensors and control system (CS) are electric. The interconnection of these technologies is achieved on the main directional control valve of each cylinder. Directional control valve directs the flow of the operating - pneumatic energy through the actuator. In other hand, directional control valves are actuated by electrical signal so, complete system is electro-pneumatic.

According to the control mode, the directional control valve can be bistable or monostable [1], Figure 1 a) and b). Bistable directional control valve has a two stable states and two solenoids which are used for switching between these two states. Solenoids, which are indicated by Z and Y in Figure 1*a*), are activated by electrical signals. For instance, after a short-term control pulse, the directional control valve switches to a stable state in which the output 4 is connected to the source of pressure 1. In this state the directional control valve remains until the signal Y appears and switch it to another stable state in which output 2 is connected to the pressure line 1. Simultaneous activation of both solenoids (Z and Y signals simultaneously) is not allowed, because then the state of the directional control valve is not defined. In this case, the overall system is blocked, and this signal is called a "blocking signal".

The monostable directional control valve, Figure 1*b*) has only one stable state. In this state (when the output 2 is connected to the pressure line) the directional control valve is positioned mechanically by the spring. To another state (when the output 4 is connected to the pressure line) which is not stable state the directional control valve is positioned by the solenoid activation (signal Z, Figure 1*b*). In this state the directional control valve remains while the Z signal is active. Inactive signal Z means return to a stable state by the mechanical spring. At the Figure 1 τ_1, τ_2 and τ_3 denotes delay times between electrical signals and pneumatic signal at output 4. This delay times are dependent on the directional control valve types but generally they are in range up to ten milliseconds.

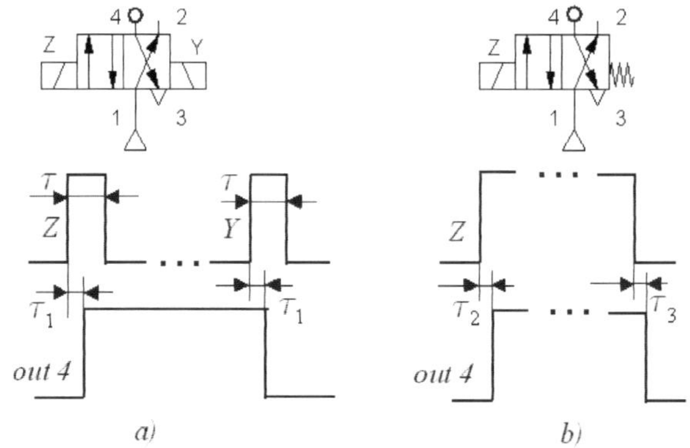

Figure 1. Bistable *a*) and monostable *b*) directional control valves with their inputs and outputs

The main task of the control system of almost all EPS is the realization of the default operation cycle. Operation cycle implies the necessary order of retractions and extractions of the piston rods of all cylinders in order to accomplish the required task. Therefore, changing of the operation task means changing of the operation cycle which make EPS very flexible. Flexibility is especially interesting nowadays, when industry 4.0 is increasingly becoming an issue of interest in underdeveloped and medium-developed countries.

However, flexibility also requires a new approach to designing control systems. Designing of CS must be automated and "designer" must be "smart machine" which, in interaction with environment, uses IoT (Internet of Things) technologies. In this way, it is achieved that one machine - the server can notify the other machine - the client that the task is finished. After that, the client can ask the server to do another task that is in accordance with the current and predetermined state of the overall system.

In this paper, algorithms that can realize the described automatic designing process are proposed. As a computer platform, WiFi microcontrollers (microcontroller with embedded WiFi interface) from the ESP8266 and ESP32 series are used, using Arduino software for their programming, [2-5]. The proposed algorithms have been implemented and experimentally confirmed on ESP8266. In doing so, the following rules and assumptions apply:

1. Double-acting pneumatic cylinders are used as actuators. Cylinders are marked with: A, B, C ... H;
2. The required operation cycle is written in the text-string form, where the extraction of the piston rod is indicated by "+" and the retraction by "-", for instance: "A+, B+, B-, A-". At the starting point, the piston rods of all the cylinders in the operation cycle are retracted, so the first movement of a cylinder is always "+". All of the cylinders must have the same number of retractions and extractions (+ and -) in one operation cycle;
3. Each cylinder is operated via a monostable directional control valve 4/2 (5/2). Solenoid actuation causes the "+" movement of the cylinder and actuation by spring causes the "-" movement. The denotes of the directional control valve's solenoids are "Z" plus cylinder. For instance, the directional control valve of the cylinder A, have the marking of the solenoid "ZA" or "zA";

4. If an operation cycle requires the holding of the piston rod of some cylinder in the retracted or extracted position prior to initiating the next movement, that task is denoted after the denote of the cylinder and the type of movement as "/ xS", where the "x" is the holding time in the tenths of a second (maximal value for x is 65535). For instance, "... F + / 65S ..." indicates that after extraction of the cylinder's piston rod F system waits 65 * 0.1 = 6.5sec before initialization movement of some another cylinder.
5. Each cylinder during movements activates two limit switches, one for the fully extraction and the other in the fully retraction position. Only one of these switches can be activated at the same time. Limit switches are designated as: cylinder mark plus "0" for the limit switch that is activated when the cylinder is retracted and cylinder mark plus "1" for the switch that is activated in the extracted position. For example, the limit switches of the cylinder F are designated as F0 (retracted position) and F1 (extracted position). Also, the f_0 and f_1 tags refer to the same limit switches.

2. Design of the control system

2.1. Design of the control system using PLC

In nowadays, the control systems for EPS are mostly realized using PLC (Programmable Logic Controller), [6-7], while for their simulation the software package FluidSim-P [8] is mainly used. It is known that the work of the PLC takes place through several so-called scan cycles: reading inputs, executing program, updating outputs and internal update. For the same application the total duration of all these scan cycles is constant and it is designated as T_{sc}. Therefore, PLC periodically with period T_{sc}, reads all inputs and updates all outputs. If there are some changes within the system that are shorter than one period T_{sc}, then these changes will not be registered, i.e. the control system behaves as like these changes did not occur. In EPS all changes are much longer than the period T_{sc}, so the period T_{sc} is negligible compared to other changes (for example, the reaction of the limit switch lasts dozens of T_{sc} periods).

Although PLC is very fast to control EPS, there are some problems with its implementation. The basic problems are in determining the logical functions that are realized within the control system. Let's explain this in one example. Let the EPS contain three cylinders A, B, and C and next requested operation cycle: A +, B +, B-, C +, C-, A-. This cycle is shown in Figure 2. Using the displacement-step diagram shown in a dark full line. The inputs to the system are information from the limit switches of each cylinder: a0, a1, b0, b1, c0 and c1 as well as ON and OFF signals, while the outputs are signals: *zA*, *zB* and *zC* which activates the solenoids of directional control valves of cylinders A, B and C, respectively. Inputs and outputs make the vectors: inputs $\boldsymbol{x}=[ON,OFF,a_0,a_1,b_0,b_1,c_0,c_1]^T$ and outputs $\boldsymbol{y}=[zA,zB,zC]^T$. From Figure 2 is obvious that the same combination of inputs in steps: 2 (B+), 4 (C+) and 6(A-) does not generate the same outputs. Thus, for the input vector $\boldsymbol{x}=[1,0,0,1,1,0,1,0]^T$ in the step 2 output vector is $\boldsymbol{y}_2=[1,1,0]^T$, in step 4 output vector is $\boldsymbol{y}_4=[1,0,1]^T$ and in step 6 output vector is $\boldsymbol{y}_6=[0,0,0]^T$. Therefore, new variables are introduced, state variables M_1 and M_2, which are together with signal $S = START$ makes state vector $\boldsymbol{m}=[S,M_1,M_2]^T$.

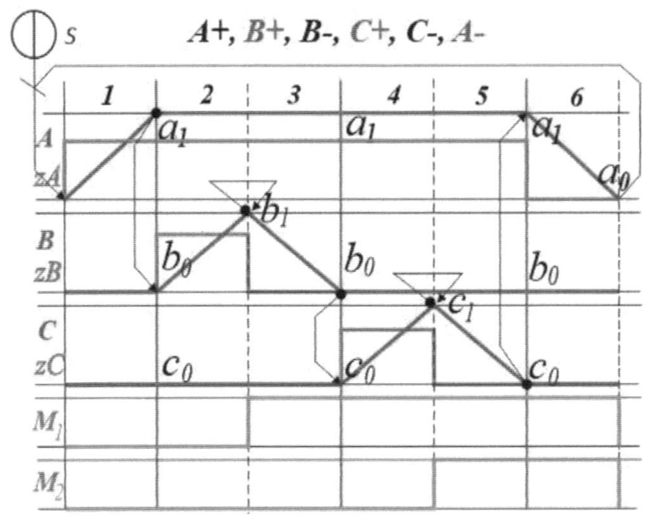

Figure 2. Displacement – step diagram with directional control valves zA, zB and zC and states M_1 and M_2

State variables M_1 and M_2 are selected in that way that together with inputs make unique combination for each generated output. Value of the state variables in some future instant $t+$, i.e. states $\boldsymbol{m}(t+)$, depends on inputs $\boldsymbol{x}(t)$ and previous states $\boldsymbol{m}(t)$ in instant t. This means that in each period T_{sc} inside PLC must be realized two logical functions: one which change states, function $\boldsymbol{g}(\square)$ and another which change outputs, function $\boldsymbol{f}(\square)$:

$$\boldsymbol{m}(t+)=\boldsymbol{g}(\boldsymbol{x}(t),\boldsymbol{m}(t)), \qquad \boldsymbol{y}(t)=\boldsymbol{f}(\boldsymbol{x}(t),\boldsymbol{m}(t)). \tag{1}$$

After one period T_{sc}, current states $\boldsymbol{m}(t)$ have the same value as previous calculated states $\boldsymbol{m}(t+)$, i.e. $\boldsymbol{m}(t)=\boldsymbol{m}(t+)$.

For the observed operation cycle, vector function $\boldsymbol{g}(\square)$ and $\boldsymbol{f}(\square)$ from (1) becomes:

$$\boldsymbol{m}(t+)=\begin{bmatrix} S(t+)=(S(t)+ON)\overline{OFF} \\ M_1(t+)=(M_1(t)+b_1)\overline{a_0} \\ M_2(t+)=(M_2(t)+c_1)\overline{a_0} \end{bmatrix}, \quad \boldsymbol{y}(t)=\begin{bmatrix} zA=S\overline{M}_2+\overline{c_0} \\ zB=a_1\overline{M}_1\overline{M}_2 \\ zC=b_0M_1\overline{M}_2 \end{bmatrix}. \tag{2}$$

At Figure 3 is shown ladder diagram for PLC for realization logical function from (2).

2.2. *Design of the control system using microcontroller – new approach*

Designing of control system of EPS using PLC in some cases can be very complex. Mainly, this is due to the PLC operational system i.e. synchronous executions of the scan cycles. In addition, for each specific operational cycle, it is necessary to re-define the logic functions and program the PLC. So, automatic programming based on a given operational cycle is either impossible or very difficult to implement.

In this section is given a proposal for the automated EPS design solution using a microcontroller. EPS control system which using the microcontroller is also given in [9], but microcontroller programming is performed for each specific cycle. Similarly, in [10] and [11], the authors deal with EPS control using a microcontroller. In [11] authors use the MATLAB/SimuLink software for designing control systems. As a link between controlled object and programs in MATLAB they use the Arduino microcontroller platform. The proposed solution is partly use the procedure given in the FluidSim-P software in the section dealing with GRAFCET programming.

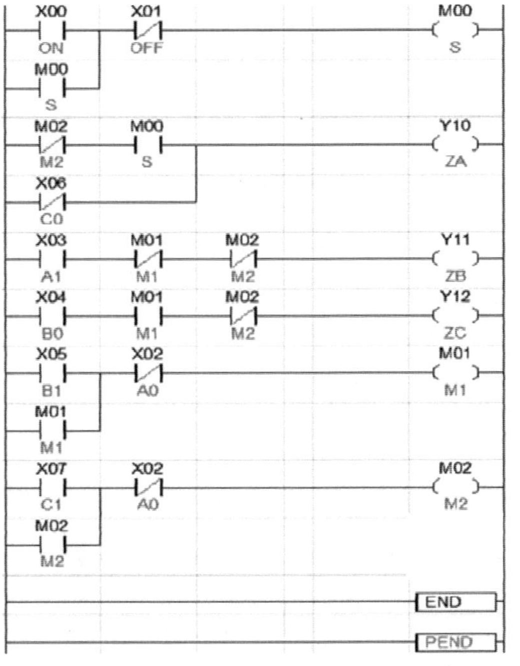

Figure 3. Ladder diagram, program for the cycle: A+,B+,B-,C+,C-,A-

The operation of almost all EPS is carried out so that the completion of the movement of one actuator-cylinder activates the next movement. This procedure continues until the last movement in the operation cycle when the start conditions are checked again. If start conditions are fulfilled the operation cycle is repeated. Therefore, the operation of these systems is asynchronous and can be divided into several steps. In doing so, the realization of the next step (movement) is conditioned by the completion of all operations (actions) in the previous step and by fulfilling certain conditions for the transition to the next step, Figure 4.

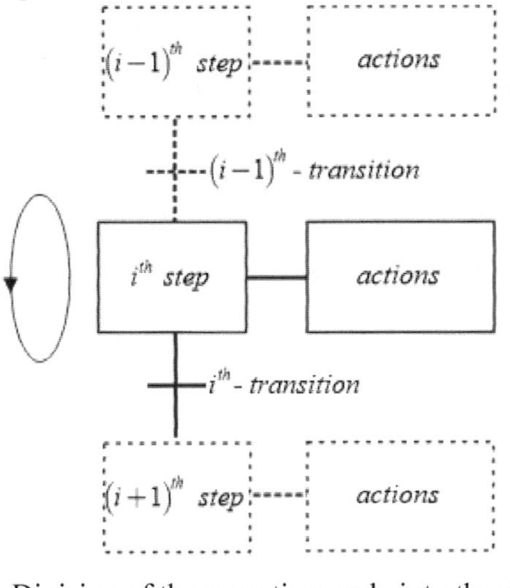

Figure 4. Division of the operation cycle into the steps with their actions and transitions

Each step is described by actions within the steps and transition conditions to the next step. Under actions in the step we consider the activation of directional control valves of all cylinders that are required in this step by default. Also, actions are considered to be the realization of the required holding time in that step. The passage conditions between two steps are determined based on the state of the limit switches of each cylinder from the operational cycle. This condition is fulfilled if all the cylinders have completed the movements required in the actions, which is signaled by the activation of the appropriate limit switches - sensors.

Actions and transitions conditions can be written in the form of three arrays, two for actions and one for transitions conditions. Let these arrays be *NIZR* and *NIZV* for actions and *NIZG* arrays for transitions conditions. In the *NIZR* array, some bits indicate the solenoids of the monostable directional control valves of each cylinder in an operational cycle. If this bit is set, then this directional control valve is activated and the cylinder piston rod is extracted (movement "+"). The *NIZV* array contains holding times in each step. The array of transitions conditions, *NIZG* is similar to the *NIZR* array, only the bits refer to the cylinder limit switches. The formation of these arrays will be explained on the previous example, i.e. for the operational cycle: A +, B +, B-, C +, C-, A-.

Operational cycle doesn't have holding times, so the all of elements of the NIZV array are zero and they will not be considered. Since the operational cycle contains three cylinders, then three bits for each solenoid are enough for the *NIZR* array. The *NIZG* array takes 6 bits, because for the three cylinders 6 limit switches are needed. Therefore, for the required operation cycle, it is necessary to form the arrays given in Table 1. The number of elements of each of the formed arrays is the same as the number of steps in the operational cycle, i.e. for each step one element of the array.

Table 1. The arrays of actions (*NIZR [6]*) and transition conditions (*NIZG [6]*) for the operational cycle: A+, B+, B-, C+, C-, A-

	actions			steps	transitions					
	NIZR[6]				*NIZG[6]*					
	zA	*zB*	*zC*		a_0	a_1	b_0	b_1	c_0	c_1
0	1	0	0	A+,	0	1	1	0	1	0
1	1	1	0	B+,	0	1	0	1	1	0
2	1	0	0	B-,	0	1	1	0	1	0
3	1	0	1	C+,	0	1	1	0	0	1
4	1	0	0	C-,	0	1	1	0	1	0
5	0	0	0	A-,	1	0	1	0	1	0

When the arrays are formed and stored in the memory of the microcontroller, then it is very easy to realize the required operating cycle. After the START signal, the first element of the action array, the *NIZR[0]* element, is sent to the output. Since in that element $zA = 1$ only the solenoid of the directional control valve of the cylinder A is activated, which further causes the extracting of the piston rod of this cylinder. Immediately after that, the microcontroller reads the state of the limit switches, i.e. reads the input status. This reading is repeated continuously until the actual state of the input is identical to the first element of the transition array, the *NIZG* element *[0]*. This state will be identical when the limit switch a_1 is activated, i.e. $a_0 = 0$ and $a_1 = 1$. Since in this step there were no movement of the cylinders B and C, the state of their limit switches are remain unchanged, i.e. $b_0 = 1$, $b_1 = 0$, $c_0 = 1$ and $c_1 = 0$ (the piston rods of both cylinders are retracted). When the actual state of the inputs identical to the element of the transition array, the step is completed. Due to the conditions for transition to the next (second) step are fulfilled.

In second step, in the array of actions (*NIZR [1]*), $zA = 1$ and $zB = 1$, so the solenoid of the directional control valve of the cylinder B is active and the cylinder B piston rod is extracted. After that, the actual state of the inputs is continuously read until the actual state of the input is identical to the other element of the transition array, the *NIZG* element *[1]*.

This procedure is repeated all the way to the last element of the action array, the *NIZR* element *[5]*. When this element is forwarded to the output, the solenoid of all directional control valves is switched off and the piston rods of the all cylinders are retracted. System waiting that all limit switches with the index "0" be activated (the corresponding bit is 1) and with the index "1" be deactivated (goes to 0). This condition is identical to the initial state of the observed EPS. After that the start state is checked. If it is active, the process is repeated and if not, the system stops in the initial position. This ensures that the system always stops in the starting position regardless of the moment when the START signal has been reset.

The problem of designing the control system in manner which is proposed is significantly simplified. Instead of control the operational cycle (like PLC), it is control of each individual step from that operating cycle. Therefore, it cannot be a problem that the same combination of inputs gives different outputs, because each step is considered separately. Since the activities in each step are the same, then the same control system can control all steps (i.e. the required operation cycle), only in certain steps the output and input values are changed, that is, the elements of arrays of the actions and conditions of transition..

In the previous section, the actions arrays and the transition condition array were formed manually based on the required operational cycle. In this chapter is described an algorithm that creates these arrays automatically. Input for proposed algorithm is operational cycle being given in a text format - a string that is written by specific syntax and rules. There are some limitations that must be respected and they are:

1. The maximum number of cylinders is 8 so that the elements of the *NIZR* array are 1-byte length = 8 bits (one bit for each directional control valve). As the 8 cylinders require 16 limit switches, then the length of the elements of the *NIZG* is 2-byte = 16 bits (one bit for each limit switch);
2. The holding time array *NIZV* is 2 bytes long. This means that it is possible to set a maximum of 65535 tenths of a second or a maximum holding time of $65535 * 0.1 / 3600 = 1.82$ hours. If the holding time are not required in the operational cycle at any step, then all the elements of the *NIZV* array are zero, and the logical variable *vremC* = *false* (0). If at least one holding time is required, then *vremC* = *true* (1);
3. The length of the string in which the requested operational cycle is written is a maximum of 256 bytes. The maximum number of steps in one operational cycle is 80.

The limitations are primarily related to the number of cylinders and limit switches (outputs and inputs), which is limited by the selected microcontroller. It is possible to use 8 cylinders for the ESP 32 microcontroller, while only 3 cylinders can be used for the ESP8266 microcontroller. Limitations related to the number of steps and the length of the string in which the operational cycle is written can be much higher. These limitations are related to the size of the operating and EEPROM memory of the microcontroller.

At Figure 5, one part of the algorithm for creating arrays is given and the function of the *praviNizove* is in Appendix. The required operating cycle of cylinders is written in the string "*sc*", which is a parameter for this function. All letters in this string were previously converted to capital letters. In the variable "*gpE*" is written the initial state of the limit switches, i.e. the state when the piston rods of all the cylinders are retracted and when the limit switches with the index "0" are activated. In the variable "*abcE*" is the initial state of the solenoids of all directional control valves, while in the variable "*abc*" the cylinder marks that can occur in any cycle.

Other variables used in this function are: "*i*" - serves as a character counter in the string "*sc*", "*masGP*" - a 2-byte length number, which serves as a mask for elimination of limit switches influence of all cylinders which could be occur but which are not be in operation cycle, "*onB*" - a logical variable that turns on when a "/" character appears in the string "*sc*" (the indication for the start of the holding time), and turns off when the "S" symbol appears (the indication for the end of the holding time). All characters between "/" and "S" are numbers that are convert to a variable "broj" that represents the holding time value, "*iR*" - the step counter in the operational cycle and "*p*" - the number which indicating the position of the cylinder from the cycle "*sc*" in the string "*abc*". For instance, the position of the cylinder A is 0, the cylinder F is 5, the cylinder H is 7, and so on.

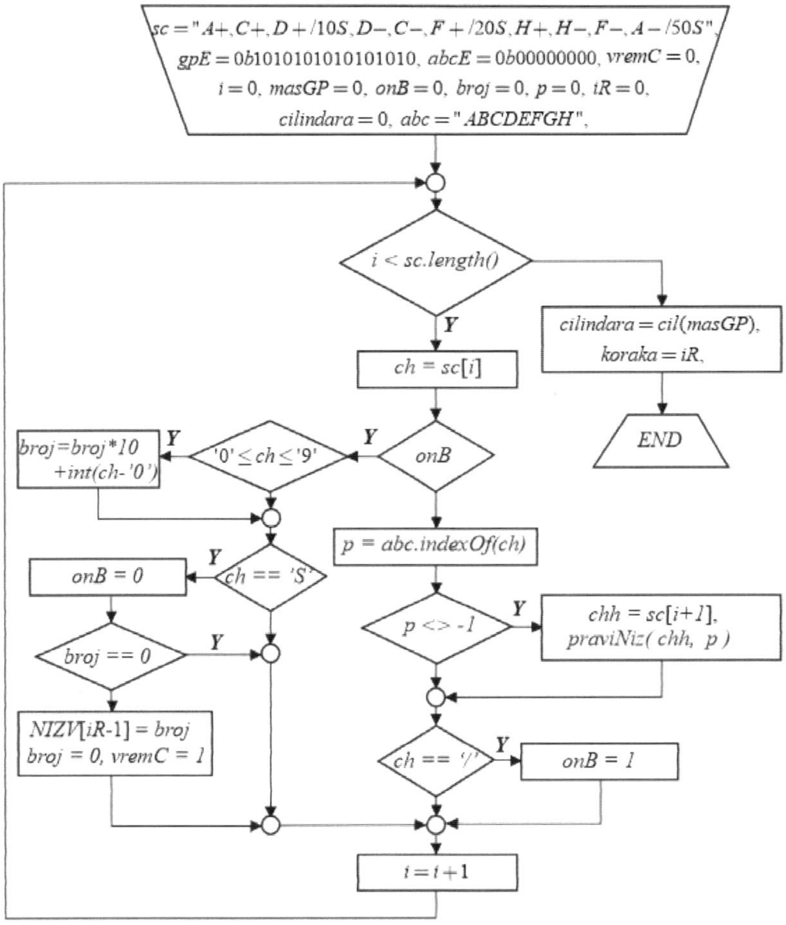

Figure 5. General algorithm for creating the arrays, function *praviNizove*

In the main program loop the characters of the operational cycle-string "*sc*" are read in sequence and are placed in the variable "*ch*". If this character is contained in the variable "*abc*" (it is checked by the command *p = abc.indexOf (ch)*), then "*p*" is the ordinal number in the string "*abc*". If p <> - 1 the character "*ch*" is contained in "*abc*", which means that this sign is a cylinder mark. This sign must be followed by a sign either "+" or "-" (i.e. extraction or retraction of the piston rod) which is read as a variable "*chh*".

The values of the character "*chh*" (sign "+" or "-") and the number "*p*" (position "*ch*" in the string "*abc*") are passed to the function *praviNiz* whose algorithm is shown in Figure 6. The function *praviNiz* creates one element of the arrays *NIZR* and *NIZG*, and the element of the array *NIZV* sets to 0. The function *praviNiz* will be explained later.

If the variable "*ch*" is character "/", then the next number is the number that represents the value of the required holding time. After that, the logical variable "*onB*" is turned on, digits in a sequence are read and the total value of the holding time are calculated. This goes all the way to the sign "*S*" when the variable "*onB*" is turned off, and the current element of the string *NIZV* is set up to value of holding time, see Figure 5.

As we have said before, Figure 6 shows the algorithm of the function (the function *praviNiz*) that creating one element of the action array *NIZR* and the transition condition array *NIZG*. To this function are assigned two parameters, the first parameter "*pm* = *chh*" is the "+" or "-" character, and the second parameter "*a* = *p*" is the ordinal number of the cylinder mark in the string "*abc*" (*abc* = ABCDEFGH). For instance, let "D+" movement from operational cycle "*sc*". Then it is *pm* = "+", and the parameter *a* = 3 (because 3 is the position of the letter D in the string "*abc*").

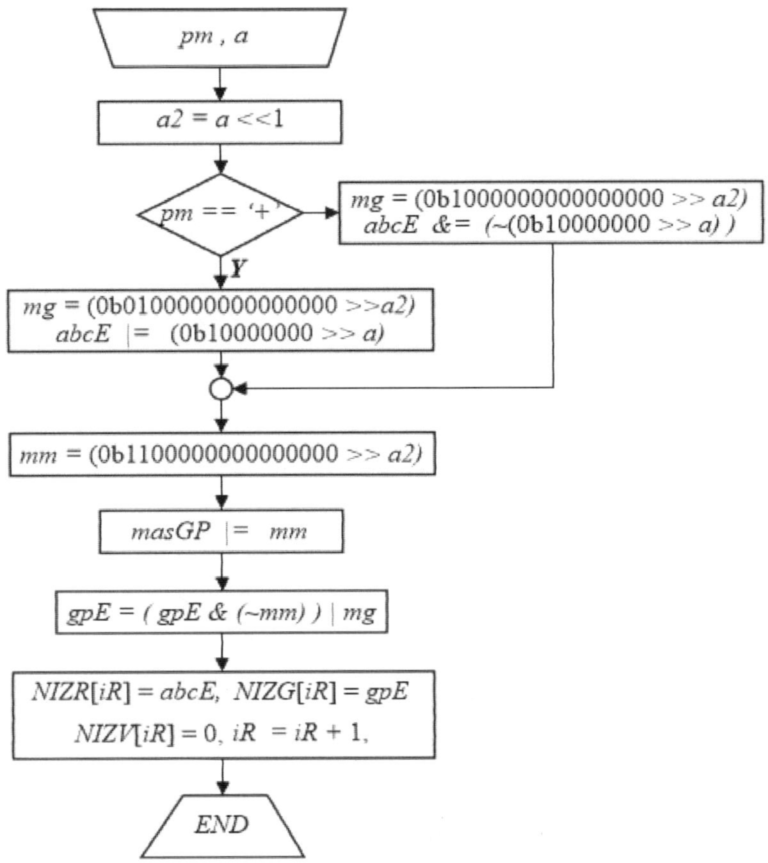

Figure 6. General algorithm for creating the arrays, function *praviNiz*

The variable *a2* is obtained as 2**a* = 6 or, which is equivalent, by moving the bits from a to one position to the left. The following masks are further defined: *m* = 0b1000 0000, *mm* = 0b1100 0000 0000 0000 and a "*mg*" mask whose value depends on the parameter "*pm*". If *pm* = "+", then *mg* = 0b0100 0000 0000 0000, and if *pm* = "-" then *mg* = 0b1000 0000 0000 0000. Using the mask *m*, via the variable *abcE* and depending on the "*pm*", is calculated the new value of the array *NIZR* element in the following way. If *pm* = "+", then *abcE* = *abcE* | (*m* >> *a*), which is the same as setting the bits *7-a* in a variable *abcE*. If *pm* = "-", then *abcE* = *abcE* & (~ (*m* >> *a*)), which is equivalent to resetting the bits *7-a* in the variable *abcE*.

Similarly, using the masks *mm* and *mg*, over the variable *gpE*, the elements of the transition conditions array are calculated (i.e. determining the state of the limit switches for which the condition of the transition to the next step is fulfilled). The variable *gpE* is calculated as: *gpE* = (*gpE* & (~ (*mm* >> *a2*))) | (*mg* >> *a2*). For a given case (*a* = 3 Þ *a2* = 6), moving the mask *mm* to 6 in the right leads to *mm* >> *a2* = 0b0000 0011 0000 0000, so that the expression (*gpE* & (~ (*mm* >> *a2*))) in the variable *gpE* resets the bits 9 and 8 (bits 15-*a2* and 15-*a2*-1 in the word *gpE*) while the other bits do not change. After that, via the OR function, i.e. the terms "| (*mg* >> *a2*) ", are replaced by bits 01 (for *pm* = "+") or by 10 (for *pm* = "-").

The variable *masGP* is also used as a mask. In it bits that are corresponding to the position of the limit switches of cylinders which are used in the required operation cycle are set to the value 1, while the other bits are zero.. In this way, the *masGP* in the logic AND function with the element of the array *NIZG* or the actual state of the limit switches distinguishes only the limit switches of cylinders which are used. For instance, let's have 6 cylinders A, B, C, D, F, and H on a machine. In one cycle we use all the cylinders, while in the second cycle we do not use, say, cylinder B, but this cylinder is physically presence of the machine and its limit switches are connected to the microcontroller. In the first case, the mask is *masGP* = 0b1111 1111 0011 0011, and in the second case, the states of the limit switches of cylinder B are ignored so that the mask has a value of *masGP* = 0b1100 1111 0011 0011.

For the operational cycle *sc*= "*A+,C+,D+ /10S,D- ,C- ,F+ /20S,H+,H- ,F- ,A- /50S*" from Figure 5, using the algorithms from Figure 5 and Figure 6, the following arrays are generated:

```
CYCLE : $A+/30S,B+,C+,D+/50S,D-,C-,B-,A-/20S,H+,H-/50S#
    Valves    Limit Switches
    ABCDEFGH  A B C D E F G H   Cy. T[s]
 0. 10000000, 0110101010101010, A+, 3.0,
 1. 11000000, 0101101010101010, B+, 0.0,
 2. 11100000, 0101011010101010, C+, 0.0,
 3. 11110000, 0101010110101010, D+, 5.0,
 4. 11100000, 0101011010101010, D-, 0.0,
 5. 11000000, 0101101010101010, C-, 0.0,
 6. 10000000, 0110101010101010, B-, 0.0,
 7. 00000000, 1010101010101010, A-, 2.0,
 8. 00000001, 1010101010101001, H+, 0.0,
 9. 00000000, 1010101010101010, H-, 5.0,
Steps = 10, Mask = 1111111100000011, Cyl. = 5
```

Figure 7. The arrays: *NIZR*, *NIZG* and *NIZV* that are generated using algorithms which are given on the Figure 5 and Figure 6 and for the given cycle

3. Remote control system

In Industry 4.0 everything is networked. In doing so, Internet technology is mainly used for the exchange and transmission of information in that network. Each unit in that network has its own unique IP address. The exchange of information takes place through the client-server system. The client sends requests, the server processes these requests and sends the answers. In this case, the client is a remote computer or mobile phone, and the server is a microcontroller ESP32 or ESP8266. All the inputs and outputs of the observed EPS are connected to the microcontroller. Besides, on this microcontroller is installed software which has the role of the web server and the EPS controller. The web server is at the same time the HTTP server, because HyperText Transfer Protocol is used to exchange information.

Operation within this system takes place as follows. After the client connects to the server, the server from the EEPROM reads the operational cycle which it is last executed. If it is reading for the first time, then the initial operational cycle is taken (the starting cycle is predefined as: "A +, B +, B-, A-"), which is then treated as the last cycle also. After that, the server offers the client a choice of two options, to enter a new operational cycle or a start / stop operational cycle. If a start / stop operational cycle is selected, then the last operational cycle on the server is taken as the operational cycle. On the basis of this cycle, arrays are created and the program continues its work as a control system.

In case the client chooses to enter a new cycle, then a window with the necessary instructions and a last operation cycle offered is opened. In the offered operational cycle, it is possible to make some changes or input a completely new cycle instead. In the case that the required cycle is not identical to the last cycle, a syntax check is performed, as well as a check that the cycle is entered under Rules 1, 2 and 4 specified in the introduction and whether the limitations given in section 2.3 of this paper have been complied with. If there is a problem, then the server informs the client about it. For instance, if the new operational cycle is entered as: "A, +, b +. , / 20s r B- qx a - ", then after checking returns the operational cycle as" A +, B + / 20S, B-, A- " without any notification of any problem, because only syntax errors are here. But if, for example, the entered operational cycle is "A, +, b +, / 20s B-qx", then the client receives a message that there is "A +", but there is no "A-" and changes are required in the entered operating cycle.

If the cycle is correct after the described checks, then the server again offers the client a new cycle or a start / stop cycle. In the case that the start / stop cycles are selected, the cycle is remembered in the server's EEPROM memory, the required arrays are created and the server continues to operate as a control system. At the same time as the server control the operation of the observed EPS, it is still ready to respond to client requests.

4. Implementation of the control system

The above suggested algorithms are realized in laboratory conditions. The Arduino Uno R3 compatible development board based on WiFi microcontroller ESP-8266EX was used. As the software environment, the Arduino IDE is used, so programming is identical to the programming of the Arduino UNO. The board includes integrated 3.3V and 5V voltage regulators, 2.4GHz antenna and required circuits that enable USB serial communication with CH340 conversion. The power supply of the board is via a USB port or an external power supply with voltages from 7V to 24V. The board is connected to the connectors: all 9 I/O ports (from D0 to D8 with GPIO addresses 0,4,5,12,13,14,15 and 16), Rx and Tx, A0 (AD converter) and RESET connectors, Figure 8. Nine I/O connectors allow operation with 3 pneumatic actuators. In addition, 6 I/O are configured as PULLUP inputs (these are inputs for limit switches) and 3 are configured as outputs (solenoids of directional control valves). To activate the solenoid, the output relay module is used (Figure 9), where the microcontroller (using the LOW logic level) activates relays with one normally open and one normally closed contact via the optocoupler and the transistor.

Figure 8. Arduino Uno R3 Development board based on ESP-8266EX microcontroller

Figure 9. Output relay module with optocouplers and four relay outputs

In the experimental setup the PULLUP inputs of microcontroller are directly connected to the limit switches, so the activation of the limit switches is directly achieved by connecting the PULLUP input to the ground, which is common to the microcontroller and inputs. The control circuits of solenoids are at one end connected to a 24V DC supply, and at the other end they are connected to the ground via the output contacts of the output relay. In Figure 10 and Figure 11, the practical realization is demonstrated, where FESTO electro-pneumatic didactic equipment was used. The pneumatic scheme of this EPS is shown in Figure 12.

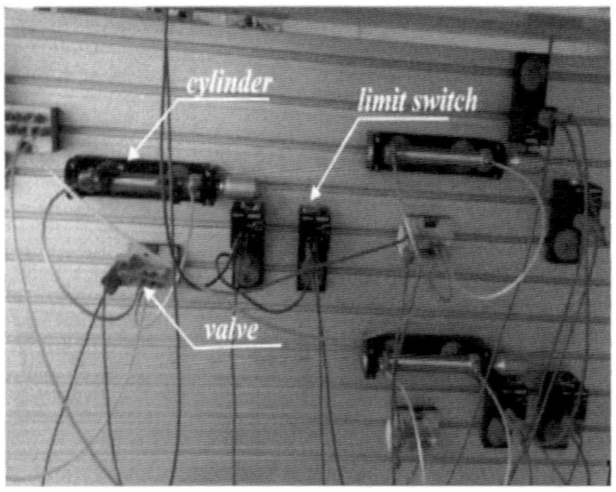

Figure 10. Three cylinders with limit switches and valves

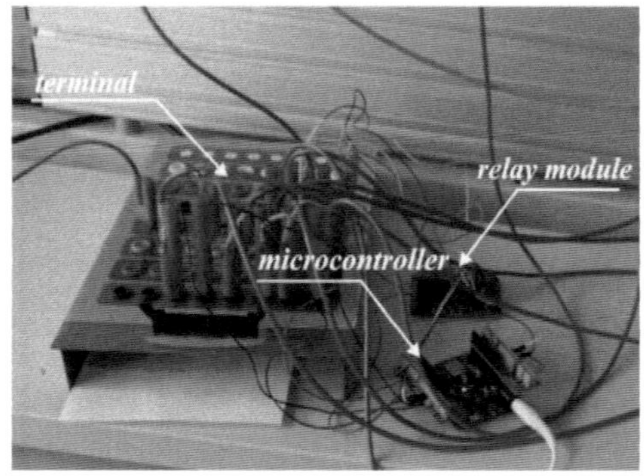

Figure 11. Microcontroller with output relay module and terminals

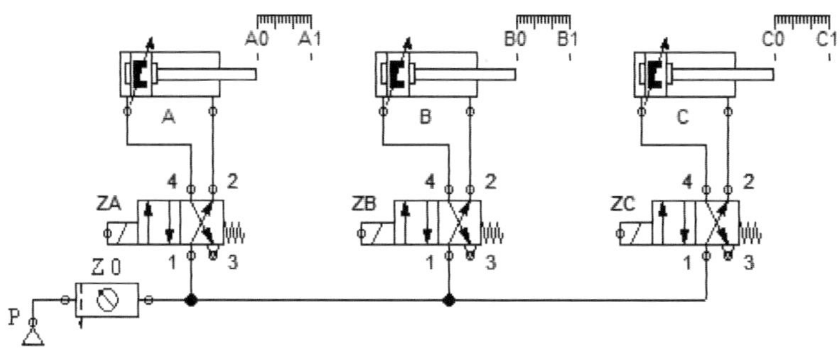

Figure 12. Pneumatic part of the considered electro-pneumatic system

In order to realize the observed EPS in industrial conditions, it is necessary galvanic separation of input module from the microcontroller as shown in Figure 13.

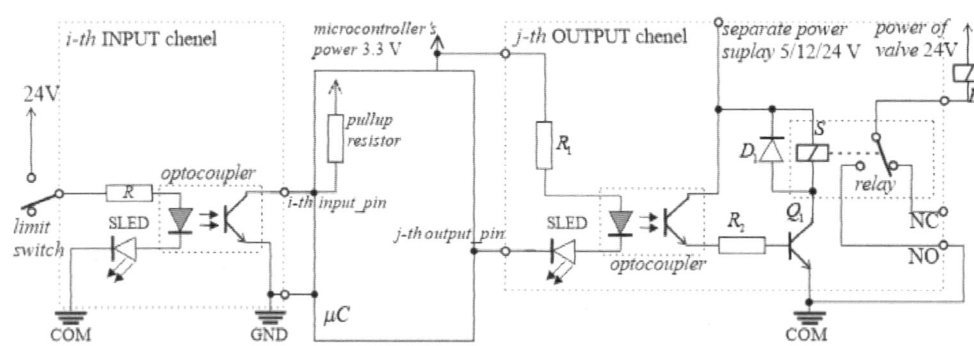

Figure 13. Input module, control module and output module of the considered electro-pneumatic system

The figure shows the galvanic separation using the optocoupler, where the voltage of the limit switch is used to activate the inputs (it is most often 24V DC). For the 8-cylinder EPS, 16 identical input channels and 8 identical output channels from Figure 13 should be used. The figure shows the relay output module. In the case of transistor outputs, in the diagram from the Figure 13, instead of the solenoid S, the solenoid EM should be connected. Power supply for the output transistors and the EM is common. In this case the relay at Figure 13 is unnecessary. The ESP32 WiFi microcontroller can be used as the control module (both microcontrollers of Chinese manufacturer Espressif Systems are) in nodeMCU or Arduino versions and with the Arduino IDE application software.

As an input module, it is possible to use ready-made modules from Figure 14 and Figure 15 as follows: two modules from Figure 14 as input and one relay module from Figure 15 as output.

Figure 14. Eight channel input module with optocouplers, high/low 3-5V output and 24V input

Figure 15. Eight channel relay output module with optocouplers, low activation and 24 V power supply

5. Conclusion

The paper deals with the design of remote control of the electro-pneumatic system, whereby the control is solved using the microcontroller ESP8266 and ESP32 with WiFi communication. The control part of the system has two functions: first, to communicate with the client in order to get tasks from it and fulfill its tasks, and second, to achieve the control function according to the assigned task in real time. Therefore, the microcontroller is configured as a web server, and the client is a computer that provides HTTP protocol and internet connection. With very small changes in the software it is possible to achieve that the client is another microcontroller, which creates all the prerequisites for direct communication between two machines (machine to machine communication), which is one of the basic requirements in the Industry 4.0.

The client sends an operation task in the form of a text string in which, according to certain rules and in a specific syntax, the operational cycle is written. Based on this operational cycle, the arrays of actions and transition conditions are generated, with one element of these arrays being able to realize all the necessary activities for one step in the operational cycle. In this way, with the realization of tasks in each step, the overall task set in the operational cycle is realized.

The fulfillment of requirements which are set for the control system was also experimentally verified. The ESP8266 microcontroller and the three-cylinder electro-pneumatic system were used. The results of the experiment confirm all the theoretical considerations which are noted and explained in the paper.

Appendix

Because of the limited scope of the paper, only some parts of the control software are provided here. Parts of the control software which are given in this paper are the functions *praviNizove* and *praviNiz*.

Flowcharts of these functions are shown at Figure 5 and Figure 6, respectively. Arduino IDE software was used to program the ESP32 and / or ESP8266 microcontroller.

```
byte praviNizove(String sc){//string sc is desired cycle
  char ch,chh;
  boolean onB = false;
 int p;
 uint16_t broj = 0;
 gpE = 0b1010101010101010; masGP = 0; cilindara = 0; iR = 0; abcE = 0; vremC = false;
 for (int i = 0; i < sc.length(); i++){// sc.length() - it is length of the string sc
  ch = sc[i];
  if (onB){
   if (ch >= '0' && ch <= '9') broj = broj * 10 + int(ch - '0');// int(ch - '0') -conversion of the character ch to int
                                                            //numeric and string between '/' and 'S' to broj
   if (ch == 'S'){
    if (broj != 0) { vremC = true; NIZV[iR - 1] = broj; }
    onB = false; broj = 0;
   }
  }
  p  = abc.indexOf(ch);
  if (p != -1) { chh = sc[i+1];  praviNiz(chh,p); }
  if (ch == '/')  onB = true;
 }
 cilindara = cil(masGP); //function cil() calculate the number of cylinders which are exist in the cycle sc
 return (iR);               // iR-number of steps
}
//
void praviNiz(char pm, byte a){//pm - it may be only '+' or '-', a-position of ch into string abc
 byte a2 = a << 1;  //multiply by 2
 uint16_t mg =  (pm == '+')?(0b0100000000000000 >> a2):(0b1000000000000000 >> a2);
 uint16_t mm = ( 0b1100000000000000 >> a2 ) ;
 masGP |= mm;
 if (pm == '+')  abcE |= ( 0b10000000 >> a) ;  //the same as bitSet  (abcE,7-a);
 else            abcE &= (~(0b10000000 >> a));  // the same as bitClear(abcE,7-a);
 gpE = (gpE & (~mm)) | mg;
 NIZR[iR]  = abcE;  NIZG[iR]  = gpE ;  NIZV[iR++]  = 0;
}
```

References

[1] *** FESTO Didactic (2000), *Fundamental of Electro-pneumatic*

[2] Javed A 2016 *Building Arduino Projects for the Internet of Things: Experiments with Real-World Applications,* Lake Zurich, Illinois, USA

[3] Schwartz M 2016 *Internet of Things with ESP8266*, Packt Publishing Ltd, Birmingham B3 2PB, UK

[4] Kolban N 2016 *Kolbon's Book on ESP32 & ESP8266*, October 2016

[5] Kolban N 2017 *Kolban's Book on ESP32*, May 2017

[6] Suyetno A, Kustono D and Sudjimat D A 2016 *Pneumatic Control System Simulation Based a PLC Micro in Mechatronics Courses,* Proceedings of the International Mechanical Engineering and Engineering Education Conferences (IMEEEC), doi: 10.1063/1.4965769,

[7] Singh R and Verma H K 2017 Development of PLC Based Controller for Pneumatic Pressing Machine in Engine-Bearing Manufacturing Plant, *Procedia Computer Science* **125** 449-458

[8] Kumar R and Rosario R *Design and Simulation of Electro-Pneumatic Motion Sequence Control*

IOP Conf. Series: Materials Science and Engineering **477** (2019) 012034 doi:10.1088/1757-899X/477/1/012034

using FluidSim, De La Salle University, Taft, Manila 1004 Philippines

[9] Swider J, Wszolek G and Carvalho W 2005 Programmable controller designed for electro-pneumatic systems, *Journal of Materials Processing Technology* **164-165** 1459-1465

[10] Kadam S P, Gaikwad V R, Bhavsar H D and Kulkarni S V 2016 Electro-Pneumatic Lift and Carry Conveying System, *International Journal of Science Technology & Engineering* **2**(10)

[11] Parikh P, Vasani R, Sheth S and Gohil J 2016 Actuation of Electro-Pneumatics System using MATLAB Simulink and Arduino Controller- A case of a Mechatronics systems Lab, *Advances in Intelligent System Research* **137** 59-64

Xilinx implementation of a serial-parallel digital converter

C D Cunțan, I Baciu and C M Diniș

Politehnica University of Timisoara, Department of Electrical Engineering and Industrial Informatics, 5 Revolution Street, Hunedoara, 331128, Romania

E-mail: corina.cuntan@upt.ro

Abstract. In the paper it is presented a serial parallel converter on 8 bit using a movement register with parallel output. The scheme created in Xilinx and implemented on a development board Nexis4DDr. The scheme contains two 4 bit counters and two registers schematically made and a frequency divider created in VHDL. The aimed sequence tact is obtained by dividing the tact frequency of 100MHz from the development board with the help of a divider circuit that allows the soft setup of the necessary frequency. The entering signal is brought to the serial input of a movement register, starting with the less significant bit. From the movement register the byte is transferred to the output register. The transfer between the two blocks is made in the moment when the 4 bits counter reaches 7 (moment which represents the closure of taking a bite) sequence which commands through a gate AND-NOT with three entrances, the command signal of the output data transfer. The output sequence can be memorized through a memorizing circuit or can be used for the command of a physical circuit. The H→L transition of the synchronizing signal, signal which is obtained also from the tact signal with the help of another 4 bit counter, determines the beginning of the conversion process. The circuit can function in a continuous way (without any reset circuit) or depending on the application it can be connected to a circuit which provides a command signal to obtain the reset signal of the implemented scheme.

1. Introduction

The most widespread shift registers application is the data conversion from parallel format in serial format and also the restoration through conversion of series data in parallel format to process or display them. A shift register is a register of n bits with the possibility of moving the stored data in it with one position of 1 bit at each clock impulse.[1] The series parallel convertor was designed using the developing software package Xilinx ISE, it's implementation being made on the reprogrammable structure FPGA of Nexis4DDR type.[2]

The probable logic circuits FPGA are probable logic gate arrays in the electric field. The circuit contains a big number of probable logic blocks (CLB) arranged in a probable connection web, surrounded by I/O blocks which are also probable.

With these types of circuits logic functions of great complexity can be obtained using both electronic schemes and codes which are written in hardware description language VHDL, Verilog or ABEL.

The software is essential for the efficient usage of the probable logic devices. The series of programs Integrated Software Environment of the Xilinx firm represents an integrated development environment for the designing of the digital systems with the help of FPGA.

The usage of programmable circuits brings a series of advantages in comparison to the usage of classic integrated circuits, among which a few are reminded:

- Reducing the size and power dissipation;
- Improving speed performance;
- Ease of debugging faulty systems;
- Decreasing the failure rate and the price of the printed circuit;[3]
- Updating and improving the implemented function by reprogramming, without any hardware change.

2. Problem Formulation

In this paper it is presented a series - parallel convertor on 8 bits using a shift register with parallel output.

The series-parallel conversion implies a data transfer from the source to destination, transfer which means using three command signals (Figure 1):

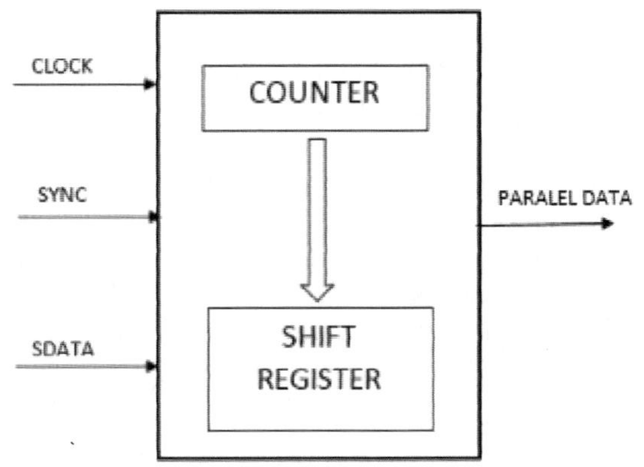

Figure 1. Block simplified scheme of the convertor

The clock signal (clock), the data series signal (SDATA) and the timing signal (SYNC). The clock signal gives the time reference needed for the transfer, defining the time period in which 1 bit is transferred. The desired frequency is obtained by dividing the frequency on the main circuit (100MHz) through a dividing circuit which allows the soft setting of the output frequency.

The data signal represents the input size for the serial - parallel convertor. (Figure 2)

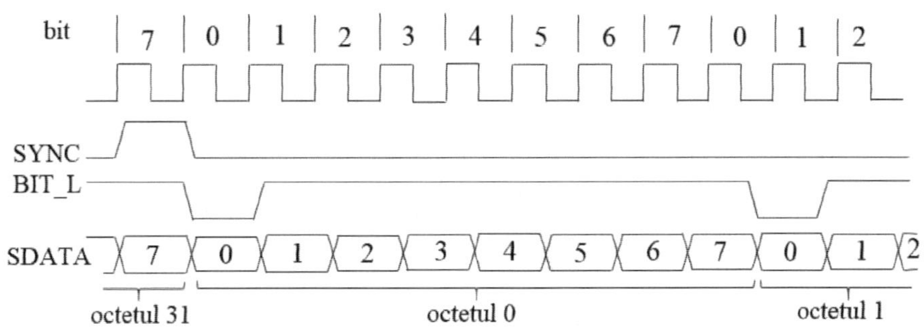

Figure 2. Temporal diagram for the parallel-series conversion

The beginning and ending of a byte is ordered by BIT_L which is obtained at the output of the 4 bits' number and which orders the data transfer from the shifting register to the parallel register. The data is transmitted starting with the most significant bit. The 32 temporal divisions are determined with the help of the synchronizing input.

The timing signal gives the time reference for defining the data format, for example the beginning of a byte or of a word from the series data row. [4]

The series - parallel conversion is realized through two shift registers. The data transfer between the registers is ordered by a counter of 4 bits which provides the transfer impulse through an identification circuit of number 7. [5] The convertor's structure contains a second counter which allows the determination of the temporal division number of the 8 bits section. The convertor's scheme realized in Xilinx is presented in Figure 3.[6]

The byte from the convertor's output is used for the transistors' command from the bridge inverter of a frequency convertor.[7]

To obtain a signal with an output frequency of 1KHz it is necessary a clock signal with a frequency of 8 KHz. This signal is obtained at the DIV1 divisor's output which divides the clock frequency of the Nexis 4 DDR module to 6250. The DIV1 divisor's output commands both the counters as well as the shift registers from the convertor's structure.

The data sequence is brought with a frequency of 4 KHz which is obtained at the DIV 2 divisor's output; this frequency results from the division of the clock frequency to 12500.

The timing signal is brought at the counter's input and determines the beginning of the counting process.

The command signal for data transfer between the registers is obtained by the first 4 bit counter. The least important 3 bits of the counter are brought at the entrance of a NAND gate with three inputs; its exit is brought through an inverter of the command pin of the data transfer in the output register.

The RCO output of the number commands the functioning validity of the next number. The most significant bit from the first counter together with the second one are used for the elaboration of the command sequence of the frequency phase convertor's transistors, as it can be observed in Figure 3.

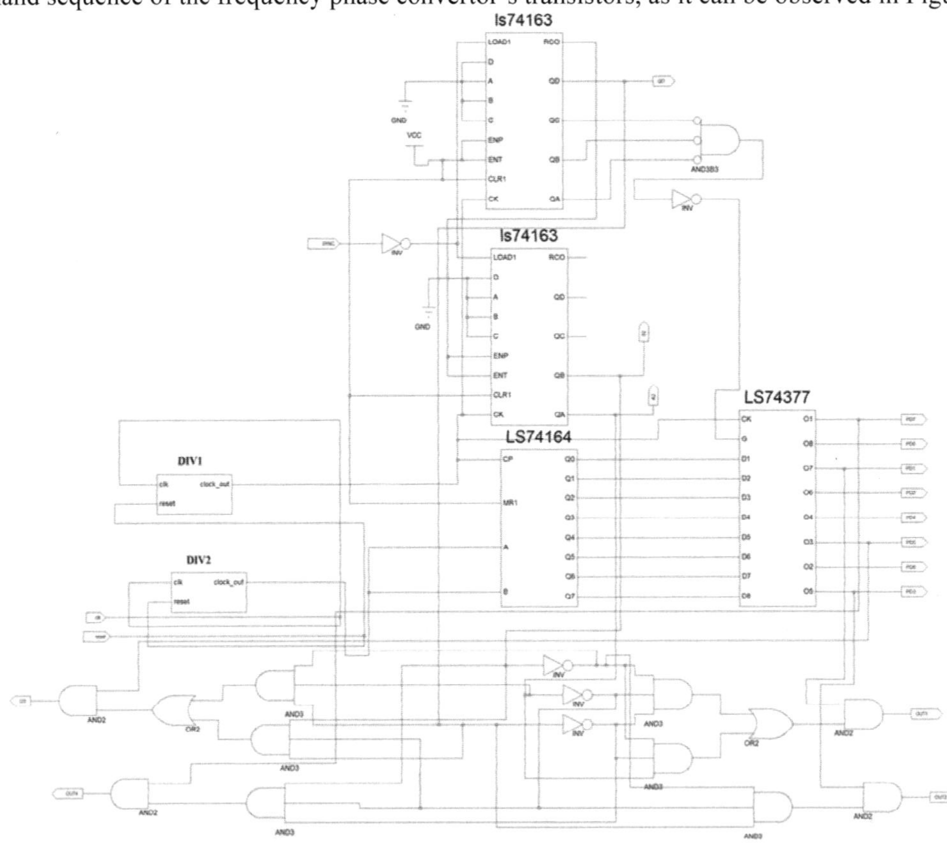

Figure 3. The series - parallel convertor's scheme realized in Xilinx

The structure of a counter (Figure 4) was realized under the form of a schematic file in Xilinx starting from the structure of a 74163 counter.

The parallel inputs are connected to the mass and the validation and reset inputs are connected to1 logic. The tact impulse for counting is received from the DIV1 divider with an output frequency of 8 KHz and the charging input is brought from the synchronizing signal through an inverting device. The outputs of the number allow obtaining the transfer command given in the parallel register and at the output.

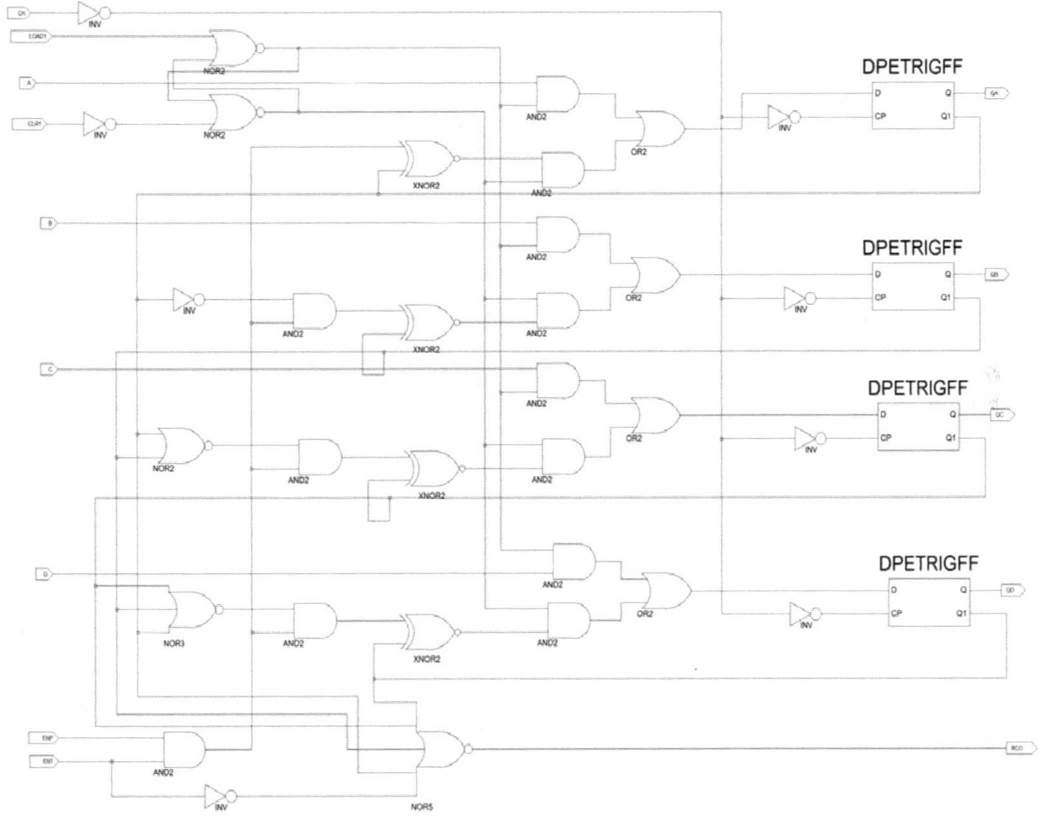

Figure 4. The structure of the counter designed in Xilinx

Also the structures of the shift register and of the parallel register were realized starting from the structure IC74164 (Figure 5) and 74377 (Figure 6).

The shifting register (Figure 5) contains 8 bi-stables of Delay type. The input data brought in a series are transferred in the parallel register at the end of 8 tact impulses. Once with the data transfer in the parallel register the serial acquisition of a new bite can begin.

The parallel register (Figure 6) has in its structure mainly 8 bi-stables of Delay type. This block receives the transfer command from BIT-L in the moment when the counter reaches number 7, transferring the output data in parallel format.

At each made block the independent functioning was checked using Nexis 4 DDR.

Finally the parallel series convertor scheme was implemented and it was verified it's functioning.

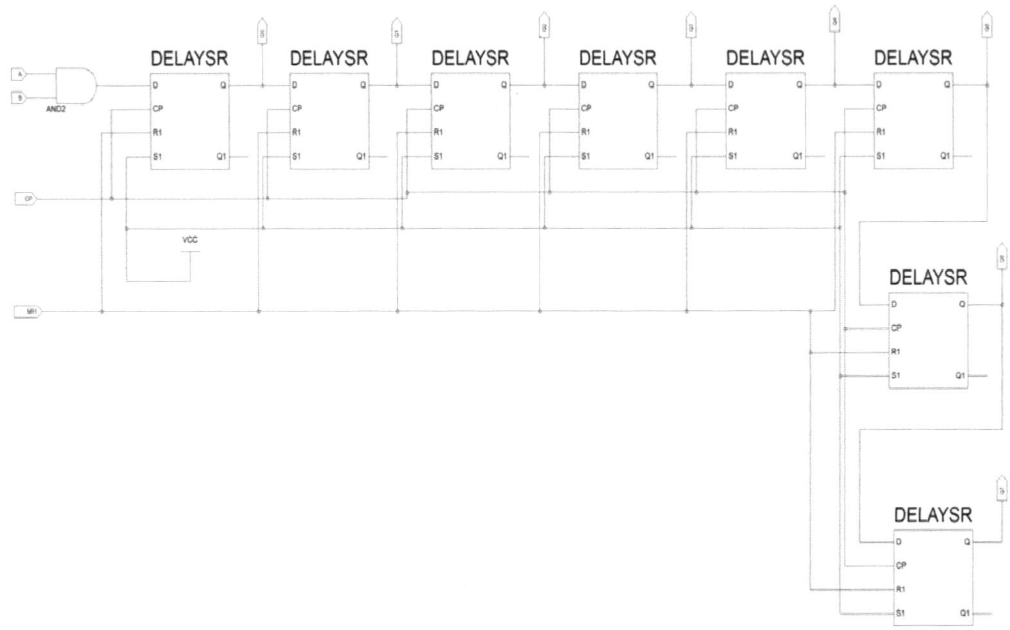

Figure 5. The structure of the shift register designed in Xilinx

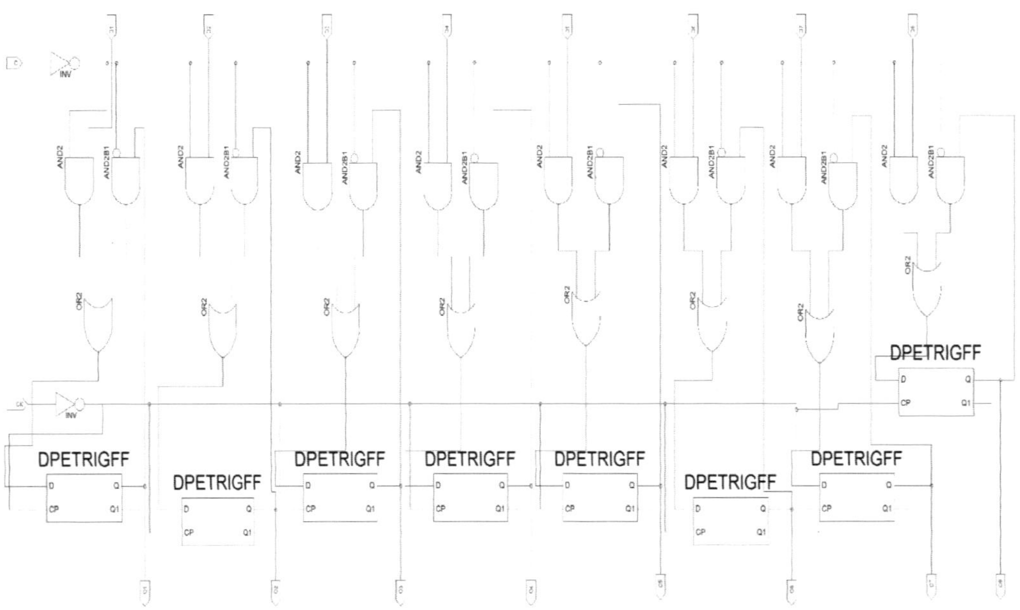

Figure 6. The structure of the parallel register designed in Xilinx

The output obtained signals go to the entrance of an phase inverter block with 4 IGBT transistors (Figure 7).[8]

The signals are brought at an optocoupler galvanic separation circuit through a MOS inverting. [9]

For the optocouplers command are used transistors (Ta1...Td1) to raise the required current for the input commands of the octocouplers. At the octocouplers' exit another group of transistors is used (Ta2, Ta3...Td2, Td3) for the command both in current as well as in the polarity of the IGBT

(T1...T4) transistors.[10] The IGBT transistors order the primary of a voltage elevated transistor resulting at the output an alternative voltage that has an alternation formed out of two voltage levels. To smooth the voltage and current jumps a L1C1 filter is used.[11]

The presentation of the functioning of the entire circuit is emphasized by the wave obtained in different points of the circuit.

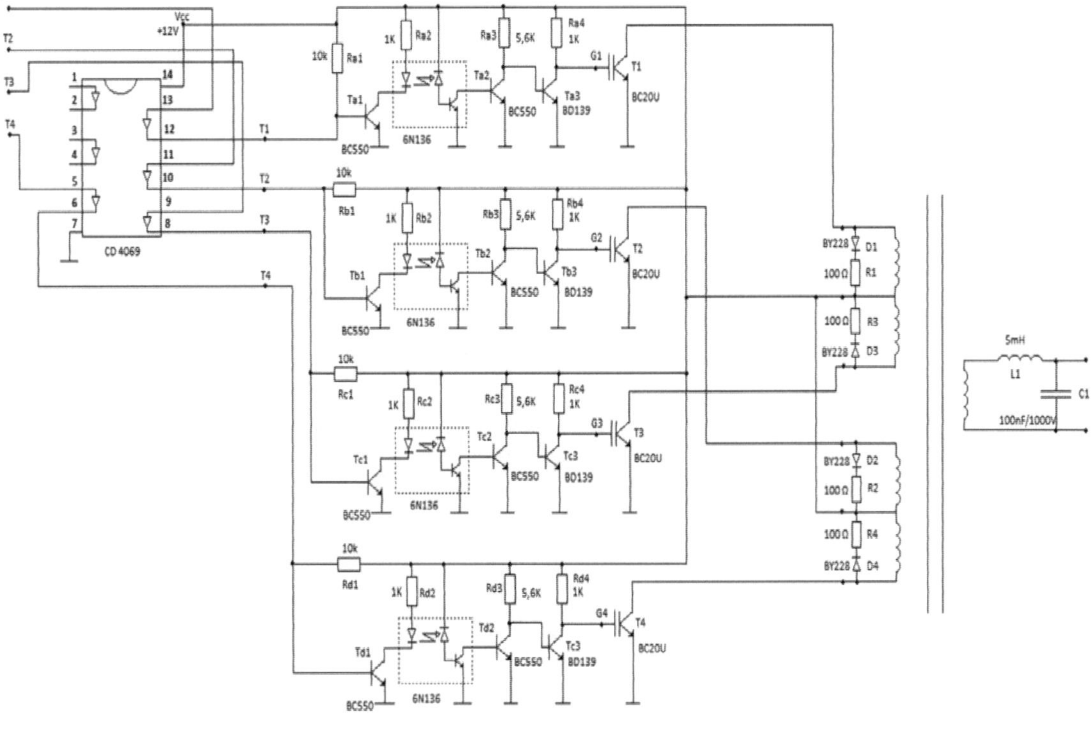

Figure 7. The single - phase inverter scheme

The waveforms were obtained with the help of a two-channel digital oscilloscope which allows the storage of the visualized signals. In this way the waveforms are presented for the command of the Ta1, Tb1 (Figure 8) and Ta1, Tc1(Figure 9).

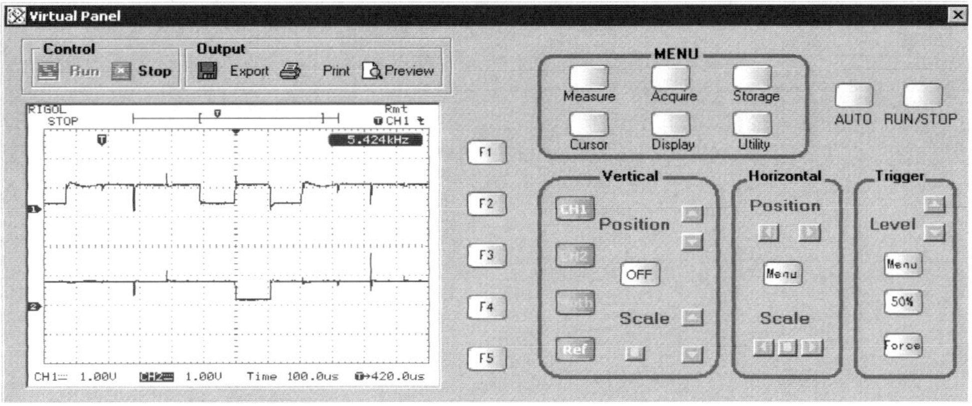

Figure 8. Command impulse for Ta1 and Tb1

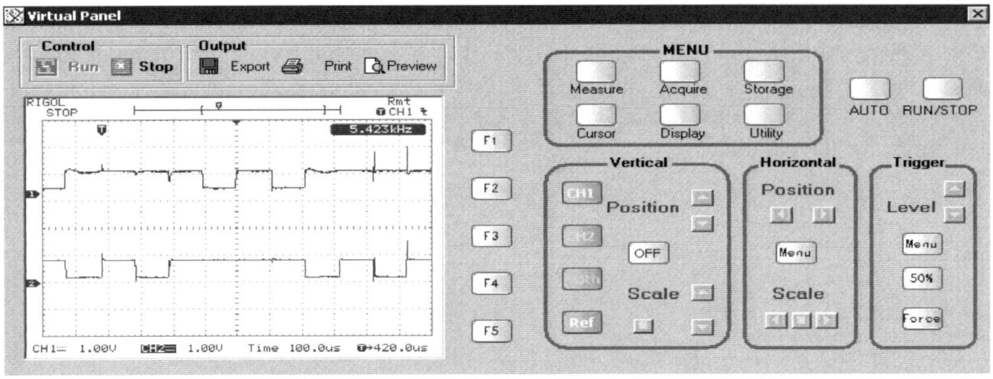

Figure 9. Command impulse for Ta1 and Tc1

These signals command the optocouplers' entrance as it follows: the optocoupler command 1,2 (Figure 10) and the optocoupler command 1,3 (Figure 11).

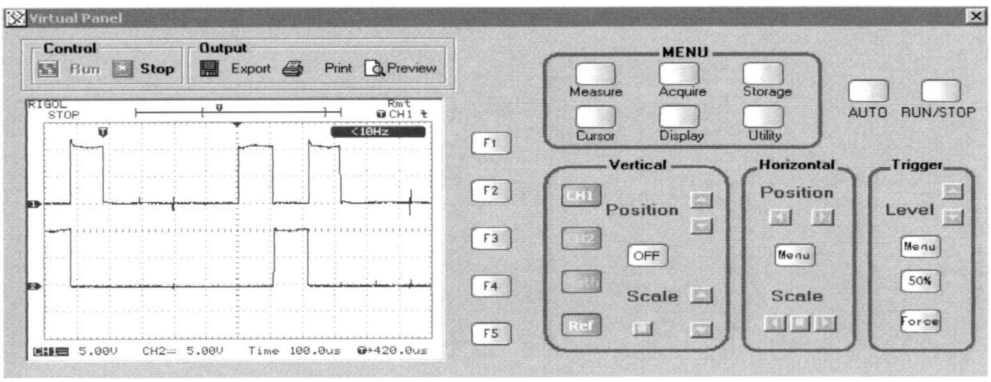

Figure 10. Command impulse for optocouplers 1 and 2

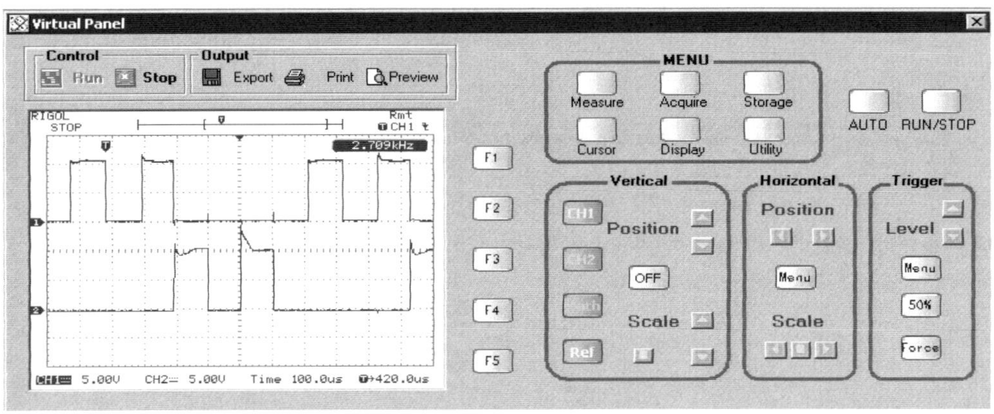

Figure 11. Command impulse for optocouplers 1 and 3

The IGBT transistor's command signals are presented in Figure 11 for T1 and T3 transistors (Figure 12) and for T1 and T2 transistors in Figure 13.

IOP Conf. Series: Materials Science and Engineering **477** (2019) 012035 doi:10.1088/1757-899X/477/1/012035

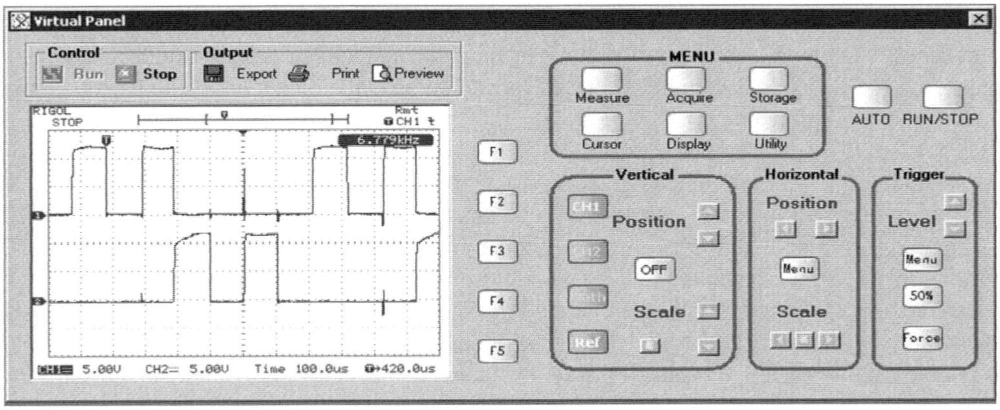

Figure 12. Command voltage for the transistors' command IGBT: T1 and T3

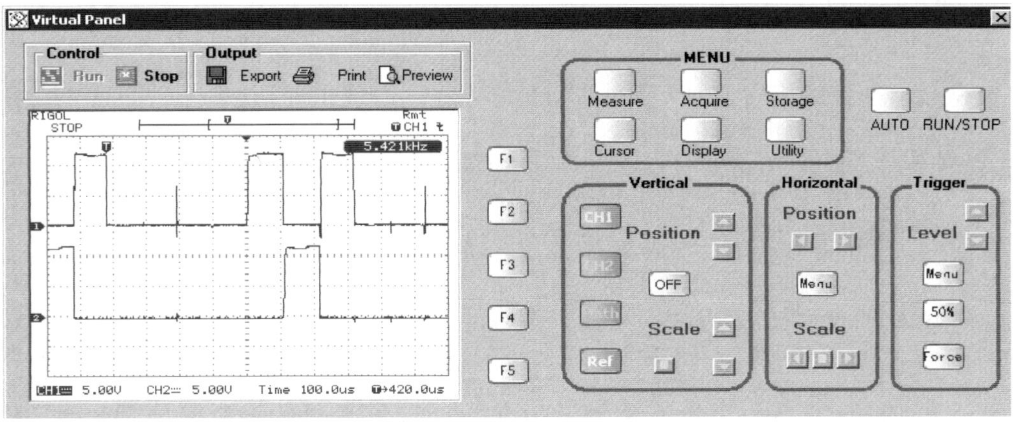

Figure 13. Command voltage for the transistors' command IGBT: T1 and T2

The output voltage (Figure 14) emphasizes the existence of three intervals on each alternation. From the figure it results that the voltage jumps corresponding to levels 1 and 2 are not equal. This fact is due to the transformer's construction.

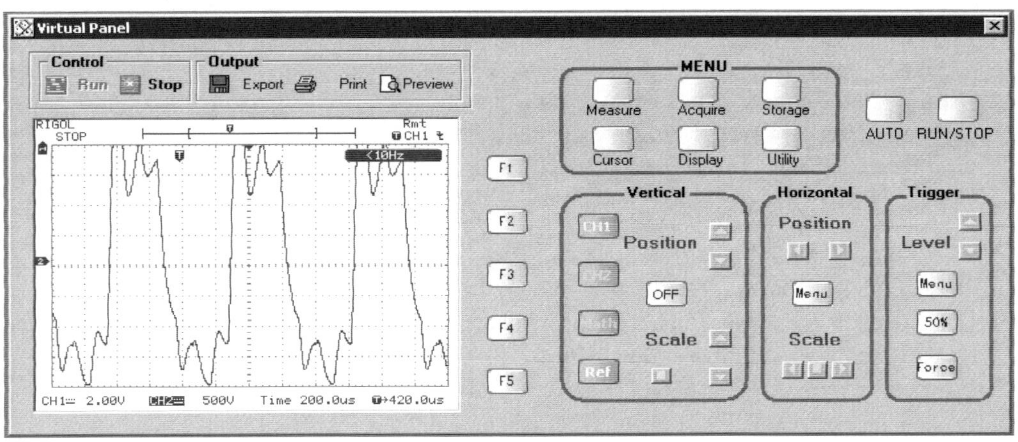

Figure 14. Output voltage

IOP Conf. Series: Materials Science and Engineering **477** (2019) 012035 doi:10.1088/1757-899X/477/1/012035

3. Conclusions

The presented waveforms correspond to a clock frequency of 1KHz which corresponds to an output frequency of 125 Hz. Using the output filter allows smoothing the jumps which reduces the number of harmonics in the output voltage.

The final IGBT transistors are protected at high-voltage by connecting an RD group on each primary winding of the transistor. Improving the functioning of the inverter can be done by using a command sequence with pauses between the transistors' commands. The usage of the parallel series convertor allows setting with ease the output frequency by changing the clock impulse. This is realized by soft modifying of the frequency divider without involving other additional circuits.

Also the output frequency can be memorized in a memory cell which can be later used independently for the inverter command.

Implementing the Xilinx circuit allows to easy change the output sequence so that other physical circuits can be commanded.

At the exist of the register from the parallel series convertor's structure it is ordered the passage on 0 in the short time of its exit. This can be realized through a logical scheme which is commanded by another frequency divider.

References

[1] Wakerly J 2002 *Digital Circuits. Principles and practices used in design*, Publishing House Teora, București, Romania

[2] Iana G, Șerban G, Ionescu L, Anghelescu P, Popescu M, Bitoleanu A and Dobriceanu M 2010 *Programming with Hardware Description Languages*, Publishing House MatrixRom, Bucuresti, Romania

[3] Nicula D 2000 *Design of digital systems implemented with programmable devices*, Technical Publishing House, Bucuresti, Romania

[4] Büchele M, Fischer H, Gorzellik M, Herrmann F, Königsmann K, Schill C and Schopferer S 2012 128-channel Time-to-Digital Converter (TDC) inside a Virtex-5 FPGA on the GANDALF module, *Journal of Instrumentation* **7** 1-7

[5] Li Y, Huang A Q and Motto K 2002 Series and Parallel Operation of the Emitter Turn-Off (ETO) Thyristor, *IEEE Transactions On Industry Applications* **38**(3) 706-712

[6] Sugimura H, Eid A M, Kown S K, Lee H W, Hiraki E and Nakaoka M 2005 *High frequency cyclo-converter using one-chip reverse blocking IGBT based bidirectional power switches*, Proceedings of the Eighth International Conference on Electrical Machines and Systems, September, Volume 2, pp 1095-1100

[7] Monmasson E and Cirstea M N 2007 FPGA Design Methodology for Industrial Control Systems—A Review, *IEEE Transactions On Industrial Electronics* **54**(4) 1824-1842

[8] Luo D F L, Hong Y and Rashid M H 2005 *Digital Power Electronics and Applications*, Elsevier Academic Press

[9] Sathyan A, Milivojevic N, Lee Y J, Krishnamurthy M and Emadi A 2009 An FPGA-Based Novel Digital PWM Control Scheme for BLDC Motor Drives, *IEEE Transactions on Industrial Electronics* **56**(8) 3040-3049

[10] Popa G N, Popa I, Diniș C M, Deaconu S I and Abrudean I C 2016 *Control of a single-phase inverter with PIC microcontroller*, The XVIth Multidisciplinary International Conference "Professor Dorin Pavel - Founder of Romanian Hydro-energetics" Science and Engineering, Vol 29, pp 447-456

[11] Deaconu S I, Babău R, Popa G N and Gherman P L 2018 Hydroelectric power plant with variable flow on drinking water adduction, *IOP Conf. Ser.: Mater. Sci. Eng.* **294** 012023

Aspects concerning the friction for the motion on an inclined plane of an axisymmetric body

F C Ciornei, S Alaci, I C Romanu, I Mihai and V C Lazar

"Stefan cel Mare" University, Mechanics and Technologies Department, Suceava, Romania, Universitatii str., no. 13, 720229 Suceava, Romania

E-mail: florina@fim.usv.ro

Abstract. The geometrical intrinsic elements defining the Hertzian point contact between two solid bodies are the common normal and the tangent plane in the contact point. To the relative motions occurring about these elements, forces and moments of friction oppose; these reactions are the elements of the friction torsor from contact. The paper presents a particular case when sliding friction and rolling friction coexist in the tangent plane. The first body is made of two identical bearing balls rigidly fastened and the second body is an inclined plane. The straight line defined by the contact points between the mobile body and the plane differs from the steepest line of the plane. By gradually increasing the slope of the plane, the situation when the body simultaneously rolls and slides with respect to the line defined by the contact points is reached. When the duration of motion, the sliding of the centre of mass and the rotation angle of the body between two marks from the plane are measured, it is concluded that the sliding velocity and the angular rolling velocity are independent. While the rotation angle is constant the sliding and time of motion vary in a relatively wide range thus confirm a significant gradient of the coefficients of sliding and rolling friction.

1. Introduction

In applied mechanics there are frequently met contacting bodies which have different boundary surfaces and therefore the theoretical contact between them takes place in a point [1]. From kinematical point of view, this situation presents the most complex case of relative motions [2]. A kinematical pair of first class is created and from the six simple motions that may occur between two free rigid bodies only the displacement along the normal in contact $\boldsymbol{n}$ is cancelled [3]. The other motions are translations in the tangent plane (T) along two perpendicular axes, rotations about these two axes and spinning rotation about the normal and they are impeded by friction forces $\boldsymbol{F}_{sl}$ and friction torques, $\boldsymbol{M}_r$ and $\boldsymbol{M}_s$ respectively, all components of the friction torsor in the contact point, according to Figure 1. The characterization of the components of the torsor of friction is a major tribological topic.

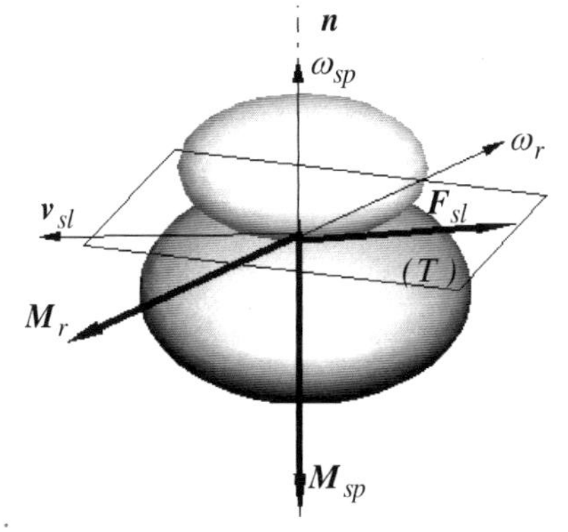

Figure 1. The components of the torsor of friction in a point contact

2. Theoretical aspects

For an actual case, the simultaneous incidence of all components of the friction torsor makes very difficult the task of establishing them [4], [5]. A typical example is the thrust ball bearing, where the hypothesis of pure rolling between the ball and the race conducts to the conclusion that spinning and rolling torques are simultaneously present. From this reason, often, considering simpler cases are preferred, where only one component [6-11] of the friction torsor is present and afterwards, for a concrete situation, the results are by some means superposed, with the necessity of stipulating in which manner they influence reciprocally. The inclined plane method is frequently used in tribology as it permits finding the characteristics of sliding and rolling friction. As principle, on an inclined plane of variable angle, initially in horizontal position, a prismatic or body of revolution is set, and then, by progressively increasing the tilting angle of the plane, the body starts moving – translation or rotation, as in Figure 2.

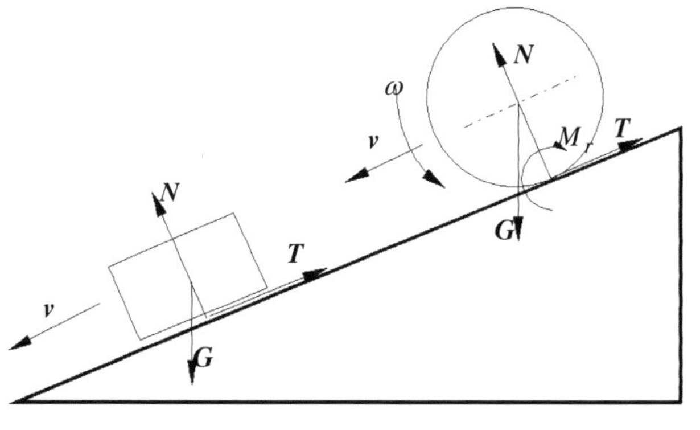

Figure 2. Bodies in rolling and sliding motions on the inclined plane

Finding the acceleration of the body for the downward motion and writing the equations of motion one can establish the coefficient of sliding friction (CSF) or rolling friction (CRF). To be mentioned that for the case of plane-parallel motion on the inclined plane of a body of revolution, this may present either sliding or rolling motion; both motions cannot simultaneously occur, Figure 2. But for spatial cases, since in the common tangent plane two perpendicular directions exist, it is possible to appear sliding on one direction and pure rolling on the other one. An illustration and the consequences emerging from this are next presented. The following coordinate systems are considered:

- the $Oxyz$ system, having the Ox and Oy axes perpendicular and contained in a horizontal plane, and the vertical axis, Oz;
- the $Ox'y'z'$ system obtained from the $Oxyz$ system by a rotation with the angle α around the Oz axis;
- the $Ox''y''z''$ system obtained from the $Ox'y'z'$ system by a rotation with the angle β around the y' axis.

An axisymmetric body is placed on the plane $Ox''y''$ such as the axis of symmetry is parallel to the Ox'' axis. The body of revolution is obtained by fastening two identical bearing balls via a cylindrical ring. This solution has the advantage that the points of application of the reaction forces between the body and the plane are well specified, different from the situation when a cylinder is placed on a plane. Another drawback of the cylindrical body consists in the edge effects occurring in the vicinity of the ends of the contact generatrix. The body presented in Figure 3 has the following motions, permitted by the two Hertzian point contacts: translations along the x'' and y'' axes, pivot rotation around the normal z'' and rolling around the x'' axis. By conveniently selecting the distance between the centers of the two balls and the tilting angles α and β it is seen, by assuming that the reaction forces from the two points of contact are identical, that the spinning motion cannot take place because the spinning torque is zero. For small values of the α angle the body rolls along the y'' axis, sliding or resting about the x'' direction.

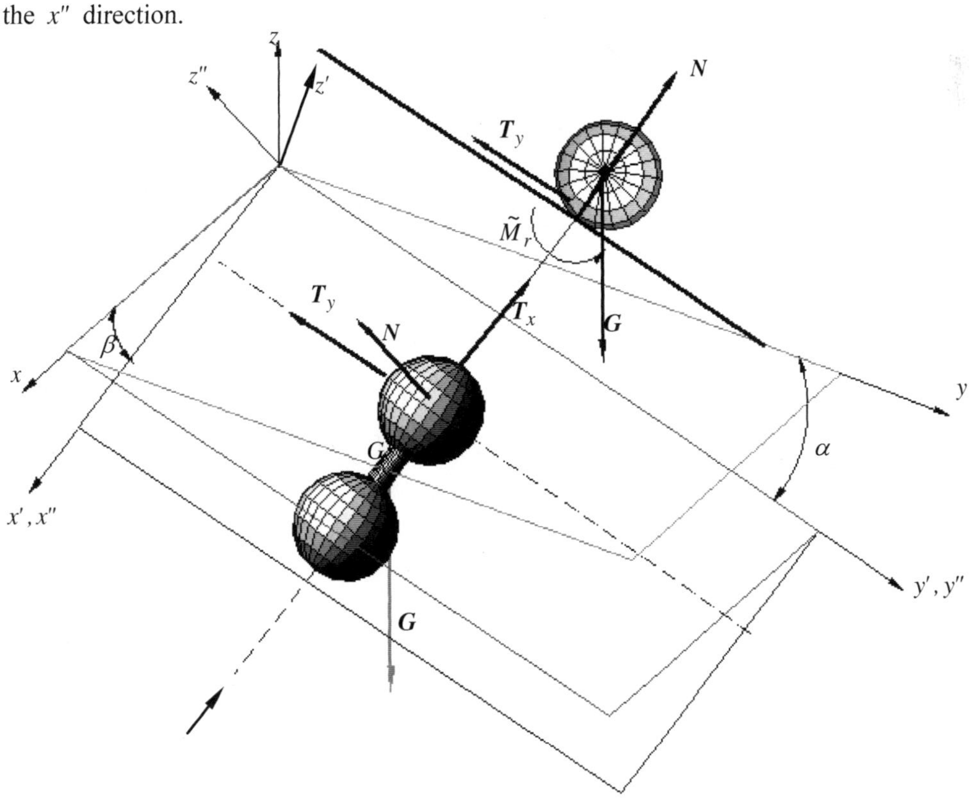

Figure 3. Axisymmetric body in sliding and rolling motions performed on two directions

In order to describe the motion of the body, the theorem of the motion of the center of mass:

$$M\ddot{\boldsymbol{r}} = \boldsymbol{N} + \boldsymbol{T}_x + \boldsymbol{T}_y - \boldsymbol{G} \tag{1}$$

and the moment of momentum theorem [12]:

$$J_x\ddot{\varphi} = T_y R - M_r \tag{2}$$

are applied. The equation (1) can be written using the projections on the axes:

$$M\ddot{x}''\,\boldsymbol{i}'' + M\ddot{y}''\,\boldsymbol{j}'' = -T_x\boldsymbol{i}'' - T_y\boldsymbol{j}'' + N\boldsymbol{k}'' - G\boldsymbol{k} \tag{3}$$

It is noticed that the unit vector $\boldsymbol{k}$ should be expressed as function of the unit vectors of the $Ox''\,y''\,z''$ system. To this end, it is applied the remark made by McCarthy [13], namely:

$$\boldsymbol{r}_1 = \boldsymbol{T}_{12}\boldsymbol{r}_2 \tag{4}$$

where $\boldsymbol{r}_1$, $\boldsymbol{r}_2$ are the position vectors of the point in the systems $''1''$ and $''2''$ and $\boldsymbol{T}_{12}$ is the displacement required to superpose the system $''1''$ over $''2''$. Considering the $Ox'\,y''\,z''$ as initial system and $Oxyz$ as final system, the displacement operator $\boldsymbol{T}_{12}$ is obtained a rotation of angle $-\beta$ around Oy'' axis followed by a rotation of angle α around the Ox' axis.

$$\boldsymbol{T}_{12} = \boldsymbol{Y}(-\beta)\boldsymbol{X}(\alpha) \tag{5}$$

where [14]:

$$\boldsymbol{X}(\alpha) = \begin{bmatrix} 1 & 0 & 0 \\ 0 & \cos\alpha & -\sin\alpha \\ 0 & \sin\alpha & \cos\alpha \end{bmatrix} \tag{6}$$

and

$$\boldsymbol{Y}(\beta) = \begin{bmatrix} \cos\beta & 0 & \sin\beta \\ 0 & 1 & 0 \\ -\sin\beta & 0 & \cos\beta \end{bmatrix} \tag{7}$$

The following decomposition of the $\boldsymbol{k}$ versor is obtained after the calculus is made:

$$\boldsymbol{k} = -\boldsymbol{i}''\cos\alpha\sin\beta - \boldsymbol{j}''\sin\alpha + \boldsymbol{k}''\cos\alpha\cos\beta \tag{8}$$

and it results that the projections of the equation(3) have the form:

$$\begin{cases} M\ddot{x}'' = Mg\cos\alpha\sin\beta - T_x \\ M\ddot{y}'' = Mg\sin\alpha - T_y \\ 0 = N - Mg\cos\alpha\cos\beta \end{cases} \tag{9}$$

The scalar equations of motion of the body have the expressions:

$$\begin{cases} M\ddot{x}'' = Mg\cos\alpha\sin\beta - T_x \\ M\ddot{y}'' = Mg\sin\alpha - T_y \\ 0 = N - Mg\cos\alpha\cos\beta \\ J_x\ddot{\varphi} = T_y R - M_r \end{cases} \tag{10}$$

The unknowns of the system of equations (10) are the components of the reaction, N, T_x T_y, M_r and the laws of motion, $\varphi = \varphi(t)$ and $x = (t)$. To the system of equations (10), must be added:

- the kinematics equation characteristic to pure rolling:

$$y = R\varphi \tag{11}$$

- the equations characteristic to sliding along Ox axis:

$$T_x = \mu N \tag{12}$$

- and the equation relating the rolling friction torque to the normal reaction force:

$$M_r = sN \tag{13}$$

where it was assumed that the torque of rolling friction is proportional to the normal load. Recent theories propose new relations where the rolling friction torque depends on the normal force raised at a power [15], [16]. The equations (9-13) complete a system of seven equations with seven unknowns and by solving it, the following expressions are found:

$$N = Mg\,cos\alpha\,co\beta \tag{14}$$

$$T_x = \mu Mg\,cos\alpha\,cos\beta \tag{15}$$

$$M_r = sMg\,cos\alpha\,cos\beta \tag{16}$$

$$\ddot{x}'' = (sin\beta - \mu cos\beta)g\,cos\alpha \tag{17}$$

$$\ddot{\varphi}_2 = Mg\frac{R\,sin\alpha - s\,cos\alpha\,cos\beta}{J_x + MR^2} \tag{18}$$

$$\dddot{y}''_2 = Mg\frac{R\,sin\alpha - s\,cos\alpha\,cos\beta}{J_x + MR^2}R \tag{19}$$

$$T_y = gM\frac{J_x\,sin\alpha + sRM\,cos\alpha\,cos\beta}{J_x + MR^2} \tag{20}$$

Besides the above equations, the inequality characteristic to pure rolling around the axis passing through the contact points, which is parallel to Ox'' axis, should be verified:

$$T_y < \mu N \tag{21}$$

An analysis upon the relation (14-20) shows that all the right members of these equations are constant values and this proves that all motions should be uniformly accelerated translation and rotation, respectively. Additionally, all the forces and moments from the contact points must be constant. Moreover, when sliding and rolling occur simultaneously with respect to the same axis Ox'', from the tangent plane, in the present situation, from the relations (15) and (16) a *proportionality between the acceleration on the sliding direction and the angular rolling acceleration should exist.*

$$\frac{\ddot{x}''}{\ddot{\varphi}_2} = \frac{(sin\beta - \mu cos\beta)cos\alpha}{M\frac{R\,sin\alpha - s\,cos\alpha\,cos\beta}{J_x + MR^2}} \tag{22}$$

IOP Conf. Series: Materials Science and Engineering **477** (2019) 012036 doi:10.1088/1757-899X/477/1/012036

If the equation (22) proves to be correct, then finding the coefficient of friction would be easier since the coefficient of rolling friction may be found by measuring the coefficient of sliding friction μ, the angular acceleration $\ddot{\varphi}_2$ and the sliding acceleration $\ddot{x}''$. Furthermore, in the two ratios from equation (22), both the numerator and the denominator are constant, and from here the conclusion that a necessary condition is:

$$\frac{\ddot{x}''}{\ddot{\varphi}_2} = const \tag{23}$$

3. Experimental set-up

A device was designed and assembled in order to test the validity of the hypothesis expressed by the equation (23). The experimental set-up, based on the schematics form Figure 3, consists in a 10 mm thick aluminum plate 1, placed on another plate made of laminated wood 2, that at its turn is placed on the top of an horizontal table. Between the table and the wood plate, on one side, and between the wood plate and the aluminum plate, on the other side, feeler gauges are introduced to accomplish the intended tilt of the aluminum plate, according to the scheme form Figure 3. The rolling body 3 is made from two identical bearing balls jointed together with the purpose to block the spinning motion and to define precisely the position of the rolling axis (the axis passing through the contact points between the balls and the plate). The rolling body is set into contact with two plates 4 made from electronic circuit bare board, insulated one from each other, but both attached firmly to a metallic prism 5. The prism 5 is on the top of aluminum plate with edges parallel to the ones of the plate. The launching position assumes that each of the balls contacts one of the electric boards and thus an electrical circuit having a source S and a small light emitting diode is closed. Therefore, when the body is set to motion, the circuit opens and the moment of start of the motion can be precisely identified. At the other end of the plate, a rubber prism 6 is positioned to stop the body and damp the motion. A ruler 7 is fixed to the rubber prism in order to observe the transversal sliding of the body. An auxiliary prism 8 was used so as to ensure the launch of the body from the same position.

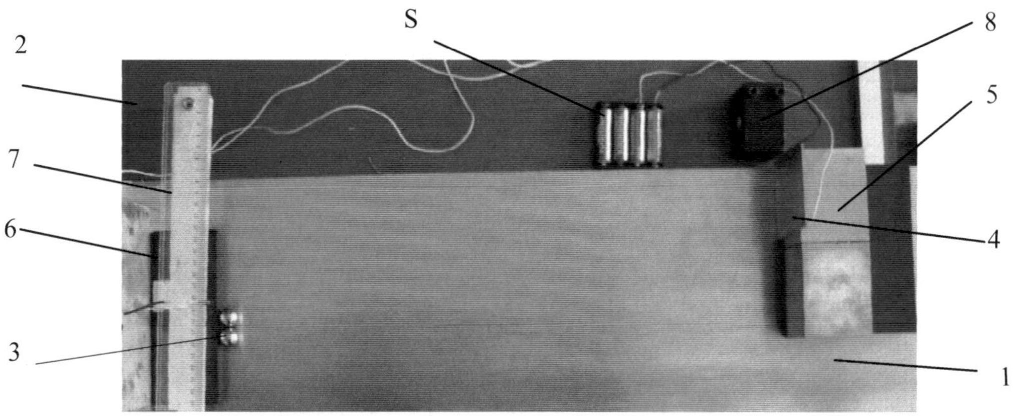

Figure 4. General view of the experimental set-up

It was concluded that the sliding motion can be estimated with a sufficient accuracy but for the rolling motion the errors in estimations are appreciable. To improve this aspect of the device, another alternative for the rolling body was selected, as presented in Figure 5. It consists in two identical bearing balls connected by a metallic cylindrical part. On the side surface of the cylinder an adhesive leaf was glued and the tracks of an axial symmetry plane 1 and normal symmetry plane 2 respectively, were traced on it. These tracks are used in finding the rotation displacement and sliding of the body, respectively.

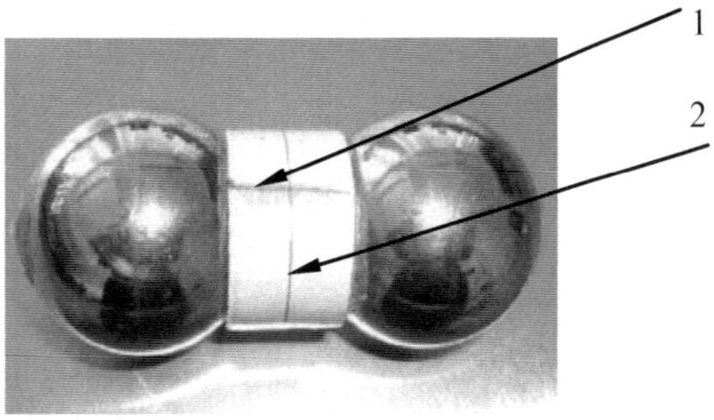

Figure 5. The new option for the rolling body

The body for the launching position and for the final position is presented in Figure 6 and Figure 7, respectively.

Figure 6. The body in the start position

Figure 7. The body at the end of motion

The motion of the body was filmed from the beginning to the moment of the collision with the rubber prism, in order to measure the time of motion. The movie was subsequently split into frames. The initial moment of the motion is precisely identified as the instant when the LED is off but the final moment required supplementary actions, for accurate timing. To this end, it was necessary that the longitudinal track is observed in the vicinity of the instant of impact with the rubber prism. To this purpose, the hypothesis of pure rolling (which was experimentally validated later on) was accepted. The distance between the starting prism and the final rubber prism was adjusted to be equal to the sum between the diameter of the ball and an integer number of circumferences and thus the centre of the ball runs the same integer number of diameters and the axial track should attain in the final position the same as in the launching one.

Concerning the accuracy of finding the final instant of the motion, in Figure 8 there are presented three successive frames from a movie: one before the impact with the rubber, one at the end and the third after impact.

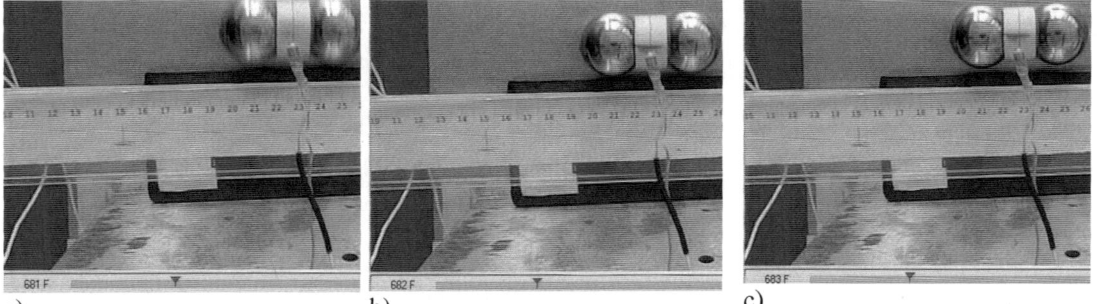

a) b) c)

Figure 8. Identification of the final moment of motion

It can be noticed that before impact, due to high rotation velocity, the track cannot be identified, Figure 8a. Quite after impact the rotation velocity decreases significantly and the track can be easily observed, Figure 8c, the body being away from the rubber prism. Therefore, the impact instant should be discovered between the two images form Figure 8a-8b, the error being equal to the velocity of frame acquisition, namely 1/30 frames/s. The position of the axial track at the end of the motion is also attested in Figure 8, being identical to the track from the launching moment.

The final positions of the body are presented for a number of nine identical launchings of the body, with the longitudinal track placed initially in the vertical plane of symmetry. It can be noticed that, practically, the final position of this track is the same and this confirms the correctness of the pure rolling hypothesis (there was no relative displacement between the plate and the body in the contact points). Concerning the sliding, a variation of it can be observed.

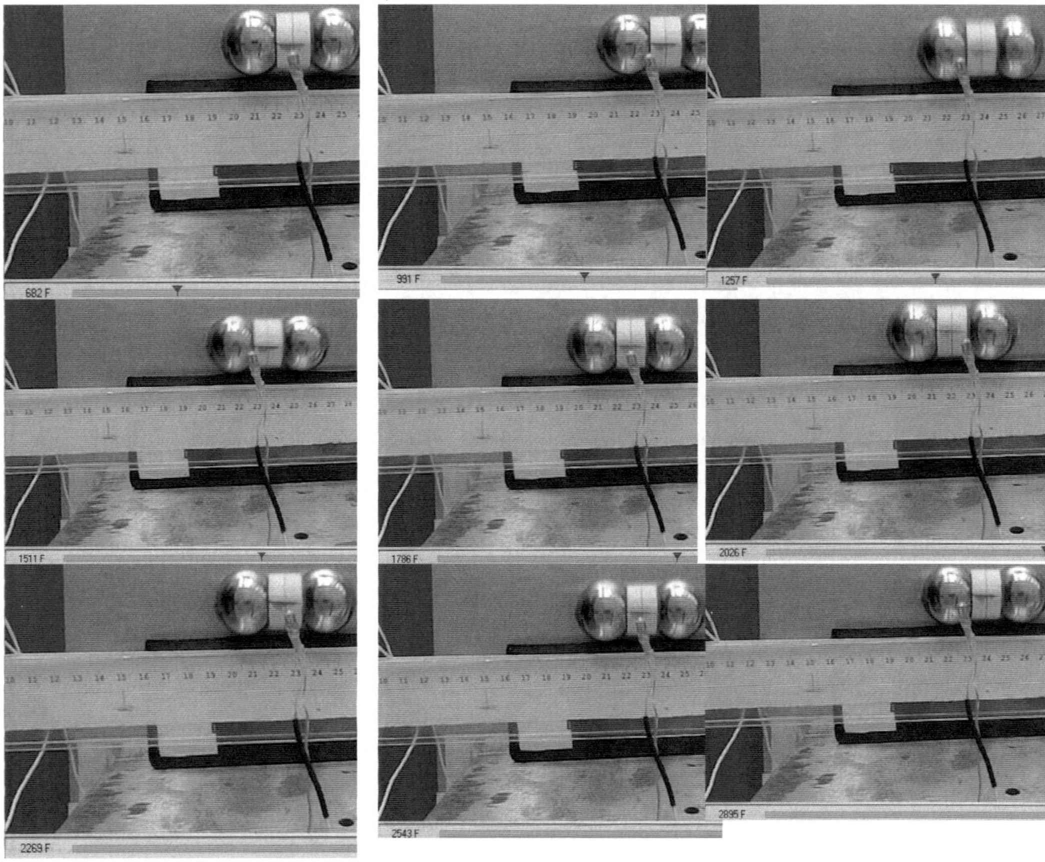

Figure 9. The final positions of the body for a series of successive launchings

For the cases from Figure 9, the variation of the running times is presented in Figure 10 and the sliding distance is presented in Figure 11.

As it can be remarked, the running times present variation within quite a wide time range. According to the above considerations, both torques M_r and T_x must be constant. Moreover, for the case of rolling motion, because the angle of rotation is practically constant for all launchings, the constant angular acceleration assumes the necessity of constant running times. This aspect directs to the conclusion that the parameters characteristic to rolling friction present a high gradient (appreciable variation from one point to another).

$$START := \begin{bmatrix} 555 \\ 871 \\ 1153 \\ 1415 \\ 1665 \\ 1911 \\ 2169 \\ 2420 \\ 2769 \end{bmatrix} \quad END := \begin{bmatrix} 682 \\ 991 \\ 1257 \\ 1511 \\ 1786 \\ 2026 \\ 2269 \\ 2543 \\ 2895 \end{bmatrix}$$

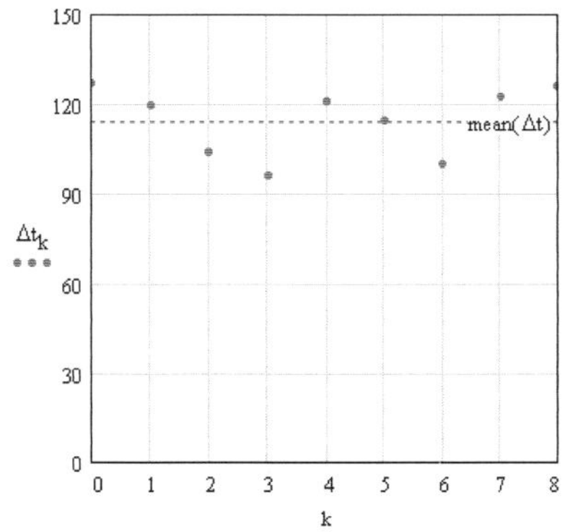

$$\frac{\max(\Delta t) - \min(\Delta t)}{\mathrm{mean}(\Delta t)} \cdot 100 = 27.035$$

Figure 10. Finding the indexes of the frames corresponding the start and the end for a series of successive launchings

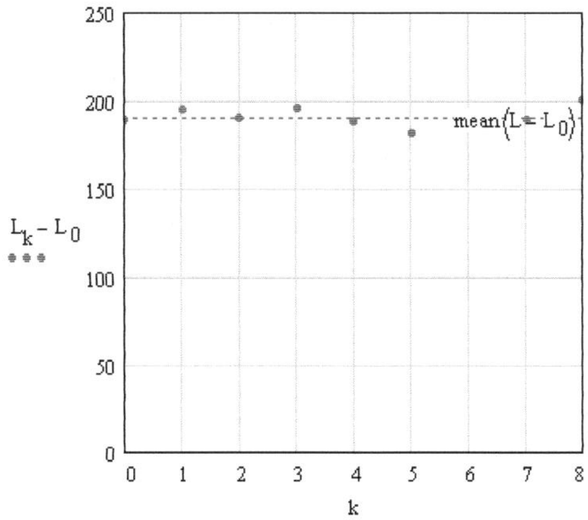

$$\frac{\max(L) - \min(L)}{\mathrm{mean}\left(L - L_0\right)} \cdot 100 = 9.965$$

Figure 11. Values of lateral sliding for the studied set of launchings

More precisely, the motion of the body is performed in a manner that ensures that the trajectories of the points of contact ball-plane are very close for two launchings, but the moments of rolling friction vary noticeably from a trajectory to another. Considering that the rolling accelerations and sliding accelerations are inversely proportional to the time of motion squared, the condition (22) requires that for φ_2 of constant value, then x_2 should be constant on its turn. But it can be easily observed that this condition is not fulfilled. From Figure 11 it is noticed that the sliding distances vary in a narrow range and therefore the relations (22) and (23) ought to be reconsidered and regarded from statistical point of view. In this situation, considering that for one launching, sliding and rolling motions take place during the same time and that the rotation is practically constant, the relation (22) can be re-written as:

$$L_{med} = \frac{(\sin\beta - \mu\cos\beta)\cos\alpha}{M\dfrac{R\sin\alpha - s\cos\alpha\cos\beta}{J_x + MR^2}}\varphi_{2max} \tag{24}$$

where L is the length of sliding distance. Now, from the above relation, the coefficient of rolling friction can be found.

4. Conclusions

The motion of an axisymmetric body obtained by two jointed bearing balls on an inclined plane is analyzed. The tilt of the plane is set in a manner that ensures pure rolling of the body with respect to the straight line passing through the contact points and sliding with respect to this line.

The equations of motion of the body are obtained and the conclusion that both rotation and translation should take place with constant accelerations is reached. This fact allows for writing a relation between the rotation acceleration and the sliding acceleration. If this relation is correct, finding the coefficient of rolling friction would be much simpler.

The experimental tests performed on a test rig designed and assembled in the laboratory revealed that for all launchings, the rotation angle was the same but the sliding distance presents noticeable variations. This fact contradicts the hypothesis of proportionality between the sliding and rolling accelerations.

The paper requires further researches in which the proportionality between the two accelerations and the parameters involved in this relation, including the coefficient of sliding friction and the coefficient of rolling friction, should be statistically considered.

Acknowledgement. This work was supported by a grant of the Romanian Ministry of Research and Innovation, CCCDI – UEFISCDI, project number PN-III-P1-1.2-PCCDI-2017-0404 / 31PCCDI / 2018, within PNCDI III.

References

[1] Johnson K L 1985 *Contact Mechanics*, Cambridge, Cambridge University Press
[2] Uicker J J Jr, Pennock G R and Shigley J E 2016 *Theory of Machines and Mechanisms*, 5th Edition, Oxford University Press
[3] Mott R L 2017 *Machine Elements in Mechanical Design*, 6th Ed, Pearson
[4] Kragelskii I V, Dobychin M N and Kombalov V S 1977 *Osnovy raschetov na trenie i iznos* (*Fundamentals of Friction and Wear Calculations*) Moscow, Mashinostroenie
[5] Goryacheva I G 2001 *Mekhanika friktsionnogo vzaimodeiistviya (Mechanics of Friction Interaction)* Moscow, Nauka
[6] Bowden F P and Tabor D 1950 *Friction and Lubrication of Solids*, Oxford, Clarendon
[7] Townsend D P, Dietrich M W and Zaretsky E W 1969 *Speed effects on ball spinning torque,* Lewis Research Center Cleveland, Ohio NASA, Washington DC
[8] Alaci S, Ciornei F C, Romanu I C and Ciornei M C 2018 Evaluation of spinning friction from a thrust ball bearing, *ModTech2018 IOP Conf. Series: Mater. Sci. Eng.* **400** 022002
[9] Dzhilavdari I Z 2003 Studies of Elasticity and Elastic Hysteresis of the Surface of Materials on the Basis of the Dynamics of the Oscillations of the Physical Pendulum with a Rolling Contact Bearing, *Trenie Iznos* **24**(1) 42-48
[10] Alaci S, Ciornei F C, Bujoreanu C, Ciornei M C and Acsinte I L 2018 *Finding the coefficient of rolling friction using a pericycloidal pendulum*, ACME International Conference TU Iasi, 6-8 June, Romania, to be published
[11] Siretean S T, Alaci S, Musca I and Ciornei F C 2018 *Determining the coefficient of rolling friction using hypocycloidal oscillations*, ACME International Conference TU Iasi, 6-8 June, Romania, to be published
[12] Ardema M 2005 *Newton-Euler Dynamics*, Springer, New York
[13] McCArthy M C 1990 *Introduction in Theoretical Kinematics*, MIT Press
[14] Fisher I 1998 *Dual-Number Methods in Kinematics, Statics and Dynamics* CRC Press
[15] Cherepanov G P 2015 The contact problem of the mathematical theory of elasticity with stck and slip ares. The theory of rolling and tribology, *Journal Applied Mathem and Mechanics* **79** 81-101
[16] Cherepanov G P 2014 Theory of rolling: solution of the Coulomb problem *Journal of Applied Mechanics and Technical Physics* **55**(1) 182-189

IOP Conf. Series: Materials Science and Engineering **477** (2019) 012037 doi:10.1088/1757-899X/477/1/012037

Kinematical analysis of a generalized Cardanic joint

F C Ciornei, S Alaci, I C Romanu, I Mihai and G A Tibu

[1]"Stefan cel Mare" University, Mechanics and Technologies Department, Suceava, Romania, Universitatii str., no. 13, 720229 Suceava, Romania

E-mail: florina@fim.usv.ro

Abstract. The necessity of transmitting the rotation motion between two shafts is commonly met in mechanical engineering. The most frequent solutions are represented mechanical systems known as couplings. The coupling solutions presented in literature depend on two main features: the relative position between the shafts and the variation of the transmission ratio. One of the most popular couplings is the Cardanic joint meant to transmit motion between intersecting axes. From technological point of view, to obtain the concurrence of all axes of the joints of the coupling it is a very difficult task. To overcome this aspect, it is assumed that the axes of the shafts are not intersecting and from the four joints of the Cardanic coupling, only the input joint remains a rotation one; the other three joints transform into cylindrical joints and thus allow for relative linear displacement of the elements besides rotation. In this manner, the new mechanism is a RCCC one. The paper presents the positional analysis of the mechanism applying the method of homogenous operators proposed by Hartenberg and Denavit. The analytical expressions of all displacements from the joints of the mechanism are obtained. The aspects referring to the obscurity occurring when inverse trigonometric functions are involved in describing the displacements and the way to avoid these are discussed.

1. Introduction

The main task of a mechanism is to transmit motion and power from the driving element to the final one under well stipulated conditions [1]. The problem of transmission of rotation motion between two shafts is often met in technology [2]. When a rigorously constant transmission ration is required, the gear mechanisms are used [3], [4]. When the constant transmission ratio is not a stringent necessity, the use of linkage mechanisms [5-7] which are more reliable and economical than gears is a handy solution.

The kinematical study of a linkage mechanism supposes establishing the position of the mechanism for a given position of the driving elements [8]. This means that for stipulated position parameters of the driving elements, all displacements from the rest of kinematical pairs of the mechanisms are required. Every method from technical literature is based on the condition of closed kinematical chain from the structure of the mechanism [9], requirement imposed by the definition of a mechanism itself. The condition of closed kinematic chain is founded on the manner of coordinate transformation of a point using a set of coordinate systems attached to the elements of the chain, chosen in such way that the first frame coincides to the last one. The condition of closed kinematic chain is expressed in a manner that depends on the structure of the mechanism and the solutions depend on the mathematical apparatus involved.

2. Methods applied in spatial kinematical analysis

For plane mechanisms [10], the condition of closed kinematic chain can be expressed in vector format or using complex numbers. For spherical mechanisms [11], the closed loop condition is formulated via 3x3 type rotation matrices. The closing equation has the form:

$$\boldsymbol{R}_{12}\boldsymbol{R}_{23}.....\boldsymbol{R}_{n,1} = \boldsymbol{I}_3 \tag{1}$$

where $\boldsymbol{R}_{k,k+1}$ is the rotation matrix that brings the frame (k) over the frame $(k+1)$ and $\boldsymbol{I}_3$ is the unit matrix of third order. The general from of a rotation matrix which describes the rotation of φ angle around an axis of $\boldsymbol{e}$ versor is, according to Angeles [12]:

$$\boldsymbol{R} = \boldsymbol{e}\boldsymbol{e}^T + (\boldsymbol{I}_3 - \boldsymbol{e}\boldsymbol{e}^T)\cos\varphi + \boldsymbol{E}\sin\varphi \tag{2}$$

where $\boldsymbol{E}$ is the anti-symmetric matrix attached to the $\boldsymbol{e}$ versor. The condition to be fulfilled by two rotation matrices in order to be identical is to have the angles and the versors of the axes, respectively, identical. From here it results the conclusion that from the nine scalar equations provided by equation (1) only three are independent. Another extremely useful instrument in expressing the spatial rotations is the quaternions algebra, [13]. For the case of spatial mechanisms the closing condition has a form identical to the relation (1) [14].

$$\boldsymbol{T}_{12}\boldsymbol{T}_{23}....\boldsymbol{T}_{n,1} = \boldsymbol{I}_4 \tag{3}$$

where $\boldsymbol{T}_{k,k+1}$ is the displacement matrix that superposes the (k) frame over the $(k+1)$ frame and having the general form:

$$\boldsymbol{T}_{k,k+1} = \begin{bmatrix} \boldsymbol{R}_{k,k+1} & \boldsymbol{d}_{k,k+1} \\ \boldsymbol{0} & 1 \end{bmatrix} \tag{4}$$

where $\boldsymbol{R}_{k,k+1}$ has the same significance as in equation (1), $\boldsymbol{d}_{k,k+1}$ is the vector that moves the origin O_k over O_{k+1} origin and $\boldsymbol{I}_4$ is the unit matrix of fourth order. It is easily noticed that the relation (3) supposes the identity between the rotation matrices and displacement vectors of the matrices from both members of equation. Thus, a system of six scalar equations actually results. The kinematic study of spatial chains was substantially simplified by Hartenberg and Denavit [15] who proved that for a kinematic chain with lower cylindrical pairs, the number of necessary parameters needed for stipulation of the relative position of two elements may be reduced from six to four by the suitable selection of the position and orientation of the axes of the frames attached to the chains' elements. To complete this operation, it is required that the axes of the cylindrical pairs should be chosen as Oz axes of the frames and each of the Ox axis should be along the common normal of two neighboring axes, as in Figure 1, [15]. With the axes chosen in this manner, the (k) coordinate system can be superposed over the $(k+1)$ frame by a roto-translation of parameters θ_k and s_k followed by a roto-translation of parameters $\alpha_{k,k+1}$ and $a_{k,k+1}$.

$$\boldsymbol{T}_{k,k+1} = \begin{bmatrix} \boldsymbol{R}^z_{k,k+1} & \boldsymbol{d}^z_{k,k+1} \\ \boldsymbol{0} & 1 \end{bmatrix} \begin{bmatrix} \boldsymbol{R}^x_{k,k+1} & \boldsymbol{d}^x_{k,k+1} \\ \boldsymbol{0} & 1 \end{bmatrix} \tag{5}$$

The two displacement vectors are expressed under the form:

$$\begin{aligned} \boldsymbol{d}^z_{k,k+1} &= [0 \;\; 0 \;\; s_k]^T \\ \boldsymbol{d}^x_{k,k+1} &= [a_{k,k+1} \;\; 0 \;\; 0]^T \end{aligned} \tag{6}$$

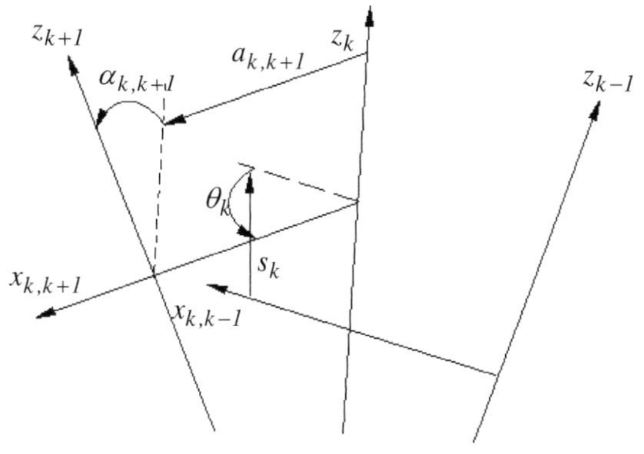

Figure 1. The Hartenberg Denavit convention

In order to obtain the rotation matrices from relation (1), this relation is applied for $\boldsymbol{e}=\boldsymbol{k}$, $\varphi=\theta_k$ and $\boldsymbol{e}=\boldsymbol{i}$, $\varphi=\alpha_{k,k+1}$ respectively, and it results:

$$\boldsymbol{R}^{z}_{k,k+1}=\begin{bmatrix} cos\,\theta & -\,sin\,\theta & 0 \\ sin\,\theta & cos\,\theta & 0 \\ 0 & 0 & 1 \end{bmatrix} \tag{7}$$

and:

$$\boldsymbol{R}^{x}_{k,k+1}=\begin{bmatrix} 1 & 0 & 0 \\ 0 & cos\,\alpha_{k,k+1} & -\,sin\,\alpha_{k,k+1} \\ 0 & sin\,\alpha_{k,k+1} & cos\,\alpha_{k,k+1} \end{bmatrix}. \tag{8}$$

The following notations are introduced for a simpler writing manner:

$$sin\,\alpha = S\alpha \qquad cos\,\alpha = C\alpha \tag{9}$$

The $\boldsymbol{T}_{k,k+1}$ operator has the matrix:

$$\boldsymbol{T}_{k,k+1}=\begin{bmatrix} C\theta_k & -S\theta_k & 0 & 0 \\ S\theta_k & C\theta_k & 0 & 0 \\ 0 & 0 & 1 & s_k \\ 1 & 0 & 0 & 1 \end{bmatrix}\begin{bmatrix} 1 & 0 & 0 & a_{k.k+1} \\ 0 & C\alpha_{k,k+1} & -S\alpha_{k,k+1} & 0 \\ 0 & S\alpha_{k,k+1} & C\alpha_{k,k+1} & 0 \\ 0 & 0 & 0 & 1 \end{bmatrix} \tag{10}$$

Another especially useful tool in kinematical analysis of spatial mechanisms is the dual numbers concept introduced by Ball and Clifford [16]. Similarly to complex numbers use, they propose the employment of numbers having the form:

$$\hat{x}=x+\varepsilon x_0 \tag{11}$$

acknowledged as dual numbers where ε is the imaginary unit defined by the Phillips equality [17]:

$$\varepsilon^2=0 \tag{12}$$

Based on the Kotelnikov transfer principle [16] that states: *all relations deduced within spherical trigonometry remain valid if the real numbers are replaced by dual numbers,* any spatial mechanism may be studied using a spherical mechanism. Founded on this remark, Yang [18] analyses the spatial mechanisms with five elements employing matrices of 3x3 type with dual numbers. It is also Yang who studies spatial mechanisms applying dual quaternions.

3. The kinematical analysis of the proposed mechanism

Next, the positional analysis of generalized Cardanic mechanism is presented. The mechanism of the Cardanic transmission presented in Figure 2 is a spherical one.

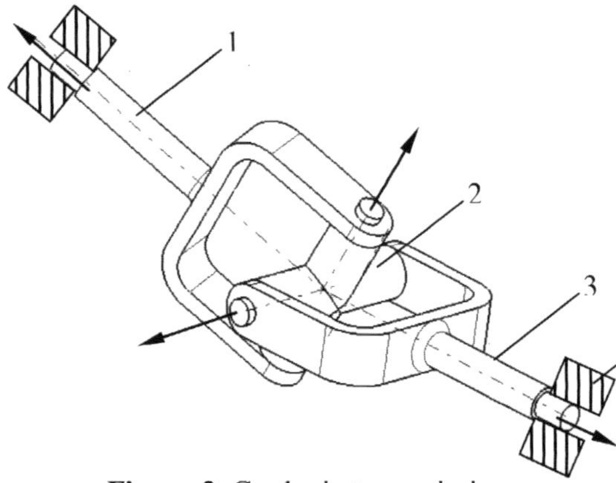

Figure 2. Cardanic transmission

The structure of the mechanism contains four rotation pairs whose axes should be theoretically concurrent in a point. This requirement is extremely restrictive and therefore, from technological point of view, special tasks must be fulfilled. To avoid these complications, the kinematical pairs of the mechanism (excluding the driving pair) are transformed into cylindrical pairs. Thus, the resulting mechanism is a *RCCC* spatial one, with the degree of mobility 1, and which doesn't require the concurrence of the axes of two pairs. For the mechanism obtained in the above manner, applying a series of particular values for the constructive parameters, the positional analysis made by way of the Hartenberg-Denavit method is presented next. The general case of the analysis of the *RCCC* mechanism was completed by Yang who applies as manner of study the dual quaternions and by Fischer who studies the kinematics of the mechanisms using matrices with dual elements. The generalized Cardanic mechanism presented in Figure 2 consists in four elements: the driving element 1, the intermediary coupling element 2 that has perpendicular (without being a strict requirement) but not intersecting arms, the final element 3 and the ground, 4. The elements 1 and 3 are identical as shape but they distinguish by the joint they have with the ground. Thus, while the element 1 and the ground 4 complete a rotation pair, R, the driven element 3 and the ground 1 make a cylindrical pair, C. The intermediate element 2 creates cylindrical pairs with the driven element and with the driving element respectively, and for this reasons the mechanism's name - *RCCC*, is, from structural point of view, justified. For the kinematical analysis of the mechanisms, the Hartenberg-Denavit methodology is applied. To this purpose, the Oz axes of the pairs are chosen with the orientation presented in Figure 2. Once the Oz axes are stipulated, the $Ox_{k,k+1}$ axes are chosen as to superpose the current axis Oz_k aver the next Oz_{k+1} axis by a screw motion with the $\alpha_{k,k+1}$ rotation in trigonometric sense, as shown in Figure 2. The motions of the mechanism will be described by the displacements from the three cylindrical pairs (three rotations θ_2, θ_3 and θ_4 together to the translations s_2 s_3 and s_4) expressed as functions of the position of the driving element θ_1. The constructive characteristics of the mechanism will be represented by:

$$\alpha_{12} = \alpha_{23} = \alpha_{34} = \pi / 2 \tag{13}$$

and

$$a_{12} = a_{34} = 0 . \tag{14}$$

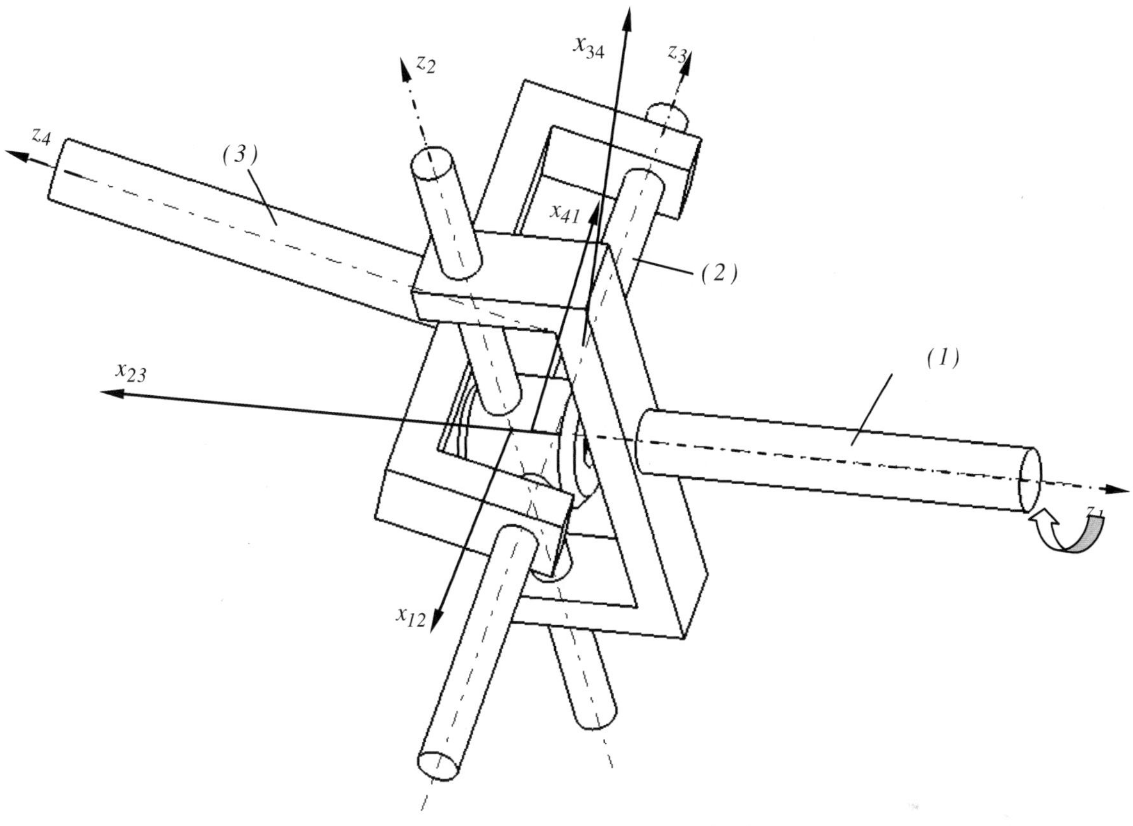

Figure 3. Generalized Cardanic mechanism

$$Z(\theta_1, s_1)X(\alpha_{12}, a_{12})Z(\theta_2, s_2)X(\alpha_{23}, a_{23})Z(\theta_3, s_3) \cdot X(\alpha_{34}, a_{34})Z(\theta_4, s_4)X(\alpha_{41}, a_{41}) = I_4 \tag{15}$$

where $Z(\theta, s)$ and $X(\alpha, a)$ are the displacements with respect to z axis and x axis respectively, applying the equations (7) and (8) and I_4 is the unit matrix of fourth order. It is obvious that by replacing the equations (1) and (2), the concrete form of the matrix equation(3) will be considerable simplified. Though, to solve this equation is not an easy task. McCarthy recommends that the equation (15) should be written under the form:

$$Z(\theta_1, s_1)X(\alpha_{12}, a_{12} \ldots\ldots = (\ldots\ldots Z(\theta_4, s_4)X(\alpha_{41}, a_{41}))^{-1} \tag{16}$$

without stipulating the point where the two terms from the initial equation must be separated. In the present case, the McCarthy method was applied by writing the equation (16) in the following manner:

$$Z(\theta_1, s_1)X(\alpha_{12}, a_{12})Z(\theta_2, s_2)X(\alpha_{23}, a_{23}) = [Z(\theta_3, s_3)X(\alpha_{34}, a_{34})Z(\theta_4, s_4)X(\alpha_{41}, a_{41})]^{-1} \tag{17}$$

where O_4 is the zero matrix of $4x4$ type. The matrix equation (15) assumes the annulations of the rotation matrix and of the displacement vector from the matrix on the left member of equation (15). The displacement vector doesn't create problems as it means the simultaneous cancellation of the first three elements from the last column of the left member matrix, but concerning the rotation matrix, it is characterized by 9 elements (the 3x3 type matrix from upper left).

$$\begin{bmatrix} C\theta_1 C\theta_2 - C\theta_3 C\theta_4 & S\theta_1 C\theta_2 + C\theta_3 S\theta_4 C\alpha_{41} - S\theta_3 S\alpha_{41} & S\theta_2 - C\theta_3 S\theta_4 S\alpha_{41} - S\theta_3 C_{\alpha_{41}} & -a_{23} - s_1 S\theta_2 - a_{41} C\theta_3 C\theta_4 - s_4 S\theta_3 \\ S\theta_1 - S\theta_3 C\theta_4 & -C\theta_1 + S\theta_3 S\theta_4 C\alpha_{41} + C\theta_3 S\alpha_{41} & -S\theta_3 S\theta_4 S\alpha_{41} + C\theta_3 C\alpha_{41} & -s_2 - a_{41} S\theta_3 C\theta_4 + s_4 C\theta_3 \\ C\theta_1 S\theta_2 - S\theta_4 & S\theta_1 S\theta_2 - C\theta_4 C\alpha_{41} & -C\theta_2 + C\theta_4 S\alpha_{41} & s_1 C\theta_2 - a_{41} S\theta 4 - s_3 \\ 0 & 0 & 0 & 0 \end{bmatrix} = \boldsymbol{O_4} \tag{18}$$

From the nine equations obtained, only three are independent. A rigorous solving assumes separation of the symmetrical parts from the two matrices:

$$\boldsymbol{R}^s = \frac{\boldsymbol{R} + \boldsymbol{R}^T}{2} \tag{19}$$

which permits finding the angle of rotation and the disjointing of the asymmetrical parts:

$$\boldsymbol{R}^a = \frac{\boldsymbol{R} - \boldsymbol{R}^T}{2} \tag{20}$$

which permits finding the versor of the axis of rotation.

For the current case, a different technique will be used consisting in the convenient selections of the three equations from which the angles θ_2, θ_3 and θ_4 should be found. The elements from the positions (3,1) and (3,2) are equaled to zero and the following system results:

$$\begin{cases} C\theta_1 S\theta_2 - S\theta_4 = 0 \\ S\theta_1 S\theta_2 - C\theta_4 C\alpha_{41} = 0 \end{cases} \tag{21}$$

that allows for the θ_4 to be found.

$$\theta_4 = atan[C\alpha_{41} C\theta_1 / S\theta_1] \tag{22}$$

The elements from the positions (3,2) and (3,3) are equaled to zero:

$$\begin{gathered} S\theta_1 S\theta_2 - C\theta_4 C\alpha_{41} = 0 \\ -C\theta_2 + C\theta_4 S\alpha_{41} = 0 \end{gathered} \tag{23}$$

and it is found:

$$\theta_2 = atan[tan\theta_4 / (S\alpha_{41} C\theta_1)] \tag{24}$$

The elements from the positions (1,1) and (2,1) are zero:

$$\begin{gathered} C\theta_1 C\theta_2 - C\theta_3 C\theta_4 = 0 \\ S\theta_1 - S\theta_3 C\theta_4 = 0 \end{gathered} \tag{25}$$

and it results:

$$\theta_3 = atan^*(tan\theta_1 / C\theta 2) \tag{26}$$

The displacements from the cylindrical pairs are obtained straightforward:

$$s_3 = s_1 C\theta_2 - a_{41} S\theta_4$$
$$s_4 = \frac{a_{41} C\theta_3 C\theta_4 - (a_{23} + s_1 S\theta_2)}{S\theta_3} \tag{27}$$
$$s_2 = s_4 C\theta_3 - a_{41} S\theta_3 C\theta_4$$

Next it will be shown that the employment of the function $atan(x)$ in the expressions of the rotations (22), (24) and (26) may generate errors in the description of the kinematics of the mechanism. Instead of function $atan(y/x)$, the use of the function $angle(x, y)$, existing in any calculus utilitarian, is proposed; this function returns the value of the angle from the domain $[0, 2\pi)$ made by the vector radius of the point of coordinates (x, y) with the positive half-axis Ox. The expressions of the rotation angles from the cylindrical pairs, as dependencies only of the angle of the driving element θ_1, are presented next:

$$\theta_2 = angle(S\alpha_{41} S\theta_1, C\alpha_{41})$$
$$\theta_3 = angle(S\alpha_{41} C\theta_1, \sqrt{S^2\theta_1 + C^2\alpha_{41} C^2\theta_1} \tag{28}$$
$$\theta_4 = angle(S\theta_1, C\alpha_{41} C\theta_1)$$

4. Discussions

The variation of the linear displacement from the pair between the driven element and the ground is presented in Figure 4, with continuous line, found by means of relations (28), with broken line, found by using the relations (22), (24) and (26). One can observe that, the two plots coincide only for the first quarter of rotation of the driving element and differ fro the rest. At a first sight it should seem that it is only a vertical translation, on portions, of the graph with continuous line.

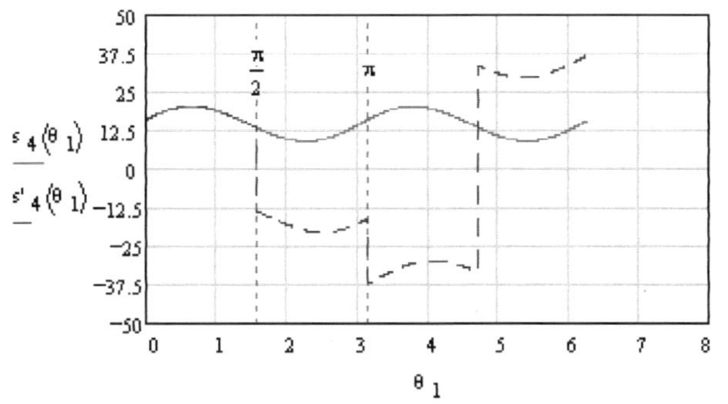

Figure 4. The translation of the driven element calculated in two ways

If the graphs from Figure 4 are parts of the same continuous function, translated vertically on intervals, their derivatives must coincide. In Figure 5 there are presented the derivatives of the two functions from Figure 4. It can be observed that the plots are identical only for the portion on which the graphs from Figure 4 coincide. This fact confirms the statement that the use of the function $atan(x)$ generates errors in the kinematical analysis of the mechanism. To clarify this aspect, one must remark that the expressions of linear displacements from the cylindrical pairs depend on the rotation angles θ_2, θ_3 and θ_4 as arguments of the trigonometric functions $cos(x)$ and $sin(x)$.

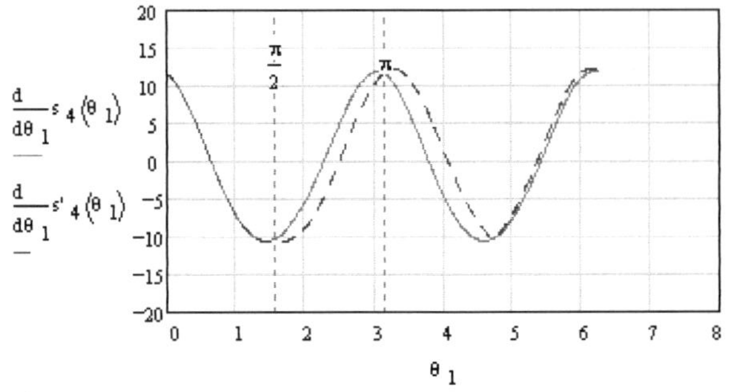

Figure 5. The derivatives of the functions from Figure 4

The following simple functions are considered next in order to explain the effect of applying the *atan(x)* function to the above mentioned ones:

$$f(\varphi) = sin(angle(cos(\varphi), sin(\varphi)) \qquad f'(\varphi) = sin(atan(sin(\varphi)/cos(\varphi))) \tag{30}$$

Theoretically, these expressions should represent the same function or at most one of them could be vertically translated on portions but in the end their derivatives should be equal.

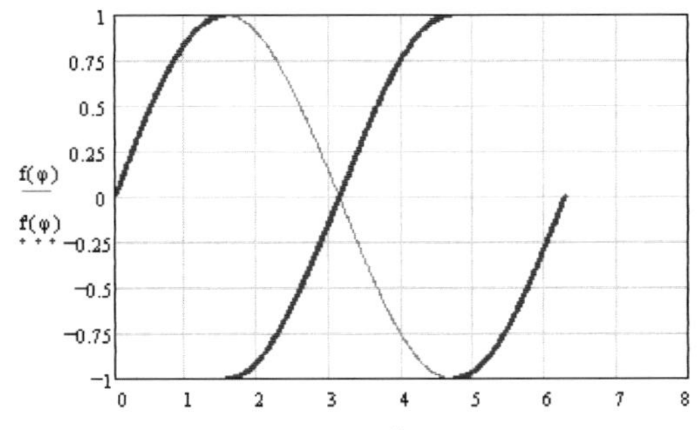

Figure 6. The effect of the functions *atan(sin(φ))* and *angle(sin(φ),cos(φ))*

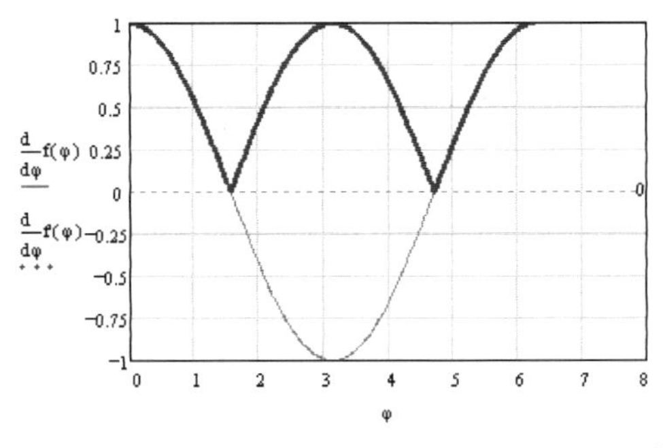

Figure 7. The effect of the functions *atan(sin(φ))* and *angle(sin(φ),cos(φ))* upon the derivative of the same function *cos(x)*

A similar situation arises for the *cos(x)* function. It can be concluded that, applying the trigonometric functions sine and cosine to the functions *atan(y / x)* and *angle(x, y)* different results are obtained. Finally, using the relations (28), the rotation velocities and displacement velocities from cylindrical pairs were found and there are shown in Figure 8 and Figure 9 respectively. Both the rotation velocity and the translation velocity from the pair between the final element and the ground are the result of the superposition of a periodic component over a constant component, the frequency of the periodic component being twice the frequency of the rotation of the driving element. The law of angular velocity variation is identical to the one of a Cardanic mechanism [2].

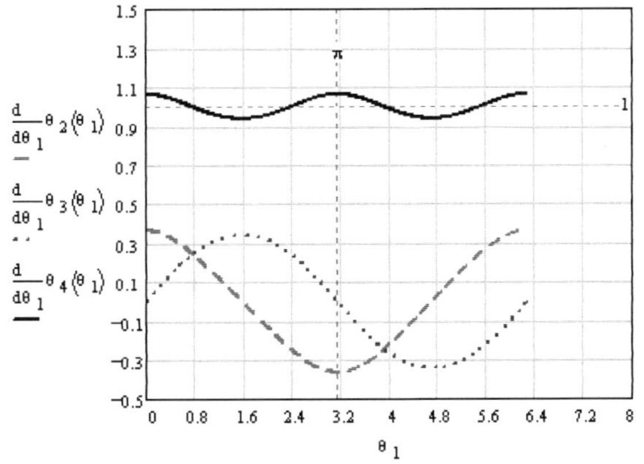

Figure 8. The rotation velocities from cylindrical pairs

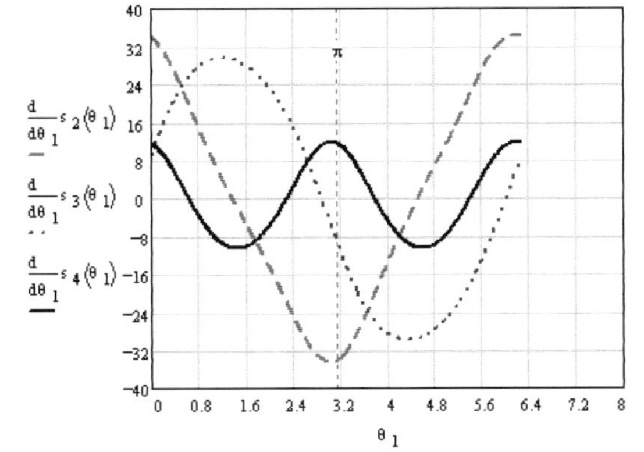

Figure 9. The linear velocities from cylindrical pairs

Concerning the motions from the intermediate pairs, from Figure 8 it possibly will say that the rotations from these pairs are the same only delayed with a quarter of period and of opposite signs. However the plots of the translation velocities disagree with this hypothesis. For a true such hypothesis, the plot $s_3 = s_3(s_2)$ should be a symmetrical shape with respect to coordinate axes. The graph from Figure 10 confirms that the motions from the two pairs present different characteristics.

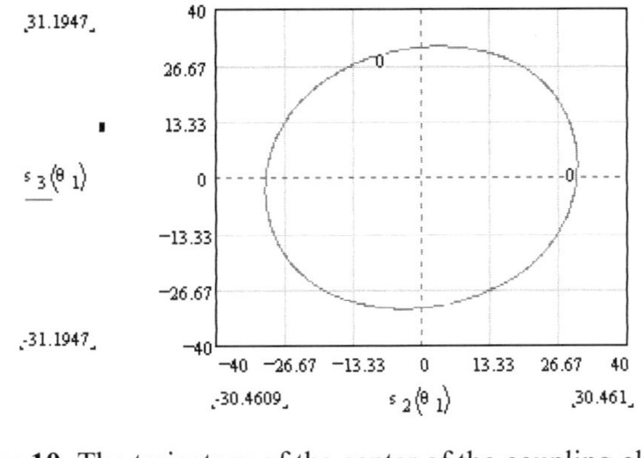

Figure 10. The trajectory of the center of the coupling element

5. Conclusions

The paper presents the kinematical analysis of the spatial mechanism obtained from a Cardanic joint, which is a spherical mechanism, by transforming the rotation pairs into cylindrical pairs. The new mechanism presents the advantage that transmits the rotation motion identically to the spherical Cardanic mechanism. But, compared to it, the condition of intersection of the axes of all pairs, which is difficult to ensure in applications, must not be strictly obeyed.

After a short review of the methods applied in spatial analysis, the method of homogenous operators proposed by Hartenberg and Denavit is chosen as technique for the analysis of the mechanism. The analytical expressions for all the displacements from the pairs of the mechanism as functions of the rotation angle of the driving element are resulting from the analysis.

In the final part of the work, the errors produced by the use of multiform function $atan(x)$ are highlighted and necessity of replacing this function with the uniform function $angle(x, y)$ is discussed.

The variations of the rotation and translation velocities from all the pairs of the mechanism are presented and the remark that in the final pair the motion is helicoidally periodic with the period half of the period of the driving element is made.

Acknowledgement. This work was supported by a grant of the Romanian Ministry of Research and Innovation, CCCDI – UEFISCDI, project number PN-III-P1-1.2-PCCDI-2017-0404 / 31PCCDI / 2018, within PNCDI III.

References

[1] Erdman A G, Sandor G N and Kota S 2001 *Mechanism Design: Analysis and Synthesis,* 4th ed, Pearson
[2] Pennock G R, ShigleyJ E and Uicker J J 2014 *Theory Of Machine And Mechanisms,* 4th ed, Oxford University Press
[3] David B. Dooner 2012 *Kinematic geometry of gearing,* Willey
[4] Litvin F L and Alfonso Fuentes A 2004 *Gear Geometry and Applied Theory,* 2nd ed, Cambridge University Press
[5] Seherr-Thoss H C, Schmelz F and Aucktor E 2010 *Universal Joints and Driveshafts: Analysis, Design, Applications,* Springer
[6] Dudita F 1974 *Cuplaje Mobile Homocientice,* Ed.Tehnica Bucuresti
[7] Dudita, F 1966 *Cuplaje Cardanice,* Ed.Tehnica Bucuresti
[8] Lung-Wen Tsai 2000 *Mechanism Design: Enumeration of Kinematic Structures According to Function,* 1st ed, CRC Press
[9] Norton R L 2008 *Design of Machinery: An Introduction to the Synthesis and Analysis of Mechanisms and Machines,* 4th ed, McGraw-Hill Higher Education
[10] Huckert J 1958 *Analytical Kinematics of Plane Motion Mechanisms,* The Macmillan Company
[11] Chiang C H 2000 *Kinematics of Spherical Mechanisms,* 2nd ed, Krieger Pub Co
[12] Angeles J 1989 *Rational Kinematics,* Springer Tracts in Natural Philosophy, Springer
[13] Brandt L 1947 *Vector and tensor analysis,* 1st ed, J. Wiley
[14] McCarthy J M 1990 *Introduction to Theoretical Kinematics*, 1st ed, The MIT Press
[15] Hartenberg R S and Denavit J 1964 *Kinematic Synthesis of Linkages,* 1st ed, McGraw-Hill Inc
[16] Dimentberg F M 1965 *The Screw Calculus and Its Applications in Mechanics*, Foreign Technology Division Translation, FTDHT
[17] Phillips J 2007 *Freedom in machinery*, Cambridge University Press
[18] Yang A T and Freudenstein F 1964 Application of dual-number quaternion algebra to the analysis of spatial mechanisms *ASME Journal of Applied Mechanics* **86** pp 300–308

Analysis of the performance of occupational health and safety in the production business system

B Vranješ[1] **and B Anđelković**[2]

[1]University of Banja Luka, Faculty of Mechanical Engineering, Stepe Stepanovića 71, Banja Luka, 78000, Bosnia and Herzegovina-Rebublic of Srpska
[2]Politehnica University of Niš, Faculty of Occupational Safety, Čarnojevića 10A, Niš, 18000, Serbia

E-mail: biljana.vranjes@mf.unibl.org

Abstract. The basic task of the occupational health and safety system in the business system is to identify all the hazards and harmfulness that occur in work processes, to assess the risks of identified hazards and harmfulness, and to design protection measures that reduce risks to an acceptable level.Monitoring and evaluating the occupational health and safety system in the business system is achieved by defining key performance indicators in the form of performance. By collecting data on the state of occupational health and safety in the business system and their analysis, we obtain information about the deficiencies of the given system. Determining the correlations between the occupational health and safety system in the business system and its performances leads to defining of the guidelines for improvement. On the basis of these data, the optimization of the given system can be performed in order to raise the level of its effectiveness and efficiency.

1. Introduction

All business processes should have defined key performance which are monitored and measured, as well as occupational health and safety processes. This approach enables the proactive measurement of the leading safety performance indicators with the goal of continuous improvement, which is one of the basic requirements of the OHSAS 18000 standard. The basic goals of the implementation of occupational health and safety measures are to reduce the risk (probability of occurrence) and to reduce the consequences of unwanted events.

Most often, only unwanted events (injuries, occupational diseases, fatalities, illnesses) are measured with all their characteristics, such as lost working hours and other costs, as performance of occupational health and safety. However, this performance covers a much wider area, for example, planning, implementing key training and training activities, participation of management in the field of occupational safety, safety culture, culture of dialogue between participants in the system of occupational health and safety (employer, trade union, workers), human resources management and the like. Occupational health and safety is an integral part of the performance of each work process and the organizational structure of the business system as a whole. Modern business and business process management tends to standardize all procedures, as well as procedures related to occupational health and safety. This approach enables the creation of additional value, competitiveness and performance of the business system. Increasing the efficiency and effectiveness of managing the occupational health and safety system means reducing the number of unwanted events and all costs associated with unwanted events, such as accidents and injuries at work, fatalities, environmental accidents, discharge of dangerous substances, production of hazardous and non-hazardous waste, energy consumption, production hazards and all other negative effects [1]. An effective system of occupational health and safety is one of the indicators of successful operation of the organization as a whole, and we can conclude that raising the level of efficiency of this system helps in achieving business goals (Figure 1).

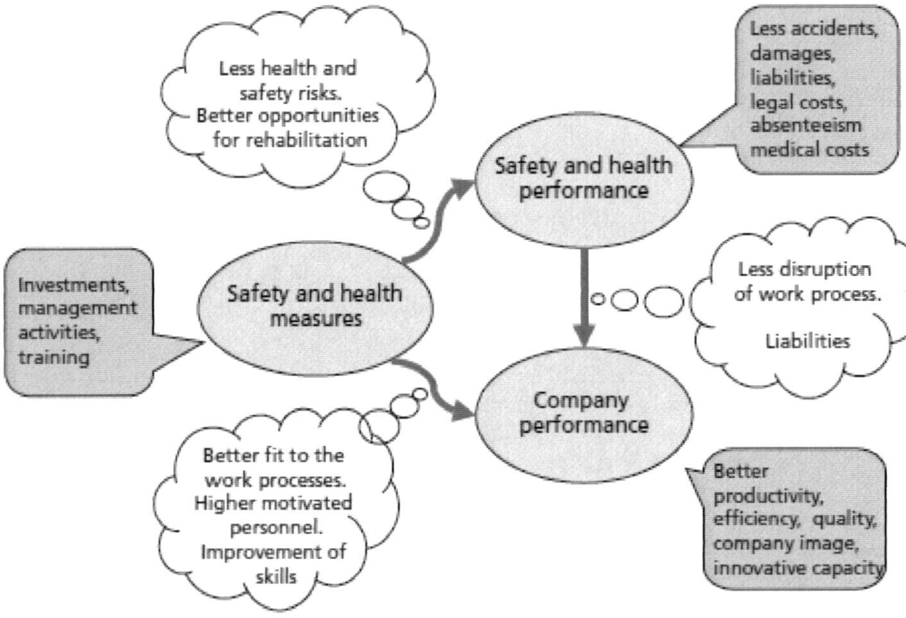

Figure 1. The advantages of applying occupational safety measures in the business system [2]

Key indicators of occupational health and safety are numerical indicators that should provide information to the management of the business system on what happened in the previous period and what are the problems and difficulties in providing occupational health and safety. Based on these information, the business system consequently adjusts its response to changes in occupational health and safety. Key indicators of occupational safety and health are used to measure progress towards certain health and safety goals or monitor trends related to business organization activities [3]. Indicators provide information and what is happening around the business system, as it stands in relation to competition in terms of occupational safety. Defined key occupational safety indicators are subject to change and may be changed. Some indicators will be excluded from monitoring, and the new ones will be included depending on the stated needs and conditions in the occupational health and safety system. Key indicators are not oriented only to the past, but also towards the future and primarily prevention. From a standpoint of the success of occupational health and safety system in the business system, prevention is a dominant goal [3].

Preventive measures include planned or undertaken measures in all work processes by the employer, with the aim of preventing or reducing the risks at work. Preventive measures, work and production procedures applied by employer based on risk assessment must ensure the highest possible degree of workers' occupational health and safety. Occupational safety must be included in all work processes of employer and at all levels of organization and management. In the process of organizing the work process and entrusting the work to the worker, the employer is obliged to consider all the capabilities of workers that can influence on the protection of his/her health and safety at work [3].

In the business system in which the research presented in this paper was carried out, the following indicators of the occupational health and safety system are monitored, Figure 2:

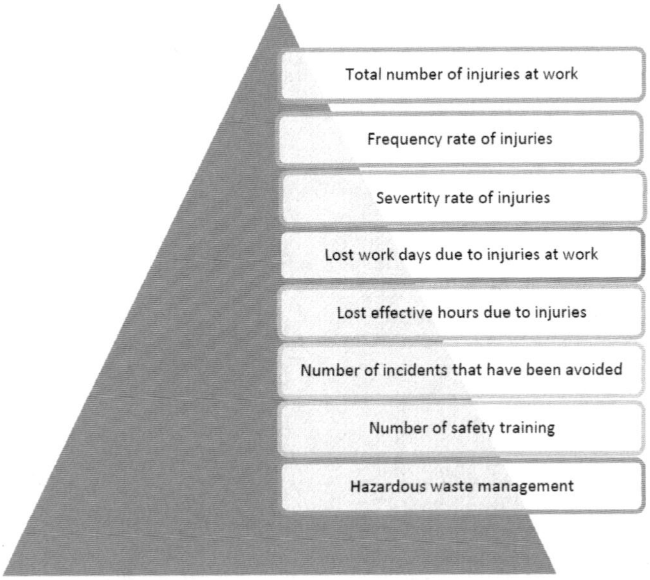

Figure 2. Indicators of occupational health and safety system in the observed business system

2. Methods of research

2.1. Research problem and goal of research
Recording the state of the existing occupational health and safety system in the production business (extraction and processing of mineral raw materials - production of iron ore) system enables the collection of data for its improvement.

The aim of the research was to get to know about the role, significance and characteristics of the occupational health and safety system in the observed production business system and based on the adopted conclusions to define the possibilities and guidelines for its optimization.

2.2. Hypothesis
Based on the research problem, the following hypotheses have been posed:
H1: By monitoring the indicators and the performances of occupational health and safety in the production business system, the level of efficiency and effectiveness of the occupational health and safety system and the business system as a whole is raised.

2.2.1. Auxiliary hypothesis. **H2:** There is a connection between the worker's (employee's) profile and the occupational health and safety system in the organization. **H3:** There is a connection between the worker's (employee's) profile and the state (performance) of occupational health and safety in the organization. **H3:** There is a connection between the occupational health and safety system in the organization and the state (performance) of occupational health and safety in the organization.

2.3. Tasks
Research tasks have been set as follows:
- A survey instrument was created: a survey questionnaire with established independent and dependent variables.

- A survey has been conducted on a determined sample.
- Data from the questionnaire were collected and statistically processed.
- The results obtained were interpreted through discussion and conclusion.

2.4. Survey method
The data were collected using a research that is descriptive-analytical, in written form, voluntary and anonymous for respondents. The questionnaire contains 43 closed-type questions, categorized into 4 groups of questions.

2.5. Research variables

Table 1. Variables of survey research (independent and dependent variables)

Independent
1. Worker profile • gender • age • level of education • total work experience
2. Profile organizational unit • number of worker in the organizational unit • the activity of the organizational unit
• The occupational health and safety system in the organization (organizational unit) • My workplace is safe in line with the organization's policies and rules of occupational safety • Direct management (supervisors, headman, etc.) understands occupational health and safety as a priority • Direct management (supervisors, headman, etc.) removes safety omission as soon as possible and provides feedback on the undertaken protection measures • Direct management (supervisors, headman, etc.) does not allow work if all protective and safety measures are not taken and provides support to workers in all matters of protection and safety (advice, additional information, teaching) • All safety incidents are reported and investigated, and information is available to all workers • Our organization has a good safety history (injuries, safety incidents, work-related disorders, occupational diseases are not common) • All means of work (machines, devices, equipment, tools) are maintained regularly, failures are immediately reported and eliminated • All protective systems and devices are always available and used • Plan (evacuation and rescue), equipment and available human resources in case of danger (natural disasters, accidents, fireworks, explosions, etc.) are satisfactory • Occupational health and safety are the main priorities of management • Management always (by its behaviour) gives a positive example when it finds itself in a work environment • Management does not blame workers in case of a safety incident • Management does not insist on productive performance if the condition of the occupational health and safety is problematic • All workers are aware of the risks of work they perform and are involved in assessing the risk of their organizational unit and workplace • The state of working environment in workplaces is good • There is exposure to potential hazards and harmfulness in workplaces • There is possibility of occurrence of unwanted event in workplaces • There is possibility of unwanted consequences of endangering human health at

workplaces
- Workplace risk assessments are audited, and workers are informed about changes
- Our organization recognizes, controls and manages risks in the workplace
- Our organization carries out regular medical examinations of all workers
- Personal protective equipment is always available, correct and in case it has been worn out replaced with a new one
- No operation is permitted without the use of personal protective equipment
- All workspaces are marked with appropriate safety signs - signs of danger, notifications and obligations (for example danger of electric current, obligation to use personal protective equipment, etc.)
- Occupational health and safety of the organization is the responsibility of all workers
- The occupational health and safety representative takes care of his/ her safety and the safety of other workers
- Measures are taken against workers who do not comply with the occupational health and safety rules
- Inspection controls are regularly implemented, and the observed shortcomings are eliminated as soon as possible

Dependent
• The state (performance) of occupational health and safety at work in the organization • Assessment of the state of occupational health and safety in the organization • Assessment of financial investments of the organization in the occupational health and safety • Assessment of management participation in the organization's occupational health and safety system • Assessment of the participation of workers in the organization's occupational health and safety system • Assessment of training and education in the organization's occupational health and safety • Assessment of the work of the trade union organization • Assessment of the work of the Occupational safety service in the organization • Dialogue/ cooperation between employers, trade unions, workers' representatives, workers and inspections in the area of Occupational health and safety • The work of the Committee on occupational health and safety in the organization (Advisory body for issues of Occupational health and safety consisting of representatives of workers, trade unions and workers)

Source: Survey Research

2.6. Research sample

Employees who are employed in the production business system (extraction and processing of mineral raw materials - production of iron ore) were defined as potential respondents in the survey i.e. about 800 potential respondents.

The expected and planned minimum number of respondents for participation in the research is:

- more than 30
- more than 10% of the total number (more than 80 respondents)
- more than the minimum representative sample determined on the basis of the variability of the dependent variables of the research (minimum scientific sample).

2.7. Statistical methods

- N – frequency
- % – percentage
- Σ – amount - sum
- M – arithmetic mean
- σ – standard deviation

- V% – coefficient of variability
- r – Pearson's coefficient of correlation-for testing the connection between independent and dependent variables

2.8. Data on survey research

The survey was organized in the period from 15.02.2018 to 15.03.2018. The research was carried out directly by answering the questionnaires by respondents in the presence of researchers. A total of 205 respondents participated, 2 survey ballots were invalid and 203 filled questionnaires were accepted as valid for statistical processing.

2.9. Testing the representativeness of the survey sample

The size of a simple random sample was obtained by the formula [4], [5]:

$$n' = \left[\frac{z_{\alpha/2} \cdot V}{G_r}\right]^2 \quad (1)$$

where is:

n' - the previous size of a simple random sample

$z_{\alpha/2}$ – confidence or reliability coefficient for the default probability of estimation ($z_{\alpha/2}$ = 1.96 for the probability of estimation 95% and $z_{\alpha/2}$ = 2.58 for the probability of estimation 99%)

V – coefficient of population variability in percentages (%)

G_r – estimation error, expressed relative in percentages

For the probability of estimating 95% i.e. $z_{\alpha/2}$ = 1,96, the selected estimation error of 10% and the value of the variability coefficient of the population of the dependent variables 23, the sample size is n_o = 20, with an additional reduction ($f = n_o/800 = 0.03 < 0.05$) we get a minimal scientific sample n_z = 20.

The set conditions of representativeness of the sample have been met:

- N(203) > 30
- N(203) > 10% n (800), N = 25.37% n
- N(203) > n_z (20)

3. Results and Discussion

Table 2. Frequency (N) and percentage (%) of independent variables

I. Worker's (employee's) profile	
1. Gender • men - 181 (**89.2%**) • women 22 - (10.8%)	**2. Age** • up to 30 years - 16 (7.9%) • from 31 to 40 - 71 (35%) • from 41 to 50 – 33 (16.3%) • from 51 to 60 - **74 (36.5%)** • over 60 years - 9 (4.4%)
3. Level of education • high school - **170 (83.7%)** • college - 9 (4.4%) • higher education (graduate engineers) - 23 (11.3%) • masters - 1 (0.5%)	**4. Total work experience** • up to 5 years - 8 (3.9%) • from 6 to 10 - 63 (31%) • from 11 to 20 - 45 (22.2%) • from 21 to 30 - 8 (4.4%) • over 30 years - **78 (38.4%)**
II. Profile organizational unit	
5. Number of worker in the organizational unit • up to 50 workers - 45 (22.2%)	**6. The activity of the organizational unit** • production and technical sector - **199 (98%)**

• from 50 to 150 workers - 37 (18.2%) • over 150 workers - **121 (59.6%)**	• economic and financial sector - 2 (1%) • sector of general, legal and personnel affairs - 1 (0.5%) • director's duties - 1 (0.5%)

Source: Survey results

Workers (employees) as respondents gave subjective assessments of raised assertions by choosing a response from 1 to 5 according to the Likert scale of intensity [4], [5]:

1. I totally disagree
2. I mostly disagree
3. I do not know (I neither agree nor disagree)
4. I mostly agree
5. I totally agree

Table 3. Arithmetic mean (M), standard deviation (σ) and coefficient of variability (V%) of independent variables of research related to The occupational health and safety system in the organization (organizational unit)

Serial number	The occupational health and safety system in the organization (organizational unit)	Lowest rating	Highest rating	Arithmetic mean - M	Standard deviation - σ	Coefficient of variability - V %
1.	My workplace is safe in line with the organization's policies and rules of occupational safety	1	5	3.99	0.91	22.81
2.	Direct management (supervisors, headman, etc.) understands occupational health and safety as a priority	1	5	3.99	0.88	22.10
3.	Direct management (supervisors, headman, etc.) removes safety omission as soon as possible and provides feedback on undertaken protection measures	1	5	3.90	0.81	23.33
4.	Direct management (supervisors, headman, etc.) does not allow work if all protective and safety measures are not taken and provides support to workers in all matters of protection and safety (advice, additional information, teaching)	1	5	3.95	0.94	23.80
5.	All safety incidents are reported and investigated, and information is available to all workers	1	5	4.15	0.99	23.86
6.	Our organization has a good safety history (injuries, safety incidents, work-related disorders, occupational diseases are not common)	1	5	4.17	0.95	22.78
7.	All means of work (machines, devices, equipment, tools) are maintained regularly, failures are immediately reported and eliminated	1	5	3.76	1.16	30.85
8.	All protective systems and devices are always available and used	1	5	4.08	0.96	23.53
9.	Plan (evacuation and rescue), equipment and available human resources in case of danger (natural disasters, accidents, fireworks, explosions, etc.) are satisfactory	1	5	4.27	0.88	20.61
10.	Occupational health and safety are the main priorities of management	1	5	4.17	0.97	23.26
11.	Management always (by its behaviour) gives a positive example when it finds itself in a work environment	1	5	4.05	0.91	22.47
12.	Management does not blame workers in case	1	5	3.75	1.07	28.53

	of a safety incident					
13.	Management does not insist on productive performance if the condition of the occupational health and safety is problematic	1	5	**3.69**	1.10	29.81
14.	All workers are aware of the risks of work they perform and are involved in assessing the risk of their organizational unit and workplace	1	5	4.14	0.77	18.60
15.	The state of working environment in workplaces is good	1	5	3.82	0.85	22.25
16.	There is exposure to potential hazards and harmfulness in workplaces	1	5	3.86	0.92	23.83
17.	There is possibility of occurrence of unwanted event in workplaces	1	5	3.76	1.05	27.93
18.	There is possibility of unwanted consequences of endangering human health at workplaces	1	5	3.84	1.01	26.30
19.	Workplace risk assessments are audited, and workers are informed about changes	1	5	3.86	0.99	25.65
20.	Our organization recognizes, controls and manages risks in the workplace	1	5	3.96	0.94	23.74
21.	Our organization carries out regular medical examinations of all workers	1	5	**4.62**	0.77	16.67
22.	Personal protective equipment is always available, correct and in case it has been worn out replaced with a new one	1	5	4.46	0.86	19.28
23.	No work is permitted without the use of personal protective equipment	1	5	4.56	0.81	17.63
24.	All workspaces are marked with appropriate safety signs - signs of danger, notifications and obligations (eg danger of electric current, obligation to use personal protective equipment, etc.)	1	5	4.43	0.80	18.06
25.	Occupational health and safety of the organization is the responsibility of all workers	1	5	4.45	0.88	19.77
26.	The occupational health and safety representative takes care of his/ her safety and the safety of other workers	1	5	4.15	0.91	21.93
27.	Measures are taken against workers who do not comply with the occupational health and safety rules	1	5	4.05	0.85	20.99
28.	Inspection controls are regularly implemented, and the observed shortcomings are eliminated as soon as possible	1	5	4.25	1.00	23.53
	Total	1	5	**4.08**	**0.93**	**23.00**

Source: Survey Results

3.1. Correlation testing

Correlations between independent and dependent variables of the study were tested by Pearson's correlation coefficient. The correlation coefficient (r) has limit values: Degrees of freedom (N-2) = 203-2 = 201.

Correlation significance levels [6]:

- **An extremely significant correlation**** - the risk of accepting the correlation between the two indicators is less than 1%, **r ≥ 0,181**
- **Significant correlation*** - the risk of accepting the correlation between the two indicators is less than 5%, **r ≥ 0,138**

Table 4. Correspondence between independent variables Worker's (employee's) profile and The occupational health and safety system in the organization (organizational unit)

Serial number	Independent variables The occupational health and safety system in the organization (organizational unit)	Independent variables Worker's (employee's) profile 1. Gender	2. Age	3. Level of education	4. Total work experience
1.	My workplace is safe in line with the organization's policies and rules of occupational safety	0.021	0.207**	0.324**	0.162*
2.	Direct management (supervisors, headman, etc.) understands occupational health and safety as a priority	0.058	0.276**	0.146*	0.217**
3.	Direct management (supervisors, headman, etc.) removes safety omission as soon as possible and provides feedback on undertaken protection measures	0.073	0.187**	0.093	0.208**
4.	Direct management (supervisors, headman, etc.) does not allow work if all protective and safety measures are not taken and provides support to workers in all matters of protection and safety (advice, additional information, teaching)	0.138*	0.237**	0.117	0.168*
5.	All safety incidents are reported and investigated, and information is available to all workers	-0.068	0.161*	0.106	0.052
6.	Our organization has a good safety history (injuries, safety incidents, work-related disorders, occupational diseases are not common)	-0.011	0.108	0.125	0.021
7.	All means of work (machines, devices, equipment, tools) are maintained regularly, failures are immediately reported and eliminated	0.141*	0.222**	0.144*	0.117
8.	All protective systems and devices are always available and used	0.019	0.116	0.062	-0.019
9.	Plan (evacuation and rescue), equipment and available human resources in case of danger (natural disasters, accidents, fireworks, explosions, etc.) are satisfactory	0.057	0.102	0.128	0.070
10.	Occupational health and safety are the main priorities of management	0.055	0.200**	0.168*	0.081
11.	Management always (by its behavior) gives a positive example when it finds itself in a work environment	0.033	0.141*	0.217**	0.078
12.	Management does not blame workers in case of a safety incident	0.037	0.240**	0.242**	0.194**
13.	Management does not insist on productive performance if the condition of the occupational health and safety is problematic	0.097	0.060	0.224**	-0.035
14.	All workers are aware of the risks of work they perform and are involved in assessing the risk of their organizational unit and workplace	0.082	0.120	0.057	0.104

15.	The state of working environment in workplaces is good	-0.037	0.190**	0.005	0.122
16.	There is exposure to potential hazards and harmfulness in workplaces	0.035	0.076	-0.111	0.055
17.	There is possibility of occurrence of unwanted event in workplaces	-0.043	-0.071	0.005	-0.071
18.	There is possibility of unwanted consequences of endangering human health at workplaces	0.023	0.010	0.001	-0.030
19.	Workplace risk assessments are audited, and workers inform about changes	0.033	0.089	0.140*	0.040
20.	Our organization recognizes, controls and manages risks in the workplace	0.016	0.122	0.128	0.084
21.	Our organization carries out regular medical examinations of all workers	-0.094	0.022	0.107	-0.014
22.	Personal protective equipment is always available, correct and in case of weariness replaced with a new one	-0.041	0.053	0.112	0.030
23.	No work is permitted without the use of personal protective equipment	0.034	0.072	-0.011	0.067
24.	All workspaces are marked with appropriate safety signs - signs of danger, notifications and obligations (eg danger of electric current, obligation to use personal protective equipment, etc.)	0.071	0.027	0.020	-0.004
25.	Occupational health and safety of the organization is the responsibility of all workers	0.091	0.061	0.171*	0.082
26.	The occupational health and safety representative takes care of his/ her safety and the safety of other workers	0.013	0.167	0.188**	0.165*
27.	Measures are taken against workers who do not comply with the occupational health and safety rules	-0.058	0.077	0.018	0.037
28.	Inspection controls are regularly implemented, and the observed shortcomings are eliminated as soon as possible	-0.088	-0.024	0.054	-0.038
	28/112	2	10	10	6

Source: Survey Results

Table 5. Statistical analysis of coefficients of independent variables correlation Worker's (employee's) profile and The occupational health and safety system in the organization (organizational unit)

1. Gender

Women give:

- higher score of variables: 4 (r = 0.138*) and 7 (r = 0.141*)

2. Age

Respondents who are older are given:

- higher score of variables: 1 (r = 0.207**), 2 (r = 0.5276**), 3 (r = 0.187**), 4 (r = 0.237**), 5 (r = 0.161*), 7 (r = 0.222**), 10 (r = 0.200**), 11 (r = 0.141*), 12 (r = 0.167*) and 15 (r = 0.190**)

3. Level of education

Respondents who have a higher level of professional education give:

- higher score of variables: 1 (r = 0.324**), 2 (r = 0.146*), 7 (r = 0.144*), 10 (r = 0.168*), 11 (r = 0.217**), 12 (r = 0.242**), 13 (r = 0.224**), 19 (r = 0.140*), 25 (r = 0.171*) and 26 (r = 0.188**)

4. Total work experience

Respondents who have more total work experience give:

- higher score of variables: 1 (r = 0.162*), 2 (r = 0.217**), 3 (r = 0.208**), 4 (r = 0.168*), 12 (r = 0.194**) and 26 (r = 0.165*)

Source: Survey Results

Table 6. Correspondence between independent variables Worker's (employee's) profile and dependent variables The state (performance) of occupational health and safety in the organization

Serial number	Dependent variables	Independent variables			
		Worker's (employee's) profile			
	The state (performance) of occupational health and safety in the organization	1. Gender	2. Age	3. Level of education	4. Total work experience
1.	Assessment of the state of occupational health and safety in the organization	-0.009	0.102	0.267**	0.008
2.	Assessment of financial investments of the organization in the occupational health and safety	0.104	-0.045	0.258**	-0.089
3.	Assessment of management participation in the organization's occupational health and safety system	-0.008	0.009	0.172*	-0.032
4.	Assessment of the participation of workers in the organization's occupational health and safety system	-0.177*	-0.027	0.024	-0.068
5.	Assessment of training and education in the organization's occupational health and safety	-0.092	0.004	0.097	-0.024
6.	Assessment of the work of the trade union organization	-0.188**	-0.070	0.051	-0.113
7.	Assessment of the work of the Occupational safety service in the organization	-0.052	0.084	0.239**	0.008
8.	Dialogue/ cooperation between employers, trade unions, workers' representatives, workers and inspections in the area of Occupational health and safety	-0.089	-0.043	0.255**	-0.130
9.	The work of the Committee on occupational health and safety in the organization	-0.069	-0.048	0.295**	-0.145*
	9/36	2	0	6	1

Source: Survey Results

Table 7. Statistical analysis of coefficients of independent variables correlation Worker's (employee's) profile and The state (performance) of occupational health and safety in the organization

1. Gender

Women give:

- lowest score of variables: 4 (r = -0,177*) and 6 (r = -0,188**)

2. Level of education

Respondents who have a higher level of professional education give:

- higher score of variables: 1 (r = 0,267**), 2 (r = 0,258**), 3 (r = 0,172*), 7 (r = 0,239**), 8 (r = 0,255**) and 9 (r = 0,295**)

3. Total work experience

Respondents who have more total work experience give:

- lowest score of variables: 9 (r = -0,145*)

Source: Survey Results

Serial number	Independent variables	Dependent variables								
	The occupational health and safety system in the organization (organizational unit)	The state (performance) of occupational health and safety in the organization								
		1. Assessment of the state of occupational health and safety in the organization	2. Assessment of financial investments of the organization in the occupational health and safety	3. Assessment of management participation in the organization's occupational health and safety system	4. Assessment of the participation of workers in the organization's occupational health and safety system	5. Assessment of training and education in the organization's occupational health and safety	6. Assessment of the work of the trade union organization	7. Assessment of the work of the Occupational safety service in the organization	8. Dialogue/ cooperation between employers, trade unions, workers' representatives, workers and inspections in the area of Occupational health and safety	9. The work of the Committee on occupational health and safety in the organization
1.	My workplace is safe in line with the organization's policies and rules of occupational safety	0.455^{**}	0.303^{**}	0.418^{**}	0.277^{**}	0.216^{**}	0.237^{**}	0.340^{**}	0.290^{**}	0.277^{**}
2.	Direct management (supervisors, headman, etc.) understands occupational health and safety as a priority	0.312^{**}	0.194^{**}	0.444^{**}	0.229^{**}	0.180^{*}	0.134	0.267^{**}	0.172^{*}	0.201^{**}
3.	Direct management (supervisors, headman, etc.) removes safety omission as soon as possible and provides feedback on undertaken protection measures	0.277^{**}	0.030	0.307^{**}	0.147^{*}	0.216^{**}	0.211^{**}	0.180^{*}	0.220^{**}	0.176^{*}
4.	Direct management (supervisors, headman, etc.) does not allow work if all protective and safety measures are not taken and provides support to workers in all matters of protection and safety (advice, additional information, teaching)	0.312^{**}	0.264^{**}	0.376^{**}	0.233^{**}	0.210^{**}	0.209^{**}	0.237^{**}	0.217^{**}	0.173^{*}
5.	All safety incidents are reported and investigated, and information is available to all workers	0.384^{**}	0.292^{**}	0.471^{**}	0.345^{**}	0.292^{**}	0.311^{**}	0.382^{**}	0.365^{**}	0.311^{**}
6.	Our organization has a good safety history (injuries, safety incidents, work-related disorders, occupational diseases are not common)	0.418^{**}	0.260^{**}	0.536^{**}	0.396^{**}	0.307^{**}	0.278^{**}	0.367^{**}	0.303^{**}	0.366^{**}
7.	All means of work (machines, devices, equipment, tools) are maintained regularly, failures are immediately reported and eliminated	0.226^{**}	0.204^{**}	0.373^{**}	0.353^{**}	0.249^{**}	0.242^{**}	0.280^{**}	0.237^{**}	0.177^{*}
8.	All protective systems and devices are always available and used	0.379^{**}	0.278^{**}	0.489^{**}	0.405^{**}	0.261^{**}	0.244^{**}	0.404^{**}	0.325^{**}	0.305^{**}
9.	Plan (evacuation and rescue), equipment and available human resources in case of danger (natural disasters, accidents, fireworks, explosions, etc.) are satisfactory	0.381^{**}	0.281^{**}	0.433^{**}	0.356^{**}	0.335^{**}	0.142^{*}	0.357^{**}	0.218^{**}	0.312^{**}
10.	Occupational health and safety are the main priorities of management	0.422^{**}	0.266^{**}	0.453^{**}	0.379^{**}	0.242^{**}	0.250^{**}	0.419^{**}	0.373^{**}	0.360^{**}
11.	Management always (by its behavior) gives a positive example when it finds itself in a work environment	0.442^{**}	0.296^{**}	0.406^{**}	0.342^{**}	0.314^{**}	0.183^{**}	0.392^{**}	0.370^{**}	0.405^{**}
12.	Management does not blame workers in case of a safety incident	0.331^{**}	0.183^{**}	0.490^{**}	0.271^{**}	0.220^{**}	0.287^{**}	0.288^{**}	0.330^{**}	0.274^{**}
13.	Management does not insist on productive performance if the condition of the occupational health and safety is problematic	0.417^{**}	0.300^{**}	0.439^{**}	0.333^{**}	0.290^{**}	0.245^{**}	0.340^{**}	0.464^{**}	0.324^{**}
14.	All workers are aware of the risks of work they perform and are involved in assessing the risk of their organizational unit and workplace	0.341^{**}	0.262^{**}	0.376^{**}	0.262^{**}	0.257^{**}	0.049	0.312^{**}	0.216^{**}	0.253^{**}
15.	The state of working environment in workplaces is good	0.341^{**}	0.226^{**}	0.327^{**}	$0,.96^{**}$	0.230^{**}	0.147^{*}	0.295^{**}	0.228^{**}	0.063
16.	There is exposure to potential hazards and harmfulness in workplaces	0.076	0.115	0.028	-0.005	0.025	-0.050	0.033	-0.014	0.112
17.	There is possibility of occurrence of unwanted event in workplaces	0.106	0.118	-0.069	-0.069	-0.029	-0.078	0.025	-0.037	0.032
18.	There is possibility of unwanted consequences of endangering human health at workplaces	0.122	0.147^{*}	-0.069	0.342^{**}	-0.036	-0.129	0.014	-0.066	0.303^{**}
19.	Workplace risk assessments are audited, and workers inform about changes	0.326^{**}	0.312^{**}	0.347^{**}	0.273^{**}	0.198^{**}	0.131	0.334^{**}	0.258^{**}	0.361^{**}
20.	Our organization recognizes, controls and manages risks in the workplace	0.363^{**}	0.307^{**}	0.383^{**}	0.193^{**}	0.205^{**}	0.165^{*}	0.454^{**}	0.297^{**}	0.227^{**}
21.	Our organization carries out regular medical examinations of all workers	0.288^{**}	0.181^{**}	0.247^{**}	0.344^{**}	0.183^{**}	0.145^{*}	0.232^{**}	0.203^{**}	0.330^{**}
22.	Personal protective equipment is always available, correct and in case of weariness replaced with a new one	0.379^{**}	0.282^{**}	0.434^{**}	0.327^{**}	0.288^{**}	0.292^{**}	0.388^{**}	0.267^{**}	0.243^{**}
23.	No work is permitted without the use of personal protective equipment	0.313^{**}	0.180^{*}	0.369^{**}	0.272^{**}	0.311^{**}	0.208^{**}	0.288^{**}	0.252^{**}	0.304^{**}
24.	All workspaces are marked with appropriate safety signs - signs of danger, notifications and obligations (eg danger of electric current, obligation to use personal protective equipment, etc.)	0.355^{**}	0.235^{**}	0.386^{**}	0.199^{**}	0.257^{**}	0.133	0.319^{**}	0.154^{*}	0.227^{**}
25.	Occupational health and safety of the organization is the responsibility of all workers	0.280^{**}	0.153^{*}	0.313^{**}	0.264^{**}	0.231^{**}	0.063	0.231^{**}	0.092	0.163^{*}
26.	The occupational health and safety representative takes care of his/ her safety and the safety of other workers	0.444^{**}	0.300^{**}	0.348^{**}	0.208^{**}	0.268^{**}	0.156^{*}	0.409^{**}	0.258^{**}	0.327^{**}
27.	Measures are taken against workers who do not comply with the occupational health and safety rules	0.284^{**}	0.171^{*}	0.244^{**}	0.364^{**}	0.190^{**}	0.214^{**}	0.310^{**}	0.216^{**}	0.246^{**}
28.	Inspection controls are regularly implemented, and the observed shortcomings are eliminated as soon as possible	0.341^{**}	0.140^{*}	0.315^{**}	0.364^{**}	0.352^{**}	0.310^{**}	0.356^{**}	0.327^{**}	0.353^{**}
	219/252	25	25	25	25	25	20	25	24	25

Figure 3. Statistical analysis of coefficients of independent variables correlation The occupational health and safety system in the organization (organizational unit) and dependent variables The state (performance) of Occupational health and safety in the organization

4. Conclusion

By collecting data on state of occupational health and safety in the production business system, from direct participants - employees, we obtain concrete data that can be used to improve and optimize that system.

The mean average assessment of the independent variables of the research "System of occupational health and safety in the organization (organizational unit)" given by workers (employees) (4.08) is in the positive area and quite high. Based on the indicators that covered this variable, the conclusion may be drawn that the occupational health and safety in the organization is very good. This result should not stop the process of continuous monitoring and measurement of occupational health and safety indicators, as well as the improvement of the overall system of occupational health and safety in the organization.

Independent variable "Our organization carries out regular medical examinations of all workers" obtained a peak average score of 4.62. It means that the workers (employees) recognized the commitment of the employer to preserve the health of employees. In the organization which was the subject of the research, regular medical examinations are carried out based on the Risk Assessment Act, systematic reviews (every 5 years) of all employees, as well as targeted reviews as needed. The organization in direct production has a permanent medical team engaged, and the Occupational Safety and Health Service, in cooperation with the Workers' medical care, organizes lectures on medical topics according to the established program.

The independent variable "Management does not insist on productive performance if the condition of the occupational health and safety is problematic" has the lowest score of 3.69. Thus it can be concluded that situations are heppening when production continues even if it does not fully ensure adequate occupational safety.

Hypothesis: There is a correlation between The worker's profile and The occupational health and safety system in the organization (organizational unit) is not fully confirmed, because out of possible 112 correlations, correlation of 28 variables has been established, a quarter of possible correlations, 25%.

Hypothesis: There is a correlation between The worker's profile and The state of occupational health and safety in the organization, is not fully confirmed, because out of possible 36 correlations, correlation of 9 variables has been established, a quarter of possible correlations, 25%.

Hypothesis: There is a correlation between The occupational health and safety system in the organization and The state (performance) of occupational health and safety in the organization, is confirmed, as out of 252 possible correlation of variables, correlation of 219 variables were confirmed i.e. 86,9%.

The confirmation of this hypothesis indicates that the improvement of all components constituting an independent variable „System of occupational health and safety in the organization" is the way to influence the state of occupational health and safety in the organization for the purpose of raising the level of efficiency and effectiveness of this system and a the business system as a whole.

Based on the previous knowledge and experience of the researchers, within this research, selected components of the occupational health and safety system and its performance were observed. However, this does not mean that some other indicators and performance that are considered significant for further optimization of the occupational health and safety system in the business system cannot be included by the research.

References

[1] Petras M, Begović I and Palačić D 2014 Key indicators of the success of the process of health, safety at work and environmental protection, *Journal of Practical Management* **V**(1) 68-78

[2] ***European Agency for Safety and Health at Work 2002 *Inventory of socioeconomic costs of work accidents*, Prepared by: Topic Centre on Research - Work and Health: Jos Mossink, http://europa.eu.int

[3] Baldauf P E J 2010 *Measuring Safety Performance: What are KPIs?,* EHS Journal – Practical

Solutions for Environmental, Health and Safety Professionals
[4] Vranješ B 2017 *Knowledge management in protection at work through newspapers new Law on Occupational safety in the Republic of Srpska (professional exam, risk assessment),* Proceedings of the 12th International Conference on Management and Safety M&S 2017, Neum and Mostar, Bosnia and Herzegovina, June 9-10, pp 113-124
[5] Aranđelović I, Mitrović Z and Stojanović V 2011 *Probability and Statistics*, Belgrade Institute for Textbooks, Belgrade, Serbia
[6] Dizdar D 2006 *Quantitative Methods*, Faculty of Kinesiology, University of Zagreb, Zagreb, Croatia

IOP Conf. Series: Materials Science and Engineering **477** (2019) 012039 doi:10.1088/1757-899X/477/1/012039

Investigation of harmonic pollution produced by power tools

A Iagăr, G N Popa and C M Diniş

Politehnica University of Timisoara, Department of Electrical Engineering and Industrial Informatics, 5 Revolution Street, Hunedoara, 331128, Romania

E-mail: angela.iagar@fih.upt.ro

Abstract. The paper analyzes the influence of power tools operating mode on the power quality, especially on the harmonic pollution. The measurements were carried out with the CA 8334B three-phase portable AC power quality analyzer. The low utilization factor of power tools (due to the short duration of operations) introduces complexity in the study of their impact on power quality. Although the absorbed powers are small, the cumulative effect produced by a large number of power tools can be substantial.

1. Introduction

Number and variety of nonlinear equipment used in the residential and commercial domains had a constant increase in the last decades, these being the main sources of harmonic distortions in low voltage distribution systems [1-6]. Simultaneous, has increased the use of equipment very sensitive to electromagnetic disturbances, in industrial automated processes, digital communications, computer networks and multimedia. Under these circumstances, power quality assurance becomes more and more complicated [7-11].

The most common types of power quality problems are: voltage dips, very short interruptions, long interruptions, voltage spikes, voltage swells, harmonic distortions, voltage fluctuations, noise and voltage unbalances [1-3].

To estimate the economic impact of harmonics is much more difficult than other power quality costs, such as voltage interruptions and dips. Harmonic distortion effects on the distribution system equipment and connected loads can be short and long term. Short-term effects include equipment malfunction (e.g., malfunction of electronic equipment based on voltage zero crossing detection or sensitive to wave shape; errors in measures; mis-operation of protective relays), loss of efficiency, electromagnetic interference with communication systems and power losses caused by harmonic currents and voltages [12], [13]. Long-term effects include thermal losses (in cables and equipment), premature aging and reduced life span of equipment [12], [13].

Among the most sensitive equipment at harmonic distortion are computers and programmable controllers, which typically require AC sources with a total harmonic distortion factor (THD) of 5%, the highest harmonic level being at most 3% of the fundamental voltage [2].

Electronic equipment can be disturbed by harmonics due to conduction coupling (via the equipment power supply) and magnetic coupling [1], [2].

The emission and propagation of harmonics through the network are influenced by multiple factors, such as network impedance, voltage distortion, time-variation of number and type of connected equipment, which complicate the analysis of harmonics effects in the network [5], [6], [14-18].

The situation becomes even more difficult in the case of small harmonic sources, because of their random nature and time variability [16], as well the propagation and partial cancellation of harmonics due to phase angle diversity (diversity effect) [15]. Also, for small distributed harmonic sources, the design of mitigation measures is very complex, as and the optimal location of capacitors for power factor correction, due to possible resonances [3].

Although it is very difficult to estimate the cumulative effect produced by a large number of small harmonic sources, it is essential to identify and analyze the sources of the harmonics and their effects. The purpose of this work is to determine, through measurements, the harmonic contributions of power tools.

2. Experimental measurements

The laboratory measurements were carried out using the power quality analyzer CA 8334B. To measure the current was used current probes MN 93A (5 A) [19].

Power quality analyzer CA 8334B gave an instantaneous image of the main characteristics of power quality for the analyzed power tools. The main parameters measured by CA 8334B analyzer were: True RMS AC phase voltages and True RMS AC line currents; active, reactive and apparent power per phase; harmonics for voltages and currents up to the 50th order; Fresnel diagrams. This analyzer provide numerous calculated values and processing functions in compliance with EMC standards in use (EN 50160, IEC 61000-4-15, IEC 61000-4-30, IEC 61000-4-7, IEC 61000-3-4). The values computed by the CA8334B analyzer were: total harmonic distortion (THD) of voltages and currents; power factor (PF) and displacement power factor (DPF). This analyzer allows the recording, time stamping and characterization of electromagnetic disturbances (swells, dips and interruptions, overrun of power and harmonic thresholds), detection of transients and recording of the associated waveforms.

Below are presented the results of measurements for various types of grinders, electric saws, drilling machines, milling and finishing machines, operating in an isolated mode.

2.1. Angle grinder

It was analyzed an angle grinder (AG 115 Faust) with the following technical characteristics: rated active power 580 W, rated (no load) speed 10,000 rpm, rated voltage 230 V, frequency of the supply voltage 50 Hz.

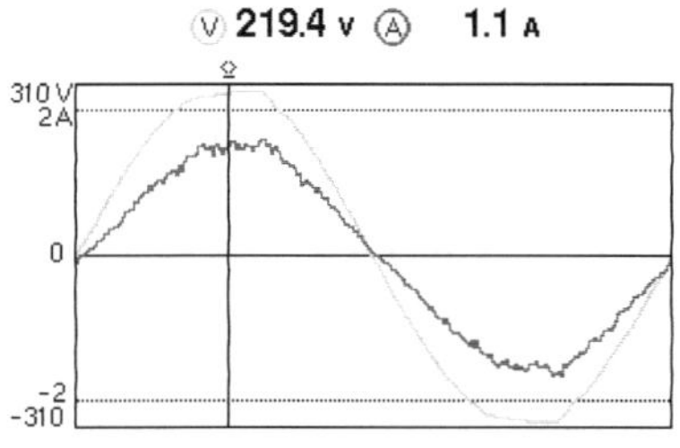

A

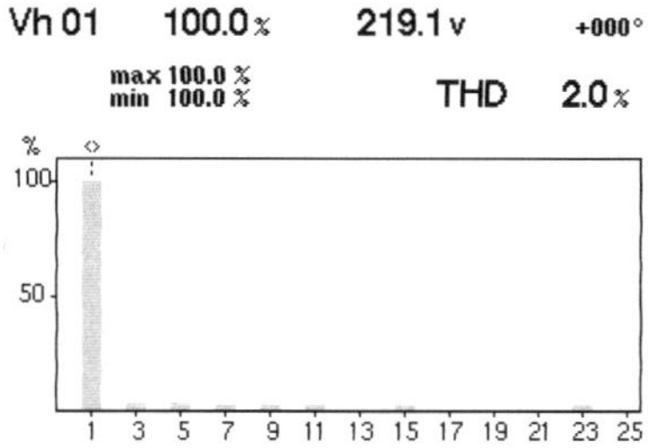

b

Figure 1. Voltage and current waveforms (a) and harmonic spectrum of supply voltage (b) for angle grinder (idle operation)

The waveforms of voltage and current (Figure 1.a) are very close to the sine wave, in idle operation of angle grinder.

The voltage harmonics are insignificant, within the EMC limits [20], [21] (Figure 1.b), and total harmonic distortion factor of voltage is very low (THD_U=2%).

Harmonic analysis of current (Figure 2) in idle operation of angle grinder shows that its distortion is within the EMC limit (THD_I=4.7%).

It is noticed that the reactive power is much lower than the active power absorbed by the network, when the angle grinder is in idle operation (Table 1). For this reason, power factor is very good, above the neutral value (PF=0.996, inductive).

Because the harmonic pollution of current is very low, DPF and PF differs very little (DPF=0.997) in this case.

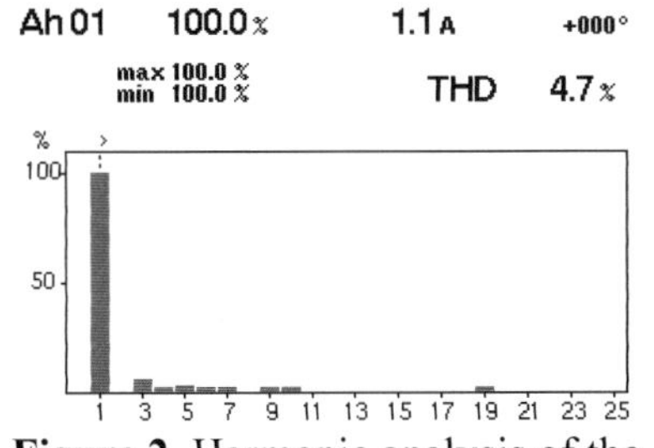

Figure 2. Harmonic analysis of the current drawn by angle grinder in idle operation

Table 1. Angle grinder

Operation mode	*Idle operation*	*Load operation*
Active power, P[W]	234.2	296.3
Reactive power, Q[VAR]	21.9	55.1
Apparent power, S[VA]	235.2	301.4
Power factor, PF[-]	0.996	0.983
Displacement power factor, DPF[-]	0.997	0.972

When the angle grinder is in load operation, the waveform of supply voltage is almost sinusoidal (Figure 3.a), but the current waveform is distorted. Voltage harmonics do not exceed the EMC limits (Figure 3.b), but total harmonic distortion factor of supply voltage is slightly higher in load operation compared to idle operation of angle grinder (THD_U=2.3%).

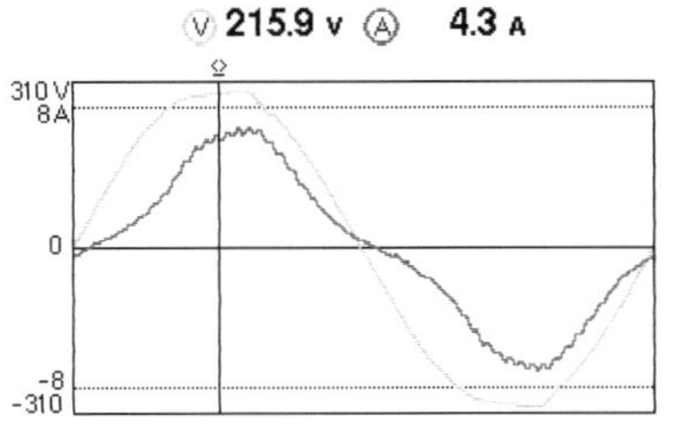

A

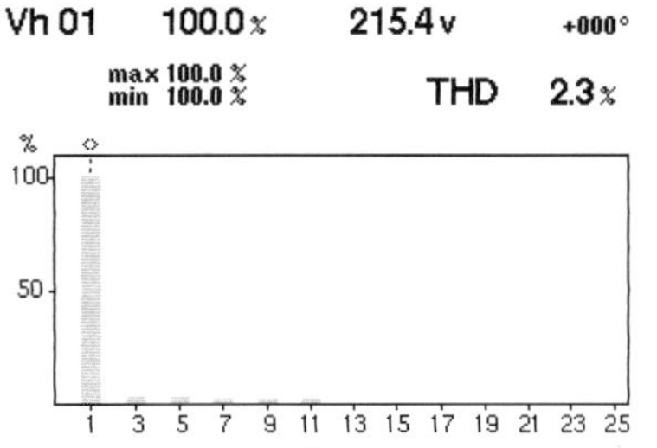

b

Figure 3. Voltage and current waveforms (a) and harmonic spectrum of supply voltage (b) when angle grinder is in load operation

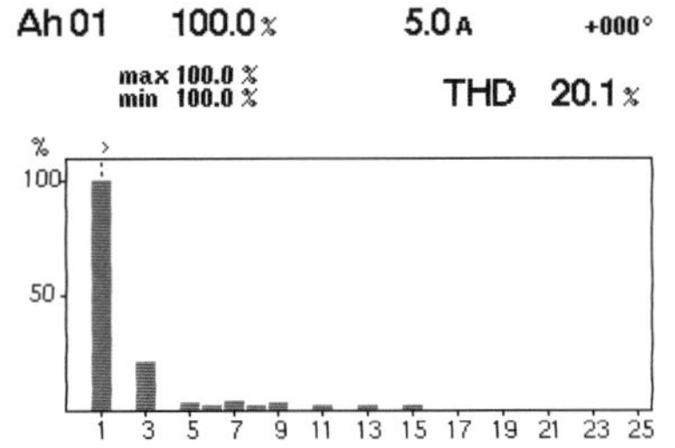

Figure 4. Harmonic analysis of the current drawn by angle grinder in load operation

It is observed (from the harmonic analysis of current, Figure 4) that the 3rd harmonic exceeds the compatibility limit and total harmonic distortion factor of current is much higher in load operation compared to idle operation of grinder (THD_I=20.1%), exceeding very much the EMC limit.

In load operation of angle grinder, current RMS is almost 4 times higher compared to idle operation. Active and reactive powers also increase compared to idle operation.

PF and DPF exceed the neutral value, but are lower in this case (PF=0.983 inductive, DPF=0.972), because the reactive power doubles and the harmonic pollution is much higher in load operation, compared to idle operation of angle grinder.

2.2. Bench grinder

Bench grinder (Jupiter) had the following technical characteristics: rated input power 370 W, grinding wheel diameter 150 mm, no load speed 3000 rpm, rated voltage 230 V, frequency of the supply voltage 50 Hz.

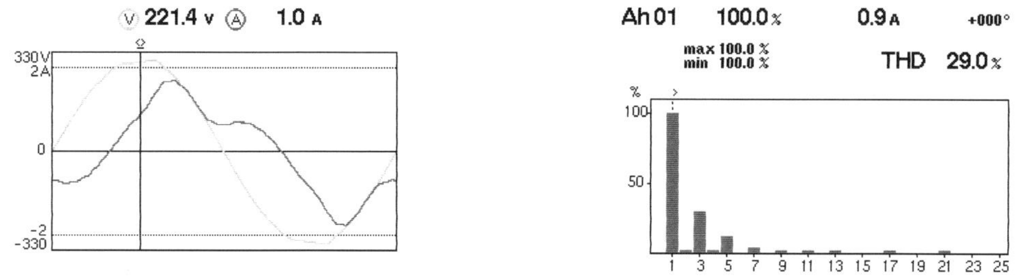

A b

Figure 5. Voltage and current waveforms (a) and harmonic spectrum of current (b) for bench grinder (idle operation)

The supply voltage is almost sinusoidal in idle operation of bench grinder, but the current waveform is distorted (Figure 5.a). The 3rd and 5th harmonics of current and its total harmonic distortion factor exceed the EMC limits (THD_I=29%, Figure 5.b).

Table 2. Bench grinder

Operation mode	*Idle operation*	*Load operation*
Active power, P[W]	132	211.6
Reactive power, Q[VAR]	160.2	146.6
Apparent power, S[VA]	207.6	257.4
Power factor, PF[-]	0.636	0.822
Displacement power factor, DPF[-]	0.659	0.836

The reactive power is larger than the active power in the case of idle operation of bench grinder (Table 2), and power factors are of similar values, much less than the neutral value (PF=0.636 inductive, DPF=0.659).

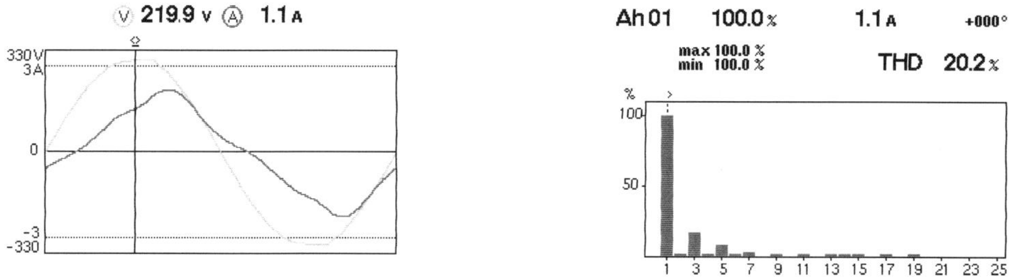

A b

Figure 6. Voltage and current waveforms (a) and harmonic spectrum of current (b) for bench grinder (load operation)

If bench grinder is in load operation (6 mm flat plate was polished), the supply voltage remains sinusoidal and the current is less distorted than in the previous case (Figure 6.a). The 3rd and 5th current harmonics exceed the limits set by EMC standards, but total harmonic distortion factor (THD_I=20.2%) is lower than in the idle operation.

In this operating mode, the active power doubles almost compared to the idle operation of the bench grinder (Table 2), but the reactive power decreases by 13.6 VAR. In these conditions, an increase of DPF and PF was observed (PF=0.822 inductive, DPF=0.836), although their values remain below the neutral value.

2.3. *Chainsaw*

In idle operation of chainsaw (PKS 1840/1 Einhell, with technical characteristics: rated active power 1800 W, blade length 400 mm, chain speed 9 m/s, rated voltage 230 V, frequency of the supply voltage 50 Hz), the waveform of supply voltage is sinusoidal, but the waveform of current is distorted (Figure 7.a).

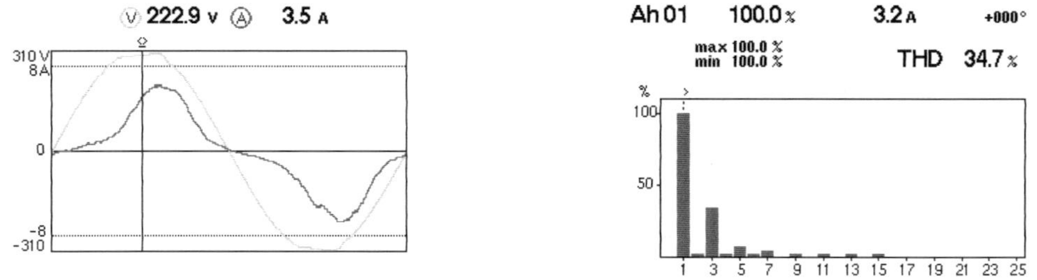

A b

Figure 7. Voltage and current waveforms (a) and harmonic spectrum of current (b) for chainsaw (idle operation)

Harmonic analysis of current (Figure 7.b) reveals that 3rd harmonic level is very high (above 30%), and total harmonic distortion factor of current exceeds very much the compatibility limit (THD_I=34.7%).

Table 3. Chainsaw

Operation mode	*Idle operation*	*Load operation*
Active power, P[W]	649.9	870.6
Reactive power, Q[VAR]	319.1	411.8
Apparent power, S[VA]	764.6	963.1
Power factor, PF[-]	0.909	0.904
Displacement power factor, DPF[-]	0.962	0.955

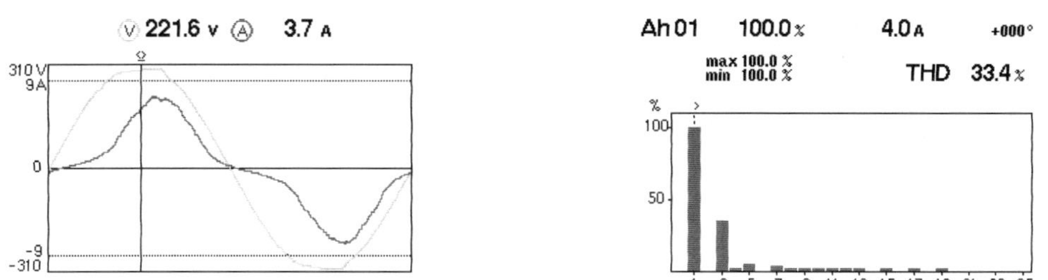

A b

Figure 8. Voltage and current waveforms (a) and harmonic spectrum of current (b) for chainsaw (load operation, chain of 4 mm thickness)

In idle operation of electric chainsaw, real power factor is PF=0.909 (inductive), slightly lower than the neutral value (Table 3). It is observed that displacement power factor (DPF=0.962) is higher than power factor (PF) due to harmonic pollution generated by the electric chainsaw.

The voltage waveform remains sinusoidal and the current waveform is still distorted (Figure 8.a) in load operation of chainsaw. It is observed that the 3rd harmonic current exceed the EMC limit and total harmonic distortion factor of current is slightly lower than in the idle operation (THD_I=33.4%).

Active and reactive power consumption increases on load operation compared to idle operation, and power factor is slightly lower (PF=0.904 inductive) than on the idle operation (Table 3). Displacement power factor remains greater than power factor (DPF=0.955), due to harmonic pollution of current absorbed by chainsaw.

2.4. Percussion hammer drill

Was analyzed a percussion hammer drill (BMH 750) with the following technical data: rated active power 750 W, no load speed 800 rpm, rated voltage 230 V, frequency of the supply voltage 50 Hz, maximum diameter of the drill 26 mm.

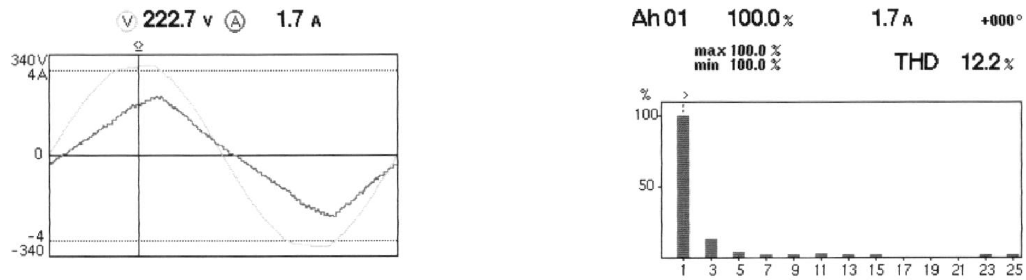

A b

Figure 9. Voltage and current waveforms (a) and harmonic spectrum of current (b) for percussion hammer drill (idle operation)

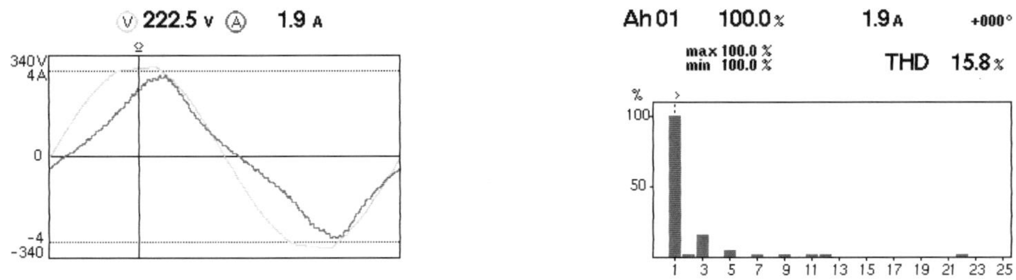

A b

Figure 10. Voltage and current waveforms (a) and harmonic spectrum of current (b) for percussion hammer drill in load operation (6 mm drill, in stone)

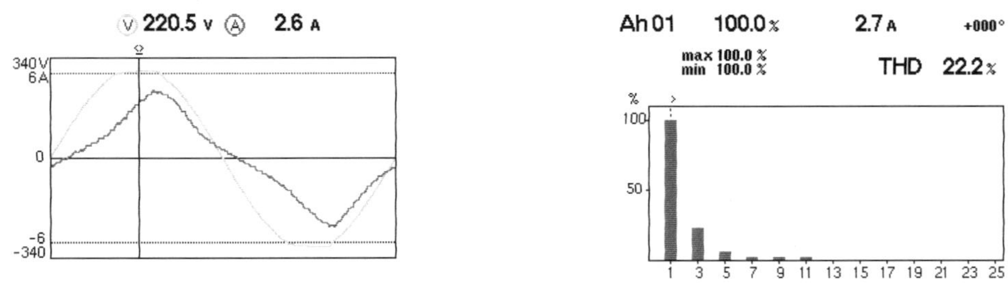

A b

Figure 11. Voltage and current waveforms (a) and harmonic spectrum of current (b) for percussion hammer drill in load operation (18 mm drill, in stone)

Total harmonic distortion of voltage is much below the EMC limit, but it is observed a significant harmonic distortion of current in all operating conditions of the hammer drill (Figures 9-11).

If the load is higher, it can be noted the increase of RMS value and harmonic distortion of current absorbed by hammer drill (Figures 10, 11). In all analyzed situations, the 3rd current harmonic and total harmonic distortion of current exceed the compatibility limits.

There is a similitude between the electrical parameters measured in idle operation and low load operation of hammer drill (6 mm drill, in stone, Table 4). Also, it is found very small differences between the measurements done in high load operation of hammer drill (with 12 mm drill, respectively 18 mm drill, in stone, Table 4).

Reactive (inductive) power is double in the latter two cases, compared to situation where the 6 mm drill was used. Also, in these situations, total harmonic distortion factor of current is about 1.4 times greater, due to the increase of 3rd and 5th harmonics.

Power factor and displacement power factor decrease with increasing of load, but have higher values than the neutral power factor. As the harmonic distortion increases, there are increasing differences between displacement power factor and real power factor.

Table 4. Percussion hammer drill

Operation mode	*Idle operation*	*Load operation*		
		6 mm drill in stone	*12 mm drill in stone*	*18 mm drill in stone*
Active power, P[W]	371.3	386.4	572.3	569.7
Reactive power, Q[VAR]	106.6	111.7	228.9	233.1
Apparent power, S[VA]	386.3	402.2	616.4	615.5
Power factor, PF[-]	0.961	0.961	0.929	0.926
Displacement power factor, DPF[-]	0.969	0.969	0.951	0.940

2.5. Column drilling machine

Analyzed column drilling machine (TB13/5) had the following technical data: rated active power 350 W, motor speed 1420 rpm, mandrel speed 515-2580 rpm, rated current 1.5 A, maximum drilling depth 50 mm, rated voltage 230 V, frequency of the supply voltage 50 Hz.

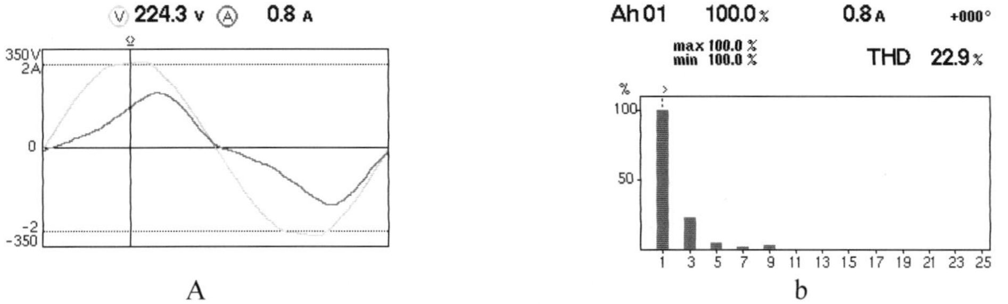

Figure 12. Voltage and current waveforms (a) and harmonic spectrum of current (b) for column drilling machine (idle operation)

Table 5. Column drilling machine

Operation mode	*Idle operation*	*Load operation*		
		3 mm drill in wood	*5 mm drill in wood*	*10 mm drill in wood*
Active power, P[W]	161.5	168.8	180.3	183.0
Reactive power, Q[VAR]	67.1	62.2	59.5	58.6
Apparent power, S[VA]	174.9	179.9	189.9	192.1
Power factor, PF[-]	0.923	0.938	0.950	0.952
Displacement power factor, DPF[-]	0.945	0.959	0.968	0.970

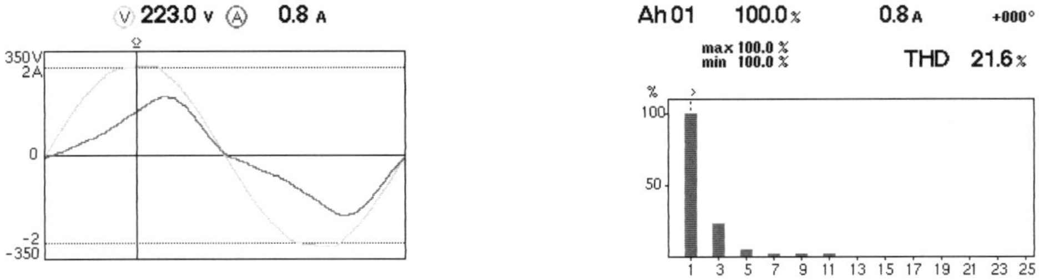

A b

Figure 13. Voltage and current waveforms (a) and harmonic spectrum of current (b) for column drilling machine in load operation (3 mm drill, in wood)

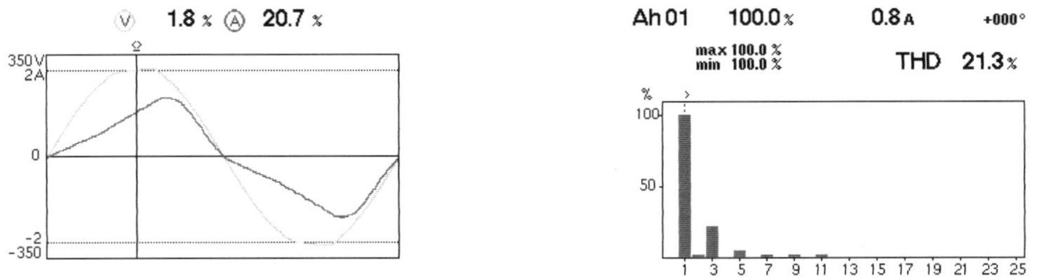

A b

Figure 14. Voltage and current waveforms (a) and harmonic spectrum of current (b) for column drilling machine in load operation (5 mm drill, in wood)

A b

Figure 15. Voltage and current waveforms (a) and harmonic spectrum of current (b) for column drilling machine in load operation (10 mm drill, in wood).

Electrical parameters of column drilling machine are not influenced too much by its operating mode (Table 5). The current (RMS value) absorbed by column drilling machine is about the same in all analyzed cases (Figures 12-15).

Active power and apparent power increase slightly when the load increases (drilling with a 5 mm drill and with a 10 mm drill in wood), but reactive (inductive) power decreases slightly (Table 5).

The values of power factor and displacement power factor are much higher than the neutral value in all analyzed cases, and increase slightly as the load increases (Table 5). Harmonic distortion of voltage is extremely small (THD_U=1.8%) in all operating modes of this column drilling machine, but current waveform is highly distorted (THD_I exceeds 20%). The level of 3rd harmonic, and, hence, total harmonic distortion factor of current, decrease when the load increases (Figures 12-15).

2.6. *Pendulum jigsaw*

Analyzed pendulum jigsaw (PSTS 710) had the following technical characteristics: rated active power 710 W, no load speed 1200-3000 rpm, saw blade length 80 mm, rated voltage 230 V, frequency of the supply voltage 50 Hz.

The experiments were carried out at a set speed of 2000 rpm. The operation of pendulum jigsaw greatly deteriorates the indicators of power quality, in terms of current harmonic pollution and reactive power consumption.

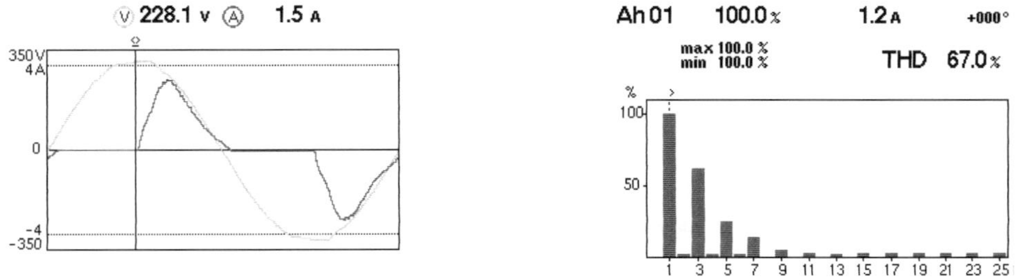

A b

Figure 16. Voltage and current waveforms (a) and harmonic spectrum of current (b) for pendulum jigsaw (idle operation)

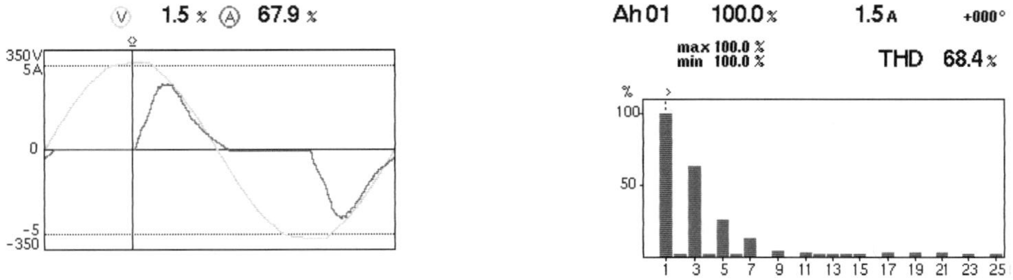

A b

Figure 17. Voltage and current waveforms (a) and harmonic spectrum of current (b) for pendulum jigsaw (load operation, in wood)

For idle operation of pendulum jigsaw (Figure 16) it is noted that the voltage waveform is approximately sinusoidal, having a small total harmonic distortion factor (THD_U=1.5%). The phase angle deviation between voltage and current is very high and inductive.

Harmonic current analysis indicates a high odd harmonic content. The 3rd, 5th and 7th harmonics are the dominating components, their levels far exceed the limits imposed by EMC standards (3rd harmonic level exceeds 60%, 5th harmonic level exceeds 25% and 5th harmonic level exceeds 14%). In these conditions, total harmonic distortion factor of current is extremely high (THD_I=67%) at idle operation of pendulum jigsaw.

In load operation of pendulum jigsaw (wood), the voltage waveform remains sinusoidal and the current waveform is strongly distorted (Figure 16). Total harmonic distortion factor of current has a higher value in this operating mode of jigsaw (THD_I=68.4%), due to the increase of 5th harmonic level.

Table 6. Pendulum jigsaw

Operation mode	*Idle operation*	*Load operation*
Active power, P[W]	213.4	247.9
Reactive power, Q[VAR]	264.2	321.1
Apparent power, S[VA]	339.6	405.7
Power factor, PF[-]	0.628	0.611
Displacement power factor, DPF[-]	0.752	0.734

IOP Conf. Series: Materials Science and Engineering **477** (2019) 012039 doi:10.1088/1757-899X/477/1/012039

Pendulum jigsaw behaves as a strong inductive load both in idle and in load operation. In idle operation, the reactive power is greater than the active power of approximately 1.24 times and increases 1.21 times in load operation (Table 6).

Consequently, the values of power factor and displacement power factor are much lower than the neutral value in both operating modes of pendulum jigsaw (in idle operation: PF=0.628, DPF=0.752, and in load operation: PF=0.611, DPF=0.734). Increasing of reactive (inductive) power consumption and current harmonic distortion leads to worsening of displacement power factor and power factor, at load operation of jigsaw (Table 6).

2.7. Circular saw

Was analyzed a circular saw (DMC 1400) with the following technical data: rated active power 500 W, no load speed 3200 rpm, saw blade diameter 140 mm, rated voltage 230 V, frequency of the supply voltage 50 Hz.

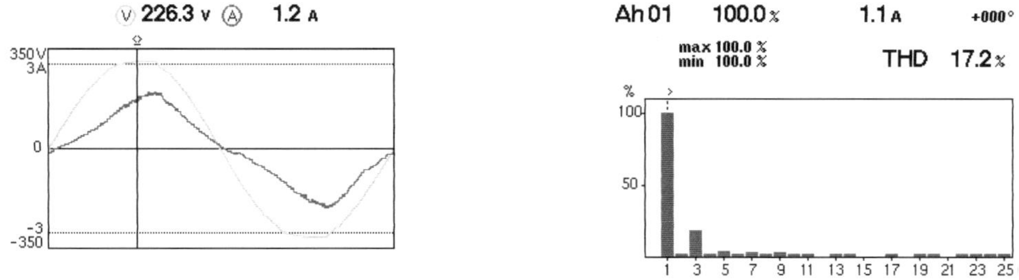

A b

Figure 18. Voltage and current waveforms (a) and harmonic spectrum of current (b) for circular saw (idle operation)

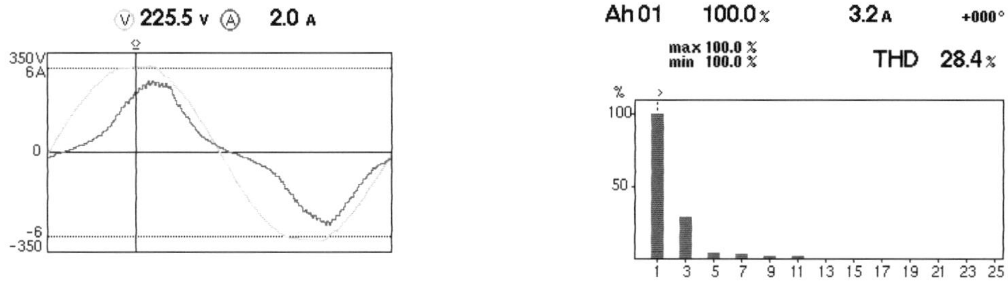

A b

Figure 19. Voltage and current waveforms (a) and harmonic spectrum of current (b) for circular saw (load operation)

Table 7. Circular saw

Operation mode	*Idle operation*	*Load operation*
Active power, P[W]	245.6	761
Reactive power, Q[VAR]	63.5	334
Apparent power, S[VA]	253.7	831.1
Power factor, PF[-]	0.968	0.916
Displacement power factor, DPF[-]	0.983	0.953

Although the harmonic pollution is much lower in the case of circular saw (Figures 18 and 19), compared to the case of jigsaw, total harmonic distortion factor of current exceeds the limit imposed by EMC standards in all operating modes.

Total harmonic distortion factor of current increases 1.65 times in load operation, compared to idle operation, due to the increase of 3rd harmonic level, in particular. Voltage harmonic distortion is extremely low in both operating modes of circular saw.

Reactive power consumption is much lower at this type of saw, which results in much higher values of power factor and displacement power factor (DPF exceeds the neutral value in both analyzed regimes, Table 7).

In load operation (wood, 40 mm thick), active power increases 3 times, reactive (inductive) power increases more than 5 times, and apparent power increases more than 3.2 times, compared to idle operation (Table 7). Consequently, displacement power factor and power factor decrease in this operation mode.

2.8. Milling machine

Milling machine (ASIST AE4F102) had the following technical parameters: rated active power 1020 W, no load speed 11500-30000 rpm, maximum stroke of the milling 40 mm, rated voltage 230 V and frequency of the supply voltage 50 Hz.

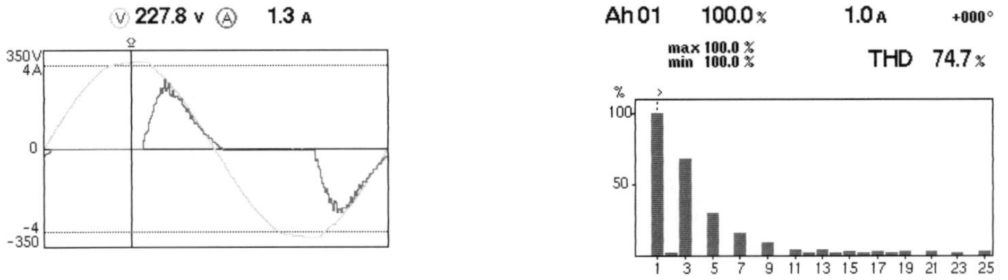

A b

Figure 20. Voltage and current waveforms (a) and harmonic spectrum of current (b) for milling machine (idle operation)

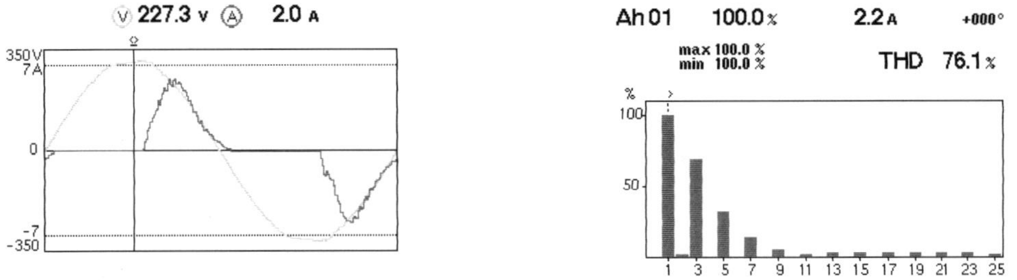

A b

Figure 21. Voltage and current waveforms (a) and harmonic spectrum of current (b) for milling machine (load operation)

The supply voltage is almost sinusoidal, with a very low total harmonic distortion factor (max. 1.5%) in both operating modes of the milling machine (Figures 20 and 21). Instead, the current waveform is pulsating and total harmonic distortion factor has an extremely high value (over 74%) both in idle operation (Figure 20) and in load operation (was used a 7 mm milling, the cutting depth being 20 mm, Figure 21).

The levels of 3rd, 5th and 7th harmonics greatly exceed the compatibility limits (3rd harmonic level exceeds 70%, 5th harmonic level exceeds 30% and 7th harmonic level exceeds 13%).

It is found that reactive (inductive) power is greater than active power of about 1.48 times at idle operation of milling machine, and reached more than 1.56 times at load operation.

Consequently, the power factor and displacement power factor have extremely low values (well below neutral) in both analyzed situations (Table 8).

Table 8. Milling machine

Operation mode	*Idle operation*	*Load operation*
Active power, P[W]	168.7	325
Reactive power, Q[VAR]	249.9	507.8
Apparent power, S[VA]	301.5	602.9
Power factor, PF[-]	0.56	0.539
Displacement power factor, DPF[-]	0.695	0.674

2.9. Belt sander

Belt sander (BS 75) had the following technical parameters: rated active power 760 W, idle speed 420 m/min, sanding belt 76 mm x 533 mm, rated voltage 230 V and frequency of the supply voltage 50 Hz.

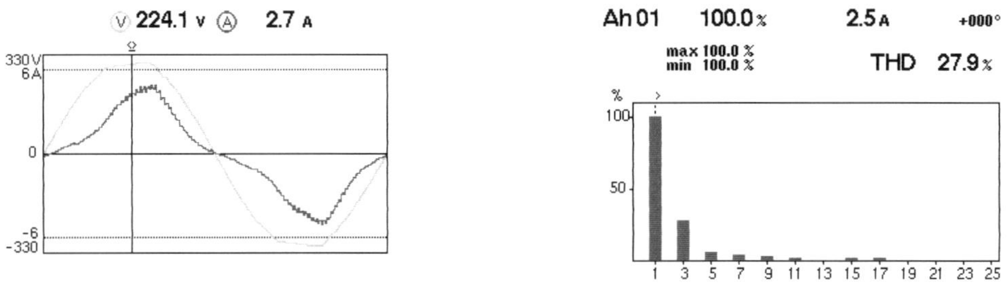

A b

Figure 22. Voltage and current waveforms (a) and harmonic spectrum of current (b) for finishing machine (idle operation)

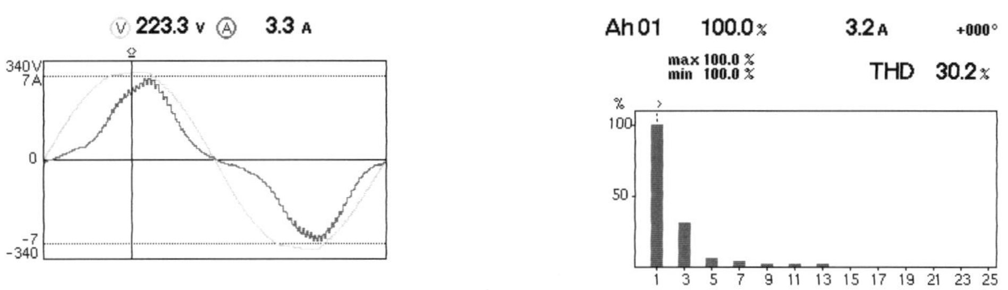

A b

Figure 23. Voltage and current waveforms (a) and harmonic spectrum of current (b) for finishing machine (load operation, wood finishing)

Table 9. Belt sander

Operation mode	*Idle operation*	*Load operation*
Active power, P[W]	549.2	680.3
Reactive power, Q[VAR]	179.5	242.5
Apparent power, S[VA]	577.8	722.2
Power factor, PF[-]	0.951	0.942
Displacement power factor, DPF[-]	0.987	0.985

It is noted that the waveform of supply voltage is sinusoidal, but the waveform of current is distorted, having a total harmonic distortion factor of over 27% both in idle operation (THD_I=27.9%,

Figure 22), as well as in load operation (wood finishing, THD_I=30.2%, Figure 23) of belt sander. In both operation modes of this belt sander, the 3rd harmonic level exceeds the compatibility limit.

The reactive (inductive) power is less than the active power of about 3 times in idle operation, respectively 2.8 times in load operation. For this reason, the values of power factor and displacement power factor are above the neutral value, in both analyzed situations (Table 9). The differences between these power factors are due to the harmonic distortion of the current, being slightly more pronounced in the load operation.

2.10. Planer

The analyzed planer (EH 616) had the following technical specifications: rated active power 600 W, no load speed 16000 rpm, maximum planing width 82 mm, maximum planing depth 2 mm, rated voltage 230 V, frequency of the supply voltage 50 Hz.

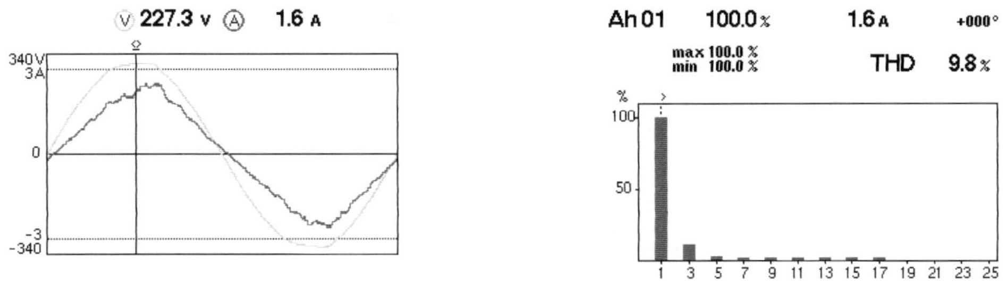

A b

Figure 24. Voltage and current waveforms (a) and harmonic spectrum of current (b) for planer (idle operation)

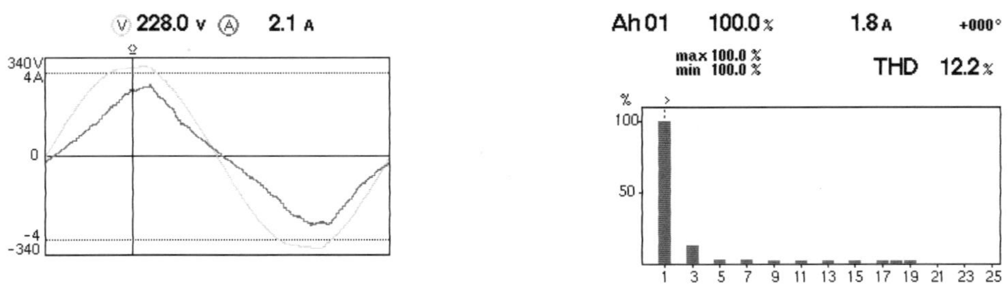

A b

Figure 25. Voltage and current waveforms (a) and harmonic spectrum of current (b) for planer (load operation, wood finishing)

Table 10. Planer

Operation mode	*Idle operation*	*Load operation*
Active power, P[W]	358.6	396.3
Reactive power, Q[VAR]	58.2	71.3
Apparent power, S[VA]	363.3	402.6
Power factor, PF[-]	0.987	0.984
Displacement power factor, DPF[-]	0.992	0.991

Harmonic pollution produced by the operation of planer is lower compared to the previously analyzed power tools. Supply voltage is sinusoidal, both in idle operation and in load operation. Total harmonic distortion factor of current exceeds the compatibility limit in both operating modes of this planing machine, due to the 3rd harmonic (the level of 3rd current harmonic does not fall within the compatibility limit).

Reactive (inductive) power is less than active power of about 6.16 times in idle operation, respectively 5.56 times in load operation (Table 10).

Because of low harmonic pollution and low reactive power consumption, the values of power factor and displacement power factor are much higher than the neutral value, in both operating modes of the planer; the differences between the power factor and displacement power factor are relatively small in this case (Table 10).

3. Conclusions

In order to maintain the harmonic distortion within the limits imposed by electromagnetic compatibility standards, it is necessary to know the sources of electromagnetic disturbances and how the disturbances levels vary depending on time, location and the connected loads.

Analyzing the impact of low-power nonlinear loads on harmonic distortion is very important, because new nonlinear loads (some of low-power i.e., LED lamps) will be progressively incorporated to power networks, with increasing of negative effects for power quality indicators.

The analyzed power tools are characterized by highly distorted waveforms of absorbed currents. Total harmonic distortion factor of the current were influenced by the operation mode (idle or load operation) of power tools, but also depends on their constructive characteristics. Most of the analyzed tools contain single-phase electric motors. Harmonic distortion in this case depends on the characteristics of the motor and varies with the voltage level. Because the 3rd harmonic exceeds the compatibility limits, neutral conductor overheating may occur, which is an important risk of fire.

Harmonic distortion caused by multiple low-power nonlinear loads has mainly negative impacts, such as reducing the life-span of various equipment of the distribution network (transformers, conductors, protective systems, etc.).

The random nature and low utilization factor of power tools (because of the short duration of operations) introduces complexity in the study of their impact on power quality.

Although the absorbed powers are small, the cumulative effect produced by a large number of tools can be substantial, because the currents THD exceeds 20% and can reach 76%. Due to high harmonic distortion, tools should be used only with network filters.

References

[1] Dugan R C, McGranaghan M F, Santoso S and Beaty H W 2003 *Electrical Power Systems Quality*, Second Edition, McGraw-Hill, New York, USA

[2] Arrillaga J, Watson N R and Chen S 2000 *Power System Quality Assessment*, John Wiley and Sons, New York, USA

[3] Masoum M A S and Fuchs E F 2015 *Power Quality in Power Systems and Electrical Machines*, Second Edition, Academic Press, Elsevier, USA

[4] Emanuel A E, Janczak J, Pileggi D J, Gulachenski E M, Root C E, Breen M and Gentile T J 1994 Voltage distortion in distribution feeders with nonlinear loads, *IEEE Transactions on Power Delivery* **9**(1) 79-87

[5] Wang Y J, O'Connell R M and Brownfield G 2001 Modelling and prediction of distribution system voltage distortion caused by nonlinear residential loads, *IEEE Transactions on Power Delivery* **16**(4) 744 - 751

[6] Tokić A, Jukan A and Uglešić I 2011 *Power quality degradation due to low power electronic loads*, 21st International Conference on Electricity Distribution, Frankfurt, Germany, 6-9 June, pp 1-4

[7] Caicedo J E, Romero A A and Zini H C 2017 Assessment of the harmonic distortion in residential distribution networks: literature review, *Ingeniería e Investigación* **37**(3) 72-84

[8] Farooq H, Zhou C, Allan M, Farrag M E, Khan R A and Junaid M 2011 Investigating the Power Quality of an Electrical Distribution System Stressed by Non-Linear Domestic Appliances,

Renewable Energy & Power Quality Journal **1**(9) 283-288
[9] Yong J, Chen L and Chen S 2010 Modeling of Home Appliances for Power Distribution System Harmonic Analysis, *IEEE Transactions on Power Delivery* **25**(4) 3147-3155
[10] Purushothama R N, Narender R N and Santosh A 2012 Modeling and Harmonic Analysis of Domestic/Industrial Loads, *International Journal of Engineering Research and Applications* **2**(5) 485-491
[11] Nicholson G, Gosbell V J and Parsotam A 2007 *Analysis of Harmonic Distortion Levels on a Distribution Network*, Australasian Universities Power Engineering Conference (AUPEC), Brisbane, Queensland, Australia, 25-28 September, pp 186-191
[12] Duarte C H and Schaeffer R 2010 Economic impacts of power electronics on electricity distribution systems, *Energy* **35**(10) 4010-4015
[13] Ghorbani M J and Mokhtari H 2015 Impact of Harmonics on Power Quality and Losses in Power Distribution Systems, *International Journal of Electrical and Computer Engineering* **5**(1) 166-174
[14] Salles D, Jiang C, Xu W, Freitas W and Mazin H E 2012 Assessing the collective harmonic impact of modern residential loads-part I: Methodology, *IEEE Transactions on Power Delivery* **27**(4) 1937–1946
[15] Gil-de-Castro A, Rönnberg S K and Bollen M H J 2014 *A study about harmonic interaction between devices*, 16th International Conference on Harmonics and Quality of Power (ICHQP), Bucharest, Romania, 25-28 May, pp 728-732
[16] Barros J, Diego R I and De Apraiz M 2013 A discussion of new requirements for measurement of harmonic distortion in modern power supply systems, *IEEE Transactions on Instrumentation and Measurement* **62**(8) 2129 - 2139
[17] Ye G, Nijhuis M, Cuk V and Cobben J F G 2017 Stochastic Residential Harmonic Source Modeling for Grid Impact Studies, *Energies* **10**(3) 1-21
[18] Dobrić G, Žarković M and Šošić D 2015 Fuzzy-based Monte Carlo simulation for harmonic load flow in distribution networks, *IET Generation, Transmission&Distribution* **9**(3) 267–275
[19] *** CA8334, Three Phase Power Quality Analyser, technical handbook, Chauvin Arnoux, France, 2007
[20] *** IEC 61000-3-2:2009, Electromagnetic Compatibility (EMC) – Part 3–2: Limits – Limits for Harmonic Current Emissions (Equipment Imput Current ≤ 16 A per Phase).
[21] *** PE 143/2014, Technical norm for limiting deforming regime in transmission and distribution networks (in romanian)

IOP Conf. Series: Materials Science and Engineering **477** (2019) 012040 doi:10.1088/1757-899X/477/1/012040

The impact of electrode material in the treatment of landfill leachate by electrocoagulation process

B N Malinovic[1], T Djuricic[1] and D Zoric[2]

[1]University of BanjaLuka, Faculty of Technology, Stepe Stepanovica 73, 78000 Banja Luka, Bosnia and Herzegovina
[2]EFT Mine and Thermal Power Stanari Ltd., Stanari , Bosnia and Herzegovina

E-mail: borislav.malinovic@tf.unibl.org

Abstract. This paper presents the results of the electrocoagulation treatment of leachate by the application of iron and aluminum as an electrode material. It was examined the influence of the current density and electrolysis duration on the removal efficiency of total nitrogen (N). Duration of electrolysis was 30 minutes, and based on the experimental results it was found that the optimal current density is 25 mA/cm^2. The highest removal efficiency of total nitrogen, achieved under these conditions, amounts 29.58% by the application Fe-Fe electrode pair, while removal of 7.70% is achieved by application Al-Al electrode pair.

1. Introduction

In addition to the municipal and industrial wastewater generated in various technological processes, landfill leachate require special attention and treatment because of the enormous impact on the quality of the environment. Disposal of waste in the form of landfill is still the cheapest and most widely used waste treatment in the world [1].

Sanitary landfill can be defined as a method of disposing of refuse on land without creating nuisances or hazards to public health or safety, by utilizing the principles of engineering to confine the refuse to the smallest practical area, to reduce it to the smallest practical volume, and to cover it with a layer of earth at the conclusion of each day's operation or at such more frequent intervals as may be necessary [2]. Leachate is formed by draining rainwater through the landfill body, in which occurs extraction of soluble colloidal and suspended matter from the waste. While sanitary landfills are often economically and environmentally friendly methods for the disposal of solid waste, the leachate water that emerges poses a threat to the environment, as it can carry toxic pollutants into groundwater and surrounding land [3].

Leachate from landfills are one of the most toxic type of wastewaters, as well as in terms of the selection of appropriate techniques for their purification. It is particularly difficult to process due to its complex and variable composition. All waters generated during landfill operations according to the EU Directive 1999/31/EC on Waste Disposal [4], as well as the Waste Directive 2006/12/EC, should be collected and treated before any discharge into the final recipient [5].

Research has shown that each landfill is a separate system, and in this sense, the composition and the amount of leachate depend only on the characteristics of the landfill [6]. For the treatment of landfill leachate there are a variety of technical and technological solutions. The treatment strategy generally depends on the characteristics of the leachate. Young landfill leachates are generally easier to treat compared with old ones. Leachate from the old landfills contain less organic biodegradable matter and more toxic components, so the purification of the water should be used a combination of several different processes. The quality and quantity of the filtrate, the size of the landfill and the lifetime, the required quality of the effluent with economic analysis are the basic elements that define the selection of the method for treatment of these waters [7].

1.1. Electrochemical methods (literature review)

Electrochemical processes, such as electrocoagulation (EC), electrooxidation (EO), electroreduction and electro-photo-oxidation, are often used for treatment of wastewaters from various industries such as textile, leather, oil industry, and recently have been used in the treatment of leachate.

The EC process involves the *in situ* generation of coagulants through the electro-dissolution of a sacrificial anode, which is usually made of iron or aluminium. EC is one of the electrochemical methods used for treatment many types of wastewater. Numerous studies describe the application of this method for treatment of wastewater, including leachate.

Mussa *et al.* performed decolorization of leachate by researching the efficiency of EC. Several different parameters were tested, such as electrode material (Ni, Ag, Cu, Pd, Pt, Stainless Steel, C), experiment duration, voltage and chloride concentration. Good color removal (70%) was achieved using carbon in 120 minutes of treatment. It has been found that the electrochemical method of EC can be used as a pre-treatment for the leachate treatment [8]. Norma *et al.* applied a combination of electrochemical techniques, EC and EO. They found that these techniques are cost-effective for treatment of leachate from sanitary landfills [9]. Electrochemical methods have long been used to remove not only suspended solids, but also the chemical oxygen demand (COD) and ammonia nitrate (NH_3-N), and can provide good color removal [8]. Ilhan *et al.* researched the application of EC for the treatment of leachate using aluminum and iron electrodes. It has been shown that EC has higher purification effects from the chemical coagulation process and can be successfully used [10]. Xiangdong *et al.* examined the EC efficacy to remove NH_3-N and COD from leachate. The treatment was preform in a batch electrochemical reactor. The experimental conditions were as follows: current density 4.96 mA/cm^2, chloride concentration 2319 mg/L, electrolysis duration of 90 minutes. Treatment was carried out with iron electrodes, where the highest removal efficiency of COD and NH_3-N was achieved, 49.8% and 38.6% respectively [11]. In their research, Bouhezila *et al.* used EC to remove COD, total nitrogen (N_{tot}), color, turbidity and bacteria from leachate, characterized by high COD, high nitrogen concentration and dark color. It was found that increasing the current density (125-500 A/m^2) significantly accelerates the process, with a distance between the electrodes of 2.8 cm and a stiring speed of 150 rpm. The removal efficiencies of COD, N_{tot}, color and turbidity was 68%, 15%, 28% and 16% respectively, when iron electrodes are used. The consumption of electricity was 0.0196 kWh/L [12]. Similar to the research of Ilhan *et al.*, in the research of Yadav *et al.* also was used iron and aluminum electrodes for EC treatment of leachate. The aim of the paper was to optimize parameters such as current density, electrolysis duration and pH value to achieve maximum COD removal. By using iron electrodes, COD removal efficiency was 56% at a current density of 466 A/m^2. Optimum electrolysis time was 90 minutes and a pH 6. The iron electrode improved higher COD removal efficiency in comparance with an aluminum electrode (47%). EC was assessed as a good pre-treatment [13].

2. Material and methods

The experimental part of the research consists in the application of EC treatment for purification of leachate. Sample of leachate was taken from the regional sanitary landfill - "JP Regional Landfill DEP-OT," Ramici, Banja Luka. Leachate drainage into pipes leading into the tank (Figure 1), and then in the reverse osmosis plant for treatment of landfill leachate.

Figure 1. Drainage pipe and landfill leachate tanks (regional sanitary landfill Ramici)

Experimental setup is shown on the Figure 2. The batch reactor for EC of 500 cm^3 capacity made from polypropylene with constant mixing (400 rpm) was used, combined with two electrodes of the same area surface. The dimensions of the electrodes are 40x40x1 mm. The total useful surface was 95.76 cm^2, and the distance between electrodes 3 cm. Electrodes were connected to digital power source (Atten, APS3005SI; 30V, 5A).

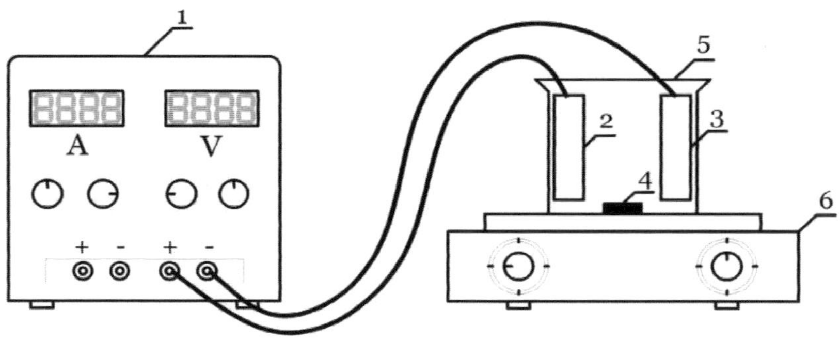

Figure 2. Schematic view of electrochemical reactor (1 – electric power source; 2 – anode; 3 – cathode; 4 – magnetic stir bar; 5 – electrochemical cell; 6 – magnetic stirrer)

All the experiments were performed at an ambient temperature, with the leachate initial volume of 400 cm^3 and with concentration of supporting electrolyte γ_{NaCl}=5 g/L. As an electrode material was used iron, Fe (steel - EN 10130-91; max. 0.08% C, max. 0.12% Cr, max. 0.45% Mn, max. 0.60% Si) and aluminum, Al (aluminum - Al 99.5/ENAW-1050 A; max. 0.25% Si, max. 0.40% Fe, max. 0.05% Cu, max. 0.05% Mn, max. 0.05% Mg, max. 0.05% Ti, max. 0.07% Zn, min. 99.50% Al), which meet the required standards. After each EC treatment leachate was collected and filtered. The samples of leachate before and after treatment were analyzed on the content of N_{tot} in accordance with standard method [14]. The determination of total nitrogen was carried by the persulphate method in standard cuvettes (Lovibond 53 55 50, Germany).

3. Results and discusion

In this paper has shown a part of results of case study for the Ramici regional landfill in Banja Luka, Bosnia and Herzegovina [15]. Results of the electrochemical treatment of landfill leachate are shown through removal efficiency, *E* (%), whose ratio could be described by following equation:

$$E = \frac{\gamma_i - \gamma_f}{\gamma_i} \times 100[\%] \quad (1)$$

where are γ_i and γ_f the initial and the final value of parameter which efficiency is determined.

The following figures shows the impact of current density on removal efficiencies of N_{tot} for Fe-Fe electrode pair (Figure 3) and Al-Al electrode pair (Figure4). In this experiment was used current densities of 5, 10 i 25 mA/cm^2.

Figure 3 shows that the highest removal efficiency is achieved at a current density of j=25 mA/cm^2 (E=29.58%). Although it was expected that the increase in the current density would lead to a constant increase in efficiency, the figure shows a decrease in efficiency. The efficiency achieved at a current density of 10 mA/cm^2 (E=10.28%) is lower than when the current density is 5 mA/cm^2 (E=16.40%).

For case of aluminum electrodes the highest removal efficiency was achieved at the lowest current density, j=5 mA/cm^2 and amounts 22.41%, and at j=25 mA/cm^2 amounts 7.70% (Figure 4). It can be seen that the iron electrode has a slightly greater removal efficiency compared to the aluminium electrode for 30 minutes of treatment. In further experiments it was applied current density of 25 mA/cm^2.

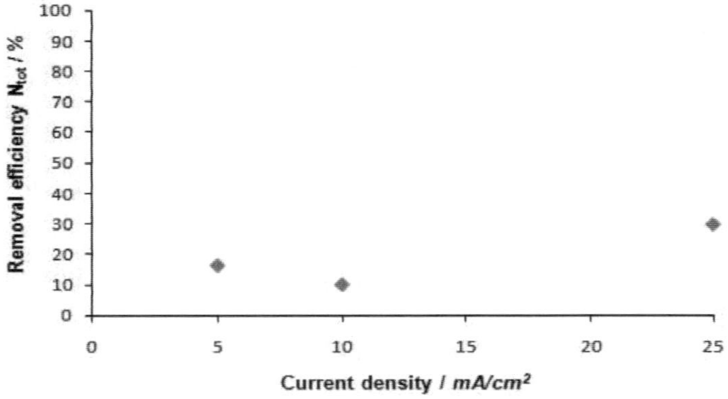

Figure 3. Removal efficiency in relation to the current density by using iron electrodes (γ_{NaCl}=5 g/L, t=30 min, stirring 400 rpm)

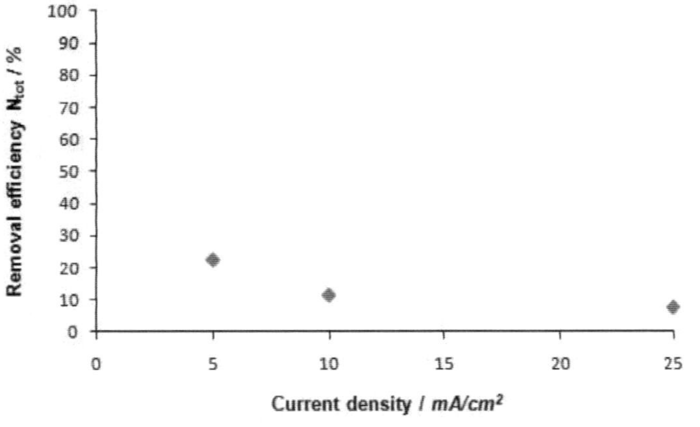

Figure 4. Removal efficiency in relation to the current density by using aluminum electrodes (γ_{NaCl}=5 g/L, t=30 min, stirring 400 rpm)

Figures 5 and 6 shows the impact of electrolysis duration (10; 20; 30 min) on the N_{tot} removal efficiency. The highest removal efficiency for iron electrodes was achieved at the electrolysis duration of 20 minutes and amounts 40.59% (Figure 5). For aluminium electrode by extending of electrolysis duration, the removal efficiency decreases. The highest removal efficiency for was achieved at 10 minutes of electrolysis and amounts 16.39% (Figure 6).

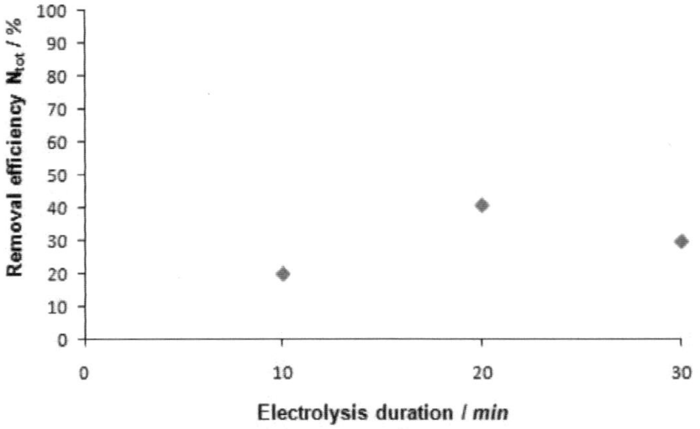

Figure 5. Removal efficiency in relation to the electrtolysis duration by using iron electrodes (γ_{NaCl}=5 g/L, j=25mA/cm^2, stirring 400 rpm)

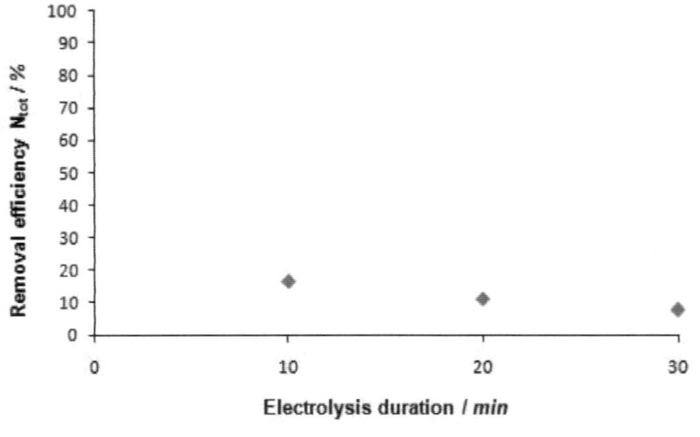

Figure 6. Removal efficiency in relation to the electrtolysis duration by using aluminum electrodes (γ_{NaCl}=5 g/L, j=25mA/cm^2, stirring 400 rpm)

4. Conclusions

It can be concluded that at lower current densities removal efficiency of N_{tot} was higher for aluminum electrodes, while at higher current density and longer electrolysis duration the removal efficiency is greater for the iron electrode. From the results we can also see that with relatively low current density and short treatment time, good purification results have been achieved. Therefore, electrocoagulation can be successfully applied as a pre-treatment for landfill leachate. In further studies it is necessary to consider the application of post treatment, such as electro-oxidation.

Acknowledgement
This work was supported in part by the Ministry of Science and Technology of the Republic of Srpska under Project 19/6-020/961-19/15.

References
[1] Ismail S and Manaf L A 2013 The challenge of future landfill: A case study of Malaysia, *Journal of Toxicology and Enviromental Health Sciences* **5**(6) 86-96
[2] Raghab S M, El Meguid A M and Hegazi H A 2013 Treatment of leachate from municipal solid waste landfill, *HBRC journal* **9** 187-192
[3] Llabanowski J, Pallier V and Feuillade-Cathaliufad G 2010 Study of organic matter during coagulation and electrocoagulation processe: Application to a stabilized landfill leachate, *Journal of Hazardous Materials* **179** 166-172
[4] ***Council Directive 1999/31/EC European Union (1999)
[5] ***Directive on waste 2016/12/EC European Union (2006)
[6] Ehrig H and Robinson H 2010 *Landfilling: Leachate treatment, in Solid Waste technology & Menagement,* Volume 1 & 2 (ed T H Christensen), Chichester, UK: John Wiley & Sons, Ltd.
[7] Fudala-Ksiazek S, Luczkiewicz A, Quant B and Olanczuk-Neyman K 2011 *The effectiveness of nitrification and denitrification with CO-treatment wastewarer with landfill leachate*, Thirteenth International waste managment and lanfill symposium, Cagliary, Italy: IWWG-International waste working group, pp 321-329
[8] Mussa Z H, Othman M R, Abdullah P and Nordin N 2013 Decolorization of landfill leachate using electrochemical technique, *International Journal of Chemical Science* **11**(4) 1636-1646
[9] Norma D, Fernandes A, Pacheco M J, Ciriaco L and Lopes A 2012 Electrocoagulation and Anodic Oxidation Intregated Process to Treat Leachate from Portuguese Sanitary Landfill, *Portigalie Electrochimica Acta* **30**(3) 221-234
[10] Ilhan F, Kurt U, Apaydin O and Talha Gonullu M 2008 Treatment of leachate by electrocoagulation using aluminum and iron electrodes, *Journal of Hazardous Materials* **154** 381-389
[11] Xiangdong L, Junke S, Jiandong G, Zhichao W and Qiyan F 2011 Landfill leachate treatment using electrocoagulation, *Procedia Environmental Sciences* **10** 1159-1164
[12] Bouhezila F, Hariti M, Lounici H and Mameri N 2011 Treatment of the OUED SMAR town landfill leachate by an electrochemical reactor *Desalination* **280**(1) 347-353
[13] Yadav J S and Dikshit A K 2017 Stabilized Old Landfill Leachate Treatment Using Electrocoagulation, *Enviroment Asia* **10**(1) 25-33
[14] ***APHA. Standard Methods for the Examination of Water and Wastewater, New York: American public healt association, 1999
[15] Malinovic B N, Pavlovic M G, Djuricic T, Bjelic D and Neskovic Markic D 2018 *Electrochemical treatmaent of leachate: case study for sanitary landfill "Ramici", Banja Luka*, XX International Conference YuCorr, Tara Mountain, Serbia, May 21-24, pp 16-27

IOP Conf. Series: Materials Science and Engineering **477** (2019) 012041 doi:10.1088/1757-899X/477/1/012041

Third approach in work loss phenomenon

P Petrović, M Kotur and D Kardaš

University of Banja Luka, Faculty of Mechanical Engineering, Stepe Stepanovića str., no.71, 78 000 Banja Luka, Bosnia and Herzegovina

E-mail: danijela.kardas@mf.unibl.org

Abstract. One of the most widespread irreversible processes in technique is the heat transfer from the higher to the lower body temperature. The main reason for this is due to the necessity of significant temperature difference in order to achieve a satisfactory process speed. Maximum work for the same heat amount (the same temperature level) will achieve reversible (ideal) system. The heat exergy represents maximum part of the heat that can be converted into work. In the system in which real processes take place, there is an increase of entropy and loss of work (exergy). This is the main reason for the low efficiency of the system for the transformation of heat energy into work. This phenomenon can be proven by different approaches. The first law of thermodynamics is mainly used for the analysis of energy utilization. The second law of thermodynamics and exergy analysis as its consequence, is more relevant for qualitative aspects of used or consumed energy. This paper presents third approach how exergy losses can be defined and its impacts on energy efficiency of the system.

1. Introduction

Exergy analysis is known as a useful tool that can be utilized to evaluate the performance of any engineering system [1]. Energetic and exergetic assessment of a process deals with the first and second law of thermodynamics, respectively. Eventhough, first law of thermodynamics is mainly used, it doesn't posess ability to analyze qualitative aspects of used or consumed energy. Exergy, as a consequence of second law of thermodynamics, can be used for such a analyzes.

Exergy is the maximum theoretical useful word (shaft work or electrical work) obtainable from a system as this is brought into thermodynamic equilibrium with the reference environment while the system interacts only with this environment. Alternatively the exergy can be defined as the minimum theoretical work required to form a quantity of matter from substances present in the reference environment and to bring this matter to a specific thermodynamic state. The exergy is a measure of the departure of the state of a system from the state of the reference environment [2].

Exergy analysis [3-10] is a methodology that uses the conservation of energy principle (embodied in the first law of thermodynamics) together with non-conservation of entropy principle (embodied in the second law) for the analysis, design and improvement of energy and other systems. Since exergy is always a measure of the approach to ideal, exergy analysis enables design more efficient energy systems by reducing inefficiencies.

Energy efficiencies do not define how nearly performance aproaches ideality but exergy efficiency does. There are now many papers introducing principles and methods of exergy analysis [1-11]. Exergy destruction and irreversibilities which is similar in a sense of loss during the process gives more realistic system losses than energy losses [11]. Exergy analysis can, therefore, indicate the possibilities for thermodynamic improvement. The term exergy was first mentioned at the conference in Lindau in 1953 by Z. Rant. The term exergy originates from the "ex" before "ergy", and the meaning is "from" and "work", and characterizes the work that can be obtained from the system [12].

Heat (as a form of energy) can be divided into into one part that can produce work (its Exergy), and one part that can not be used to produce work, which is commonly referred to as Anergy [13]. For heat Q at temperature T, the following decomposition can be made with reference to the ambient temperature T_o:

$$Energy = Exergy + Anergy$$

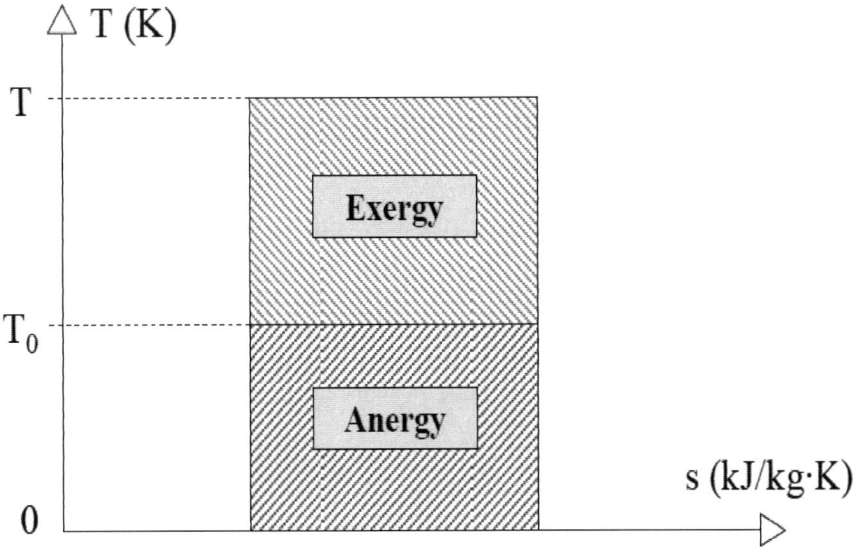

Figure 1. Decomposition of Energy (Heat) into Exergy and Anergy [13]

First law of thermodynamic says that amount of energy that enters the system is the same one that comes out of the system in accordance with the law of conservation of energy. Second law says that the amount of entropy that comes out of the system is greater than the one that enters the system in accordance with the law of increasing entropy.

One of the most widespread irreversible processes in technique is the heat transfer from the higher to the lower body temperature. The main reason for this is due to the necessity of significant temperature difference in order to achieve a satisfactory process speed. All technical systems in which heat is converted into work (motors, thermal power plants, etc.) still are considerably behind ideal ones. All these processes are irreversible. The heat exergy represents maximum part of the heat that can be converted into work. In the system in which real processes take place, there is an increase of entropy and loss of work (exergy). This is the main reason for the low efficiency of the system for the transformation of heat energy into work. The maximum work for the same amount of heat (the same temperature level) will achieve the ideal (reversible) system.

The classification of exergy can be done in the way similar to classification of energy, with some exceptions. In this paper, emphasis is on thermodynamic and heat converted into work relations so thermo-mechanical exergy can be classified as it's shown on Figure 2.

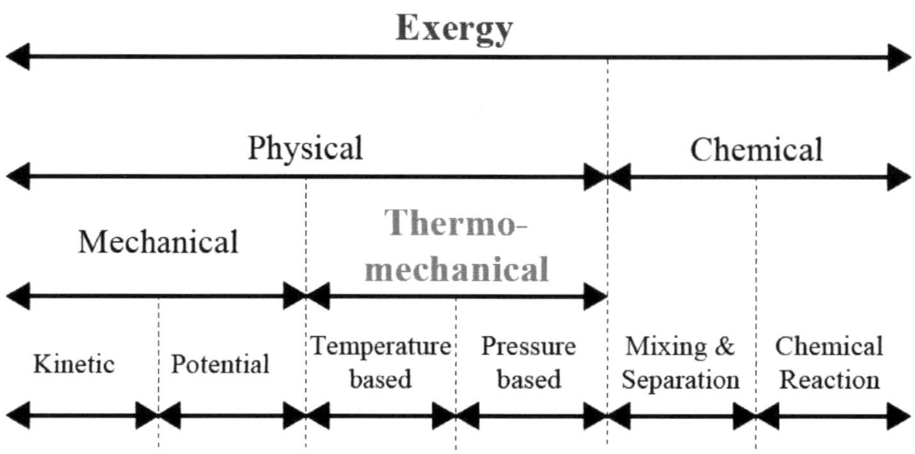

Figure 2. Classification of exergy for pVT systems [13]

The exergy consumption during a process is proportional to the entropy created due to irreversibilities associated with the process. Exergy is also a measure of the potential to cause change, as a consequence of not being in equilibrium relative to the reference environment [14]. It can be said that exergy is a concept that connects the universal concept of entropy to the conditions on Earth through temperature. The loss of exergy (work) is determined as a product of ambient temperature and entropy increment within the system's control boundary: This term in the literature is called the "Gouy - Stodola rule" according to French physicist Gouy and the Slovak engineer Stodola.

This is analitical approach for exergy analyze. Taking into consideration ambient temperature as one of the criteria for thermodynamic analyzes of systems which are operating in the temperature range "far from the ambient temperature" led some researchers to doubts about the validity of the exergetic method as a tool for thermodynamic analysis. A more extensive critique of exergetic analysis is presented in [15] and [16].

2. Analitical approach

Entropy of an isolated system grows if even the smallest part of its process is irreversible. This is always the case in the conditions on Earth, where the changes of the state are irreversible, and the consequence of this is the so-called "degradation" of energy. What really means energy "degradation"? This is to say that the heat as a form of energy has a lower quality than other forms (mechanical, electric, etc). This is in cases where heat participates in the process and there is irreversibility and losses (conversion various forms of energy into heat at the ambient temperature) [17]. In fact, there is no purely mechanical process because always with every mechanical movement friction occurs. Duo to that one part of the mechanical energy is transformed into a heat and such a mechanical process (with friction, with the appearance of heat) is not reversible. So, a purely mechanical process would be without friction, and in that case it would be completely reversible. Friction is a cause of irreversibility, a cause of energy "degradation", and hence itself makes a non-return process. A process is reversible if, after the process is completed, all system and all its elements and surroundings can be restored to their respective initial states. When there is a friction in process, mechanical energy (spent work) turns into heat:

$$W = Q. \tag{1}$$

Based on well-known equation for the Carnot cycle efficiency of an engine operating between two fixed temperatures:

$$\eta_c = \frac{Q - Q_o}{Q} = \frac{W_k}{Q} = \frac{T - T_o}{T}, \tag{2}$$

It is possible get maximum work in reversible process:

$$W_{max} = W_k = \eta_c Q = Q\frac{T - T_o}{T}, \tag{3}$$

where W_k is obtained work which is also maximum work of the system because it is obtained with Carnot cycle; T_o is ambient temeprature and T temeprature of the system. Difference between spent work duo to friction (W) and maximum work (W_{max}) which could be obtained from heat (Q) with Carnot cycle is:

$$W - W_{max} = Q - Q\frac{T - T_o}{T} = Q\left(1 - \frac{T - T_o}{T}\right) \tag{4}$$

$$\Delta W = W - W_{max} = Q\frac{T_o}{T} = \frac{Q}{T}T_o = \Delta S T_o \tag{5}$$

From this it can be seen that the difference in work (i.e. loss in work due to irreversibility) is product of the ambient temperature and the increase in entropy (due to friction). So, it can be said that the work loss due to irreversibility is proportional to the entropy increase. Expression 5 is also known as a Gouy-Stodola theorem for the lost work. Hence, lost work is proportional to the entropy generation rate.

3. Graphical approach

In order to prove the necessity of ambient temperature in the criteria for thermodynamic evaluation of systems which are operating in the temperature level "far from" the ambient temperature or in a technological process when they do not have a direct connection with the environment, in [18], [19] is developed graphical interpretation of exergy losses in T-s diagram. Figure 3 represents heat exchange between flows of the same substance (same mass flows) which change a phase. In second stream is process of condensation and in first is evaporation. The temperature difference is due to different pressures in flows.

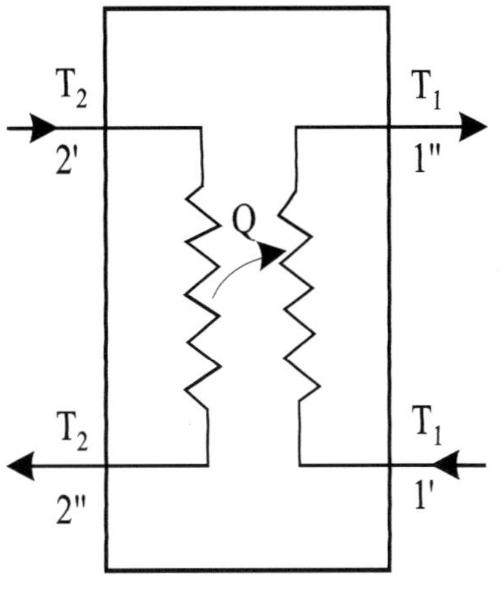

Figure 3. Exchange of heat between flows of the same substance

If in the T-s diagram (Figure 4) there is a change state of both flows, where is $T_2 > T_1 > T_0$, it follows that the surface 2'2"a'b" represents the heat taken from 1kg of flow two, and the surface 1'1"c'a' presents heat brought to 1kg of flow one. Due to the equality of heat, the equality of surfaces is also followed (the sum of 2'2"a'b'= $Q_2 = Q_1$ = surface 1'1"c'a"). The equality of these surfaces follows from the applied structures of the rectangle a'c'e2". On the same heat exergy diagram, the surfaces are represented by flow 2 (E_2 = surface 2'2"ab) and flow 1 (E_1 = surface1'1"ca). The loss of work during heat transfer from a higher to a lower temperature level is equal to the difference between the exergy, that is, the surface 2'2"ab minus the surface1'1"ca. Graphical construction applied in Figure 4 also shows that the difference in these surfaces is equal to the surface of bccb'. From the same image, it is seen that this surface is equal to the product of the ambient temperature and the increase in entropy in the system, which is the Gouy-Stodola rule.

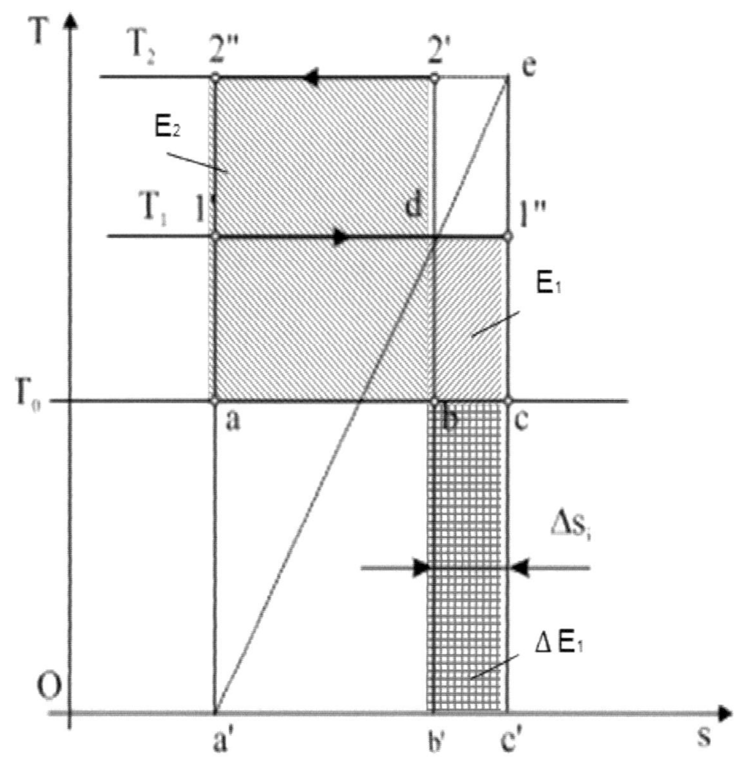

Figure 4. *T-s* diagram of heat exchange of the same substance flows at $T_2 > T_1 > T_0$

4. New graphical approach

For this approach, we can use the same logic applied in the second approach. If the heat taken from 1 kg of the substance of flow 2 (surface 2'2"a'b') is shown as the sum of exergy and anergy,

$$Q_2 = E_2 + A_2 \tag{6}$$

it follows that the surface 2'2"a'b" is equal to the sum of the surfaces 2'2"ab plus the surface of abb'a' where the heat anergy is equal to the surface of abb'a' = A_2. Anergy is a part of the heat energy that can not be converted into work and which is passed on to the environment. In analogy, the heat received by 1 kg of flow 1 (surface 1'1"c'a") can be represented as the sum of exergy and anergy, and consequently the heat anergy is equal to the acc'a' surface = A_1.

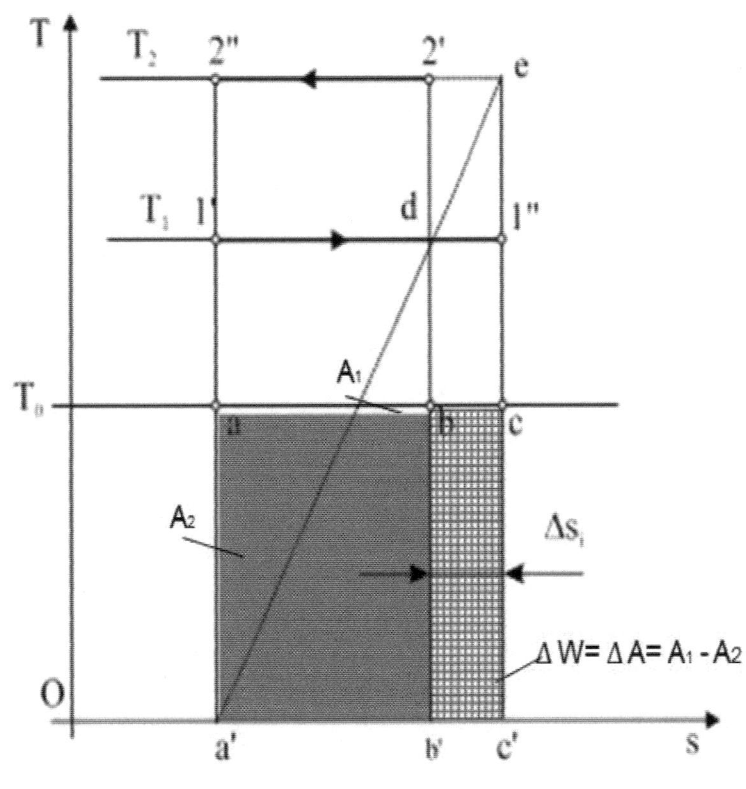

Figure 5. New graphical approach: *T-s* diagram of heat exchange of the same substance flows at $T_2>T_1>T_0$

As work loss due to irreversibility can be expressed as difference between the exergy input and output exergy, it follows that:

$$\Delta E = \Sigma E_u - \Sigma E_i = E_2 - E_1 \tag{7}$$

Exergy can be written as:

$$E_1 = Q_1 - A_1 \tag{8}$$

$$E_2 = Q_2 - A_2. \tag{9}$$

So it follows that:

$$\Delta E = Q_2 - A_2 - (Q_1 - A_1). \tag{10}$$

Work loss i.e. exergy loss can be written using anergy:

$$\Delta E = \Delta W = A_1 - A_2 \tag{11}$$

When some amount of heat (equal to the sum of its exergy and anergy) passes from a higher to a lower temperature level, then its exergy decreases and anergy increases. The difference in the heat anergy of the lower and higher temperature levels (after and before the process) is equal to the loss of work (exergy) due to the irreversibility of the process.

This follows that:

$$\Delta E = A_1 - A_2 = T_0 \Delta S_S \tag{12}$$

which is in T-s diagram (Figure 5) the surface acc'a'=A_1 minus the surface abb'a'=A_2. If we use Gouy – Stodola rule that is the product of the ambient temperature and the increase in the entropy of the system.

In both areas, under and above of ambient temperature, work loss due to irreversibility in heat exchange with temperature differences, it can be determined analytically by the expression:

$$\Delta W = T_0 \frac{T_2 - T_1}{T_1 T_2} Q \tag{13}$$

Showing exchange of heat in graphical methods given in Figures 4and 5 provides comprehensive interpretation of this phenomenon. By bringing closer temperatures T_2 and T_1 to the absolute zero, losses of work are inclined to infinity.

5. Conclusion

Some natural phenomena can be defined by different approaches. Today in thermodynamics we use different definitions of second law: Carno, Clausius and Thomson. This is almost possible with analysis of work loss due to irreversibility. So far, there is the Gouy - Stodol's rule ($\Delta E = T_0 \Delta S_s$), then graphic interpretation of Gouy - Stodol's rule which has been published in [18], and in this paper is presented the new graphic method. Gouy - Stodo 's rule is presented through the difference of anergy, before and after the process ($\Delta E = A_1 - A_2$) when the heat exchange is in the temperature range above ambient temperature. In fact, it is a slightly simpler approach than the approach presented [18]. In the temperature range below the ambient temperature, the graphical approach presented in [18], it can not be simplified with an exposed logic as it has been done for a temperature range above the ambient temperature. In both areas, under and above of ambient temperature, work loss due to irreversibility in heat exchange with temperature differences, it can be determined analytically (13).

References

[1] Boroumand Jazi G, Saidur R, Rismanchi B and Mekhilef S 2012 A review on the relation between the energy and exergy efficiency analysis and the technical characteristic of the renewable energy systems, *Renewable and Sustainable Energy Reviews* **16** 3131-3135

[2] Tsatsaronis G and Cziesla F 2009 *Exergy and Themodynamic Analysis*, in Exergy, Energy System Analysis and Optimization (ed C A Frangopoulos), EOLSS, vol 1

[3] Al-Sulaiman A F, Hamdullahpur F and Dincer I 2011 Greenhouse gas emission and exergy assessments of an integrated organic Rankine cycle with a biomass combustor for combined cooling, heating and power production, *Appl. Therm. Eng.* **31**(4) 439-446

[4] Ozturk M 2014 Energy and exergy analysis of a combined ground source heat pump System, *Appl. Therm. Eng.* **73** 362-370

[5] Çakır U and Çomaklı K 2016 Exergetic interrelation between an heat pump and components, *Appl. Therm. Eng.* **105** 659-668

[6] Ezzat M F and Dincer I 2016 Energy and exergy analyses of a new geothermal–solar energy based system, *Solar Energy* **134** 95-106

[7] Yamankaradeniz N 2016 Thermodynamic performance assessments of a district heating system with geothermal by using advanced exergy analysis, *Renewable Energy* **85** 965-972

[8] Saloux E, Sorin M and Teyssedou A 2016 Modeling the exergy performance of heat pump systems without using refrigerant thermodynamic properties, *Energy and Buildings* **112** 69-79

[9] Wall G 2003 Exergy tools, *Proc. Inst. Mechanical Engineers, Part A: J. Power and Energy* **217** 125-136
[10] Hepbasil A 2012 Low exergy (LowEX) heating and cooling systems for sustainable buildings and societies, *Renewable and Sustainable Energy Reviews* **16** 73-104
[11] Suleman F, Dincer I and Agelin-Chaab M 2014 Energy and exergy analyses of an integrated solar heat pump system, *Appl. Therm. Eng.* **73** 559-566
[12] Baehr H D 1965 *Energie und Exergie; die Anwendung des Exergiebegriffs in der Energietechnik*, VDI-Gesellschaft Enegietechnik Düsseldorf
[13] Gundersen T 2009 *An Introduction to the concept of exergy and energy quality*, Norwegian University of Science and Technology Trondheim, Norway
[14] Rosen M A and Bulucea C A 2009 Using Exergy to Understand and Improve the Efficiency of Electrical Power Technologies, *Entropy* **11** 820-835
[15] Alefeld G 1988 *Probleme mit der Exergije*, BWK 40 Nr.3, s.72/87
[16] Alefeld G 1988 *Energie I,II Hauptsatz der Thermodynamik*, BWK 40 (1988) Nr.11
[17] Malić D 1977, *Termodinamika i Termothenika*, University in Belgrade, Belgrade, Serbia
[18] Petrović P 2000 *Prikazivanje gubitaka rada u T-s dijagramu (Grafička interpretacija zakona Goui-Stodola)*, XXXI Kongres o KGH, Belgrade Serbia
[19] Petrović P, Tica G and Kotur M 2010 Graphical presentation of losses of work in T-s diagram, *Annals of Faculty of Engineering Hunedoara* **VIII**(3) 355-356

IOP Conf. Series: Materials Science and Engineering **477** (2019) 012042 doi:10.1088/1757-899X/477/1/012042

Linear regression equation of mining terrain curvatures caused by hard coal excavation from a few seams and their approximated values by the use of smoothed spline

J Orwat[1,2]

[1]Silesian University of Technology, Department of Mining and Geology, Akademicka str., no. 2, 44100 Gliwice, Poland
[2]Silesian University of Technology, Department of Construction, Akademicka str., no. 5, 44100 Gliwice, Poland

E-mail: justyna.orwat@polsl.pl

Abstract. An article presents the own research results of curvatures shaping on the mining terrain surface. The curvatures have been caused by the hard coal exploitation which was conducted within five years in two coal seams situated at the great depth. The curvatures were observed throughout surveys carried out on the points of measuring line established parallel to the edges of mining pits. There was used a longwall excavation system with the cave-in of roof rocks. At the observational line were performed the geodetic measurements of the points heights and the distances between these points, at the time intervals amounted from three to four months. It has been selected such measuring cycle which represents the excavation influences coming from an exploitation of two coal beds. For curvatures measured after the termination of exploitation in these coal seams, it has been designated their average graph by the use of a mean-squared approximation and the smoothed spline functions. Then, it has been determined the equation of linear regression for the measured values of curvatures and their average values. On the base of this equation, it's possible a making of usefulness assessment of the smoothed splines to designation of average courses of measured curvatures.

1. Introduction

Underground mining exploitation of hard coal seams can cause the discontinuous deformations of terrain surface e.g. ground steps [1], sinkholes [2], [3] and/or the continuous deformations of terrain surface such as: subsidence, inclinations, curvatures, horizontal displacements and horizontal strains.

It should be emphasized that there is a some group of objects which are especially sensitive to the mining terrain curvatures. The linear and large surface objects belong to it, for instance highways, railway lines, pipelines, production and storage halls, and many others.

Builders in the mining areas, which design the objects like these and the protections on the dynamic influences of mining exploitation, take into account the values of deformations indicators (the curvatures especially [4]) forecasted by the use of selected theory of exploitation influences, e.g. the geometric-integral theories [5], [6], the cellular automata theory [7], [8].

To check the protections correctness and the credibility of made forecast, they compare the prognosis results with the average values of real deformations indicators, which are determined from the results of geodesic surveys.

IOP Conf. Series: Materials Science and Engineering **477** (2019) 012042 doi:10.1088/1757-899X/477/1/012042

The obtaining way of the average graphs of measured deformations indicators can be different. So far the orthogonal polynomials were used in an approximation of average graphs of measured subsidence, inclinations, horizontal strains [9] and curvatures [10]. The basic problems with their use are:

- an unknown, optimum and proper degree of polynomial within a deformation indicator;
- a division necessity of whole graph of deformation indicator into some parts;
- the improper results in case of curvatures.

For these reasons, in the article was shown an obtaining method of the average graphs of observed curvatures [11], [12] by the use of smoothed spline functions [13]. The measured curvatures have been determined from the results of geodesic measurements, which were conducted before and after hard coal exploitation in two coal seams, at the great depths.

To check, how much the curvatures average graph obtained from an approximation done by the use of smoothed splines corresponds their real graph, it was determined the linear regression equation between these graphs of curvatures.

2. Observed graph of curvatures

This section presents:

- an instance of hard coal exploitation which was conducted in Poland, in the Upper Silesian Coal Basin, in several coal seams, at the great depths;
- a characteristic of geodesic surveys which were carried out at the observational line established above the exploitation;
- an obtainment method of curvatures measured graph.

2.1. Hard coal exploitation

Some mining plant, located in the southern part of Poland and in the Upper Silesian Coal Basin, conducted an exploitation of hard coal deposits in two seams numbered *338/2* and *358/1* by the use of longwall system with the roof rocks cave-in.

This extraction system is regarded as the most effective in deposit obtaining and optimal in costs. Its appropriate configuration allows also to protect the building objects from the excessive mining damages [14].

Inclination of these hard coal seams was amounted to 6° in the south-eastern direction. Extraction was conducted in the years 2001 – 2006 at the large depths.

The first coal seam *338/2* was exploited in the years 2001 – 2006 by the use of five longwalls: *B-1*, *B-2*, *B-3*, *B-4* and *C-1*. This hard coal seam was located at the depth of 600 m and has the thickness till 3.0 m.

Exploitation in the second coal seam *358/1* was carried out in the years 2002 – 2006 by the use of seven longwalls: *B-1*, *B-2*, *B-3*, *B-4*, *B-5*, *B-6* and *B-7*. The *358/1* hard coal seam was located at the great depth of 1000 m and has the thickness till 2.9 m.

2.2. Geodesic surveys

Mining terrain curvatures, caused by an exploitation of two hard coal seams *338/2* and *358/1*, were observed by the geodesic surveys conducted at the measuring line *№ 8*. This line consisted of 53 measuring points and was located perpendicularly to the runs of the *B-2* longwall in the *338/2* seam. The average length of measuring segments was amounted 35 m.

The points heights measurements were carried out in reference to the 'Kronstadt' altitude system, by the use of precise leveller with an accuracy of 0.1 mm. It was used the double geometric levelling from the middle of measuring segments.

The distances between the subsequent observational points were measured by the use of distance-meter device with an accuracy of 1 mm.

2.3. Obtainment method of curvatures observed graph

On the base of differences of points heights, which were measured before an exploitation has been stared and after an exploitation ending of two coal seams, there were calculated the subsidence of measuring points from the following formula:

$$S_i^{meas} = H_i^A - H_i^B, \tag{1}$$

where:

H_i^A – the absolute height of the *i* point after termination of exploitation [*mm*];

H_i^B – the absolute height of the *i* point before beginning of exploitation [*mm*];

S_i^{meas} – measured subsidence of the *i* point [*mm*].

The subsidence graph of measuring points, registered along the observational line *№ 8* after the termination of exploitation in the *338/2* and *358/1* coal seams, has been shown at the Figure 1.

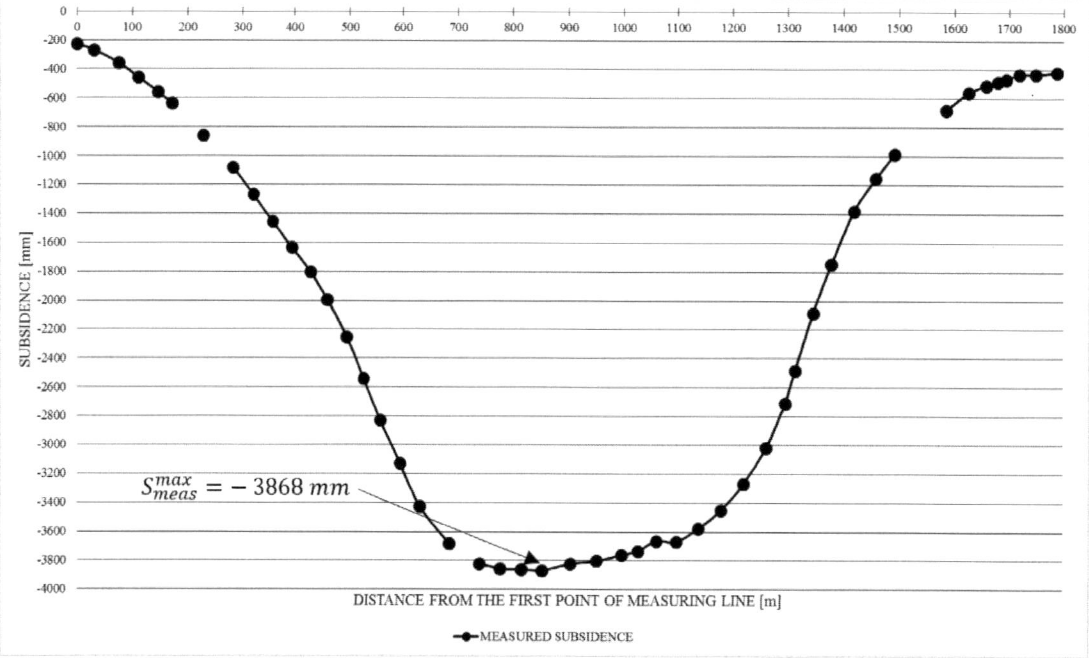

Figure 1. The subsidence graph noticed along the measuring line *№ 8* and caused by exploitation of two coal seams (*338/2, 358/1*)

The curvatures values, which were observed on two neighbouring segments of measuring line *№ 8*, have been calculated from the subsidence of three neighbouring points divided by an average length of these segments:

$$C_{i-1,i,i+1}^{meas} = 2\frac{S_{i-1}^{meas} - 2S_i^{meas} + S_{i+1}^{meas}}{L_{i-1,i}^{meas} + L_{i,i+1}^{meas}}, \tag{2}$$

where:

$C_{i-1,i,i+1}^{meas}$ – curvature measured on the *i-1,i*; *i,i*+1 segments of measuring line [$10^{-6} \cdot 1/m$];

$L_{i-1,i}^{meas}$ – measured length of the *i-1,i* segment [*m*];

S_{i-1}^{meas} – measured subsidence of the *i-1* point [*mm*];

i – measuring point.

The curvatures graph of measuring sections, registered along the observational line *№ 8* after the termination of exploitation in the *338/2* and *358/1* coal seams, has been shown at the Figure 2.

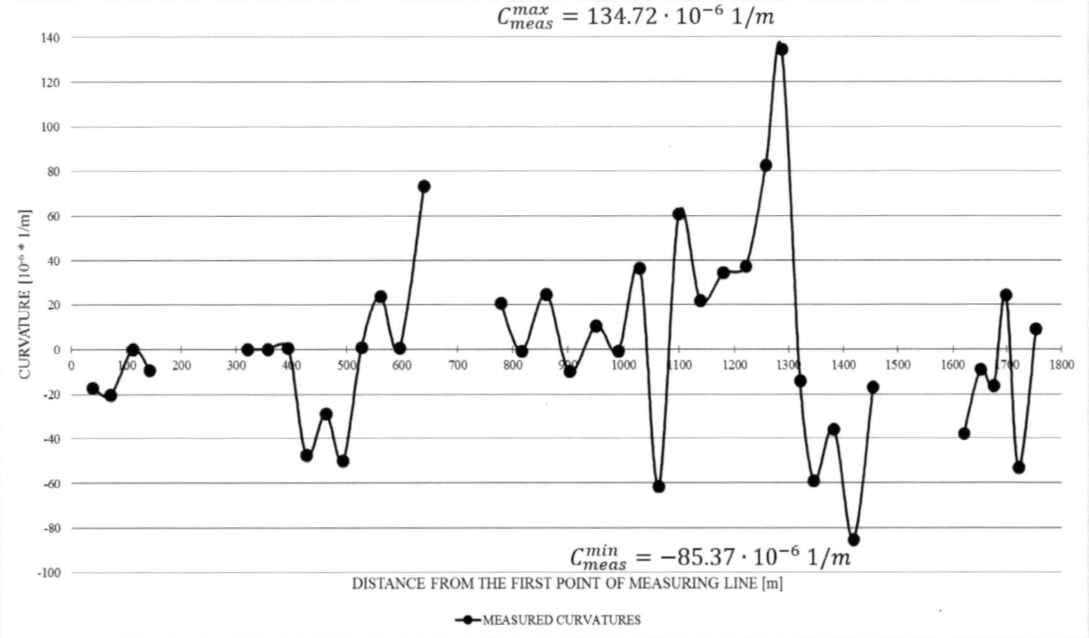

Figure 2. The curvatures graph noticed along the measuring line *№ 8* and caused by exploitation of two coal seams (*338/2*, *358/1*)

3. Average graph of curvatures

This section contains:

- an approximation of average graph of measured subsidence, which was done by the use of smoothed spline functions and minimization criterion of differences between the average and model values of subsidence;
- a manner of obtainment of curvatures average graph.

3.1. Approximation of average graph of the measured subsidence

To determine the average graph of curvatures, which were observed after the exploitation end of the *338/2* and *358/1* coal seams, first it's necessary to do an approximation of average graph of measured subsidence [15], [16].

The average values of measured subsidence were calculated by the use of smoothed splines. There were used the *R* computer programme and the *smooth.spline* function. Basic arguments of this function are:

- *df* – number of the freedom degree assuming the values from *0* to *n*, where *n* is number of measuring points;
- *spar* – smoothing parameter of approximating function assuming the values from *0* to *1*, where *0* denotes a total lack of smoothing and approximating function assumes the form of approximated function;
- λ – penalized criterion;
- *cv* – cross validation, when *cv = truth* then we deal with the leave-one-out validation, if *cv = false* then we deal with the general cross validation (*GCV*).

For the purpose of designation of proper value of the *spar* smoothing parameter [17], it was applied a minimization criterion of differences between the approximated and model values of subsidence (the least squares method). Model values of subsidence were obtained from the Bialek's formula with the values of its parameters determined from the results of surveys: $a = 0.890$; $tg\beta = 2.616$; $A_l = 0.227$.

In the Table 1 have been juxtaposed the parameters values of the *smooth.spline* function and made approximation.

Table 1. The parameters values of subsidence approximation

Parameter	Value
Df	20.95
Spar	0.41
λ	$4.53 \cdot 10^{-6}$
GCV	403.36
$\sum(S_{mod} - S_{approx})^2$ [mm²]	858891.36
$\sum(S_{meas} - S_{approx})^2$ [mm²]	6148.87

At the Figure 3 has been shown an average graph of subsidence caused by exploitation of two hard coal seams, determined on the way of approximation by the use of smoothed splines.

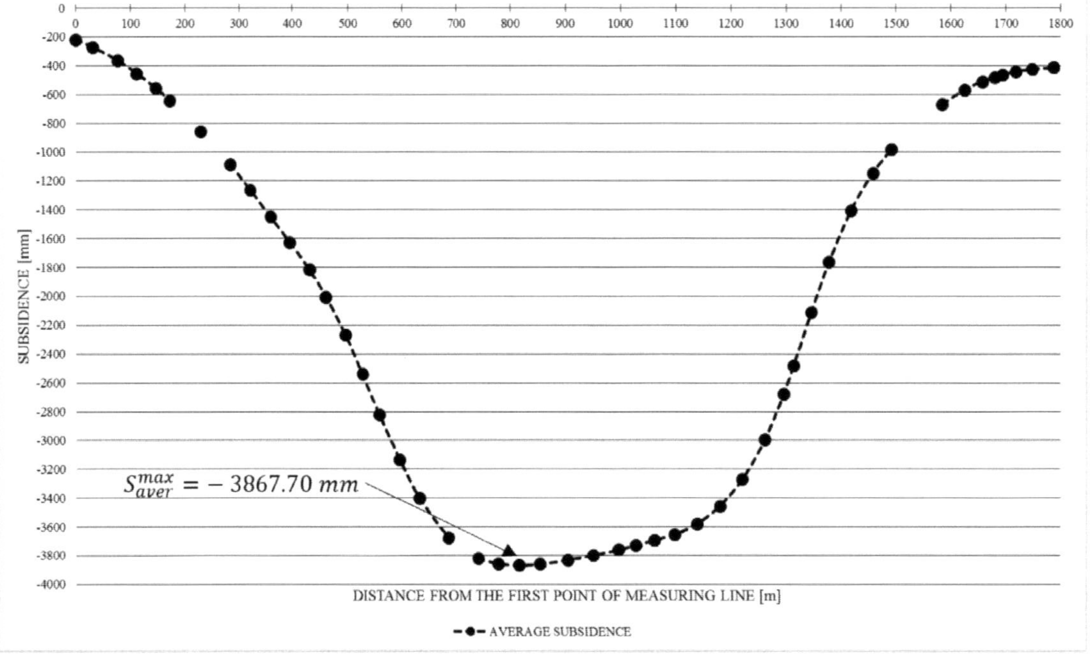

Figure 3. Approximated graph of subsidence measured at the observational line *№ 8* after the ending of exploitation in two coal seams (*338/2, 358/1*)

3.2. *Average graph of the observed curvatures*

The average graph of the observed curvatures has been determined on the base of an average graph of measured subsidence. The average values of measured curvatures have been obtained from the same formula as the measured values of curvatures, but instead the measured values of subsidence have been used the approximated values of measured subsidence:

$$C_{i-1,i,i+1}^{aver} = 2\frac{S_{i-1}^{approx} - 2S_{i}^{approx} + S_{i+1}^{approx}}{L_{i-1,i}^{meas} + L_{i,i+1}^{meas}}, \quad (3)$$

where:

$C_{i-1,i,i+1}^{aver}$ – average value of curvature measured on the *i-1,i*; *i,i*+1 segments of measuring line $[10^{-6} \cdot 1/m]$;
$L_{i-1,i}^{meas}$ – measured length of the *i-1,i* segment [*m*];
S_{i-1}^{approx} – approximated value of subsidence measured in the *i-1* point [*mm*];
i – measuring point.

Figure 4 shows an average graph of curvatures caused by exploitation of two hard coal seams, which was obtained from an average graph of measured subsidence.

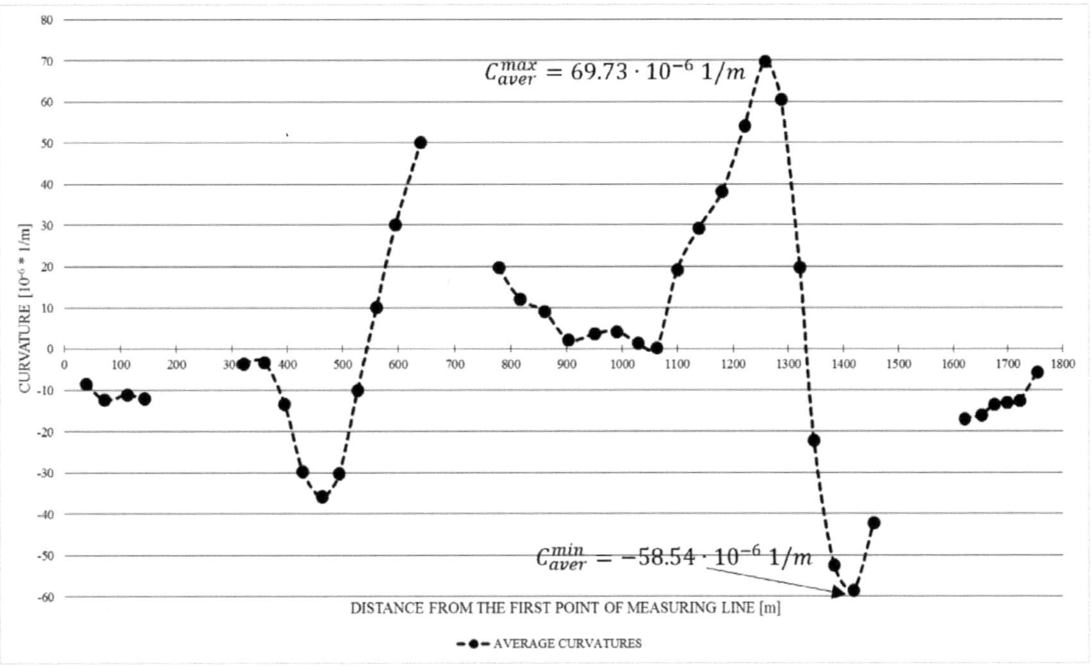

Figure 4. Average graph of curvatures observed at the measuring line *№ 8* after the ending of exploitation in two coal seams (*338/2, 358/1*)

4. Linear regression equation between the observed and average graphs of curvatures

The correlation between the measured and average graphs of curvatures has been shown in the Table 2 by the use of:

- the P_C percentages of the extreme measured values of curvatures which correspond the extreme average values of observed curvatures;
- the σ_C standard deviation between the observed and average values of curvatures;
- the R_C correlation coefficient.

Table 2. Relation between the measured and average curvatures

Parameter	Value
P_{Cmax} [%]	51.76
P_{Cmin} [%]	68.57
σ_C $[10^{-6} \cdot 1/\mathrm{m}]$	25.29
R_C	0.8122

For the purpose of relationship showing of the measured values of curvatures with their average values, it has been determined a linear regression equation. This correlation has been presented at the Figure 5.

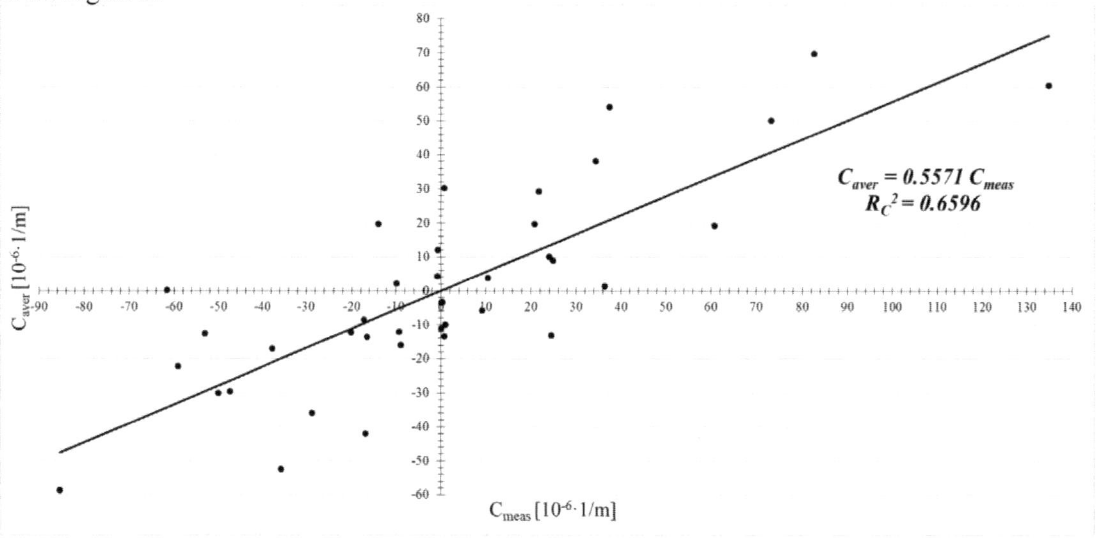

Figure 5. Linear regression between the average and observed values of curvatures at the measuring line *№ 8* after the ending of exploitation in the *338/2* and the *358/1* coal seams

5. Conclusions

Figures 1 and 3 indicate that the maximum value of average subsidence (-3867.70 mm) is almost equal the maximum measured subsidence (-3868.00 mm) and has been occurred almost in the same point (point number 819b, about 816 meters from the beginning of measuring line) like the maximum value of measured subsidence (point number 820, about 853 meters from the beginning of measuring line). Maximum subsidence has been observed above middle of the *B-2* longwall in the *338/2* coal seam.

Figures 2 and 4 show that the minimum values of observed and average curvatures have been occurred in the same places (point number 831, about 1418 meters from the line beginning, the run end of the *B-2* longwall in the *338/2* coal seam).

However, the maximum value of average curvatures has been occurred in the different place than the maximum value of observed curvatures (point number 827, about 1258 meters from the first point of measuring line instead point number 828, about 1287 meters from the first measuring point). The maximum curvatures have been noted near the southern run of the *B-4* longwall in the *358/1* seam.

From the data juxtaposed in the Table 2 results that the smoothed splines functions better reflect a minimum observed curvature (almost in 69 %) than a maximum observed curvature (only in 52 %).

The value of standard deviation between the observed and average curvatures equals to almost 40 % of the extreme values of average curvatures.

The linear regression equation between the measured and average curvatures has a form: $C_{aver} = 0.5571\ C_{meas}$. It means that the average values of observed curvatures are about 44.29 % smaller than the measured values of curvatures. Nevertheless, the value of correlation coefficient should be considered as high (more than 0.81). That means, that relationship of the average curvatures with the measured curvatures is rather strong and definitely better than in case of exploitation of three hard coal seams [18].

Acknowledgement

The results which have been presented in this article are the effect of work financed by the Ministry of Science and Higher Education in Poland, by means of a transfer of funds designated for the statutory activity of the Silesian University of Technology in Gliwice, the Faculty of Mining and Geology (project no. BKM-535/RG5/2018 [06/050/BKM18/0064]).

References

[1] Strzałkowski P and Ścigała R 2017 The Causes of Mining Induced Ground Steps Occurrence – Case Study from Upper Silesia in Poland, *Acta Geodynamica et Geomaterialia* **14**(3) 305-312

[2] Strzałkowski P 2017 The Proposal of Predicting Formation of Sinkholes with an Exemplary Application, *Journal of Mining Science* **53**(1) 53-58

[3] Strzałkowski P 2015 Mathematical Model of Forecasting the Formation of Sinkhole Using Salustowicz's Theory, *Archives of Mining Sciences* **60**(1) 63-71

[4] Szojda L and Kapusta Ł 2017 The Influence of a Variable Ground Curvature at the Passing of an Exploitation Front on the Technical Condition of a Multi-family Residential Building, *Mining Review* **73**(12) 63-74

[5] Orwat J 2017 Evaluation of Forecasts of Area Deformation Caused by Multideposit Exploitation of Hard Coal Made by the Budryk-Knothe Theory – Case Study, *Conf. Proc. of SGEM 2017; Informatics, Geoinformatics and Remote Sensing – Geodesy and Mine Surveying* **17**(22) 289-297

[6] Orwat J 2017 Verification of Forecasts of Subsidences, Inclinations and Curvatures of Area Caused by Multideposit Mining Conducted by Including Mining Margin – Case Analysis, *Conf. Proc. of SGEM 2017; Science and Technologies in Geology, Exploration and Mining – Exploration and Mining* **17**(13) 979-986

[7] Sikora P 2016 Simulation of Rock Mass Horizontal Displacements with Usage of Cellular Automata Theory, *Archives of Mining Sciences* **61**(4) 749-763

[8] Mielimąka R and Sikora P 2017 Simulation of Rockmass Subsidence Caused by Frequent Mining Exploitation with the Use of Cellular Automata Theory, *Mining Review* **73**(12) 84-92

[9] Stoch T, Niedojadło Z, Sopata P and Moskała S 2014 Functions Approximating the Average Course of Deformation Indicators for the Determination of the Random Dispersion Parameters of the Post-Mining Dislocations and Land Surface Deformations, *Mining Review* **70**(8) 113-123

[10] Popiołek E, Ostrowski J and Milewski M 1995 Appraisal Attempt of Usefulness of Subsidence Trough Profile Curvature as an Object Threat Indicator Under the Influences of Underground Mining Exploitation (Case Study of the LGOM Deposit), *Mining Review* **1**(11) 1-7

[11] Orwat J and Mielimąka R 2017 Approximation of Average Course of Measured Curvatures of Mining Area with Reference to Their Forecast Values by Bialek's Formulas, *AIP Conf. Proc.* **1863** 130003-1 – 130003-4

[12] Mielimąka R and Orwat J 2017 Approximation of Average Course of Measured Curvatures of Mining Area with Reference to Their Forecast Values by Knothe's Formulas, *AIP Conf. Proc.* **1863** 130005-1 – 130005-4

[13] Orwat J 2018 Evaluation of Application Possibilities of Smoothed Splines to Designation of the Average Values of Terrain Curvatures Measured after the Termination of Hard Coal Exploitation Conducted at Medium Depth, *IOP Conf. Ser.: Mater. Sci. Eng.* **294** 012030

[14] Ścigała R 2013 Mining Extraction System with Advancing Longwall as a Method of Protecting – Technical Objects and an Element of Rational Deposit Management, *Mineral Resources Management* **29**(3) 191-208

[15] Orwat J 2017 Approximation of Average Course of Measured Subsidences of Mining Area by Smooth Splines, *AIP Conf. Proc.* **1863** 130004-1 – 130004-4

[16] Orwat J and Mielimąka R 2017 Average Course Approximation of Measured Subsidence and Inclinations of Mining Area by Smooth Splines, *Journal of Sustainable Mining* **16**(1) 8-13

[17] Orwat J and Mielimąka R 2018 Smoothing Parameter as Shape Parameter of Function Approximating the Average Course of Terrain Surface Subsidence, *AIP Conf. Proc.* **1978** 390005-1 – 390005-4
[18] Orwat J 2018 Determination of Equation Describing the Measured and Average Curvatures Graphs Observed as a Result of Multideposit Exploitation at the Great Depth, *Conf. Proc. of SGEM 2018; Informatics, Geoinformatics and Remote Sensing – Geoinformatics, Geodesy and Mine Surveying* **18**(2.2) 685-691

IOP Conf. Series: Materials Science and Engineering **477** (2019) 012043 doi:10.1088/1757-899X/477/1/012043

Relation between the theoretical and average observed curvatures of mining terrain

J Orwat[1,2]

[1]Silesian University of Technology, Department of Mining and Geology, Akademicka str., no. 2, 44100 Gliwice, Poland
[2]Silesian University of Technology, Department of Construction, Akademicka str., no. 5, 44100 Gliwice, Poland

E-mail: justyna.orwat@polsl.pl

Abstract. In the article have been shown the obtaining methods of average graph of terrain curvatures caused by underground mining exploitation and their theoretical graph calculated on the base of subsidence forecasted by the use of the geometric–integral theory of mining exploitation influences. The average graph of curvatures measured after the ending of excavation in the 338/2 and 358/1 hard coal seams, was designated using the smoothed splines and the least squares method. First it has been done an approximation of the subsidence measured values (with help of the R-project computer program) and then, it has been determined an average graph of the observed curvatures. The choice criterion of proper value of smoothing parameter of approximating function was applied for the subsidence. Right value of smoothing parameter occurs when the standard deviation between the average and predicted graphs of subsidence reaches the minimal value. Average values of measured curvatures were compared with the theoretical values of curvatures. They were calculated on the base of subsidence forecasted by the use of the EDN-OPN computer program and the Bialek's formula with the values of its parameters designated from the results of geodesic measurements. It was determined a relation between the theoretical values of curvatures and their average measured values. This relationship allows to evaluate the mathematical model which was used to calculation of the curvatures theoretical values.

1. Introduction

Every mining plant, which strives for a concession to an extraction of the useful ore, has to prepare a forecast of mining exploitation influences on terrain surface, buildings and technical infrastructure objects.

These forecasts can be made by the use of the various theories of exploitation influences, e.g. the geometric-integral theories [1], [2], a cellular automata theory [3], [4], and with the different values of these theories parameters [5], [6].

Credibility of made predictions can be only checked when the excavation will be completed. Then, the proper geodesic surveys are being done and their results are comparing with the surveys results done before an exploitation has begun. On the base of this, there are determined the deformations indicators values, for example of subsidence, inclinations, curvatures, horizontal strains.

The forecasted graphs of deformations indicators shouldn't be compared with the deformations indicators graphs directly received from the results of geodesic measurements (they are irregular), but with their average graphs obtained on the way of approximation. This irregularity is mainly caused by the measurements errors, a randomness of deformation process and the cracks of superficial layer of rock mass [7].

The average graphs of measured deformations indicators can be obtained by the use of the various, approximating functions, e.g. the orthogonal polynomials [8]. Unfortunately, these functions aren't flexible (especially in case of curvatures) and require a division of the whole graph of deformation indicator into the sections.

Because of that, in the article were applied the smoothed splines functions to determination of an average graph of terrain curvatures observed as a result of underground mining exploitation conducted in two hard coal seams.

Then, there have been compared the theoretical graph of curvatures with the average graph of curvatures measured after the ending of hard coal extraction from the 338/2 and 358/1 seams.

The theoretical curvatures have been calculated from the subsidence forecasted by the use of a geometric-integral theory of mining exploitation influences which takes into account an existence of the operating margin [2], namely the Bialek's formula with the values of its parameters calculated from the geodesic measurements results.

To evaluate the credibility of made forecast, it was determined the equation of linear regression between the theoretical and average measured curvatures [9].

2. Average graph of observed curvatures

In this section have been presented an example of operation of hard coal which was carried out by some mining plant in Poland, in two coal seams located at the large depth; a description of the geodesic measurements which have been taken on the observational points stabilized in the ground above the exploitation field; the obtainment ways of the observed and average graphs of curvatures.

2.1. Mining exploitation case

Some hard coal mine, which is located in the Upper Silesian Coal Basin in Poland, in the years 2001 – 2006 realized an extraction of hard coal from the *338/2* and *358/1* seams. Exploitation was conducted at the great depth and by the use of longwall system with cave-in of the roof rocks. Slope of coal layers was small and amounted to 6°. Hard coal deposit was inclined in the south-eastern direction.

Coal extraction from the *338/2* seam was conducted at the depth of 600 m and at the height of 3.0 m. Within five years (2001 – 2006), there have been exploited the following longwalls: *B-1*, *B-2*, *B-3*, *B-4*, *C-1*.

Hard coal seam *358/1* was exploited at the depth of 1000 m, in the years 2002 – 2006. Exploitation of seven longwalls (from the *B-1* to the *B-7*) was running at the height of 2.9 m.

2.2. Geodesic observations

Shaping of terrain surface curvatures, under the exploitation influence of two hard coal seams, has been observed by the use of the geodesic measurements. These measurements were carried out at 53 points of the observational line *№ 8*. This line was established above the *B-2* longwall in the *338/2* seam, perpendicularly to its run. The average length of line segments was amounted 35 m.

Surveys of the points heights have been taken by the use of precise leveller. Accuracy of measurements was amounted 0.1 mm. The heights of points were defined in the Kronstadt altitude system. It was used a geometric levelling from the middle with the double targeting.

The sections lengths have been measured by the use of the distance-meter device with an accuracy of 1 mm.

2.3. Graph of the observed curvatures

Curvatures observed on the neighbouring sections of the measuring line *№ 8* have been obtained from the subsidence values measured on three neighbouring points and an average length of these sections. It presents the formula (1).

IOP Conf. Series: Materials Science and Engineering **477** (2019) 012043 doi:10.1088/1757-899X/477/1/012043

$$C_{i-1,i,i+1}^{meas} = 2\frac{S_{i-1}^{meas} - 2S_{i}^{meas} + S_{i+1}^{meas}}{L_{i-1,i}^{meas} + L_{i,i+1}^{meas}}, \tag{1}$$

where:
$C_{i-1,i,i+1}^{meas}$ – curvature observed on the *i-1,i*; *i,i*+1 sections of measuring line [$10^{-6} \cdot 1/m$];
$L_{i-1,i}^{meas}$ – measured length of the *i-1,i* section [*m*];
S_{i-1}^{meas} – measured subsidence of the *i-1* point [*mm*];
i – measuring point.

The curvatures, noted on the measuring sections of the observational line *№ 8* after the exploitation end in the *338/2* and *358/1* seams, have been presented at the Figure 1.

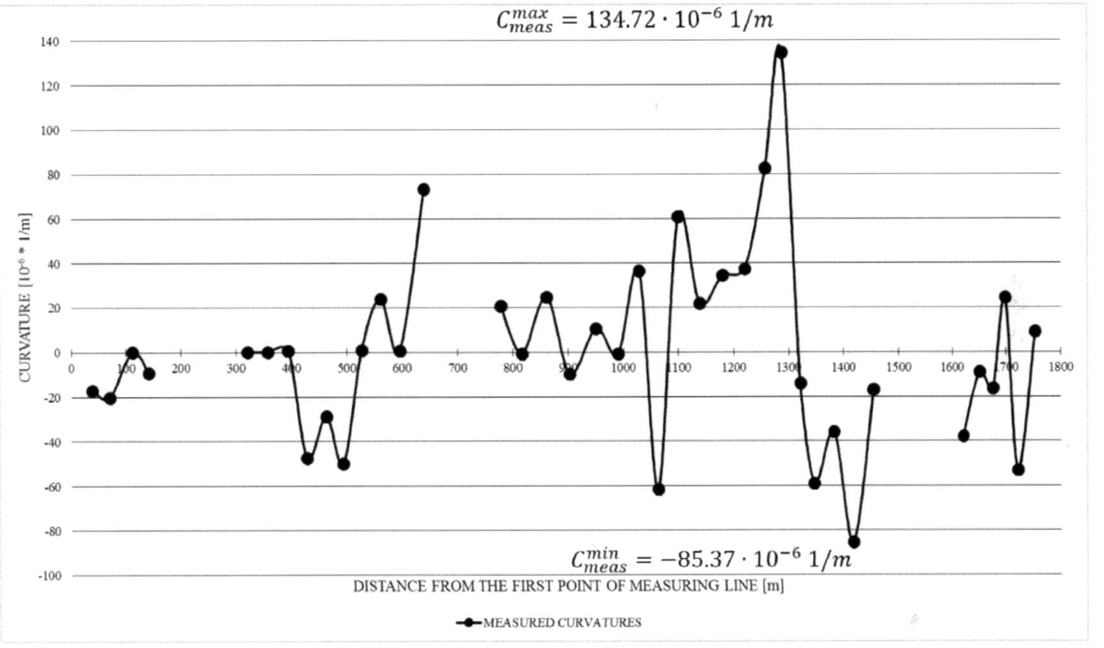

Figure 1. The curvatures observed along the measuring line *№ 8* after an extraction ending in the *338/2* and *358/1* hard coal seams

Subsidence of measuring points have been calculated from the points heights differences. Heights of points have been measured before the beginning of exploitation and after the termination of exploitation of two hard coal seams. Subsidence were calculated from the formula:

$$S_i^{meas} = H_i^A - H_i^B, \tag{2}$$

where:
H_i^A – absolute height of the *i* point after the exploitation termination [*mm*];
H_i^B – absolute height of the *i* point before the exploitation beginning [*mm*];
S_i^{meas} – measured subsidence of the *i* point [*mm*].

2.4. Average graph of the observed curvatures

Average values of the observed curvatures [10], [11] have been calculated from the approximated values of measured subsidence and the average length of two neighbouring segments, from the below formula [12]. This formula is analogous as the formula (1):

$$C_{i-1,i,i+1}^{aver} = 2\frac{S_{i-1}^{approx} - 2S_i^{approx} + S_{i+1}^{approx}}{L_{i-1,i}^{meas} + L_{i,i+1}^{meas}}, \quad (3)$$

where:

$C_{i-1,i,i+1}^{aver}$ – average value of curvature observed on the *i-1,i*; *i,i*+1 sections of measuring line $[10^{-6} \cdot 1/m]$;

$L_{i-1,i}^{meas}$ – measured length of the *i-1,i* section [*m*];

S_{i-1}^{approx} – approximated value of subsidence measured in the *i-1* point [*mm*];

i – measuring point.

Figure 2 presents the average values of curvatures observed along the measuring line *№ 8* after the end of exploitation in the *338/2* and *358/1* coal seams.

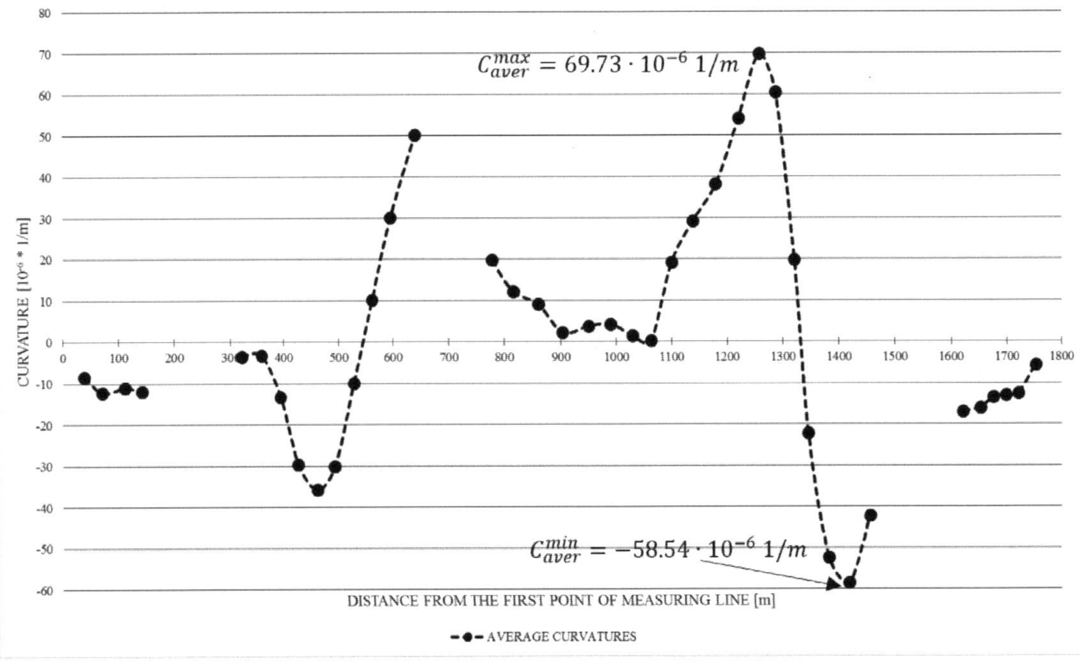

Figure 2. The average curvatures observed on the measuring line *№ 8* after the termination of exploitation of the *338/2* and *358/1* coal seams

Approximation of the average values of measured subsidence has been conducted by the use of the smoothed splines functions [13], [14] and the least squares method. It has been applied the *smooth.spline* function from the *R* computer program. This function is based on the following arguments:

- *df* – number of the freedom degree assuming the values from *0* to *n*, where *n* is number of measuring points;
- *spar* – smoothing parameter of approximating function assuming the values from *0* to *1*, where *0* denotes a total lack of smoothing and approximating function assumes the form of approximated function;
- λ – penalized criterion;
- *cv* – cross validation, when *cv* = *truth* then we deal with the leave-one-out validation, if *cv* = *false* then we deal with the general cross validation (*GCV*).

To choose the right value of the smoothing parameter *spar* [15], it was used a residuals minimization criterion between the approximated and forecasted values of subsidence [10], [11].

Predicted values of subsidence have been calculated by the use of the Bialek's formula with the values of its parameters obtained from the surveys results.

In the Table 1 have been presented the parameters values of the *smooth.spline* function and an approximation [16].

Table 1. The parameters values of subsidence approximation

Parameter	Value
Df	20.95
Spar	0.41
A	$4.53 \cdot 10^{-6}$
GCV	403.36
$\sum(S_{fore} - S_{approx})^2$ [mm^2]	858891.36
$\sum(S_{meas} - S_{approx})^2$ [mm^2]	6148.87

3. Theoretical graph of curvatures

Theoretical values of curvatures have been calculated on the base of the forecasted values of subsidence by the use of the Bialek's formula with the parameters values specified from the results of geodesic measurements:

$$C_{i-1,i,i+1}^{theor} = 2\frac{S_{i-1}^{fore} - 2S_{i}^{fore} + S_{i+1}^{fore}}{L_{i-1,i}^{meas} + L_{i,i+1}^{meas}}, \qquad (4)$$

where:

$C_{i-1,i,i+1}^{theor}$ – theoretical value of curvature on the *i-1,i*; *i,i*+1 sections of measuring line [$10^{-6} \cdot 1/m$];

$L_{i-1,i}^{meas}$ – measured length of the *i-1,i* section [*m*];

S_{i-1}^{fore} – forecasted value of subsidence by the use of the Bialek's formula in the *i-1* point [*mm*];

i – measuring point.

At the Figure 3 has been shown the theoretical graph of curvatures along the observational line *№ 8* after the exploitation ending in the *338/2* and *358/1* coal seams, obtained from the subsidence forecasted by the use of the Bialek's formula.

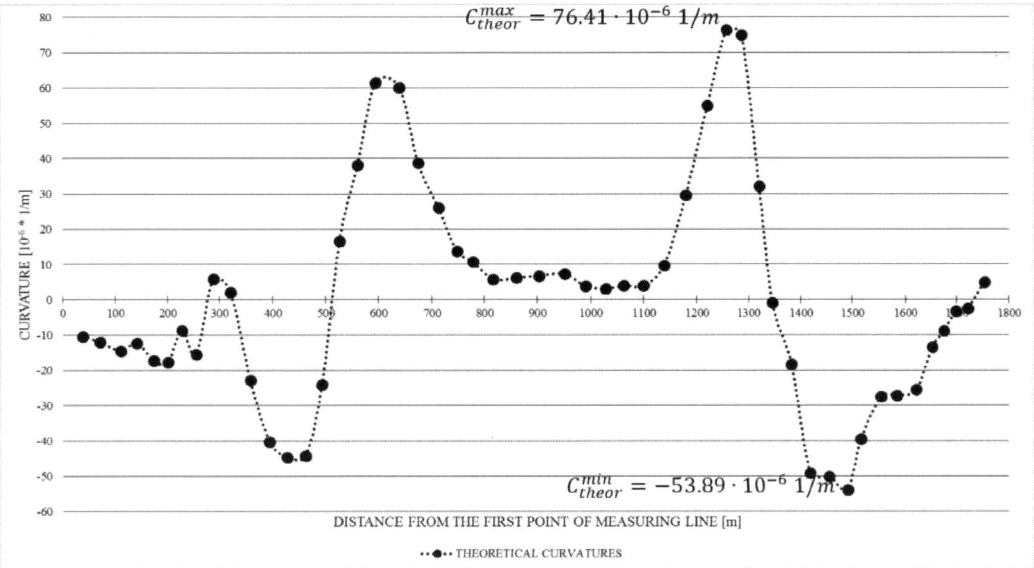

Figure 3. The theoretical curvatures along the measuring line *№ 8* after the operation end in the *338/2* and *358/1* seams

Subsidence forecast has been made by the use of the *EDBJ1* computer program. In calculations was assumed the Bialek's formula which is classified as the geometric-integral theory of mining exploitation influences. This formula takes into account: the far influences (the β-angle), an exploitation margin (the A_l parameter), an old goaf reactivation, a desymmetrization of subsidence trough profile relative to an exploitation edge (the r_1, r_2-radius) and the exploitation influences coming from several coal seams (the a parameter). The parameters values of this formula have been determined from the subsidence measured after the termination of exploitation in two hard coal seams, namely: $a = 0.890$; $\mathrm{tg}\beta = 2.616$; $A_l = 0.227$.

$$S_f = (1 - a_s)s(r_1) + a_s s(r_2) - A_1\left(2 + \frac{A_3}{2}\right)\frac{s(r_1)[r_1\gamma(r_1)]^2}{A_3\left[\frac{s(r_1)}{2} + \frac{s(r_2)}{2}\right]^2 + [r_1\gamma(r_1)]^2}, \tag{5}$$

where:
β – range angle of main influences [°];
γ – simplified octahedral strain [mm/m];
A_l – dimensionless multiplier which takes into account an asymmetry of subsidence trough profile;
$A_3 = 6.667$;
S_f – final subsidence [mm];
a_s – coefficient constant for each subsidence trough, it defines what part of final subsidence was calculated by the use of the r_i radius ($a_s = 0.4 \div 1.25A_l$);
h – exploitation depth [m];
r_i – radius of influences dispersion for each subsidence trough [m];
$r_l = \frac{h}{tg\beta}F_r(A_l)$;
$r_2 = 2r_l$;
$s(r_l)$, $s(r_2)$ – subsidence calculated using the Knothe's formula with the use of two different radiuses of influences dispersion r_l, r_2 [mm].

4. Relation between the theoretical and average observed curvatures

Relationship the theoretical graph of curvatures with the average observed graph of curvatures has been shown by the use of a linear regression equation and statistics published in the Table 2. There have been calculated: the P_C percentages of the extreme average values of observed curvatures which correspond the extreme theoretical values of curvatures, the σ_C standard deviation between the average observed and theoretical values of curvatures, the R_C correlation coefficient between these curvatures.

Table 2. Relation between the average observed curvatures and the theoretical curvatures

Parameter	Value
P_{Cmax} [%]	109.58
P_{Cmin} [%]	92.06
σ_C [$10^{-6} \cdot 1/m$]	14.04
R_C	0.9030

The equation of linear regression between the theoretical curvatures and the average observed curvatures is shown at the Figure 4.

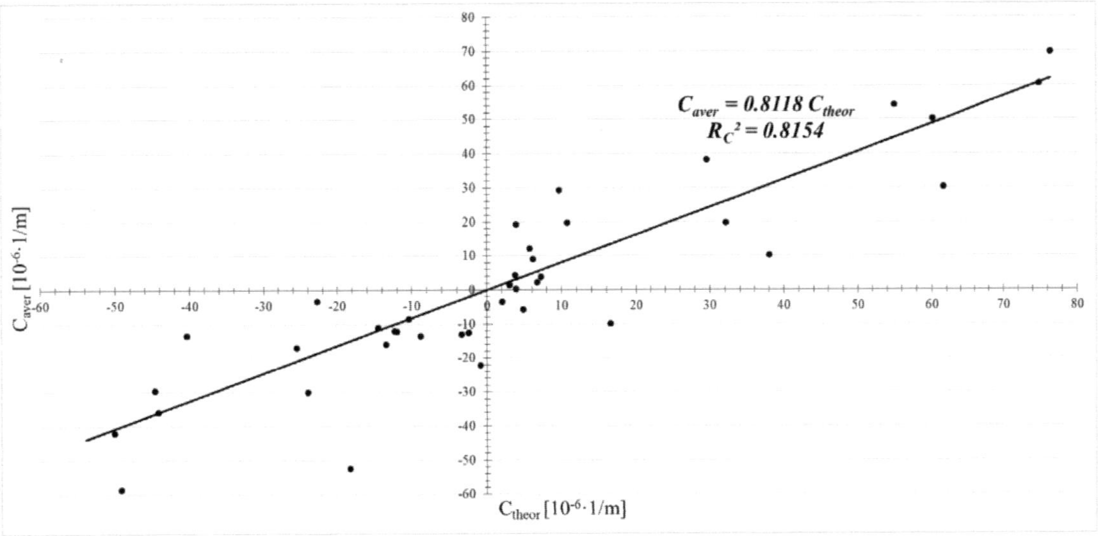

Figure 4. Linear regression equation between the theoretical and average observed curvatures at the measuring line *№ 8* after the extraction termination in the *338/2* and the *358/1* hard coal seams

5. Conclusions

Figures 1 and 2 indicate that the minimum values of measured and average curvatures occur in the same point number 831, which is located about 1418 meters from the first point of line and at the end of run of the *B-2* longwall in the *338/2* seam. But the maximum value of average curvatures occurs in another point than the maximum value of observed curvatures, namely in point number 827 (about 1258 meters from the beginning of observational line) instead in point number 828 (about 1287 meters from the line beginning). The maximum curvatures occur nearby the southern run of the *B-4* longwall in the *358/*1 hard coal seam.

From the Figures 2 and 3 result that the minimal value of curvatures theoretical graph occur in the different place than the minimal value of curvatures average graph (in the 832 point, about 1491 m from the first measuring point and outside of exploitation field of the *B-2* longwall in the *338/2* coal seam, instead in the 831 point). From these figures can be concluded that the maximal values of theoretical and average curvatures have been noted in the same point number 827.

Data juxtaposed in the Table 2 show that the theoretical graph of curvatures, obtained on the base of subsidence forecasted by the use of the Bialek's formula with the determined values of its parameters, fits very well to the average graph of observed curvatures, obtained from the measured subsidence approximated by the use of the smoothed splines functions. It's confirmed by the high values of the R_C correlation coefficient (above 0.9), the P_C percentages of the extreme average values of observed curvatures which correspond the extreme theoretical values of curvatures (both close to 100 %) and rather by the small value of the σ_C standard deviation (about $14 \cdot 10^{-6}$ 1/m).

The equation of linear regression between the theoretical and average observed curvatures (C_{aver} = 0.8118 C_{theor}) indicates that the theoretical values of curvatures are over 23 % larger than the average values of observed curvatures.

Taking into account the correlation coefficient value, it can be seen that the relation between the theoretical and average observed curvatures is stronger than in case of this relation determined for an operation of three coal seams [9].

Due to the above arguments, it can be said that the forecasts of mining exploitation influences on terrain surface, made by the use of the Bialek's formula with the values of its parameters (a, tgβ, A_I) designated from the surveys results, are credible and fairly accurate.

Acknowledgement

The results which have been presented in this article are the effect of work financed by the Ministry of Science and Higher Education in Poland, by means of a transfer of funds designated for the statutory activity of the Silesian University of Technology in Gliwice, the Faculty of Mining and Geology (project no. BKM-535/RG5/2018 [06/050/BKM18/0064]).

References

[1] Orwat J 2017 Evaluation of Forecasts of Area Deformation Caused by Multideposit Exploitation of Hard Coal Made by the Budryk-Knothe Theory – Case Study, *Conf. Proc. of SGEM 2017; Informatics, Geoinformatics and Remote Sensing – Geodesy and Mine Surveying* **17**(22) 289-297

[2] Orwat J 2017 Verification of Forecasts of Subsidences, Inclinations and Curvatures of Area Caused by Multideposit Mining Conducted by Including Mining Margin – Case Analysis, *Conf. Proc. of SGEM 2017; Science and Technologies in Geology, Exploration and Mining – Exploration and Mining* **17**(13) 979-986

[3] Sikora P 2016 Simulation of Rock Mass Horizontal Displacements with Usage of Cellular Automata Theory, *Archives of Mining Sciences* **61**(4) 749-763

[4] Mielimąka R and Sikora P 2018 Simulation of Rock Mass Subsidence Due to Subsequent Mining Extraction Using the Theory of Cellular Automata, *Management Systems in Production Engineering* **7**(1) 457-469

[5] Strzałkowski P 2010 Forecasts of Mine-Induced Land Deformations in Consideration of the Variability of the Parameter Describing the Process Kinematics, *Archives of Mining Sciences* **55**(4) 865-872

[6] Ścigała R 2013 The Identification of Parameters of Theories Used for Prognoses of Post Mining Deformations by Means of Present Software, *Archives of Mining Sciences* **58**(4) 1347-1357

[7] Kowalski A and Jędrzejec E 2015 Influence of Subsidence Fluctuation on the Determination of Mining Area Curvatures, *Archives of Mining Sciences* **60**(2) 487-505

[8] Stoch T, Niedojadło Z, Sopata P and Moskała S 2014 Functions Approximating the Average Course of Deformation Indicators for the Determination of the Random Dispersion Parameters of the Post-Mining Dislocations and Land Surface Deformations, *Mining Review* **70**(8) 113-123

[9] Mielimąka R and Orwat J 2018 Correlation Between the Average and Forecasted Curvatures Caused by Hard Coal Exploitation Conducted at the Great Depth in Several Coal Seams, *Conf. Proc. of SGEM 2018; Informatics, Geoinformatics and Remote Sensing – Geoinformatics, Geodesy and Mine Surveying* **18**(2.2) 669-676

[10] Orwat J and Mielimąka R 2017 Approximation of Average Course of Measured Curvatures of Mining Area with Reference to Their Forecast Values by Bialek's Formulas, *AIP Conf. Proc.* **1863** 130003-1 – 130003-4

[11] Mielimąka R and Orwat J 2017 Approximation of Average Course of Measured Curvatures of Mining Area with Reference to Their Forecast Values by Knothe's Formulas, *AIP Conf. Proc.* **1863** 130005-1 – 130005-4

[12] Orwat J 2018 Evaluation of Application Possibilities of Smoothed Splines to Designation of the Average Values of Terrain Curvatures Measured after the Termination of Hard Coal Exploitation Conducted at Medium Depth, *IOP Conf. Ser.: Mater. Sci. Eng.* **294** 012030

[13] Orwat J 2017 Approximation of Average Course of Measured Subsidences of Mining Area by Smooth Splines, *AIP Conf. Proc.* **1863** 130004-1 – 130004-4

[14] Orwat J and Mielimąka R 2017 Average Course Approximation of Measured Subsidence and Inclinations of Mining Area by Smooth Splines, *Journal of Sustainable Mining* **16**(1) 8-13

[15] Orwat J and Mielimąka R 2018 Smoothing Parameter as Shape Parameter of Function Approximating the Average Course of Terrain Surface Subsidence, *AIP Conf. Proc.* **1978**

390005-1 – 390005-4
[16] Orwat J 2018 *Linear Regression Equation of Mining Terrain Curvatures Caused by Hard Coal Excavation from a Few Seams and Their Approximated Values by the Use of Smoothed Spline*, International Conference of Applied Sciences 2018 (ICAS 2018), May 09-11, Banja Luka, Bosnia and Herzegovina

IOP Conf. Series: Materials Science and Engineering **477** (2019) 012044 doi:10.1088/1757-899X/477/1/012044

Determination of Reliability for Dacia Solenza 1,9D

A Budiul Berghian, T Vasiu and C Birtok Baneasa

Politehnica" University of Timisoara, Department of Engineering and Management, 5 Revolution Street, Hunedoara, 331128, Romania

E-mail: adina.budiul@fih.upt.ro

Abstract. This paper analyses running and failure of Dacia Solenza 1,9D. The study comprises plotting the Pareto diagram in accordance with the data collected by observation of the vehicle running/failure and determination of its reliability using the Weibull++8 specialized software of the company ReliaSoft, conclusions being presented in graphic form within this study.

1. Introduction

In the system of economic efficiency, which points all aspects and components of the production activity, products quality and its continuous improvement have a significant place.

Reliability is one of the prominent elements of product quality and according to SR-EN-ISO 8402-1991 represents: *Ability of a product to accomplish a specified function in fixed conditions and for a certain period of time* [1], [2], [5].

From a certain point of view, quality can be considered as a "static" feature to satisfy some requirements at a certain moment, while reliability is "dynamic", as manifestation in time of the quality.

2. Pareto diagram

The analysis of the data collected from the observation of the functioning/drawback of the Dacia Solenza 1,9D car allows the construction of so called Pareto diagram (Figure 1). This is a method of analysis of fault distribution and of evaluation allowing in the same time the identification of fault categories upon which one must operate with priority, in case of faults with unknown causes.

3. Determination of reliability for Dacia Solenza 1,9D

Information on reliability for Dacia Solenza 1,9 have been obtained observing the running of a lot of vehicles of the same generation and fabrication, on a period of a six months. This control programme comprises data on the number of kilometers from which intervene changes in status or behavior of the vehicle, moment of down time, moment of re-starting up after removal of failure causes.

After processing the obtained data from the exploration of the car Dacia Solenza with the specialized soft Weibull ++ 9.0.9, the following results present graphically in Figures 2-8 were obtained.

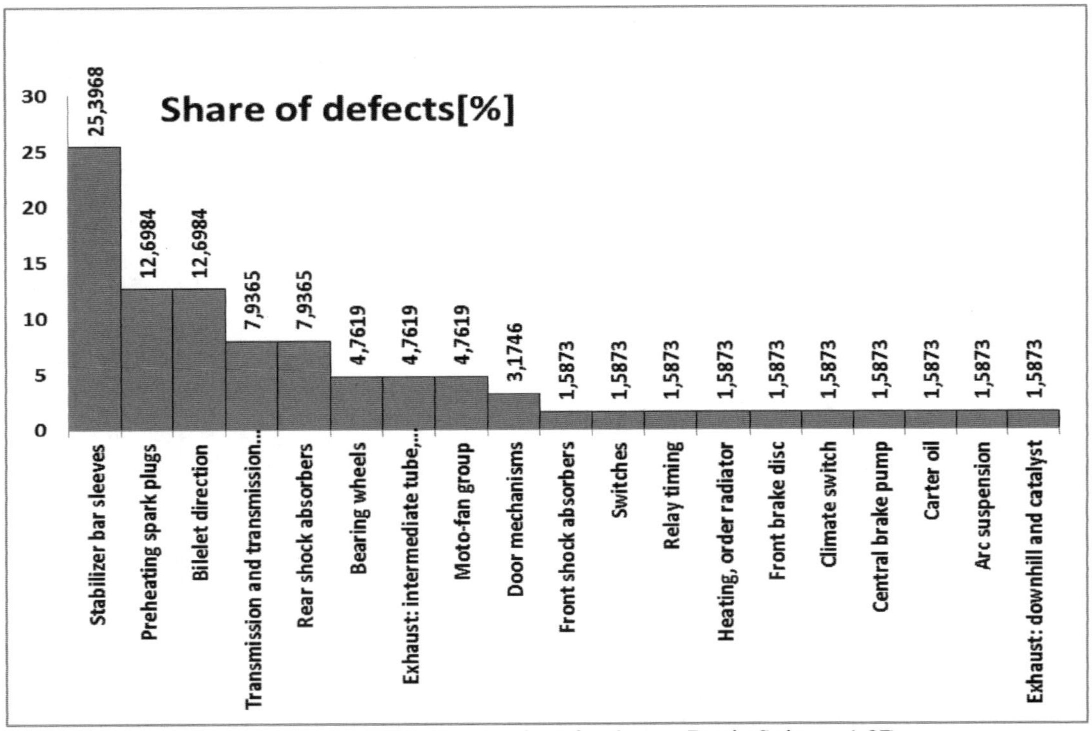

Figure 1. The Pareto chart for the car Dacia Solenza 1,9D

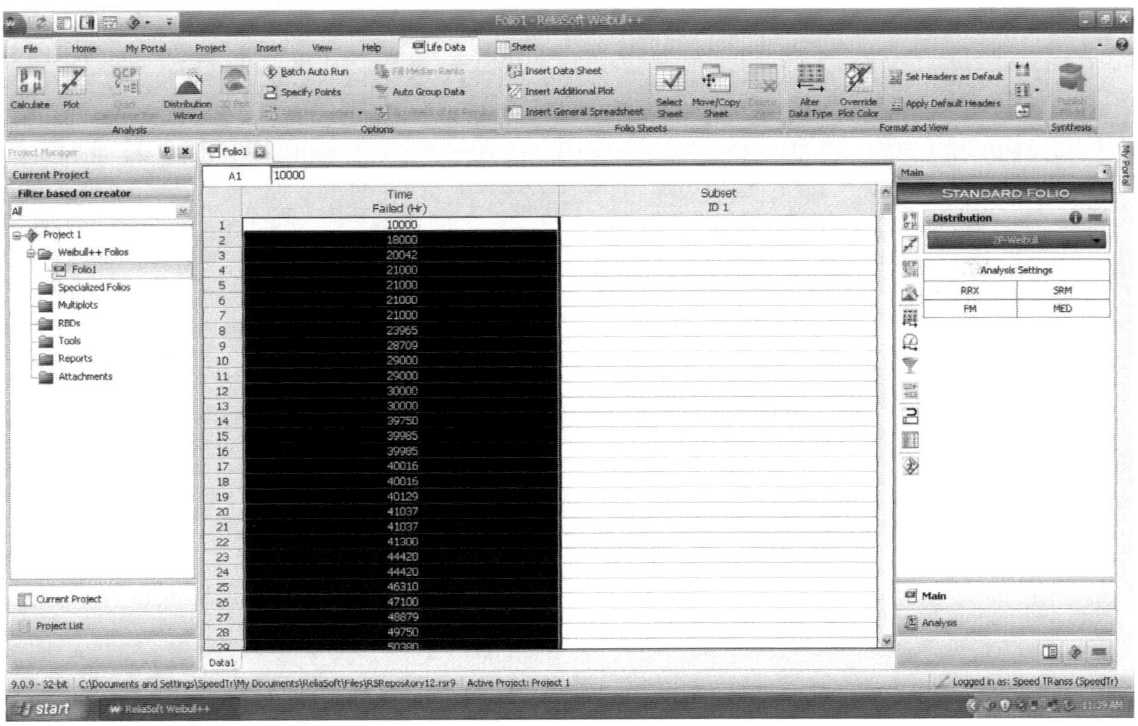

Figure 2. Interface for determing the Weibull placement parameters

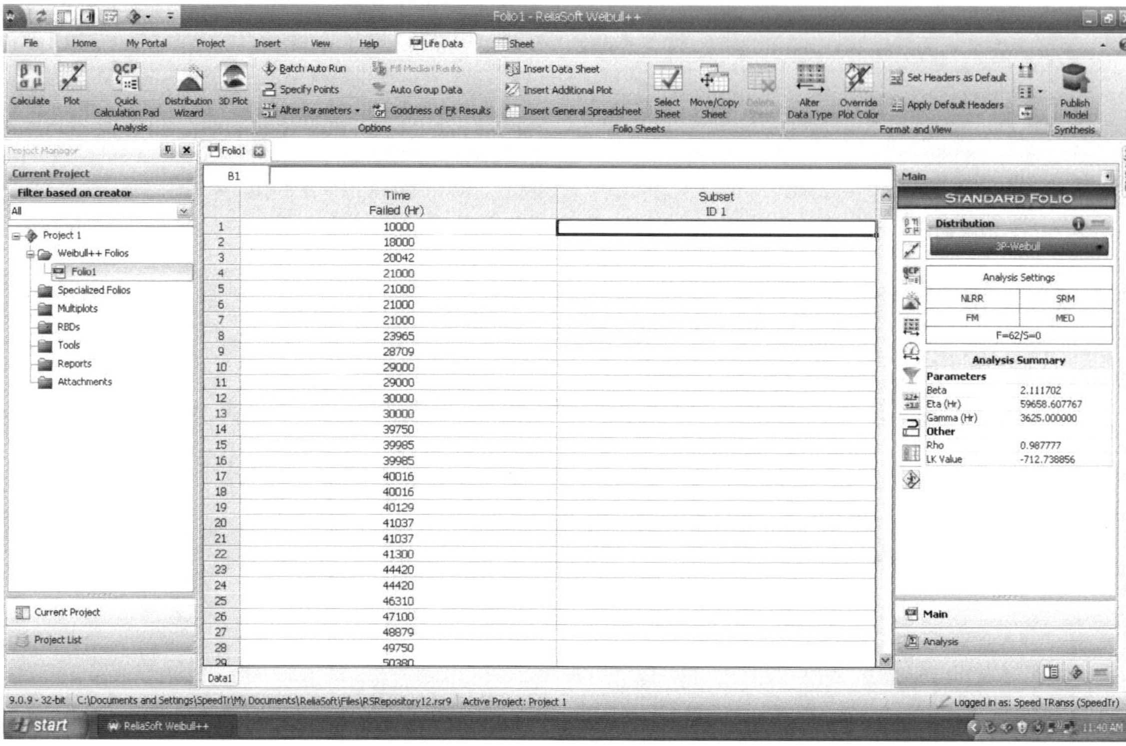

Figure 3. Procesing the results with the help of the Weibull ++ 9.0.9 soft

The certification of the fact that the real reliability of the studied vehicle respects the Weibull placement law can be seen in Figure 4, in which the Alain Plait diagram is presented where the double-logarithmic coordinates show the time independence of the cumulated relative frequency. [3]

The calculated values of the parameters of Weibull's law are, with the correlation coefficient ρ=0,987777 : β=2.11702, η=59658.607 and γ=3625,00 [4], [5]. In this way, the mathematical expression of the law of the reliability's placement of Dacia Solenza, 1,9D is:

$$R(t) = e^{-\left(\frac{t-\gamma}{\eta}\right)^{\beta}} = e^{-\left(\frac{t-3625}{59658.607}\right)^{2.11702}} \quad (1)$$

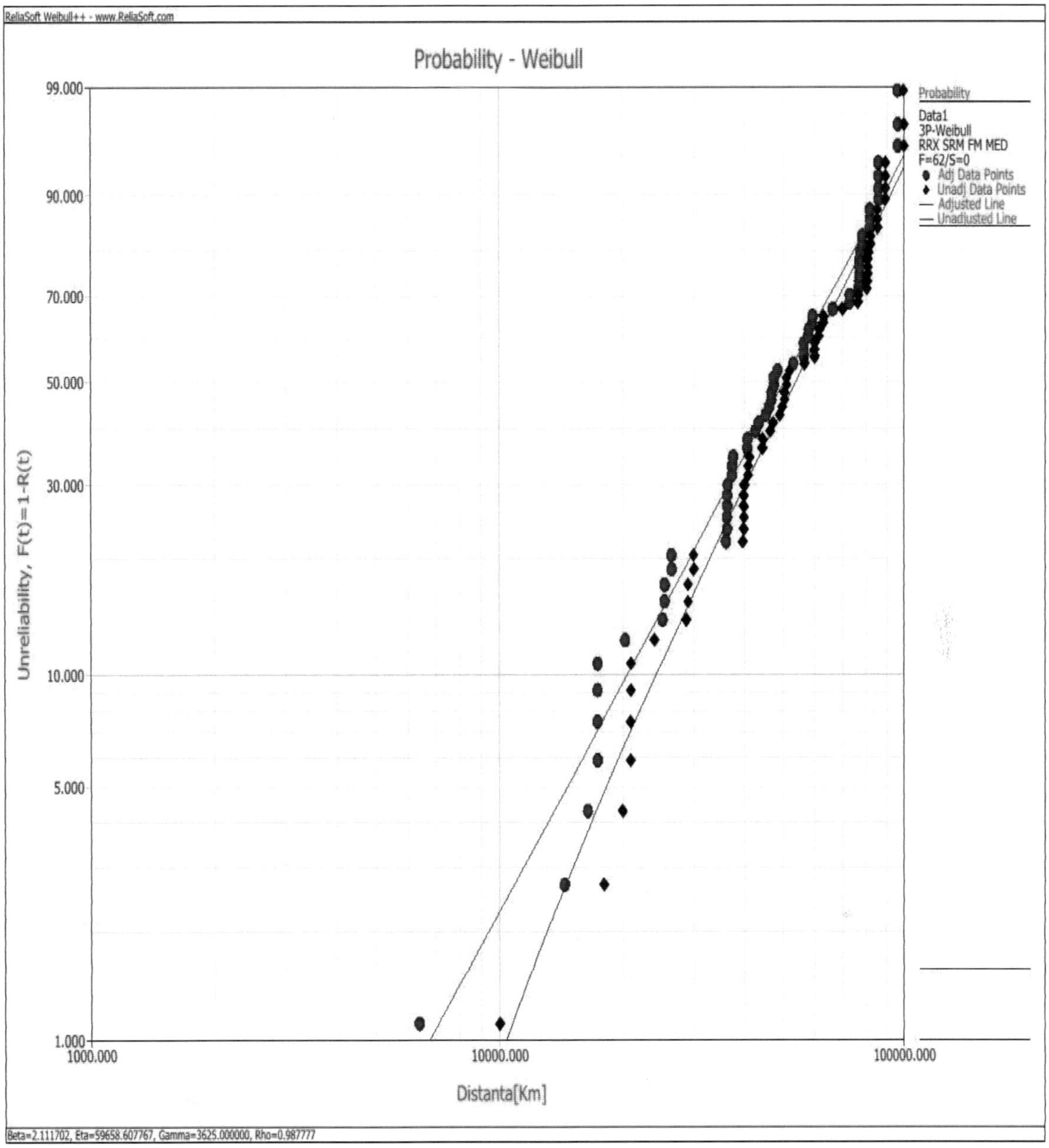

Figure 4. Alain Plait diagram

IOP Conf. Series: Materials Science and Engineering **477** (2019) 012044 doi:10.1088/1757-899X/477/1/012044

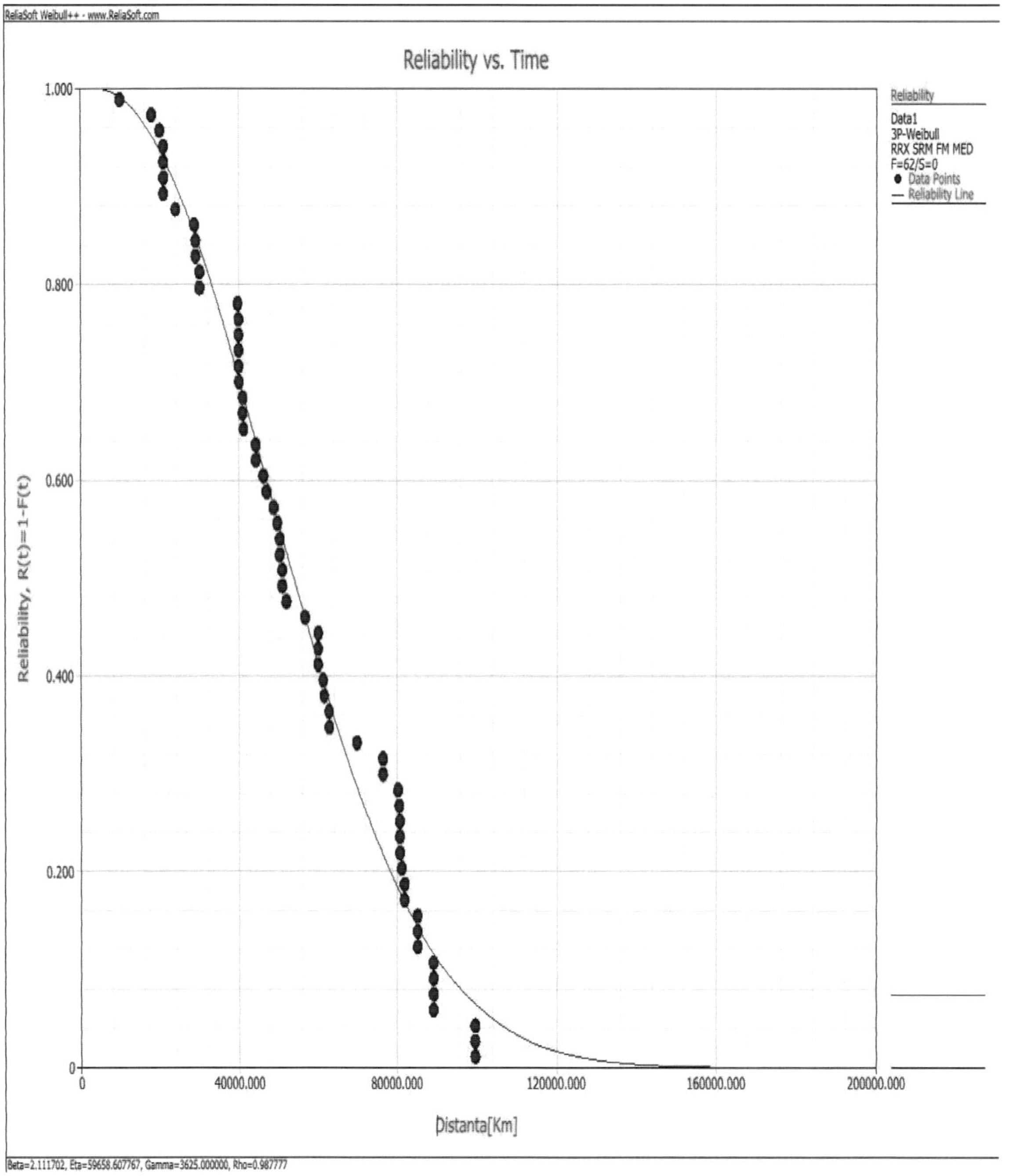

Figure 5. Reliability graphic according to km

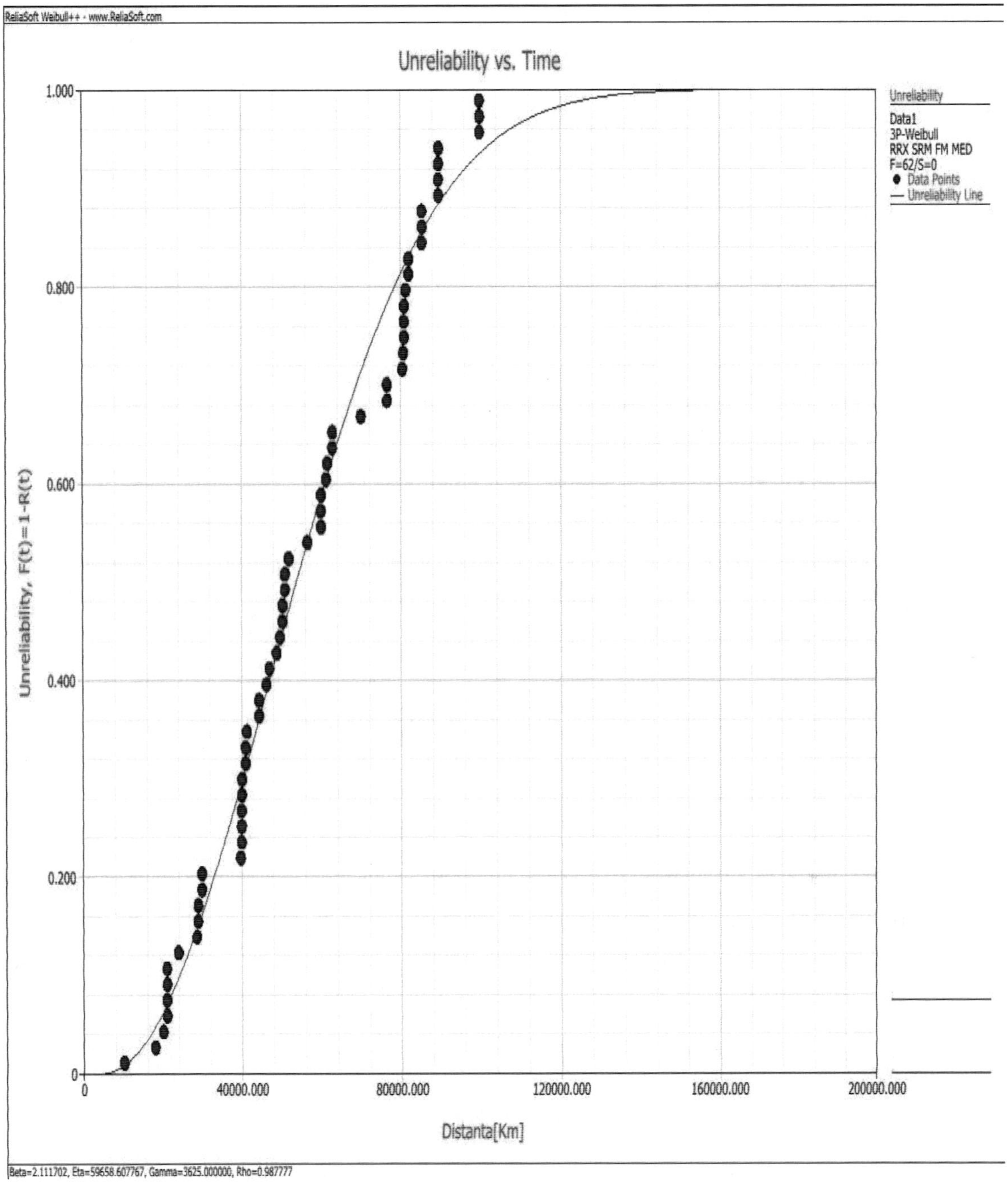

Figure 6. Non-reliability graphic according to km

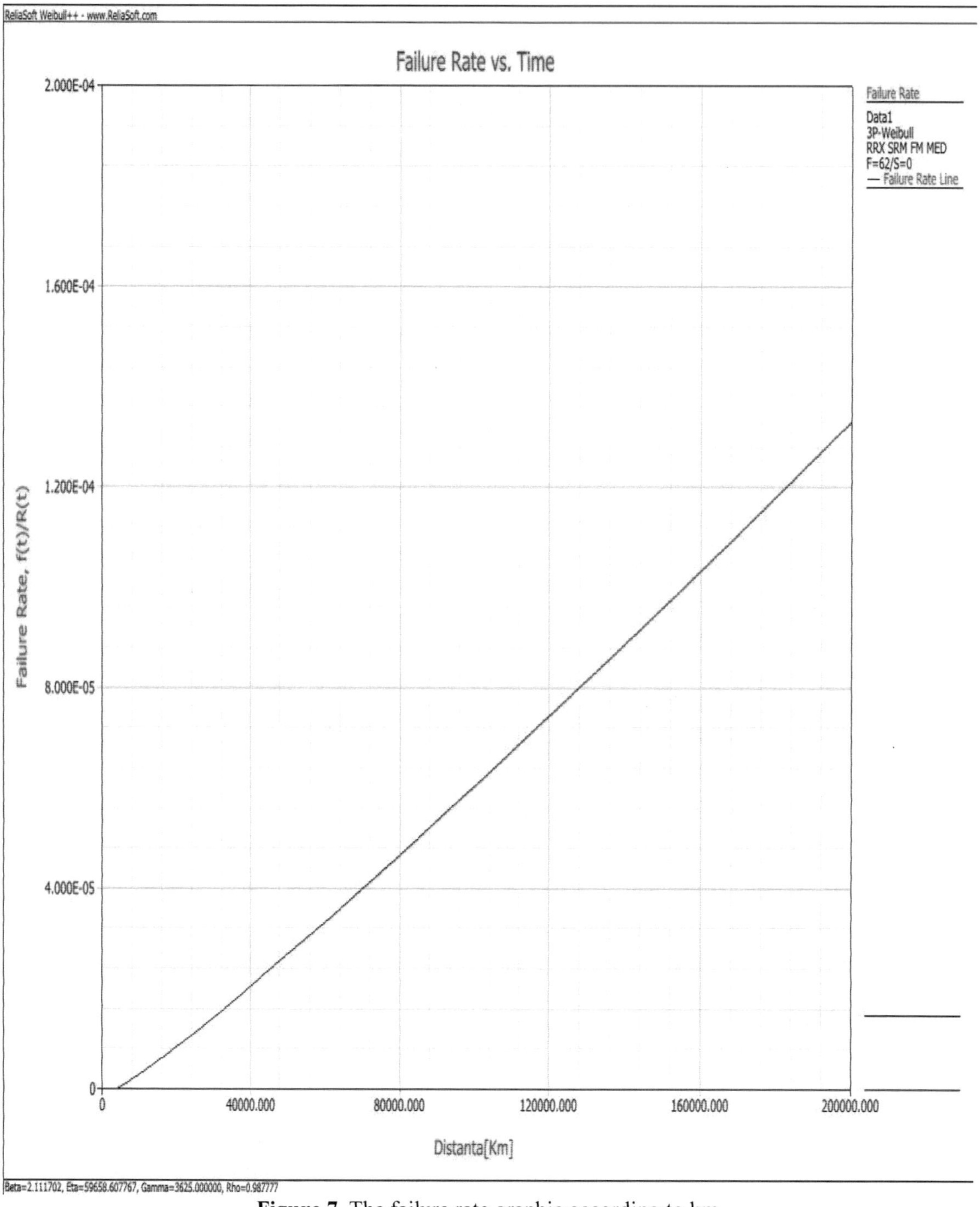

Figure 7. The failure rate graphic according to km

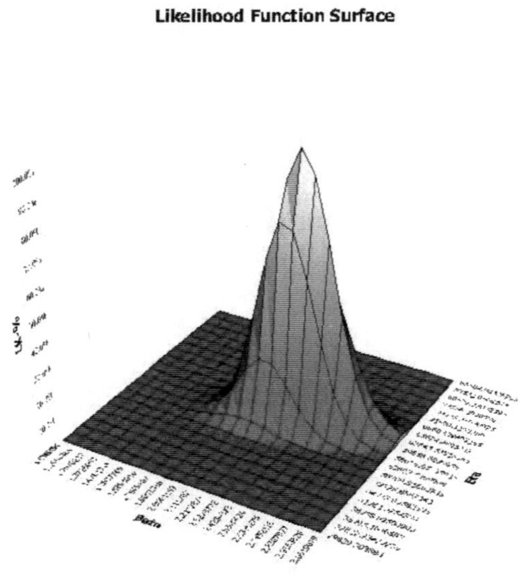

Figure 8. Graphic represention of the Likelihood function

4. Conclusions

From Figure 1 it is found out that by reducing or eliminating the first four categories of faults, stabilizer bar sleeves 25,3968; pre-heated spark plugs 12,6984; bilelet direction 12,6984; transmission and drive shaft 7,9365, the number of accidental stops is reduced with approximately 65%.

From Figure 5 one can observe that the reliability has an accentuated decrease from the moment the car starts. This means that the mechanical problems emphasized through Pareto diagram (Figure 1) occur very fast. The major milestone that contribute to the appearance of the flaws must be analysed with big responsibility in what concerns both the quality of the material and the processing, and their fitting.

Not respecting these observations makes the reliability of the car to drop below 0,2 after only 80.000Km, and the failure rate has a rapid growth in this distance (Figure 7); this means that the car's warranty period, when free maintenance operations are made, has bigger costs for the producer. Also, low reliability imposes a higher maintenance which duration brings frustrations to the owner of the car.

References

[1] ***STAS 8174/1-77, Reliability. Terminology

[2] ***STAS ISO 9001/1991, *Quality Systems. Model of Quality Insurance in Project /Development, Production, Assemblage and Service*

[3] Baron T 1976 *Quality and reliability of products*, Publishing House: Editura Didactică şi Pedagogică, Bucharest

[4] Budiul Berghian A and Vasiu T 2008 *Reliability and maintainability of industrial entities. Theory and applications,* Publishing House: Editura "INFOMIN" Deva

[5] Budiul Berghian A and Vasiu T 2015 Determination of Reliability for Dacia 1304, 1,9 D utility vehicle, *IOP Conf. Ser.: Mater. Sci. Eng.* **85** 012007

Big Data as the basis for the innovative development strategy of the Industry 4.0

T Latinovic[1], D Preradović[2], C R Barz[3], A Pop Vadean[3] and M Todić[1]

[1]University of Banja Luka, Faculty of Mechanical Engineering, Banja Luka, Bosnia and Herzegovina
[2]University of Banja Luka, Faculty of Economics, Banja Luka, Bosnia and Herzegovina
[3]Technical University of Cluj Napoca, Department of Mechatronics and Machine Dynamics, Cluj Napoca, Romania

E-mail: tihomir.latinovic@mf.unibl.org

Abstract. The aim of the paper is establishing a link between Big Data connected with Industry 4.0. We are developing a modern economy which is based on supply chains and logistics. Within each logistics are IT solutions. Some of the world's solutions are compiled in this paper and guidelines are set in the development of such solutions. Industry development 4.0, and in particular the development of Big data, gives us the possibility to introduce new ways of the industry. In the development of industry, maximizing production and reducing costs is extremely important. Real-time data and advanced algorithms give us the possibility to obtain real-time operational industry. Big data and their analysis are of great use in the industry because they improve the accuracy of the decision-making process. We must organize a structure that can design, manage and develop projects based on IoT and Big Data / Analytics technologies. In the industry, good sensor technology provides access to information. The paper has a focus on the managing the database with Big data in the Industry 4.0.

1. Introduction

Data processing with computers, network and internet give us a big benefit. One of the advantages was that information could easily be generated and easily exchanged between people, from people to machines and machines to machines. All this data that generates people and machines, and when a man manages a machine and when we have a collaborative work of man and machine, generate huge amounts of information called "Large Data". Companies then use these large data, through the so-called. "Data Mining", to achieve a better service to its customers and achieve greater profit. It is much important to know both software and hardware for various ways of generating large data, how to memorize them, and to study applications for commercial use.

Data becomes Big Data depending on the capacity of an organization. Something Big Data is for one organization, for the other it is not. There is some opinion that is data transfer over 1 TB. In a real system in production, service, weather, etc., a different large number from different types of devices sometimes are smart and transmit data. Such data has a high speed, a large amount and variety (unstructured data type).

For a better insight into BD size, we set some examples in Table 1 for some of the events that result in a large amount of data.

Table 1. Table with event and description about the amount of data [1]

Event	Description
Flight of Air Bus	10 TB of data is generated every 30 minutes of flight
Smart meters	In 2009, is about 76 million digital electricity meters, which produced 350 million transactions a year. It is estimated if it goes with the current tempo by 2020, the number of these devices amount to 800 million
Cell phone cameras	In 2013 in the world was about 5 million phones cameras. Most of them have GPS. This number increasing from day to day, and also increasing the number of photos and videos that users create
Web users	At the end of 2014, the number of internet users was nearly 3.08 billion which contributed to the volume of Internet traffic measured in ZB
Blogs	Data from 2013 indicate that there are 200 million registered blogs
Email	About 300 billion emails are sent every day
RFID	The use of RFID tags (chips) is increasing at the global level, 2005 was 1.5 million tags, and for 7 years later, 30 billion tags
Twitter	Twitter generates about 12 TB of data per day, with 200 million users and 230 million tweets per day (97,000 tweets per second)
NYSE	New York Stock Exchange daily generates about 1 TB of data

2. Digital Manufacturing

The digitization process consists of simulation, three-dimensional representation and various collaborative tools for creating products through appropriate production processes. Production has been enhanced by the introduction of improved design such as DFM, Computer Integrated Manufacturing (CIM), flexible production, slim production and other production systems that use an information technology the entire production process [2]. 3D tools that handle production from a point of view are developed (Orientation according to demand, Data-driven and Digitally executed). Various strategies are in place to guide the life cycle of products. Production is becoming global, multidisciplinary, based on innovation and customer-oriented [3].

Consequently, Big Data should be understood not only as data sets but also as a set of techniques for processing and analyzing them. Below are the main attributes of Big Data [4]:

- volume – this feature refers to the quantity and means the high dynamics of data growth requiring advanced technologies for efficient processing,
- velocity – data is streaming delivered to the organization in real time, which requires sufficient computing power to provide fast response
- variety –data, the lack of structuring, in different formats, which is generated by various sources,
- variability – Change in the intensity of data during a time,
- complexity – Is a variety way of complex data and unstructured data (eg. documents, video, photos),
- value – refers to the ability to capture unique information from large data sets that significantly influence organizational efficiency.

2.1. Big data
Big data is recently being arrived and applied into the dataset, big data growth is very large and even they don't work with traditional database management system. According to the IDC Big Data is a new generation technology and architecture designed to efficiently extract value from the huge volume of a wide variety of data. Big data is a term for massive data sets having the large, varied & complex type of structure with the challenges of analyzing, Storing and visualizing for extra processes or results. [5]

2.2. Traditional data
Traditional data Traditional data is basically stored in the database or is in the data-warehousing. Example of the traditional database is RDBMS, DBMS, SQL and more. Data Warehouse is a set of tools and techniques to enable the collection of data from the operational system, integration and management of that data into a central database and then the analysis, visualization and other operation can perform in a dashboard. machine. [6],[7]. In Table 2 we see a comparative study of traditional and Big Data.

Table 2. Comparative Study of Traditional vs. Big Data

	Traditional Data	**Big Data**
1.	Data generated in Enterprise level or it is including tradition al data	Data generated in Outside and Enterprise level.
2.	Traditional data sources are include ERP transaction data, CRM data, Web transaction data, Financial data	Non-traditional data sources are including Social Media, Log data, Device data, Sensor data, Video, Images
3.	Data store in gigabytes or terabytes	Data store in Petabytes, Zettabytes, Exabyte's
4.	Data managed in the centralized form	Data manage in physically distributed form
5.	By default, stable and interrelationship	Unknown Relationship
6.	Specialized high-level software as well as hardware used	Inexpensive commodity boxes in cluster mode

3. Industry 4.0
In industry 4.0 products and production systems such as machines, warehouses and operational resources are enhanced for Cyber-Physical Systems (CPS) and are connected to the world's manufacturing networks (Kagermann et al., 2013) [8]. These intelligent entities in production have the ability to interpret data, encourage actions, and have the ability to autonomy self-control and self-optimization (Lee, 2008) [9]. Intelligent products in the future will be identified, know their history, status and how they are produced. Intelligent manufacturing systems are linked with intelligent business processes. This gives us the ability to control and optimize the value chain, starting from the production to the final delivery of the product (Kagermann et al., 2013) [8]. A gradual approach to the physical world and the digital world with CPS systems enables autonomous and decentralized production (Brettel et al., 2014, Monostori, 2014) [10].

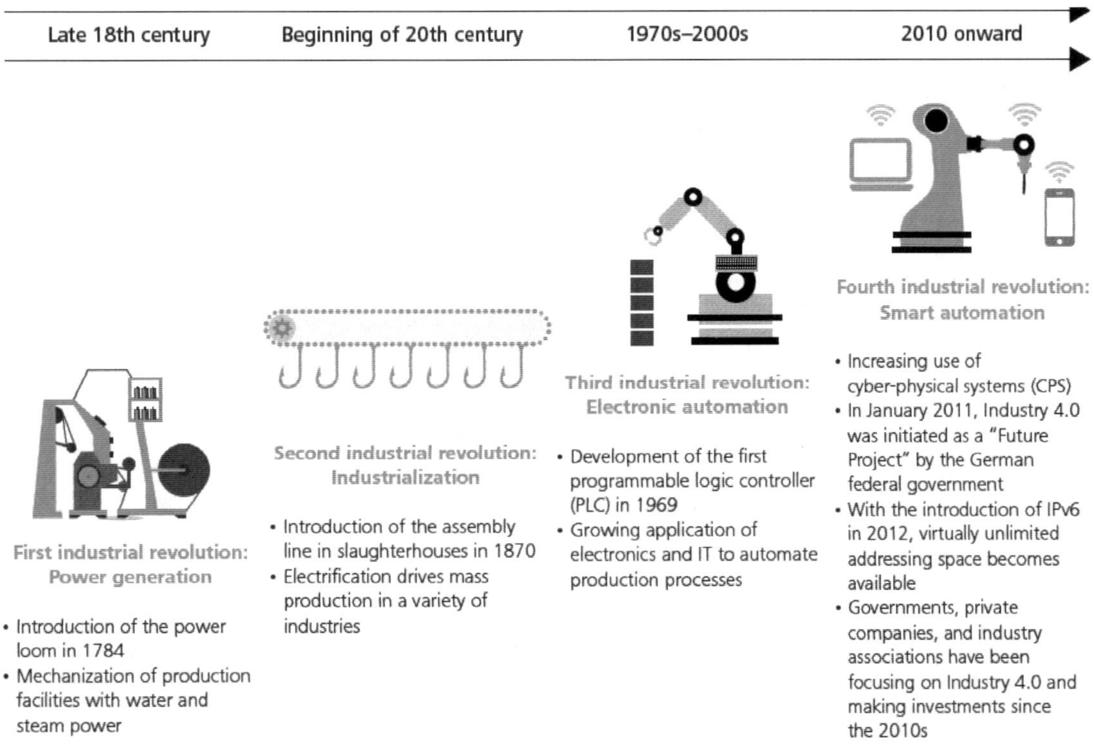

Figure 1. History of Industry 4.0 Graphic: Deloitte University Press | DUPress.com [11]

3.1 Industry 4.0.

Industry 4.0 has six design principles:

- **Interoperability**: With the Internet of things humans communicate with Smart Factories
- **Virtualization**: We working on a virtual copy of the Smart Factory created by 3D software and linked with real sensor data from physical processes.
- **Decentralization**: CPS systems inside Smart Factories have the possibility to make decisions
- **Real-Time Capability**: real-time collecting data
- **Service Orientation**: Service must be oriented on the **Internet of Services**
- **Modularity**: Smart Factories must be flexible adaptation with the module and working islands.

4. Big Data software solutions

There are software solutions that are directly implemented for Big Data in Industry 4.0. There are Hadoop, Cassandra, Scylla, Hypertables, Accumulo, Amazon SimpleDB, Clouds, MonetDV, Apache Flink, Elassandra.

For memory files are: Elastic, MongoDB, Cloud Datastore, Azure DocumentDB, RethinkDB, ToroDBm IBN Cludant etc.

For graphical bases are: Neo4J, Arango DB, Sparksee, GraphBase, Trinity, etc.

NoSQL term used by Carlo Strozzi is the type of open source relational database data that do not have standard SQL interface, but are still relational [12]. There are several approaches to classify the NoSQL database, there are different categories and subcategories, which are subject to type some of them overlapping. Below is provided a basic classification of the database according to the model data, with examples:

1. The key-value: Redis, Riak, OrientDB
2. Document: CouchDB, MongoDB, eXistDB
3. Graph: InfiniteGraph, Neo4J, Stardog
4. Column: Vertica, Druid, HBase
5. Multi-model: CortexDB, MarkLogic, FoundationDB

Semi-structured data are actually a form of structured data, but not in accordance with the formal structure of the data model associated with relational databases. It could be said that the semi-structured data describing themselves (Self-describing), in general, refers to the fact that no scheme. Semi-structured data have appeared all higher use of the Internet where the network is through this type of data, but also appeared the need for the semi-structured data model.

OEM (Object Exchange Model) is one of the standards for presentation of semi-structured data. The key-value databases have the simplest data model. The data is organized so that the associative array notation. A unique key is used to identify a unique record and can be used with three available operations: delete the default key, change the record that is related with that key and enters a new key for the given file. The document database is semi-structured databases, implemented without tables. Each tuple can contain any number of fields and no size limitations of the same. Even if the use in a structured way, there are no limits or limiting scheme database according to a pre-defined structure. Which means that application developers with ease can create dynamic data. This type database can be implemented as a layer through a relational or object database. Another option is to implement directly as semi-structured file formats, such as JSON or XML. Traditional concepts like and the index keys are designed in terms of function as in relational databases data. Using this, we should achieve results of the same performance as in the system implemented by tables.

Table 3. Capabilities and characteristics of representative Big Data software solutions [13]

BIG DATA SOFTWARE SOLUTIONS				
Capabilities /Characteristics	Hadoop HDFS & MapReduce	Cassandra	MongoDB	SimpleDB
Data model	File system	Column	Document	Key-Value
Batch processing / distributed computing	Yes	No	No	No
Real-time queries	No	Yes	Yes	Yes
Random access	No	Yes	Yes	Yes
Horizontal scaling	Yes	Yes	Yes	Yes
Strength	Data processing	Write	Read	Full Indexing
Architecture type	Master-Slave	Peer-to-Peer	Master-Slave	Web Service / Cloud Computing
CAP theorem	Consistency, Partition Tolerance	Availability, Partition Tolerance	Consistency, Partition Tolerance	Availability, Partition Tolerance

5. Conclusion

BD technology has recently been widely used. A large number of scientific papers and projects confirm the necessity of adopting BD tools for everyday work. There are still no projects in Bosnia and Herzegovina for the introduction of Big Data, any forums or blogs on this topic. Also, great data is not mentioned in the curriculum in higher education at the University of Banja Luka.

Many colleagues dealing with computer science do not get involved in this area, because they do not generate large data in the industry in the Western Balkans, but only in the field of Telecommunication.

Big companies almost do not exist, and therefore the need for innovations in the BD and Industries 4.0 industries. Economic areas that have the greatest potential to exploit BD in order to increase profitability, increase efficiency and reduce costs are primarily in the information sector, followed by the financial sector, insurance, management, and retail.

This paper outlines the guidelines for the potential use of Big Data in companies, where they are created, and the software tools that need to work with BD. The paper also lists digital production as the basis of Industry 4.0. When we digitally handle some of it, we send it to digitized production and it produces it, without any worker. Information technologies fully assume the role of service providers and products. These are created by BD, who must process them to overcome digital production. This paper presents the basis of Industry 4.0 as well as most software tools distributed by the type of data they are processing. In further research, we will deal with Bosnia and Herzegovina and where there is no industry development strategy 4.0 because we are still dealing with small and medium-sized enterprises with Industries 3.0 and below. One of the possible applications is the Tax Administration and Telecom Srpske and BiH. The world is slowly shifting from Big to Fast Data and the ability to exploit data generated in real-time and based on their response time. The need for speed is indispensable because it increases profitability and customer satisfaction. The primary goal is to satisfy the client's needs when he wants, where he wants and how he wants, otherwise, he will take over the competition.

Contemporary companies face a number of difficulties when it comes to applying information technology. Business systems are becoming more complex and require significant human resources, so staff in IT departments has a difficult task to maintain and improve the system, both because of technology advancement and difficulties their acceptance by employees, especially when it comes to large companies. Large companies, for as opposed to small and medium enterprises, have a difficult situation because they are less agile and flexible, so they cant quickly respond to change and adapt. Business software and information technology have become integral and the inseparable part of every business entity. Many companies are thanks to innovation in the field of information systems and business applications have succeeded in expanding the current or emerging markets. It can be assumed that modern business is highly dependent on the information systems and software solutions they use. Because of that tracking contemporary trends in the business software market becomes so important that any inertness can lead to the loss of competitive advantage, and in some cases may even endanger survival companies.

Further research can be linked to more specific technical requirements, infrastructure and BD architecture, exploring the possibilities of introducing new exploitation technologies, exploring new methods of accessing fast data, as well as extreme data.

References

[1] Latinović T S, Preradović D M, Barz C R, Latinović M T, Pop P P and Pop-Vadean A 2016 Big Data in Industry, *IOP Conf. Ser.: Mater. Sci. Eng.* **144** 012006

[2] ***http://www.plm.automation.siemens.com/en_us/plm/digital-manufacturing.shtml

[3] ***https://ieeexplore.ieee.org/abstract/document/6857843/

[4] Gierej S 2017 Big data in the industry – overview of selected issues, *Management Systems in production engineering* **25**(4) 251-264

[5] Suppiah Y, Tariqi Raja R M and Mohamad Amran M F 2016 A Study on Social Data Analytics and Privacy Concern among Social Media Users, *International Journal of Computer Applications* **149**(5) 46-49

[6] ***https://gcn.com/blogs/reality-check/2014/01/hadoop-vs-data-warehousing.aspx

[7] Tapan P Gondaliya, Hiren D Joshi and Hardik J Joshi 2015 *NewBig Things in Era of DigitalData : 'BigData'&BigDataChallenges with its Solution Using Different Tools*, 10th

International Conference CALIBER2015, March 12-14, Shimla, India

[8] Kagermann H, Wahlster W and Helbig J 2013 *Umsetzungsempfehlungen für das Zukunftsprojekt Industry 4.0 –Abschlussbericht des Arbeitskreises Industry 4.0*, Forschungsunion im Stifterverband für die Deutsche Wissenschaft, Berlin, Germany, pp 51-55

[9] Lee L C, Su W S and Tang K 2014 *Scheme of handover self-optimization using big data analytics*, The 16th Asia-Pacific Network Operations and Management Symposium, 17-19 September, Hsinchu, Taiwan

[10] Monostori L 2014 *Cyber-physical Production Systems: Roots, Expectations and R&D Challenges*, CIRP Conference on Manufacturing Systems, Vol 17, pp 9-13

[11] Sniderman B, Mahto M and Cotteleer M J 2014 *Industry 4.0 and manufacturing ecosystems*, July 1, National Academy of Science and Engineering

[12] Raghav Mehra, Nirmal Lodhi and Ram Babu 2015 Column Based NoSQL Database, Scope and Future, *International Journal of Research and Analytical Reviews* **2**(4)

[13] Gölzer P, Cato P and Amberg M 2015 *Data processing requirements of industry 4.0 – use cases for big data applications*, ECIS 2015 Proceedings, Twenty-Third European Conference on Information Systems (ECIS), Münster, Germany, pp 3

Optimal solution of the Guarini puzzle extension using tripartite graphs

A E Iordan

Politehnica University of Timisoara, Department of Electrical Engineering and Industrial Informatics, 5 Revolution Street, Hunedoara, 331128, Romania

E-mail: anca.iordan@fih.upt.ro

Abstract. In this paper there will be presented a software implemented in the C# programming language using .Net Frameworks platform which allows an optimal solution of a puzzle obtained through Guarini puzzle extension. The generalization of Guarini puzzle consists in considering 2·n knights placed on the chessboard of 4·n dimension. Half of these knights are white and are positioned on the first line of the chessboard, and the other half are black and are positioned on the last line of the chessboard. The aim of this puzzle is to move the knights through a minimum number of moves so on the first line we have all the black knights, and on the last line we have all the white knights. This puzzle belongs to a category of problems which can be efficiently resolved by use of tripartite graph. Numbering the chessboard squares these will represent the vertices of the graph, and the possible moves of the knights will represent the edges of the graph. Modelling this problem associated with this puzzle, it can be noticed that the corresponding vertices of the squares from the first and last line form the first vertices subset from the tripartite graph, the ones corresponding to the second line of squares form the second vertices subset from the tripartite graph, and the ones corresponding to the third line of squares form the third vertices subset from the tripartite graph.

1. Guarini puzzle generalization

The generalization of Guarini puzzle [1], [2] consists in considering 2·n knights placed on the chessboard of 4·n dimension: the n white knights are at the first line, and the n black knights are at the last line. The scope of this puzzle is to exchange the knights to obtain the position represented on the right side of figure 1 (particular case n=3) in the minimum number of knight moves, not permitting more than one knight on a square at any time.

This puzzle belongs to a class of problems which can be efficiently resolved by use of tripartite graph [3]. Numbering the chessboard squares these will represent the vertices of the graph, and the possible moves of the knights will represent the edges of the graph. Modelling this problem associated with this puzzle, it can be noticed that the corresponding vertices of the squares from the first and last line form the first vertices subset from the tripartite graph, the ones corresponding to the second line of squares form the second vertices subset from the tripartite graph, and the ones corresponding to the third line of squares form the third vertices subset from the tripartite graph.

2. Tripartite graphs

In graph theory a multipartite graph is a graph whose vertices can be divided into n different independent sets [4]. In other words, it is a graph that can be colored with n colors, so that no two endpoints of an edge have the same color.

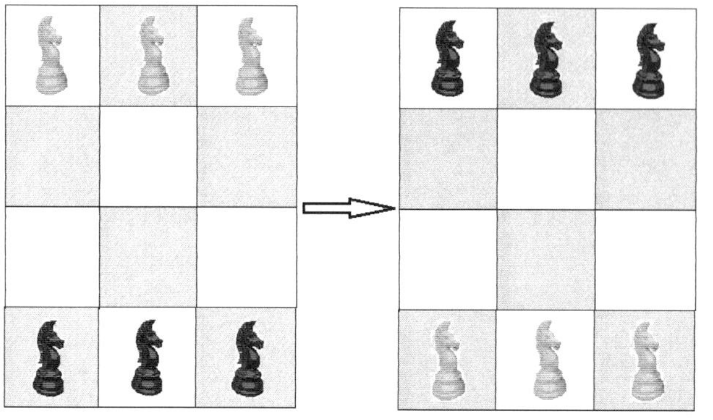

Figure 1. Start state and goal state for the six knights puzzle

When n = 2 these are the bipartite graphs [5] (Figure 2), and when n = 3 they are called the tripartite graphs (Figure 3) [6]. A complete multipartite graph [7] is a multipartite graph in which there is an edge between every pair of vertices from different independent sets. These graphs are described by notation with a capital letter K subscripted by a sequence of the sizes of each set in the partition. For example, $K_{2,2,2}$ is the complete tripartite graph of a regular octahedron (Figure 4), which can be partitioned into three independent sets each consisting of two opposite vertices.

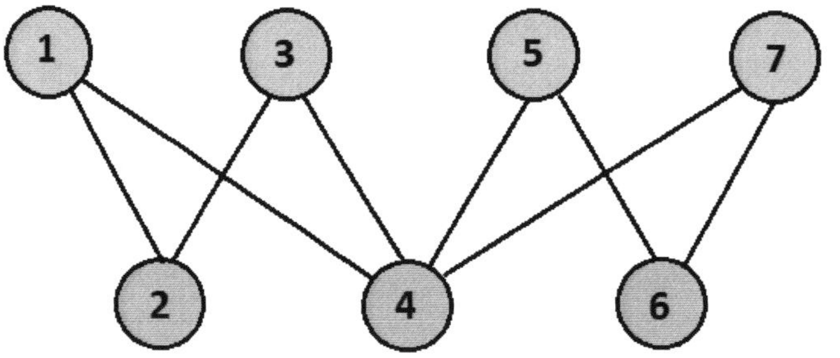

Figure 2. Example of a bipartite graph

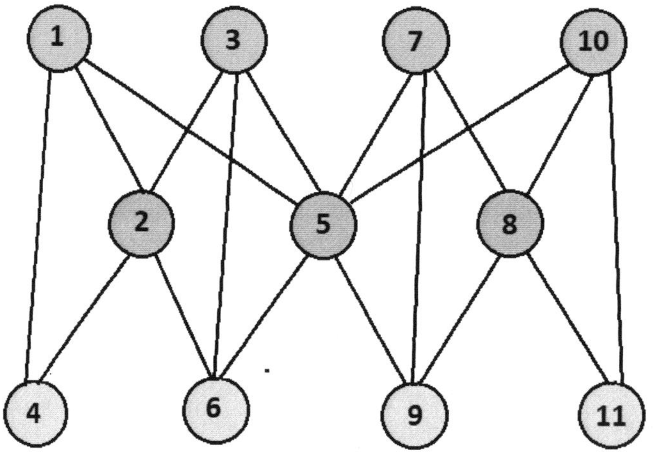

Figure 3. Example of a tripartite graph

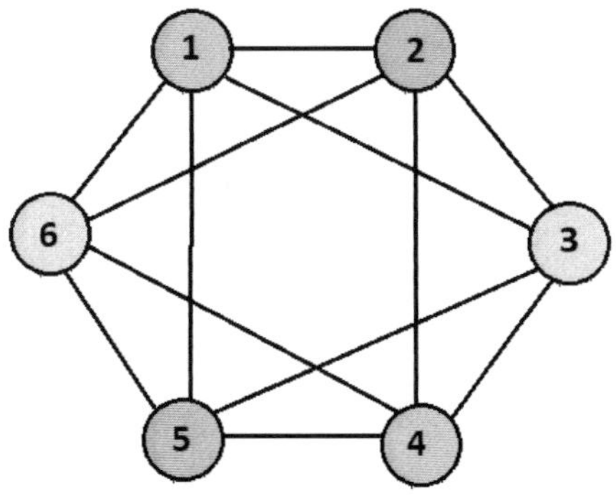

Figure 4. Complete tripartite graph $K_{2,2,2}$

3. The analysis of the software

The scope of this paper is to present a software which permits an optimal solution of the puzzle presented in the previous paragraph. The mathematical modelling of this puzzle with the aim of an optimal resolving can be used to teach tripartite graph notion [8].

From the perspective of unified modelling language [9], the analysis of didactical game application includes the representation of the use-case diagram [10]. The use-case diagram offers simplified and graphical representation of what the didactical software must really do. Use-case diagram (Figure 5) is based upon functionality and thus will focus on the "what" offers the didactical software and not how will be realized.

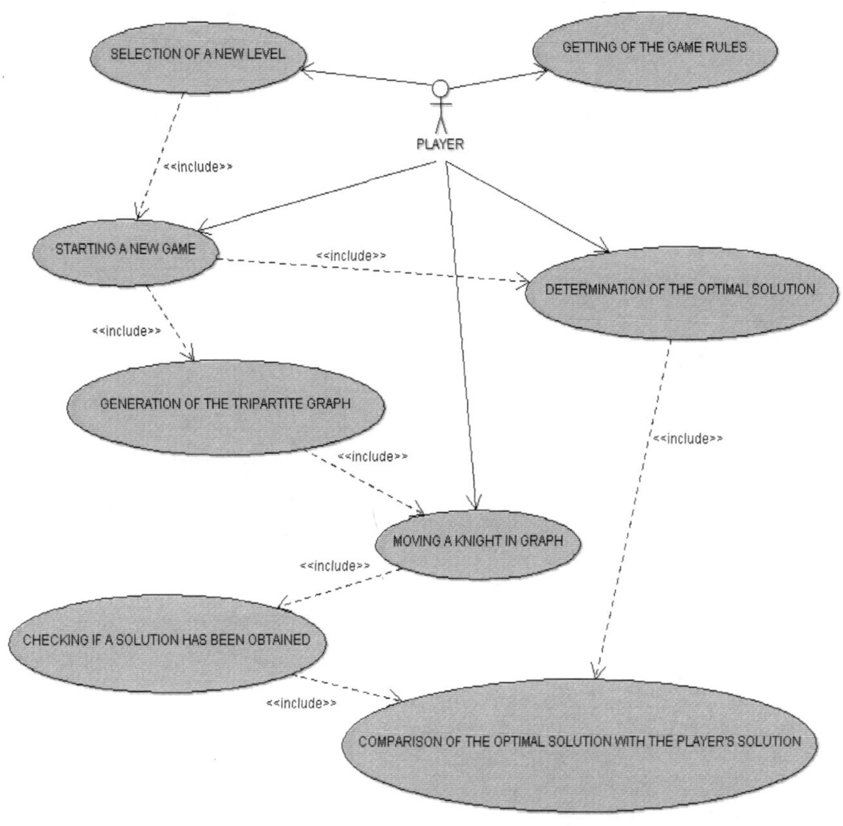

Figure 5. Use-case diagram

The use-case diagram corresponding to this didactical software includes:

- One actor - the user who is external entity with that the software interacts;
- Eight use-cases which describe the performance of the software;
- Association relationships between player and use-cases, and dependency relationships between use-cases.

4. The design of the software

Conceptual modelling allows identifying the very important elements for the didactical software application. Class diagram [11] is represented (Figure 6) in order to be observed the connection mode between the classes and the interfaces that are used and also the composition and aggregate relationships between instances.

After identifying the specific elements of the didactical software, they were been implemented 3 interfaces and 9 classes grouped into three packages.

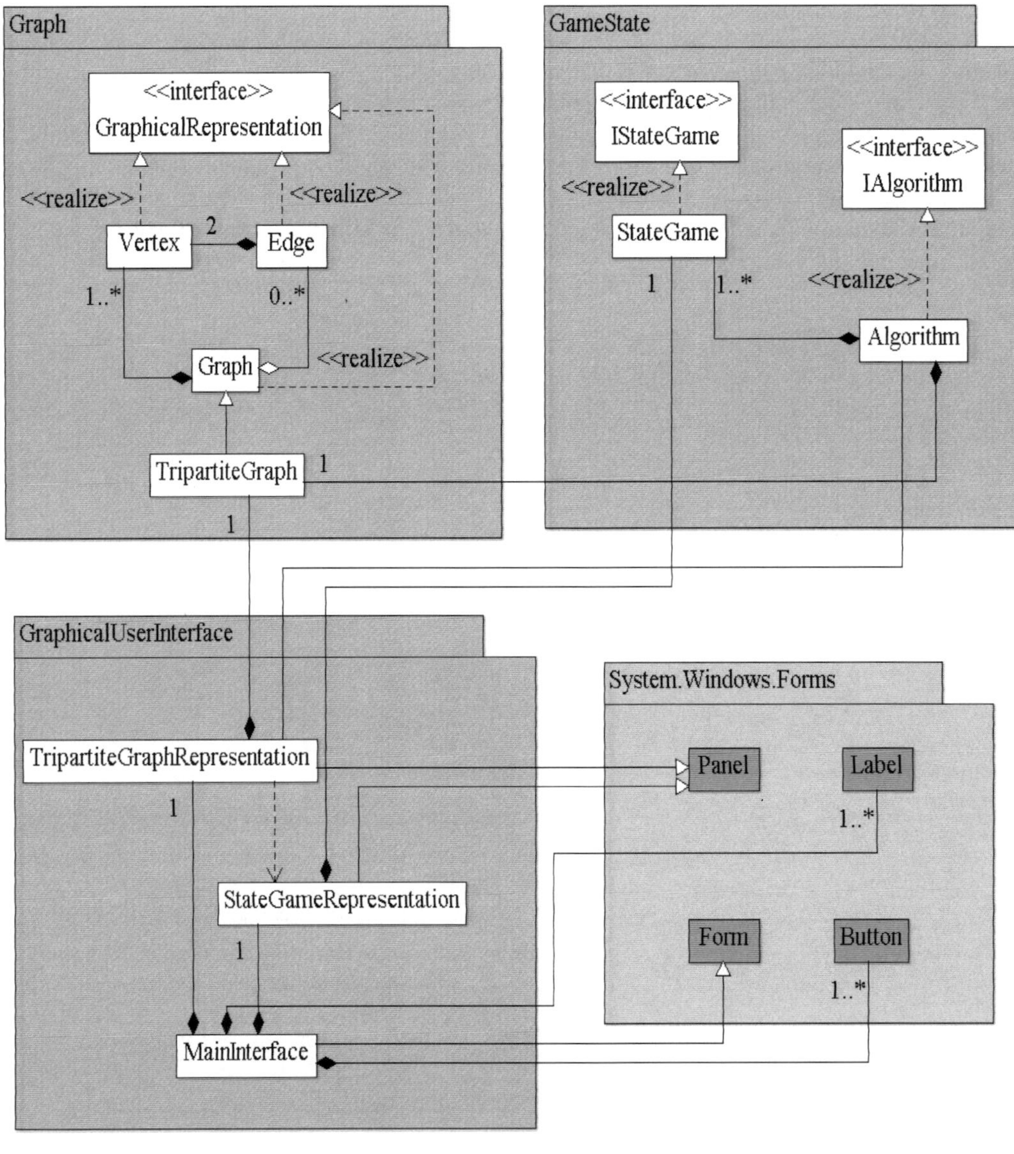

Figure 6. Class diagram

Package “GameState” brings together concepts that correspond to the algorithm corresponding to the tripartite graph, consisting of 2 classes and 2 interfaces. For memorizing a state configuration, there was implemented “StateGame” class which implements “IStateGame” interface.

The algorithm corresponding to tripartite graph is implemented by using “Algorithm” class which implements “IAlgorithm” interface. An instance of this class is composed by one instance of “StateGame” class and one instance of “TripartiteGraph”, according to composition relationship which exists in diagram.

Package “GraphicalUserInterface” brings together concepts that correspond to the graphical user interface corresponding to the didactical software, consisting of 3 classes. For designing the graphical interface of the software, it was implemented the „MainInterface” class.

An instance of this class is composed by one instance of “TripartiteClassRepresentation” class and one instance of “StateGameRepresentation” class, according to composition relationship which exists in diagram. The “StateGameRepresentation” class was implemented for show the current state of the puzzle, and the “TripartiteGraphRepresentation” class was implemented to present the tripartite graph associated to the current state of the puzzle.

Package “Graph” (Figure 7) brings together concepts that correspond to the graph used to solve the puzzle, consisting of 4 classes and one interface. The concept of a vertex from a graph is implemented by using “Vertex” class which implements “GraphicalRepresentation” interface. The concept of an edge from a graph is implemented by using “Edge” class which implements “GraphicalRepresentation” interface. An instance of this class is composed by one instance of “Vertex” class.

The concept of graph is implemented by using “Graph” class which implements “GraphicalRepresentation” interface. An instance of this class is composed by several instances of “Vertex” class and several instances of “Edge” class, according to composition relationship which exists in package diagram. The concept of tripartite graph is implemented by using “TripartiteGraph” class which inherits from “Graph” class.

In the design phase, it also performs statechart diagrams that specify the dynamic behavior of class instances. Figure 8 shows the statechart diagram associated with an instance of the “Algorithm” class.

5. Graphical user interface

The software is accomplished using the C# object-oriented programming language [12], [13] on the .Net Framework platform [14], [15]. Given that specified requisites in uses cases diagram (Figure 5) it was designed graphical user interface.

After the level of difficulty was selected, it is generated the tripartite graph associated to the start state (Figure 9), and the player can begin the game. The application permits moves of any knight between two adjacent vertices, such that any vertex of the tripartite graph does not contain more than one knight at any time.

After each move, the current state of the puzzle is updated (Figure 10) and it is checked whether the goal state has been reached.

When the goal state is reached (Figure 11), it is checked whether the solution obtained is optimal.

6. Conclusions

Because the graph theory is a difficult field of computer science, the computer games can be of helpful in the teaching/learning process of manifold aspects regrading to graphs. The utilization of didactical software signifies an efficient way of stimulation and development of the student motivation.

The didactical software presented in this paper contributes to the qualitative assimilation of the knowledge and to the competences development related to the tripartite graphs, having great educational opportunities, because it shows the most accessible and the most interesting type of activity for students.

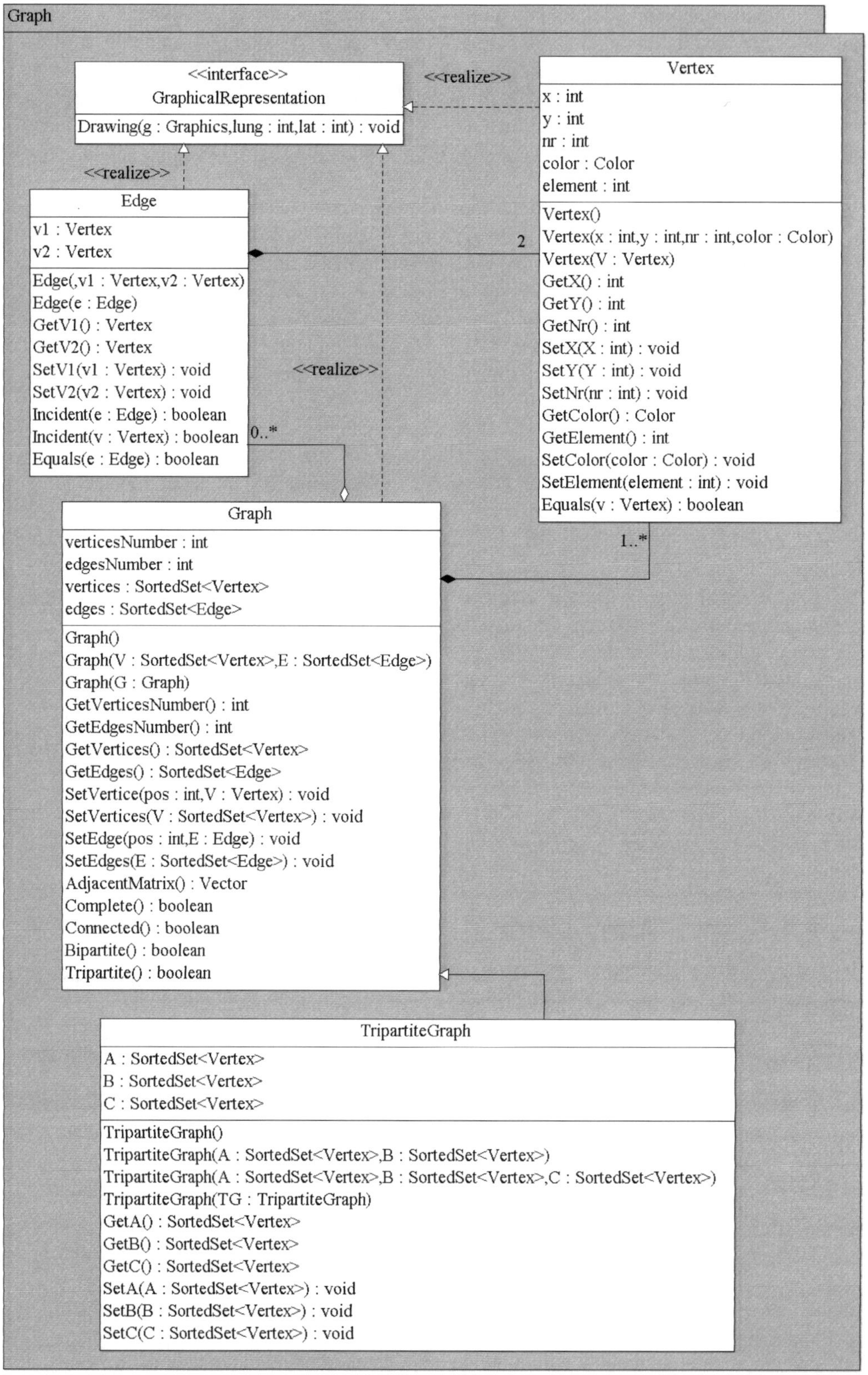

Figure 7. Package diagram

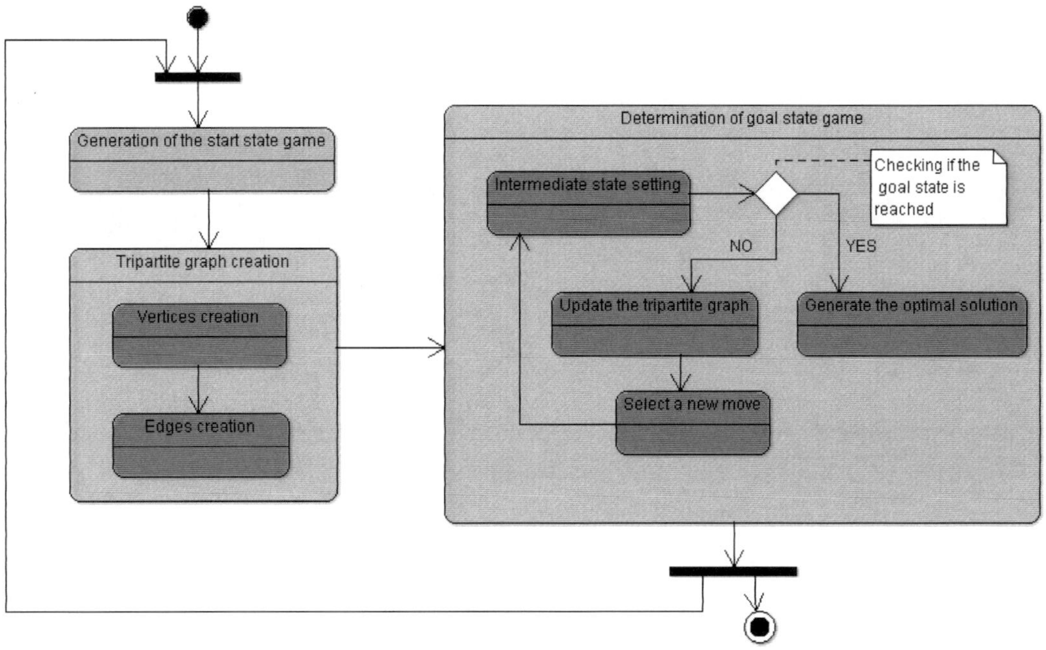

Figure 8. Statechart diagram for an instance of the "Algorithm" class

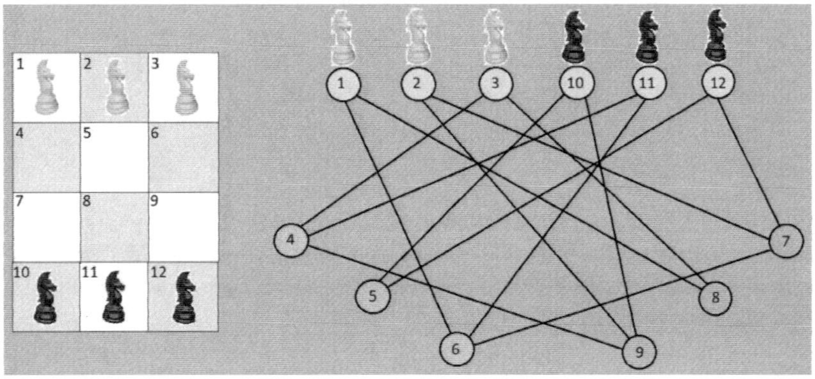

Figure 9. Graphical user interface corresponding to the start state of the 6 knights puzzle

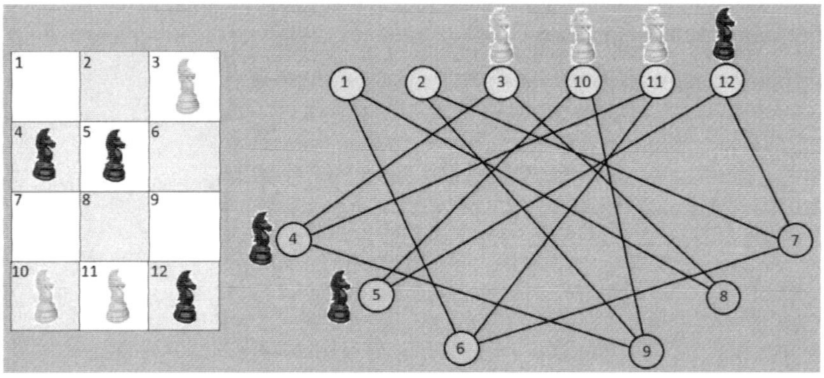

Figure 10. Graphical user interface corresponding to an intermediate state

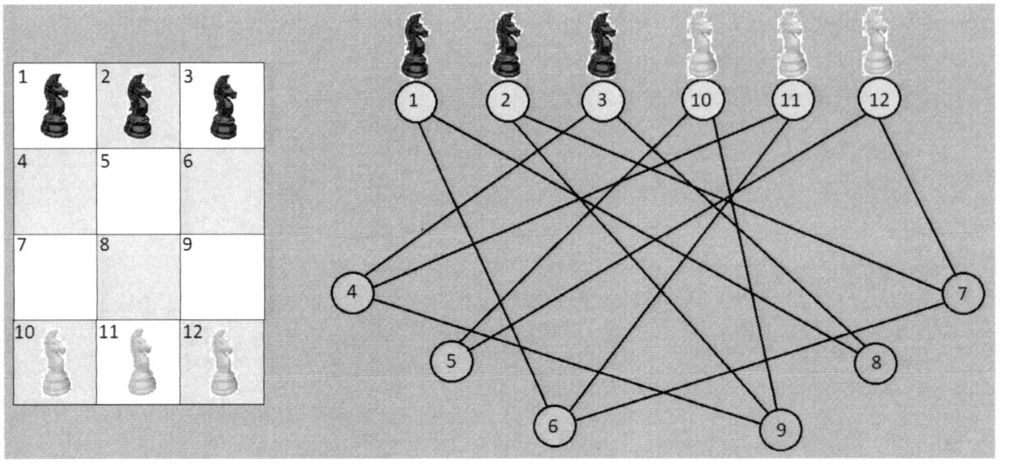

Figure 11. Graphical user interface corresponding to the goal state of the 6 knights puzzle

References

[1] Levitin A and Levitin M 2011 *Algorithmic Puzzles*, Oxford University Press, New York, USA
[2] Grabarchuk S 2005 *The New Puzzle Classics: Ingenious Twists on Timeless Favorites*, Sterling Publishing Co Inc., pp 204-206
[3] Rajasekaran G and Sampathkumar R 2015 Optimal Orientations of Some Complete Tripartite Graphs, *Mathematics Subject Classification* **29**(8) 1681-1687
[4] Even S and Even G 2012 *Graph Algorithms*, Cambridge University Press, New York, USA
[5] Muscalagiu I, Vidal J and Cretu V 2007 *Experimental Analysis of the effects of agent synchronization in Asynchronous search algorithms*, Kes International Symposium on Agent and Multi-Agent, Wroclaw, Poland, May 31- Jun 1, pp 619-636
[6] Diestel R 2017 *Graph Theory*, Springer Verlag, New York, USA
[7] Rob R, Panoiu C and Panoiu M 2015 Voltage and Current Signals Processing in Functioning of Nonlinear Loads, *Soft Computing Applications* **2** 1171-1184
[8] El-Zohny H and El-Morsy H 2012 Shortest Path Algorithm for some Graphs before and after Folding, *International Journal of Pure and Applied Mathematics* **75**(2) 141-148
[9] Bruegge B and Dutoit A 2013 *Object Oriented Software Engineering Using UML*, Prentice-Hall, Inc. Upper Saddle River, NJ, USA
[10] Dennis A, Wixom B H and Tegarden D 2012 *Systems Analysis and Design with UML*, John Wiley & Sons Ltd, Hoboken, NJ, USA
[11] Rumpe B 2016 *Modeling with UML*, Springer International Publishing AG, Switzerland
[12] Doyle B 2013 *C# Programming From Problem Analysis to Program Design*, Cengage Learning, Australia
[13] Purdum J 2012 *Beginning Object Oriented Programming with C#*, John Wiley & Sons, Inc., Indianapolis, Indiana, USA
[14] Gittleman A 2011 *Computing with C# and the .Net Framework*, Jones and Bartlett Publishers, Burlington, MA 01803, USA
[15] Thuan T and Hoang L 2015 *.Net Framework Essentials*, O'Reilly & Associates, Inc. Sebastopol, CA, USA

IOP Conf. Series: Materials Science and Engineering **477** (2019) 012047 doi:10.1088/1757-899X/477/1/012047

Experimental test bench for studying the boiling dynamics

V Stoica, M Ilie, A Borborean, A Cioabla and D Lelea

Politehnica University of Timisoara, Faculty of Mechanical Engineering, Mihai Viteazu str., no. 1, 300222 Timisoara, Romania

E-mail: virgil.stoica@upt.ro

Abstract. The need of reducing the fuel consumption and pollution in case of road vehicles force the researchers to find solutions. Since the improvements of the internal combustion engine by design are on a rather flat trend, the researchers are looking to recover the lost energy. About 70% of the fuel's potential energy is released into the atmosphere as heat. This energy can be recovered in different manners like Rankine cycle, thermoelectric Peltier modules, absorption refrigerator cycle, and so on. The steam engine might be a promising candidate. Depending on the operating regime, the fuel economy can be up to 31.7%, accordingly to some re-ports from the open literature. The scope of the present research is to make an experimental research on a steam engine where the boiling takes place directly in the engine cylinder. So the preliminary investigations are required for this purpose. As a first stage, an experimental test bench for studying the vaporization in an enclosure is presented in the paper. The experimental test bench will be equipped with a controllable heat source at adjustable volume and a complete automatization for process monitoring and control.

1. Introduction

The energy recovery in case of the internal combustion engine is one of the possibilities to improve the global efficiency of it. Therefore, it is proposed for study a steam engine at which the boiling will take place directly within the cylinder eliminating in this way, the boiler which is an additional system to care of and also dangerous in some situations. For the proposed engine, the liquid water would be sprayed on an in-cylinder hot wall. The vaporization will take place leading to a pressure increase and finally to the piston movement. Of course, the efficiency would be rather small if the vapor does not superheat meanwhile. For doing a feasibility study of such an engine, it is important to know how the vaporization takes place in a closed container, where the pressure will be not constant due to vaporization and increase in vapor amount.

The vast majority of research that is somewhat connected to the proposed topic relates to spray cooling. In this studies, the pressure at which the process takes place is kept constant and also there is a continuous spraying.

Liang and Mudawar [1] have done a review of the spray cooling and they pointed out the influence of different parameters on the spray cooling process and vaporization regime. Mudawar and Estes [2] developed a correlation for predicting the critical heat flux (CHF) for different nozzles flow rates and subcooling. They pointed out, that there is an optimum distance between the nozzle and the surface that maximizes CHF. Visaria and Mudawar [3] have also done a systematic study concerning the influence of different parameters like nozzle type, subcooling, flow rates, orientation on the CHF. Horacek et al. [4] concluded that the CHF is more influenced by the length of the contact line than the wetted area fraction. During this research, they pointed out the influence of the dissolved gas on the

CHF. Gilberto Moreno Jr [5] analyzed the influence of the heater type (coated, non-coated), the distance between the nozzles and heater, coolant type and subcooling on heat transfer characteristics. Matthias Winter [6] studied also the influence of the working fluid, subcooling, saturation conditions and the surface topography. Rui Zhao et al [7] presents a mathematical model for the spray cooling and several experimental results. The model end experiments include the vaporization effect but they are not focused on this aspect.

From of all these studies, we can conclude that there are many parameters that are influencing the vaporization process during the spray cooling. We will focus, for the beginning, on analyzing the influence of only a few parameters such as: the amount of liquid injected, the temperature of the heater, the subcooling, and the volume of the test cell, to determine the heat transfer coefficients required for the analysis of the operation of a steam engine with built-in boiler.

2. Test bench description

The experimental test bench used to determine the heat transfer coefficients and to analyze the boiling dynamics is depicted in Figure 1. The boiling process takes place within a cylinder closed at the bottom by a heater and at the top by a wall with an adjustable position. This solution, with an adjustable wall on the top, was chosen in order to make the experiments for different volumes of the cylinder. The bottom electrical plate heater is commercially available, mostly used for small boilers.

The heater is powered by a variable power supply to control the dissipated heat, and its temperature is monitored via two thermocouples. On the cylinder side, the water injector is mounted. The pressurized water is ensured from a reservoir through a sub-merged pump and liquid heater. The water pressure and temperature are monitored with the two sensors mounted in front of the injector. The steam exhaust is ensured through the exhaust pipe, controlled by an electromagnetic valve (steam valve). The wall and steam temperatures are monitored with thermocouples, and the pressure within the cell is monitored with a pressure sensor mounted on the cylinder side. For data acquisition and actuators control, NI data acquisition system is used together with a signal conditioning box. Data analysis and visualization are done using a desk-top PC and LabView software.

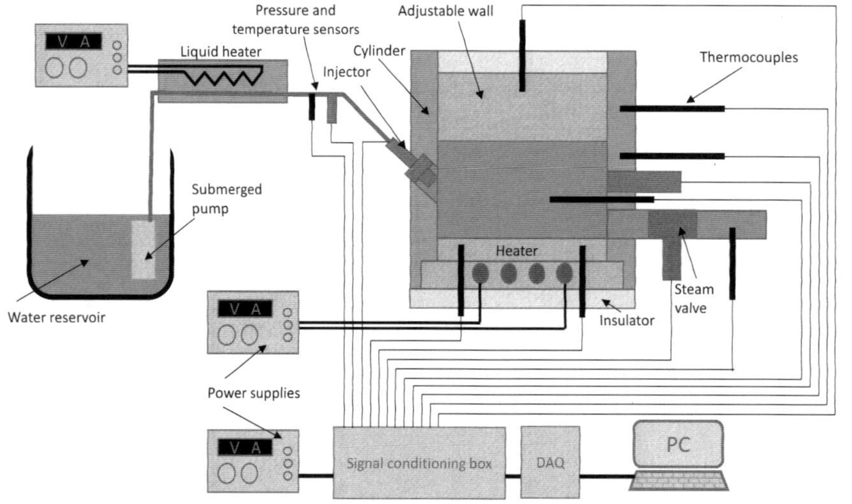

Figure 1. The proposed experimental test bench

3. Data analysis

The processes that take place within the cell are not steady state, therefore, the analysis of the time evolution of temperatures and pressures are not straightforward. The main goal of the presented experimental set-up is to determine the heat transfer coefficients between the liquid-wall, liquid-vapour and vapour-wall, to identify or estimate the injection law and to see if the proposed mathematical model can be suitable for analysing the functionality of a steam engine at which the boiling takes place directly in the cylinder.

The heat and mass transfer within the cylinder are schematically presented in Figure 2. As can be seen, two independent subsystems are considered, vapour and liquid respectively. Based on this arrangement a mathematical model was developed [8].

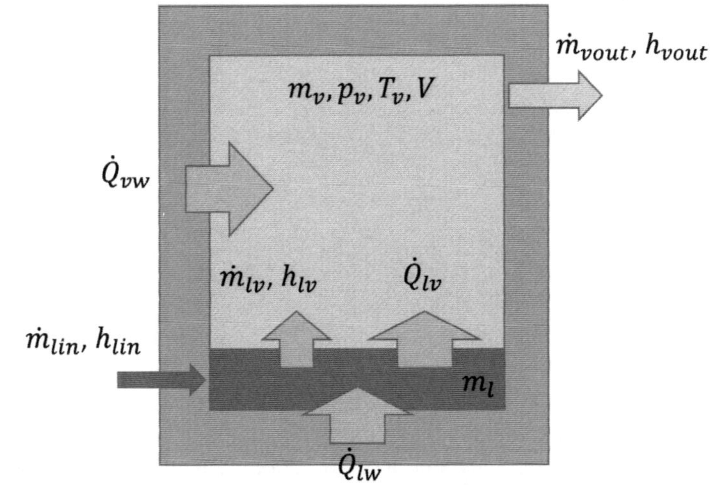

Figure 2. Schematic of the heat and mass fluxes trough experimental cell

After the injection takes place over the heater, a liquid film forms and it will re-ceive heat from the wall. The liquid temperature is computed based on the energy balance:

$$\alpha_{lw} \cdot A \cdot (T_l - T_w) + \dot{m}_{lin} \cdot h_{lin} - \dot{m}_{lv} \cdot h_{lv} = \frac{d(m_l \cdot c_{vl} \cdot T_l)}{dt} \tag{1}$$

where: T_l is the liquid temperature, T_w is the wall temperature, A is the liquid-wall contact area (heater area), α is the heat transfer coefficient, $\dot{m}_{lin} \cdot h_{lin}$ is the input water enthalpy, c_{vl} is the specific heat of water and m_l is the liquid mass within the cell.

When the liquid temperature achieves the boiling point, it will start to vaporize and the mass transfer is computed based on a simplified formula obtained from the Schrage's theory.

$$\dot{m}_{lv} \cdot \Delta H_v = \frac{1}{R} \cdot (T_l - T_{vs}) \tag{2}$$

Once the vaporization starts, the vapour mass within the cell will increase therefore the pressure will increase and leads to changes in the saturation temperature which can be computed with the Clausius-Clapeiron equation:

$$ln\left(\frac{p_v}{p_o}\right) = -\frac{\Delta H_v}{R} \cdot \left(\frac{1}{T_{vs}} - \frac{1}{T_o}\right) \tag{3}$$

where: p_v is the pressure of the vapour within the cylinder, p_o is a reference temperature, $\Delta H_v$is the latent heat of vaporization, R is the gas constant and T_o is the boiling temperature at pressure p_o.

Since there is a vapour incoming within the vapour system, for computing the pressure and temperature the first principle of thermodynamics for open systems, the equation of state and the mass conservation are used:

$$\dot{U} = \dot{Q} - \dot{W} + \sum \dot{m}_{in} \cdot h_{in} - \sum \dot{m}_{out} \cdot h_{out} \tag{4}$$

where the internal energy is:

$$\dot{U} = \frac{dU}{dt} = \frac{d(m \cdot c_v \cdot T)}{dt}, \tag{5}$$

the equation of state

$$p \cdot V = m \cdot R \cdot T, \tag{6}$$

and the mass conservation:

$$m_l = m_{in} - m_{lv} \tag{7}$$

$$m_v = m_{lv} - m_{vout}. \tag{8}$$

The water injection rate is calculated based on the following formula:

$$m_{in} = m_0 \cdot \left(1 - e^{-a \cdot \left(\frac{t - t_0}{\Delta t}\right)^{m+1}}\right) \tag{9}$$

where m_0 is the total injected mass, a and m are shape parameters, t is the current time, t_0 start of injection time, and Δt is the duration of the injection.

After the vaporization is complete and the pressure and temperature reach a constant level, the exhaust valve opens and the steam is removed from the test cell. Then a new experimental cycle resumes. This is repeated until the measured data repeats, indicating that at each start of the measurement cycle in the test cell either we have only gaseous water at atmospheric pressure or we always have the same mixture of gaseous water and air. It is indicated that the entire amount of air is completely removed from the test cell.

All the above equations have been implemented in a MatLab Simulink.

4. Results and discussions

Based on the presented model and the conditions presented in Table 1, some numerical results are presented.

Table 1. Parameters used for the calculations

Parameter	Value	Units
Test cell volume	2	l
Heating wall temperature	673	K
Injected liquid	6	g
Heat transfer coefficient (wall-vapour)	50	W/mK
Diameter of the cell	0.12	m

In Figure 3 the evolution of the liquid and vapour phase is presented. The liquid mass achieves its maximum value at about 26.8 seconds and the values are much smaller than the injected one which is 6g. This is happening because during the injection the injected water starts immediately to vaporize as can be seen in Figure 3, the vapour mass starts to increase once the injection starts.

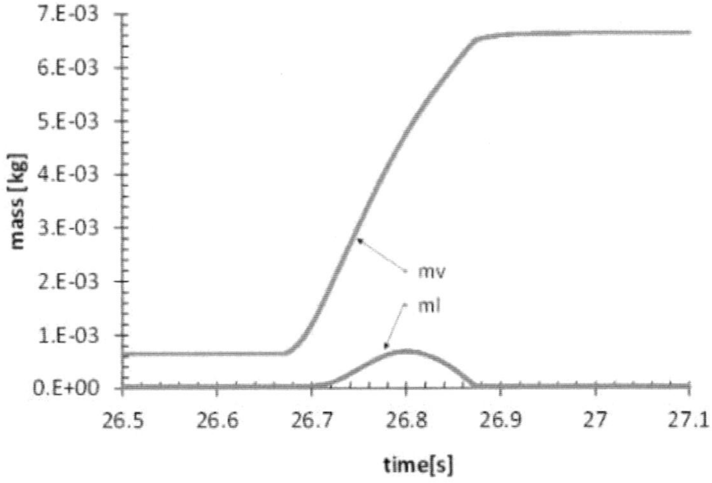

Figure 3. Liquid and vapour masses within the cell. The injection starts at 26.66 seconds

The scope of the presented model and the proposed experiments is to determine, in dynamic conditions, the heat transfer coefficients for wall-liquid, wall-vapour and liquid-vapour contact surfaces. The wall liquid heat transfer coefficient can be estimated based on the pressure variation within the cell, during the vaporization process. A higher heat transfer coefficient will lead to a faster vaporization and a faster change of the vapour pressure Figure 4. As can be seen in Figure 5, the change in temperature is rather small and cannot be used to get some useful information.

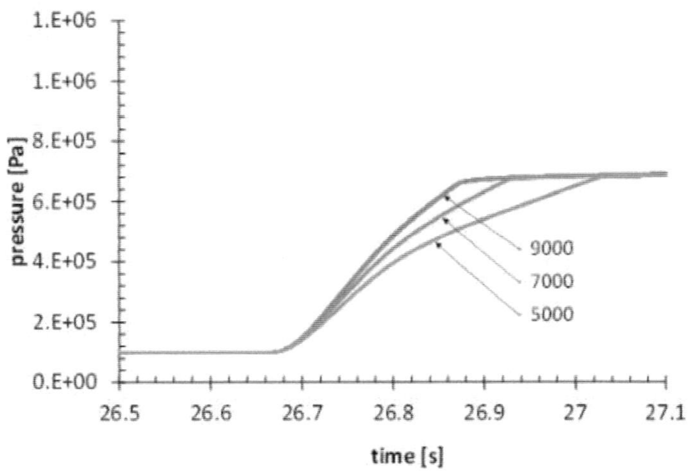

Figure 4. Pressure evolution within the cell for different heat transfer coefficients between the heating wall and the liquid phase

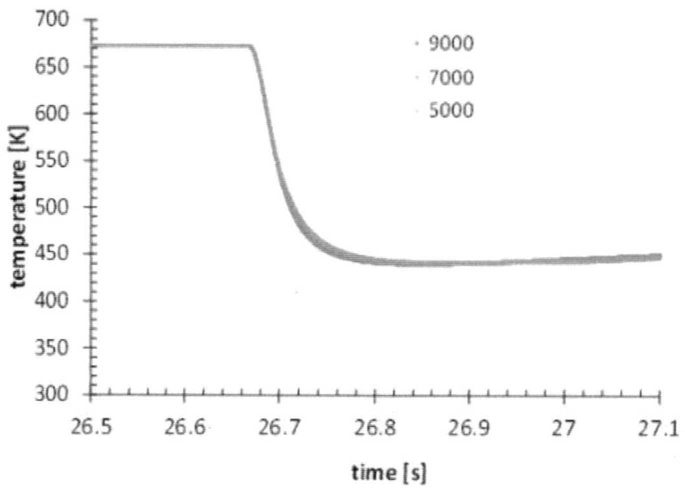

Figure 5. Temperature evolution within the cell for different heat transfer coefficients between the heating wall and the liquid phase

The pressure and temperatures are similar at the end of the vaporization process, regardless the heat transfer coefficient between the wall and liquid. Besides, the evolution is influenced by the wall-vapor heat transfer coefficients and adsorbed liquid film and vapor. From Figure 6 and 7, it can be seen that a higher liquid-vapor heat transfer coefficient leads to a faster change in pressure and temperature, although, after a long enough time, the values are going to be similar.

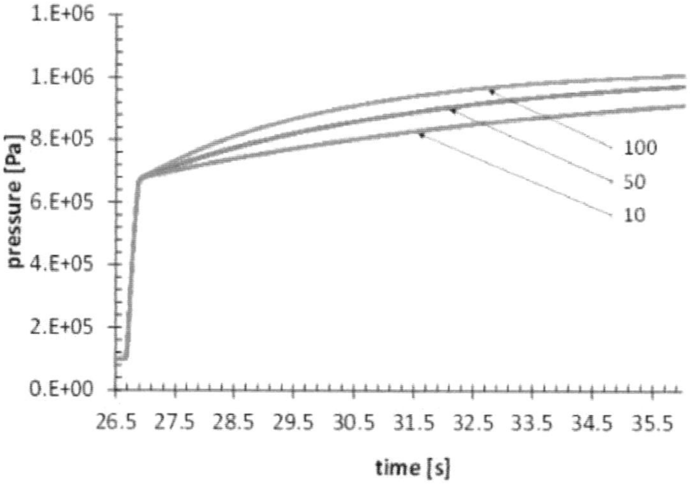

Figure 6. Pressure evolution within the cell for different heat transfer coefficient between the liquid and vapour phases

Some other parameters can be changed (test cell volume, heating wall temperature, the quantity of the injected water, the temperature of the injected water and so on) in order to obtain the complete set of unknown parameters.

The final goal of the experiment is to see if such a simplified model can describe well enough the processes that take place within the experimental cell and to use it for designing a steam engine where the boiling takes place directly within the cylinder.

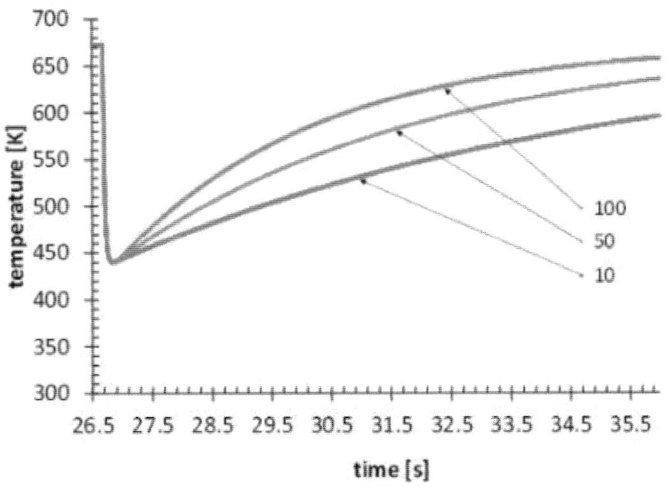

Figure 7. Pressure and temperature evolution within the cell for different heat transfer coefficient between the liquid and vapour phases

5. Conclusions

The experimental test bench and a simplified mathematical model for vaporization data analysis are presented in the paper. The experimental test bench has been designed in order to allow analysis of different parameters like the test cell volume, the quantity of the injected mass, the heat dissipated by the heater and the temperature and pressure of the injected liquid.

The simplified mathematical model used to describe the behavior of the temperature and pressure inside the test cell revealed that the evolution of these parameters is influenced by heat transfer coefficients (wall-liquid, wall-vapor, liquid-vapor). The final goal is to estimate these heat transfer coefficients by correlating the experimental results with the numerical ones.

Acknowledgment

This work was supported by research grants PCD-TC-2017 no.42 financed by the University Politehnica Timisoara.

References

[1] Liang G and Mudawar I 2017 Review of spray cooling – Part 1: Single-phase and nucleate boiling regimes, and critical heat flux, *International Journal of Heat and Mass Transfer* **115** 1174-1205

[2] Mudawar I and Estes K A 1996 Optimizing and Predicting CHF in Spray Cooling of a Square Surface, *Journal of Heat Transfer* **118** 672-679

[3] Visaria M and Mudawar I 2008 Effects of high subcooling on two-phase spray cooling and critical heat flux, *International Journal of Heat and Mass Transfer* 51 5269-5278

[4] Horacek B, Kiger K T and Kim J 2005 Single nozzle spray cooling heat transfer mechanisms, *International Journal of Heat and Mass Transfer* **48** 1425-1438

[5] Moreno G Jr 2009 *Experimental investigation of flow boiling and spray cooling on enhanced surfaces in FC-72*, University of Texas at Arlington, PhD Thesis

[6] Winter M 2015 *Heat Transfer Mechanisms during Spray Cooling of Electronic Devices*, E-Publishing-Service at Technische Universität Darmstadt, Darmstadt, PhD Thesis

[7] Zhao R, Cheng W, Liu Q and Fan H 2010 Study on heat transfer performance of spray cooling: model and analysis, *Heat Mass Transfer* **46** 821-829

[8] Stoica V, Trif-Tordai G, Ilie M and Calinoiu D 2018 *Flash boiling steam engine,* International Conference of Applied Science ICAS2018, May 9-11, Banja Luka, Bosnia and Herzegovina

IOP Conf. Series: Materials Science and Engineering **477** (2019) 012048 doi:10.1088/1757-899X/477/1/012048

Flash boiling steam engine

V Stoica[1], G Trif-Tordai[1], M Ilie[1] and D Calinoiu[2]

[1]Politehnica University of Timisoara, Faculty of Mechanical Engineering, Mihai Viteazu str., no. 1, 300222 Timisoara, Romania
[2]Politehnica University of Timisoara, Department of Fundamental of Physics for Engineers, Vasile Parvan str., no. 1-2, 300223, Timisoara, Romania

E-mail: virgil.stoica@upt.ro

Abstract. Increasing the efficiency of internal combustion engines and reducing their pollution is the key issue nowadays. As a result, researchers have focused their attention on energy recovery either thermal or kinetic. In terms of thermal energy recovery, this can be done with promising results, especially as most of the potential energy of the fuel is lost as heat. Energy recovery can be done either using the Clausius-Rankine cycle or the Organic Rankine Cycle, the use of thermoelectric generators that would convert thermal energy directly into electricity, Sterling engines or even steam engines. In this paper, we propose a steam engine in which water vaporization takes place directly in the cylinder. The engine, unlike the classic locomotive engines is working in a closed cycle. The condensate is preheated and re-injected into the engine. Such an engine has the advantage of a known, cheap manufacturing technology and simple maintenance. A steam engine efficiency of only 10% of the energy dissipated on the exhaust would provide a global increase in engine efficiency by about 3%, taking into ac-count that about 30% of the potential energy of the fuel is dissipated on the exhaust.

1. Introduction

In our days researcher try by any way to increase the internal combustion engines efficiency, to reduce the fuel consumption and pollution.

In addition to optimizing engine operation, there is an increasing interest focused on recovering thermal energy dissipated by the engine in the atmosphere. Thus, this energy can be recovered with a higher or lower efficiency using thermogenerators, Stirling engines, engines based on the Clausius-Rankine cycle or cooling cycles.

R. Zhao et al. [1] presents an energy recovery solution for turbo-charged internal combustion engines where the exhaust gas energy is recovered with an additional turbine in which the exhaust gases mixt with water vapor expand. The water vapor is formed by injecting liquid water into the exhaust stream. The study shows that the adopted solution can reduce specific consumption by 6.0-11.2% in various operating regimes when the steam injection also occurs at the inlet of the super-charger's turbine. Gonca [2] presents the advantages of steam injection in the cylinder of an internal combustion engine with spark ignition. The results obtained from thermodynamic calculations reveal an increase in power and thermal efficiency by up to 5%. Experimental trials [3] with saturation water injection in the admission valve gate, in the case of a transformed engine to run after the Miller cycle, reveal a reduction of effective power by 6.5% and thermal efficiency by 10%, but which is offset by supercharging, power and efficiency increasing by 18 and 12%, respectively. However, it is not specified whether this overcharging is assured by the engine or an external source. G. Kokkülünk et al [4], have made both numerical and experimental attempts [5] to see what happens in the case of steam injection in recirculated flue gas (EGR). The results highlight the drastic reduction of NOx emissions but they are accompanied by a slight increase in specific consumption.

Experimental attempts are, however, limited because a classic engine with mechanical control is used without the possibility to change parameters such as injected doses and the moment of diesel fuel or saturation water injection. The saturation water injection and injection system is presented briefly without making any reference to the components used or to its operating algorithm.

Attempts to inject water into the valve gate were also carried out by Daniel Busuttil [6]. The experimental stand used is limited in the possibility of controlling different parameters such as water injection timing, which is done using an electronic timing circuit. However, an increase in engine torque of up to 16% was observed when the air at the entrance of the engine was at ambient temperature.

Bosch, in collaboration with one of the cooperating company, presented at the Vienna Motor Symposium in 2015 [7] the water injection solution they are working on. They want to inject water directly into the engine cylinder in order to reduce both the maximum cycle temperature and the exhaust temperature thus reducing the emissions of nitrogen oxides. Besides, this approach reduces the risk of detonation without the need to reduce the timing advance, consequently increasing the efficiency of the engine at high load and speed.

Also, BMW tests the water injection solution and achieves a fuel saving of 8% [8].

The energy recovery solution by generating steam, by means of exhaust gases, which is introduced into the cylinder during engine operation is presented in the STIPE (Steam Injected Piston Engine) report. The report shows that in the case of an internal combustion engine with spark ignition, it is possible to increase the thermal efficiency from 37% to 48%. This is possible, on the one hand, by increasing the compression ratio, the detonation phenomenon being avoided precisely by steam injection, and on the other hand by reintroducing a part of the exhausted into the engine cycle [9].

In the present work, we propose another water based solution for internal combustion engine exhaust gases heat recovery. More specifically, at the presented steam engine the boiling takes place directly in the cylinder since overheated and pressurized water is injected onto the surface of a hot wall. So, a separate boiler is no longer needed. The piston steam engine becomes much safer in operation. The manufacturing technology is known being similar to that of two stroke internal combustion engines. It works on a water basis and therefore does not pose any risk to the environment.

2. Proposal and numerical approach

The proposed steam engine arrangement is presented in Figure 1. The exhaust gases from the internal combustion engine passes through the main heat exchanger 1 for steam super heating, followed by a secondary heat exchanger 2 for liquid water heating. The hot liquid water is injected with the injector 4 into the engine cylinder 3 where vaporizes and expands pushing the piston 7 from the top dead center to the bottom dead center. At the bottom dead center, the piston opens the exhaust port 10, and the steam is evacuated. The evacuated steam condenses in the condenser 5, and the obtained liquid water is pumped towards the injector with the pump 6, and the circuit is restarted. The linear movement of the piston is transformed into a circular one, with the connecting rod 8 and the crankshaft 9.

The thermodynamic analysis is done taking in account some simplifications and hypothesis based on the model depicted in Figure 2. In the cylinder, two independent subsystems are considered, liquid (l) and vapor (v), in between which exist a heat and mass transfer noted as Q_{lv} and m_{lv}. It is also considered that: the vapor phase be-haves like an ideal gas, the heat capacities are constant, and the volume of the liquid phase is neglected. The mixing of new-come fluid is done in the bulk and instantaneously and the processes are taking place in time but not in space.

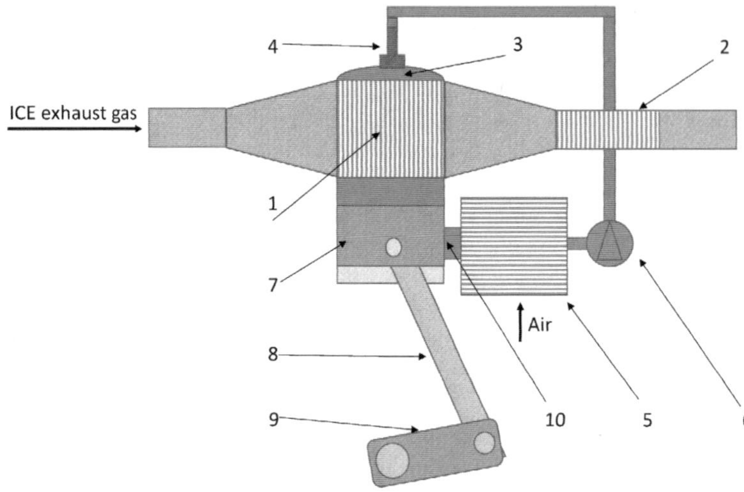

Figure 1. Proposed steam engine for heat recovery

For estimating the performances of the engine, the liquid and steam evolution through the engine is computed considering a constant angular speed of the crank-shaft.

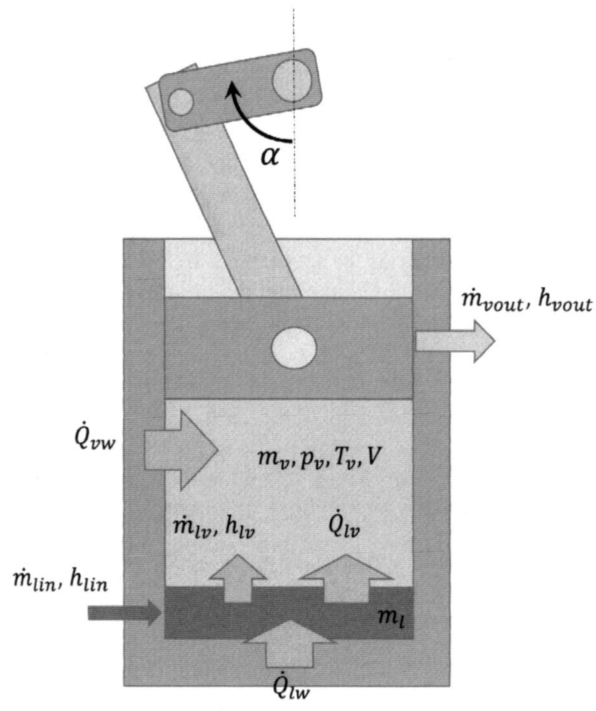

Figure 2. Schematic of the heat and mass fluxes trough the engine

For computing the boiling temperature at different pressures the Clausius-Clapeiron equation is used:

$$ln\left(\frac{p_v}{p_o}\right) = -\frac{\Delta H_v}{R}\cdot\left(\frac{1}{T_{vs}} - \frac{1}{T_0}\right) \quad (1)$$

where: pv is the pressure of the vapor within the cylinder, po is a reference temperature, ΔH_v is the latent heat of vaporization, R is the gas constant and To is the boiling temperature at pressure po.

The mass of liquid that vaporizes m_{lv} is estimated based on the Schrage theory, which after mathematical approximations and simplifications can be written as:

$$\dot{q}_i = \frac{1}{R_{vl}} \cdot \left[T_{vi} - T_v \cdot \left(1 + \frac{p_v - p_l}{\Delta H_v \cdot \rho_l}\right)\right] \quad (2)$$

where $\dot{q}_i$ is the heat flux at the interface, Rlv is the interface thermal resistance, Tvi is the temperature of the vapor at the interface, Tv is the vapor temperature in the bulk, pv is the vapor pressure pl is the liquid pressure and $\rho_l$is the liquid density

Considering the vapor temperature at the interface equal with the liquid temperature, and the vapor pressure equal with the liquid pressure, then it can be written:

$$\dot{q}_i = \dot{m}_{lv} \cdot \Delta H_v = \frac{1}{R} \cdot (T_l - T_v) \quad (3)$$

The liquid-wall, liquid-gas and gas-wall heat transfer is calculated based on the convective heat transfer equation:

$$\dot{Q}_{lw} = \alpha_{lw} \cdot A \cdot (T_l - T_w) \quad (4)$$

The temperature and pressure of the vapor within cylinder is computed based on the first principle of thermodynamic for open systems, the equation of state and the mass conservation:

$$\dot{U} = \dot{Q} - \dot{W} + \sum \dot{m}_{in} \cdot h_{in} - \sum \dot{m}_{out} \cdot h_{out} \quad (5)$$

where the internal energy is:

$$\dot{U} = \frac{dU}{dt} = \frac{d(m \cdot c_v \cdot T)}{dt}, \quad (6)$$

the equation of state:

$$p \cdot V = m \cdot R \cdot T, \quad (7)$$

and the mass conservation:

$$m_l = m_{in} - m_{lv} \quad (8)$$

$$m_v = m_{lv} - m_{vout}. \quad (9)$$

The produced work L and heat inputs Q are computed by integration over one cycle, one crank shaft rotation respectively.

$$L = \int_0^{2\pi} p_v dV \quad (10)$$

$$Q = \int_0^{2\pi} \dot{Q}_{lw} \cdot \omega \cdot d\alpha \quad (11)$$

The thermal efficiency is:

$$\eta = \frac{L}{Q} \quad (12)$$

The exhaust flow rate is computed considering the exhaust port a convergent nozzle:

$$m_{vout} = A \cdot \rho_{vout} \cdot \sqrt{2 \cdot \frac{\chi}{\chi-1} \cdot \frac{p_v}{\rho_v} \cdot \left[1 - \left(\frac{p_{vout}}{p_v}\right)^{\frac{\chi-1}{\chi}}\right]} \tag{13}$$

where χ is the adiabatic coefficient.

The water injection rate is calculated based on the following formula:

$$m_{in} = m_0 \cdot \left(1 - e^{-a\left(\frac{\alpha-\alpha_{soi}}{\Delta\alpha}\right)^{m+1}}\right) \tag{14}$$

where m_0 is the total mass that is going to be injected, a and m are shape parameters, α is the current position of the crankshaft, α_{soi} start of injection, and $\Delta\alpha$ is the duration of the injection.
All the above equations have been implemented in a MatLab Simulink model presented in Figure 3.

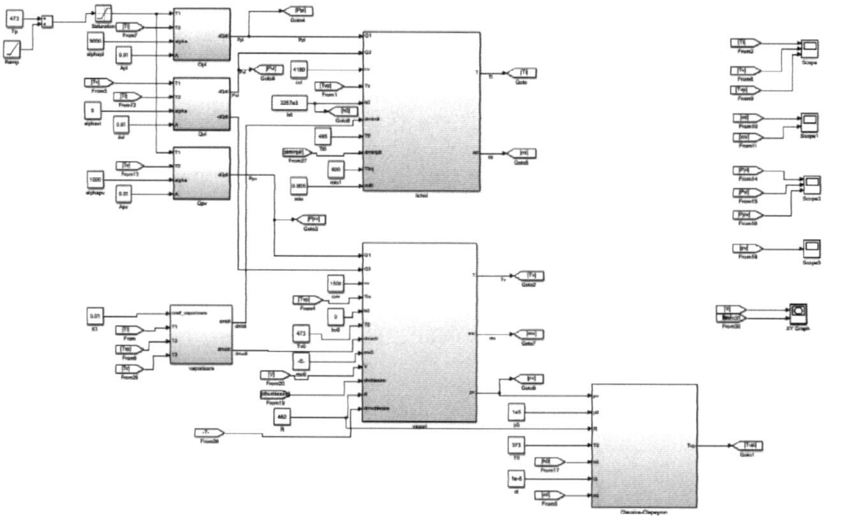

Figure 3. MatLab Simulink model

3. Results and discussions

Based on the mentioned model it was possible to estimate the fluid evolution with-in the engine cylinder. The functional and geometrical characteristics used for the calculations are presented in Tables 1, 2 and 3.

The most important diagram for evaluating the engine performances is the p-V diagram presented in Figure 4. From this diagram was calculated the produced work based on the formula 10. Two important points can be identified: the start of injection (SOI), before TDC, is the point where the diagram presents an inflexion point and the pressure starts to increase faster due to the vaporization process, and the opening of the exhaust port, before BDC, where the pressure starts to decrease achieving the ambient pressure considered before the condenser.

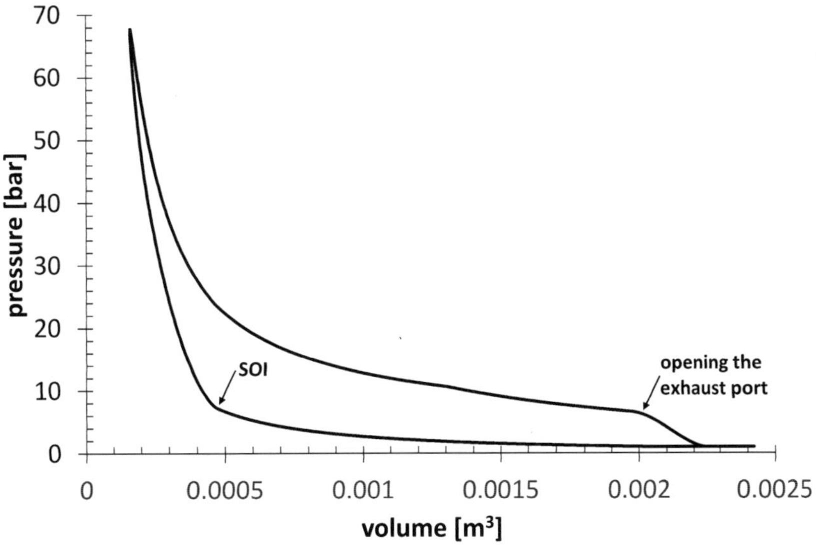

Figure 4. pV diagram

In Figure 5, the temperature evolution of the fluids are presented. The temperature of the liquid phase is most of the time equal with the temperature of the heating wall considered constant. This is happening because it was assumed that there is a small quantity of liquid adsorbed by the wall, which will not vaporize.

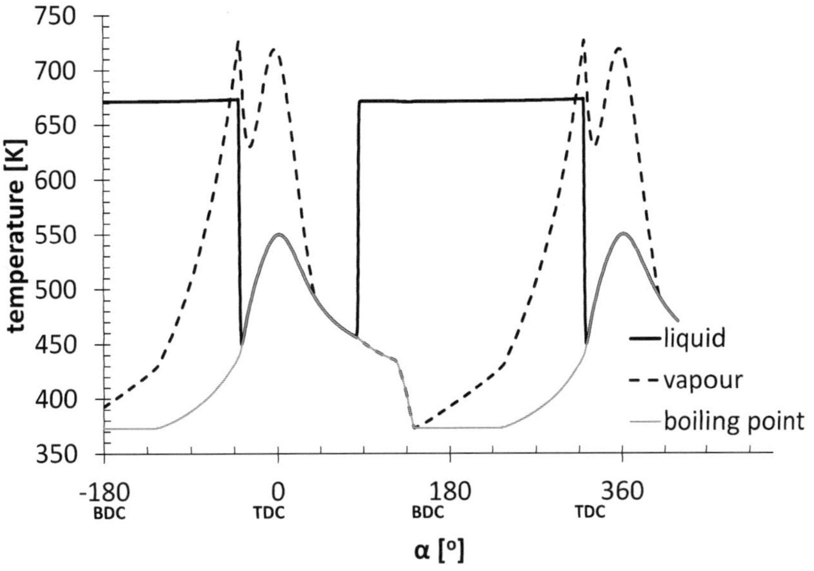

Figure 5. Temperature evolution of the fluids within the cylinder

The liquid temperature decreases during water injection as it can be seen in Figure 5, due to the low temperature of the injected water and to its vaporization. The temperature follows the change of pressure and increases suddenly at the end of the vaporization process. Vapour temperature decreases during injection due to freshly arrived vapours at a lower temperature. The temperature continues to decrease during vapour expansion and the vapours start to condensate when achieving the saturation

temperature. The pressure decreases more when the exhaust port opens. As it is considered that all the condensate is evacuated, only saturated vapour exist within the cylinder when the compression starts.
Its temperature will increase during this stroke due to compression.

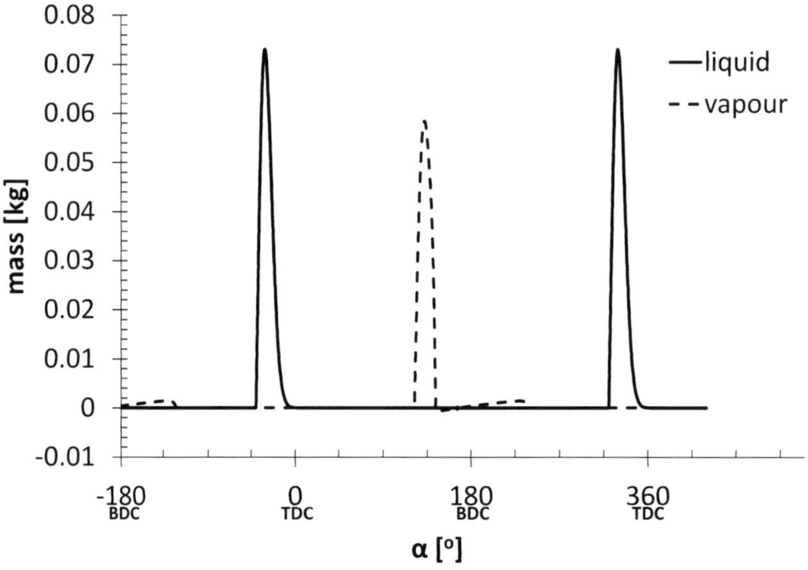

Figure 6. Mass fluxes trough the engine

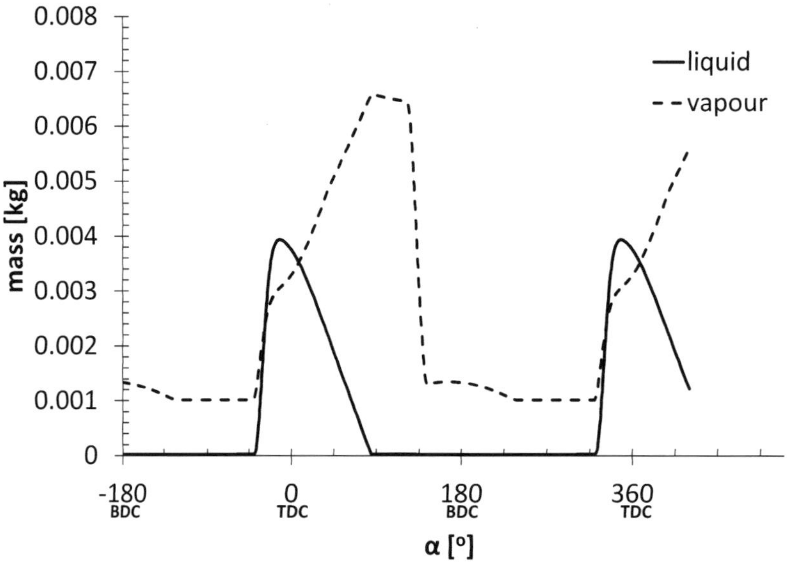

Figure 7. Evolution of the liquid and vapour phases

In Figure 6, the mass fluxes of liquid and vapour are presented. The liquid is injected before TDC and the gas is evacuated during the opening of the exhaust port around BDC. The mass of vapour within cylinder starts to increase after injection due to vaporization as can be seen in Figure 7. After vaporization process finish, when there is no more liquid that can vaporize, the mass of vapour achieves its maximum and stays at this level till its temperature achieves the saturation temperature

IOP Conf. Series: Materials Science and Engineering **477** (2019) 012048 doi:10.1088/1757-899X/477/1/012048

and the condensing process starts. As can be seen, the vapour mass slightly decrease during this process. The evolution of the vapour mass continues with an abrupt decrease due to the opening of the exhaust port.

The estimated efficiency of the presented engine is 19% before any other analysis and optimization. An increase of the efficiency is expected after a numerical optimization.

Table 1. Geometrical characteristics of the engine

Parameter	Value	Units
Bore	0.24	m
Crank radius	0.025	m
Connecting rod length	0.12	m
Compression ratio	15	-
Exhaust port height	0.01	m
Exhaust port length	0.04	m

Table 2. Water injection characteristics

Parameter	Value	Units
Injected mass	0.006	kg
Start of injection	40	°BTDC
Injection duration	30	°
Temperature	465	K

Table 3. Functional parameters

Parameter	Value	Units
Engine speed	30	rpm
Temperature of the hot wall	673	K
Heat transfer coefficient wall-liquid	9000	W/m^2K
Heat transfer coefficient wall- vapor	50	W/m^2K
Heat transfer coefficient liquid-vapor	50	W/m^2K
Pressure in front of condenser	1	bar

4. Conclusions

Within the present paper a steam engine was proposed as a possible candidate for heat recovery in case of internal combustion engine. The advantage of the presented engine consists in eliminating the boiler and all the problems related to it. The vaporization process is taking place directly in the cylinder. For its performance analysis, a mathematical model is proposed and, some numerical results are presented.

The mathematical model has been developed to analyse the operation of such an engine in different configurations and operating parameters. The final goal is to identify the optimal solution to design and execute it.

For the verification of the model and, for the determination of some heat transfer coefficients, some preliminary experiments must be carried out. So we will propose an experimental stand for the study of heat and mass transfer and, vaporization and condensation processes in enclosed enclosures.

Acknowledgment

This work was supported by research grants PCD-TC-2017 no.42 financed by the University Politehnica Timisoara.

References

[1] Zhao R, Li W, Zhuge W, Zhang Y and Yin Y 2017 Numerical study on steam injection in a turbocompound diesel engine for waste heat recovery, *Applied Energy* **182** 506-518

[2] Gonca G 2017 Effects of engine design and operating parameters on the performance of a spark ignition (SI) engine with steam injection method (SIM), *Applied Mathematical Modelling* 1-21

[3] Gonca G and Sahin B 2017 Effect of turbo charging and steam injection methods on the performance of a Miller cycle diesel engine (MCDE), *Applied Thermal Engineering* **118** 138-146

[4] Kokkülünk G, Parlak A, Ayhan V, Cesur I and Gonca G 2014 Theoretical and experimental investigation of steam injected diesel engine with EGR, *Energy* **74** 331-339

[5] Kökkülünk G, Gonca G, Ayhan V, Cesur I and Parlak A 2013 Theoretical and experimental investigation of diesel engine with steam injection system on performance and emission parameters, *Applied Thermal Engineering* **54** 161-170

[6] Busuttil D 2015 Experimental investigation on the effect of injecting water to the air to fuel mixture in a spark ignition engine, *MM Science Journal* 585-590

[7] Brooke L 2015 Bosch developing new water-injection system for production engines, *SAE International* 02 June 2015

[8] Collie S 2015 BMW water injection system offers improved performance and fuel economy, *NewAtlas*, 02 July 2015

[9] Pedersen T H, Koch T, Sørensen E B and Kjellström B 2002 *STIPE (STeam Injected Piston Engine)*,TK Energi A/S Denmark, Luleå University of Technology Sweden

Aluminium matrix composites: Processing of matrix from re–melted aluminium wastes in micro–melting station

C Bulei[1], M P Todor [1], I Kiss[2] and V Alexa[2]

[1]University Politehnica Timisoara, Faculty of Engineering Hunedoara, Doctoral School, Timisoara / Hunedoara, Romania
[2]University Politehnica Timisoara, Faculty of Engineering Hunedoara, Department of Engineering & Management, Hunedoara, Romania
E–mail: imre.kiss@fih.upt.ro

Abstract. This assessment focuses on the composites with metal matrix, the development of these composites having an important innovation in area of materials engineering over the last years, having in view that these type of composites offer several attractive advantages over the traditional engineering materials due to their superior properties. Therefore, the metal matrix composite become economical alternatives to the monolithic alloys due to their improved specific strength, stiffness and wear resistance combined with better physical properties such as low density and low coefficient of thermal expansion. Even though the specific strength of composites is not up to the mark, advantages like low processing cost and ease of processing make them as potential candidates for various innovative applications. In the meantime, the recycling of any waste is as such a worldwide phenomenon. In this context application of aluminium wastes for development of some value added product has been thought of in this work. This paper presents an overview of aluminium matrix composite systems on aspects relating to processing of matrix from re–melted aluminium wastes.

1. Introduction

The efficient use of materials and the materials substitution are strongly interrelated factors.[1–5] The threat of materials substitution has encouraged all producers to apply new technologies aimed at reducing the amount of material required to meet consumer needs.[1–3], [6–9] Therefore, materials substitution significantly affects the trend toward more efficient use of the monolithic materials.[1–3], [7–8] The increasing use of alternative materials in many technological applications has motivated the industry to provide metal matrix composites. The motivation for materials substitution has mainly been the opportunity to achieve weight savings.[1–3], [6–9] To develop these materials, the metallurgy has undergone some actions:

— the efficient use of materials in technological processes, devoting significant research resources to develop lighter weight alloys and metal matrix composites and increased process technologies that require less metal.
— the materials substitution, knowing that when a new material can offer a cost (or performance) advantage over the current material in an established industrial application, will begin to displace the old in that application.

Reducing the weight of the vehicle increases the fuel efficiency, which results in lower life–cycle costs.[1–3], [6–9], [13–15] Significant reductions in weight and increases in strength are desired to meet fuel efficiency needs. Moreover, the industry experts have projected that substantial weight savings could be achieved along with the property improvements.[1–3], [6–8] The many applications of these materials do not all have the same cost and performance requirements.[1–5], [8], [13–15] Accordingly, the investment criteria of user companies specializing in different product areas are different and the barriers to investment are highest in cost–sensitive areas, wherein expensive new materials must compete with cheap, well–established conventional materials. The relatively high cost of these composites will probably prevent their extensive use in some commercial applications. However, they may well be used in specific mechanical components such as pumps, bearings, gears, springs and suspensions.[10–15]

2. Development of metal matrix composites

Metal matrix composites with ceramic particulates – having a high volume fraction of the hard, brittle and elastically deforming ceramic reinforcement in the soft and plastically deforming metal matrix – have in recent years grown in use in various applications. In particular, these composites are noted for their high strength and wear resistance, coupled with enhanced high temperature properties.[15–20]

Discontinuously–reinforced aluminium alloy–based metal matrix composites (MMCs) containing particulate, whisker or short fibre reinforcements in an aluminium alloy metal matrix have in the time period spanning the last decades, engendered considerable scientific and technological interest for use in a spectrum of high performance markets.[15–20] Besides, the discontinuously–reinforced aluminium alloy–based metal matrix composites based on particulate reinforcements are attractive primarily because they can be made to have properties that are isotropic in three dimensions, or in a plane, when compared one–on–one with the continuously–reinforced metal matrices.[16–20]

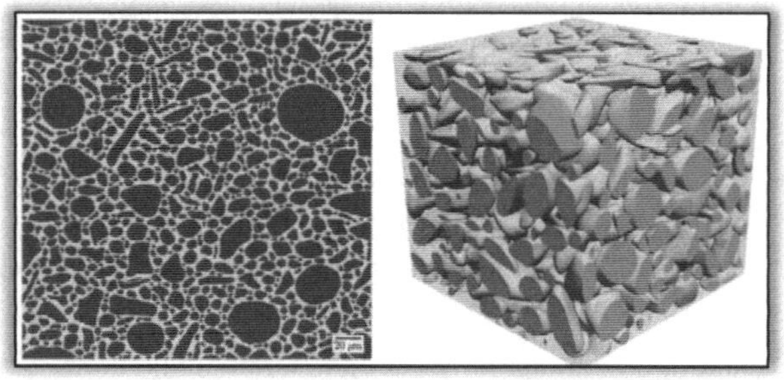

Figure 1. Aluminium alloy–based metal matrix composites with particle reinforcements

In order to make metal matrix composites more attractive, several research priorities should receive attention in automotive sector. The priorities relative to the metal matrix composites are grouped in several categories of descending priority, which reflect the special needs to be done in order to promote the development of these materials in advanced sectors, including the automotive industry:[13–15]

— research in the area of development of lower cost manufacturing technologies, first in importance being the need for the low–cost and the highly–reliable manufacturing processes.
— research in the area of development of lower cost materials, including the search for all–type fibers or particulates and better matrices and continued development activities on existing materials is important to lower costs as well.
— implementing a rigorous control over the matrix, the reinforcement and reinforcement–matrix interphase in the metal matrix composites, which is critical to both the performance and cost of the composites.

In area of research and development of lower cost manufacturing technologies, the actions are oriented to develop high efficient manufacturing processes, research should concentrate on optimizing the industrial and laboratory processes, while in area of research and development of new and cheaper materials all actions need to increase the metal matrix composites production experience with present reinforcements and matrices, in the best route to lower costs, development of cheaper materials (reinforcements and matrices) being a major need.[1–4], [13–15]

Last, but not least, the priorities for implementing a rigorous control over the matrix, the reinforcement and reinforcement–matrix interphase in metal matrix composites can optimize the reinforcement–matrix interface and can prevent deleterious chemical reactions between the reinforcement and the matrix which weaken these composite.[4], [13–15], [20] Therefore, perhaps the most important element in the development of metal matrix composites will be an understanding of the relationships among the constituent properties (reinforcements and matrices), microstructure, and the macroscopic properties of the composite's structure.

Metal matrix composites generally consist of lightweight metal as matrix (alloys of aluminium, magnesium or titanium), reinforced with ceramic particulate, whiskers, or fibres.[3–5], [16–20] In this sense, the reinforcement is very important because it determines the mechanical properties, cost, and performance of a new engineered material.[13–20] Reinforcements for metal matrix composites have a dual demand profile, which is determined by the composites processing and by the matrix system of the material, which can be achieved only by using non–metal inorganic reinforcement components.[15–20]

Aluminium metal matrix composites consist of aluminium alloy as the main constituent, which serves as matrix phase. The second constituent is embedded in the matrix and serves as reinforcement, which is usually non–metallic and generally ceramic such as silicon carbide, graphite and aluminium oxide. Sometimes boron carbide, aluminium nitride, silicon nitride, titanium carbide and boron nitride can also be considered.

Properties of these composites can be modified by varying the nature of constituents and their volume fractions. Generally, based on the type of the reinforcement, aluminium metal matrix composites are classified as fibre reinforced, whisker (short fibre) reinforced and particulate reinforced and illustrated in Figure 2. The imbedded phase is most frequently one of the shapes illustrated in Figure 2: fibers, particles, or flakes.

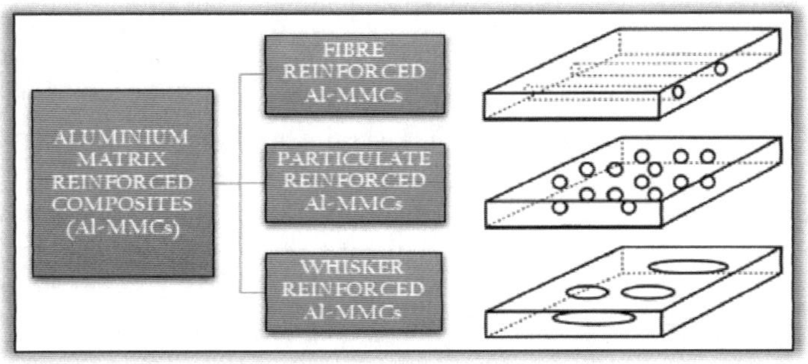

Figure 2. Possible shapes of the reinforced phases in a composite materials: (a) long fiber (aligned continuous, aligned discontinuous, random), (b) particle (spherical, cubic or platelets), and (c) short fiber (whiskers) or flake

3. Recycling the aluminium beverage cans

Aluminium is not surprisingly one of the most recycled materials today.[4], [5], [8], [10], [12,] Also, it is also the only packaging material that completely covers the cost of its own collection and processing by recycling. [4], [5] It is widely recognized that the recovering aluminium for recycling is not only economically viable, but energy efficient and ecologically sound.[1–5], [8], [10–12] The cost of recycling aluminium, which can be used an infinite amount of times, is less than producing new aluminium with raw materials.[1–5], [8], [10–12] The aluminium is one of the few materials – along with the copper and the steel – for which recycling costs are exceeded by the sales price of the recycled product. Aluminium is worth recycling it from all points of view: it is easy to carry, recyclable, infinitely reusable, it does not rust. Therefore, aluminium is rarely lost.[1–7], [10–12] In this sense, aluminium have a great potential, it is often re–used for various applications and could offer practical solutions to recycling. The use of synthetic composite materials like metal matrix composites allows to create components with certain mechanical, physical, chemical properties, imposed from the design stage, making it possible to reduce material consumption, respectively to reduce the quantity of waste resulting from the manufacturing procedures.[13–20].

Recycling is the reprocessing of materials in new products being a modern concept in the waste management, emerged as one of the possibilities to limit waste and to use resources more efficiently.[1], [2], [4], [6–8], [10–12] Moreover, the importance of recycling is increasing and the secondary materials are essential to the industry's survival.[1], [2], [6–12] Generally, the recovered metal are melted down in a furnace, cast into ingots, which are either used in the foundries.[1], [2], [6–8], [10–12] Thus, in order to carry out the recycling process of collected aluminium waste, melting facilities are required. This enables cost efficiency and faster recycling of waste from aluminium.[4], [5], [10–12]

Aluminium cans are an important and extremely important source of secondary aluminium. Once the beer or juice is consumed, if the empty can is properly collected and subsequently recycled, it can be revalued indefinitely without losing anything. Thus, aluminium cans can become a very important raw material (Figure 3). Aluminium cans are recycled in a closed circuit after the recycling process, and aluminium can be re–injected again or can be used for other products. Recyclable aluminium can come from a wide range of sources, including industries and private households. They also include metal containers used as packaging by large juice and beer beverage manufacturers.[4], [5], [9]

Figure 3. Aluminium cans recyclability by melting

We propose a micro–station for melting of aluminium wastes from the beverage cans (Figure 4). The objective of the research is to develop some recycling facilities for melting of aluminium wastes by using charcoal briquettes into a micro–station, designed and manufactured in laboratory. The main activities described here are carried out during the experimental phase of the research and they are oriented towards the recovering aluminium for recycling. [4], [5] In this research we experimented a reusable backyard foundry that melts aluminium soda and beer cans, easily and safely.

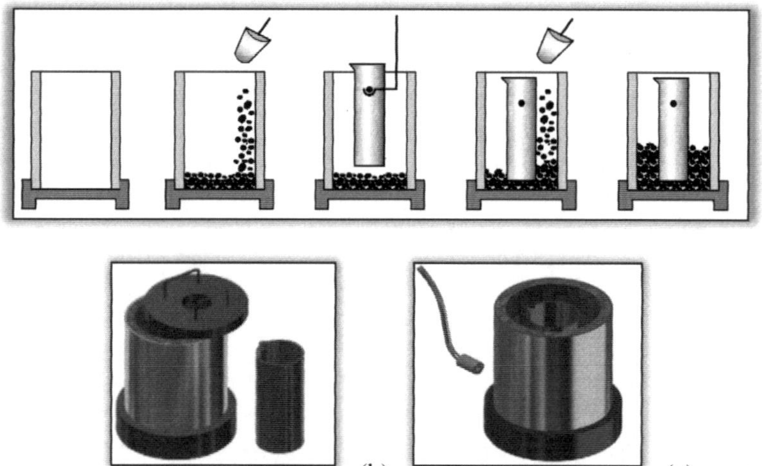

(a)

(b) (c)

Figure 4. Melting of aluminium beverage cans: (a) preparing the micro–furnace (b) the melting micro–furnace with the feeder and the crucible; (c) the assembled melting aggregate;

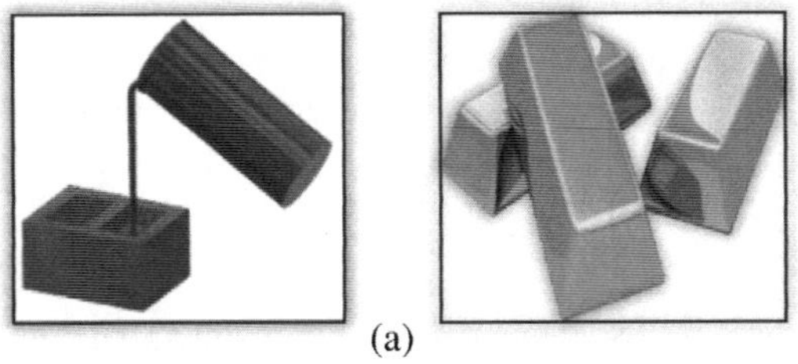

(a) (b)

Figure 5. Casting of aluminium from the beverage cans: (a) aluminium ingots casting; (b) aluminium ingots

A lot of charcoal briquettes are placed around the crucible until they are filled.[4], [5] The charcoal is ignited and when the crucible is hot, the aluminium cans are introduced. Melting takes place at a temperature of 680–750°C. Thus, small ingots or secondary aluminium chips are obtained.

Depending on the end use, aluminium can be simply cast into the finished part, or cast into an intermediate form, such as an ingot, which can be re–used in industry for different utilities, depending on the market demand. Also, this ingots can be used as matrix for a large diversity of metal matrix composites.

4. Concluding remarks

In a world which is increasingly demanding sustainability, recycling of the non–ferrous metal has become a very important practice. Opting for recycling processes does not only mean being responsible towards the environment, it is also a very reasonable industrial process that rely on using non–renewable resources. Also, materials substitution and efficient use of materials are, as we mentioned above, priorities of the several industries.

With the worldwide volume of wastes processed increasingly sourced from consumer and light industrial waste streams, the percentage of valuable nonferrous metals has dramatically increased. This trend, coupled with environment protection legislation and ever increasing waste minimization, has driven the need for complicated, integrated nonferrous recovery plants. Non–ferrous metals, including aluminium do not degrade during the recycling process and thus can be recycled an infinite number of times. Thus, the non–ferrous metals recovery and any recycling process has become increasingly important – both domestically and globally. The trends are continually increasing resource recovery rates with a particular focus on the reduction of metal losses to and diversion from landfills relatively low cost and isotropic properties especially in those applications not requiring extreme loading or thermal. The general trend is for improvements to be sought in every area possible – new ways of applying existing technologies are sought to gain improvements along with the development of new technologies for solutions to existing and emerging applications. The beverage cans metal recyclers are also focusing on increasing both their material recovery rates and the quality of the recovered material.

The aluminium can is the most recycled beverage packaging. Obviously, this is only possible if they are collected and (re)introduced into the recycling circuit. An aluminium can goes through various processes before it is fully recycled and in combination with other recycled aluminium cans turned into a melt aluminium which is then made into aluminium matrix for metal matrix composites.So far, extensive studies have been done for the production of aluminium matrix composites and now these are being manufactured commercially for numerous industrial applications. In light metal matrix composites, aluminium is mostly used as a base metal matrix phase.

Aluminium matrix composites constitute an important category of weight–efficient materials and their processing has the vast development in the various research on advanced materials to overtake the need of low cost, light–weight and high–technological properties. Their relatively low cost and properties especially in those applications not requiring extreme loading or thermal conditions. Also, the processing problems and commercial difficulties associated with continuously reinforced aluminium matrix composites are contributory to the recent interest in their particulate composites (commonly ceramic particles such as SiC).

References

[1] ***U.S. Congress – Office of Technology Assessment 1988 *Advanced Materials by Design, Chapter 1: Executive Summary*, Washington, U.S. Government Printing Office https://www.princeton.edu
[2] ***U.S. Congress – Office of Technology Assessment 1988 *Advanced Materials by Design, Chapter 6: Impacts of Materials Substitution on the Basic Metals Industries* Washington U.S. Government Printing Office https://www.princeton.edu
[3] ***U.S. Congress – Office of Technology Assessment 1988 *Advanced Materials by Design, Chapter 4: Metal Matrix Composites* Washington U.S. Government Printing Office https://www.princeton.edu
[4] Bulei C, Todor M P and Kiss I 2018 Metal matrix composites processing techniques using recycled aluminium alloy *IOP Conference Series: Materials Science and Engineering* **393** 1–6
[5] Bulei C, Todor M P, Heput T and Kiss I 2018 Recovering aluminium for recycling in reusable backyard foundry that melts aluminium cans, *The 7th International Conference on Advanced Materials and Structures (AMS 2018)*, 28–31 March 2018, Timisoara, Romania
[6] ***Centrul Naţional pentru Dezvoltare Durabilă / National Centre for Sustainable Development 2007 *Stategia Naţională pentru Dezvoltare Durabilă a României: Orizonturi 2013–2020–2030 / National Sustainable Development Strategy for Romania: Horizons 2013–2020–2030* Bucuresti http://www.mmediu.ro
[7] Capră G S and Stan A–C 2013 *Strategia Naţională de Gestionare a Deşeurilor 2014–2020 / National Waste Strategy 2014–2020*, Bucuresti http://www.mmediu.ro
[8] Ganea O, Simionescu A, Boța I and Popa M 2011 *Reciclarea deșeurilor, o piață emergentă – Raport privid piața muncii în domeniul gestiunii deșeurilor în România* București www.eco–social.ro
[9] Niero M and Olsen S I 2016 Circular economy: To be or not to be in a closed product loop? A Life Cycle Assessment of aluminium cans with inclusion of alloying elements *Resources, Conservation & Recycling* **114** 18–31
[10] Wernick I K and Themelis N J 1998 Recycling metals for the environment *Annual Review of Energy and the Environment* **23**(1) 465–497
[11] Das S K and Yin W 2007 The worldwide aluminium economy: The current state of the industry, *JOM*, **59**(11) 57–63
[12] Millbank P 2004 Aluminium recycling vital to global supply chain, *Aluminium International Today* **16**(5) 44–49
[13] Mallick P K 2010 *Advanced materials for automotive applications: An overview. Materials, design and manufacturing for lightweight vehicles* Woodhead Publishing Series in Composites Science and Engineering 1–32
[14] Das S 2010 *Recycling and life cycle issues for lightweight vehicles, Materials, design and manufacturing for lightweight vehicles*, Woodhead Publishing Series in Composites Science and Engineering 309–331
[15] Mouritz P A 2012 *Introduction to Aerospace Materials – Chapter 16: Metal matrix, fiber–metal and ceramic matrix composites for aerospace applications – Introduction to metal matrix composites* Woodhead Publishing Limited 394–410
[16] Manigandan K, Srivatsan T S, Ren Z and Zhao J 2015 *Influence of reinforcement content on tensile response and fracture behavior of an aluminum alloy metal matrix composite, Advanced Composites for Aerospace, Marine, and Land Applications: Metal–Matrix Composites*, The Minerals, Metals & Materials Society, John Wiley & Sons, New Jersey
[17] Park S J, Seo M K 2011 Production and processing of metal matrix composites *Interface Science and Technology* **18** 501–629
[18] Gawdzińska K, Chybowski L and Przetakiewicz W 2016 Proper matrix–reinforcement bonding in cast metal matrix composites as a factor of their good quality *Archives of Civil and Mechanical Engineering* **16**(3) 553–563
[19] Bahrami A, Soltani N, Pech–Canul M I and Gutiérrez C A 2015 *Development of metal–matrix composites from industrial/agricultural waste materials and their derivatives*, Critical Reviews in Environmental Science and Technology, Taylor & Francis, 1–66

[20] Meena K L, Manna A, Banwait S S and Jaswanti D 2013 An analysis of mechanical properties of the developed Al/SiC–MMC's, *American Journal of Mechanical Engineering* **1**(1) 14–19

Application a machine vision system in labelling machines

R Łabudzki and J Leopold

Poznan University of Technology, Faculty of Mechanical Engineering and Management, Poznan, Poland

E-mail: leopoldjedrzej@gmail.com

Abstract. Machine Vision systems are becoming more and more significant recently. Set of cameras connected with a PC computer have a great potential in solving many problems that can occure in the industry. Labeling machnes are reasonable application example of these systems. Within this work the construction of labeling machine was presented and concept of using cameras to check label quality, provide help with pre-setting of the machine and monitor the whole labeling process was discussed. This paper describes the proposal of the solution, that will be universal and possibile to apply on every labeling machine with the similar construction to one described below. Moreover, this application could be considered as an inspiration to develope new vision system based structures with other kind of machines, by using same solutions as those shown in this paper.

1. Introduction

Nowdays almost every product has it's own label. It provides the possibility to locate the brand name, logo and basic informations about the particular item. Sometimes producent is even obligated to do this, like in the case of the information about some food product's composition. For this reason, labelling is inherent production process. However, the industry requirements are often very strict, so the label must be applied very precisely, even if the shape of the product or the label itself is extraordinary. This generates many problems, what results with the need of the individual approach in some cases.

It is difficult to determine the duration of the labelling machine calibration. It is highly related with the exact shape of the container and label. For simple cases (rectangular shaped label and plain surface of the label place) it can take up to two hours, but not less than an hour. However if the surface of the object is irregularly curved (in different directions and with variable radii) the pre-settings can last few days, even for the operator, who knows the construction of the labelling machine.

The defects, that are unacceptable, can be categorized. It is easy to determine the type of the defect that has occurred. recognizing the defect type is crucial in determining the setting which needs to be changed for compensating this defect.

In this article, we would like to propose the machine vision system structure, that will fit every labelling machine with the similar construction to the one described below. It will control the quality of the whole labelling process, and recognize the type of the defect. Moreover, knowing the connections between some particular settings and the defects, it is possible to prepare some tips for the operator.

2. Construction of the labelling machine

To fully understand the whole machine vision system described below, it is important, to learn about construction of the device that this system is prepared for. The labelling machine that is shown below, is the simple version, that can be extended by, for example, additional labelling head, so it could apply labels on the both sides of the product. In this section however is described only basic version of the machine, because it contains all elements and it is enough to understand the construction of the extended versions.

The construction of the labelling machine is shown in the Figure 1.

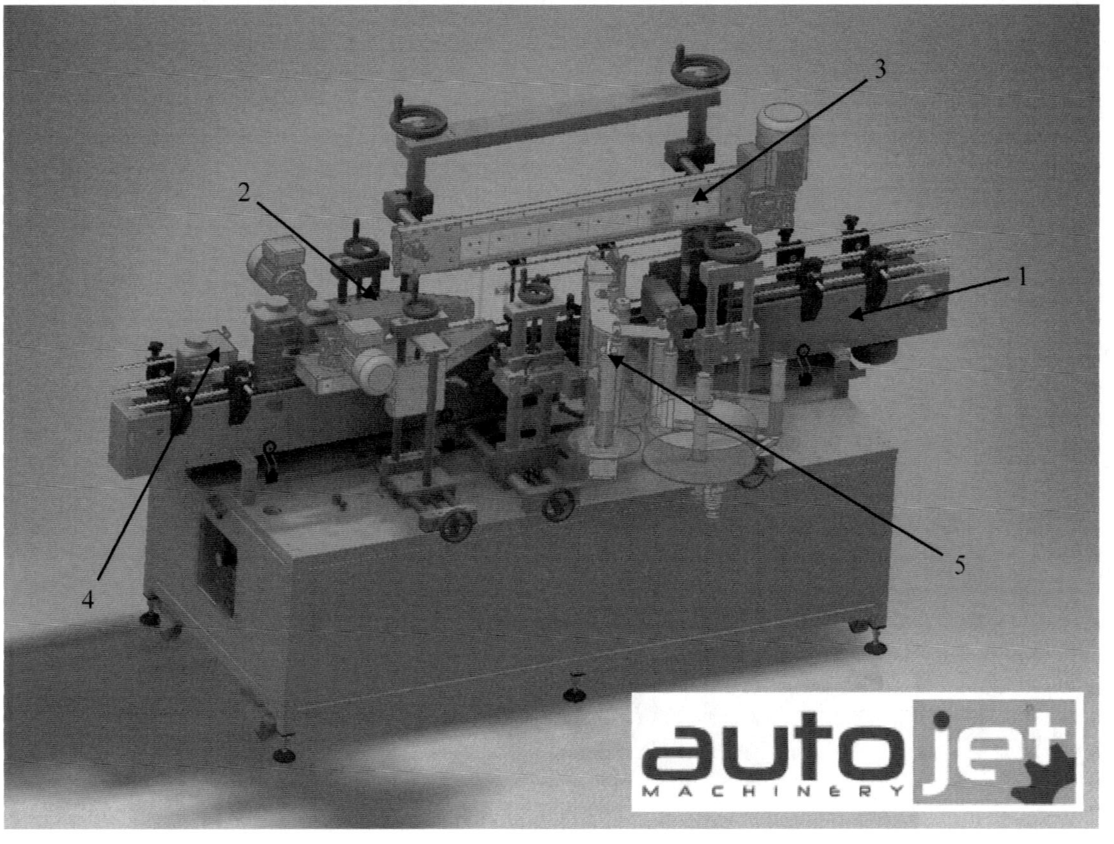

Figure 1. The construction of the labelling machine. 1-Transporter; 2-Side positioners; 3-Top positioner; 4-Separator; 5-Labeling head. This is the model of the labelling machine produced by the Autojet company. For more information, please check [1]

Principal of operation in this machine is very simple. Items that are going to be labeled are driven by the transporter (1), that is moving constantly from the left to the right side. Containers don't have to be placed at regular intervals, because separator (4) provides regular gaps between the products. If there will be two items placed one right after the other, the separator will hold the second one of them for a while, and release after previously determined time. While an item goes further, it gets between the side positioners (2), that are fixing it in the right position until they reach the top positioner (3). It presses the products to the transporter, what is necessary to ensure stability during the labelling process. Next step is labelling itself. Item is detected by sensors and then labeling head (5) is giving the label. while the item passes next to the label, it sticks to him. For better adhearing, just applied label is pressed to the container with a brush Then, already labeled containers are leaving from under the top positioner, and can be collected as the final products.

3. Calibration of the labelling machine

After assembling whole labeling machine and uploading the program, there is a need to calibrate it. This procedure is necessary to accomplish required performance and quality of label application. Skipping this step will result with incorrectly applied labels (with defects described in section 4) in best case, but it can also damage the products or the machine itself.

The labelling machine with featured pre-setting gear is shown in the figure 2 below. Marked elements are the basics instruments for calibrating the labelling machine, but that is not all that could be changed. However, beside those described below, there are settings that are usually made once and there is no need to change them during the exploitation of the machines, which is why They have been omitted.

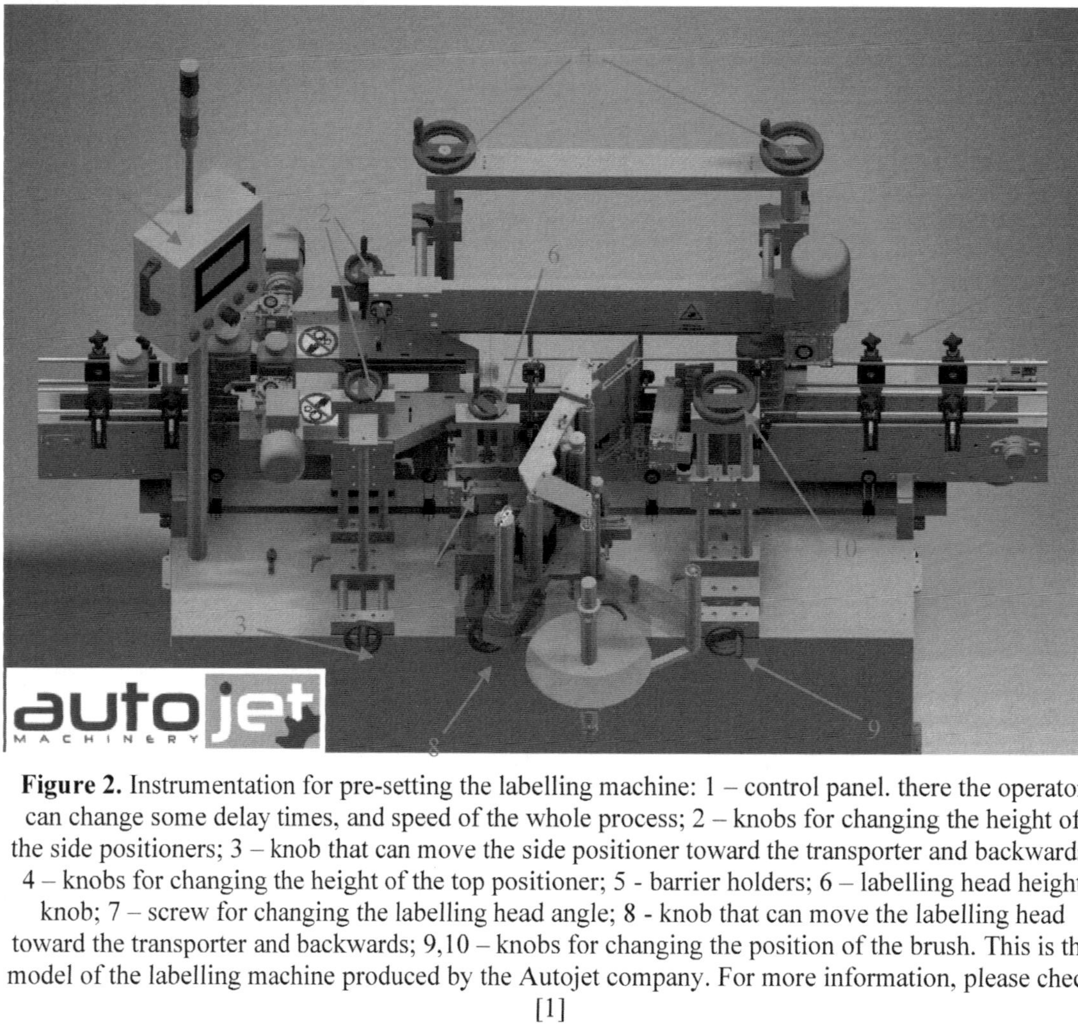

Figure 2. Instrumentation for pre-setting the labelling machine: 1 – control panel. there the operator can change some delay times, and speed of the whole process; 2 – knobs for changing the height of the side positioners; 3 – knob that can move the side positioner toward the transporter and backwards; 4 – knobs for changing the height of the top positioner; 5 - barrier holders; 6 – labelling head height knob; 7 – screw for changing the labelling head angle; 8 - knob that can move the labelling head toward the transporter and backwards; 9,10 – knobs for changing the position of the brush. This is the model of the labelling machine produced by the Autojet company. For more information, please check [1]

4. Labeling Defects Classification

According to what was mentioned in the introduction, in this section, all labeling defects are presented and categorized. All defects that can occure due to the incorrect labeling process are mentioned below, however, defects made during other processes that the item or label were subjected (like for example stains of the ink on the label caused by damaged printer), do not appear in this section.

Figures 3-8 depict all types of labeling defects. Possible causes of each defect and tips helpful with compensating it are posted in the figure descriptions.

Figure 3. Label absence – unsuccessful labelling process. Item did not get at least one of the labels it should get. Possible cause: empty label roll, labelling head too far from the product. Compensation: 1. Check label content; 2. move labelling head towards the transporter

Figure 4. Angle error – applied label is twisted. Possible cause: incorrect labelling head angle. Compensation: 1. change the labelling head angle. (angle of the label is varying accordingly to the labelling head angle)

Figure 5. Folded label – The label is creased. Possible cause: unsynchronized speeds of label and the transporter, incorrect brush setup, incorrect labeling head setup. Compensation: 1. check the transporter and labeling speed (synchronise them if necessary); 2. monitor the process of labelling to figure if the label is being folded during it's application or when pressing with a brush, then change adequate equipment setup

Figure 6. Mechanical damage – label is torn, or some part of it is missing. Possible cause: any element of the machine is interfering with the product path. Compensation: 1. Monitor the process of labelling to find the exact spot, where label is being damage. 2. move the equipment so i won't interrupt

Figure 7. Inaccurate adhering – label does not adhere to the product with all of its surface. Possible Cause: position of the item is not fixed, incorrect brush setup, incorrect labelling head setup. Compensation: 1. check if the top positioner is holding the item in required position. (if necessary, move the top positioner down so it reach the container and press it toward the transporter) 2. move the brush towards the product, so it could stronger press the label (try different angles). 3. move the labelling head toward the transporter

Figure 8. Displacement – label position is incorrect (too high in this case). Possible cause: incorrect labelling head setup. Compensation: 1. change the position of labelling head accordingly to the displacement

Defects described above often occure simultaneously. However they should be consider separately, so the compensating procedures proposed above will help in such cases as well. The situation is more complicated if the defects are not recurrent. If some defect occure once, there is high probability that it wasn't caused by the labeling machine, but by the faulty product or label. In those cases it is necessary to take off incorrectly applied label and check if the problem appears repeatedly for the same product. This will allow to distinguish and eliminate faulty items. If the problem is not connected with particular item, it can be caused by low quality labels.

5. Machine vision system

Machine vision (system vision) it is an using computer vision in industry. While computer vision is focused mainly on image processing at the level of hardware, machine vision most often requires the use of additional hardware I/O (input/output) and computer networks to transmit information generated by the other process components, such as a robot arm. One of the most common applications of machine vision is inspection of the products such as microprocessors, cars, food and pharmaceuticals. Machine vision systems are used increasingly to solve problems of industrial inspection, allowing for complete automation of the inspection process and to increase its accuracy and efficiency. As is the case for inspection of products on the production line, made by people, so in case of application for that purpose machine vision systems are used digital cameras, smart cameras and image processing software.

A typical machine vision system consists of several components of the following (Figure 9):

- one or more digital or analogue camera (black and white or colour) with optical lenses,
- interface the camera to digitize the image (the so-called frame grabber),
- processor (this is usually PC or embedded processor such as DSP),

(In some cases, all the elements listed above are included in the one device, the so-called smart cameras),

- device I/O (input/output), or communication links (e.g. RS-232) used to report the results of system,
- lens for taking close-ups,
- adapted to the system, specialized light source (such as LEDs, fluorescent lamps, halogen lamps, etc.),
- software to the imaging and detection of features in common image (image processing algorithm),
- sync-sensor to detect objects (this is usually an optical or magnetic sensor), which gives the signal for the sampling and processing of image,
- the regulations to remove or reject products with defects.

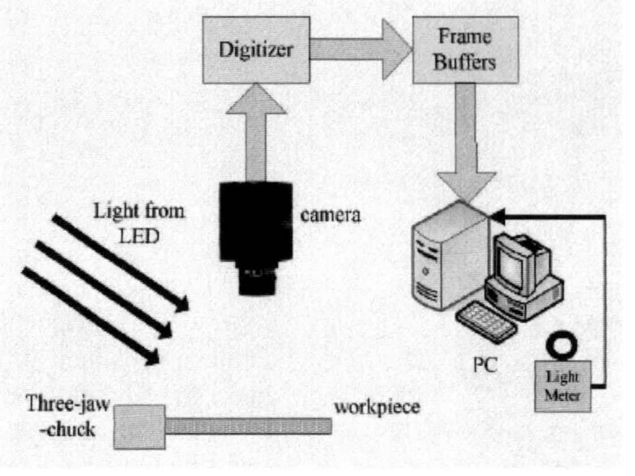

Figure 9. A typical vision system operation

Software machine vision system for both commercial and open-source consists of many different image processing techniques, such as:

- counting pixels (counting the light and dark pixels),
- binarization (colour conversion from the shades of gray in two colours: white and black,
- segmentation (used to locate and/or counting objects,
- "hard" to identify the image (to locate an object that can be rotated, partially hidden by another object or change its size),
- barcode reading (decoding of bar codes, read or scanned by machines),
- text recognition (automatic reading of text - letters and numbers, such as serial numbers),
- measurement (measuring the size of the object),
- edge detection (edge detection object),
- template matching (finding, matching, and/or counting specific patterns),
- others.

In most cases, the machine vision system uses a combination of these processing techniques in order to perform a complete inspection. For example, a system that reads bar codes can also check the surface of an object to detect scratches or other damage, and to measure the length and width of the manufactured component.

Figure 10 shows general structure of the program with sample result windows using Adaptive Vision Studio software for labelling defects. One of the steps of realizing algorithm of recognition the label shows Figure 11.

There are many video programs available on the market, and the selection of the right software significantly affects the reduction of integration time, its costs and - in the future - the possibility of performing modifications. The main factor determining the program selection is the complexity of the vision application. For simple applications, such as bar code reading, simple measurements or checking the completeness of parts, software with graphical interface is sufficient. Sometimes, however, even for such applications it is necessary to use video libraries, because with variable lighting conditions, background or uniqueness of code execution (often occurs when knocking out the DataMatrix code on metal parts) only the use of specialized algorithms allows correct execution of the application.

The end user is the second factor determining the choice of software. If you want to be able to modify video procedures yourself, add new products to check, the most convenient solution is also to use a graphical interface program. Thanks to the ability to create access levels, no unauthorized person will be able to interfere in the program, and after the training, maintenance staff will be able to implement modifications themselves.

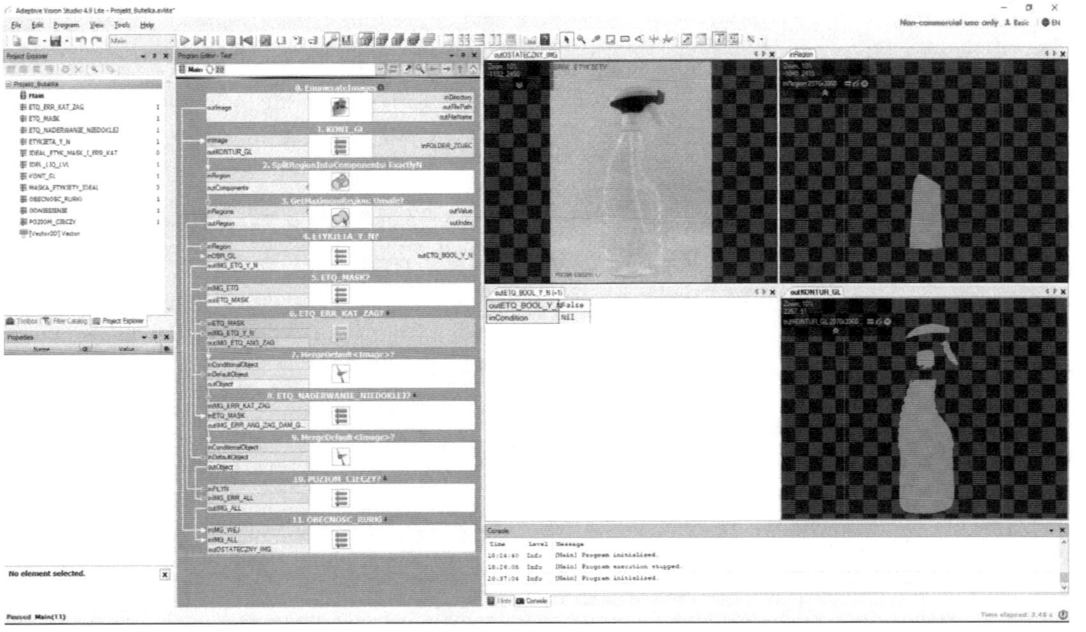

Figure 10. The general structure of the program with sample result windows

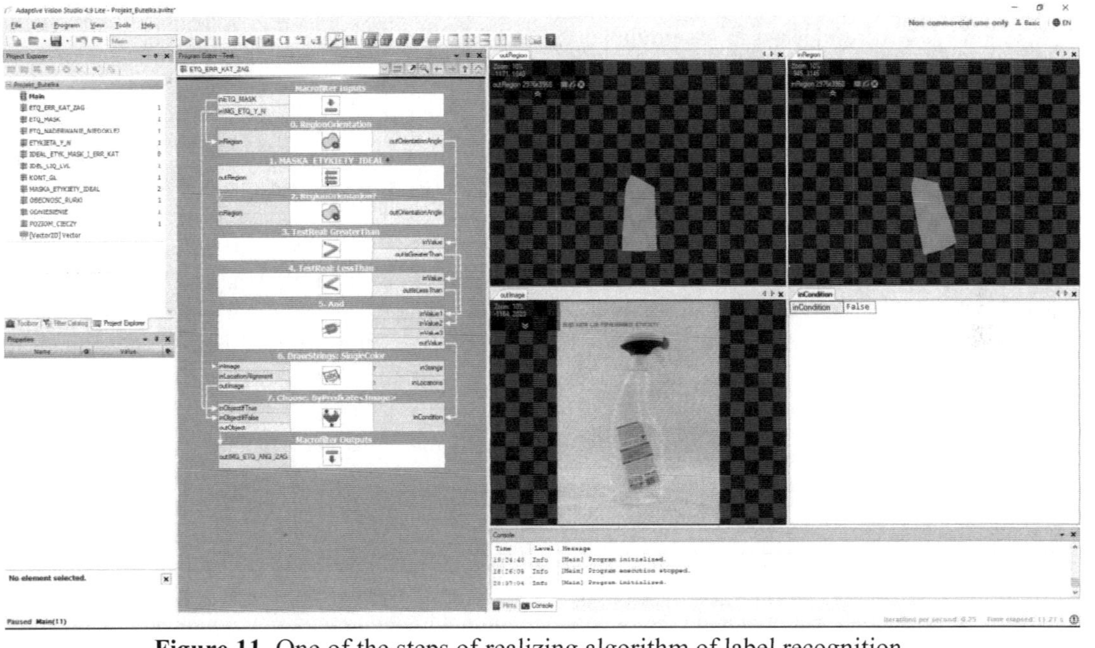

Figure 11. One of the steps of realizing algorithm of label recognition

References

[1] ***http://www.autojet.com.pl/

IOP Conf. Series: Materials Science and Engineering **477** (2019) 012051 doi:10.1088/1757-899X/477/1/012051

Application of genetic algorithms to generate the geometry of elements with a given stiffness coefficient

F Sarbinowski, R Labudzki and R Talar

Poznan University of Technology, Faculty of Mechanical Engineering and Management, Piotrowo 3, 61-102 Poznan, Poland

E-mail: filip.j.sarbinowski@doctorate.put.poznan.pl

Abstract. Work presents the optimization process based on genetic algorithms of two mechanical systems . The first of these is the system of two springs, connected serially by a rigid medial element whose movement is forced in the direction transverse to the original axis of the springs. The goal of the optimization is to obtain a given characteristic describing the dependence of displacement of the medial element on the excitation force. The second part is devoted to multi-criteria optimization of geometry of statically indeterminate system of three rods, which is to ensure the required stiffness of the system while meeting the condition of mass minimization.

1. Introduction

The dynamic increase in the computing power of computers allows for developing new computational algorithms and adapt existing ones to solve common problems. An example of this trend may be genetic algorithms, presented for the first time in [1]. Since then, they have found application in sociology [2], economics [3] or many fields of technology, such as electronics [4], logistics [5] or mechanics [6].

The essence of genetic algorithms is to imitate the optimization process that occurs among living organisms [7]. The process begins with the generation of the a predetermined number of initial population of solutions with random features (parameters) presented numerically, often in binary form. At this stage, the key is to ensure the greatest possible diversity of individuals. Then each of the individuals is evaluated - the degree to which the characteristics of a given individual meet the goal function is defined, therefore an appropriate adaptation function is required, assigning a given set of features (individual) a fitness score from 0 to 1. The best-adapted individuals have the best chance to take part in the reproduction - a process aimed at identifying those between which the exchange of features will occur. It can be implemented, by using, among others, the roulette method - percentage expression of chances for the participation of the individual in the reproduction process is obtained by calculating the ratio of his fitness and the sum of fitness of all individuals. Such randomly selected solutions are coupled in random pairs in which crossover occurs - the creation of new individuals (children) on the basis of the exchange of some traits between individuals - parents (Figure 1).

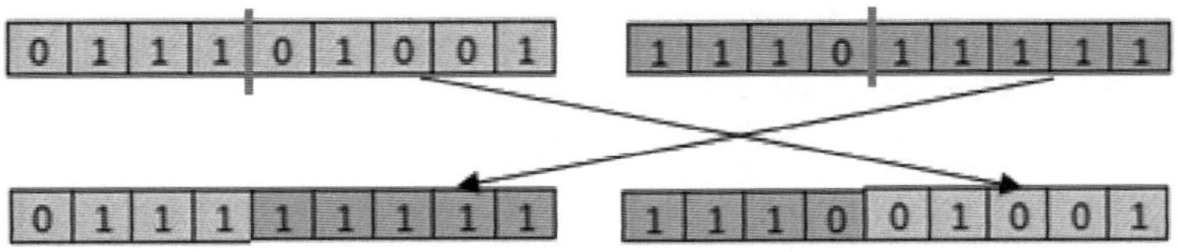

Figure 1. Single point crossover

In order to allow occurrence in the current population of features that did not appear in the previous population, the mutation process is implemented to the algorithm - it is determined with a certain probability of occurrence of a phenomenon in which random feature of a randomly selected individual is modified. In this way a new population is created. It can has the same number of population as the previous one or a different one, defined by the program. The new population is re-evaluated and the algorithm is repeated until the program termination condition, usually associated with obtaining a specific minimum match value or reaching the maximum number of iterations, is met. The diagram of genetic algorithms used in this paper is presented in Figure 2.

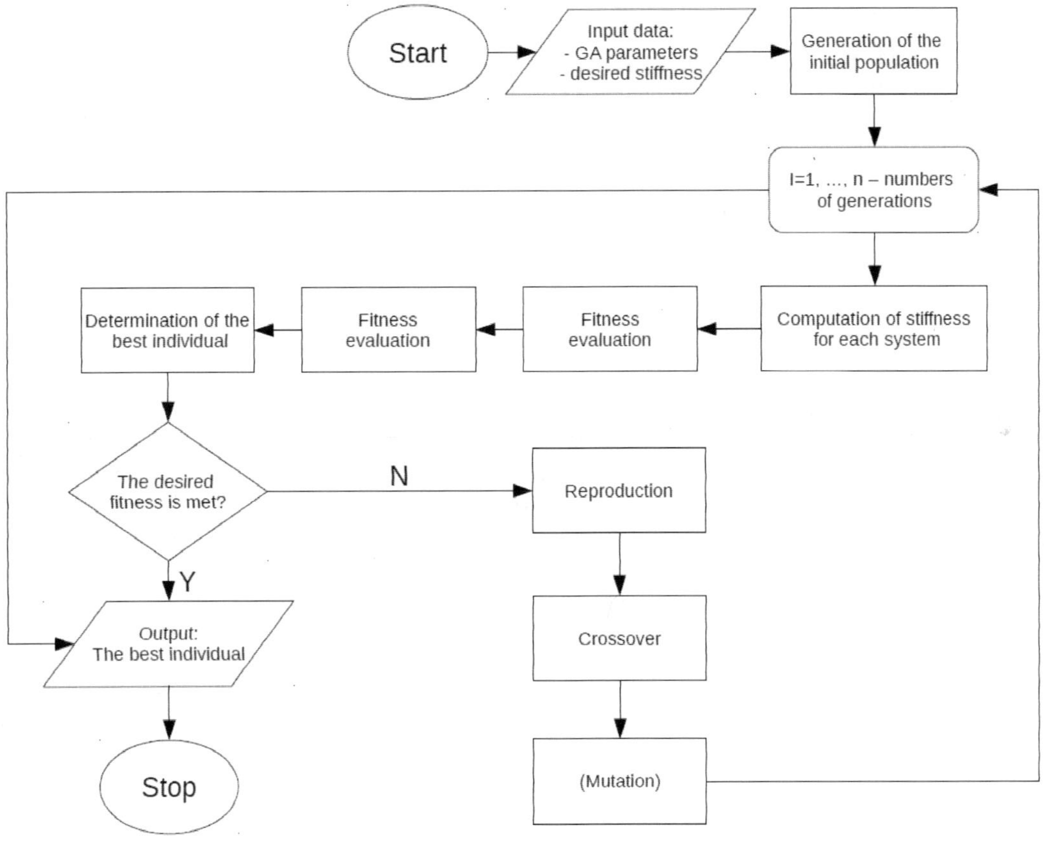

Figure 2. Diagram of genetic algorithm

2. Optimization tasks

Genetic algorithms, as numerical methods, are a convenient tool for solving problems that are difficult or impossible to solve with analytical methods, which may occur in the case of nonlinear models and problems described by many dependent variables. Such problems include the task of optimizing the stiffness of the structures presented in this section – simple but nonlinear spring system and rod system with mass minimization criteria.

2.1. Spring system

The system shown in Figure 3 will be optimized. It consists of a central plate and springs with stiffness k and initial length L_0, which engage the plate with the foundation and keep it in an equilibrium position. Despite the use of springs with linear characteristics, displacing the plate horizontally is strongly non-linear - the stiffness of the system increases with its deformation.

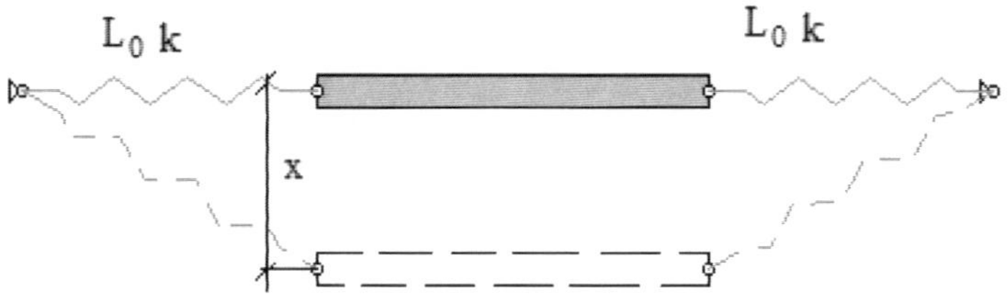

Figure 3. Optimized system

In devices such as bumpers, dampers or springboards the model of which may represent the presented system, the key is to choose the spring parameters so that the values of the forces that cause the individual strains are precisely defined. Let us assume that we are looking for a system with the following relationship between deformations and their stiffness (Table 1).

Table 1. Characteristic points of desired stiffness curve

i	x_i [mm]	K(x) [N/mm]
1	5	0.02
2	30	0.16
3	70	0.20

Obtaining the desired effect with analytical methods, by approximating the function may turn out to be a very demanding process, therefore the problem will be solved using a genetic algorithm whose goal will be to find such parameters k and L_0 that the system stiffness curve will go through the points indicated in Table 1. Stiffness is defined as a derivative of the force after the deformation that this force caused. The function of force dependence on horizontal deformation is:

$$\begin{cases} P(x) = 2k(\sqrt{L_0 \cos(\alpha)^2 + (L_0 \sin(\alpha) + x)^2} - L_0)\cos(\alpha) \\ \alpha = \arctan(\frac{L_0}{x}) \end{cases} \quad (2.1)$$

hence the vertical stiffness of the system is therefore equal to:

$$K(x) = \frac{\delta P(x)}{\delta x} = \frac{2kx \cos(\alpha)}{\sqrt{{L_0}^2 + x^2}} \quad (2.2)$$

For the purposes of calculations, 100 stiffness curves were generated with random values of $k \in (0; 50)$ and $L_0 \in (0; 30)$ parameters, indicated with an accuracy of 0.01. The fitness to the goal function by each of them was calculated on the basis of equation (2.3):

$$fit_i = \sum_{i=1}^{3} \frac{1 - \left| \frac{K(x_i) - K(p_i)}{K(p)} \right|}{3} \quad (2.3)$$

IOP Conf. Series: Materials Science and Engineering **477** (2019) 012051 doi:10.1088/1757-899X/477/1/012051

where $K(p_i)$ means the stiffness of the currently tested pair of parameters k and L_0. This definition of match means that each of the three points is treated equally and the full convergence of stiffness in one of the three required points guarantees a match at 0.33.

The next operation is to select individuals that will create a new population. This was done using the roulette method described in section 1, in such a way that the number of individuals did not change. The crossover process is carried out on the newly formed population. Individuals are selected in random pairs in which the random number of traits is exchanged. It was assumed that during crossover there is a five-percent probability of occurrence of a mutation - change of the value of a randomly selected parameter to a random value, selected from a previously set interval. For each individual of the newly created population, the match is recalculated and the entire algorithm is repeated until a 0.95 fitness is obtained.

Achieving the results with the assumed accuracy required seventeen iterations of the program. The objective function has been met with a fitting degree of 0.96 for parameters $k = 0.1$ and $L_0 = 20.1$. The graphical representation of the result is shown in (Figure 4). Interestingly, after reducing the population to 25 individuals, the algorithm returned the exact same result after the same number of iterations, but it is worth noting that the execution time of one iteration was significantly shortened.

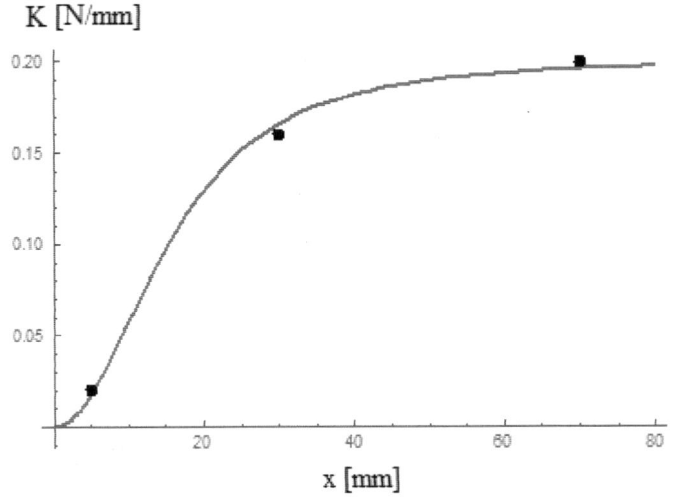

Figure 4. The obtained function of stiffness and strain dependence for k = 0.1 N/mm and L_0 = 20.1 and the goal points

2.2. Rod system

Genetic algorithms are perfect for optimizing more complex structures. Figure 5 shows a convergent system of steel rods connected with a joint and loaded with horizontal force P = 1000 N at point O. The aim of optimization is to choose such geometric parameters of the structure (cross-sections A_1 A_2 A_3 and angles α, β, γ) so that the force P does not cause displacement point O in the horizontal direction by a value greater than 0.7 mm and in the vertical direction by a value greater than 0.2 mm, while maintaining the condition of minimizing the mass and constant height h = 1m of the system.

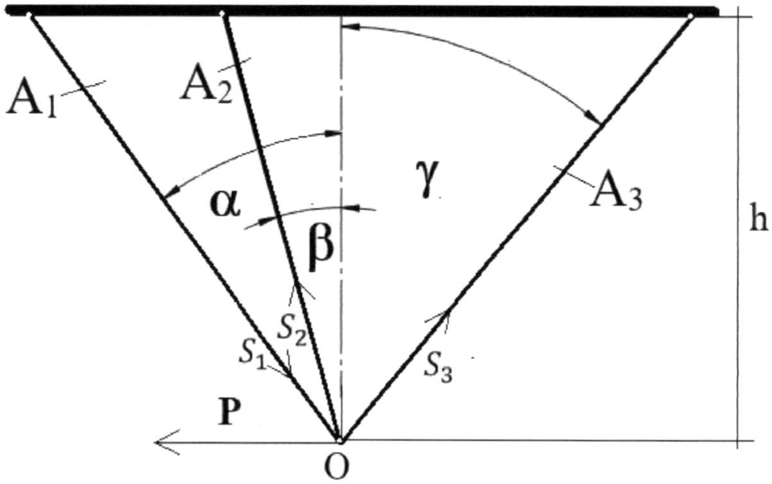

Figure 5. Scheme of considered rod system

The system meets the following equations of equilibrium:

$$\begin{cases} \Sigma F_x = 0; \; P + S_1 \sin(\alpha) - S_2 \sin(\beta) + S_3 \sin(\gamma) \\ \Sigma F_y = 0; \; -S_1 \cos(\alpha) + S_2 \cos(\beta) + S_3 \cos(\gamma) \end{cases} \tag{2.4}$$

where S_1, S_2, S_3 are the internal forces of the respective bars. The analyzed system is a statically indeterminate design, so to calculate the displacement of the node O of the structure, in addition to the equilibrium equations, a geometric displacement equation will be required (Figure 6).

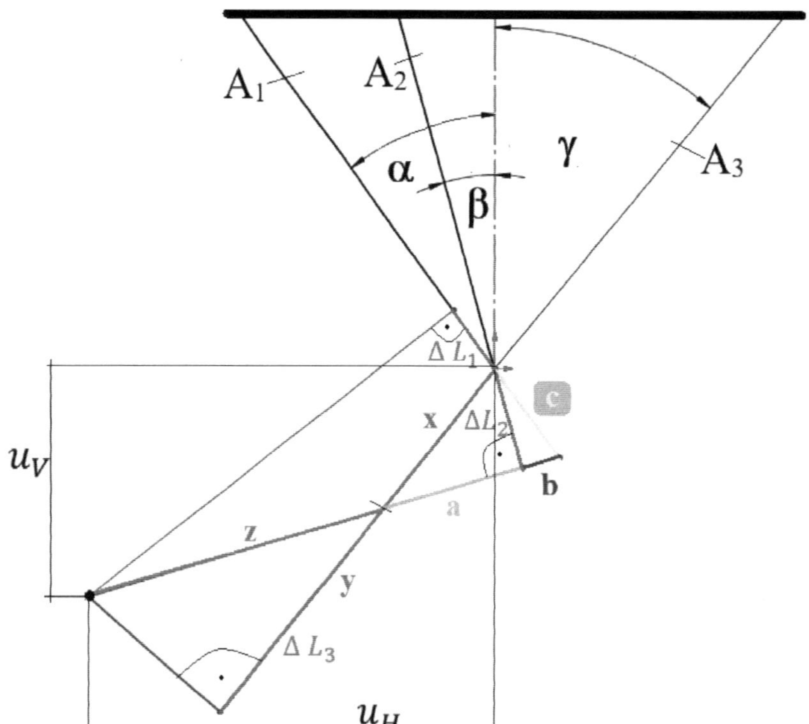

Figure 6. Geometric displacement scheme

$$\Delta L_3 = x + y \tag{2.5}$$

where:

$$x = \frac{\Delta L_2}{\cos(\beta+\gamma)} \qquad a = \Delta L_2 \tan(\beta + \gamma)$$

$$y = z\cos(90 - (\beta + \gamma)) \qquad c = \frac{\Delta L_2}{\cos(\alpha-\beta)}$$

$$z = u - a \qquad b = \Delta L_2 \tan(\alpha - \beta) \tag{2.6}$$

$$u = \frac{\Delta L_1 + c}{\sin(\alpha-\beta)} - b \qquad \Delta L_i = \frac{S_i L_i}{E A_i}, i = 1,2,3$$

$$L_1 = \frac{h}{\cos(\alpha)}, \; L_2 = \frac{h}{\cos(\beta)}, \; L_1 = \frac{h}{\cos(\gamma)}$$

The values of vertical and horizontal displacements are then:

$$u_H = \Delta L_2 \cos(\beta) + (z + a)\sin(\beta)$$
$$u_V = x\sin(\gamma) + z\cos(\beta) \tag{2.7}$$

The first operation performed by the algorithm was to generate 100 sets of searched parameters meeting the condition of maximum displacements of the *O* point and additional conditions: $\alpha > \beta$; β, $\gamma >= 0$. For each of them, the mass m_i of the system and the value of the fit were calculated. As the task of the algorithm is to minimize the mass, and not to obtain a specific value, the fit is not a standardized parameter – it is a qualitative assessment referring to a given population:

$$fit_i = \frac{\sum m_i - m_i}{\sum m_i} \tag{2.8}$$

According to the roulette method, one hundred individuals were randomly selected, which were crossed with a mutation probability of 1%. Those of new individuals that meet the condition of maximum displacement of the O-point of the system are subjected to the assessment of the fit due to the mass and the algorithm is repeated until the mass of the system in ten consecutive iterations will not change by more than 1%.

Calculated geometric parameters, for which the system have the desired stiffness while maintaining the criterion of mass minimization, are presented in the Table 2. For these parameters, the appropriate displacements are $u_H = 0.69$ mm, $u_V = 0.17$ mm, and its total mass is equal $M = 6.67$ kg. Convergence of algorithm has been obtained after 374 iterations.

Table 2. Obtained results

α [°]	β [°]	γ [°]	A_I [mm^2]	A_I [mm^2]	A_I [mm^2]
31,0	16,7	24,2	18,1	22,8	12,6

In order to confirm the correctness of calculations, the optimized system was modeled, in the COMSOL Multiphysics 5.1 environment and its load was simulated. Figure 7 demonstrate the obtained displacement map - it shows a large convergence with the obtained results.

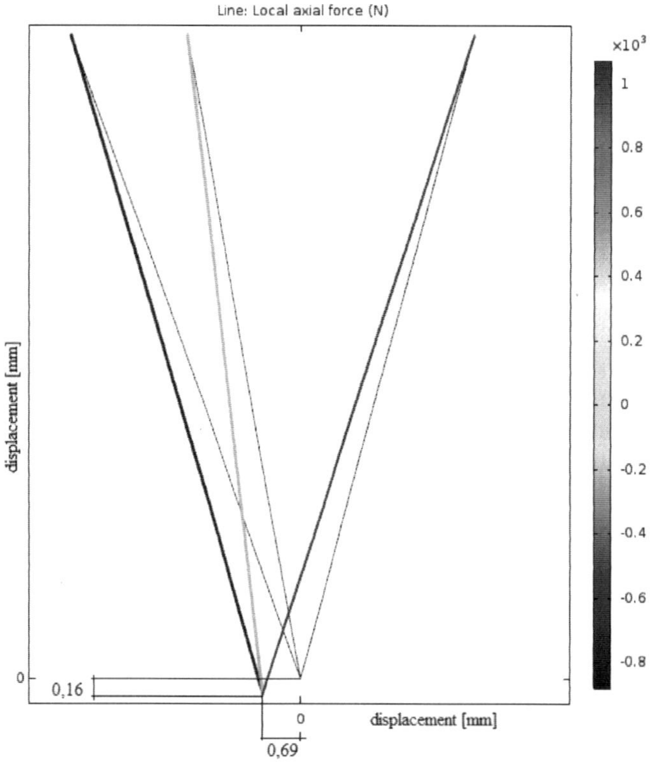

Figure 7. Results obtained by FEM analysis

As stated in [8], the size of the initial population may influence the number of iterations of the algorithm and the same calculation time. This phenomenon has been examined by re-performing the above calculations with different population sizes of 20, 50, 100, 150, 200, 300, 500 and 1000 individuals respectively. The obtained dependence is presented on Figure 8.

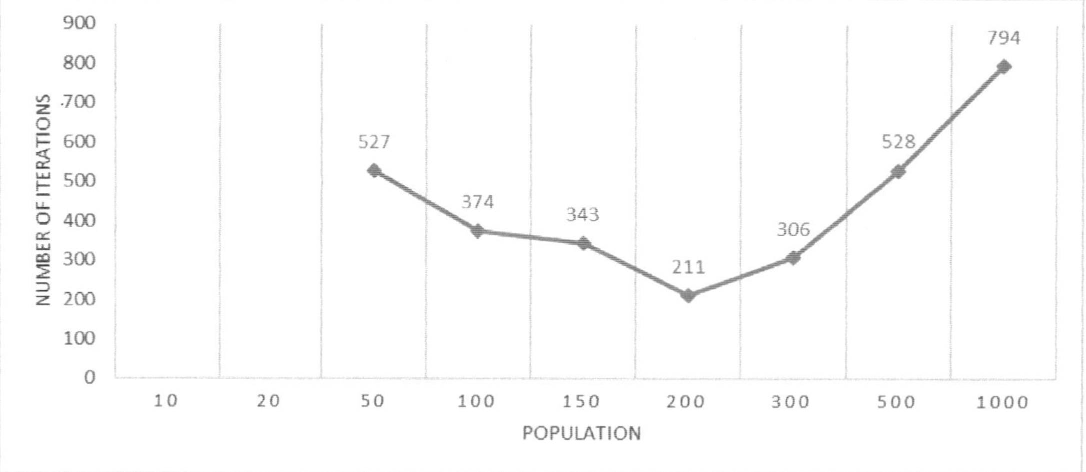

Figure 8. Dependence of the number of algorithms iterations on the initial population

3. Conclusions

Genetic algorithms are a considerable tool for optimizing mechanical structures. The calculations carried out showed that the efficiency and accuracy of the obtained results goes hand in hand with their simplicity. Both examples indicated that an important feature of this type of algorithms is the appropriate selection of input parameters, including the size of the initial population. The analysis of the influence of the latter on the time of calculations of the second task showed, that there may be some optimal starting population number, for which the number of iterations of the algorithm required to achieve the set goal function is the shortest. In addition, for too small population sizes, convergence may not be possible - for its values of 10 and 20, the goal function has not been achieved. It is also worth noting that - as both examples show - the size of the population does not have a significant impact on the result of the calculation.

References

[1] Holland J H 1975 *Adaptation in Natural and Artificial Systems,* MIT Press Cambridge, MA, USA
[2] Moretti S 2002 What Contribution?, *Computer Simulation in Sociology* **Febuary 1** 43-57
[3] Tapia G S and Coello C A 2007 *Applications of multi-objective evolutionary algorithms in economics and finance: A survey,* IEEE Congress on Evolutionary Computation, Singapore, Singapore, September 25-28, pp 532-539
[4] Pedapenki K K and Swathi G 2017 Application of Genetic Algorithm in Electrical Engineering, *International Journal of Pure and Applied Mathematics* **114**(8) 35-43
[5] Takeyasu K and Kainosho M 2014 Optimization technique by genetic algorithms for international logistics, *Journal of Intelligent Manufacturing* **23**(5) 1043-1049
[6] Starosta R 2006 Application of genetic algorithm and fourier coefficients (GA-FC) in mechanism synthesis, *Journal of theoretical and applied mechanics* **26**(2) 395-411
[7] Goldberg D E 1975 *Genetic Algorithms in Search, Optimization, and Machine Learning,* Addison-Wesley Publishing Company, Inc.Reading, Massachusetts, USA
[8] Mora-Melià D, Martínez-Solano F J, Iglesias-Rey P L and Jimmy H, 2017 *Population size influence on the efficiency of evolutionary algorithms to design water networks,* XVIII International Conference on Water Distribution Systems Analysis, **186**, Cartagena, Columbia, July 24-28, pp 341-348

Convergence of V2X communication systems and next generation networks

A Costandoiu[1] and M Leba[2]

[1]University of Petrosani, System Engineering Doctoral School, Petrosani, Romania
[2]University of Petrosani, Department of Computer and Electrical Engineering, Petrosani, Romania

E-mail: alincostandoiu@upet.ro

Abstract. The main goals of the new generation of Vehicle-to-everything (V2X) communication systems is to increase the public safety trying to achieve the accident-free milestone, increase traffic efficiency and lower the carbon emissions for a cleaner environment, improve traffic flow in intersections and on the highways, more efficient parking system, improve fuel efficiency and ultimately enabling self-driving vehicles. Due to multiple research and standardization institutions in Europe and US, two different technologies have emerged, namely C-ITS (Europe – ITS-G5) and DSRC (US – WAVE) based on 802.11p. Both are fully standardized, tested and ready for mass deployment. As the science evolves at an always accelerating pace, a new technology, C-V2X (LTE/Cellular based V2X) is being developed and it started taking the spotlight from already mature and well tested technologies but with less convergence towards 5G. The foundation for connected and automated cars is an optimal technology that is scalable and can evolve in the years to come. C-V2X allows its users to communicate leveraging the existing LTE infrastructure and in future, the 5G mobile network. In this paper we will make a brief introduction on existing technologies, we will present and analyze the standardization landscape and the capabilities of existing technologies. Furthermore, we discuss the proposals for spectrum harmonization followed by pros and cons of legacy and new V2X technologies highlighting the pressure of different automotive and industry associations on steering the development and deployment of these V2X communication systems.

1. Introduction

The present article gives a good overview of standards for cellular and legacy V2X technologies as well as potential future directions. First section is an introduction into V2X landscape and necessity of standardization. The other sections of the article are organized as follows: Section 2 presents the standardization landscape, analyzing key organizations and the development status, Section 3 gives an overview of spectrum usage harmonization attempts, while Section 4 provides trends and directions for future releases, the road towards cellular V2X communication in Europe and in the U.S. Finally, Section 5 offers a conclusion of the article.

1.1. State-of-the-art

Either it takes the form of combat sports or two technology confronting each other vying to become an industry standard, "Everyone loves a fight!" says the president of Ultimate Fighting Championship (UFC). This is the main purpose of this article, to give the best overview of these two technology

confrontation, between Wi-Fi-based 802.11p and cellular C-V2X, as they vie to represent vehicle-to-everything (V2X) communications on a global scale.
The 802.11p standard also defines how to exchange data through a link without the need to establish a basic service set (BSS), without the need of association and authentication procedures to exchange data.

DSRC uses IEEE 802.11p, which is an approved amendment to the IEEE 802.11 standard and adds wireless access in vehicular environments (WAVE). It offers low-latency (down to 2-ms) communication of safety messages between vehicles and between vehicles and infrastructure like roadside units.

C-ITS is the IEEE 802.11p equivalent protocol stack in Europe, covering PHY and MAC, named also ITS-G5. Similar to DSRC, it operates in the 5.9-GHz band as well. DSRC and C-ITS facilitates the communication for vehicle-to-infrastructure (V2I) and vehicle-to-pedestrian (V2P). This gives the vehicle additional awareness on top of what's provided by advances in RADAR, LIDAR and advanced camera systems. A strong challenger to DSRC and C-ITS for V2X has emerged from the cellular industry, namely Cellular-V2X. Similar to 802.11p based communications, C-V2X direct communications supports active safety and enhance situational awareness by detecting and exchanging information using low-latency transmission in the 5.9-GHz ITS band for vehicle-to-vehicle (V2V) as well as V2I and V2P scenarios. C-V2X can also function without the assistance of network and has a higher range than DSRC/C-ITS, that exceeds one mile, even in areas where mobile network connection is not available.

C-V2X can combine the communication capabilities of roadside units (RSUs) with the cellular network to help improve safety and support autonomous driving. The roadside units will use a high-throughput connection, as they are static with other cars on the road to build dynamic high quality maps using cameras and many data sensors on board, and share them with nearby vehicles as needed [1].

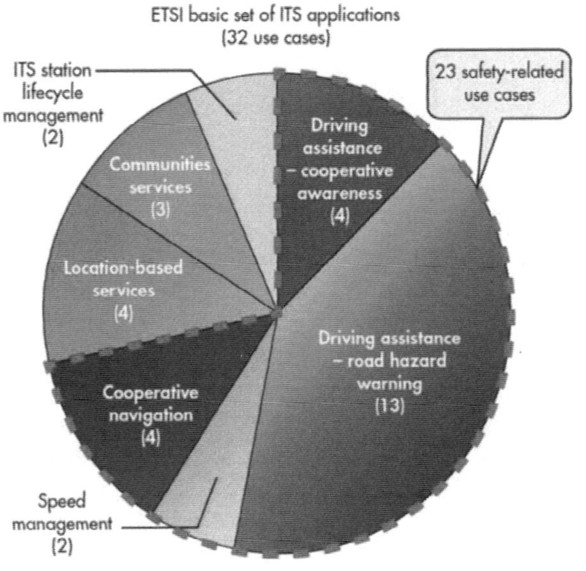

Figure 1. ETSI basic set of ITS applications

Figure 1 give a brief overview of the use cases covered. V2X is using wireless connectivity to enable connected vehicles to achieve "Cooperative Awareness" for the realization of the "Basic Set of ITS Applications" as defined by ETSI (European Telecommunications Standards Institute) [2].

1.2. Necessity of standardization

A standardization process for wireless vehicular communication ensures, as in any other domains, not only interoperability but also supports regulations and legislation in the legal framework, and creates larger markets for global products.

An essential requirement for the worldwide deployment of V2X communication systems are global standards, which provide regulations to ensure interconnectivity among V2X sub-systems and components as well as interoperability of implementations from different vendors. Additionally, global standards also serve for many other purposes, like: open standards create trust for customers in products and services, create global markets compared to proprietary systems, help to lower development and maintenance costs, and increase competition between vendors helping to lower the production costs.

2. Standardization landscape

Historically speaking, standardization in Europe and in the U.S. has developed in parallel with many overlaps, mainly because the activities were supported by different SDOs, research and development programs, as well as promoted by different stakeholders. In the end, these different paths led to different sets of standards. These two approaches are C-ITS (Cooperative Intelligent Transportation Systems) in Europe and DSRC (Dedicated Short Range Communication) in the U.S. ITU (International Telecommunication Union) and 3GPP (3rd Generation Partnership Project) have initiated first standardization studies. This article focuses on the most relevant standards planned for deployment in the next years, and namely the standardization of C-ITS in Europe, DSRC in the U.S. and C-V2X with PC5 interface as a global standard [2].

2.1. Global overview

Policy and rule makers in many countries around the globe have been preparing policies on V2X for some time, reflecting the fact that technologies for cooperative driving are continuously evolving, and automated vehicles are already being introduced in the market. More work is still needed in the areas of security/privacy, conformance, public awareness and economic sustainability for investors, since these kinds of deployment require involvement from many parties, like the state, regional and local governments and road authorities together with the private sector.

In Europe, the modernization and digitalization of transport and the use of intelligent transport systems (ITS) has increasing interest from the European Committee, due to the many benefits, like increase of road safety, address gas emission and traffic congestion issues, and support jobs and economic growth in the transportation sector.

Back in August 2008, the EC published the Decision 2008/671/EC [3] to allocate the frequency band 5875–5905MHz (30MHz) for safety-related applications of ITS, and adopted the ITS Action Plan COM(2008)886 [4], with the initiatives to accelerate and coordinate the deployment of ITS systems on roads and in auto vehicles in the EU.

In July 2010, a legal framework, Directive 2010/40/EU [5], was adopted to accelerate the deployment of ITS across Europe, with the V2I link defined as a key priority area.

The major SDOs active in Europe in the C-ITS development area are ETSI (European Telecommunications Standards Institute) and CEN (European Committee for Standardization) with their Technical Committees (CEN TC 278 – Intelligent Transport Systems) [6]. CEN works closely with ISO to produces joint specifications. Supported by a mandate of the European Commission (EC) [7], ETSI and CEN have created a stable and consistent set of standards for a minimal deployment of V2X communication system, which was supposed to be taken as basis for European deployment. ETSI focused on specifications and requirements for the communication system and V2V applications while CEN mainly focused on standards for V2I applications. The European standardization efforts are going in parallel with activities of the Car-2-Car Communication Consortium (C2C-CC) [8], an industry consortium of automobile manufacturers, equipment suppliers, and research organizations like ERTICO [9], a European organization of public and private stakeholders, and ETSI CTI (Center for Testing and Interoperability). The C2C-CC has developed a customized profile of the European C-ITS Release 1 that is restricting the large set of developed standards and complements missing specifications.

In 2013, automobile manufacturers in C2C-CC signed an agreement for the adaptation of the V2X C-ITS system. Deployment plans were developed in the Amsterdam Group, a strategic alliance with stakeholders from C-ITS ecosystem in Europe, with the objective to facilitate joint deployment of cooperative ITS in Europe. The Amsterdam Group includes umbrella organizations like CEDR–ASECAP–POLIS, representing stakeholders of the ITS infrastructure on highways, cities and traffic management, and the C2C-CC. There were many pilot deployment projects supporting the system introduction and concept demonstration. A trilateral C-ITS corridor that interconnects Vienna–Frankfurt–Rotterdam is planned to be equipped with roadwork protection systems for highways until end of 2018 [10] [11] [12].

In November 2014, the EC launched a C-ITS Deployment Platform, a multilateral framework involving national authorities and C-ITS stakeholders (not including the cellular mobile sector) to identify the remaining barriers and to propose solutions for the commercial deployment of C-ITS systems in the EU zone.

The initial phase of the C-ITS Platform (2014–2016) has focused on the development of a common vision on the interoperability deployment of the system in the EU, including the identification of key technical, legal and commercial issues and the development of policy recommendations to address these issues. The first phase has ended with publication of an technical and strategical expert report in January 2016 [13], which was complemented by a “cost vs. benefit” analysis and a public consultation on this topic.

In July 2016, the C-ITS Platform second phase started. The main objective was to further develop a shared vision on the interoperable deployment of C-ITS systems in the EU, by defining the common technical and legal framework to address key issues on security, data protection, compliance assessment and hybrid communication identified in the first phase, and by investigating further the benefits that C-ITS could bring in terms of autonomous driving. The 5GAA (5G Automotive Association) attended two C-ITS Platform meetings as guests during year 2017. The second phase ended with the publication of an expert report in September 2017 [14].

In parallel, the EC has adopted and is open to various other policies which are relevant for connected, cooperative and automated mobility within EU zone. In particular, “5G for Europe: An Action Plan” – COM(2016)588 [15] calls for the availability of 5G along main European transportation paths.

In October 2017, the Radio Spectrum Committee (composed of the EC and representatives of regulators from EU Member States) approved a new mandate for CEPT (European Conference of Postal and Telecommunications) to study an extension of the ITS-safety related spectrum band 5.9GHz, with the possibility to extend the dedicated ITS spectrum to 50MHz in bandwidth (from 30MHz currently). The document recognizes recent developments in relation to Cellular (LTE) based V2X specification for ITS, which could underpin the path to 5G connectivity for the automotive and road transport sectors. The EC also recognizes that the two V2V radio systems (IEEE 802.11p and C-V2X PC5) cannot inter-communicate with each other due to different radio access techniques. CEPT shall deliver a final report on the topic of harmonization and coexistence by March 2019 [16].

U.S. was the first to assign a dedicated spectrum in the 5.9GHz band for DSRC-based V2X services, but unfortunately there has been a huge delay between the spectrum being assigned, and being put into use. In October 1999, the United States Federal Communications Commission (FCC) allocated 75 MHz of spectrum in the 5.9 GHz band to be used by intelligent transportation systems (ITS).

In the U.S., the relevant SDOs for V2X communications are IEEE (Institute of Electrical and Electronics Engineers) and SAE (Society of Automotive Engineers). Specifically, the 802.11 Wireless LAN and the 1609 Standards were developed by working groups from IEEE, while the DSRC was developed by the technical committee from SAE. IEEE has developed the IEEE 1609 standard family of protocols [17] on top of the IEEE 802.11 PHY and MAC layers. This combination of IEEE 802.11 and 1609 standards is also known as WAVE (Wireless Access for Vehicular Environment). Additional layers coming above the protocol stack, like V2X message sets and related performance requirements, are specified by SAE. Altogether, the WAVE standards, V2X message sets and performance requirements build a consistent set of standards ready for basic functionality commercial deployment.

In January 2017, the US DoT (Department of Transportation) gave a Notice of Proposed Rulemaking (NPRM) that require all new light vehicles being sold on US market to be capable of V2V communications. It tries to mandate the IEEE 802.11p as the communication technology standard for V2V, and require automotive OEMs to start implementing these specifications and requirements two years after the final rule is adopted, this being around 2019-2020 (although it gave a three-year period to accommodate automotive OEM production lines). The NPRM attracted lots of attention and hundreds of responses, especially from the C-V2X supporters and promoters, but the US DoT has given no indication on when it will proceed.

In parallel, the US DOT's Federal Highway Administration Infrastructure Deployment Guidelines were temporarily published but then withdrawn from public access. These guidelines give specifications about V2I communications, in order to help transportation planners to integrate the technologies that allow their vehicles to communicate with roadside infrastructure.

Unlike in Europe, where deployment is industry-driven and voluntarily, a regulatory decision was expected in the U.S. until 5GPP, Qualcomm and GSMA started to push rule makers for technology neutrality approach. This rulemaking process was planned to be initiated and to make DSRC mandatory, indicating a deployment at the beginning of 2020 [2].

In China no dedicated spectrum was allocated initially for ITS services, but the Ministry of Industry and Information Technology (MIIT) allocated 50MHz in the 5.9GHz band for ITS systems, on a license-exempt basis, like in all other regions.

In November 2016, the Chinese government announced the dedicated allocation of 20MHz (5905–5925MHz) for C-V2X trials in six major cities. It is possible that China will be the first country to launch C-V2X, due to bigger opening to new technologies. Research being conducted for this study suggests that C-V2X commercial launch in China could be in the second half of 2019.

In Japan, the 5.9GHz band was used by a Japanese wireless technology called Electronic Tolling Collection (ETC). Reports show that this band is considered for use to provide V2V and V2I communications, but it is unclear whether a decision will be made soon. Until now, Toyota already deployed a proprietary "ITS Connect" solution in 12 cities.

In September 2016, Ministry of Communications in South Korea allocated 70MHz of spectrum in the 5855–5925MHz band for ITS services on a license-exempt basis, and finally on a technology-neutral basis.

Last year, in September 2017, Australia's telecommunications regulator, ACMA, closed a consultation on the allocation of dedicated spectrum in 5.9GHz band for ITS application, proposing rules to follow a technology-neutral approach [18].

2.2. Wi-Fi based V2X (802.11p)

2.2.1. Dedicated Short Range Communication (DSCR) standards in U.S. DSRC is based on the widely deployed WLAN standard defined in 2012, IEEE 802.11, which defines the physical transmission (PHY) and medium access control (MAC) (Figure 2 shows the overall DSCR protocol stack). PHY and MAC are derived from the former IEEE 802.11a standard, and adapted to the requirements for V2X communication.

Similar to IEEE 802.11a, DSRC operates in the 5 GHz band (U-NII band), but it was shifted from the regular Wi-Fi channels to some dedicated DSRC channels. These channels range from 5.855 GHz to 5.925 GHz, commonly referred as the "5.9 GHz band". The spectrum is further subdivided into 10 MHz channels (see Figure 3).

DSRC uses the Orthogonal Frequency Division Multiplexing (OFDM) scheme, a state-of-the-art widely used multi-carrier transmission scheme which is proven to be robust against interference and fading, and also re-uses the same preamble and the pilot design for synchronization as well as channel estimation. Comparing to the normal usage in Wi-Fi, OFDM operates with “half clock”, which reduces the commonly used 20 MHz channel spacing to 10 MHz and doubles the time parameters, particularly the OFDM symbol duration with the cyclic prefix. These changes attribute to the characteristics of the wireless channel in vehicular environments as it can adapt to the inter-carrier interference caused by the Doppler spread due to fast moving vehicles [19].

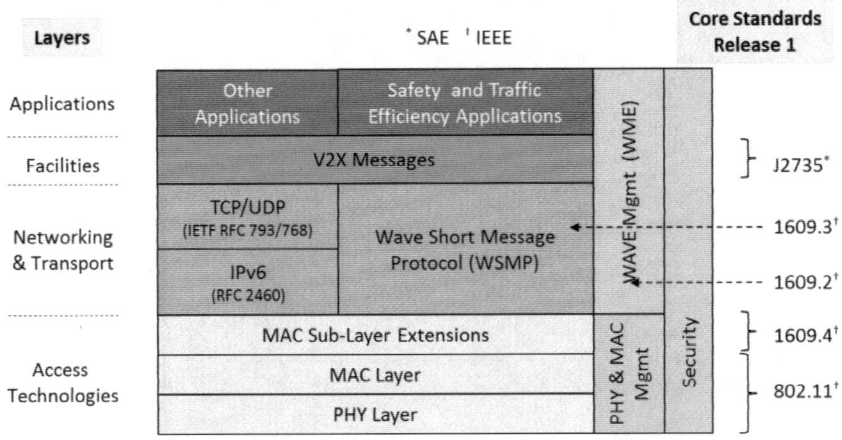

Figure 2. DSRC Protocol stack and related standards in the U.S.

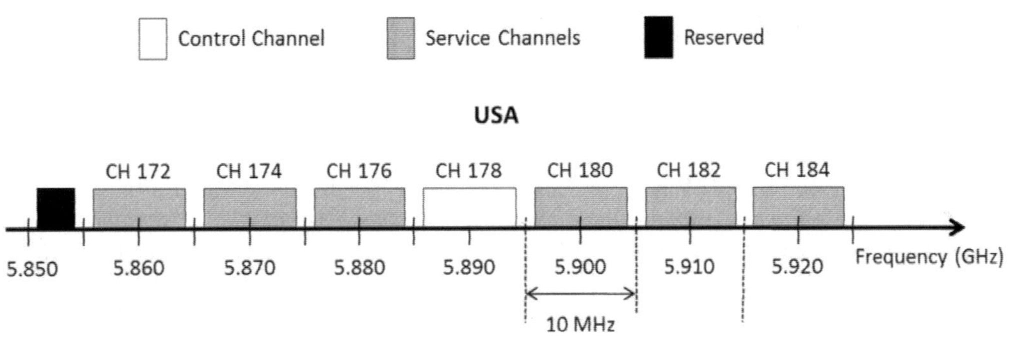

Figure 3. IEEE 802.11p DSRC Channel Frequency Band in U.S.

2.2.2. Cooperative – Intelligent Transportation System (C-ITS) standards in Europe. The parallel research and development of V2X communications by different SDOs has led to a different protocol stack in the U.S. and Europe. This section presents the C-ITS standards in Europe in comparison to the DSRC standards. Figure 4 describes the overall protocol stack and the corresponding core standards behind it, keeping the same structure of horizontal layers for access technologies, networking & transport, facilities and V2X messages, applications, and vertical management and security entities for all layers, same as in Figure 2.

ITS-G5 is the IEEE 802.11p DSRC equivalent in the C-ITS stack covering PHY and MAC layers. The last two letters in naming indicate that it operates in the 5 GHz band. It operates in the 5.9 GHz band in channels of 10MHz each, whereas the European spectrum allocation is sub-divided into part A to D (see in Figure 5). ITS-G5A with 30 MHz is the primary frequency band that is dedicated for safety and traffic efficiency applications, ITSG5B has 20 MHz for non-safety application, and ITS-G5C is shared with the RLAN band, while ITS-G5D is reserved for ITS future use [20].

A specific requirement for Europe is that the ITS-G5 spectrum must minimize the interferences to the 5.8 GHz band. However, the key technology features of IEEE-802.11 for DSCR and ITS-G5 are similar. On PHY layer, it applies OFDM with the same parameter set "half clocked" compared to IEEE 802.11a, with an adapted spectrum masks. On MAC layer, ITS-G5 also employs EDCA with CSMA/CA, identical to DSRC, and access categories allow for prioritization of data traffic.

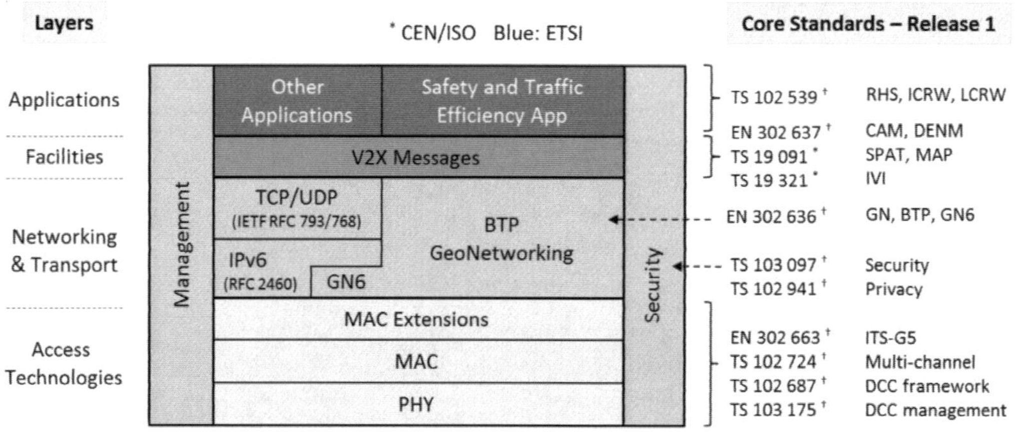

Figure 4. Protocol stack and related core standards for C-ITS in Europe

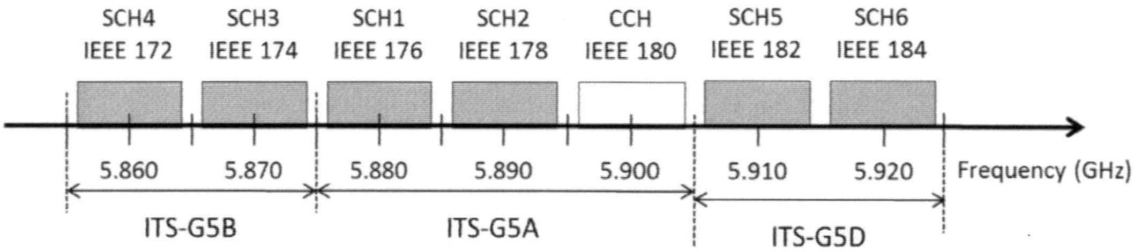

Figure 5. IEEE 802.11p ITS-G5 Channel Frequency Band in Europe

Standards for Networking & Transport and V2X messages also rely on the IP protocol. While the usage of TCP/UDP and IP version 6 is similar, C-ITS specifies a totally different ad-hoc routing protocol with additional function for multi-hop communication, named GeoNetworking described by the ETSI EN 302 636 standard series. Main feature is the utilization of geographical coordinates for addressing and forwarding of messages. The geographical addressing has the advantage of making the packet delivery independent from the communication range of a single wireless hop (which can vary from several 10 meters up to 1 km under line-of-sight conditions found on highways). It enables much more efficient multi-hop routing with low protocol overhead for establishment and maintenance of network routes in a highly dynamic environment with frequent topology changes. This feature provides more technical opportunities in application support, but has the disadvantage of an increased protocol complexity and overhead.

Standards at the Facilities and V2X messages layer define more application-related functionalities specific for European market. The most relevant V2X messages are: Cooperative Awareness Message (CAM) (ETSI EN 302 637-2) [21]. CAM periodically conveys critical information about the state of the vehicle in support of safety and traffic efficiency application. Vehicles receiving CAM messages can track other vehicles positions and movement. It can be considered as an equivalent to the BSM in DSRC protocol stack. Distributed Environmental Notification Message (DENM) (ETSI EN 302 637-3) [22] helps to disseminate safety information in a specific geographical region. Comparing with the

CAM, which is sent periodically by all vehicles, the DENM message transmission has to be triggered by an application, which offers more flexibility.

For V2I (vehicle-to-infrastructure) communications, additional services are defined to communicate with road users, to control of roadside infrastructure for priority access, preemption in case of emergency and to provide information from the vehicles to the infrastructure (see Table 1).

Table 1 Overview of V2I services based on V2X communication (ETSI TS 103 301 [23])

Service name		Description
IIS	Intersection information service	Provides dynamic information about status of an intersection, such as traffic light state, residual time until traffic light changes, right of way for allowed maneuvers, public transport prioritization
TPS	Topology service	Offers static topology information for intersection or road segment, and paths for pedestrian crossings, for vehicles, public transportation
IVI	In-vehicle information service	Gives mandatory and advisory road signage information inside the vehicle, including static, variable and virtual signs; examples: such as contextual speeds and road works warnings.
SCS	Signal control service	Controls traffic lights for prioritization of public transport and preemption of public safety vehicles
INS	Infrastructure notification service	Informs vehicles and pedestrians about traffic situation, hazards, road works warnings, and other events
IAS	Infrastructure awareness service	Informs vehicles about existence of infrastructure (e.g., for interference management near tolling zones); estimation of traffic situation and flow based on messages from vehicles in vicinity

These services define dedicated messages (see Figure 4), namely the Signal Phase & Timing (SPAT) message for IIS, the MAP message for TPS, and the In-Vehicle Information (IVI). In the signal control service message are bidirectional exchanged. DENM and CAM are also used for infrastructure-related services like INS and IAS.

Similar to the DSRC standards, C-ITS applications are not fully standardized. Only minimum functional and performance requirements for three groups of applications are defined:

- Road hazard signaling (RHS): includes use cases such as emergency vehicle approaching, hazardous location and emergency electronic brake lights;
- Intersection collision risk warning (ICRW): refer to potential vehicle collisions at intersections;
- Longitudinal collision risk warning (LCRW): refer to potential vehicle rear-end/head-on collisions.

2.3. Cellular based V2X (4G and future 5G)

V2V communications are based on device-to-device (D2D) communications defined as part of ProSe services in Release 12 and later in Release 13 of the 3GPP specifications. Sidelink aims to enable D2D communications within legacy cellular-based LTE radio access networks.

In legacy LTE communications, two UEs communicate through the Uu (Over the air) interface and data are always going through the network LTE eNB. Differently, Sidelink enables the direct communication between multiple UEs in proximity of each other using the newly defined PC5 interface, and data does not need to go through the eNB (see Figure 6).

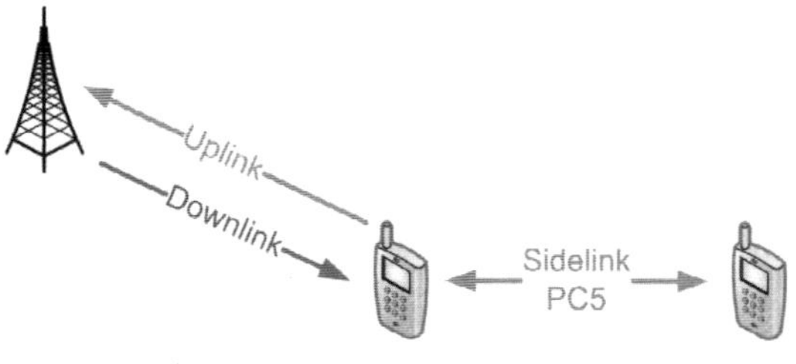

Figure 6. Model of Sidelink PC5 interface

Resources assigned to the PC5 interface are taken from the Uplink resource pool. The reasons for this is, first, the UL subframes are usually less occupied; second, most of the Downlink subframes are never really empty so that they can be reused. Usually there are always at least the cell specific reference signals (CRS) transmitted in all DL subframes.

Based on fundamental link and system level changes, in Cellular V2X communication there are two possible high-level deployment configurations currently defined and illustrated in Figure 7.

In both cases GNSS (Global Navigation Satellite System) is used for time synchronization. This can be seen as a drawback for the C-V2X systems while operating in tunnels or area where GNSS signal is missing.

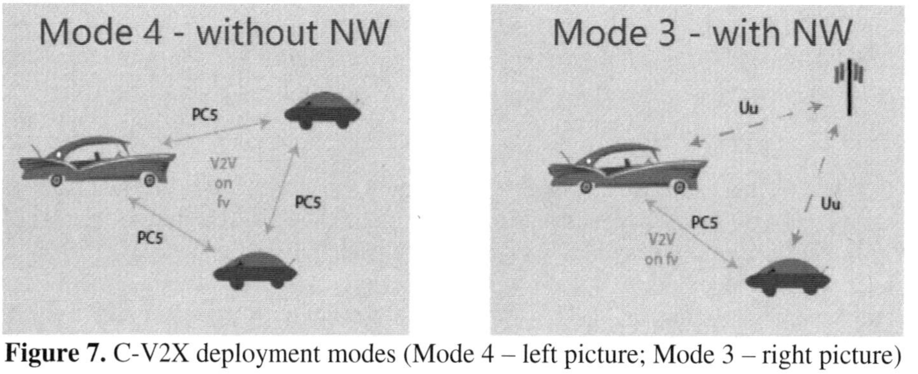

Figure 7. C-V2X deployment modes (Mode 4 – left picture; Mode 3 – right picture)

In "Mode 4", scheduling and interference management are supported based on distributed algorithms implemented between the vehicles. As specified earlier the distributed algorithm is based on sensing with semi-persistent transmission.

In "Mode 3", scheduling and interference management are managed by the eNBs via control signaling over the Uu interface. The eNodeB will assign the resources for V2V signaling dynamically [24].

According to the requirements, ProSe communication has to work in regions, where network coverage is missing. Therefore, ProSe communication is specified for the following scenarios (see Figure 10):

- In the "in coverage" case, the network controls the resources used for ProSe so interferences with the cellular traffic are avoided and in addition the ProSe communication may be optimized;
- For the "out-of-coverage" case such a control is not possible. The UE uses resources which are preconfigured, either in the mobile device or in the USIM of the UICC card;
- A special case is given in the "partial coverage" case. Coordination between the network and the preconfigured values is necessary in order to enable communication and to limit the interferences to other UEs at the cell boundary.

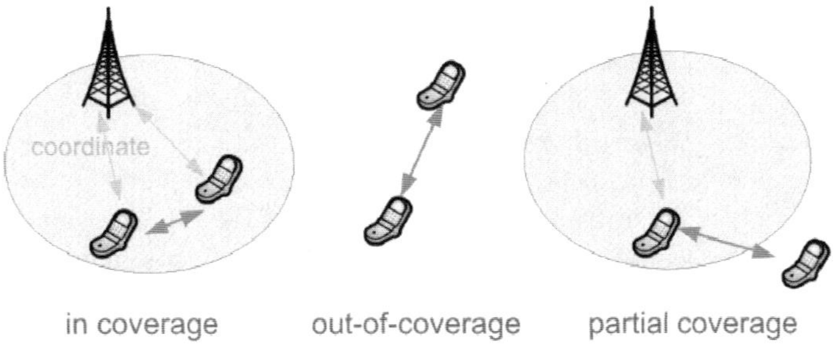

Figure 8. Coverage scenarios

The PC5 air interface, in ProSe communication system, is connectionless. There is no equivalent to the RRC connection procedures from legacy LTE system. Messages are created on the application level of the UE and transmitted with the first opportunity. If a connection is required, it has to be done within the application. For transmission and reception of the associated data packets, the following protocol stack is used (see Figure 9).

LTE-Uu interface can be unicast, multicast or broadcast transmitted via MBMS (Multimedia Broadcast/Multicast Services) [14].

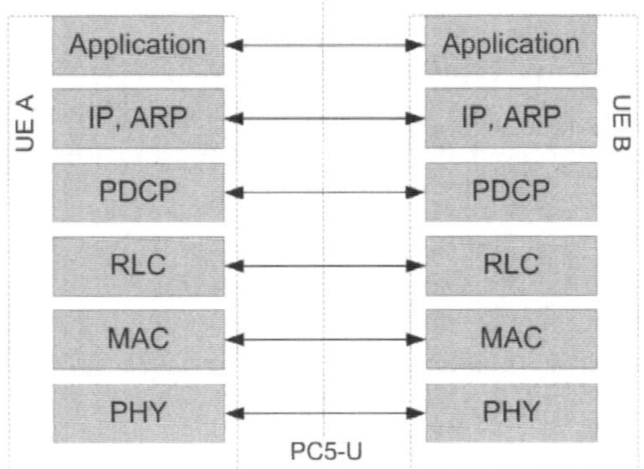

Figure 9. PC5 air interface protocol stack

3GPP Release 14 including C-V2X is a key step to the next generation of cellular technology, namely 5G. Naturally C-V2X is already on a backwards compatible evolution path with enhancements being specified beginning with 3GPP Release 15.

3. Harmonization of spectrum usage

3.1. Technology-neutrality of the regulatory bodie

Considering that FHWA (Federal Highway Administration) in US and European Committee had preferred 802.11p V2X as main technology for enabling C-ITS, the 5GAA association and GSMA asked both regulatory bodies to keep and respect technology neutrality. Regulators shall facilitate interoperability regulations between technologies while the market should decide which one is more suited and optimal for a strong foundation in order to remain sustainable for years to come.

Created in September 2016, 5G Automotive Association (5GAA), a relatively new cross industry association, is committed to support further enhancements of this future proof technology, to make sure that the full potential of the C-V2X technology is reached. This includes leading efforts to address key technical and regulatory issues, as well as integrating vehicle platforms with advanced cellular connectivity, networking and computing solutions [25].

The technology neutral nature of spectrum regulations in Europe means that both LTE-V2X and 802.11p have equal rights to operate in the 5.9 GHz band, subject to compliance with the relevant regulatory technical conditions. 5GAA is a supporter of LTE-V2X as a platform to evolve towards 5G technologies.

3.2. Spectrum harmonization efforts in Europe

C-V2X and 802.11p use different physical layers and medium access control protocols. Therefore, the operation of the two technologies in the 5.9 GHz band and in the same geographic area without an agreed coexistence solution would result in mutually harmful co-channel interference. These interferences can be avoided in the short term by allocating so-called "safe harbor" channels to C-V2X and 802.11p in the 5875-5905 MHz band (see Figure 10). This will encourage stakeholders to further enhance this proposal and come to speedy agreements on this topic for the benefit of the European ITS ecosystem as a whole.

The below regulatory measures all refer to the ETSI Harmonized Standard EN 302 571 [26] which defines requirements for operation of ITS equipment in 5855-5925 MHz, covering the essential requirements of article 3.2 of the Radio Equipment Directive (2014/53/EU). According to ECC (European Communication Committee), ITS equipment complying with EN 302 571 are exempt from individual licensing for operating in this band.

The band 5855-5925 MHz is subject to the following harmonization measures in Europe:

- The European Commission has harmonized the band 5875-5905 MHz for safety-related applications of Intelligent Transport Systems in the European Union via the legally binding Commission Decision 2008/671/EC [3].
- The same harmonization is applied by the ECC via ECC Decision (08)01 [27], which additionally indicates that CEPT administrations shall consider within a future review of this Decision the designation of the frequency sub-band 5905-5925 MHz for an extension of ITS spectrum.
- ECC also recommends, via ECC Recommendation (08)01 [28], that CEPT administrations should make the frequency band 5855-5875 MHz available for ITS non-safety applications.

The proposed spectrum partitioning shall happen in three steps. Finally, partitioning might be complemented by additional technical mechanisms which would allow each of C-V2X and 802.11p to access the remaining 20 MHz in a fair manner, to reduce the risk of harmful co-channel interference.

Harmonization of safety-related applications is shown in Figure 10 below, where the two technologies are referred to as Technology A and Technology B, where the equipment is tuned to operate at 5875-5885 and 5895-5905 MHz, respectively avoiding co-channel interference between the two V2X technologies.

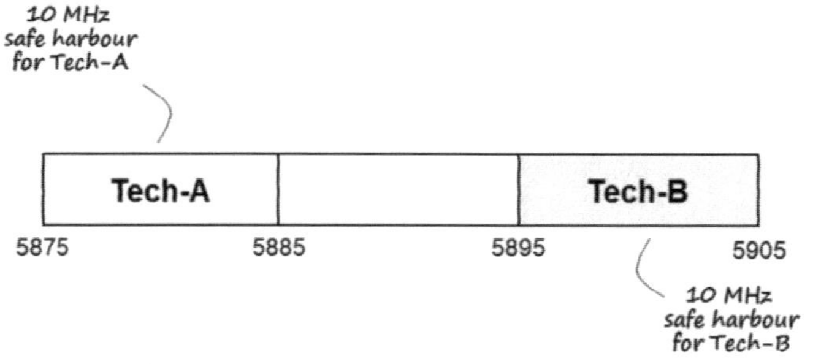

Figure 10. Step1: Partitioning of 5.9 GHz

Figure 11 illustrates Step 2, with a detect-and-vacate solution for C-V2X and 802.11p coexistence. Technology A equipment should operate without any special measures in 5875-5885 MHz. If Technology A equipment needs to transmit in 5885-5895 MHz, then it would need to monitor activity on the relevant channel, and proceed with transmissions if and only if Technology B transmissions are not detected in that channel. A symmetrical procedure would apply to Technology B.

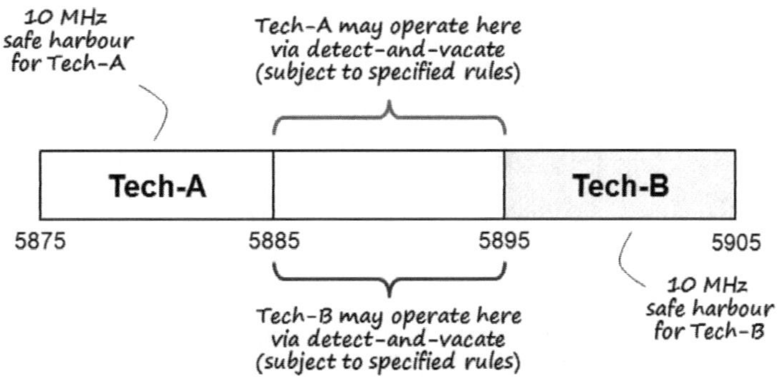

Figure 11. Step2: Partitioning of 5.9GHz with mutual "detect and vacate" techniques for middle 10MHz

Figure 12 illustrates Step 3, with an extended detect-and-vacate solution for C-V2X and 802.11p coexistence. Technology A equipment should operate without any special measures in 5875-5885 MHz. If Technology A equipment needs to transmit in 5885-5905 MHz, then it would need to monitor activity on the relevant channel, and proceed with transmissions if and only if Technology B transmissions are not detected in this channel. A symmetrical procedure would apply to Technology B.

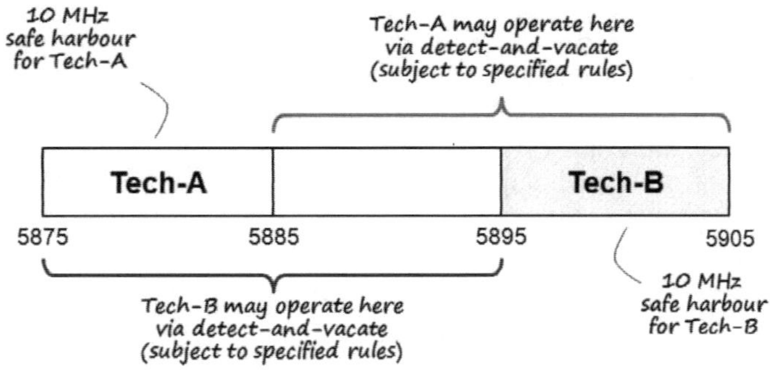

Figure 12. Step3: Partitioning of 5.9GHz with enhanced mutual "detect and vacate" techniques for middle 10MHz

ITS industry to will further develop, and cope with this agreement on the spectrum partitioning, with a view to expedite the successful deployment of C-V2X and 802.11p in the 5.9 GHz band [29].

4. Road to Cellular V2X

C-V2X started to gain momentum once release of 3GPP Rel.14 was published. Following this, studies about the readiness of 802.11p technologies were published and pushed to regulators. These studies were highlighting how C-V2X is only in the early stages of development and it will take many years, by 2020-2021, until readiness for commercial deployment.

In 2016, a European strategy on C-ITS was published, setting out priorities for the deployment of C-ITS, with a goal of deployment by 2019. At that time, the technology being specified was based on the IEEE 802.11p standard, although using C-V2X for V2I, rather than deploying dedicated roadside infrastructure, was recognized as a means of accelerating infrastructure penetration across all roads from the very first launch of C-ITS services. Since then, the mobile industry has developed C-V2X specifications within Release 14 of the 3GPP specifications for LTE/4G, in the expectation that 5G-based C-V2X will be analyzed in 3GPP specifications from Release 16 onwards. Initial 5G deployment, providing enhanced mobile broadband services, will be specified starting from Release 15 [18].

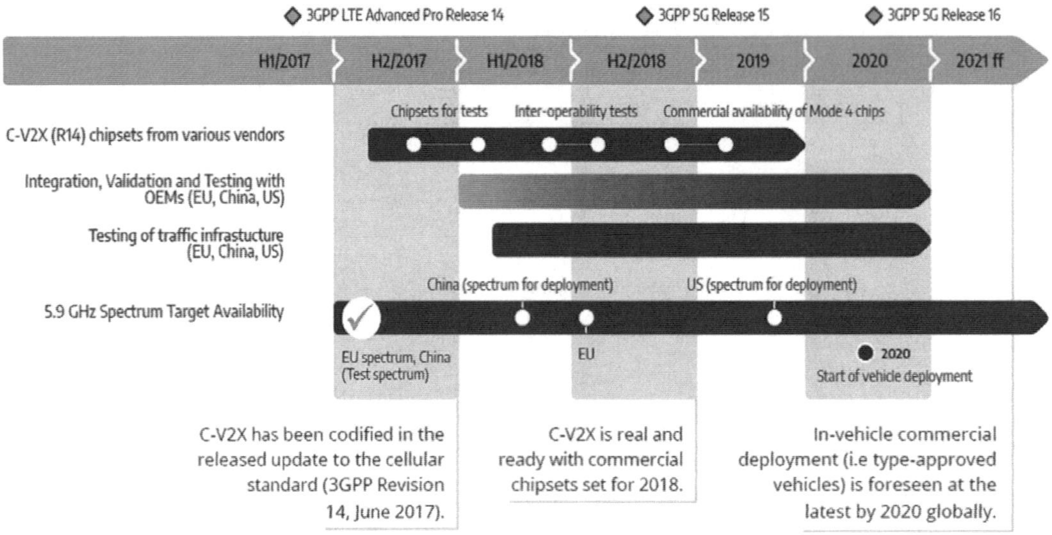

Figure 13. Scenario for C-V2X deployment [18]

The 4 major chipset vendors Intel, Qualcomm, Huawei and Samsung which are members of 5GAA provide almost all of the chipsets used in connected cars today. These vendors are committed to provide C-V2X chipsets and stopped supporting 802.11p standard. The availability of these chipsets will coincide with planned validation and testing of the communication modules by vehicle manufacturers including supportive vendors Audi, Continental, Ford, Nissan, PSA, SAIC and Bosch.

This testing and validation aims to confirm the C-V2X functional capabilities, and V2V communications in particular. These chipset vendors look to build on the successfully tested pre-standard versions from 2015 and 2016, which showed a superior technical advantage in all safety relevant parameters such as vehicle speed, braking distance, scalability and reliability, compared to similar tests performed over IEEE 802.11p based protocols. These validation tests are an essential prerequisite to ensure a successful launch in the second half of 2018 of the first commercial chipsets from Qualcomm [30].

An additional safety benefit of C-V2X is the support for Vulnerable Road User (VRU) collision avoidance, with the integration of V2N and V2P (vehicle to pedestrian) functionality. It ensures that a VRU becomes visible to vehicles via V2N utilizing smart phone applications. It will not require the integration of smartphones with C-V2X functionality in the first step of deployment. this feature will be implemented as a second step, for direct V2P communication between vehicles and VRUs. These enhancements will not possible in other V2V and V2I technologies such as DSRC or ITS-G5.

C-V2X also allows for further enhancements by future architecture and design decisions including network slicing, quality of service, edge computing and innovation, leveraging the C-V2X foundation. Rohde & Schwarz, a leading supplier of test and measurement equipment, is enabling vehicles to move to the next level of autonomy by providing initial signaling testing support for cellular vehicle-to-everything (C-V2X). Rohde & Schwarz and Qualcomm have collaborated by testing a pre-commercial Qualcomm 9150 C-V2X chipset, supporting 3GPP Release 14 specifications for PC5 based direct communications with the R&S CMW500 wideband radio communication tester. The 9150 C-V2X chipset is a product of Qualcomm Technologies and operates in bands 46D and 47 for 5.8/5.9GHz ITS spectrum [31].

Both technologies, C-V2X and IEEE 802.11p, have the potential to increase the safety and efficiency in transport. However, with the currently defined C-V2X technology communication, combined with LTE cellular networks for V2N, has the potential to bring additional benefits, including:

- better coverage for V2N, by exploiting existing cellular network coverage

IOP Conf. Series: Materials Science and Engineering **477** (2019) 012052 doi:10.1088/1757-899X/477/1/012052

- reduced infrastructure deployment costs and improved service reliability, by using existing mobile infrastructure
- other telematics services in vehicles (e.g. infotainment) to be provided via cellular interface
- increased deployment flexibility, with coverage for both short-range and wide-area applications
- integration with smart-city and other connected-transportation initiatives using cellular
- enhanced security, by usage of mobile subscriber identity module (SIM) cards
- future proof evolution to 5G, facilitating earlier deployment and after-market deployment.

5. Conclusions

By leveraging existing cellular infrastructure and continued industry investment, deployment of C-V2X technology can be achieved at a much lower cost compared to other competing V2X technologies. Consequently, C-V2X technology can save taxpayer money that otherwise would be used for new deployment, maintenance, and upgrade of roadside units utilized by non-cellular technologies.

C-V2X benefits from a higher spectrum efficiency and supports higher ranges compared to non-cellular technologies. Additionally, with support from the existing cellular ecosystem, C-V2X is integrated into the latest cellular standards with a clear upgrade path for future enhanced safety features and migration towards 5G (see Figure 14).

On the other side, the IEEE 802.11p is a ready to deploy technology that could save lives and make the traffic more efficient from today if mass deployed [32]. It has the advantage of many years of research, development and testing [33].

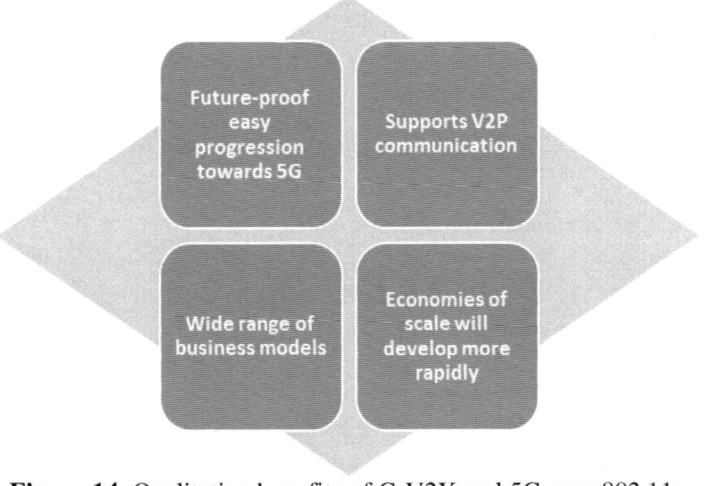

Figure 14. Qualitative benefits of C-V2X and 5G over 802.11p

The path to fully automated vehicles will require coexistence for a period of time between vehicles with no active control systems, different levels of automated vehicles and fully automated vehicles. C-V2X technology will help to enable this coexistence by ensuring that fully automated vehicles and drivers of other levels of automated vehicles can maintain complete awareness surrounding vehicles and other road hazards. C-V2X technology is well positioned to implement the full slate of V2X benefits highlighted in this paper.

Qualcomm was initially a big enabler of DSRC and offered also second-generation 802.11p DSRC chips. Currently their focus switched to cellular V2X, announcing the interruption of 802.11p enabled chipsets.

As until now, the trend is to adopt both 802.11p and C-V2X technologies and let the market decide which one is more suited for commercial deployment and future development and evolution towards 5G.

A "Analysis Mason" study from December 2017 [18] indicates that C-V2X (PC5) outperforms 802.11p by reducing serious injuries and fatalities on the EU roads. This is due to a combination of the superior performance of C-V2X at the radio link level for ad-hoc direct communications between road users, together with the market conditions which better places the deployment of C-V2X in commercial vehicles and smartphones. For these reasons, it is essential that EU regulations remain neutral and do not obstruct the deployment of C-V2X in favor of 802.11p for the provision of direct vehicle communications and between vehicles and VRUs (vulnerable road users).

The absence of interoperability measures at radio link level between C-V2X and 802.11p is unlikely to present a serious barrier to the reduction of accidents on the EU roads in the short to medium term. The low penetration of C-ITS technologies in vehicles in the second half of this decade means that a vehicle equipped with C-V2X or 802.11p is far more likely to interact on the roads with a vehicle that is not equipped with any C-ITS technologies at all. Only starting from the middle of the next decade, when penetration rates are expected to reach a significant level which results in visible impacts on lowering the accident rates. Any regulations from rule-making-bodies which mandate that C-V2X to be backward compatible with 802.11p standards, will only have a limited effect in the early stages of deployment pre-2025. Nevertheless, such regulations will run the risk of unnecessarily obstructing full potential of C-V2X development in favor of 802.11p, thereby obstructing the smooth and rapid adoption of C-V2X and resulting in a higher rate of road injuries and fatalities on the longer term.

References

[1] Slovick M 2017 DSRC vs. C-V2X: Looking to Impress the Regulators, *Electronic Design* [Online] Available: http://www.electronicdesign.com/automotive/dsrc-vs-c-v2x-looking-impress-regulators

[2] Festag A 2015 Standards for vehicular communication – from IEEE 802.11p to 5G, *Elektrotechnik & Informationstechnik* **132**(7) 409-416

[3] ***2008/671/EC: Commission Decision of 5 August 2008 on the harmonised use of radio spectrum in the 5875 - 5905 MHz frequency band for safety-related applications of Intelligent Transport Systems (ITS), *Official Journal of the European Union*

[4] Ven T and Wedlock M 2014 ITS action plan – framework service contract TREN/G4/FV-2008/475/01,[Online]Available: https://ec.europa.eu/transport/sites/transport/files/themes/its/studies/doc/2014-07-its-action-plan-d5-action-b.pdf

[5] ***European Commission 2010 Framework for the Deployment of Intelligent Transport Systems Directive 2010/40/EU, [Online] Available: https://ec.europa.eu/transport /sites/transport/files/themes/its/road/doc/2010_09_15_its_directive_2010-40-eu-standard.pdf

[6] ***European Committee for Standardization *CEN/TC 278 Intelligent transport systems*, [Online]. Available: http://www.itsstandards.eu/tc278

[7] ***European Commission 2016 M/546 Commission implementing decision C(2016)808, [Online].Available:http://ec.europa.eu/growth/tools-databases/mandates/index.cfm?fuseaction=search.detail&id=568#

[8] ***Car-2-Car Communication Consortium, [Online]. Available: http://www.car-2-car.org

[9] ***ERTICO, ERTICO - ITS Group, [Online]. Available: http://ertico.com/history

[10] ***Amsterdam Group, [Online]. Available: http://www.amsterdamgroup.eu

[11] ***InterCor Joint Deployment 2017 *Cooperative ITS Corridor*, [Online]. Available: https://itscorridor.mett.nl/English/default.aspx

[12] ***European Comission 2016 *Amsterdam Declaration – Cooperation in the field of connected and automated driving* [Online]. Available: https://english.eu2016.nl/binaries/eu2016-en/documents/publications/2016/04/14/declaration-of-amsterdam/2016-04-08-declaration-of-amsterdam-final-format-3.pdf

[13] ***C-ITS Platform, European Comission 2016 *C-ITS Deployment Platform – Final Report*, [Online]. Available: http://ec.europa.eu/transport/themes/its/doc/c-its-platform-final-report-january-2016.pdf

[14] ***C-ITS Platform, European Comission 2017 *C-ITS Platform Phase II Final report*, [Online]. Available: https://ec.europa.eu/transport/sites/transport/files/2017-09-c-its-platform-final-report.pdf

[15] ***European Comission 2016 *5G for Europe: An Action Plan – COM (2016) 588, footnote 29*, [Online]. Available: http://ec.europa.eu/newsroom/dae/document.cfm?doc_id=17131

[16] ***CEPT Electronic Communications Committee 2018 *9th ETSI ITS Workshop - CEPT/ECC Activities on ITS*, [Online]. Available: https://cept.org/files/10898/ECC%20presentation%20ITS%20-%20Thomas%20Weilacher%20and%20TW%20March18.pptx

[17] ***IEEE 2001 1609.x family of standards [Online]. Available: http://odysseus.ieee.org/query.html?qt=1609.&charset=iso-8859-1&style=standard&col=sa

[18] ***Analysys Mason SBD 2017 *Final Report for 5GAA - Socio-economic benefits for cellular V2X,* [Online]. Available: http://5gaa.org/wp-content/uploads/2017/12/Final-report-for-5GAA-on-cellular-V2X-socio-economic-benefits-051217_FINAL.pdf

[19] Mecklenbräuker C F 2011 Vehicular channel characterization and its implications for wireless system design and performance, *Proc. IEEE* **99**(7) 1189-1212

[20] Campolo C, Molinaro A and Scopigno R 2015 From today's VANETs to tomorrow's planning and the bets for the day after, *Vehicular Communications* **2** 158-171

[21] ***ETSI EN 302 637-2 V1.3.2 2014 *Intelligent Transport Systems (ITS); Vehicular Communications; Basic Set of Applications; Part 2: Specification of Cooperative Awareness Basic Service*, [Online]. Available: http://www.etsi.org/deliver/etsi_en/302600_302699/30263702/01.03.01_30/en_30263702v010301v.pdf

[22] ***ETSI EN 302 637-3 V1.2.2 2014 *Intelligent Transport Systems (ITS); Vehicular Communications; Basic Set of Applications; Part 3: Specifications of Decentralized; Environmental Notification Basic Service,* [Online]. Available: http://www.etsi.org/deliver/etsi_en/302600_302699/30263703/01.02.02_60/en_30263703v010202p.pdf

[23] ***ETSI TS 103 301 V1.1.1 2016 *Intelligent Transport Systems (ITS); Vehicular Communications; Basic Set of Applications; Facilities layer protocols and communication requirements for infrastructure services,* [Online]. Available: http://www.etsi.org/deliver/etsi_ts/103300_103399/103301/01.01.01_60/ts_103301v010101p.pdf

[24] ***3GPP C-V2X 2016 *Initial Cellular V2X standard completed,* [Online]. Available: http://www.3gpp.org/news-events/3gpp-news/1798-v2x_r14

[25] ***5G Automotive Association 2018 *Timeline for deployment of LTE-V2X,* [Online]. Available: http://5gaa.org/wp-content/uploads/2018/02/5GAA_Timeline-for-deployment-of-LTE-V2X_FINAL.pdf

[26] ***ETSI EN 302 571 V2.1.1 2017 *Harmonised Standard covering the essential requirements of article 3.2 of Directive 2014/53/EU,* [Online]. Available:

https://www.etsi.org/deliver/etsi_en/302500_302599/302571/02.01.01_60/en_302571v020101p.pdf

[27] ***CEPT/ECC 2015 *ECC Decision (08)01: The harmonised use of the 5875-5925 MHz frequency band for Intelligent Transport Systems,* [Online]. Available: https://www.ecodocdb.dk/download/b470d271-048b/ECCDEC0801.PDF

[28] ***CEPT/ECC 2015 *ECC Recommendation (08)01: Use of the band 5855-5875 MHz for Intelligent Transport Systems*, [Online]. Available: https://www.ecodocdb.dk/download/798c1836-20c6/REC0801.pdf

[29] ***5G Automotive Association 2017 *Coexistence of C-V2X and 802.11p at 5.9 GHz*, [Online]. Available: http://5gaa.org/wp-content/uploads/2017/08/5GAA_Position_Paper.pdf

[30] ***Slovick M 2017 Qualcomm Doubles Down on Cellular Chips for Wireless Vehicle Communications, *ElectronicDesign*, [Online]. Available: http://www.electronicdesign.com/automotive/qualcomm-doubles-down-cellular-chips-wireless-vehicle-communications

[31] ***2017 *Rohde & Schwarz supports 3GPP Cellular V2X device testing for vehicle-to-vehicle connectivity*, [Online]. Available: https://www.rohde-schwarz.com/de/unternehmen/news-und-presse/pressebereich/pressemitteilungen-detailseiten/rohde-schwarz-supports-3gpp-cellular-v2x-device-testing-for-vehicle-to-vehicle-connectivity-pressemitteilungen-detailseite_229356-505411.html

[32] ***Auto-Talks; NXP 2016 *IEEE802.11p ahead of LTE-V2V for safety applications*, [Online]. Available: https://www.auto-talks.com/wp-content/uploads/2017/09/Whitepaper_LTE-V2V_USletter__05.pdf

[33] Filippi A, Moerman K, Daalderop G, Alexander P D, Schober F and Pfliegl W 2017 *Why 802.11p beats LTE and 5G for V2x*, [Online]. Available: https://www.siemens.com/content/dam/webassetpool/mam/tag-siemens-com/smdb/mobility/road/connected-mobility-solutions/documents/its-g5-ready-to-roll-en.pdf

IOP Conf. Series: Materials Science and Engineering **477** (2019) 012053 doi:10.1088/1757-899X/477/1/012053

Dry sliding wear research on C45 carbon steel, 41Cr4 alloyed steel and X3CrNi13-4 martensitic stainless steel

L Nedeloni[1,2], P C Pedrali[3], L Cîndea[1], A V Petrica[1], A M Budai[1], I L Conciatu[1] and A Băra[1]

[1]Eftimie Murgu University of Reşiţa, Faculty of Engineering and Management, Traian Vuia Square, No. 1-4, 320085, Resita, Romania
[2]Westphalian University of Applied Sciences, Department of Materials Science and Testing, Gelsenkirchen, Germany
[3]Regional University of Northwestern Rio Grande do Sul, Department of Exact Sciences and Engineering, Rio Grande do Sul, Brazil

E-mail: l.nedeloni@uem.ro

Abstract. In this paper, the experimental research refers to the dry sliding wear behaviour of three type of steels namely carbon steel (C45), alloyed steel (41Cr4) and martensitic stainless steel (X3CrNi13-4). From the X3CrNi13-4 martensitic stainless steel, three different batches with varying percentages of chromium and nickel were analyzed. For all the steels, the research was made using a tribometer (CSM Tribometer) through to the pin-on-disk method (POD), where the friction coefficient and wear rate were determined. All these tests were made according to the ASTM G99 standard, with the same working conditions for all five samples, only for the two batches of the X3CrNi13-4 steel, the motor speed was varied. The obtained results were presented through characteristic graphs for the friction coefficient and through images acquired with a digital and a laser microscope respectively with a specialized software regarding the wear track profiles. It can be concluded that the C45 carbon steel showed the good tribological behaviour with the highest dry sliding wear resistance.

1. Introduction

As a type of wear, sliding wear can occur in many industrial applications which can be part especially of the automotive and energy industry [1-9]. For example in Hydro Power Plants (HPPs) where hydraulic machines operate, beside the abrasive, adhesive, erosion wear (hydro-abrasion / cavitation erosion) respectively corrosion, the sliding wear is also present. For example, for the radial bearing, crank or pin crank of a Kaplan turbine respectively for the needle spear of a Pelton turbine, the sliding wear is the wear mechanism [10-23].

Due to the dynamic interaction of the mating components, also regardless of the materials from which they are manufactured and respectively the operating environment, different components can be more or less affected by sliding wear. Depending on the material properties and also depending of the used materials pairs, the sliding wear resistance can be high or low [24-35].

Different materials like metallic materials (iron alloys) and non-metallic materials can be used for manufacturing of machine parts in the industry. In this regard, it is fundamental to evaluate the resistance of these materials, resistance which can be estimated by testing the materials with professional apparatus [36-47].

The Fe alloys, due to their alloying elements, can possess good mechanical proprieties. Many studies have been conducted with regard to investigate the carbon steels, alloyed steels and stainless steels. For example, samples of C45 carbon steels, 41Cr4 alloyed steels and respectively X3CrNi13-4 martensitic stainless steels have been tested by several authors were investigated for their wear and corrosion resistance and it was found out that some steels which have a good wear resistance can have a poor corrosion resistance and vice versa. Also, martensitic stainless steels which widely are used in industrial processes due to their excellent mechanical properties and corrosion resistance in various media, are continuously investigated [48-52].

Regarding their sliding wear resistance of these steels mentioned above it can be concluded that this is strongly affected by increasing the parameters values of the testing machine (a tribometer or a tester) especially like applied load and sliding speed. On the other hand, through a higher carbon content and by applying different treatments, the sliding wear behaviour of the materials can be also improved [53-55].

In this regard, also this paper is dedicated to present and to compare the experimental results (by means of friction coefficient and wear rate) of the dry sliding wear research on a C45 carbon steel, a 41Cr4 alloyed steel and an three different batches with varying percentages of chromium and nickel of the X3CrNi13-4 martensitic stainless steel, all this by using a tribometer through to the pin-on-disk method with the experimental procedure described below and with the authors knowledge in the term of wear from preview studies.

2. Experimental procedure

Dry sliding wear research was done by using a tribometer (CSM Instruments) through to the pin-on-disk method (POD). For the tested samples the counterpart was a 100Cr6 steel ball (6 mm in diameter) introduced in a pin holder. According to the ASTM G99 Standard [56], the temperature was set at 20°C and the atmosphere environment was the air with the humidity set to 50%. Before and after the tests, the samples (with dimensions Φ 16 x 10 mm) and the counterparts were cleaned with acetone.

The applied load was set only at 10 N and the sliding distance was stop at 500 m. The other main parameters for this dry sliding wear research are shown in Table 1.

Table 1. Parameters for the dry sliding wear research

Samples	Sliding speed ($mm\ s^{-1}$)	Motor speed (rpm)	Radius (mm)	Test duration (s)
C45	170	400	4.05	2940
41Cr4	170	400	4.06	2940
X3CrNi13-4 (1)	191.2	450	4.06	2620
X3CrNi13-4 (2)	170	400	4.06	2940
X3CrNi13-4 (3)	148.8	350	4.06	3370

By using the tribometer software (InstrumX) with an acquisition rate of 10 Hz, the friction coefficient was registered as graphic representation for all the tests and regarding the wear rate, this was calculated with the help of equation 1 [56]:

$$K = \frac{2\pi r \cdot h\left(3h^2 + 4s^2\right)}{6 \cdot L \cdot d \cdot s} \tag{1}$$

where: K is the wear rate; h – the wear track depth; s – the wear track width; L – the applied load and d is the sliding distance, which is only for 500 m.

Also, for the results exposure the samples mass was measured by weighing before and after the tests with the help of a Sartorius digital balance.

The chemical composition of the tested samples (C45 as a carbon steel, 41Cr4 as an alloyed steel (which was typical heat treated for armament industry) and X3CrNi13-4 as martensitic stainless steel in three different batches) is reported in Table 2.

Table 2. The chemical composition of the tested samples (wt.%)

Samples	C	Si	Mn	P	S	Cu	Ni	Cr	Mo	Fe
C45	0.47	0.24	0.68	0.010	0.004	0.14	0.09	0.14	0.01	97.49
41Cr4	0.394	0.202	0.59	0.005	0.016	0.23	0.272	0.95	0.03	97.17
X3CrNi13-4 (1)	0.07	0.41	0.56	0.027	0.014	0.16	5.17	11.15	0.35	82.08
X3CrNi13-4 (2)	0.05	0.62	0.65	0.021	0.012	0.25	3.62	12.95	0.36	81.46
X3CrNi13-4 (3)	0.06	0.43	0.42	0.015	0.009	0.07	3.81	12.5	0.32	82.36

3. Results and discussions

The friction coefficient evolution for the five tested samples is shown in Figure 1.

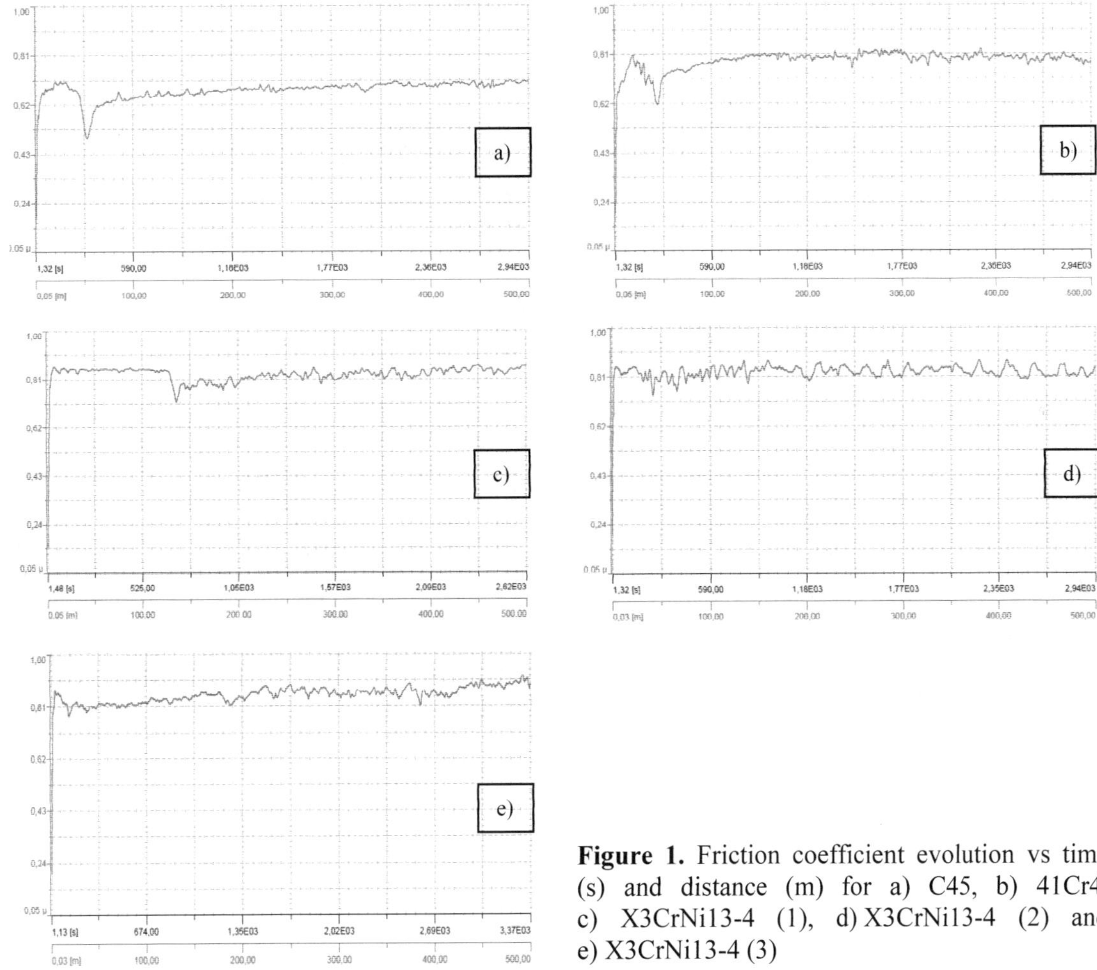

Figure 1. Friction coefficient evolution vs time (s) and distance (m) for a) C45, b) 41Cr4, c) X3CrNi13-4 (1), d) X3CrNi13-4 (2) and e) X3CrNi13-4 (3)

The friction coefficient reach values between 0.50 - 0.90 for all samples. For the C45 and 41Cr4 samples, the values are the lowest and for the X3CrNi13-4, the values are among the highest and their curves are almost the same. It can be seen that the materials pair (C45 with 100Cr6 counterpart) has the best tribological behaviour and after 100 m the steady state [57] is reached between the values 0.62 - 0.70 with small deviations compared to the other curves (Figures 1 b), c), d) and e).

For a better comparison between the friction coefficient evolutions of the samples, in Table 3 the main friction coefficient values are shown (it can see that the lowest mean value belongs to C45 steel).

Table 3. Comparison between friction coefficients values

Samples	Start	Min	Max	Mean	Std. Dev.
C45	0.193	0.177	0.712	0.673	0.039
41Cr4	0.202	0.202	0.832	0.785	0.041
X3CrNi13-4 (1)	0.163	0.152	0.863	0.827	0.042
X3CrNi13-4 (2)	0.216	0.206	0.873	0.827	0.030
X3CrNi13-4 (3)	0.205	0.205	0.920	0.851	0.037

Furthermore, through images acquired with a digital and with a confocal 3D laser scanning microscope (Keyence), the wear track of the 100Cr6 counterparts and of the samples are shown in Figure 2 respectively in Figure 3 and Figure 4.

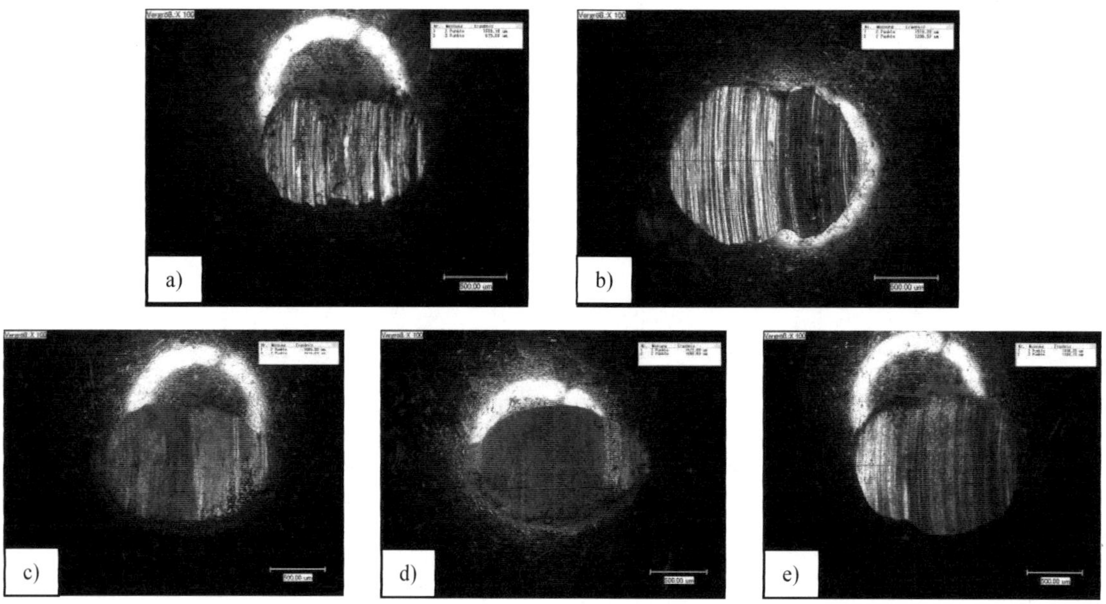

Figure 2. 100Cr6 counterparts wear track (magnification 100 X – 500 μm) for a) C45, b) 41Cr4, c) X3CrNi13-4 (1), d) X3CrNi13-4 (2) and e) X3CrNi13-4 (3)

From these images the length and the height, are the following:
- C45: (1) 1338.19 μm and (2) 815.88 μm;
- 41Cr4: (1) 1518.28 μm and (2) 1208.57 μm;
- X3CrNi13-4 (1): (1) 1485.92 μm and (2) 1016.03 μm;
- X3CrNi13-4 (2): (1) 1572.66 μm and (2) 1083.69 μm;
- X3CrNi13-4 (3): (1) 1439.23 μm and (2) 1164.73 μm.

From these measured values one can observe that the counterpart of C45 sample was the smallest values, which mean that the eroded area was less compared to the other areas of the samples. From the images in Figure 2, the abrasive wear it is observed for the all samples. Besides that, in the case of counterpart for C45 sample, also galling or the pull-out phenomenon is occurred and in the case of the X3CrNi13-4, the abrasive wear is combined with the adhesion wear.

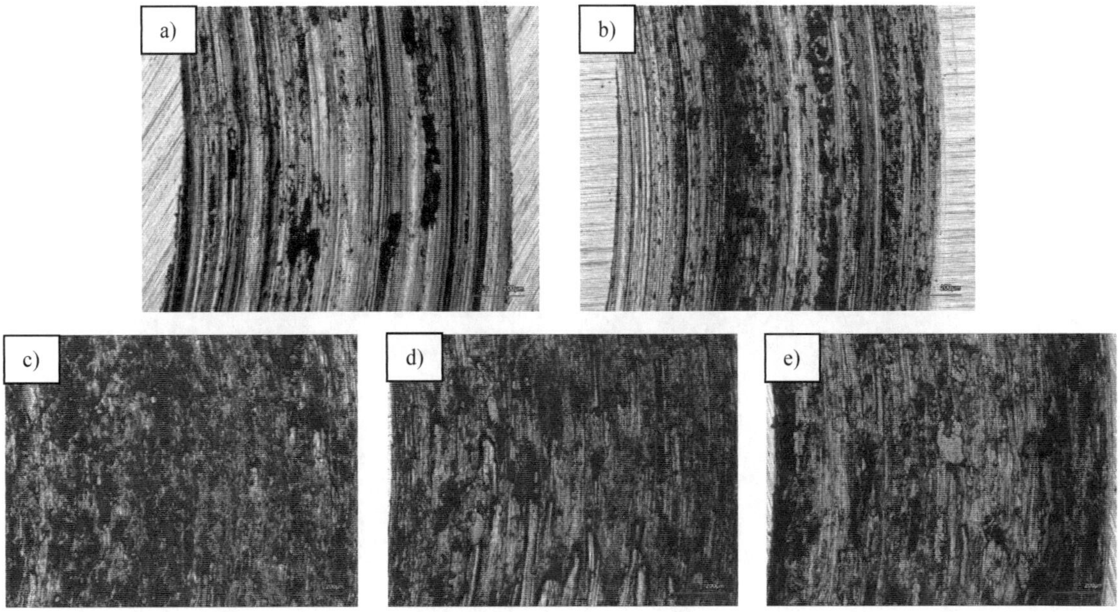

Figure 3. Samples wear track profiles (magnification 100 X – 200 μm) for a) C45, b) 41Cr4, c) X3CrNi13-4 (1), d) X3CrNi13-4 (2) and e) X3CrNi13-4 (3)

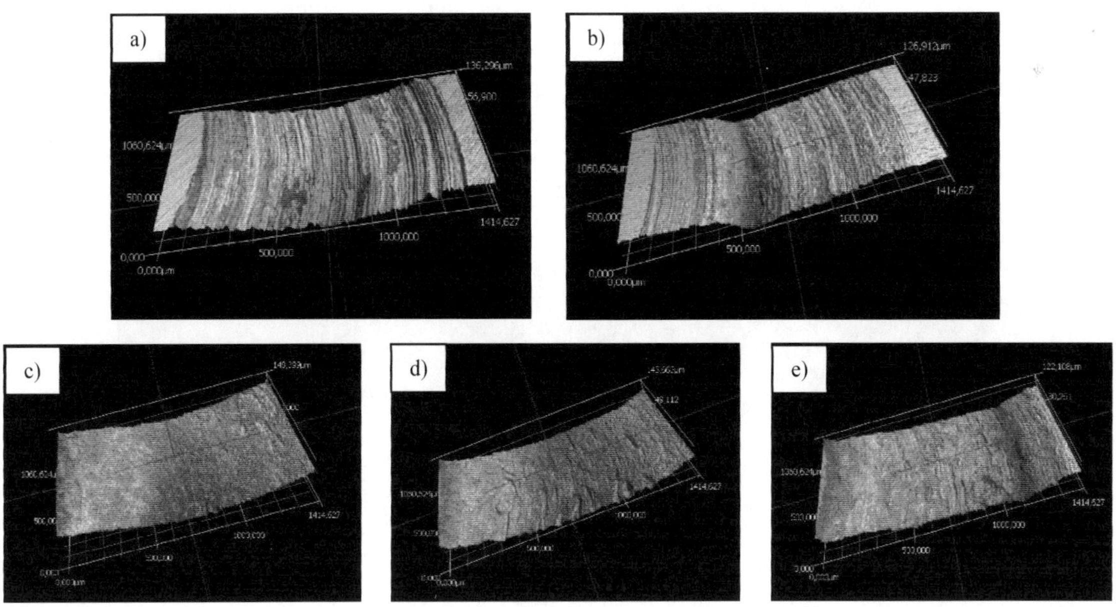

Figure 4. 3D images of samples wear track profiles (height magnification – 200%) for a) C45, b) 41Cr4, c) X3CrNi13-4 (1), d) X3CrNi13-4 (2) and e) X3CrNi13-4 (3)

Also, from Figure 3 can observe the abrasive wear (in the case of the C45 sample and 41Cr4 sample) respectively the abrasive wear and adhesion wear (in the case of X3CrNi13-4 samples). The C45 and 41Cr4 steels have the smallest wear track profiles compared to those of the X3CrNi13-4 stainless steel (Figure 3 and Figure 4).

The 3D images processing with MultiFileAnalyzer (which, like other software presents some advantages [58-61]) show interesting shapes of the wear profiles, especially regarding the wear depth illustrated with blue color. To know exactly which steel has the lowest wear rate values, through MultiFileAnalyzer, in Figure 5 these measurements will be shown by mean of wear track depth and wear track width, and through Table 4 and Figure 6, the wear rate values will be also presented.

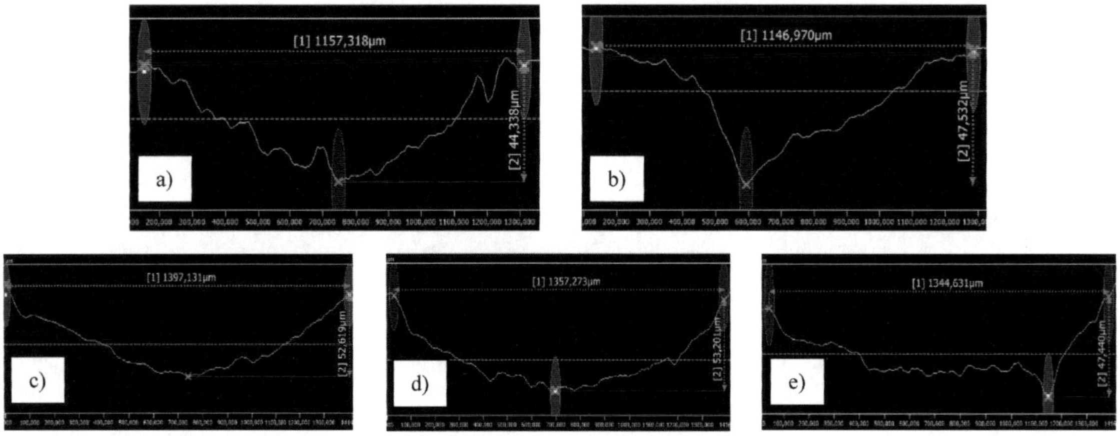

Figure 5. Depth and width wear track profiles with values for a) C45, b) 41Cr4, c) X3CrNi13-4 (1), d) X3CrNi13-4 (2) and e) X3CrNi13-4 (3)

Table 4. Wear rate values

Sample	Wear track depth, h (µm)	Wear track width, s (µm)	Cross section area, A (µm^2)	Volume loss, V (mm^3)	Wear rate, K (mm^{3}N^{-1}m^{-1})
C45	44.338	1157.318	34246.43	0.8715	$1.74\cdot10^{-4}$
41Cr4	47.532	1146.970	36392	0.9284	$1.86\cdot10^{-4}$
X3CrNi13-4 (1)	52.619	1397.131	49062.56	1.2516	$2.50\cdot10^{-4}$
X3CrNi13-4 (2)	53.201	1357.273	48194.32	1.2294	$2.46\cdot10^{-4}$
X3CrNi13-4 (3)	47.440	1344.631	42565.90	1.0858	$2.17\cdot10^{-4}$

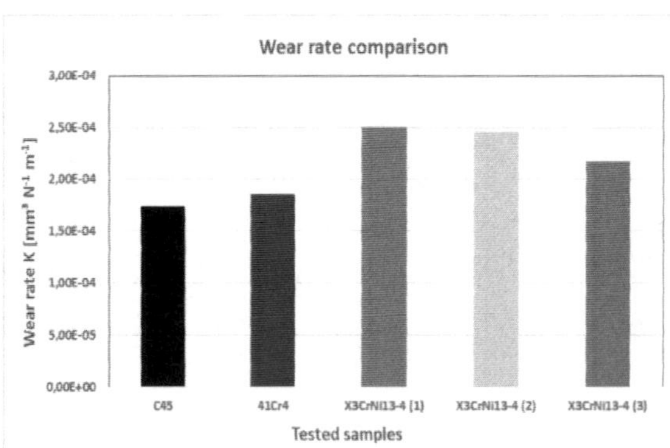

Figure 6. Comparison between the wear rates

It can be seen that the lowest wear rate value is attributed to the C45 carbon steel followed by 41Cr4 alloyed steel. The same is regarding the worn material volume.

In this case of the three X3CrNi13-4 samples (with highest wear rate values), these have the lowest sliding wear resistance compared with the other two samples. The X3CrNi13-4 martensitic stainless steel samples wear resistance was the following X3CrNi13-4 (3) by 350 rpm, X3CrNi13-4 (2) by 400 rpm and X3CrNi13-4 (1) by 450 rpm. From these assessments, it can be seen that a higher sliding speed can produce a more severe material wear.

According with [62], the authors also tested these materials regarding their cavitation erosion resistance through the stationary specimen method, the material mass losses were as following: C45 - 5.39 mg, 41Cr4 - 2.42 mg, X3CrNi13-4 (1) - 2.22 mg, X3CrNi13-4 (2) - 2.67 mg and respectively X3CrNi13-4 (3) - 4.47 mg. This confirms that an alloy which has a good cavitation erosion resistance due can have a poor sliding wear resistance [63].

Furthermore, according to [64] and [65], with an increasing of the material hardness especially after heat treatment, cavitation erosion resistance can also increase. This is true also from our side, regarding the hardness of these materials from some preview research of the authors, where it can be found that steels like 41Cr4 alloyed steel (which was heat treated) has the highest hardness (around 400 HV) compared to other four samples of this research. We mentioned that because even if in this research, the 41Cr4 steel has a good sliding wear resistance (very close to that of the C45 steel), but the abrasive wear caused to the 100Cr6 counterpart was the highest (as is highlighted in Figure 2 b).

4. Conclusions

The authors pointed out in introduction the presents of sliding wear which occur in many industrial applications due to the dynamic interaction between the mating components and make a connection with other studies in the literature regarding the C45 carbon steels, 41Cr4 alloyed steels and X3CrNi13-4 martensitic stainless steels tribological behaviour and success to make reproducibility with similar researches and their own preview work.

In this way, the main conclusions of the present paper can be summarized as follows:

- the friction coefficient of the tested samples reach values between 0.50 - 0.90 mostly for the C45 steel, which was having the lowest values. So, for the total distance of 500 m, the materials pair (C45 steel and 100Cr6 ball counterpart) showed the best tribological behaviour. After the first 100 m, the steady state was reached and friction coefficient curve was enough stable compared to the other curves, this due to the highest carbon content and formation of the tribofilm.
- the abrasive wear occurred between the tested samples and the 100Cr6 ball counterparts, where in the case of C45 sample sporadic areas of galling were observed and in the case of the X3CrNi13-4 samples, the abrasive wear was combined with the adhesion wear.
- the lowest wear rate value respectively worn material volume was attributed to the C45 carbon steel followed by 41Cr4 alloyed steel and by three X3CrNi13-4 samples especially by X3CrNi13-4 (3), where due to a lowest sliding speed the wear was less.
- the C45 carbon steel showed the good tribological behaviour with the highest dry sliding wear resistance, this confirmed also from the smallest wear track profile. Also the 41Cr4 alloyed steel showed a good sliding wear resistance but as a materials pair, this was produced the highest abrasive wear for 100Cr6 ball as counterpart compared the other tested steels.

References

[1] Nová I and Machuta J 2016 Monitoring of the diffusion processes during carburizing automotive steel parts, *Manufacturing Technology* **16**(1) 225-232

[2] Korka Z I 2017 Reduction of vibration and noise generated by multipoint sources of dynamic contact processes, *Romanian Journal of Acoustics and Vibration* **14**(1) 2

[3] Palásti-Kovács B, Néder Z, Czifra A and Váradi K 2004 Microtopography Changes in Wear Process, *Acta Polytech Hung* **1**(1) 108-119

[4] Gillich G R et al. 2017 Assessing corrosion damage from the natural frequency changes,

Romanian Journal of Acoustics and Vibration **14**(2) 63-68
[5] Drăgoi C, Pîrăianu V F and Pîrăianu C 2012 Evaluating the lifetime of hydro-mechanical equipment for low head hydropower plants, *UPB Scientific Bulletin, Series D: Mechanical Engineering* **74**(1) 73-82
[6] Coman L, Hamat C, Pittner A and Ianici S 2009 *Experimental research regarding the hydraulic forming of revolution shape parts*, Annals of DAAAM and Proceedings of the International DAAAM Symposium, pp 141-142
[7] Ruja I, Frunzaverde D, Marta C, Suciu L, Cziple F and Roşu M 2010 Theoretic and experimental studies on the casting of manganese steel balls for the ore grinding mills, *Metalurgia International* **15**(12) 13-18
[8] Csaki I et al. 2018 Nickel based coatings used for erosion-corrosion protection in a geothermal environment, *Surface & Coatings Technology* **350** 531-541
[9] Chen L et al. 2018 Erosion-corrosion behavior and electrochemical performance of Hastelloy C22 coatings under impingement, *Applied Surface Science* **456** 985-998
[10] Marta C, Suciu L and Nedelcu D 2009 Simulation of the Kaplan rotor hub casting and comparison with the experimental results, *Metalurgia International* **14**(10) 33-38
[11] Frunzaverde D, Câmpian V, Nedelcu D, Gillich G R and Mărginean G 2010 Metallographic and numerical methods investigations about failure of a kaplan turbine runner blade, *WSEAS Transactions on Fluid Mechanics* **5**(3) 122-131
[12] Pola A, Gelfi M, Eleonora Depero L and Roberti R 2008 Study of annealing temperature effect on stress-corrosion cracking of aluminum brass heat-exchangers tubes by microdiffraction experiments, *Engineering Failure Analysis* **15**(1-2) 54-61
[13] Miclosina C O, Campian C V, Frunzaverde D and Cojocaru V 2011 Analysis of an outer bearing bush of a hydropower plant kaplan turbine using finite element method, *Recent Researches in Environment, Energy Planning and Pollution* 221-224
[14] Frunzaverde D et al. 2010 *Influence of welded layers thickness on the cavitation erosion resistance*, Proceedings of the 6th WSEAS International Conference on Energy, Environment, Ecosystems and Sustainable, pp 316-320
[15] Budai A M et al. 2017 Lifetime estimations of the operating mechanism of kaplan turbine runner blades, *Technical Gazette* **24**(2) 271-277
[16] Pîrăianu V F and Ilinca C 2012 *Practical solutions for small hydro power plant design*, 18th European Concurrent Engineering Conference, pp 59-62
[17] Ilinca C D and Pîrăianu V F 2017 *Developing maintenance program for small hydroelectric power plant using vibration measurements and simulation*, International Multidisciplinary Scientific Geo Conference, pp 183-190
[18] Pittner A M, Campian C V, Nedelcu D, Frunzaverde D and Cojocaru V 2010 *Stress concentration factors for pin lever of runner blade mechanism from Kaplan turbines*, International Conference on Geography and Geology, pp 181-185
[19] Kastawan I M W and Rusmana 2017 Voltage generation of three-phase double sided internal stator axial flux permanent magnet (AFPM) generator, *IOP Conf. Ser.: Mater. Sci. Eng.* **180** 012105
[20] Nedelcu D, Cojocaru V, Ghican A, Peris-Bendu F and Avasiloaie R 2015 Considerations regarding the use of polymers for the rapid prototyping of the hydraulic turbine runners designed for experimental research on the model, *Materiale Plastice* **52**(4) 475-479
[21] Miclosina C O, Balint D I, Campian C V, Frunzaverde D and Ion I 2012 A method to combine hydrodynamics and constructive design in the optimization of the runner blades of Kaplan turbines, *IOP Conference Series: Earth and Environmental Science* **15**(3) 032015
[22] Glăvan D O and Babanatsas T 2018 Tool machinery vibrations frames comparison concerning welded or moulded manufacturing structures, *MATEC Web of Conferences* **121** 1-6
[23] *** 2011 Surfaces for longer life and higher energy efficiency, Sulzer Technical Review **3**
[24] Korka Z I, Bara A, Clavac B and Filip L 2017 Gear pitting assessment using vibration signal

analysis, *Romanian Journal of Acoustics and Vibration* **14**(1) 44-49

[25] Cîndea L et al 2016 The influence of thermal field in the electric arc welding of X60 carbon steel components in the CO2 environment, *Appl Therm Eng* **103** 1164-1175

[26] Vencl A et al. 2014 Structural, mechanical and tribological characterization of Zn25Al alloys with Si and Sr addition, *Materials and Design* **64** 381-392

[27] Rozing G, Alar V and Marušić V 2015 Study of stainless steel resistance in conditions of tribocorrosion wear, *Interdisciplinary Description of Complex Systems* **13**(3) 461-471

[28] Afronie E M, Manescu T S and Ianici S KOD 2012 *Calculus of metalic and steel-concrete mixed structures at fire actions*, 7th International Symposium on Machine and Industrial Design in Mechanical Engineering (KOD 2012) Hungary, May 24-26, pp 515-518

[29] Păduraru L et al. 2018 Investigations on dry sliding wear and corrosion resistance of thermal sprayed molybdenum coatings, *IOP Conf. Ser.: Mater. Sci. Eng.* **416** 012027

[30] Glăvan D O, Babanatsas T, Babanatis Merce R M and Glăvan A 2018 Comparative study of tool machinery sliding systems; comparison between plane and cylindrical basic shapes, *IOP Conf. Ser.: Mater. Sci. Eng.* **294** 012068

[31] Montesano L, Pola A, Gelfi M and La Vecchia G M 2015 Wear Behavior of Zn-15Al-1Cu-Mg Alloy after Aging, *Procedia Engineering* **109** 228-233

[32] Hu J 2016 Development of a Composite Technique for Preconditioning of 41Cr4 Steel Used as Gear Material: Examination of Its Microstructural Characteristics and Properties, *Science and Technology of Nuclear Installations* **2016** 1-6

[33] Zurowski W 2012 Structural factors contributing to increased wear resistance of steel friction couples, *Maintenance and Reliability* **14**(1) 19-23

[34] Pola A, Gelfi M and La Vecchia G M 2016 Comprehensive numerical simulation of filling and solidification of steel ingots, *Materials* **9**(9) 1-13

[35] Suwanpinij P, Massaro P, Pola A and Sricharoenchai P 2016 *Simulation of precipitation in V-containing hsla steel for the strengthening enhancement*, Proceedings of the 7th European Congress on Computational Methods in Applied Sciences and Engineering, pp 7757-7764

[36] Utu D, Marginean G, Pogan C, Brandl W and Serban V A 2007 Improvement of the wear resistance of titanium alloyed with boron nitride by electron beam irradiation, *Surface and Coatings Technology* **201**(14) 6387-6391

[37] Kumar N and Kumar P 2016 Influence of machining parameters on surface roughness and dry friction, *Engineering Solid Mechanics* **4** 109-116

[38] Ivan A, Ivan M and Both I 2010 Comparison of FEA and experimental results for a steel frame connection, *WSEAS Transactions on Applied and Theoretical Mechanics* **5**(3) 187-96

[39] Both I Wald F and Zaharia R 2016 Benchmark for numerical analysis of steel and composite floors exposed to fire using a general purpose FEM code, *Journal of Applied Engineering Science* **14**(2) 275-284

[40] Rigou I V, Mărginean G, Frunzăverde D and Câmpian C V 2012 Silver based composite coatings with improved sliding wear behaviour, *Wear* **290-291** 61-65

[41] Marušić V et al. 2018 The possibility of extending work time of tribocouple welding improved stainless steel with improved carbon steel, *MATEC Web of Conferences* **184** 01014

[42] Nedeloni M D, Birtărescu E, Nedeloni L, Ene T, Băra A and Clavac B 2018 Cavitation erosion and dry sliding wear research on X5CrNi18-10 austenitic stainless steel, *IOP Conf. Ser.: Mater. Sci. Eng.* **416** 012028

[43] Birtărescu E et al. 2018 Some laboratory tests regarding the X20Cr13 martensitic stainless steel behaviour, *IOP Conf. Ser.: Mater. Sci. Eng.* **416** 012025

[44] Cojocaru V et al. 2010 *Laboratory tests concerning the influence of surface hardening on the cavitation erosion resistance*, 3rd WSEAS International Conference on Engineering Mechanics, Structures, Engineering Geology, pp 210-213

[45] Nogoud Y A, Abudarag S, Mansor S and Jonker A 2018 Dynamic Analysis and Simulation of a Horizontal Slider Crank Mechanism using Spreadsheets, *Journal of Engineering and*

Computer Science **19**(1) 13-17
[46] Haris N A et al. 2016 Abrasion and erosion wear properties of surface deformed stainless steel, *Journal of Engineering and Applied Sciences* **11**(12) 7717-7720
[47] Chirila C B, Crescenzo P, Lahire P, Pescaru D and Ţundrea E 2004 Factoring Mechanism of Reverse Inheritance, *Periodica Politechnica* **49**(63) 1-6
[48] Nedelcu D, Cojocaru V, Nedeloni M, Peris-Bendu F and Ghican A 2015 Failure analysis of a Ti-6Al-4V ultrasonic horn used in cavitation erosion tests, *Mechanika* **4** 272-276
[49] Rozing G, Alar V, Marušić V and Samardžić I 2016 Tribocorrosion wear of austenitic and martensitic steels, *Metalurgija* **55**(3) 441-444
[50] Milinović A, Marušić V and Samardžić I 2016 Research into boride layers growth kinetics on C45 carbon steel, *Metalurgija* **55**(4) 671-674
[51] Nedelcu D, Nedeloni M D and Lupinca C I 2014 Cavitation erosion research on the X3CrNi13-4 stainless steel, *Materials Science Forum* **782** 263-268
[52] Li Y, He Y, Xiu J, Wang W, Zhu Y and Hu B 2017 Wear and corrosion properties of AISI 420 martensitic stainless steel treated by active screen plasma nitriding, *Surface & Coatings Technology* **329** 184-192
[53] Nedeloni L, Korka Z I, Pascal D T, Kazamer N and Nedeloni M D 2018 Comparative Study on Dry Sliding Wear Resistance of Carbon Steel, Alloyed Steel and Cast Iron, *IOP Conf. Ser.: Mater. Sci. Eng.* **416** 012026
[54] Krol S, Zalisz Z and Hepner M 2005 Comparison of the friction and wear properties of titanium and oxidised titanium in dry sliding against sintered high speed steel HS18-0-1 and against C45 carbon steel, *Journal of Materials Processing Technology* **164-165** 868–875
[55] Marušić V, Nedić B and Stoić A 2009 Application of Tribometer Measurements for Evaluation of Machinability, *Metalurgija* **51**(1) 1-6
[56] *** 2000 Standard Test Method for Wear Testing with a Pin-on-Disk Apparatus ASTM G99
[57] Cozza R C 2013 A study on friction coefficient and wear coefficient of coated systems submitted to micro-scale abrasion tests, *Surf Coat Tech* **215** 224-233
[58] Lahire P, Parigot D, Courbis C, Crescenzo P and Ţundrea E 2004 An attempt to set the framework of Model-Oriented Programming, *Periodica Politechnica* **49**(63) 1-6
[59] Hatiegan C et al. 2015 Finite element analysis of thin plates clamped on the rim of different geometric forms. Part I: Simulating the vibration mode shapes and natural frequencies, *Romanian Journal of Acoustics and Vibration* **12**(1) 69-74
[60] Nedelcu D et al. 2011 *The aerodynamic force calculus for a plate immersed in a uniform air stream using Solidworks Flow Simulation module*, Proceedings of the 4th WSEAS International Conference, pp 98-103
[61] Nedelcu D et al. 2011 *The hydrodynamic characteristics calculus for isolated profile Go428 using Solidworks Flow Simulation module*, Proceedings of the 4th WSEAS International Conference, pp 92-97
[62] Hamat C O, Nedeloni M D, Haţiegan C, Ciubotariu R C and Pădureanu I 2015 Cavitation erosion research on C45 carbon steel. Part I: multiple tests of 180 minutes, *Annals of „Constantin Brâncuşi" University of Târgu Jiu* **3** 127-132
[63] Gottardi G, Tocci M, Montesano L and Pola A 2018 Cavitation erosion behaviour of an innovative aluminium alloy for Hybrid Aluminium Forging, *Wear* **394-395** 1-10
[64] Petrogalli C et al. 2015 Improvement of Fatigue Resistance of a Tool Steel by Surface Treatments, *Procedia Engineering* **109** 154-161
[65] Pola A, Montesano L, Gelfi M and La Vecchia G M 2016 Comparison of the sliding wear of a novel Zn alloy with that of two commercial Zn alloys against bearing steel and leaded brass, *Wear* **368-369** 445-452

IOP Conf. Series: Materials Science and Engineering **477** (2019) 012054 doi:10.1088/1757-899X/477/1/012054

Necessary measures and calculation for dimension of coaxial heat exchanger for deep boreholes

V Stefanović[1], B Drobnjaković[2] and S Pavlović[1]

[1]Faculty for Mechanical Engineering Niš ,Aleksandra Medvedeva Street,18000, Niš, Srbija
[2]Institute for Mining and Metallurgy Bor, Zeleni bulevar Street, 19210 Bor, Srbija

E-mail: bojan.drobnjakovic@irmbor.co.rs

Abstract. This paper aims to illustrate the complexity of the approach to modeling a potential heat exchanger envisaged for deep drilling. The problem was presented in one example of the installation study for the heating and cooling of the University building in the city of Aachen, Germany, with a deep coaxial heat exchanger, located in a depth well of 2500m (BHE). Prior to any modelling of the heat exchanger, the geological analysis of the wall layers along the well was to be approached; to conduct complex field and laboratory geophysical measurements and ultimately to approximate the modeling of heat processes in the heat exchanger. Also, this study was aimed at assessing the feasibility of such an installation for heating and cooling space, as well as clarifying operational characteristics and performance (in the long run). Direct heating of the building in the winter period requires a temperature of 40^0C. In the summer period, for the cooling of the university building, an adsorption cooling unit is used, which requires a temperature of at least 55^0C. Dredged walls up to a depth of 2500m have extremely low permeability and porosity, less than 1%. Their thermal conductivity varies between 2.2W/(mK) and 8.9W/(mK), more values are associated with quartz sandstone. The maximum temperature in the well is 850C at a depth of 2500m, corresponding to a mean specific heat flow of 85mW/m2-90mW/m2. The experience of twenty years of work has indicated that only in a short period of time, the borehole can supply the desired temperature for the operation of the absorption refrigerant. In winter, the heat exchanger can also supply a building with sufficient heat.

1. Introduction

Much before 2004, it is planned that in the framework of the Aachen unification, a deep well will be built in which a heat exchanger (BHE) will be installed, for the use of geothermal energy for heating and cooling the new building of the student center, which is classified into the energy class "Super C". In the summer period, a large roof area should provide shade, while in winter it is necessary to provide passive heating by solar radiation through glass surfaces. Solar cells on the roof should generate electricity for the needs of a circulation pump for a heat exchanger [1].

A deep coaxial heat exchanger is a closed system where cold water flows through the outer tube, heats up, enters the inner tube and flows upwards. The borehole under the official name RWTH-1 has a depth of 2500m and is intended to provide heat for heating and cooling through BHE. The outlet water temperature from the inner tube of the exchanger should have a temperature of at least 55^0C-80^0C. In order to operate an absorption cooling system during the flight, a minimum temperature of 55^0C is

IOP Conf. Series: Materials Science and Engineering **477** (2019) 012054 doi:10.1088/1757-899X/477/1/012054

required. A period of 30 to 40 years is foreseen. In winter, the building will be heated directly by geothermal heating water. The well was drilled in the spring of 2004, before the construction of the university building (Figure 1). The building is located in the urban part in the center of Aachen. The drilling operation was monitored from the petrographic side and the temperature side. The bore is crossed by a wall series dominated by sandstone and gravel. Organic gravel and thin coal layers were locally found in the upper half to 1016 m as part of the carbonic delta cycles. From limestone 1016m to 1440m, limestone walls from the period of the Upper Devon dominate. In the deeper part of the well, dominant shale and series of sandstones date from the period of the lower devon [2]. During the drilling, material from the drilling rig was sampled (Figure 2), length 1m. Cores samples were available for 3 sections along the well (1391.5m-1515.7m, 2128.2m-2142.8m and 2536.8m-2544.5m) with a total length of 150m samples of the core. No fluid flow into the well was detected during drilling. The most important information needed on the walls for long-term simulation of the heat exchanger's performance in deep wells is the thermal conductivity of the wall material, then the specific heat capacity, porosity, permeability and density [3].

The building is located in the urban part in the center of Aachen. Borehole is in the red circle.

Figure 1. Building "Super C" energy class

2. Necessary measures

This section will show the required measurements.

2.1. Laboratory Measurements

Necessary physical measurements are measurements of density and thermal conductivity of layers; 57 samples were selected at intervals of 50m in depth, from the walls considered to vary in geophysical terms. The density of the ρm core was measured (Figure 2) with helium pycnometer [4]. The device calculates the density of the matrix over the amount of helium that saturates the pores in the dried core sample. The density varies from 2640 kg/m^3 to 2840 kg/m^3, the mean value is 2780 kg/m^3. The accuracy of the device for the increase of 1 kg/m^3 is 0,1kg/m^3. Bulk density can not be measured because only core samples from the most stable, massive parts of drilled walls were available.

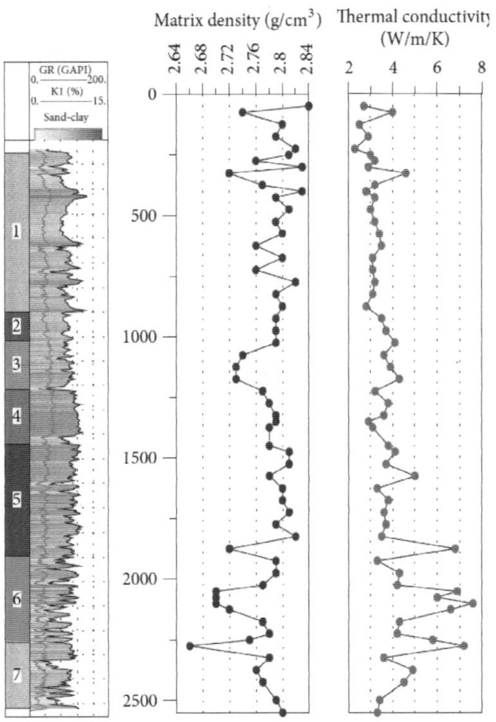

Figure 2. Density and heat measurements conductivity in the borehole zones

The core density is measured with a multisensor logger, and on some cores with a helium gas pycnometer. The multisensor logger uses the source of gamma radiation and measures the intensity of gamma radiation. The density is calculated from the absorbed gamma radiation. The process requires careful calibration of the measuring device. The density of helium gas pycnometer measured varies between 2391kg/m^3 and 2897kg/m^3, on average 2820kg/m^3. The density of gamma measurements varies between 2576kg/m^3 to 3233kg/m^3, on average 2830kg/m^3.

Thermal conductivity of the core (Figure 2) was measured using a TK04 needle probe (TeKa Geophysical Instruments product) in a mixture of the core with water. The thermal conductivity of the samples varies from 2.2W/(mK) to 8.9W/(mK), on average 3.8W/(mK). High values can be explained by a high percentage of quartz. Cemented quartz sands often appear in the deeper part of the well below 1895m [2].

The thermal conductivity of the core was measured using the Thermal Conductivity Optical Thermal Scanner (TCS) manufactured by Lippmann & Rauen GbR [5]. The thermal conductivity scanner moves along the sample, detects the heat radiation of the sample, and calculates the heat conductivity of the sample with an error of ± 3% at measured temperatures before and after scanning. The thermal conductivity of the core varies between 2.3W/(mK) and 4.9W/(mK) with minimum values of 2.0W/(mK) and a maximum of 5.9W/(mK). The arithmetic mean is 3.02W/(mK), as well as the geometric mean 2.99W/(mK), [6].

Additional sizes measured on the core with a multisensor logger are density, porosity, magnetic sensitivity, sound velocity and natural gamma radiation. Due to their low porosity, no differences in the saturated and dry samples of the core were detected between them. The porosity (φ) of the three selected core samples was calculated according to the quantity of water that evaporates from the sample. Porosity is almost zero, with a mean value of 0.012%.

2.2. Measurements in the borehole

The measured temperature logs are shown in Figure 3. The first measurements of temperature in April 2005 show higher temperatures in the upper half of the wells and lower temperatures in the lower half of the well compared to the June 2006 measurements. This is probably caused by the influence of drilling fluid; which heats the upper part and cools the bottom of the well during drilling. In the first 50m, the temperature anomaly occurs in both temperature curves. This is unusual, but can be explained by the thermal losses of drilling water in higher permeable sediments at the top of the well due to the cooler wall mass. Measurement of density with radioactive source was not done because it was the center of the city of Aachen. The sound velocity V_p, electrical resistance, and gamma radiation are shown in Figure 4. Logs of the measured gamma radiation are used for geological interpretation in the section. Gamma radiation gives an indication of the clay content in the wall. Therefore, these data are used to determine thermal conductivity and radiological heat generation. V_p data (logging data) are compared with core data (black line versus blue line "log /Core V_p" in Figure 3). Data "in city" and data from the laboratory fit perfectly. The results best suit the limestone part up to 1440m and in the cores from the deepest part of section 1. Between 1460m and 1490m a slight deviation can be observed, which can be attributed to an anisotropic environment.

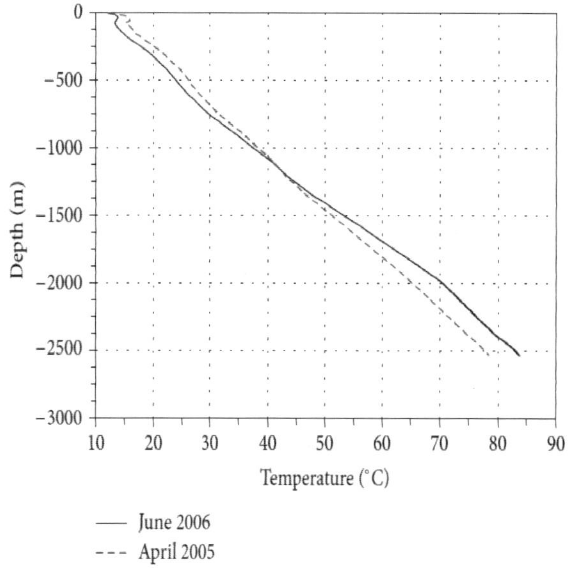

Figure 3. Temperature diagram in the borehole in different periods of time

2.3. *Interpretation of log records*

Based on the recorded geological data, the well profile is divided into three larger zones (Figure 4): (I) zones of lower limestone delta layers, (II) zones of the upper devon layers, clay stone, limestone sandstone and limestone and zones (III) with a fairly homogeneous succession of sand and clay stone of the lower Devonian era (Figure 4).

The value of the logarithm of electrical resistance reaches values of up to 1000Ω for limestone and quartz cement sand, while the organic shards have a resistance lower than 100Ω.

2.4. *Heat Characteristics*

It is not possible to measure the thermal conductivity directly in the well, so that it is necessary to find existing logs for determining thermal conductivity. The content of clay volume was derived from the contribution of potassium as a whole to gamma radiation. As described in the previous section, the well was divided into 7 zones.

IOP Conf. Series: Materials Science and Engineering **477** (2019) 012054 doi:10.1088/1757-899X/477/1/012054

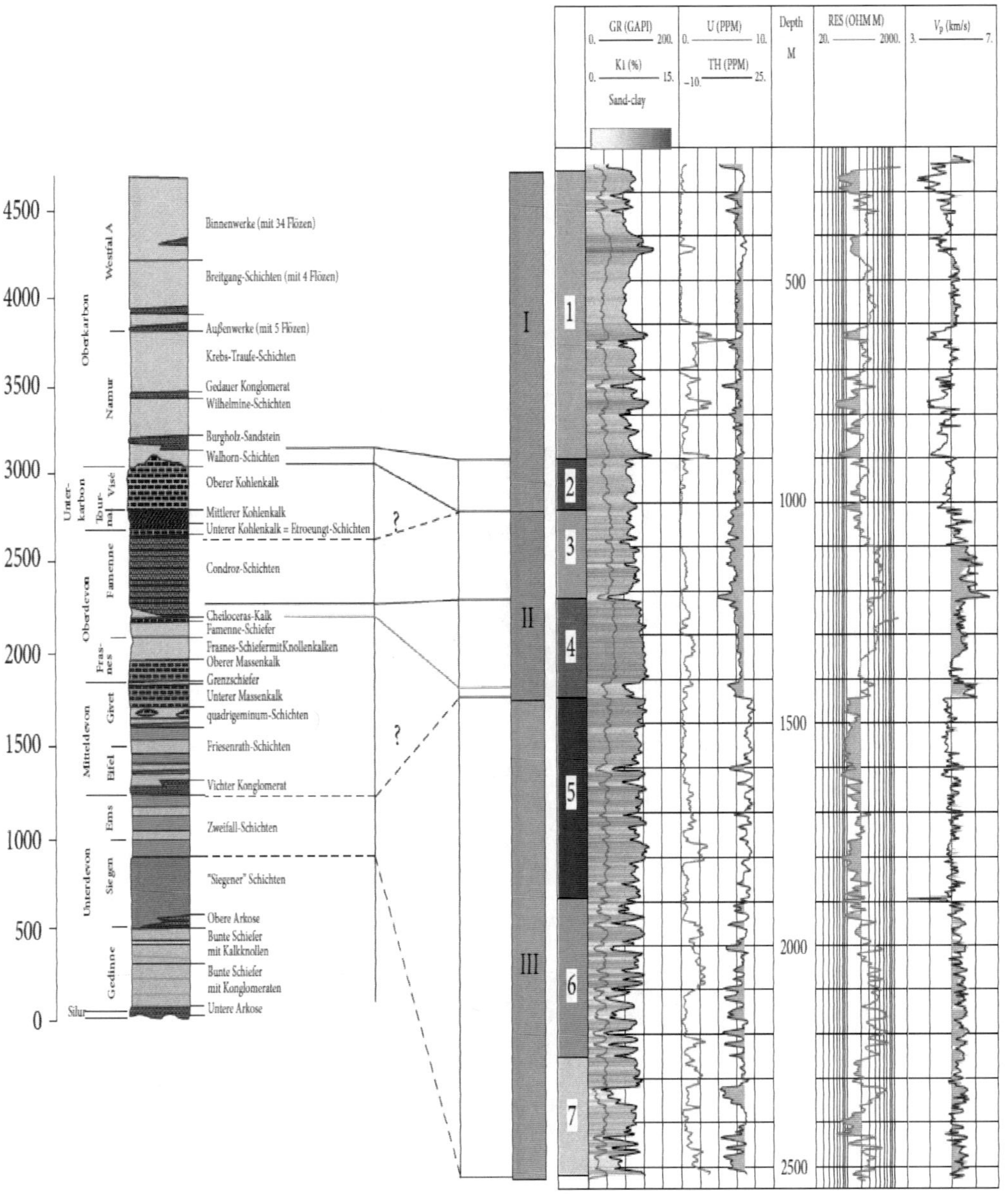

Figure 4. Geological zoning of the borehole [2]

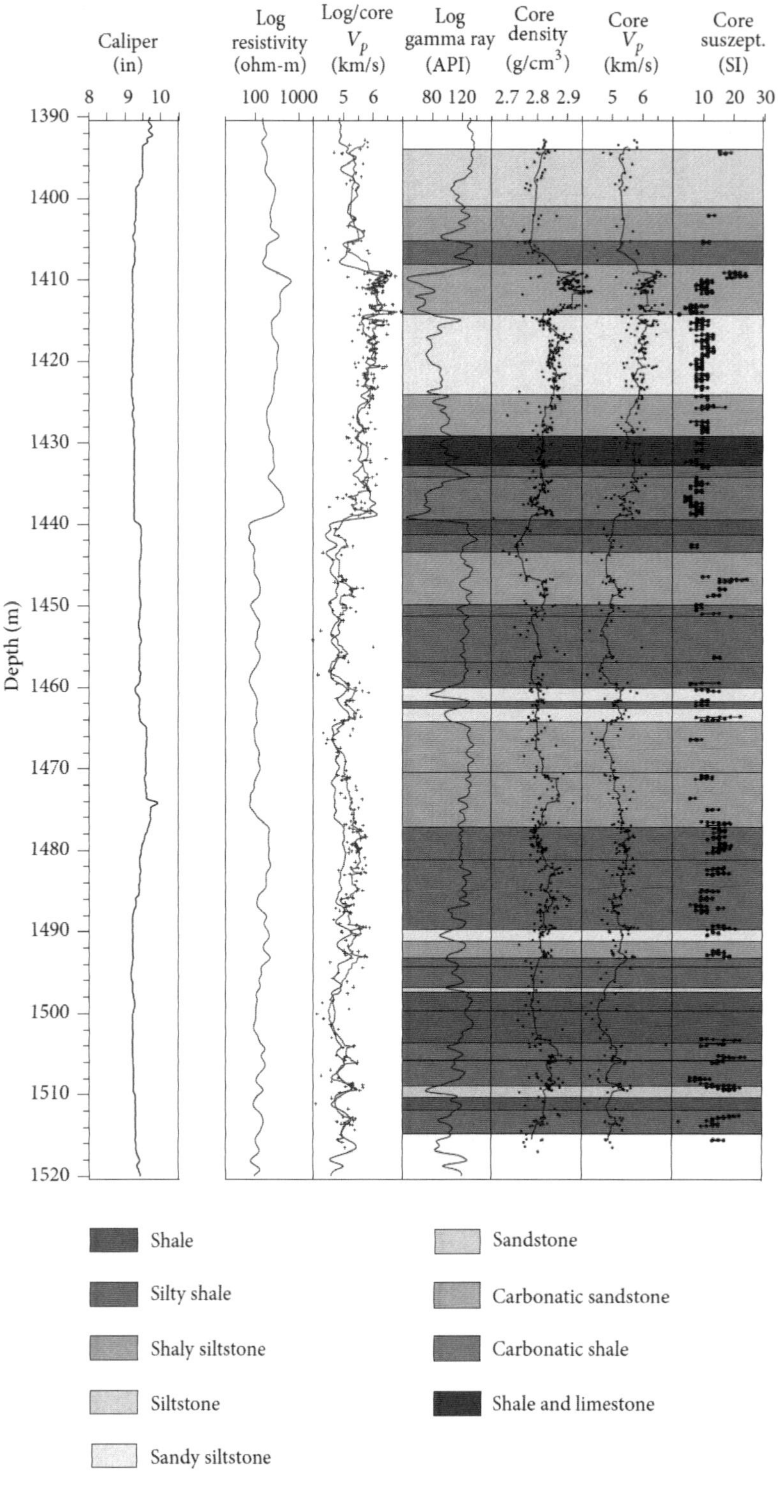

Figure 5. Petrography data on cores by zones. Colors indicate the type of wall according to the petrography description of the core [2]

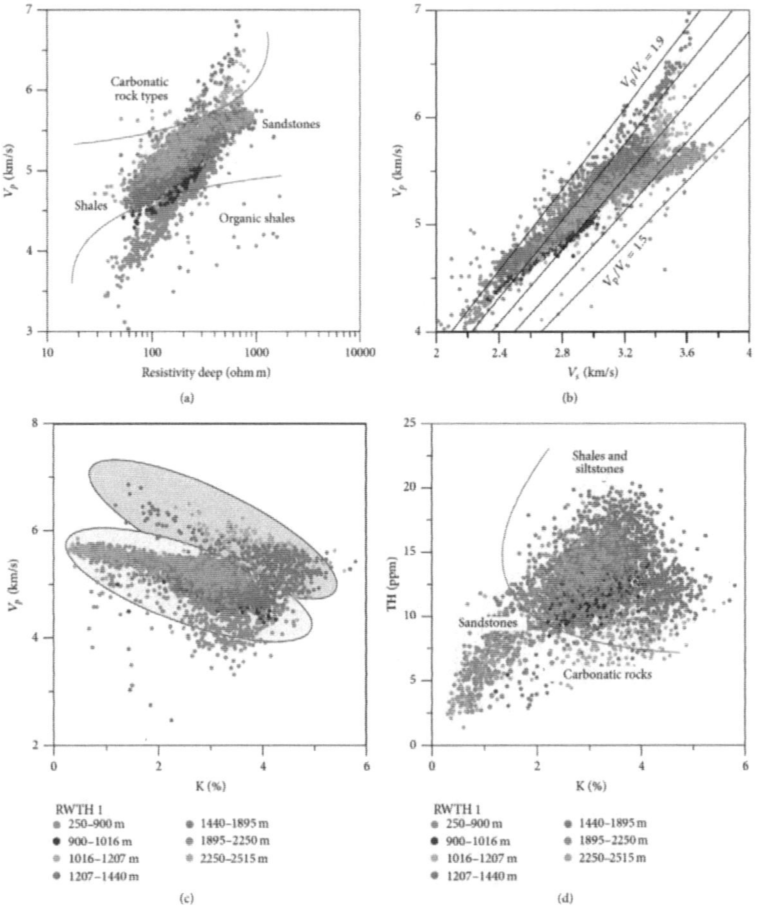

Figure 6.

Because of the paleo compensation and post-experimental processes, the ratio of natural gamma activity to relative volume of clay differs between these areas. This segmentation allows the continuous thermal conductivity of the entire profile along the depth of the well, using a natural gamma log for each individual zone and calibrating with thermal conductivity measured on a charged core [6].

The rate of compression and the effect of gamma radiation of the potassium substance were used to identify 7 zones with a defined content of 100% clay. Assuming that clay is the only source of relevant potassium radiation, 0% of the clay is assigned to the lowest total value. Identification of thermal conductivity λ1 on a scale with clay content and thermal conductivity λ2 on quark scale with the rest of the wall gives the thermal conductivity of $\lambda_{\log}$. The log (log) has a vertical resolution of 0.076m and the values are harmonized at an interval of 2m. In order to calculate the heat conductivity, the arithmetic (AM) and geometric (GM) mean were used:

$$\lambda_{\text{AM}} = \lambda_1 V_{clay} + \lambda_2 (1 - V_{clay}) \tag{1}$$

$$\lambda_{GM} = \lambda_1^{V_{clay}} \lambda_2^{(1-V_{clay})} \tag{2}$$

For each pre-identified zone, the thermal conductivity is conditioned by the clay content in the rest of the wall. For each measured value on the sample, the corresponding log value (log) was identified.

Calculated λl_{log} were adjusted to the measured changes λ1 and λ2 [6]. This resulted in thermal conductivity logs (Figure 7). "TC AM" is based on the arithmetic mean of λ_{log} calibrated by minimizing the variance between λlog and the measured value. "TC GM" is based on the geometric calibration method λlog by minimizing the minimum square deviation between λ_{log} and the measured values. In addition, the continuous profile of heat production through the entire well was derived from the sound record [7], [8] and the spectral gamma log [9]. The results are shown in Figure 7.

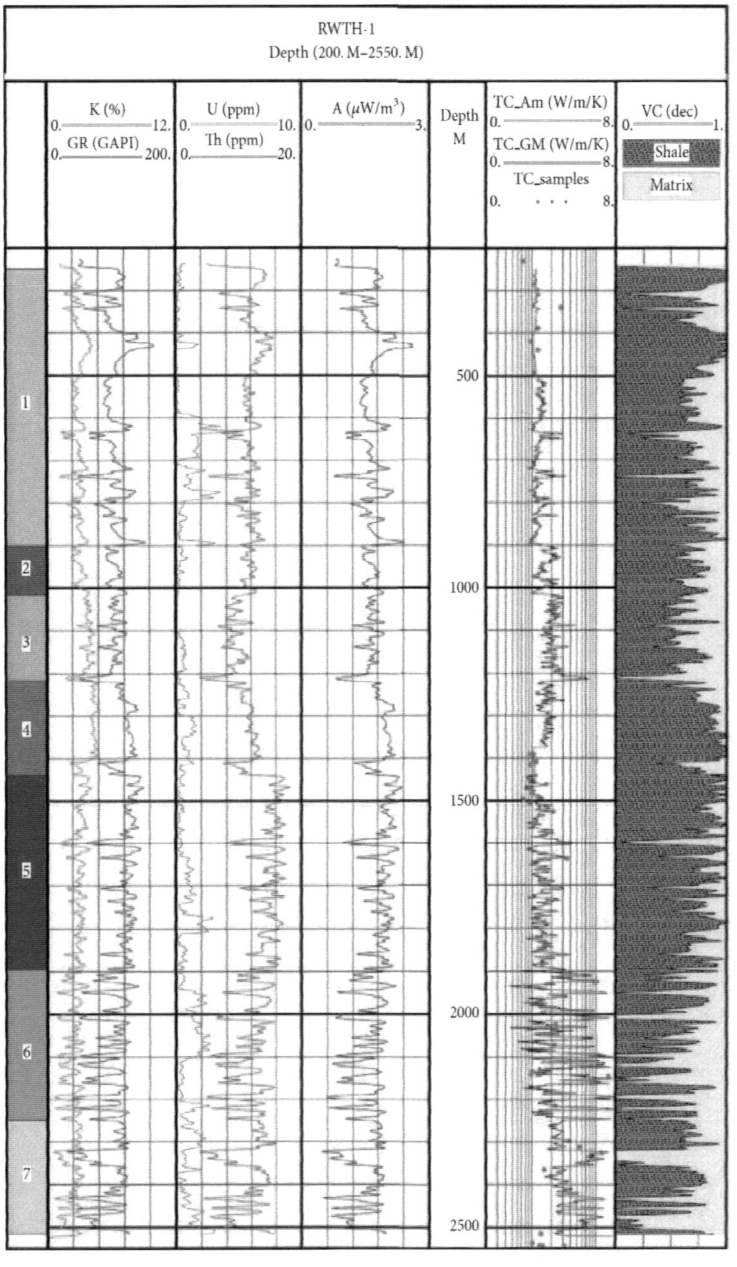

Figure 7. Thermo conductivity TC AM (arithmetic mean) and TC GM (geometric mean) and heat generation A are derived from the volume of fractions based on total (GR) and spectrum (K, U, Th); Colors indicate different borehole zones

3. Numerical cylindrical model

For the simulation of the influence of different parameters on the output heat exchanger from the heat exchanger and consequently on its thermal power, a final difference method (FD) was used called SHEMAT (Simulator for HEat and MAss Transport) [10]. Due to the fact that the permeability of the layers is very low and the natural flows of groundwater are not detected, a cylindrical symmetrical model can be used. Due to symmetry, only 2D networks are sufficient (Figure 8). Constant properties of zone materials are specified for each network block. The properties of different geological zones are marked on the right [2]. The different properties of the heat exchanger zone are based on the geometry of the heat exchanger, as shown on the left side of Figure 8. The heat exchanger has four enclosures of different lengths. The total 2D network model includes a cylindrical body with a radius of 99.5m and a depth of 2973.5m. It is assumed that the geological zones are almost horizontal due to the low inclination of the layers in relation to the well of about 100. Fluid and heat flows are calculated via the interface by dividing the blocks (access to the hidden network).

It turned out that the heat exchanger affects the soil cooling in a radius of about 10m around the borehole [3]; therefore the radial environment is more than sufficient. The minimum size of a measuring cell of 5mm in each zone of special interest (input, output, housing) was used because of the required high accuracy at these locations.

The size of the network of the conveyor zone from the well to the surrounding walls increased with a factor of 1.5 which resulted in a maximum cell size of 25m. The number of cells is 91×201=18291 cells. For each budget step, it takes about 1 second. The image shows the model in comparison with the geological profile. The lithology column is simplified on units of 25m (maximum cell size). The radius and depth of the casing, as well as changes in lithology, are well represented.

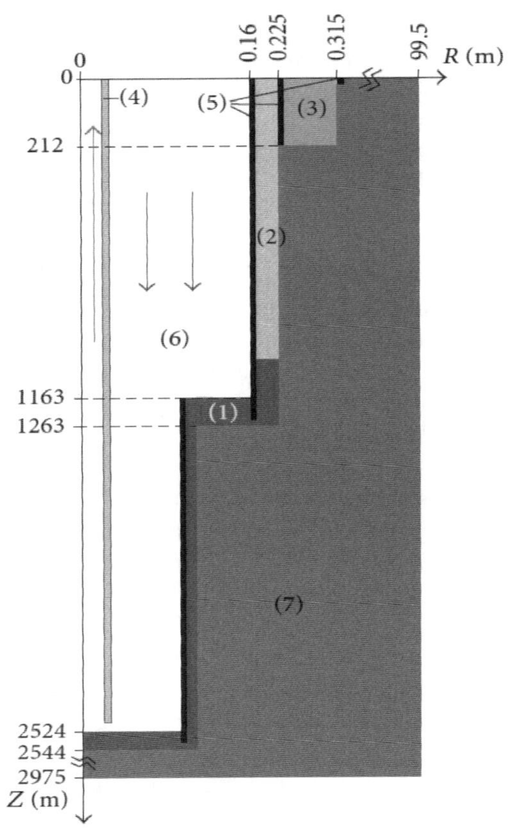

Figure 8. Used 2D cylindrical grid (radius 99.5m and depth 2973.5m) of the heat exchanger in different wall zones of the well. Thermal conductive cement (1), insulation cement (2), normal cement (3), inner tube (4), steel pipe (5), water (6) and surrounding borehole (7)

3.1. Model parameters

The wall permeability in the model was set to zero because of the extremely low porosity and the absence of water inflow during drilling and sampling. The flow of water in the inner tube, as well as between the inner and outer pipes is completely defined by constant volume flow. Since there are no layers with high specific heat capacity, the average value of specific heat capacity; 2.2MJ/(m^3K) ± 20% was used for surrounding walls (Figure 8) in simulations [11], [12]. Three outer pipes of steel tubes (Figure 8: (5)) have a specific heat capacity of 3.51MJ/m^3K). The specific values of the thermal capacity of the cement are 1.4MJ/(m^3K) for insulation cement up to the depth of 1025m, 1.58MJ/(m^3K) for thermal conductive cement from a depth of 1025m to the final depth of the well, and 1.65MJ/(m^3K) for normal cement. The thermal conductivity of the cell's cells varies with the depth, according to the average thermal conductivity of the determined log (Figure 7). The inner tube, separating the cold water flowing from the hot water flowing upwards, needs to be insulated and have a realistic thermal conductivity of 0.1W/(mK) unless otherwise stated. The inlet pipe has a thermal conductivity of 50W/(mK) steel (Figure 8). The values for cement were given by Karat: 0.52W/(mK) for insulation cement, 2.02W/(mK) for thermal conductive cement and 1.21W/(mK) for normal cement, under assuming that the shutdown is error-free [6].

As shown in Figure 7, heat generation A is less than 3μW/m^3 , on average only about 0.5μW/m^3 , and is therefore neglected in the model.

Based on measurements [13], a constant surface temperature from 100^{0}C to 35^{0}C is used in the model as the upper boundary condition of the surface of the surface. Constant specific heat flow is set as the lower limit state. He's welded. The German average of 69mW/m^2 [14], 72mW/m^2 is near the nearby Konzen bore [15]. The values specifiec het flow of 110mW/m^2 at a depth of 1312m and 90mW/m^2 between 1791m-2291m are values for Peer and Soumagne wells [16].

3.2. Simulations and results

Here, the simulations and the resulting obtained temperature Tout from the heat exchanger placed in the well and the time interval for which a temperature greater than 55^{0}C is reached which is necessary for the operation of the absorption air conditioner are described and analyzed. The size of the time step has a great effect on the duration of the simulation and the accuracy of the rapid temperature change; therefore, the simulation of the operation of the operation requires short-term steps for sufficient accuracy. This results in an extremely large countdown time (weeks).

3.2.1. The effects of heat conductivity of the inner tube. The thermal conductivity of the inner tube λ_{pipe} has a significant effect on the achieved output temperature and the corresponding thermal power P_T. In principle, the importance of isolation increases with decreasing the volume flow. This effect is important because of the longer time required to transfer heat from the water in the outer tube to the water in the inner tube. As shown in Figure 9, at a flow rate of 4m^3/h, a reduction in thermal conductivity from 0.005W/(mK) to 0.0001W/(mK) insulation results in an increase in the maximum temperature T_{max} of less than 20K; i.e. 0.005W/(mK) is the optimum value. On the other hand, an increase of 0.05W/(mK) reduces the time t (T> 55^{0}C) to achieve the required temperature greater than 55^{0}C. According to [17], GRP pipes usually have a conductivity of 0.36W/(mK). In Weggis (Switzerland), a steel pipe with double walls is in use, between which there is a vacuum of 0.02MPa. Its minimum thermal conductivity λpipe is 0.09 W/(mK), [18]. The latest GRP pipes have a thermal conductivity of 0.36 W/(mK). On the other hand, even with the best available insulation with a thermal conductivity of λpipe of about 0.1W/(mK), the temperature losses are 19% to 31%. Therefore, as shown in Figure 9, only insulation with less than 0.1W/(mK) is sufficient to maintain the required temperature for operation of the air conditioner. As shown in Figure 10, in the case of realistic insulation of 0.1W/(mK), the temperature falls below the required minimum of 55^{0}C, but after an uninterrupted operating time of ½ days.

3.2.2. Effects of flow velocity. The water in the heat exchanger in the well may circulate with different flows F(m^3/h). The water retention time inside the pipe and the corresponding total circulation time are 1,0h and 11,7h for a flow of 10m3/h and 1,9h and 23,4h for a flow of 5m^3/h. Flow values F are directly related to the thermal power P_T of the heat exchanger of the well and are calculated as:

$$P_T = c_{pw}\,\rho_w\,F_w\,(T_{out} - T_{in}) \quad (3)$$

where c_{pw} is indicated with the thermal capacity of water, with ρw water density and with the T output and the input temperature of the heat exchanger, respectively. In order to see the effect on the output temperature, a volume flow change is simulated: 1m^3/h, 4m^3/h, 6m^3/h, and 10m^3/h. Isolation of 0,01W/(mK) provides even 1/10 better thermal conductivity than the real heat conductivity of the inner tube.

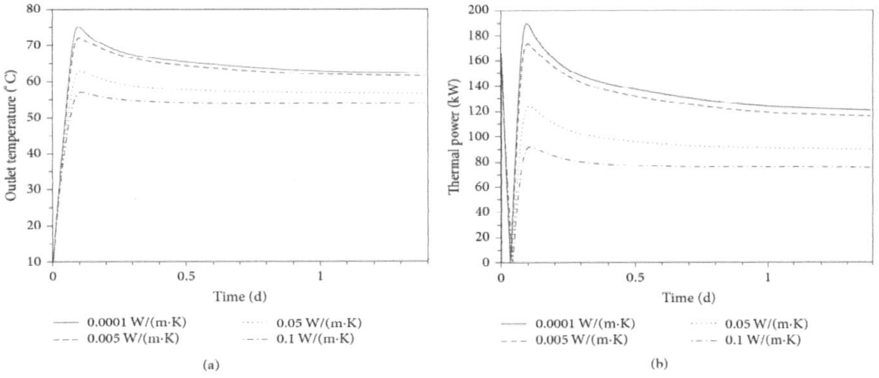

Figure 9. Variation of the temperature of the outgoing water and the heat energy of the heat exchanger in the well with time, for the different heat conductivity of the inner tube and for the continuous operation of the heat exchanger

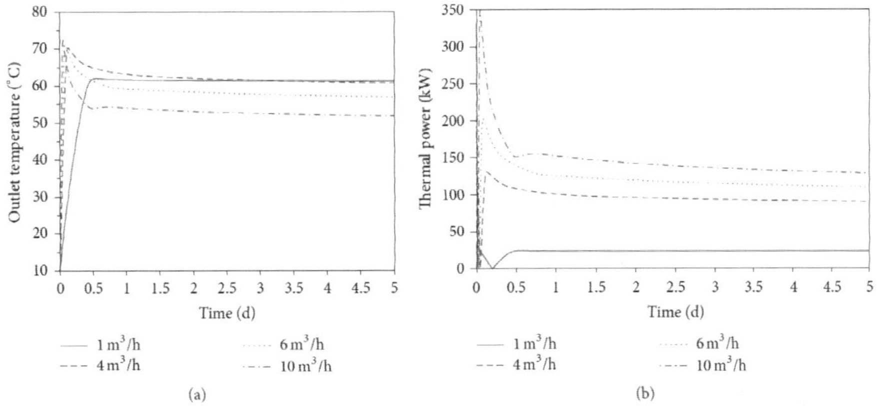

Figure 10. Behavior of the temperature of the outgoing water and heat energy of the heat exchanger with time, for different flows of water and for continuous operation

As shown in Figure 10, with a drop in insulation of 1/10, the temperature drop below is minimum required from 55^{0}C and if the flow increases. After 5 days of continuous operation, the temperature and heat output are 52^{0}C, 57^{0}C, 60.5^{0}C, and 61^{0}C and 120kW, 110kW, 90kW and 20kW, respectively.

3.2.3. Thermal capacity effect tubing. The effect of the specific heat capacity of the inner tube, the specific heat capacity of c_p "windows", "district heating networks", "glass fiber reinforced plastics" and other materials were used in the simulation. These variations had no effect on T_{max} and t ($T>55^0C$).

3.2.4. Operational Cycle Effects. The previously described simulations have shown that the continuous operation of the heat exchanger will not bring the temperature needed for the operation of the absorption heat sink. Cyclical operations with heat extraction in the daytime and heat recovery were simulated to maximize the output temperature (Figures 11 and 12). The water flow is $3m^3/h$; a period of 6 hours of heat extraction is shifted with a period of 18 hours of heat recovery. Shorter extraction periods were ineffective, as it took 1h to remove cooled water from the inner tube. A low flow value has been selected to maintain a high temperature over a longer period of time. Large flow would have the opposite to cause too much cooling of the wall. Moreover, it turned out that even with the best insulated tubes, the temperature drop would occur during stopping. In continuous or cyclical operation, the temperature and heat output are about 49^0C and 34^0C to 51^0C and 29kW and 20kW to 35kW, respectively (negative values are due to the start of cold water after restarting the system).

Figure 11 shows the output temperature and the corresponding thermal power after 20 years of operation for cyclic and continuous flow of $3m^3/h$ and for a four-day interval of operation. Figure 12 shows a long-term simulation over a period of 10 to 20 years with a flow of $3m^3/h$ for continuation of operation as well as for a daily working cycle consisting of 6h pump operation and 18h recuperation.

The maximum and minimum temperatures in cyclic operation are displayed. It is shown that even after 20 years of operation, the heat exchanger in the well did not stop the loss of efficiency. In the first years, the reduction in thermal power and temperature is the highest, and then remains stable after 20 years of operation. The water temperature is sufficient for direct heating in the winter. It is important to consider that the constant input temperature is 40^0C. The temperature difference between the maximum cyclic and non-cyclic temperature is 20^0C. The time required is increased until hot water flows from the inner tube. Different flows are simulated for cyclical operations consisting of 6h pumping, followed by 18h - recovery with real thermal conductivity of inner tube $\lambda_{pipe}=0.1W/(mK)$.

The cycle curves in Figures 11 and 12 show that the required minimum temperature of 55^0C has never been achieved.

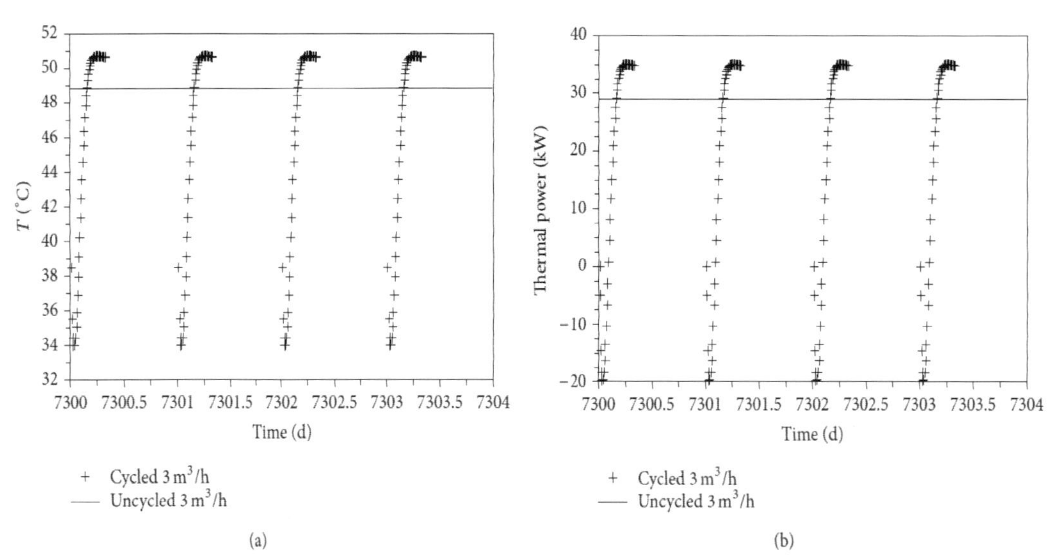

Figure 11. Output water temperature and associated heat output relative to the time after 20 years of operation for cyclic and continuous flow of $3m^3/h$

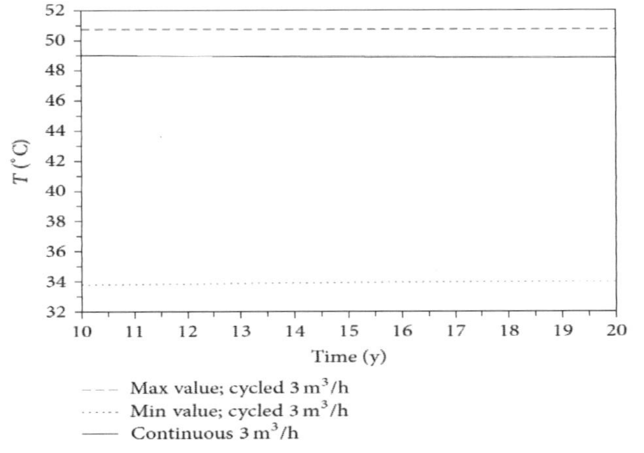

Figure 12. Maximum and minimum temperatures for cyclic and continuous work, with a constant flow of $3m^3/h$ and a time interval between 10 and 20 years of operation

3.2.5. Simulation of the Crypto Control Unit for the Super C category building. It is clear from the above figures that it is impossible to achieve, given the currently available insulation for the inner tube, the temperature of 55^0C required for the operation of the adsorption chiller. Even in the case of better insulation in the future, flow rates are limited to reach the required temperature.

In addition to the adsorption refrigerant, two compressor cooling units are planned with a capacity of 130kW. The adsorption chiller works with a coefficient of effect (COP) of 0.38-0.63 for lower temperatures, 55^0C-85^0C followed [19]. Therefore, in order to cool the building with 1kW, the adsorption chiller should use a more/less 1/0.38=3kW geothermal *energy* resulting in 4kW of additional heat, which contributes to global warming of the environment. Therefore, even with the future improvement of the isolation material, we can wonder if this is a sustainable way of cooling, it is probably not. In any case, according to German regulations, the maximum cooling power required is limited to the ambient temperature of 28^0C. In Aachen, this temperature will be 37 hours a year. Also, the need for cooling above 20^0C is 580 hours a year in Aachen [19]. In case the adsorption refrigeration unit can not meet, it can only operate 50% of the total cooling power, which corresponds to a cooling period of 190 hours per year.

As for the heating of the building, the total heat transfer coefficient (U-value) in the building is 7.9285kW/K [19]. With this value, it is possible to calculate the maximum required heating power to maintain the room temperature of 18.3^0C, at an outside temperature of 12^0C. Installed heating bodies are: standard radiators (70^0C-35^0C), floor heating (32^0C-30^0C) and concrete core (28^0C-25^0C). This high-temperature acceptance of the temperature allows the storage of large amounts of heat in the building and the reduction of the return temperature of the return water from the building to 25^0C.

Air temperatures recorded at Euro planetarium Genk (60km from Aachen) in 2005, every 30 minutes, were taken to simulate the required thermal power over time for the Super C building (Figure 13 (1)). The required power (Figure 13 (4)) and the temperature of the outgoing water from the building (25^0C) are used as input parameters for the simulation of the heat exchanger in a deep well. Figure 13 (2) shows the required flow value for a semi-annual simulation. The flow through the heat exchanger is changed every 30 minutes until the required power is reached. It is clear that the output temperature (Figure 13 (3)) slowly decreases over time, but the heat output is satisfactory (Figure 13 (4)).

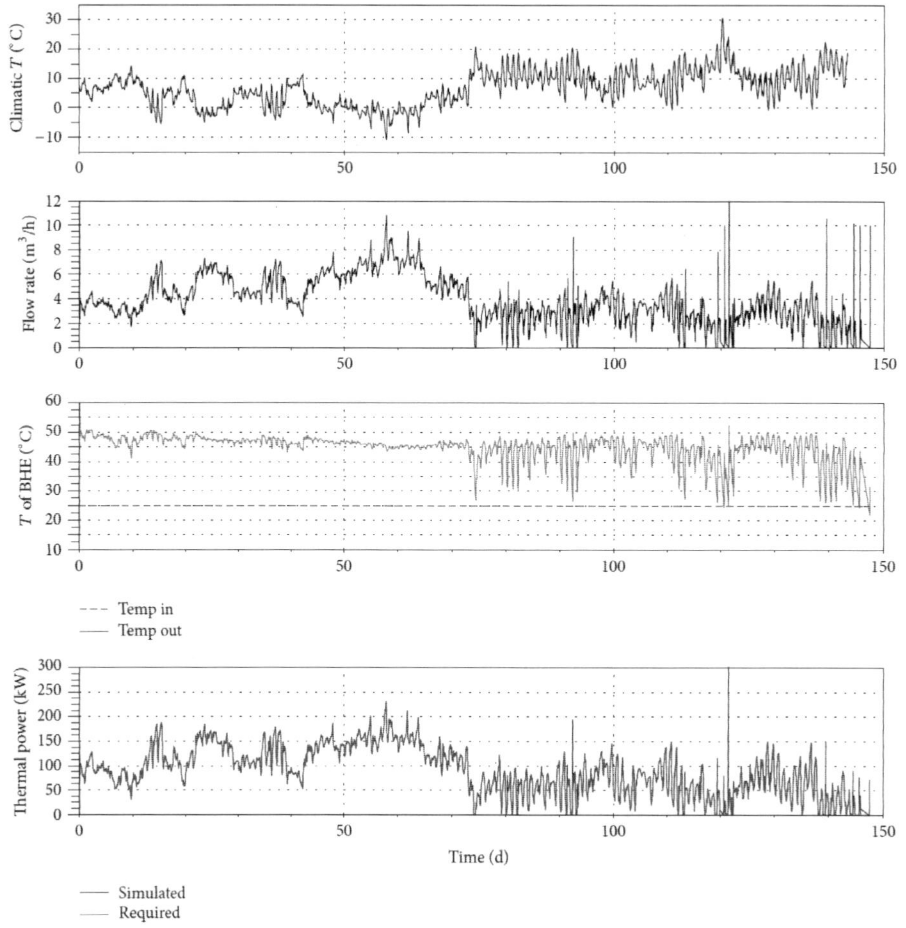

Figure 13.

With actual temperature data in the first half of 2005, the need for heating and cooling for the Super C building was calculated, using a mean value of U 7.9285kW/K [19]. The assumption is that the heating bodies in the building return water with a constant temperature of 25^{0}C which will be the input water temperature in the BHE. Output temperature and water flow are simulated in BHE. The required heat power in the first half of the year is compared with the calculated thermal energy required by the building.

4. Final result

In 2008, the building of the Super C class (Figure 1) was officially connected to the city heating, but with the intention of using the borehole for geothermal heating in the future. At the time, there were still problems in finding suitable insulation material for the inner tube. Finally, in 2011, RWTH Aachen decides that the operation of a deep-water drill shaft will not be economical due to the high additional costs of the thermal insulation of the inner tube.

5. Conclusion

In order to heat or cool the A-Class building in the energy class "Super C", in the city center of Aachen, Germany, it is necessary to have 40^{0}C for direct heating of the building in winter, while cooling airs for the building with adsorption cooling unit requires a temperature of minimum 55^{0}C. Measurements and calculations for deep-coaxial heat exchanger (CBHE) depth of 2500m, with a maximum temperature in a well at 85^{0}C at that depth of 2500m, reveals the contradiction between the maximum required temperature for the adsorption air conditioning unit and the maximum thermal power. Numerical calculations indicate that only in a short period of time in a cyclical operating mode, the borehole can supply at the same time the required temperature of 55^{0}C and a heat output of 120kW, with an input temperature of 40^{0}C and a flow of 3m^3/h. However, after 20 years of operation, it has been shown that the temperatures are too low for the correct operation of the adsorption cooling unit. During the winter, the heat exchanger can supply the building with sufficient amount of heat, with temperatures from 25^{0}C to 55^{0}C and throughput to a maximum of 10m^3/h. Also, numerical calculations indicate the great importance of isolating the inner tube. This requires a very expensive inner tube, which in turn results in the fact that the total heat exchanger for this well in Aachen is economically unsustainable.

References

[1] Dijkshoorn L, Speer S and Pechnig R 2013 Measurements and Design Calculations for a Deep Coaxial Borehole Heat Exchanger in Aachen, *International Journal of Geophysics* 916541

[2] Oesterreich B, Trautwein U and Lundershausen S 2005 Die geothermiebohrung RWTH-1, VAG—Vereinigung Aachener Geowissenschaftler e.V. Infoblatt

[3] Signorelli S 2004 *Geoscientific investigations for the use of shallow low-enthalpy systems*, ETH Zürich University, Zürich, Switzerland, Doctoral Thesis

[4] ***1997 *Akkupyc Manual*, Micromeritics Instrument Corporation, Norcross, Ga, USA

[5] ***2005 *Instrument Manual TCS—High Precision Thermal Conductivity Measurement*, Lippmann and Rauen GbR, Schaufling, Germany

[6] Speer S 2005 *Design calculations for optimising of a deep borehole heat-exchanger*, Institute of Applied Geophysics, RWTH Aachen University, Aachen, Germany, M.S. Thesis

[7] Çérmak V, Bodri L, Rybach L and Buntebarth B 1990 Relationship between seismic velocity and heat production: comparison of two sets of data and test of validity, *Earth and Planetary Science Letters* **99**(1-2) 48-57

[8] Rybach L 1978 The relationship between seismic velocity and radioactive heat production in crustal rocks: an exponential law, *Pure and Applied Geophysics* **117** (1-2) 75-82

[9] Bücker C and Rybach L 1996, A simple method to determine heat production from gamma-ray logs, *Marine and Petroleum Geology* **13**(4) 373-375

[10] Clauser C 2003 *Numerical Simulation of Reactive Flow in Hot Aquifers*, Springer, Berlin, Germany

[11] Beck A 1998 *Methods for determining thermal conductivity and thermal diffusivity, in Handbook of Terrestrial Heat Flow Density Determination*, (R Haenel, L Rybach and L Stegena, Eds.) Kluwer Academic Publishers, Norwell, Mass, USA

[12] Vosteen H D and Schellschmidt R 2003 Influence of temperature on thermal conductivity, thermal capacity and thermal diffusivity for different types of rock, *Physics and Chemistry of the Earth* **28**(9-11) 499-509

[13] Clauser C 1984 A climatic correction on temperature gradients using surface-temperature series of various periods, *Tectonophysics* **103**(1-4) 33-46

[14] Schellschmidt R, Hurter S, Forster A and Huenges E 2002 *Atlas of Geothermal Resources in Europe*, Office for Official Publications of the European Communities, Brussels, Belgium

[15] Karg H 1995 *Untersuchung des Temperatursfeldes im Untergrund der Region Aachen-Maastricht-Lüttich*, Applied Geophysics and Geothermal, Aachen University, Aachen, Germany, Master Thesis

[16] Verkeyn M 1995 *Bepaling van de Warmestroomdichtheid in Belgiëeen verkenning naar de mogelijkheden en de beperkingen* (Detemination of the Specific Heat Flow in Belgium with possibilities and limitations), University Leuven Belgium, Master Thesis

[17] Summa F, Benner L and Otto F 2005 *Tagung für Ingenieurgeologie Erlangen, Geothermie unter geotechnischen und wirtschaftlichen aspekten*
[18] Kohl T, Brenni R and Eugster W 2002 System performance of a deep borehole heat exchanger, *Geothermics* **31**(6) 687-708
[19] ***2004 *Superc entwurfplannung, unterlagen zur hubau*, Tech. Rep. 33239, Arup GmbH, New York, NY USA

IOP Conf. Series: Materials Science and Engineering **477** (2019) 012055 doi:10.1088/1757-899X/477/1/012055

Recycling of textile wastes into textile composites based on natural fibres: the reinforcement type and the architecture

M P Todor [1], C Bulei[1], I Kiss[2] and V G Cioată[2]

[1]University Politehnica Timisoara, Faculty of Engineering Hunedoara, Doctoral School, Timisoara / Hunedoara, Romania
[2]University Politehnica Timisoara, Faculty of Engineering Hunedoara, Department of Engineering & Management, Romania

E–mail: imre.kiss@fih.upt.ro

Abstract. The textile composite materials market has witnessed an important growth in the past decades and is estimated to continue during the forecast period, according all statistics and trends. Since the last decades, the usages of various textiles composite materials in different applications have increased steadily, penetrating and conquering new markets, constantly. While textile composites are already proven to be weight–saving and high strength materials, the current challenge is to make these materials cost–effective by reducing the cost of different types of reinforcements (fibres, yarns or fabrics). The efforts to manufacture economically beneficial composite textiles have resulted in several innovative production techniques, currently being used in the area of the composites industry. Applications vary significantly in size and complexity, loading and surface quality, operating temperature, suitable production volumes and peculiar added value. The expanding choice of raw materials, in terms of the reinforcement type together with the matrix type, followed by many production techniques, gives impressive flexibility in manufacturing of textile reinforced composites. The aim of this work was investigation of the possibilities of reusing of textile wastes, especially the textiles derived from the packaging industry, as reinforcement in production of new added–valued composite materials.

1. Introduction

The increased ecological consciousness has generated the concept of the sustainable development of environmental material resources with improved economic activities.[1–7] Also, the waste legislation increased the pressure on solving the waste management through reuse or recycling processes.[1–7] In this sense, the recycling of textiles is an issue that requires immediate attention in order to address the management of textiles derived from various scraps generated throughout textile manufacturing processes, as well as household waste. One possibility for recycled fibres and different fabrics could be to employ them as reinforcement in the composites, which in turn could be used in several applications.[1–9] Natural based reinforced composite materials have enormous potential due to their light weight and ability to be manufactured in many complex shapes and structures. The steady increase in the use of composite materials has brought benefits in many areas.[6–10]

At the same time, the growing interest in research of natural based composite materials is mainly due to its low weight associated with moderate mechanical properties when are compared with the synthetic fibres.[8–13] Also, natural based materials (fibres or fabrics) are biodegradable, low in cost and relative easy to obtain. These properties are already being explored by some industries sectors, like the automotive, which is already using natural based composites in car components and few structural devices.[6–10] In this sense, the bast fibres (like flax, hemp, kenaf, sisal and jute) are a promising reinforcements for use in natural reinforced composites on account of its low cost, low density, high specific strength, no health risk, easy availability, renewability and much lower energy requirement for processing.[8], [9], [12–26] Therefore, environmental concern has resulted in a renewed interest in area of bio–based materials, among them, the bast fibres being perceived as an environmentally friendly substitute to the glass fibres for reinforcement.[2], [4], [8–10], [17–19]

Several research projects have been carried out to develop the recycling methods and seek ways of effectively using recycled materials (including wastes) in new applications.[1], [4], [12–14] Currently, the use of composites has had a great impact in the advanced industry areas, such the automotive industry and construction and building materials.[8], [9], [14], [23] Mainly, those are related with the sustainability and the environment aspects.[1–7] Textile materials have been reused in the production of add–values products, crushed or cut, for manufacturing noise reduction and temperature insulation materials, and as reinforcements of natural based composites. [6], [8], [9], [15–26]

The textile wastes are part of the solid wastes generated by the population and contains, mainly, used clothing or packaging materials.[1–9], [26] The textile recycling should be taken as a means to obtain few environmental or/and economic benefits for various reasons, as it contributes to the reduction of spaces required for landfills, reduces the need to produce virgin materials and reduces pollution.[1–4], [26] The use of recycled materials is primarily aimed at environmental preservation, but creation raw materials for the industrial area is a topical issue, with a promising industrial future. In the past decades, special attention has been focused on the used clothes or packaging bags, which represent a source of raw materials environmentally responsible and economically profitable.[3–6], [16], [22–26] Wastes from the production of textiles are undesirable but are inevitable by–products in many manufacturing processes, which are often not given the recognition and economic value that they actually have. Recycling textiles are an excellent opportunity to reduce the environmental impact of commonly used products.[1–7], [11–14], [26] In this sense, the textile materials recycling involves:

— the re–use of used fibrous materials, mainly from the population, and
— the re–use of clothing manufacturing production losses.[3–7], [14], [17–19], [26]

The textile recycling strategies are based on the knowledge of their classification:[2], [4–6], [26]

— post–production textiles (yarns or fabrics), which are losses of production process, and
— post–consumption textiles (clothing or carpets), which are discarded once its useful life expires.

Recovery from the textile waste stream includes the following activities:

— re–use of the textile waste in its original form, the largest volume of goods being sorted for second hand clothing markets, or
— conversion to new products, in which the fibre is then re–engineered into value–added products (car components or building materials such as insulation and roofing felt).[8], [18], [22–26]

The traditional textile's recycling is based on introducing them back to a productive stream.[20–22] Upcycling is the process of converting wastes or useless products into a better quality's or a higher environmental valued products.[1–7] The idea of using textile waste in applications that do not involve a new industrial process results in a novel recycling idea, due to the economic and environmental benefits that can be achieved.[1], [4], [11–13], [26] In fact, the existing materials are given more value, not less and the process is essential to minimize environmental impact and decrease the amount of waste that goes into landfills.[1–7] So finding possibilities to re–use the already created resources is a way to create an endless supply of materials without further depleting natural resources. Moreover, upcycling the textile waste is definitely a necessary step in closing the loop on textile manufacturing, getting the excess and waste under control and eliminate it whenever possible.[13], [22]

2. About recycling of textile fibres and fabrics

The experts estimates that 95% of the textiles we use could probably be recycled, but only 25% actually are. Much of the remainder is incinerated or sent to landfill.[1–7], [10], [15] Therefore, a solution is needed to recycle some of this end–of–life textile waste into a valuable resource, by

IOP Conf. Series: Materials Science and Engineering **477** (2019) 012055 doi:10.1088/1757-899X/477/1/012055

rethinking its use of resources for new materials like reinforcements in composites. It can be an alternative to a variety of existing materials depending on the application. In designing the new added–value materials, must be focused on making a composite that can replace or even bring additions to existing material offerings and thereby be an alternative.[4–7], [11–14], [26] Therefore, recycling of these solid textile waste may be viable alternative for industries. Textile wastes such as from fibre, textile and clothing can be sourced from the community, manufacturing industry and consumers. These are also known as pre–consumer, post–consumer and industrial textile waste. Meanwhile, the post–consumer textiles, often known as "dirty waste", are collected mixed with other household items.[1–7], [11–14]

Material selection in the new added–values products is made during a complex and lengthy product development process. Actual selection of materials for a specific application is primarily driven by the trade–off between the cost of the material (including the purchase price and the processing costs) and its performance attributes (such as properties, quality and so on). In this sense, the use of textile fibres and fabrics, in specific applications, may solve several problems, namely:[8], [22], [26]
— reduction of spaces required for landfills,
— reduces pressure on virgin resources, using efficiently the resources, and
— promoting of an alternative material as reinforcing materials.

Recycling is a key concept of modern waste management, consisting in the reprocessing of waste materials into new or reusable products.[1–7], [11–14], [25], [26] The recycling of textile waste can helps to conserve resources and reduce the use of landfills, as we mention before. In fact, recycling has become a necessity, to protect the environment, sending textile wastes to the landfills being the last alternative in an integrated waste management system. But, in the light of a fierce need for raw materials, can be a solution to promote of an alternative material as reinforcing materials into natural based composites.

3. Comparison of properties: E–glass vs. bast fibres

Composite materials have at least two components in which properties are clearly different between them.[8–10], [14–26] Separately, these constituents maintain their characteristics, but when are mixed the combination of the two materials results in a "composite material", which is superior in both form and function to that of the constituent materials separately.[8–10], [14–26] Therefore, the properties of the composites can be considered as a combination between the properties of the reinforcement and the matrix, in addition to the properties of interfaces between the reinforcement and matrix.

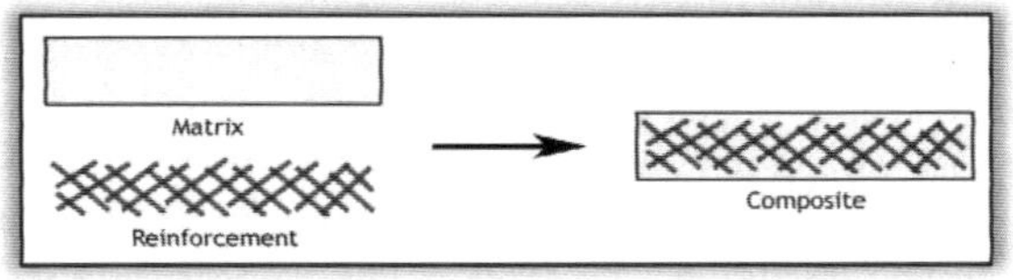

Figure 1. The main components of the composite materials

In general, in the polymer matrix composites, the fibres provide the strength of composite while the polymer, which is sometimes referred to as the matrix, provides dimensional stability and the transfer of shear stresses between the reinforcement. The role of the reinforcement is fundamentally one of increasing the mechanical properties of the matrix.[8–10], [14–26] The result is a strong functional material which can be used in a wide variety of applications.

Fibre reinforced polymer composites are composed of a polymeric medium such as polyester or epoxy, and a fibrous material – such as carbon fibre, glass fibre, aramid fibre or natural fibre. Natural fibre reinforced polymers have been identified as a potential low impact alternative to synthetic fibre reinforced polymers.[8–10], [14–26] Therefore, the replacement of glass fibre with natural textile for reinforcement in polymer composites appears to be a modern approach. Bast, leaf and fruit fibres have all been identified as potential replacement of glass in fibre reinforced polymer composites, bast fibres receiving the most attention.[16–26] Flax, jute and hemp have been identified as very important fibres for use as reinforcement in polymer composites due to their proven mechanical performance in comparison with glass fibre. For most other applications, the fibres need to be arranged into a fabric, to make handling possible. Therefore, the reinforcement with natural textile fabrics (plain weaves, unidirectional fabrics and non–crimp fabrics) can be a solution. Subsequently, different types of textile fabrics were produced and evaluated as reinforcement in composites manufactured by well–established manufacturing techniques.[8–10]

Available material properties of natural textile used as reinforcements are reviewed here. In general, these materials are categorized into fibre, yarn and fabric forms. Bast fibres has been employed for centuries as a packaging material, but in recent times it is found to be a valuable aid to sound environmental management, bag cloth industry having the biggest consumer of these fibres available in the markets.[8–10], [14–25] Among all bast fibres, along with kenaf, hemp, flax and ramie, jute is one of the most significant and versatile fibres of commercial and technical importance, being one of the cheapest vegetable fibres and it is second only to cotton in the amount produced and the variety of its uses.[8–10], [19–24] Natural fibre bags have gained an advantage as being an eco–friendly option instead of both of non–biodegradable polymer bags and paper bags.

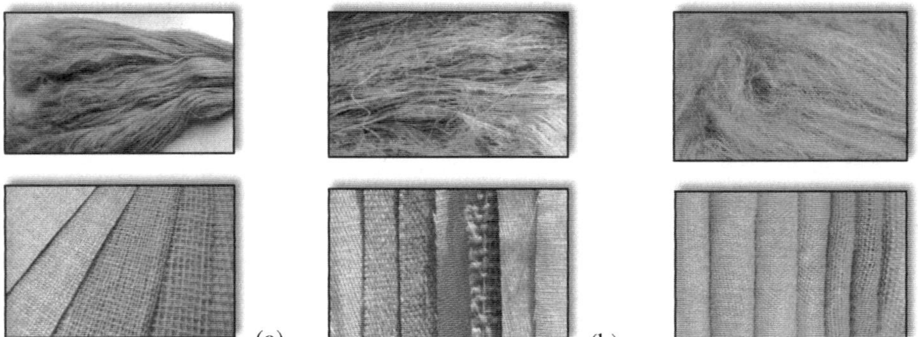

(a) (b) (c)

Figure 2. Long fibres and fabrics: (a) flax; (b) hemp; (c) jute

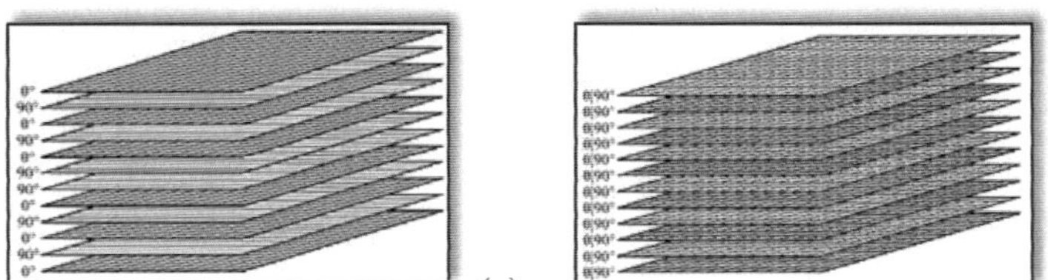

(a) (b)

Figure 3. Fabric constructions: (a) 0/90 cross–ply layers; (b) 0/90 plain weave fabrics

In polymeric composite terms, a textile fabric is defined as a manufactured assembly of long fibres of carbon, aramid, glass or natural origin, or a combination of these – the term hybrid refers to a fabric that has more than one type of structural fibre in its construction –, to produce a flat sheet of one or more layers of fibres.[8–10] Woven fabrics are produced by the interlacing of fibres in a regular pattern or weave style. The large variety of possible textile preforms presents a challenge for natural composite materials science.[8–10]

Most fabric constructions offer more flexibility for layup of complex shapes than straight unidirectional tapes offer.[8], [9] The plain weave construction results from each fibre alternating over and then under each intersecting strand. Composites using textiles have drawn attention, since requirements and survivability are becoming increasingly important in different applications.[8], [9] Textile composites now in abundant production and their contribution has been landmarked by fibres used than fabric structure and fabric types.

4. Results & Discussions

Three different natural fibres were investigated in the present study (flax, jute and hemp) in comparison with E–glass. Density, modulus values, and specific modulus (modulus divided by the density of the material) and elongation to failure for the flax, jute and hemp fabric destined to fibre–resin composite fabrications were presented in Figure 4. Since compared with the bast fibres, glass fibre do not absorb the moisture greatly, therefore, only the moisture content of the flax, jute and hemp fabric were presented in Figure 4.

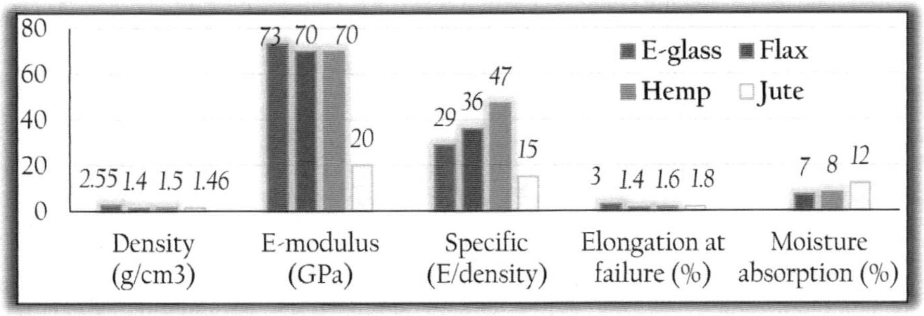

Figure 4. Comparison of properties: E–glass vs. bast fibres (hemp, jute and flax)

The density of natural fibres is generally lower than that of E–glass: typically hemp fibre is 1.5 g/cm³, flax is 1.4 g/cm³, jute is 1.46 g/cm³, while glass can be as much as 2.55 g/cm³. Hence, the specific properties like specific modulus sometimes achieve a better ratio between E–Modulus and density than E–glass reinforced composites. The density of a material is a key consideration in application as a lower weight material can reduce costs in some areas of applications. While applications for natural fibre composites could be similar to glass fibre composites, several research has indicated that composites consisting of natural fibres have comparably low intrinsic mechanical properties compared with their synthetic fibre composites.

The application of natural based composites has increased and is gaining preference over glass fibre and carbon fibre, having in view that the natural based composites excel in most parameters except strength. Therefore, total global natural based composite materials market is expected to grow in the next years, the efforts must being carried out to find new uses for natural fibres or fabric resource wastes, including its utilization as reinforcement in polymer composites.

5. Concluding remarks

A significant amount of different wastes from the textile industry and post–consumer product are disposed worldwide. This is not only a cause for environmental concern but also represents a waste of useful resources. In fact, turning them into useful materials serves a dual function:

— elimination of textile wastes, and

— development of a new added–value product.

Last decades, natural based reinforced polymer composites have been investigated extensively. At the same time, the use of these composites has continuously increased during recent years due to their low cost and environmental friendliness. Natural fibres as a substitute for synthetic fibres such as glass fibre in composite components have gained renewed interest during recent years due to their properties, renewability, and also, their abundance. This has led to the application of various natural fibres, like bast fibres in many polymer reinforced composites in a wide range of applications, products made from textile–reinforced composites having the same advantages.

Considerable growth has been seen in the use of natural based reinforced composites in many industrial applications over the past decade, most developments being focused on the random discontinuous fibre composite systems. The development of continuous fibre reinforced composites is, however, essential for manufacturing materials. The current work aims to develop high–performance natural fibre composite systems using continuous textile reinforcements like woven fabrics. Textile fabrics are characterized by plaiting, good elasticity and fine porosity, can be processed slightly into fabrics, and they can retain suspended particles by absorption. The type, porosity and thickness of the fabric, further treatment give plies a wide variety and adaptability in finding new applications.

References

[1] Sule A D and Bardhan M K 2001 Recycling of textile waste for environment protection – An overview of some practical cases in the textile industry, *Indian Journal of Fibre & Textile Research (IJFTR)* **26**(1–2) 223–232

[2] Sandin G and Peters MP 2018 Environmental impact of textile reuse and recycling – A review, *Journal of Cleaner Production* **184** 353–365

[3] Farrant L, Olsen S I and Wangel 2010 A environmental benefits from reusing clothes, *International Journal of Life Cycle Assessment* **15**(7) 726–736

[4] Pensupa N, Leu S Y, Hu Y, Du C, Liu H, Jing H, Huaimin W and Lin C 2017 Recent trends in sustainable textile waste recycling methods: Current situation and future prospects, *Topics in Current Chemistry* **375**(5) 76

[5] Domina T and Koch K 1999 Consumer reuse and recycling of post–consumer textile waste, *Journal of Fashion Marketing and Management: An International Journal* **3**(4) 346–359

[6] Wang Y 2010 Fibre and textile waste utilization, *Waste and Biomass Valorization* **1**(1) 135–143

[7] Cuc S and Vidovic M 2011 Environmental sustainability through clothing recycling, *Operations And Supply Chain Management* **4**(2/3) 108–115

[8] Todor M P, Bulei C, Heput T and Kiss I 2018 Researches on the development of new composite materials complete / partially biodegradable using natural textile fibres of new vegetable origin and those recovered from textile waste, *IOP Conf. Ser.: Mater. Sci. Eng.* **294** 012021

[9] Todor M P, Bulei C and Kiss I 2018 Composite materials manufacturing using textile inserts with natural origins fibres, *IOP Conf. Ser.: Mater. Sci. Eng.* **393** 012088

[10] Saxena M, Pappu A, Sharma A, Haque R and Wankhede S 2011 *Composite materials from natural resources: Recent trends and future potentials, advances in composite materials – Analysis of natural and man–made materials*, InTech, http://www.intechopen.com/

[11] Hawley J M 2000 Textile recycling as a system: A micro/macro analysis, *Journal of Family and Consumer Sciences* **92**(4) 40–43

[12] Bartl A, Hackl A, Mihalyi B, Wistuba M and Marini I 2005 Recycling of fibre materials, *Process Safety and Environmental Protection* **83**(4) 351–358

[13] Bhatia D, Sharma A and Malhotra U 2014 Recycled fibres: An overview, *International Journal of Fibre and Textile Research* **4**(4) 77–82

[14] Mishra R, Behera B and Militky J 2014 Recycling of textile waste into green composites: Performance characterization, *Polymer composites* **35**(10) 1960–1967

[15] Nickel J and Riedel U 2003 Activities in biocomposites *Materials todays* **6**(4) 44–48

[16] Elisa Zini E and Scandola M 2011 Green composites: An overview, *Polymer Composites* **32**(12) 1905–1915

[17] Faruk O, Bledzki A, Fink H and Sain M 2012 Biocomposites reinforced with natural fibres, 2000 – 2010 *Progress in Polymeric Science* **37** 1552–1596

[18] Faruk O, Bledzki A, Fink H and Sain M 2014 Progress report on natural fibre reinforced composites, *Macromolecular Materials and Engineering* **299**(1) 9–26

[19] Bledzki A K and Gassan J 1999 Composites reinforced with cellulose based fibres, *Progress in Polymer Science* **24** 221–274

[20] Summerscales J, Dissanayake NPJ, Virk A S and Hall W 2010 A review of bast fibres and their composites. Part 1 – Fibres as Reinforcements, *Composites Part A* **41** 1329–1335

[21] Summerscales J, Dissanayake NPJ, Virk A S and Hall W 2010 A review of bast fibres and their composites. Part 2 – Composites, *Composites Part A* **41** 1336–1344

[22] Pickering K L, Efendy M G A and Le T M 2016 A review of recent developments in natural fibre composites and their mechanical performance, *Composites Part A* **83** 98–112

[23] Sen T and Reddy H N 2011 Various industrial applications of hemp, kenaf, flax and ramie natural fibres, *International Journal of Innovation, Management and Technology* **2** 192–198

[24] Cheung H Y, Ho M P, Lau K T, Cardona F and Hui D 2009 Natural fibre–reinforced composites for bioengineering and environmental engineering applications, *Composites Part B: Engineering* **40**(7) 655–663

[25] Dicker M, Duckworth P, Baker A, Francois G and Hazzard M 2014 Green composites: a review of material attributes and complementary applications, *Composites Part A* **56** 280–289

[26] Pena Pichardo P, Martinez-Barrera G and Martinez-Lopez F 2018 *Chapter 6: Waste and Recycled textiles as reinforcements of building materials*, InTech, http://www.intechopen.com/

IOP Conf. Series: Materials Science and Engineering **477** (2019) 012056 doi:10.1088/1757-899X/477/1/012056

Mitigation of climate change from the aspect of controlling F-gases in the field of cooling technology

G Tica[1] and A R Grubic[2]

[1]University of Banja Luka, Faculty of Mechanical Engineering, Department of Thermal Engineering, Bosnia and Herzegovina
[2]Ministry of Foreign Trade and Economic Relations of Bosnia and Herzegovina, Department of Environmental Protection, Bosnia and Herzegovina

E-mail: gordana.tica@mf.unibl.org

Abstract. The main objective is to identify, analyze and deliver data on F-gases in Bosnia and Herzegovina and identify potential mitigation measures. Bosnia and Herzegovina became a member of the United Nations Framework Convention on Climate Change (UNFCCC) on December 6, 2000. In its status as non-Annex I country within the UNFCCC, Bosnia and Herzegovina is obliged to undertake the following reporting activities, such as: to calculate annual emissions of greenhouse gases using a defined methodology and reporting to the Conference of the Parties (COP), introduce and implement measures to mitigate the effects of climate change by regulating anthropogenic emissions and adaptation measures to climate change. Also cooperates in the development and transfer of technology, methods and processes that lead to limitation, reduction and stabilization of greenhouse gas emissions, include an impact assessment of climate change and appropriate economic development strategies and policies aimed at minimizing the negative impacts of climate change on the economy, the environment and the health of the population; conduct systematic observation and research, exchange data and exchange information on climate and climate change in order to improve scientific discoveries on the causes and consequences of climate change.

This project not only supports the country of Bosnia and Herzegovina in fulfilling the above obligations as a signatory to the UNFCCC, but will also serve as an important strategic document for sustainable development. Also, this project promotes relations between entities and business sectors, establishing links between the environment and other sectors. The data on the emission of F-gases in Bosnia and Herzegovina, including sources of these gases, are provided. In this paper, potentially intrusive measures are suggested. Ozone is a gas located in the earth's atmosphere in a layer of 15 to 35 km above the Earth and has a special significance for the planets and the entire life on it. It is very important his role for the protection of life on Earth, and the disrupted occurrences of ozone flows through the atmosphere can have harmful effects on the health of man and the environment as a whole. The amount of ozone in the tropospheric (in the first lower layer above the Earth) and the tratospheric (upper-upper layer) was in the past in a natural balance. Uncontrolled human activity, in particular the development of the refrigeration industry, and the accelerated increase in the emissions of individual industrial gases, there was an increase in ozone levels in the tropospheric layer and in the stratospheric reduction. Concentration of ozone in the stratosphere and troposphere is limited by the balance between the chemical processes produced by ozone and the processes that destroy it. This equilibrium of the process is determined and depends on the amount of gases (involved in the reaction) and the speed (or efficiency of individual reactions), depending on the intensity of sunlight, location and area in the atmosphere, temperature and other natural climate factors. Global ozone has declined steadily over the last few decades. The Earth's stratosphere has increased the amount of reactive gases, containing chlorine and bromine. The biggest impact on climate change is the carbon dioxide emission CO_2, due to the far-reaching emissions (measured in billions of tonnes per year). Among the "greenhouse gases" with the far-reaching "global warming potential" (GWP) are the so-called "Ozone gases" or substances that damage the ozone layer CFC and HCFC. Fortunately, there is a far smaller amount of these gases that reaches the atmosphere than CO_2.

1. Refrigerant

The cooling sector is one of the most important areas of energy in which ozone depleting substances are used. Namely, the most widespread working tools since DuPont introduced them to the market in the twenties of the last century are chlorofluorocarbons (hereinafter referred to as CFCs) and chlorofluorocarbons (hereinafter HCFCs), which appeared on the market a little later, are also used.

Significant changes in the use of refrigerants have occurred in the last decade, mainly in response to environmental issues, "hole in the ozone layer" and "global warming or greenhouse effect". Previously, there was not much dilemma in choosing a refrigerant because most of the needs could be met with well-known and well-tested fluids. The cooling sector is unique in that the ozone-depleting substances (ODS) are not only used in new devices but also in servicing existing equipment. Research on the field is carried out in relation to working assets, that are still in the existing correct systems that are in operation. They are replaced in accordance with the limits related to the time limit for their use. By evacuating from use, they are replaced with new working assets, which are thermodynamically suitable for them. Refrigerants is divided into: primary and secondary.

Primary refrigerants are those that are directly used in the cooling process .The primary refrigerants are grouped into the following four groups, shown in the table below. The scope of interest in this report is the group of F- gases.

Table 1. Distribution of primary refrigerants [1-3]

No	Groups of primary refrigerants
1.	Halogenated carbonate refrigerants (so-called F-gases)
2.	Azeotropic refrigerants
3.	Inorganic refrigerants
4.	Hydrocarbon refrigerants

1.1. Halogenated carbonate refrigerants

Halogenated carbonate refrigerants belong to a group of artificial, synthetic refrigerants. These are halogenated derivatives of saturated hydrocarbons (CFC, HCFC and HFC). The trade name for them is freons or F-gases. The basic information about them is given in the table below.

Table 2. List of Controlled and Alternative Substance [1-3]

No	Label	Hemical name	Properties
1	Aneks A		
	CFC	Group I – chlorofluorohydrocarbons	
		Group II– Halons	
	Aneks B CFC	Group I – chlorofluorohydrocarbons	Other fully halogenated saturated hydrocarbon

			derivatives
		Group II – carbon tetrahydrous	
		Group III – trichloroethane	
	Aneks C		Fully halogenated saturated hydrocarbon derivatives
	HCFC	hydrochlorofluorocarbons	
	HBFC	Bromo Fluoro Coal	
	Aneks E	Bromochloromethane	
		Group I-Methyl bromide	
2.	Mixes controlled substances, the containers they contain:	Bromohlorodifluormetan Bromotrifluormetan Dibromotetrafluoroetan	
	HCFC+PFC+HFC		
	Mixtures they containes:		
	Mixture they contains HBFC		
	Mixture they contains HCFC (with or without PFC+ HFC+CFC)		
	Mixture they contains carbon tetrachloride		
	Whith 1,1,1 trichloroethane (metal hloroform)		
	Whith bromometan (metal bromide / bromochloromethane		
3.	New controlled substances		
4.	Alternative substances		
	SF6	sulfur hexafluoride	
	HFC	Hydrofluorocarbons	
	PFC	Perfluorocarbons	
	Mixture alternative substances		

IOP Conf. Series: Materials Science and Engineering **477** (2019) 012056 doi:10.1088/1757-899X/477/1/012056

2. Actual working substances and their control

In order to mitigate measures in the state of Bosnia and Herzegovina it is mandatory to implement operational programs and projects, control and reporting systems within the Montreal Protocol. Phase of the permit for the import and use of substances harmful to the ozone layer of CFC (R-11, R-12, R-502 ...) has been completed, and the next steps are being implemented.

2.1. Measures to implement the Montreal Protocol and deadlines for the abolition of CFC consumption

The Montreal Protocol specifies measures and deadlines for the gradual reduction and final exclusion, the abolition of production and consumption of ODS (ODS). For CFCs, these are the following deadlines:

1. For developed countries – 31 December 1999 / (January 2000),
2. For developing countries, the period of 10 years is still prolonged - December 31, 2009 / (January 2010).

2.2. Measures to implement the Montreal Protocol and deadlines for the elimination of HCFC consumption

HCFCs are substances that are less harmful to the ozone layer than CFCs. Contains hydrogen and chlorine. Perhaps the R-22 is the most commonly used in this group. The next phase is still in progress, import, use and implementation treatment in devices is controlled. Monitoring is carried out by relevant and responsible state institutions.

2.3. Measures to implement the Montreal Protocol. Research on new working substances

In addition to explosive (butane, propane), HFC and their mixtures are constantly on the market. The most well-known working substance from the HFC group is the one-component working substance R-134 a, which is mainly used as a replacement for R-12. HFCs are new environmentally friendly non-ozone-depleting substances (R-134a, R-404a, R-407 C, R-410A). In smaller home coolers today, the R-22 working substances are often replaced with the R-600a hydrocarbon. Substitutes for R-22 are the most commonly used working substance mixtures that can be azeotropic and zeotropic (R-407C, R-410A). Azeotropic working substances R-502 (CFC/HCFC), used in the past for lower cooling temperatures, are now most frequently replaced by a pseudo-zeotropic mixture of R-404a or an azeotropic mixture of R-507.

By the term "consumption" of working mediums, it is assumed that the total annual quantity (in ODP tonnes) of all substances in one group (eg CFC) is calculated in such a way that when the total annual quantity produced in the country concerned accrues the imported amount of these substances during the reporting period calendar year and take away the total exported amount of the same from the country. Therefore, consumption = production + import-export.

Since Bosnia and Herzegovina has no own production of any of the ODS, this means that for Bosnia and Herzegovina the "consumption" of each "control substance" is equal to the total amount of its annual imports.

3. Data on F-gases, emissions in Bosnia and Herzegovina, including sources of these gases

Ozone-Depleting Substances (ODSs) are chemicals / bromine-containing chemicals that contain chlorine or bromine (not in nature, but produced by humans), which have the potential to react to ozone molecules (chemical label-O_3) in the stratosphere. This ability of these chemicals, when they arrive at the atmosphere, the Earth's planetary stratosphere, and under the influence of ultra-violet sun rays, are themselves disintegrating, releasing chlorine or bromine, in their free state, break down-destroy ozone molecules in the stratosphere, officially, according to the Montreal Protocol, is called

IOP Conf. Series: Materials Science and Engineering **477** (2019) 012056 doi:10.1088/1757-899X/477/1/012056

the degradation potential ozone (Ozone Depleting Potential - ODP) or often, in our domestic practice, called ozone depletion factor.

The reference substances for determining ODP, or the most commonly used substances in various sectors of industrial production, are CFC- compounds CFC-11 and CFC-12, with potential ODP = 1. Bosnia and Herzegovina is a member of the Vienna Convention and the Montreal Protocol and all amendments other than Beijing. Implementation of the Montreal Protocol and the gradual exclusion program from the use of ODS takes place through the Ozone Unit of Bosnia and Herzegovina at the Ministry of Foreign Trade and Economic. Relations, where public activities are undertaken.

Strategic documents were drafted on the basis of which Bosnia and Herzegovina fulfills its obligations for gradual exclusion from the use of ozone-depleting substances. But certain progress has been made, as many investments now use new equipment without ozone depleting supstances. Many investment projects were made, through which the business entities received new equipment that did not use ODS and destroyed the old ones in their production processes among others:

- The depletion program for methyl bromide (the application of new technologies contributed to reducing the use of methyl bromide at 1,700 kg per year in 2001 to 100 kg in 2006),
- Project for the management and control of halons in Bosnia and Herzegovina - used in fire extinguishers in Bosnia and Herzegovina, the establishment of a control system for the import and consumption of halons, and training user's staff on international standards in handling halon equipment and protection against uncontrolled release, the use of alternative halon replacement substances, and the establishment of a halon center where halon will be collected which is no longer used for further recycling or destruction,
- RMP Program (Refrigiration Management Plan) Management Program in the Refrigeration Service Sector,
- Project for training and certification of service workers in the refrigeration,
- The project of national reclaiming centers for freon recycling (CFCs, HCFCs and their polymers).

The legal basis for the implementation of the Montreal Protocol on ODS, as an international agreement, accepted by Bosnia and Herzegovina on the basis of the succession of the obligations of the former joint state of SFR Yugoslavia (original) named:

- "Sl. list SFRJ MU ", No. 16/90,
- "Fig. list Republic Bosnia and Herzegovina ", No. 13/94,
- the "Montreal Protocol on Substances that Deplete the Ozone Layer" ("Official Gazette" of SFRJ MU", No. 16/90.

Official Gazette of the Bosnia and Herzegovina), as its international obligation, is based on:

- The Law on the Procedure for the Conclusion and Execution of International Treaties of Bosnia and Herzegovina ("Official Gazette of Bosnia and Herzegovina No. 29/00)
- The Law on the Council of Ministers of Bosnia and Herzegovina ("Official Gazette of Bosnia and Herzegovina " No. 30/03 and 42/03), the decisions of the Council of Ministers on the conditions and manner of implementation of the Montreal Protocol and the gradual exclusion from the use of the ODS:

In particular, the decision of the Council of Ministers determines:

- Conditions for import and export of ODS, (licensing system, approval of quotas for import and issuance of import licenses),
- Conditions for importing and exporting devices, which contain as built-in or functionally use ODS.

Monitoring and controlling the import and export_of these substances and products that contain or use the substances referred to in the preceding paragraphs and the reporting of relevant domestic and international institutions, organizations or authorities on import / export and annual consumption of these substances.

This decision also establishes the Program and Deadlines for the gradual exclusion from the use of ODS in Bosnia and Herzegovina, ie the most significant substances classified by the basic groups defined in the annexes of the Montreal Protocol and the decisions of the Conference to the Parties to the Montreal Protocol, which bind Bosnia and Herzegovina as a party to this international multilateral agreement / contract.

The dynamics of the gradual exclusion of ODS from use in Bosnia and Herzegovina, that is permitted quantities of imports / consumption of ODS in Bosnia and Herzegovina, by years and indicators of realized realization, are monitored on the basis of the official annual reports of Bosnia and Herzegovina to the authorities of the Montreal Protocol on Import and Consumption of ODS in Bosnia and Herzegovina.

4. Legislative law on Bosnia and Herzegovina entity

Implementation of the Montreal Protocol and the protection of the ozone layer, was also challenged by the entity laws. The Law on Air Protection (Federation of Bosnia and Herzegovina, "Official Gazette of the Federation of Bosnia and Herzegovina ", number: 33/03) Law on Air Protection (Republic of Srpska, Official Gazette of RS No. 53/02). By-laws in both entities (the Regulation of the Federal Ministry of Physical Planning on the gradual exclusion of ozone-depleting substances (ODP) and the Republic of Srpska Government Regulation on the gradual exclusion of ozone-depleting substances regulates the operational implementation of Bosnia and Herzegovina programs of gradual exclusion from the use of substances that damage the ozone layer ODS, especially in the following areas of activity:

1. The control and inspection of the devices in operation, which contain the ODS and the work of the service provider include,
2. Collecting ODS from the existing cooling units in operation,
3. Disposal of old or defective refrigeration units to landfills,
4. Establishment of the ODS Import / Export Register and Distribution in Bosnia and Herzegovina,
5. Obligations and manner of reporting of the importer ODS,
6. Obligations and rights of the competent entity institutions / ministries in the process of licensing and issuing licenses for import / export of ODS in Bosnia and Herzegovina.

Deadlines for the prohibition of import of key ODSs and equipment, which contains or use ODS obligations of Bosnia and Herzegovina according to the Montreal Protocol and its own (National) program for the gradual exclusion of ODS from use in Bosnia and Herzegovina.

Table 3. Time limits for exclusion, which Bosnia and Herzegovina is obliged to respect, according to the Montreal Protocol [2], [3]

No	Time limits - exclusion	
1.	Methyl bromide	From January 1, 2006..
2.	Halons	From January 1. januara 2007.
3.	Methyl chloroform	From January 1. januara 2007.
4.	Equipment and devices that contain or use ODS	From July 15, 2007.

5.	A group of CFC compounds	From January 1, 2010.
6.	A group of HCFC compounds	From January 1, 2020

The table below shows the values recorded in the past, long-term period, by 2010.Allowed/ planned and realized quantities (in tons) of imports/consumption of ODS in Bosnia and Herzegovina were shown for the period 2003-2010. [5]

Table 4. Permitted (planned) and realized quantities of imports / consumption of ODS in Bosnia and Herzegovina for the period 2003-201, [kg] [2], [3]

Groups ODS		2003	2004	2005	2006	2007	2008	2009	2010
CFC	allowed	235,30	167,00	102,10	63,00	43,00	23,00	3,00	0
	realized	230,00	187,90	51,24	32,70				
Halons	allowed			0,68	0,68	0	0	0	0
	realized	0,38	0,38	0	0				
Methylchloroform (trichloroethane)	allowed	17,00	17,00	11,00	5,00	0	0	0	0
	realized	36,00	4,0	6,84	0				
Methyl bromide	allowed		19,65	8,90	8,90	0	0		0
	realized	16,40	12,60	1,29	0				

4.1. Regulation of imports / exports of ODS in Bosnia and Herzegovina

Regulation of imports / exports of ODS in Bosnia and Herzegovina is done through licensing, quota and licensing systems for export / import, distribution and sale of ODS in the Bosnia and Herzegovina market. The most important procedures and procedures for:

- Submission of requests from economic organizations / legal entities and institutional competencies in Bosnia and Herzegovina for the implementation of the procedure for issuing licenses for import and export of ODS,
- Submission of requests from economic organizations / legal entities that have received the ODS import / export license for determining the quota (quantity) of annual ODS imports in Bosnia and Herzegovina,
- Submission of application and procedure for issuing licenses for import / export of ODS, for each individual contingent,
- Regulation of the import of equipment and devices, which contain or use ODS.

Import into the country or export from the country of renewed / recycled ODS, according to the Montreal Protocol, is not permitted. Keeping records (users) of the user / consumer ODS, as a rule, is done as follows:

- Keeping records of imports and exports of ODS by the Importers,
- Keeping records of distribution and sale of ODS in the Bosnia and Herzegovina market, by the Importers,
- Keeping records of consumption by ODS by producer or service economic organizations,
- Monitoring and control of import, export, sale and consumption of ODS, with inspection supervision (competent institutions of the entity and Brčko District)

- Preparation of data and information on import / export, distribution / sale and inventories of ODS for annual and occasional reporting of certain instances / institutions of Bosnia and Herzegovina.

The system of reporting on the import / export, consumption and distribution of ODS to the Bosnia and Herzegovina market - is done through reporting forms. In the system of compulsory reporting by various domestic and international factors, Bosnia and Herzegovina, as a party to the Vienna Convention for the Protection of tthe grant-funding he Ozone Layer and the "Montreal Protocol on ODS" and beneficiaries of the Multilateral Fund for the Implementation of the Montreal Protocol, (was and was still) obligatory to introduce and implement in practice reporting systems for at least the following areas of activities and activities:

- The annual report on import / re-export, distribution / sale on the Bosnia and Herzegovina market and ODS stocks, as of December 31 of the previous year, shall be submitted by each legal entity / company importer / exporter to the competent authority of the Entity, Brcko Ditrix. Reports are submitted on the prescribed forms (attached to the DECISION of the Council of Ministers),
- Annual Report on the Registration / Registration of ODS Imports in Bosnia and Herzegovina and Exports from Bosnia and Herzegovina is submitted by the Customs Sector Bosnia and Herzegovina - the relevant service - Ozone Unit of Bosnia and Herzegovina: Copy of the Reports: Entity Entities and Statistics Bureau of the Brčko District. Reports are submitted on the standardized forms of the Customs Service,
- Every annual report on the consumption of ODS in the production process and / or servicing of existing devices in operation shall be submitted by each economic entity / company / service workshop to the relevant entity institution or institution of the Brčko District,
- The Annual Report on Import / Export and Consumption of ODS in Bosnia and Herzegovina is submitted by the Bosnia and Herzegovina Ozone Unit - the Secretariat of the Montreal Protocol (Ozone Secretariat). Copies of reports: MLF, UNIDO, Bosnia and Herzegovina and Entity competent institutions,
- Reports shall be submitted on the standardized format of the Ozone Secretariat.

4.2. *Implementation of the Montreal Protocol and the National Program for the Phasing-out of ODS in Bosnia and Herzegovina*

Implementation of investment projects for technological restructuring of production capacities, with the transition from the use of controlled ODSs in the production process to the permitted substances (excluding ODS), as well as training projects and awareness raising on the need for protection of the ozone layer (approved and financed by MLF) Bosnia and Herzegovina Ozone Unit - MLF, at least once a year and at the end of each individual project, in particular.

Annual Report to the Multilateral Fund for the Implementation of the Montreal Protocol (MLF) is submitted by the Bosnia and Herzegovina Ozone Unit, on behalf of the Government of Bosnia and Herzegovina. Annual report in the usual format, and completion report, on standard MLF format. Annual Report on Import / Export and Consumption of ODS in Bosnia and Herzegovina, submitted by the Ozone Unit of Bosnia and Herzegovina to the Secretariat of the Montreal Protocol (Ozone Secretariat).

Copies of reports: Secretariat of the Montreal Protocol, UNIDO and relevant Bosnia and Herzegovina institutions for the field of environmental protection and international relations. Copies of the reports are submitted to the: Secretariat of the Montreal Protocol,UNIDO and the Council of Ministers (through the MVTEO). The following are the reports, officially registered numerically and submitted to the Ozone Secretariat for the period from 2010 to 2014 on a standardized format by importers. Of the estimated amount of cooling and air conditioners, up to 50% of the same uses the old non-ecological freon R-22 (HCFC group) and R-12 (from the CFC group).

Table 5. List of devices in Bosnia and Herzegovina using CFC and HCFC, [psc] [3]

Entity	Freeze	Freezers	Air - conditioners
Republic of Srpska	187,000 psc	187,500 pcs	50,000 pcs
Federation	350,000 pcs	350,000 psc	125,000 psc
Bosnia and Herzegovina	537,000 psc	537,000 psc	175,000 psc

Table 6. The amount of old-non-ecological freons R-22 and R-12 in refrigerated and air-conditioned, [kg], [3]

Entity	Freeze	Freezers	Air - conditioners	Total
Republic of Srpska	187,500 pcs x 0,2 kg=37,500 kg	187,500 pcs x 0,3 kg=56,250 kg	50,000 x 1 kg=50,000 kg	143,750 kg
Federation	350,000 pcs x 0,2 kg=70,000 kg	350,000 pcs x 0.,3 kg=105,000 kg	125,000 pcs x 1 kg=125,000 kg	300,000 kg
Bosnia and Herzegovina	537,500 pcs x 0,2 kg=107,500 kg	537,500 pcs x 0,3 kg=161,250 kg	175,000 pcs x 1 kg=175,000 kg	443,750 kg

Table 7. Lists the quantities of controlled freons R-22 and R-12 in chains, cooling cabinets, cooling cabinets, cooling chambers in hospitals, dental clinics, cafes, pastry shops, butcher shops, bakeries, markets, various organizations, grocery stores, hypermarkets and the like, [pcs]

Entity	Showcases	Cooling cabinets	Sanks	Cooling chambers
Republic of Srpska	100,000 pcs	100,000 pcs	10,000 pcs	5,000 pcs
Federation	250,000 pcs	250,000 pcs	25,000 pcs	10,000 pcs
Bosnia and Herzegovina	350,000 pcs	350,000 pcs	35,000 pcs	15,000 pcs

The quantity of old freons is by the free estimation still cc 30% for the above mentioned cooling devices, Table 8.

Table 8. The amount of old freons found in the above mentioned refrigerant devices, [kg]

Entity	Showcases	Cooling cabinets	Sanks	Cooling chambers
Republic of Srpska	30,000 pcs x 1 kg=30,000 kg	30,000 pcs x 1 kg=30,000 kg	3000 pcs x 1,5 kg= 4,500	1500 pcs x5 kg= 7,500 kg

Federation	75,000 pcs x 1kg=75,000 kg	75,000 pcs x 1 kg=75,000 kg	7,500 pcs x 1,5 kg=11,250	3,000 pcs x 5 kg= 15,000 kg
Bosnia and Herzegovina	105,000 pcs x 1 kg=105,000 kg	105,000 pcs x 1 kg=105,000 kg	10,500 pcs x 1,5 kg= 5,750	4,500 pcs x 5 kg= 22,500 kg

Table 9. Total consumption of old freons found in the above mentioned cooling devicesle, [kg]

Entity	Total Consumption
Republic of Srpska	72,000 kg
Federacija	248,250 kg
Bosnia and Herzegovina	320,250 kg

Table 10. Total cooling and air conditioners in ships, boats, trains, buses, trucks-refrigerators, truck trucks, dumpers, etc ,(pcs)

Entity	Boats	Boats	Train	Buses	Refrigerated trucks	Trucks
Republic of Srpska	200 pcs	300 pcs	100 pcs	1000pcs	500 pcs	1000 pcs
Federation	400 pcs	500 pcs	150 pcs	2500pcs	1000 pcs	2000 pcs
Bosnia and Herzegovina	600 pcs	800 pcs	250 pc	3500pcs	1500 pcs	3000 pcs

The amount of old freons R-22 and R-12 in the aforementioned means of transport and according to the free estimation cc 30%, showed in Table 11.

Table 11. The quantity of old freons R-22 and R-12 by the means of transport, [kg]

Entity	Boats	Boats	Train	Buses	Refrigerated trucks	Trucks
Republic of Srpska	60 pcs x 20= 1200 kg	90 pcs x 2 = 180 kg	30 pcs x 15= 1500 kg	300 pcs x 15= 4500 kg	150 pcs x 20= 3000kg	300 pcs x 7= 2100 kg
Federation	120 pcs x 20= 2400 kg	150 pcs x 2= 300 kg	45 pcs x 50= 2250 kg	750 pcs x 15= 11250 kg	300 pcs x 20= 6000 kg	600 pcs x 7= 4200 kg
Bosnia and Herzegovina	180 pcs x 20 3600 kg	240 pcs x 2= 480 kg	75 pcs x 50= 3750 kg	1050 pcs x 15= 15750 kg	450 pcs x 120= 9000 kg	900 pcs x 17= 6300 kg

Table 12. Total quantity (consumption) of old freons R-22 and R-12 by the means of transport, [kg]

No	Entity	Total , [kg]
1	Total Consumption in Republic of Srpska	12, 480 kg
2	Total Consumption in Federation	26,400 kg
3	Total Consumption in Bosnia and Herzegovina	38, 880 kg

5. Recapitulation of old non-ecological freons R-22 and R-12 found in cooling and air-conditioning units by entities in Bosnia and Herzegovina

Table 13. Recapitulation of old non-ecological freons R-22 and R-12 found in cooling and air-conditioning units by entities in Bosnia and Herzegovina

Name of refrigeration device	The amount of freon in the cooling device in Republic of Srpska, [kg]	The amount of freon in the cooling device in Federacija, [kg]	The amount of freon in the cooling device in Bosnia and Herzegovina, [kg]
Fridges	37,500 kg	70,000 kg	107,500 kg
Freezer	56,250 kg	105,000 kg	161,250 kg
Air-conditioners	50,000 kg	125,000 kg	175,000 kg
Refrigerated display	30,000 kg	75,000 kg	105,000 kg
Refrigerating cabinets	30,000 kg	75,000 kg	105,000 kg
Sanks	4,500 kg	11,250 kg	15,750 kg
Refrigerating chambers	7,500 kg	15,000 kg	22,500 kg
Ships	1,200 kg	2,400 kg	3,600 kg
Boats	180kg	300 kg	480 kg
Trains	1,500 kg	2,250kg	3,750 kg
Buses	4,500kg	11,250 kg	15,750 kg
Refrigerated trucks	3,000 kg	6,000 kg	9,000 kg
Truck Trailers	2,100 kg	4,200 kg	6,300 kg
Total	228,23 kg	502,650 kg	730,880 kg

Since we do not have a real population census, and with the number of 4,300 inhabitants in Bosnia and Herzegovina (Republic of Srpska 1,500 and Federation 2,800), one assessment of household-refrigerated appliances, air conditioners, which still use old freons can be performed.

6. Conclusion

Recapitulation of old non-elastic freon R-22 and R-12 which are found in cooling and air-conditioning devices according to entities and Bosnia and Herzegovina. How much and which cooling and air

IOP Conf. Series: Materials Science and Engineering **477** (2019) 012056 doi:10.1088/1757-899X/477/1/012056

conditioners use the old non-ecological freons (NEF) R-22 and R-12 is very difficult to estimate, and especially precisely, because there is no relevant institution in the Federation as well as in the Republic of Srpska, or Bosnia and Herzegovina, this issue. From long-time workers in this field, who are present in the field and who are continuously in practice, certain personal data have been obtained from their personal experience, which are not necessarily adequate to the actual situation. Due to war events in this region, and also after the war, this problem has not been seriously addressed, ie (non) use of non-ecological freons that destroy the ozone layer.

Europe, the United States and Australia were the first to deal with this problem, because highly developed and polluted the atmosphere the most, creating ozone holes, which forced them to ban in their territory the use of inadequate freons. They quickly switched to the use of ecological freons R-134a; R-404a; R-410; R-407C; R-507 to try to save the planet and the ozone layer in the atmosphere. Since 2000., it has been prohibited to use non-ecological freons in the aforementioned territories, but in the former Yugoslavia it has not been implemented in life-practice.

Newly emerged countries Croatia and Slovenia joined very quickly in Europe and prohibited the use of old non-ecological freons through their Laws on Ecology, while Bosnia and Herzegovina, Serbia, Macedonia and Montenegro still use the freons R-22 and R-12 much heavily.The situation in these countries was impossible to prohibit the use of cooling and air conditioners on old-age freons overnight, as almost 70% of the cooling and air-conditioning systems are still used. Once a free estimate, which is perhaps the most relevant at this time, is to take some sort of average at the level of the Republic of Srpska, the Federation, or Bosnia and Herzegovina.

References

[1] ***Code of good practice in dealing with substances that damage the ozone layer, Ministry of Environmental Protection, Physical Planning and Construction, Croatia, APO d.o.o. environmental protection services, SEI - Stockholm Environment Institute Faculty of Mechanical Engineering and Naval Architecture

[2] Trott A R and Welch T 2010 *Refrigeration and Air-Conditioning: Third edition*, Oxford Auckland Boston Johannesburg Melbourne New Delhi

[3] ***2016 *Preparation of a National Survey on Alternatives to ODSs*, UNIDO 2

IOP Conf. Series: Materials Science and Engineering **477** (2019) 012057 doi:10.1088/1757-899X/477/1/012057

Mitigation of climate change in terms of the use of renewable energy sources in Bosnia and Herzegovina and mitigation scenarios defined

G Tica

University of Banja Luka, Faculty of Mechanical Engineering, Department of Thermal Engineering, Bosnia and Herzegovina

E-mail: gordana.tica@mf.unibl.org

Abstract. Many projects that were done in Bosnia and Herzegovina are evident and they aimed at defining mitigation scenarios with measures that will lead to CO2 reduction by 2025, and by 2040 already defined. Those are as follows: Scenario I (business as usual) In this scenario, it is defined that GHG emissions are growing in proportion to the increase in production / energy consumption (with the implementation of energy efficiency measures). This is the most destructive environmental scenario, where there is no inclusion of renewable energy sources (biogas, geothermal energy, solar energy) Next is Scenario II (reference scenario) This scenario implies the implementation of projects from entity strategies for the development of the electricity sector, with the increase of energy efficiency.And next Scenario III (advanced scenario). This scenario implies the reduction of energy consumption, with the emphasis on the following: increasing the efficiency of (power) plants, next intensive exploitation of the potential of renewable energy sources, and intensive using renewable energy sources; all RES have been taken into account, especially the use of solar energy.

1. Renewable energy sources (RES) in Bosnia and Herzegovina

In paper is showed development of scenarios, where is included overview of existing legislations, acttivities and data. Overview of basic events and situation in the sector, including legislation, projects, etc, which has direct links or can be associated with "emission reduction".

In paper is showed active legislation, adoption of laws and by-laws in the framework of the mentioned Law Bosnia and Herzegovina:

- Energy Sector Study for Bosnia and Herzegovina, March 2008 Republic of Srpska Energy Development Plan Republic of Srpska (2012.) until 2030,
- Law on Spatial Planning and Construction (Official Gazette of Republic of Srpska, No. 40/13);
- Law on EE (Official Gazette of Republic of Srpska, No. 59/13),
- The Law on RES and Efficient Cogeneration (Official Gazette of Republic of Srpska, No. 39/13) Federation of Bosnia and Herzegovina,
- Law on Spatial Planning and Land Use at the Federation Bosnia and Herzegovina level (Official Gazette Federation Bosnia and Herzegovina, No. 2 / 06,72 / 07,32 / 08,4 / 10,13 / 10,45 / 10),
- Energy efficiency Act (draft, May 2012),
- Law on the use of renewable energy sources (RES) and efficient cogeneration (Official Gazette of Federation Bosnia and Herzegovina No. 70/13).

Projects that involve the use RES for Heat Production: energy from the potential of biogas, solar energy, geothermal energy. Projects that were in the jurisdiction of the city and the mayor's competence: SEAPs, DELTER projects, USAID PROJECTS 3E, Projects of the Department of Environmental Protection Fund.

1.1. Scenarios related to the use of biogas (BG) in Bosnia and Herzegovina

The S1 scenario is a scenario without taking migration measures, which means that no increase in the use of energy from modern forms of biomass such as biogas (BG) is expected, as energy prices from these sources are still non-competitive in relation to technologies using conventional energy sources. This scenario does not imply the introduction of any changes in the existing trends in the rate of increase in the number of animals and the input of nitrogen fertilizers on arable land. A significant feature of this scenario is the relatively low level of interest and activities of state and entity institutions in this energy subsector.

The S2 scenario, given the very small existing biogas share, is based on activities that focus on implementing the system for its collection and combustion in farms in Bosnia and Herzegovina. The most important features of this scenario are: gradual introduction of new technologies (orientation to the eyes, increased use of RES and biogas), planning of production and energy consumption on farms to meet the needs for space heating, drying of hay, grain, vegetables etc., production of biogas in simple plants and the use of biogas to cover a significant proportion of household energy needs even in a small number of livestock (use of mini-plants).

The S3 scenario is based on a high level of activity to mitigate climate change that is being implemented at different levels of government - from state to entity. More intensive use of biogas for the production of heat and electricity is expected, which will prove to be very profitable thanks to the improvement of the equipment used for this purpose. Biogas from agriculture (livestock breeding) is a significant source of energy in the scenario with measures S2 and S3. These are the cogenerations that are supposed to be efficient locating (electricity generation and heat usage). All electricity generation is transferred to the network, so that its share cannot be divided into categories as a whole. The heat from agricultural plants is given to district heating of rural areas. The total installed capacity of the cogeneration plant for biogas from agriculture (cattle breeding) in the Republic of Srpska is defining two times larger forces for five years. By analogy, the plans have been defined for the Federation of Bosnia and Herzegovina, ie in the following Table 1, the total values at the level of the entire Bosnia and Herzegovina are presented.

Bosnia and Herzegovina, ie in the following Table 1, the total values at the level of the entire Bosnia and Herzegovina are presented.

Table 1. Scenario S3. Total installed power in agriculture of cogeneration plants on biogas (BG) in Bosnia and Herzegovina, and evaluation of events by 2050, [1-3]

Bosnia and Herzegovina	BIOGAS PLANTS	2015	2020	2025	2030	2035	2040	2045	2050
Intalled power	[MW]	2,5	5	7,5	10	12,5	17,5	20,5	23
	GWh/year	5,7	11,54	17,31	23,00	28,8	34,57	41,5	47,5
Production	$GgCO_2$/year	6,98	13,97	20,95	27,84	34,86	41,85	48,87	55,85
Of	PJ/year.	0,02	0,04	0,06	0,08	1,0	1,02	1,05	1,07
electricity	$GgCO_2$/year	0,024	0,05	0,07	0,096	1,21	1,23	1,25	1,27
	GWh/year	11,17	22,34	33,51	44,68	55,85	67,02	77,19	89,36
Production	$GgCO_2$/year	13,52	27,04	40,56	54,09	67,61	81,13	94,64	108,1

heat	PJ/year	0,04	0,08	0,12	0,16	0,2	0,2	0,24	0,28
energy	$GgCO_2$/year	0,048	0,096	0,145	0,113	0,242	0,242	0,288	0,336

In line with scenario 3, we can see Table 1 with installed power in agriculture of cogeneration plants on biogas in Bosnia and Herzegovina, and evaluation events by 2050.

In next figure, we can see total installed capacity in agriculture of cogeneration plant on biogas in Bosnia and Herzegovina and total emissions of $GgCO_2$ / year, and assessment of events until 2050.

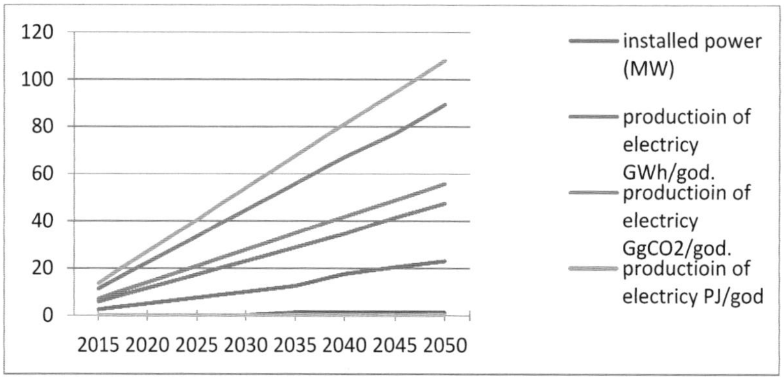

Figure 1. Total installed capacity in agriculture of cogeneration plant on biogas in Bosnia and Herzegovina and total emissions of $GgCO_2$ / year, and assessment of events until 2050

1.2. Measures to encourage increased use of biogas plants in Bosnia and Herzegovina

Investing in agricultural production is a prerequisite for encouraging the use of biogas. It is generally expected that through the policy of rural development, the application of such measures will result in energy savings in heating wood, electricity, heat energy supplied by district heating and, in particular, fossil fuel energy. On the other hand, the use of biomass (and biogas) in cogeneration production of heat and electricity is increased. If biomass (including biogas) is predicted as a fuel in the industrial sector, improvements through energy measures are manifested through: reducing the intensity of electricity consumption and useful heat, increasing the efficiency of heat generation technologies, increasing cogeneration in the production of heat and electricity.

Energy savings as a result of the application of measures in industry are shown in Table 2. Since the circumstances in the incentives are negligible, and consequently the production of electricity and heat from biogas, a sharper shift and significant energy savings in industry are expected from the year 2020. Whatever the developed scenarios. For households, for now, there are individual uses on several farms. However, these are too small installations, low power and impact on savings, or almost insignificant when it comes to describing the degree of savings.

According to the scenario with measures - S3, the final energy consumption in the industry would be 9% lower than in S2 by 2025. in the same period, electricity consumption would decrease by about 7%. Energy savings in industry [PJ] using biomass (with biogas) should be estimated to be at least twice as high.

Table 2. Energy savings [PJ] in the industry using biomass (with biogas) for cogeneration for the defined values of the emission reduction of $GgCO_2$

Savings	2010	2015	2020	2025	2030	2035	2040	2045	2050
Energy savings	0.049	0.608	1.114	1.216	1.26	1.31	1.364	1.413	1.46

[PJ]									
Energy savings in industry [$GgCO_2$]	176.4	2188.8	4010.4	4377.6	4536	4734	4910.4	5086.8	5263

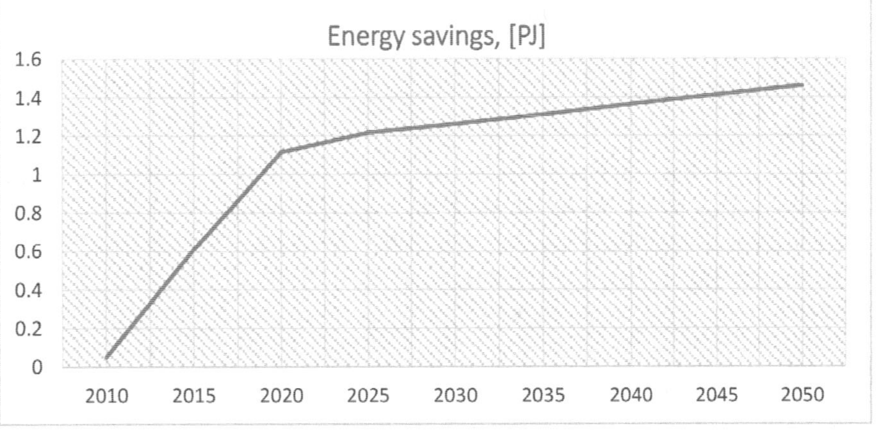

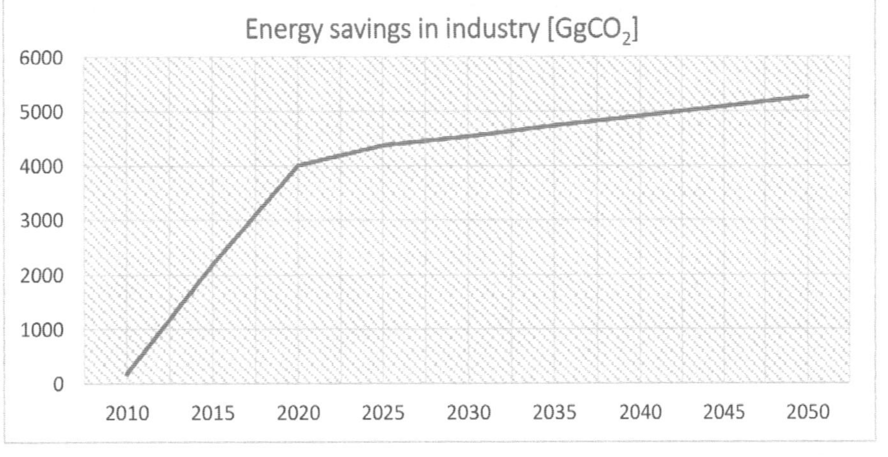

Figure 2. Energy savings according to scenario S3 [PJ] in the industry using biomass (with biogas - BG) for cogeneration by 2050, and saving emissions [$GgCO_2$] by using biogas

2. Sun energy (SE)

Next RES which was conserved was sun energy. It is showed current situation and changes up to 2050. The results of the research on the possibility of using solar energy for the production of heat – using solar collectors for 15 cities in Bosnia and Herzegovina, as well as for the production of electricity, shows justification based on already initiated initiatives, but unfortunately poor statistics and archiving in these cities. From all the collected data and conducted analyzes, it can be concluded that there is a great potential for the application of solar energy in the territory of Bosnia and Herzegovina, which amounts to 70,5 [mil.Gwh] of the total energy of total solar radiation per year. The technical potential is 685 [PJ], which is about three times more than the total primary energy needs in the energy balance of Bosnia and Herzegovina.

According to the results of the calculation of the possible degree of covering the heat demand for the preparation of hot water for the average household, about 74% of the thermal needs for the preparation of hot water in the Republic of Srpska and 78% in the Federation Bosnia and Herzegovina can be covered from solar collectors. The degree of coverage of heating needs depends on the thermal insulation of the building, but on average it is about 30%. Estimates are that solar energy could settle about 5% of energy needs in B&H. In summer, 80% of hot water needs could be provided, and in the winter between 35 and 50%. Estimates are that in Bosnia and Herzegovina there were about 7,000 square meters of installed collectors, with an annual increase of around 28%. It is estimated that the construction and use of solar collectors, in incentive and co-financing households, as well as public facilities will be proportionally increased.

If we talk about public facilities, these are, above all, hospitals, retirement homes, and sports facilities if they are used throughout the year. In private facilities, these are family houses that have at least five members of the household, whether existing or new buildings. In multi-story buildings, the application of this technology is cost-effective only in the case of a centralized system for the preparation of consumable hot water. If there is an individual preparation, subsequent work and installation in each apartment will greatly increase this technology, which leads to its inefficiency.

High interest and increased use of solar collectors in all sectors can be noticed. A lot of projects have been launched, there is especially significant public sector activity (eg school roofs, hospitals, etc.), where electricity is produced and part of the energy needed to cover heat needs. Activity and growth in all three sectors should be emphasized.

2.1. Scenarios related to the use of solar energy (SE) in Bosnia and Herzegovina

S1 scenario does not imply the introduction of significant changes in relation to the current trend of solar energy use, that is, the state according to which solar energy will not be used in a higher degree. S2 script and S3 script. In Bosnia and Herzegovina, an initiative for the use and production of equipment has begun, and consequently a more intensive and active analysis of the cost-effectiveness, sustainability and energy efficiency of solar systems. Prices are based on those from imports, from Western countries and the Far East. If we look at the standard system of 4 m², the price of a complete system with installation is estimated at around 3,500 to 4,000 euros, it is expected that the installed surface of solar collectors until 2040. it grows up to 100,000 m², which would be about 25 m² per 1,000 inhabitants. If co-financing measures are applied in view of the existing potentials, we can expect coverage of around 200,000 m² by 2025, i.e. about 42,000 households, which is about 11% of the total number of households, and proportionally to 2040 growth. The application of this technology, considering its cost-effectiveness, is most desirable in buildings that are used throughout the year in all aspects of their stay, ie, using consumable hot water. Developed scenarios S2 and S3, by 2040, with a five-year display threshold in a diagram, followed by a commentary on the results obtained.

Table 3. Energy savings in major areas of consumption, using solar energy, for 3 sectors, [PJ] and saving emissions [GgCO_2] for defined savings values) displayed on a diagram, followed by a commentary on the obtained results

No	Saving energy	2010	2015	2020	2025	2030	2035	2040	2045	2050
1	SE – in industy [PJ]	0.00	0.011	0.029	0.047	0.065	0.083	0.101	0.119	0.137
	[Gg CO_2]	0	39.6	104.4	169.2	234	238.8	363.6	488.4	613.2
2	SE - in households	0.00	0.009	0.056	0.103	0.15	0.197	0.244	0.291	0.338
	[Gg CO_2]	0	39.4	201.6	370.8	540	709.2	828.4	947.6	1066.8
3	SE - in	0.001	0.002	0.015	0.028	0.041	0.054	0.067	0.080	0.093

services [Gg CO_2]	3.6	7.2	54	100.8	147.6	194.4	241.2	288.0	336.8

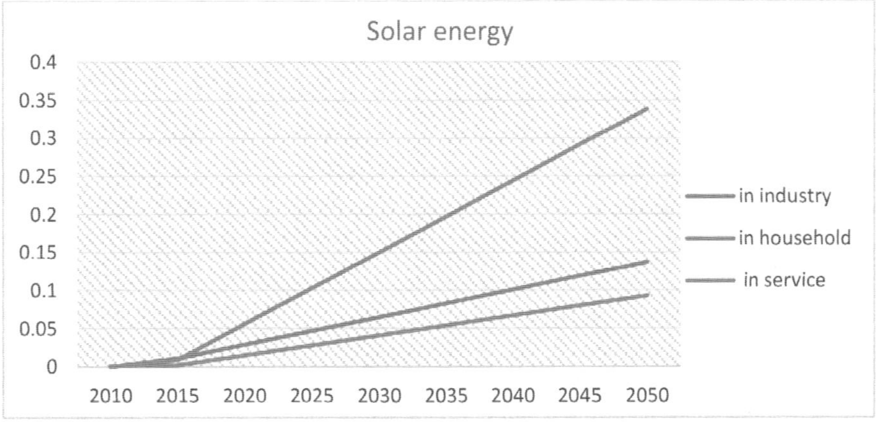

Figure 3a. Diagram showing energy savings, using solar energy (SE) for S3 consumption sectors [PJ]

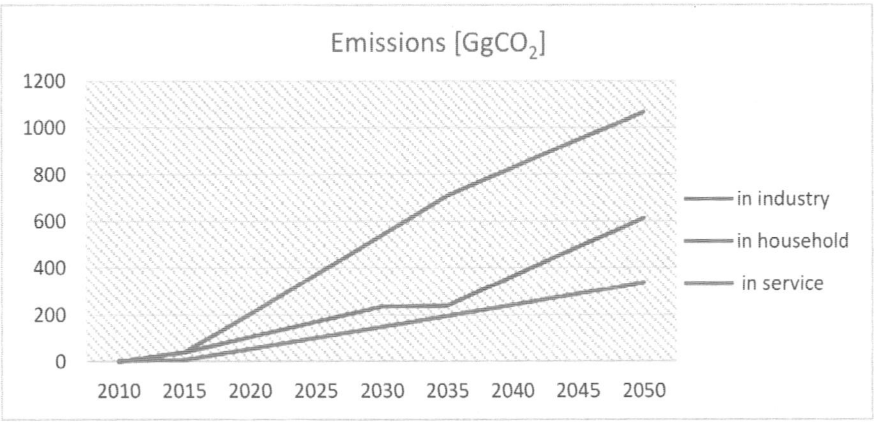

Figure 3b. Diagram showing emission savings, using [GgCO_2]

Compared to the baseline scenario, the use of solar energy for heat needs by 2050 is also increased: in industry by 10%, in households by about 25%, and in the service sector in the amount of 6,7% of useful needs.

2.2. Measures to encourage increased use of solar energy in Bosnia and Herzegovina

The initial step in the measures to encourage the use of the SE, as all other forms of RES, is to give impetus to the launch of SE production. In Bosnia, for now, there are initial - real possibilities i.e. capacity for production. A greater possibility is expected to be encouraged by applying a number of different measures. When it comes to the use of SE solar cells for electricity production in Bosnia and Herzegovina, this has just started to be analyzed and applied, but without significant figures, percentages that could be said in terms of already implemented savings. Solar collectors are analyzed for central heating systems for fuel oil, electricity, LPG (liquefied petroleum gas) - tank and natural gas, where the collector replaces the classic electric boiler for Pvt.

Due to the investment in solar collectors, the total annual costs of the observed system are increased, with the largest increase in the annual cost (in houses with central heating in the cold zone) a combination of boilers for central heating and Ptv on electricity and collectors (5,1%) , followed by natural gas and collector (5,0%), LPG tank and collector (3,0%) and fuel oil and collector (2,9%). The highest savings were obtained in the combination of boilers for central heating and Ptv on electricity and collectors (5,4%). More intensive cell construction can be expected by the end of 2040, the installed power of photovoltaic systems per capita in Bosnia and Herzegovina, i.e. a total of 60 MW installed system power. Technological development of solar cells is moving in two directions - the development of solar cells with high efficiency of transformation of solar radiation in electricity and development so-called *Low-cost solar cells*. The application of these measures results in energy savings of heating wood, electricity, heat supplied by district heating, and fossil fuels. On the other hand, the use of solar energy is increased (Table 27). According to the scenario with measures - S3, the final energy consumption in industry would be 9% lower than in the base scenario by 2025 - S1. The use of solar energy (SE) in households by 2025 should increase by five times. In B&H, there is a sensible planned project activity, with different types of domestic and foreign financing, in all three sectors.

3. Geothermal energy (GE)

The third RES considered was geothermal energy. The geothermal resource of Bosnia and Herzegovina is of a triple shape: hydrothermal systems, geographical areas and warm dry rocks. These areas cover mainly the central and northern part of Bosnia and Herzegovina. Of the three types of resources mentioned above, the most attention is drawn to hydrothermal systems, because their exploitation is the most developed and cheapest in comparison to the other two forms. By compiling the potentials of Republic of Srpska and Federation Bosnia and Herzegovina, the total thermal power and energy of geothermal phenomena in Bosnia and Herzegovina were calculated. The total possible installed capacity of geothermal sources at 42 locations is 9,25 MWt, if only the possibility of space heating, i.e. 90.2 MWt is considered, if geothermal energy is observed for space heating and recreational and balneological needs. With the use of all mentioned sources with a 0,5-point recovery factor, it is possible to produce 145,75 TJ energy in one year only for space heating, i.e. a total of 1.42175 TJ energy if one looks at the space heating and bathing together. The conducted research shows that a large part of the Republic of Srpska is perspective in terms of the presence of geothermal waters, mostly in the area of Posavina, Semberija, Banja Luka basin and Lijevče polje. The energy potential is estimated at 1260 RJ. The greatest A step forward was made in relation to the planned, in spite of the fact that the concessional policies began to be realized. Concession realizations are happening intensively in the territory of Banja Luka, Sarajevo, Bijeljina and Doboj, as well as plans related to the implementation of deep wells in order to heat the gardens. According to the previous research it was found that about 25% of the territory of Bosnia and Herzegovina is considered a potential geothermal resource.

Description of the scenario. Each scenario in one paragraph (passage) is described descriptively, with the main features that make and characterize it. Scenarios related to the use of geothermal energy in Bosnia and Herzegovina for the use of this energy source is in aquaculture, agro-culture, and heating of settlements.

3.1. Scenarios related to the use of geothermal energy in Bosnia and Herzegovina for the use of this energy source is in aquaculture, agro-culture, and heating of settlements

The scenarios related to the use of geothermal energy are mainly based on the estimated reserves as well as the technological possibilities for its exploitation. The S1 scenario is based on existing trends in the potential use of geothermal energy without any special additional exploration of the potential and without changing the attitude towards this energy source. The highest increase in the consumption of geothermal energy is assumed in the S1 scenario, in the period 2015-2025, amounting to about 2,5%.

The S2 and S3 scenarios are based on the introduction of a support model, where the main activities are focused on the implementation of hydrothermal systems on the whole Bosnia and Herzegovina level. In both scenarios, final energy consumption indicators are given along with the average annual rates of increase or decrease in five-year periods and the shares of the form of energy in final consumption. There is no geothermal energy as a participant in the structure of electricity generators in S2 and S3, although there are conditions for this, that is, at some locations, concessionaires - concession interest and already realization are happening intensively in the territory of Banja Luka, Bijeljina and Doboj , and realization of plans for realization of deep wells in order to heat the gardens is realized.

The S3 mitigation scenario with the implemented measures is very important considering the envisaged EU accession program for the EU between 2015 and 2020 or the takeover of GHG emissions reduction commitments. According to S2 and S3, a significant representation of the use of geothermal resources by heat pumps in the household sector is assumed. Developed scenarios, by 2050, with a five-year threshold of presentation in the diagram, accompanied by a commentary on the obtained results.

Note: The energy production from Renewable energy source for $GgCO_2$ emissions is made in relation to electricity consumption, considered to be the only correct aspect, since it is not treated in heating plants, i.e. remote heating.

The electricity eligibility factor for 2014 was taken: 0,744 0,826 (Source: Production - Consumption: for 2011: 0,850 t CO_2 / MWh 0,944 for 2012: 0,730 t CO_2 / MWh 0,811 for 2013: 0,720 t CO_2 / MWh 0,800).

Table 4. Consumption of geothermal energy in Bosnia and Herzegovina according to scenarios S1, S2 and S3, and $GgCO_2$ emissions for defined scenarios

Geothermal energy. Consumption [PJ] energy	2010	2015	2020	2025	2030	2035	2040	2045	2050
S1 scenarios	0.0	0.05	0.05	0.09	0.095	0.1	0.15	0.2	0.25
Emissions [$GgCO_2$]	0	180	180	324	342	360	540	720	860
S1 scenarios	0.0	0.04	0.06	0.06	0.07	0.08	0.09	0.1	0.11
Emissions [$GgCO_2$] 4	0	0.04	216.6	216.6	252.7	288.8	324.9	361	408
S1 scenarios	0.0	0.04	0.06	0.07	0.07	0.08	0.09	0.1	0.11
Emissions [$GgCO_2$]	0	144.4	216.6	216.6	252.7	288.8	324.9	361	408

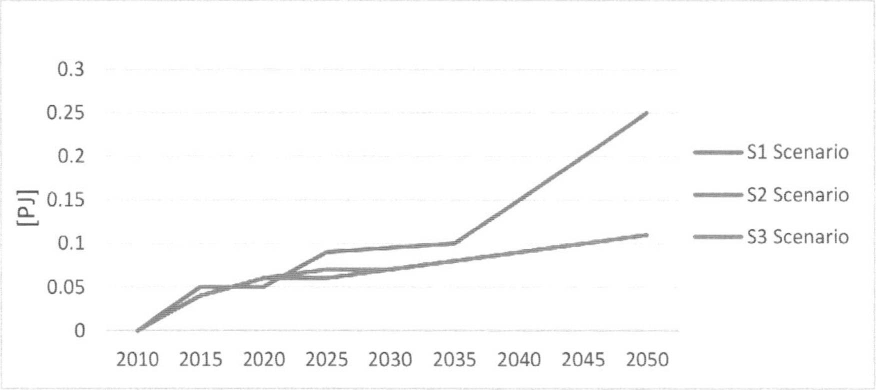

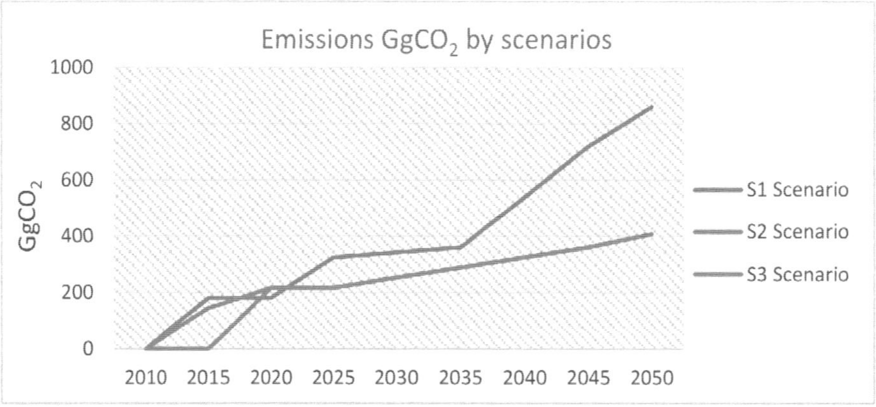

Figure 4. Diagram showing consumption of geothermal energy in Bosnia and Herzegovina ($GgCO_2$ / year). and emissions $GgCO_2$

4. Conclusion

Biogas. Based on available livestock data for 2010 and 2011, biogas production potential was estimated from 800,000 to 850,000 m^3 / day. So far, only one biogas plant in Bosnia-Herzegovina has been constructed (designed and constructed) in the territory of Srbac municipality. The second biogas plant is in the phase of completion and experimental testing at Donji Žabari near Brčko. The installed electric power of the first plant mentioned is 35 kW and a heat of 70 kW. In households, there are currently single use at several farms. However, these are too small installations, low power and little impact on savings, or almost insignificant when it comes to the degree of savings.

Sun energy. The results of exploring the possibility of using solar energy for heat generation using solar collectors for 15 cities in Bosnia and Herzegovina, as well as for electricity generation, show justifiability on the basis of already initiated initiatives. Estimates are that there are about 7,000 m^2 of installed collectors in Bosnia and Herzegovina, with an annual increase of around 28%. It is possible to notice the great interest and increase in the use of solar collectors in all sectors. A large number of projects have been launched, and particularly those in the public sector (e.g. solar roofs of schools, hospitals, etc.) where electricity is produced and part of the energy is used to cover the heat needs. The estimate is that proportionally with incentive and co-financing will increase the construction and use of solar collectors both in households and in public facilities.

Geotermal energy. The geothermal resource of Bosnia and Herzegovina is of a three-dimensional form: hydrothermal systems, geoprocessed zones and warm dry rocks. These areas mainly cover the central and northern part of Bosnia and Herzegovina. Of these three forms of resource, the greatest attention is drawn to hydrothermal systems, because their exploitation is the most developed and least

IOP Conf. Series: Materials Science and Engineering **477** (2019) 012057 doi:10.1088/1757-899X/477/1/012057

expensive in relation to the other two forms. By aggregating the potential of the Republic of Srpska and the Federation Bosnia and Herzegovina, the total heat power and geothermal energy of Bosnia and Herzegovina were calculated. The total installed capacity of geothermal sources at 42 sites is 9,25 MWt if only the heating capacity is observed, i.e. 90,2 MWt if geothermal energy is observed for space heating and recreational and balneological needs. With the use of all the mentioned sources with the utilization factor of 0,5 it is possible that in one year 145.75 TJ of energy will be generated only for space heating, ie a total of 1.42175 TJ of energy if space heating and bathing are contemplated together. The conducted research shows that a large part of the RS is prospective in terms of the presence of geothermal water, mostly in the area of Posavina, Semberija, Banja Luka valley and Lijevče fields. The energy potential was estimated at 1260 TJ. The greatest potential for using this energy source is in aquaculture, agro culture and heating of settlements. According to the previous research, it is established that about 25% of the territory of Bosnia and Herzegovina is considered as a unavailable. Still with a small share, but with a modest expansion trend, heat pump systems are applied to small and medium objects. A breakthrough has been made since concession policies have begun to be realized. Concessional realization takes place intensively in the areas of Banja Luka, Sarajevo, Bijeljina and Doboj, and plans are being made to realize the deep wells in order to heat the cities.

References

[1] ***2009 *The First National Report of Bosnia and Herzegovina in accordance with the United Nations Framework Convention on Climate Change*, Banja Luka
[2] ***2013 *Second National Report of Bosnia and Herzegovina in accordance with the United Nations Framework Convention on Climate Change*, Banja Luka
[3] ***2013 *Strategy for adaptation to climate change and low-emission development Federation Bosnia and Herzegovina*

IOP Conf. Series: Materials Science and Engineering **477** (2019) 012058 doi:10.1088/1757-899X/477/1/012058

Development of application software without programming

J Hamzabegovic

University of Bihac, Faculty of Technical Engineering Bihac, Irfana Ljubijankica str., 77000 Bihac, Bosnia and Herzegovina

E-mail: jasna.hamzabegovic@unbi.ba

Abstract. End-user programming is an increasingly popular discipline of software engineering. In order to design a development environment that allows programming by non-programmers, it is desirable to understand their mental model. Most languages designed for end-users are based on visual interaction techniques such as programming with graphically presented rules and agents. Visual techniques allow direct manipulation in the environment in such a way that users grasp and pull visual components to match the application, as well as to select specific actions on certain events. Due to its omnipresence in operating systems, desktop and mobile applications, direct manipulation is an interaction technique which non-programmers are familiar with, and therefore highly desirable in application development environments by end-users or non-programmers.

In this paper we present a review of the features and the functionality of the development environment that would allow the end-users logopedists to develop their therapeutic software without writing programming code and without any programming knowledge.

1. Introduction

In the last decade, the interaction between humans and computers has evolved from a system of easy-to-use to a simple development system. For many years, the development of new ones and the modification of existing application programs have been the privilege of educated programmers. However, today professional developers can not meet the needs of all end-users due to the shortage of this profession in the global market.

There were 1.3 million software jobs open last year, according to Trilogy Education Services. Code.org calculates 512,720 open computing jobs now, and the U.S. Bureau of Labor Statistics predicts that in 2020 there will be 1.4 million more software development jobs than applicants who can fill them [1].

2. End-user programming

In order to solve the problem of software developer deficiency, software development by the end-user (EUD[1]) is increasingly encouraged. The EUD is a "set of methods, techniques and tools that enable program system users, acting as unprofessional developers, to create, modify or extend application software for their own needs" [2]. In particular, the EUD allows end-users to design or customize the user interface and functionality of their own software. This is very important because end-users know

[1] End-User Development

IOP Conf. Series: Materials Science and Engineering **477** (2019) 012058 doi:10.1088/1757-899X/477/1/012058

their own context and needs better than anyone else and have timely information about changes and new developments in their area. Through EUD end users can set up software to suit their requirements better than would be possible without EUD.

End-users do not undergo training in programming languages, development processes, or modeling, and diagnostic notation passed by professionals. They do not have time or motivation to learn traditional techniques, because end users usually want to write code to achieve an ad-hoc short-term goal - to ease the solution of the current problem. Accordingly, the EUD supports and responds to requests by offering appropriate tools, structures and development processes that are useful, and quickly learned and easily integrated into existing practice.

The EUD overlaps two similar concepts: end-user programming (EUP) and end-user software engineering (EUSE). End-user programming allows end users to create their own programs [3]. The difference between the EUSE and the EUP is that EUSE methods, techniques and tools provide the entire life cycle of software development, including modifications and extensions of the software, and not just the "create phase". End-user software engineering is a relatively new sub-domain of the EUD, which is being developed to reduce errors that end users make in developing their own software. As the focus on quality creation, quality modification and expansion of software created by the end user, the subject of software engineering research are methods, techniques and tools that enable this quality.

3. Psychology of programming

Programming is a demanding mental action that involves designing detailed instructions for the computer to solve a problem. Therefore, understanding the intellectual background as a human skill in the field of computer programming is of great importance. This knowledge is of great benefit in designing both programming languages, as well as human-computer interfaces, and other user-friendly software systems.

Cognitive science is defined as a scientific discipline dealing with the study of human mind and intelligence [4]. It is interdisciplinary because it includes: psychology, psychiatry, linguistics, anthropology, computing and biology. The name "cognitive science" was first used by Christopher Longuet-Higgins in his commentary from 1973 [5], and in what is said in his report by Lighthill [6], who was researching the current state of artificial intelligence. The term "cognitive" in the title is "used for any type of mental operation or structure that can be studied in precise terms" [7].

Cognitive science deals with the research of the nature of mental processes such as: perception, memory, attention, locking, language, motor control and problem solving. Its goal is to understand the concepts and processes that take place in the human mind: how they form, how they evolve and how they spend, or understand how the human mind works. It is precisely this knowledge from cognitive science that is very important in computing in the construction of programming languages. The cognitive science branch dealing with psychological aspects of programming is called psychology of programming. The psychology of programming (PoP) is an interdisciplinary field that includes research on the cognitive activities of the programmer (perception, recognition, reasoning and conclusion), about tools and methods for programming, and about teaching programming skills.

It is desirable to achieve programming performance such that the created program meets its specification, it is adaptable to the future and works effectively. If the psychological aspects of computer programming are understood, one may conclude how to achieve greater programming efficiency.

Cognitive psychologists and computer experts are concerned with the study of psychology of programming. The main motivation for computer experts is to improve existing tools and develop new ones, as well as the discovery of general principles that apply to people in the context of program tasks. Psychologists, however, are interested in new theories of human knowledge applied in other areas. For them programming, as a very complex job, provides good opportunities for studying cognitive processes at a high level in a complex environment. It is precisely the dual character of programming psychology that requires a good understanding of psychology, social sciences and program engineering from the PoP researchers.

Methods of research in the field of PoP are different. The most common methods of research in the field of cognitive psychology are controlled experiments performed in laboratories. In the field of social sciences, this is field research using the technique of qualitative analysis. In the field of program engineering, the problems of beginners in programming, as well as the problems in programming teaching are studied. A new source of research materials are various so-called. open source communities that share program code among themselves, and blogs and forums where developers participate, which are publicly available on the Internet.

4. Mental model

A mental model is the thought process of a person about how he / she works, ie. understanding the surrounding world. The mental model is based on an incompletely established factual state, experience, but also on intuitive knowledge. It helps shape actions and behaviors, influence what people need to pay attention to in complicated situations, and define how people approach problems and how to solve them.

4.1. Mental model of the end-user programmer

For "end users developers", better development tools will be built if they know how they think. If the tool works in the way the user expects and if it is felt by the user, such a tool will be well accepted because it can be easily learned and easily used. Likewise, the tool can be designed to redefine the way the end-user develops the problem. In both cases it is important to know the mental model of the user.

The human mental model is shaped by his education and experience, and he develops as he learns further. The term "natural" or "naturalness" applies to software development technology that is referred to the mental model that the user possesses before learning the tool or programming language. It is very important to minimize the "mental distance" between the user's objectives and the language specification [8], [9].

4.2. Mental model of speech therapist

In the context of this work under the mental model is meant the way logos visualize therapeutic applications, and mental representations of entities and connections that make up the entire system. In order to understand the mental model of speech therapists, the results of two studies were used.

The first study as a starting point uses methods from Pane, Ratanamahatan and Myers (2001), who dealt with the question of naturalness in the context of a programming language for children, studying how children and adults use the natural language to solve program problems [10]. They used the results of these studies to design a programming environment that offers concepts close to the natural mental model of the end-user, the future developer. Following this model, a second study looked at the behavior of the target population of the speech logic in order to identify the control logic and specify the elements and actions they apply in the natural language in everyday practice, and then designed a new language and environment based on these abstractions. After this general approach, it was explored that logos describe the behavior of a future development environment that should enable them to create interactive and multimedia applications, and how these applications should look, i.e. what should they contain and what to enable. Answers to these questions were given by logos as the best connoisseurs of the mental model of children with dyslexia for whom applications would develop. The results of all these studies have been applied to the modeling and design of the desired development environment.

4.3. Researching the language and concepts of speech therapists, future programmers

The first efforts in researching the mental end-user model were focused on researching the language, concepts and ways in which logos create therapeutic exercises for children with dyslexia. They wanted to find out how they would specify and implement therapeutic applications, and what development techniques would be most natural for them. There was little concern about their lack of language skills for specifying the elements of the user interface (such as terms: toolbox, control, textbox, grid, checkbox, listbox), as well as language and diagrams for specifying the behavior of the application. But these problems are through direct communication with them and a detailed explanation is successfully solved. There was also an interest in how users would describe the properties of created objects, their behavior, and actions as a result of certain events. For the needs of the research, a simple Windows application for the matching of elements, ie, designing a future therapeutic application.

Participants received a general introduction to the research goals and were then asked to review and designate all the elements of the screen editor design (window elements and toolbars) that can be used to create workout exercises for children with dyslexia. They were presented with a way of adding objects and staging the scene. Labeling by logopeds meant the encircling of interface elements that would also use the precipitation of those elements they would not need. They were also suggested to be able to add new elements that were not foreseen by the existing proposal. This first phase of the research was carried out separately with each logoped in order to explain the functions of the offered tools.

The aim of this research was to obtain relevant information about elements of a future visual language that the target population will use to create therapeutic software interfaces. The results showed the great need for letters and words to be used with sound and / or image due to the visual perception inherent to dyslexics, which, in contrast to the rest of the population, think in the pictures. Participants also insisted that visual components are directly manipulated in a way that they capture and pull visual components when they interface the user interface of the application, and that they can select specific actions on certain events. Because of its omnipresence in mobile applications, direct manipulation is an interaction technique with which logos are introduced and therefore highly desirable in the development environment for the development of applications by the logopedic-programmer.

After this phase, during which they proposed elements that they consider natural and intuitive, logopeds were asked to show how the program should work. They received an empty sheet of paper where they needed to create a simple exercise from therapeutic practice, using elements from the design editor of the proposed development environment. They should have defined the behavior of the application in different possible situations and based on different requests. We wanted to see what end-users consider as a specification of behavior. They wrote down their requests and remarks on a white sheet of paper. Describing the behavior of the application by the participants helped to study and understand their natural mental model.

When describing the behavior of the application, participants are inclined to combine procedural steps and declarative statements. They use declarative statements to determine constraints on behavior (eg, "all fields must be filled in"). Procedural statements often represent a test and result (for example, "if a devoted letter is not what is expected, this field will leave blank"). Of course, none of the participants in the behavior specifications mentions conventional program constructions such as: variable, conditional statement or loop, because they do not know the terminology and concepts. Where loops are needed (for example, when filling in the field, that is, all the letters of one word or all the words of one sentence), participants indicate an iteration, thinking that it is expected to be applied, i.e. repeat as much as needed. Of course, none of the participants were able to describe what is happening behind the app.

5. Visual programming

In computer science, the visual programming language (hereinafter referred to as "VPL") allows users to create programs by manipulating more graphic elements of the program than text files. VPL allows programming with visual expressions, spatial layout of text and graphic symbols that are used either as elements of syntax or secondary writing. Visual programming languages also make it possible to create a more natural programming environment that will be familiar and familiar to the user. In order to be used by non-programmers, visual programming languages must use intuitive metadata and specific to the area they are intended to, reduce the cognitive load of end-users in their learning and adoption, to enable the writing of a readable and understandable code, and They are simple and fun.

As scientists point out, in this type of programming more than one dimension is used to represent the semantics [11]. Examples of such additional dimensions are reflected in the use of multidimensional objects, the use of spatial relationships, and the time dimension by typing, so-called. "before-after" semantic relations. Any such potentially significant object or relationship is a sign, and the collection of one or more of them is an image expression. Examples of visual expressions used in visual programming include diagrams, idle sketches, icons, or demonstrations of actions using graphic objects. If in some programming language the syntax includes visual expressions, then this programming language is visual. Multidimensionality is the important difference between VPLs and strictly textual languages.

5.1. Visual programming and parametrization

Visual programming and parameterization tools allow the creation of a program logic by stacking and linking visual elements and their spatial layout instead of writing textual commands of the programming language [12]. Semantic and syntax rules, as well as textual programming languages, are still used, but the language elements in this case are represented by graphical symbols. Although these are tools for developing applications without programming, they still need to be familiar with the programming principles. With the "right" visual programming tools there is no need to write program commands or other texts, except in cases where it is necessary to designate program elements (variables, labels, blocks, etc.) for transparency purposes. Research has shown that visual programming and parameterization tools can be classified into four sub-categories: graphical user interface tools, high-level parameterization software programs, tools that use diagrams for describing data flows or process flows, and graphical representation of program logic tools. Typical representatives of the graphical editing tool for the user interface are MS Visual Studio and MS Expression Blend.

The basic advantage of the visual programming language is that it requires little initial knowledge in order to be able to start efficiently using it. This is because all the elements of the programming language, its semantics and syntax are visually represented, and the use of context dependencies dynamically narrows the choice of elements that can be used and in this way the user facilitates the programming process. All tools in this group are, in fact, software tools, and of the classic programming tools they differ only in that programming is not carried out by writing textual commands of the programming language, but by stacking graphic elements.

6. Component software development

It's the way software is developed with the help of ready, reusable components that appeared in the late 90s of the 20th century.

The concept of a component in the hardware industry has been present for a long time. Building a computer does not happen from scratch, but it folds out of the finished components. The use of finished components carries numerous advantages, such as: the ability to use proven correct components, faster computer construction, easy maintenance, low cost, and the inclusion of non-professionals in assembling the computer. That is precisely one of the reasons for the rapid progress of this branch of industry and the large supply of low-cost computers. A similar concept is also conceived for building software and can substantially help alleviate its deficit. Even if the components that were built were intuitive and close to non-programmers, experts from other areas, and if they were offered an easy way to compose them, the application software deficit would certainly be lower. Most people think about software components similar to hardware, thinking that program engineers compile them into software applications in the same way that hardware engineers compile a computer from hardware components. Various researchers attempted to define software components by keeping this attitude, which resulted in several differently descriptive informal models.

One of the most popular component definitions is provided by the ECOOP Working Group (European Conference on Object-Oriented Programming) which states that "... the software component is a unit of composition by contract of a particular interface and explicit contextual dependence." The software component can be distributed independently and be a subject for a third-party composition. " Others, however, require the extension of this definition, taking into account three different aspects. The aspect of the specification that should describe what component works and how it can be used, the aspect of the packaging perspective that views the component as a delivery unit, and the aspect of the integrity perspective that it sees as a kind of encapsulation. It follows that the component is a software package that offers its various services through interfaces. In accordance with the above, the definition of the component given by D'Souza and the authors. According to them: "The component is a coherent software package that can be developed independently and delivered as a unit, which offers interfaces that can be connected to other components and participate in building a larger system." [13]. It is an independent software package that offers some functionality. It's very important to define its interface well. It can be output through which component provides functionality for other components or input through which component gets services from other components.

All these definitions emphasize the tendency of the nature of the components in the form of a "black box" so that the developer can use them to create a larger system without any knowledge of how the components themselves are implemented. Component models allow program engineers to use components from different languages and platforms together. End users also benefit from this technology. Thus, for example, spreadsheets, word processing, drawing, and database applications often use a component model to incorporate edited data from one application into files created and managed by another application. Currently, the most popular commercial component models are: CORBA[2] from Object Management Group, COM[3] from Microsoft, and JavaBeans from Sun Microsystems.

6.1. Features and benefits of component software development

Development based on software components represents the process of software development by simply stacking or installing already finished components. The basis of this development is a component model, which defines the components and mechanism of their composition. The idea is to facilitate software development by selecting and integrating existing software components into a previously defined software architecture [14]. Software components can be developed by different developers and using different languages and development platforms. With this approach to software development, the costs and time necessary for its appearance on the market are significantly reduced, while at the same time improving: maintenance, reliability and overall software quality [15].

As the development of component software becomes increasingly important in the IT world, numerous concepts and conventions for its construction have been introduced. A special discipline within software engineering that develops software reusable components is called component-based software engineering (CBSE). This discipline was due to the failure of the object-oriented paradigm to support the effective reuse of the code. Unlike class objects that are detailed and specific, the components are more abstract. In addition to the benefits that are reflected in the reuse of finished components, CBSE is based on the basic design principles of software engineering applied to the development of components, which imply that: components must be independent to each other, not to hinder one another, their implementation must be hidden and unmanageable the one who uses them, communication with them must take place through a well-defined and simple interface.

The life cycle and model of software in component-based software engineering differ from the life cycle and software models in traditional software engineering. These two items are exactly what focuses on component based software engineering.

7. Integrated Development Environment

The common elements of an application are: screen user interface, program logic, database management system, and reporting system. For the purpose of their realization, there are special

[2] Common Object Request Broker Architecture
[3] Component Object Model

software tools for the creation, correction and testing of these elements, and sometimes they are all integrated into the so-called integrated development environment - IDE.

Modern integrated development environments contain a multitude of visual aids that facilitate the creation of a screen user interface, modeling and linking to a database or making complex reports. However, any more demanding program logic that forms the backbone of the application must be written in the selected programming language. In this part there are tools for development of applications without programming that are aimed at maximizing the programming process, and in some cases completely avoiding it.

7.1. Development environment for component software development

The development environment is very important for the development of software based on components. Not only can the application be added through this component environment, but it can be used to correct errors and test applications. In addition, the efficiency and usability of the development environment is a key factor that affects the easier development of this type of software. In such a visual programming environment, the application assembly process is visualized, and only error correction and interactive testing.

7.2. Development environment for logopeds-developers

When building most of the most popular visual development environments, the greatest attention is paid to the editor in order to make it easier to visualize and build a user interface. In this paper, a development environment for speech logoses is proposed, in which the centralized place is occupied by the visual WYSIWYG[4] editor. Components for constructing application programs by logopeds must be intuitive and meaningful units, minimally dependent on each other so that the change of one component in the system minimally affects other components in the system. The component is usable, i.e. after being designed, implemented and tested, it can be used whenever necessary. Its internal work is hidden and isolated from the interface so that logos can build applications using components without any knowledge of how they work. Components should be reliable because they are the prerequisite for the development of high quality and reliable component-based software. They must be interoperable, i.e. support multiple platforms, executed by anyone, without the need for the source code to be available, and able to describe their public interfaces, i.e. flexible properties and workable events.

The development environment whose prototype is made is completely visual because it is intended for non-programmers. All elements are concrete objects presented with icons that will enable simple and effective handling by logopeds, with the aim of simple design and creation of application programs. Visual syntax will allow logopeds to create their own program by simply customizing components in the work window, and by stacking them together and linking them to the scene.

The Development Environment for Speech and Language Therapist (DESLT) contains the main menu through which the logoped-programmer creates a future therapeutic application as a new project.

After that the window of the development environment appears (Figure 1). This window contains a toolbar with components that allow you to visualize objects and their composition in complex designs of future applications using point-and-click and drag-and-drop techniques.

[4] What You See Is What You Get

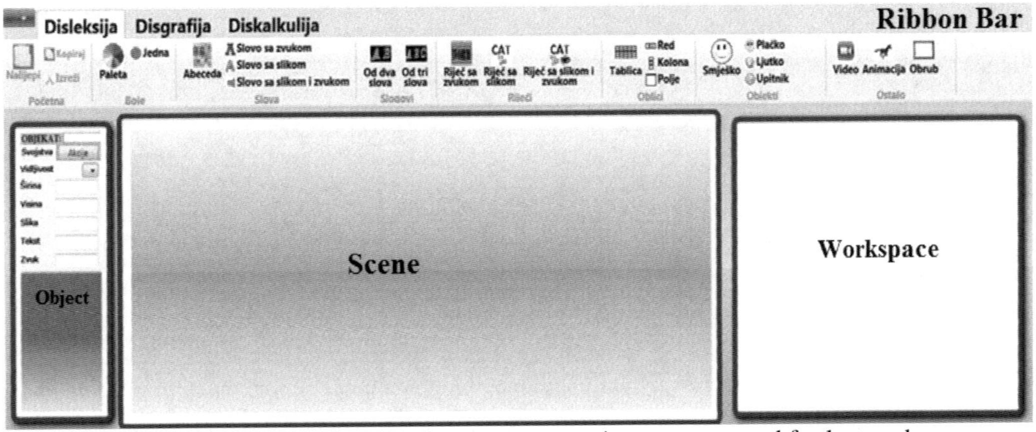

Figure 1. Design editor of a development environment created for logopeds

Programming is reduced to placing an arbitrary number of objects on the scene. Before that, the logoped-programmer instantiates them in the right-hand part of the screen (WorkSpace) and specifies their properties in the left part (Object) (Figure 1).

When creating an application after it instantiated the components in the right part of the screen (WorkSpace), they set their properties to the left (Object) and arranged them on the scene (Scene), logoped tasks, and certain rules (Figure 2).

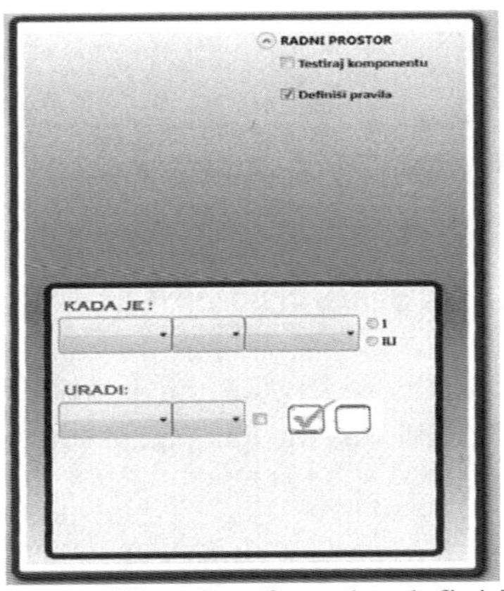

Figure 2. The plate for rules definition

While the logoped using drag & drop technique instantiated the component, in the background, this finished component was placed in the list of interface components as predicted by a software engineer. At the end the installation of the future therapeutic application, this list will contain all the instantiated components.

The rules are given by specifying the relations between the components that are selected from the list of components or directly selected on the scene. Figure 3 shows the rule: When the *Text* property of the Queue component is equal to the *Text* property of the WordWithTextImageSound component (WTIS), you rotate the component WTIS.

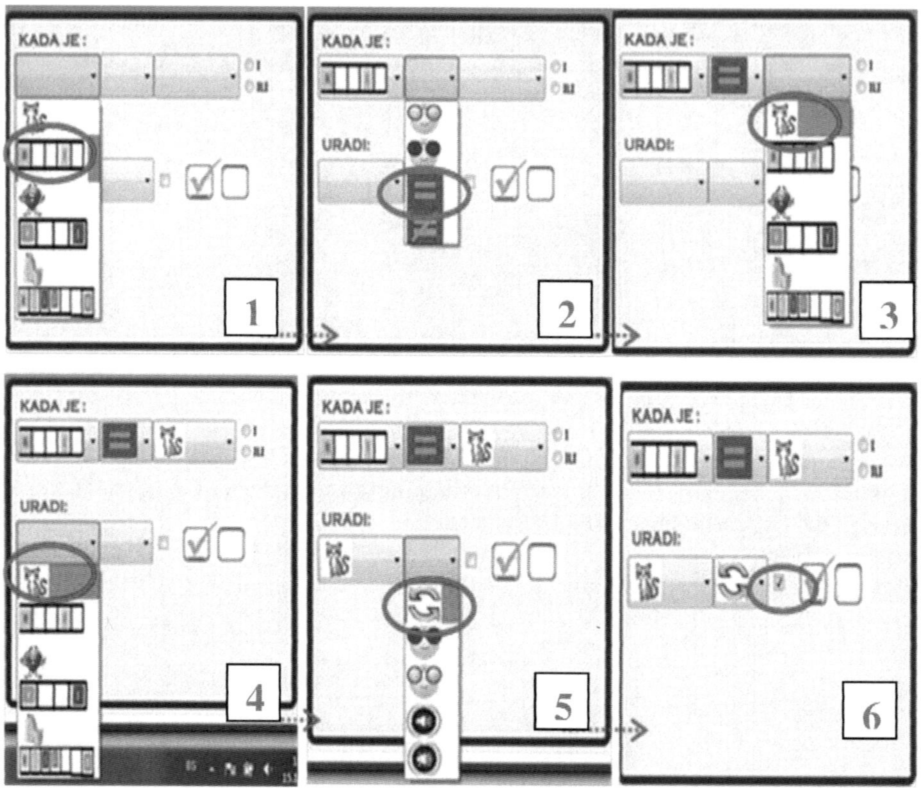

Figure 3. Defining the rules

When defining the rules the desired action is selected from the list of predefined actions (Rotate, Show, Hide, Run, Speak Off, etc.) previously programmed by the program engineer (Figure 4). These actions are represented by symbols behind which the corresponding C # code is written by a program engineer. The same actions are available to the logopedia and when defining the behavior of the component.

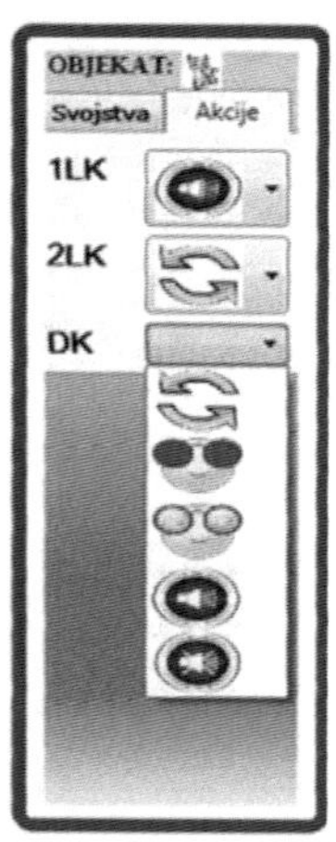

Figure 4. The list of predefined actions: Rotate, Unhide, Hide, StartSound, Mute,...

8. Conclusions

Visual programming has an ambitious vision in the future: to enable a world in which the non-programmer, i.e. someone who has no experience in writing code can quickly create flexible software solutions that work to strengthen business processes and are able to meet his real business needs. The development of component-based software is a paradigm aimed at building applications using components developed by third parties.

In the future, the rapid and easy development of therapeutic software by the logoped developer is possible using the libraries of pre-built components. By including only a small number of program engineers, the new components could be designed and distributed over the World Wide Web. Logopeds should be able to download and install them in their environments with just one mouse click.

In order for these components to be not too complex, they are expected to be small in size, and that the web transfer will take a short time. The downloaded components will be able to install themselves, so the logopeds will not face the complicated and computer-dependent installation process

Application programs for dyslexia therapy, which the logopeds will create, will feature the features of modern windows applications. A development environment with a design editor that resembles the most commonly used MS Office program for logopeds is user-friendly. Logopeds will be able to develop therapeutic software applications, as the most competent person, with previously developed training in using a specially created development environment for them. By empowering the logopeds as the ultimate beneficiaries in the development of their own application programs, it will initiate and give support to experts in other fields to become software creators for their own needs. Current software paradigms allow them that.

References

[1] ***2017 Tech: Where the jobs -- and demand – are, March 23, USA TODAY, https://www.usatoday.com/story/tech/talkingtech/2017/03/22/tech-where-jobs/99496462/

[2] Lieberman H, Paterno F and Wulf V 2006 *End User Development*, Springer

[3] Ko A J, Abraham R, Beckwith L, Blackwell A, Burnett M M, Erwig M, Scaffidi Ch, Lawrance J, Lieberman H, Myers B A, Rosson M, Rothermel G, Shaw M, Wiedenbeck S 2011 *The State of the Art in End-User Software Engineering, ACM Computing Surveys* **43**(3) 1-444

[4] Luger G 1994 *Cognitive science : the science of intelligent systems*, San Diego: Academic Press

[5] Gregory R L and Murrell J N 2006 *Hugh Christopher Longuet-Higgins. 11 April 1923 - 27 March 2004: Elected FRS 1958,* Biographical Memoirs Fell. R. Soc., pp 149-166

[6] Lighthill J 1973 *Artificial Intelligence: A General Survey*, Artificial Intelligence: a paper symposium, Science Research Council

[7] Lakoff G and Johnson M 1999 *Philosophy in the Flesh: The Embodied Mind and its Challenge to Western Thought*, New York: Basic Books

[8] Green T R G 1989 *Cognitive dimensions of notations*, People and Computers IV, Cambridge, Cambridge University Press

[9] Green T R G 1990 *Programming languages as information structures*, Psychology of Programming, London, Academic Press, pp 117-138

[10] Pane J F, Ratanamahatana C A and Myers B A 2001 Studying the language and structure in non-programmers solutions to programming problems, *International Journal of Human-Computer Studies* **54** 237-264

[11] Burnett M 1999 *Visual Programming*, Encyclopedia of Electrical and Electronics Engineering, Postscript, pp 275

[12] Wesley M, Johnston J R, Hanna P and Millar R J *Advances in Dataflow Programming Languages*, University of Ulster, World Wide Web, http://www.cs.ucf.edu/~dcm/Teaching/COT4810-Spring2011/Literature/DataFlowProgrammingLanguages.pdf

[13] D'Souza D F and Wills A C 1997 *Objects, Components, and Frameworks with UML – the Catalysis Approach*, Addison-Wesley Longman Publishing, Boston MA USA

[14] Pour G 1998 *Component - Based Software Development Approach: New Opportunities and Challenges, Proceedings Technology of Object-Oriented Languages*, Proceedings of Technology of Object-Oriented Languages and Systems TOOLS 26 (Cat. No.98EX176), Santa Barbara, CA, USA

[15] Pour G, Griss M and Favaro J 1999 Making the Transition to Component-Based Enterprise Software Development: Overcoming the Obstacles – Patterns for Success, Proceedings of Technology of Object-Oriented Languages and Systems TOOLS 29 (Cat. No.PR00275), Nancy, France, France

Overview of the main topics in wind energy systems planning

C P Chioncel[1], N Gillich[1], E Spunei[1] and O G Tirian[2]

[1]Eftimie Murgu University of Resita, Department of Electrical Engineering and Informatics, Train Vuia Street., no. 1-4, 320085 Resita, Romania
[2]Politehnica University of Timisoara, Department of Electrical Engineering and Industrial Informatics, 5 Revolution Street, Hunedoara, 331128, Romania

E-mail: c.chioncel@uem.ro

Abstract. The renewable energy production suffered very much in Romania, being practically stopped since a couple of years, because of legislative changes made "during the game". Beside the main steps of wind power plant planning, identifying locations with good wind potential, the design of the wind park layout, choosing the best wind turbine type for the given locations, proximity to the transmission/distribution network and other technical features, the legislation in the electricity production from RES will have to ensure economical feasible support scheme, in a predictive legislative framework. The present paper presents possible scenarios for the development of the Romanian RES market at the horizon of 2030, strongly related to the direction that the policy transposes in the legislation.

1. Introduction

Through the ambitious targets settled for 2030, the European Union has positioned itself as a world leader in decarbonization initiatives. [1], [2]

To achieve those targets, in the main areas of the society, certain trends and awareness must be accomplished. The population will have to support such a target, for this, among them, a better understanding of energy use and consumption must be achieved, hand in hand with cultivation of the ecological consciousness. Outgoing from those previously mentioned, also, the social behavior will have to encourage to share mobility, in the context of a growing urbanization. [3]

In favor of reaching these targets, the technologies have a rapidly developing, and in this sense, the production and exploitation costs for energy production from renewable sources are declining. [4-7] At the same time, the digitalization of the transmission and distribution network to the so-called smart electric grid as well as the lowering cost of the electric vehicles will put pressure on the need of clean and save electricity production units. [9]

From economic reasons also, the volatility of goods prices (i.e. petrol) or the increase of the cost of greenhouse gas emissions justify the accent on independent and safe energy resources (i.e. wind, sun, water). [10]

The politic will has for sure a main role in achieving this target, by creating a legislative framework, characterized by stability and predictability, as well as subsidies and incentives (i.e. electric cars), because any investment in the energy market is high and the return of investment, lasting. [11]

2. Scenarios and forecasts for Romania

2.1. Natural Potential

The natural potential to exploit the renewable energy sources (RES) in Romania, is so high, that, the entire energy consumption could be covered from those. The Romanian natural renewable energy potential, splitted on the main sources, is presented in Figure1. The diagram excludes the geothermal potential being under 1GB as well as the biomass potential, the references to this potential in Romania being, at least for the moment, unclear. [12]

From the existing potential, approximatively 32% would be enough to cover the maximum on energy from renewable sources in Romania. [12]

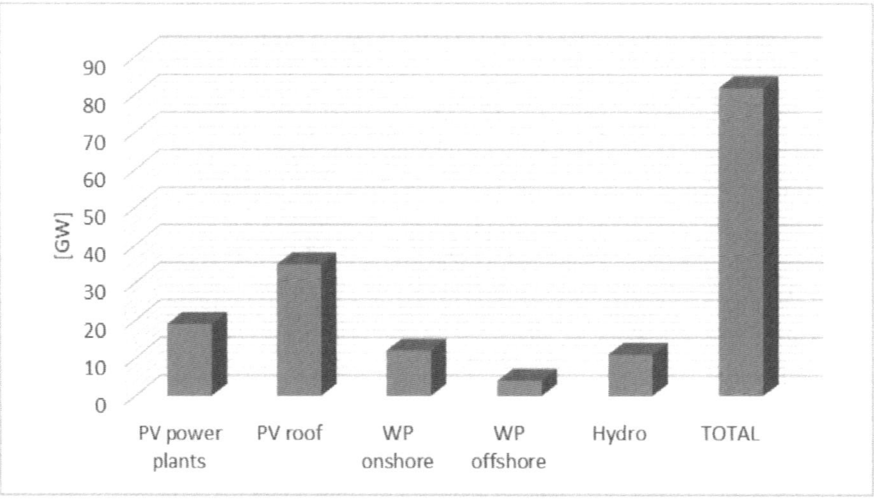

Figure 1. Romania renewable energy potential

The hypotheses considered, are based on the official date of the Romanian authorities as well as from feedback obtained from the different professional associations acting in the renewable energy sector. [12-15]

Outgoing from the market forecast, a decreasing of the technology costs (especially for photovoltaic solar panels and wind turbines), and the development of storing renewable energy capacities, make the gain of energy from renewable sources the towards competitiveness to the energy gain from conventional sources. For the period until 2030, the medium investment cost in the wind energy sector is estimated to be around 1.200 EUR/kW, while, for example, the coal-based power plants require substantial investments to comply with the pollution requirements specified through the environmental regulations. As well, the electric energy consumption will increase, considering at least two main factors, electrifying transport, to estimated 500.000 electric cars in 2030, and an increase in the residential sector, because of a higher number of households with electric cooling and heating system, increasing the number of households connected to the electric grid, in general, an improvement of the life quality. [10], [12-14]

Based on macroeconomic indicators, the evolution of the energy consumption in Romania, provides in 2020 the achievement of 45% from the EU average at the level from 2015, in 2030 this consumption will reach 54% of the EU average at 2015 level so that in 2050 the electric consumption / inhabitant will exceed the EU average from 2015. [15]

To fulfil this evolution in the macroeconomics indicator as well as in the marked forecast, the infrastructure will have to develop: transition to the smart grids through a higher level of automatization and network digitalisation; improving the balancing capacity; interconnection inside the national electricity transmission network as well as increasing interconnectivity with neighbouring countries.

2.2 Analysed scenarios

A first scenario, with the highest probability, under the current conditions, assume: expanding the life of existing nuclear facilities; two additional nuclear reactors (unity three and for) that will be commissioned in 2030 respectively 2031; gradual and natural elimination of all coal groups by around 2035. Under this scenario, the net installed capacities, broken down by sources, is: nuclear (2.000 MW), hydro (6.800), wind (4.400 MW), solar (1.800 MW), biomass (140 MW), coal (3.200 MW), natural gas (3.500 MW), masout (100 MW), Figure 2.

For this scenario, the net installed capacity from renewable energy sources will increase in 2030, comparative to 2020, with 15%: additional installed capacity, in wind power plants, from 3.000 MW to 4.400 MW, Figure 3, and PV systems, from 1.500 MW to 1.800 MW, Figure 4. Outgoing from the given scenario and based on the previous experience in the Romania market, the estimated percentage distribution of the energy by sources in 2030, is shown in Figure 5.

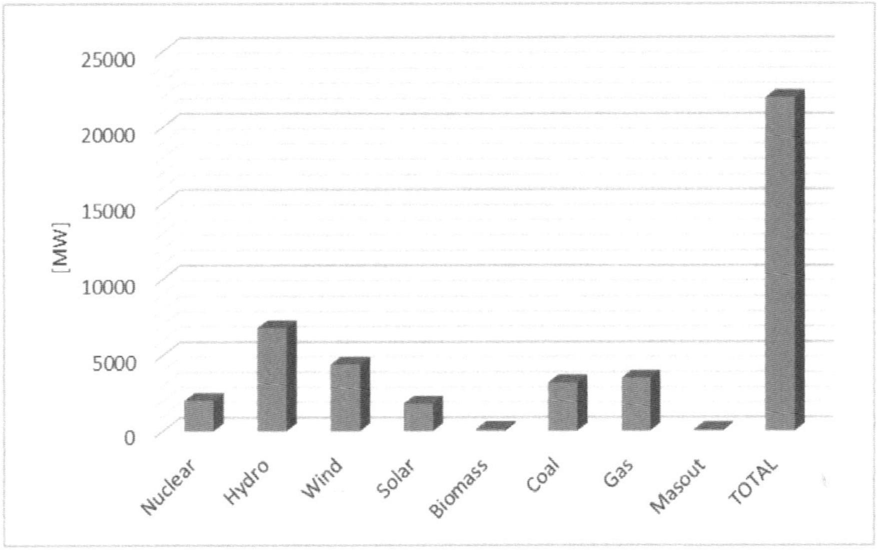

Figure 2. Net installed capacity forecast 2030

A second scenario captures a 40% increase in net installed capacity in RES compared to 2020. Under those assumptions, the nuclear energy production will increase, expanding the life of the existing nuclear facilities, reactor units one and two (U1 & U2) and gathering additional installed capacity, the units three and four (U3 and U4). In the same time, the coal capacities with theirs associated entities, will be gradual and natural, eliminated by around 2035. The wind power plants will double their installed capacity, increasing from 3.000 MW to 6.000 MW, Figure 6, as well as the installed capacity in PV systems that will also double, from 1.500 MW to 3.000 MW, Figure 7. Based on this second scenario, the estimated net energy production, assigned according to the used sources in 2030, is shown in Figure 8. It can be observed that reported to the increase installed capacity in RES, a decrease in the energy production in natural gas power plants will be assumed.

IOP Conf. Series: Materials Science and Engineering **477** (2019) 012059 doi:10.1088/1757-899X/477/1/012059

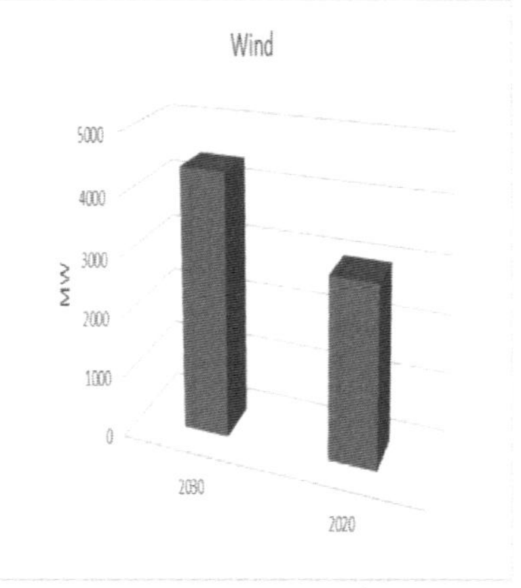

Figure 3. Additional installed capacity in wind power plants

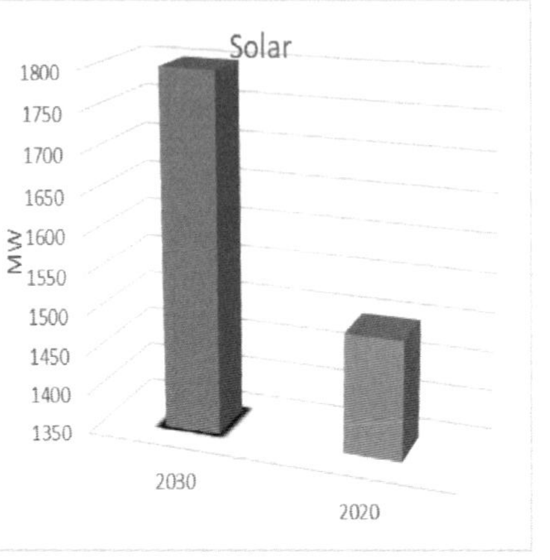

Figure 4. Additional installed capacity in PV power plants

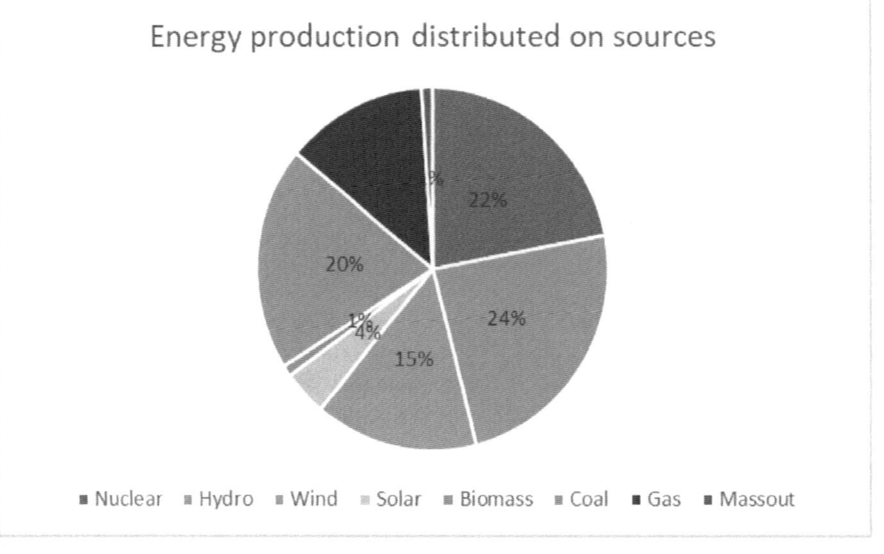

Figure 5. Estimated distribution of the energy production, by sources, in 2030

The two analyzed scenarios, both considering the initialization and completion of the related investments for the units three & four of the romanian nuclear power plant units, aim to align the RES quota. Romania will have to follow the European Union targets to ensure a fair participation in targeting common objectives of the member states.

For the first scenario, the evolution of the SRE quota is starting from the actual value of the quota, 25% in 2016, as well as the pace of deployment of existing capabilities, ≈ 1 GW annually, reaching a 30% quota by the 2030 horizon, is inertial, Figure 9. The second scenario is based on the increasing of the RES installed capacity and the decrease of the using level of the natural gas power plants, Figure 9. The final energy consumption and the electricity consumption are constant in both addressed scenarios.

IOP Conf. Series: Materials Science and Engineering **477** (2019) 012059 doi:10.1088/1757-899X/477/1/012059

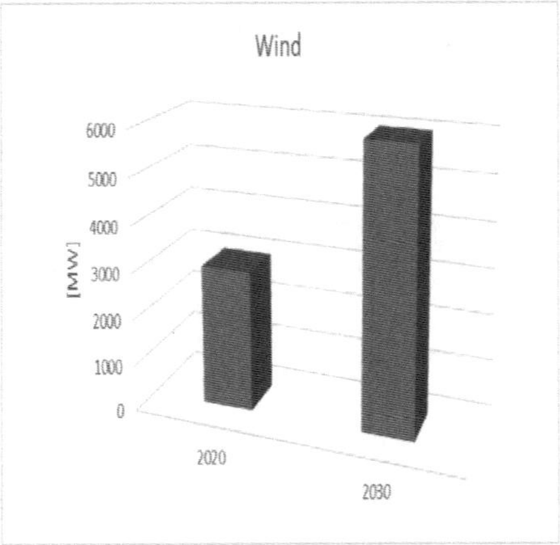

Figure 6. Additional installed wind power plant capacity in 2030 compared to 2020

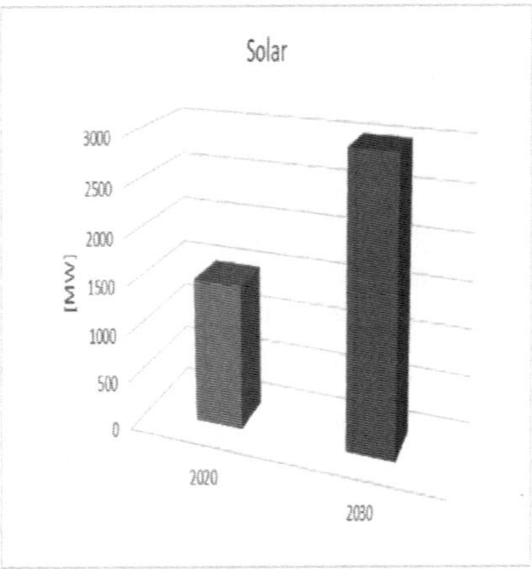

Figure 7. Additional installed PV power plant capacity in 2030 compared to 2020

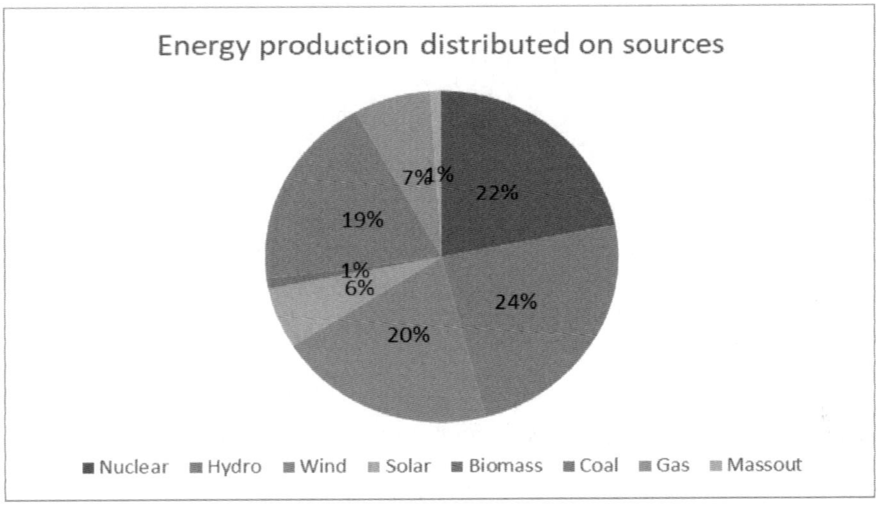

Figure 8. Estimated distribution of the energy production, by sources, in 2030

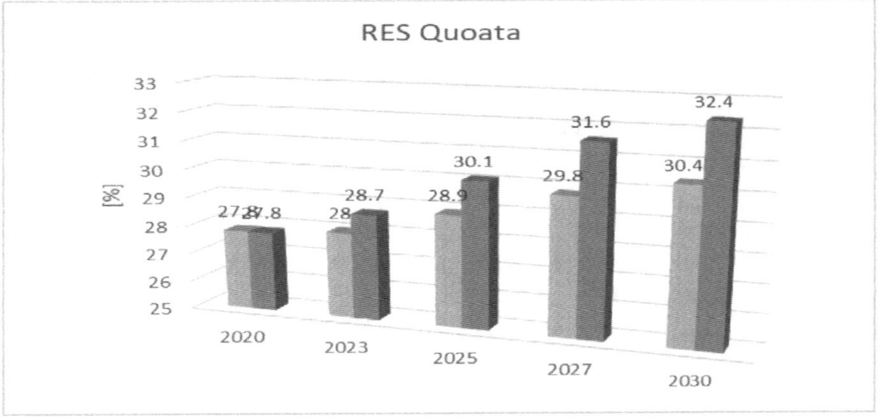

Figure 9. Evolution of the RES quota in the analysed scenarios until 2030

3. Conclusions

To be able to implement those scenarios, both, at national and European level, support schemes for RES have proven to be an effective way to favor the use of renewable energy, even if the increase of the technical performance of the equipment will reduce subsidies but will not lead to complete elimination.

General energy policies converge towards achieving the environmental targets set through the Paris Treaty, maintaining also the energy security at national level. Encouraging the development energy electricity production from renewable energy sources (RES) need to have at their disposal instruments facilitated by the authorities, preventing in the same time retroactive changes to the support scheme.

The development of the infrastructure is essential to an economically and technically increased absorption of the renewable energy, such as infrastructure, with the transmission and distribution network, smart grids and interconnections.

Romania is one of the European countries with a great potential of renewable sources, wind, water, and solar (WWS), that, with the right energy policy, will be able to transform this country in a standard of electricity produced from RES.

References

[1] ***www.ewea.org (European Wind Energy Association)
[2] ***www.windeurope.org
[3] ***Renewable Energy Policies in a Time of Transition, © 2018 IRENA, OECD/IEA and REN21
[4] Chioncel C P, Chioncel P and Babescu M 2011 Optimization of the Wind Systems, Based on the Maximum Wind Energy, *International Proceedings of Computer Science and Information Technology* **8** 190-194
[5] Chioncel C P, Gillich N, Petrescu D I and Erdodi G M 2016 *Optimal operation of an wind power system with energy storage in electric accumulators*, 9th International Conference and Exposition on Electrical and Power Engineering (EPE), Iasi, Romania, October 20-22, pp 858-861
[6] Chioncel C P, Chioncel P, Gillich N and Tirian G O 2009 Performance ratio of a photovoltaic plant, *Bulletin of Engineering, University Politehnica Timisoara, Fascicule 2* 55-58
[7] Chioncel C P, Bereteu L, Petrescu D I and Babescu M 2016 *Determination of reference mechanical angular speed for wind power systems*, Soft Computing Applications, pp 1213-1222
[8] Chioncel C P, Erdodi G M, Petrescu D I, Elisabeta S and Gillich N 2015 *Control of Wind Power Systems Imposing the Current in the Intermediate Circuit of the Converter at Variable Wind Speed*, 9th International Symposium on Advanced Topics in Electrical Engineering (ATEE), pp 923-928
[9] Chioncel C P, Tirian G O, Gillich N, Hatiegan C and Spunei E 2017 Overview of the wind energy market and renewable energy policy in Romania, *IOP Conf. Ser.: Mater. Sci. Eng.* **163** 012009
[10] ***www.gwec.net (Global Wind Energy Council)
[11] Prostean G, Vasar C, Prostean O and Vartosu A 2015 Risk analysis for renewable energy projects due to constraints arising, *IOP Conf. Ser.: Mater. Sci. Eng.* **106** 012017
[12] ***www.rwea.ro (Romania Wind Energy Association)
[13] ***www.rpia.ro (Romanian Photovoltaic Industry Association)
[14] ***www.anre.ro (Romanian Regulatory Authority for Energy)
[15] ***www.transelectrica.ro (Romanian Transmission and System Operator)

Aspects concerning the cavitation erosion and dry sliding wear behaviour of the YSn83 antifriction alloy and EN-GJS-400-15 spheroidal cast iron

M D Nedeloni[1,2], L Nedeloni[1,2], L Cîndea[1], V Iancu[1], A V Petrica[1], A-M Budai[1] and I L Conciatu[1]

[1]Eftimie Murgu University of Reşiţa, Faculty of Engineering and Management, Traian Vuia Square, No. 1-4, 320085, Resita, Romania
[2]Westphalian University of Applied Sciences, Department of Materials Science and Testing, Gelsenkirchen, Germany

E-mail: l.nedeloni@uem.ro

Abstract. In this paper, the cavitation erosion respectively the dry sliding wear behaviour of the Ysn83 antifriction alloy and EN-GJS-400-15 spheroidal cast iron was studied. As cavitation erosion aspects, the authors show some preview results of the Ysn83 antifriction alloy behaviour and continue to study and to compare also the cavitation erosion behaviour of the EN-GJS-400-15 spheroidal cast iron. In this regard, the cavitation erosion tests were made using a cavitation stand and the experimental results were presented through mass loss and cavitation erosion rate vs time curves. As dry sliding wear aspects, the tests for the two investigated materials were done with a tribometer through the pin-on-disk method where the authors varied the parameters especially the linear speed and the distance. The obtained results were presented as coefficient of friction evolutions and as wear rates. From all the tests, compared with the YSn83 antifriction alloy, the EN-GJS-400-15 spheroidal cast iron showed a superior cavitation erosion and dry sliding wear behaviour.

1. Introduction

Cavitation erosion and sliding wear are present in various industrial applications such as in the energy, shipping, machine building or automotive industry [1-6]. The cavitation erosion occurs due to the cavitation phenomenon, where the wear mechanism is through erosion and the sliding wear occurs due to the dynamic interaction of the mating components where the abrasive wear is more common [7-13].

Depends on the resistance of the used materials, the operating conditions and the environment, the wear can be more or less [14-19].

The YSn83 antifriction alloy (studied in this paper) is a nonferrous alloy. Such as alloys can be used for ingots, revolution parts (like bearings) or for coatings. Also, the YSn83 antifriction alloy (with a good shock resistance) can be used for sliding bearings manufacturing [20].

Studies regarding these nonferrous alloys and anti-friction materials were made especially concerning their casting and solidification respectively their use in the military technique and aeronautic constructions [21-23].

The other studied material in the present paper (EN-GJS-400-15 spheroidal cast iron) is part of the gray cast iron category and can be used for impact and shock-resistant parts, housings, valve bodies, pipe, cylinders or gear boxes [3], [13], [17].

Due to the good mechanical properties of the cast irons, many studies regarding especially the spheroidal cast irons wear resistance have been achieved [4], [11] and [24-28].

Also, the cavitation erosion respectively dry sliding wear process regarding the resistance / behaviour of some materials (where their mechanical properties were improved through surface and ultrasound treatments or even by nitrogen ion implanted) and coatings like Stellite 6, self-fluxing remelted coatings or through HVOF spraying have been investigated [29-45].

In this regard, the present paper is highlights the cavitation erosion and dry sliding wear behaviour of the YSn83 antifriction alloy respectively EN-GJS-400-15 spheroidal cast iron and try to make reproducibility (regarding similar results and used apparatus) with their own preview work [46-48] and similar studies mentioned above.

To compare the two investigated materials, the experimental results will be presented through micrographs respectively mass loss and cavitation erosion rate vs time curves (concerning the cavitation erosion behaviour) and as friction coefficient evolution and as wear rates (concerning the dry sliding wear behaviour).

2. Materials and methods

The materials used in this research are the YSn83 antifriction alloy and the EN-GJS-400-15 spheroidal cast iron, where the chemical composition of them is reported in Table 1 and Table 2.

Table 1. YSn83 antifriction alloy chemical composition (wt.%)

Cu	Sb	Pb	Fe	Sn
3.53	10.9	0.55	0.06	84.96

Table 2. EN-GJS-400-15 spheroidal cast iron chemical composition (wt.%)

C	Si	Mn	P	Cu	Cr	Mg	Ni	S	Fe
3.5	2.06	0.22	0.050	0.04	0.04	0.038	0.03	0.015	93.99

From these materials, two cubic samples with the edges of 16 mm were made and tested on the cavitation stand respectively on tribometer.

The cavitation stand is composed of: ultrasonic generator, electro-acoustic piezo converter, acoustic transformer, ultrasonic horn and respectively of the water container, the sample support and the tested sample.

The stationary specimen method was used and the tests were realized according to G32-10 standard [49] with the following main parameters from Table 3.

Table 3. Parameters for cavitation stand / cavitation erosion tests

Frequency (kHz)	Amplitude (μm)	Water temperature (°C)	Horn and sample distance (mm)
20 ±0.5	50	25 ±2	0.6

For these tests the total cumulated time was 150 minutes, divided into one period of 5 and 10 minutes respectively in nine periods of 15 minutes. After each period the mass loss of the samples was weighed with the help of a digital balance.

The tribometer arrangement was a pin holder with a 100Cr6 steel ball (6 mm in diameter) respectively the tested samples. For these sliding wear tests the authors varied the samples radius, linear speed, working speed respectively the distance.

The pin-on-disk method (POD) was used according to G99 standard [50] with the following main dry sliding parameters from Table 4 and Table 5.

Table 4. YSn83 antifriction alloy dry sliding wear parameters

Faces	Load (N)	Radius (mm)	Linear speed (mm s^{-1})	Working speed (rpm)	Distance (m)	Distance (laps)
Face 2 (1)	5	1.5	46.2	300	10	1090
Face 2 (2)	5	3	85.7	275	10	536
Face 2 (3)	5	4.5	117.5	250	10	355
Face 2 (4)	5	6	141.5	225	10	267
Face 1 (1)	7	1.5	30.8	200	7.5	815
Face 1 (2)	7	3	61.8	200	10	541
Face 1 (3)	7	4.5	93.6	200	12.5	446
Face 1 (4)	7	6	123.7	200	15	406

Table 5. EN-GJS-400-15 spheroidal cast iron dry sliding wear parameters

Faces	Load (N)	Radius (mm)	Linear speed (mm s^{-1})	Working speed (rpm)	Distance (m)	Distance (laps)
Face 1 (1)	7	1.5	85.1	550	100	10800
Face 1 (2)	7	3	163.9	525	100	5340
Face 1 (3)	7	4.5	234.4	500	100	3560
Face 1 (4)	7	6	297.9	475	100	2660
Face 2 (1)	10	1.5	58.7	400	75	8080
Face 2 (2)	10	3	121.1	400	100	5510
Face 2 (3)	10	4.5	186.7	400	125	4470
Face 2 (4)	10	6	249.5	400	150	4020

By using the tribometer software, the friction coefficient was monitored and the wear rate was calculated with equation 1 [50]:

$$K = \frac{2\pi r \cdot h\left(3h^2 + 4s^2\right)}{6 \cdot L \cdot d \cdot s} \qquad (1)$$

where: K is the wear rate; h – the wear track depth; s – the wear track width; L – the load and d is the distance (in meters).

Based on some preview tests, for this research the testing time for YSn83 antifriction alloy was only between 1 - 4 minutes respectively 5 - 20 minutes for the EN-GJS-400-15 spheroidal cast iron.

For the two types of tests (cavitation erosion and dry sliding wear), the samples were polished with abrasive paper in order to have a roughness around 0.8 μm and were cleaned with acetone.

3. Results and discussions

3.1. *Cavitation erosion behaviour*

Regarding the cavitation erosion YSn83 antifriction alloy behaviour, based on the reference [20], the authors present some preview results and compare them with experimental results regarding the cavitation erosion behaviour of the EN-GJS-400-15 spheroidal cast iron.

Comparatively, the main results are shown in Table 6 and in detail in Table 7 for EN-GJS-400-15 spheroidal cast iron, where the mass loss was much less than for YSn83 antifriction alloy.

Table 6. Comparison between the cavitation erosion results

Samples	Sample mass (mg)		Max. mass loss (mg)		Max. cavitation rate ($mg\cdot h^{-1}$)
	Before	After	On period	Cumulative	
	m (mg)	m (mg)	Δm (mg)	mc (mg)	Vec ($mg\cdot h^{-1}$)
YSn83	31230.37	30592.87	127.21	637.5	506.404
EN-GJS-400-15	28314.86	28301.53	1.4	13.33	11.800

Table 7. Cavitation erosion results for the EN-GJS-400-15 spheroidal cast iron

Cumu-lated time	Period (min)	Sample mass	Eroded mass loss		Cavitation rate
			On period	Cumulative	
t (min)	Δt	m (mg)	Δm (mg)	mc (mg)	Vec ($mg\cdot h^{-1}$)
0	0	28314.86	0	0	0.000
5	5	28313.7	1.16	1.16	11.800
15	10	28312.44	1.26	2.42	6.328
30	15	28311.32	1.12	3.54	4.620
45	15	28310.13	1.19	4.73	4.800
60	15	28308.92	1.21	5.94	4.560
75	15	28307.85	1.07	7.01	4.740
90	15	28306.55	1.3	8.31	5.080
105	15	28305.31	1.24	9.55	4.740
120	15	28304.18	1.13	10.68	4.760
135	15	28302.93	1.25	11.93	5.300
150	15	28301.53	1.4	13.33	5.920

Also, only for the EN-GJS-400-15 spheroidal cast iron sample, the cumulative mass loss and cavitation erosion rate vs time curves are presented in Figure 1 and 2. From these two types of curves, it can observed that the mass loss process is linearly and the cavitation erosion rate cross the characteristic stages (incubation, acceleration, maximum rate and deceleration [49]) for the first 30 minutes and then it is maintain between the range of 4 - 6 $mg\cdot h^{-1}$.

For a better comparison, in Figure 3 the eroded samples surfaces are presented through images acquired with a digital microscope (Keyence VHX-600).

From this Figure, in the case of YSn83 antifriction alloy, visible caverns (with a depth until 300 μm) can be observed and in the case of EN-GJS-400-15 spheroidal cast iron, pits (with a depth until 100 μm) toke place.

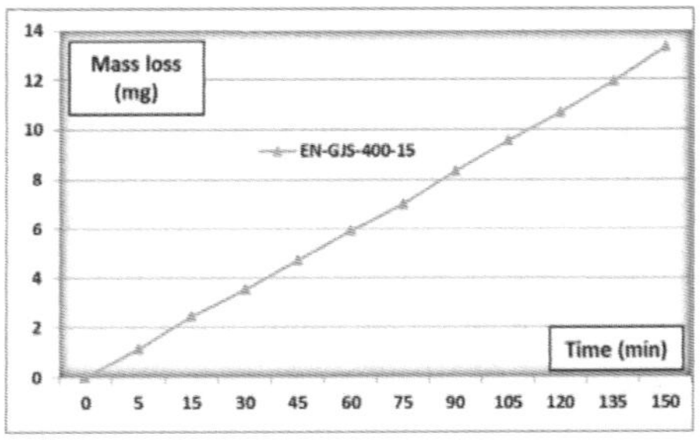

Figure 1. Cumulative mass loss vs time curve for EN-GJS-400-15 spheroidal cast iron

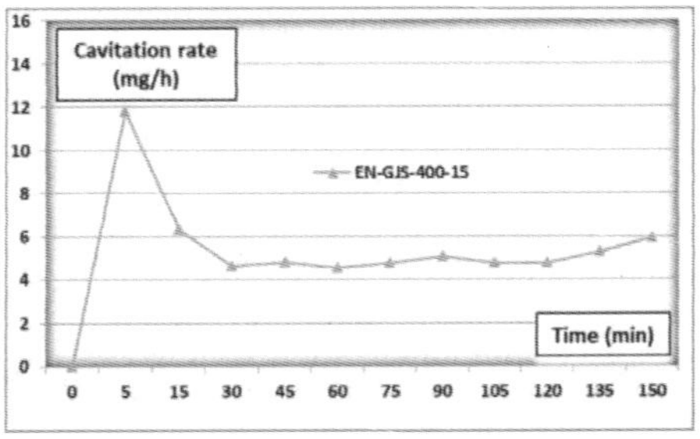

Figure 2. Cavitation erosion rate vs time curve for EN-GJS-400-15 spheroidal cast iron

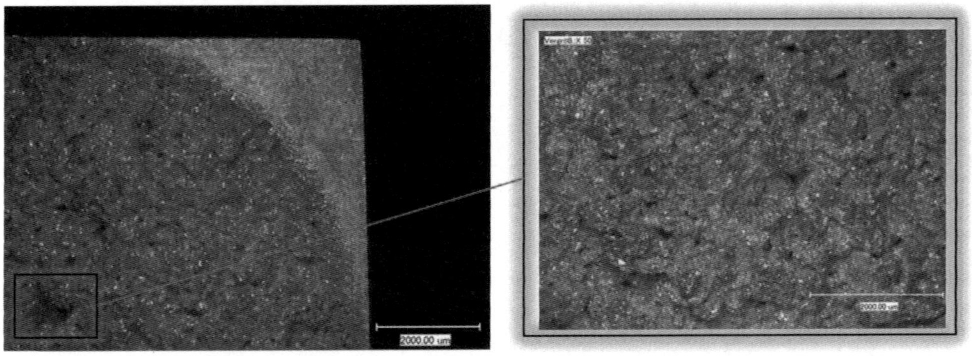

Figure 3. Details of an eroded area for the YSn83 alloy (magnification of 30X and 50X – 2000 μm)

Figure 4. Details of an eroded area for the EN-GJS-400-15 cast iron (magnification of 30X and 50X – 2000 μm)

From others results of the authors, this cast iron has a better cavitation erosion resistance than some Al alloys [51-53] but a poor once than the C45 carbon steel, some bronzes and stainless steels [54-56].

3.2. YSn83 antifriction alloy dry sliding wear behaviour

For the YSn83 antifriction alloy the main results (coefficient of friction evolutions, wear tracks profiles and the wear rate values) are presented in Figures 5 ÷ 10 and in Table 8.

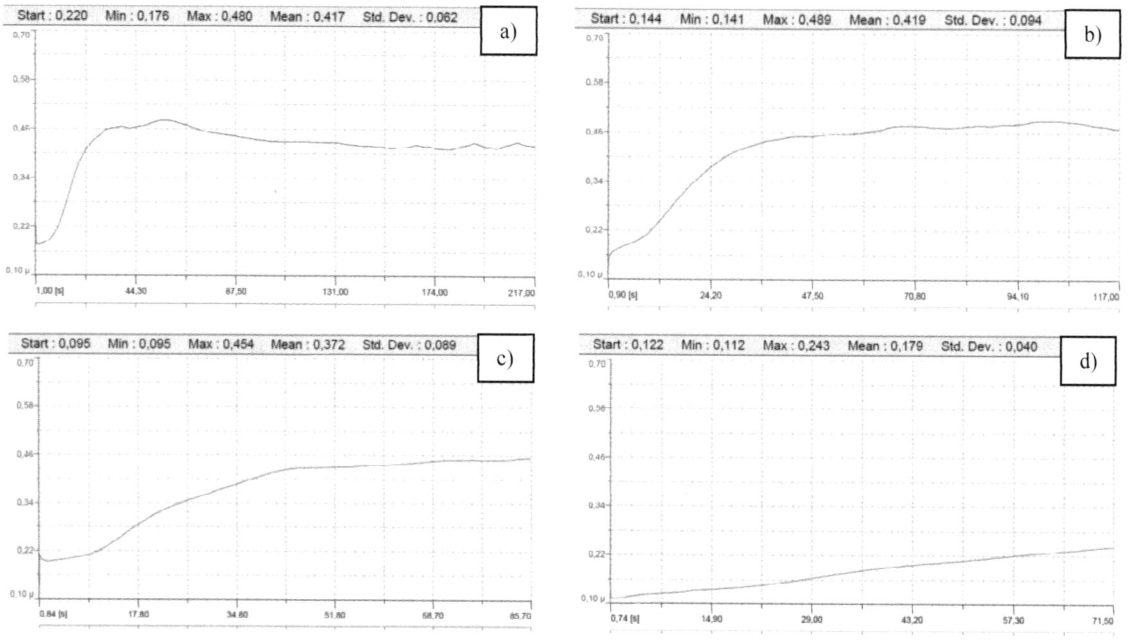

Figure 5. Coefficient of friction evolutions for the YSn83 antifriction alloy at 5N load

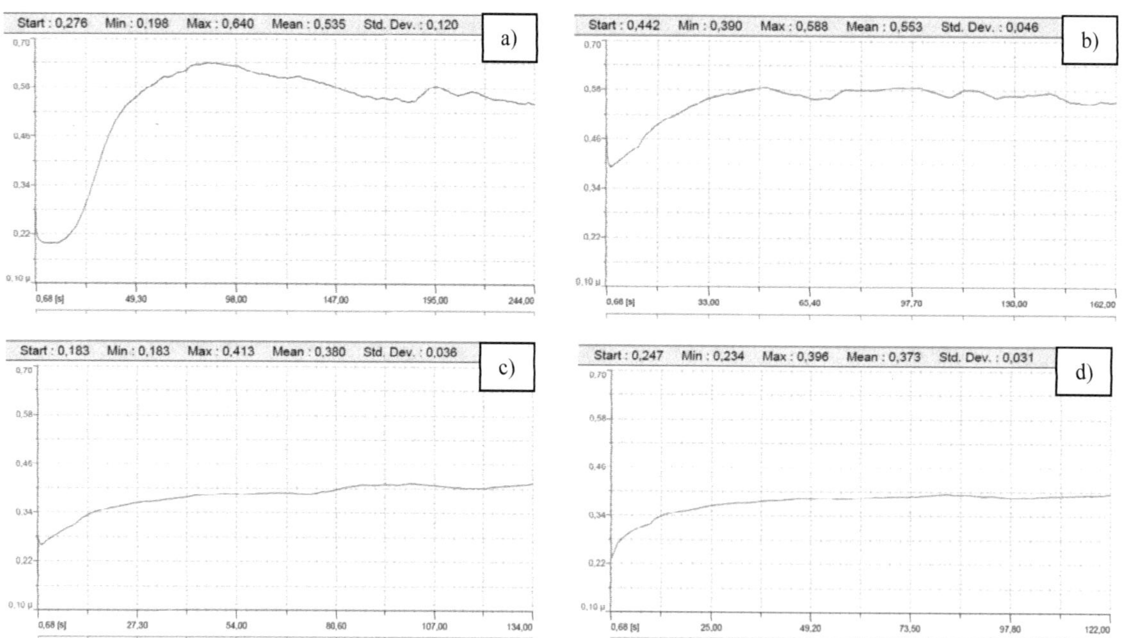

Figure 6. Coefficient of friction evolutions for the YSn83 antifriction alloy at 7N load

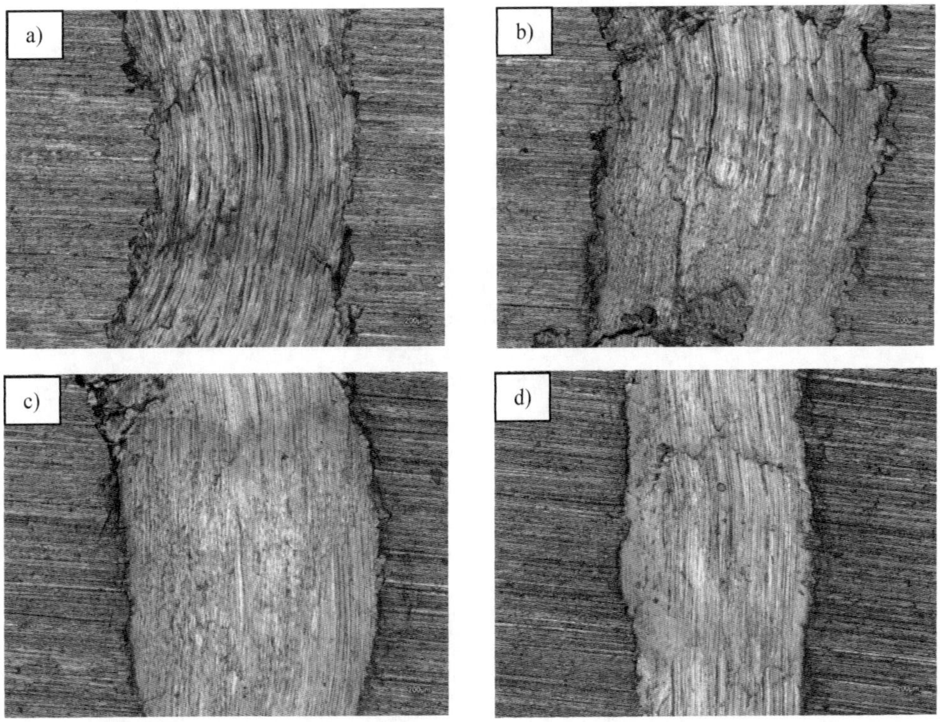

Figure 7. Wear tracks profiles for the YSn83 antifriction alloy at 5N load

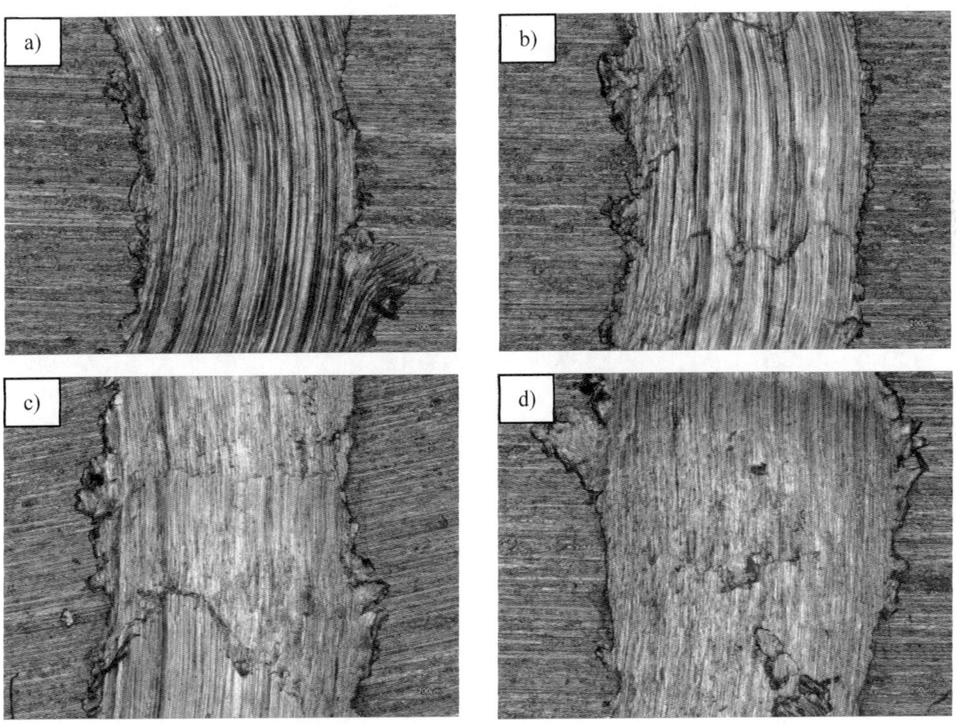

Figure 8. Wear tracks profiles for the YSn83 antifriction alloy at 7N load

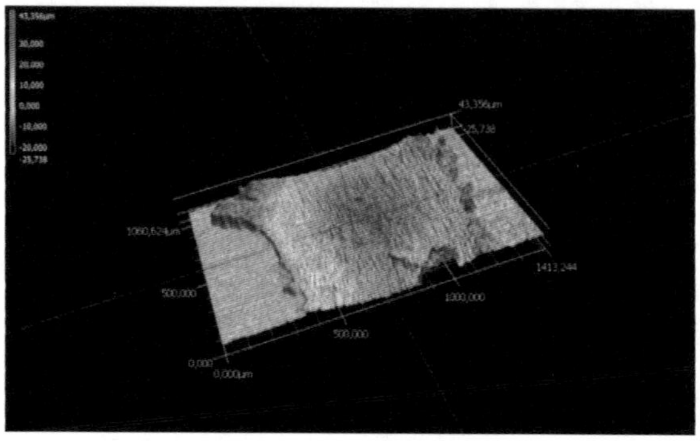

Figure 9. 3D wear track details for Face 1(4) of the YSn83 antifriction alloy

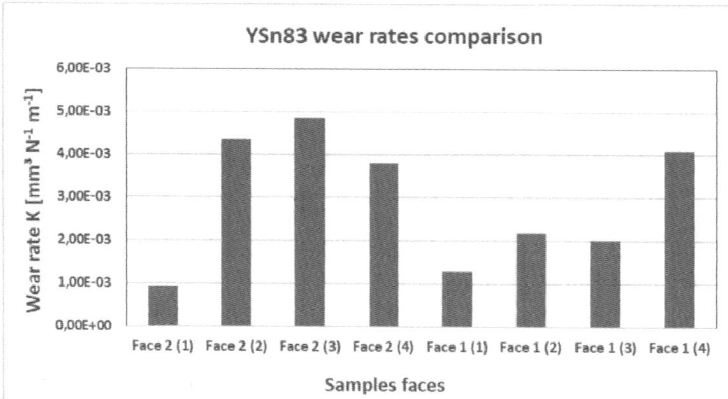

Figure 10. Comparison between the wear rates for the YSn83 antifriction alloy

Table 8. Wear rate values for the YSn83 antifriction alloy

Faces	Wear track depth, h (μm)	Wear track width, s (μm)	Cross section area, A (μm^2)	Volume loss, V (mm^3)	Wear rate, K (mm^{3}N^{-1}m^{-1})
Face 2 (1)	11.887	626.825	4968.72	0.0468	$9.37\cdot10^{-4}$
Face 2 (2)	19.831	872.150	11534.88	0.2174	$4.35\cdot10^{-3}$
Face 2 (3)	16.297	792.796	8616.19	0.2436	$4.87\cdot10^{-3}$
Face 2 (4)	12.780	590.187	5030.16	0.1896	$3.79\cdot10^{-3}$
Face 1 (1)	16.328	655.473	7138.36	0.0673	$1.28\cdot10^{-3}$
Face 1 (2)	16.632	731.329	8112.12	0.1529	$2.18\cdot10^{-3}$
Face 1 (3)	11.734	792.858	6203.28	0.1754	$2\cdot10^{-3}$
Face 1 (4)	19.149	892.882	11402.46	0.4299	$4.09\cdot10^{-3}$

It can be seen that almost all the coefficient of friction evolutions has reached the steady state stage, except one (Figure 5 d). In all these curves, the coefficient of friction mean values was between 0.179 and 0.553. The Figures 7 and 8 show that only for small distances (between 7.5 and 15 m), the YSn83 antifriction alloy presented a severe abrasion wear. Regarding the wear rates values, these tend to increase with increasing the radius and linear speed as it can be seen from Figure 10 and from the volume loss values. At small radius and linear speeds values, the wear rate is the lowest.

3.3. EN-GJS-400-15 spheroidal cast iron dry sliding wear behaviour

For the EN-GJS-400-15 cast iron the main results (coefficient of friction evolutions, wear tracks profiles and the wear rate values) are presented in Figures 11 ÷ 16 and in Table 9.

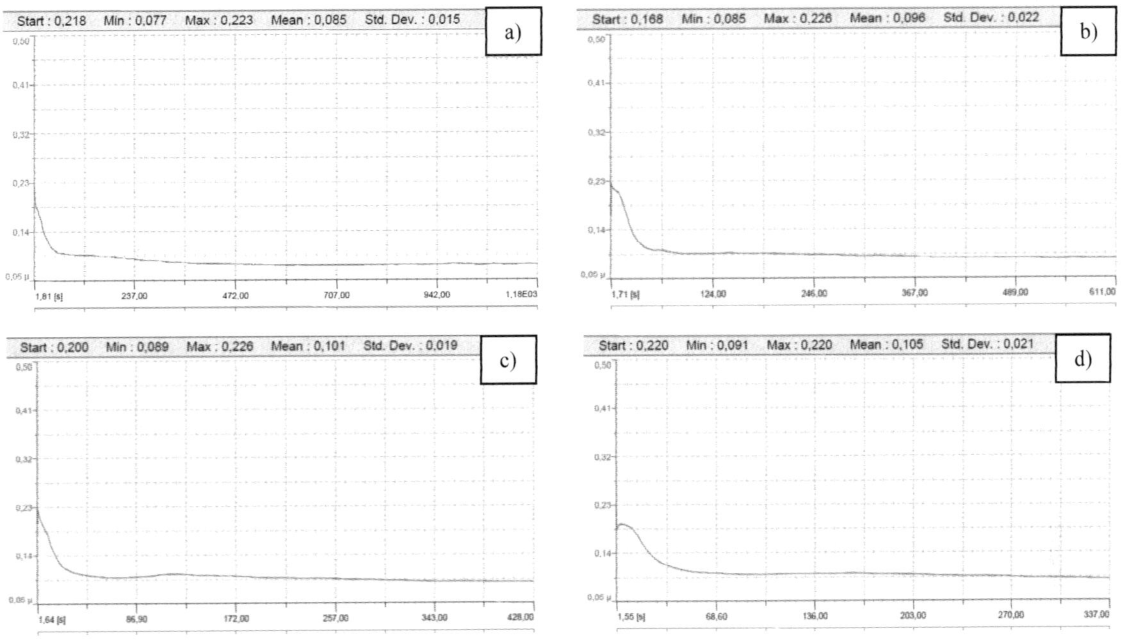

Figure 11. Coefficient of friction evolutions for the EN-GJS-400-15 cast iron at a load of 7N

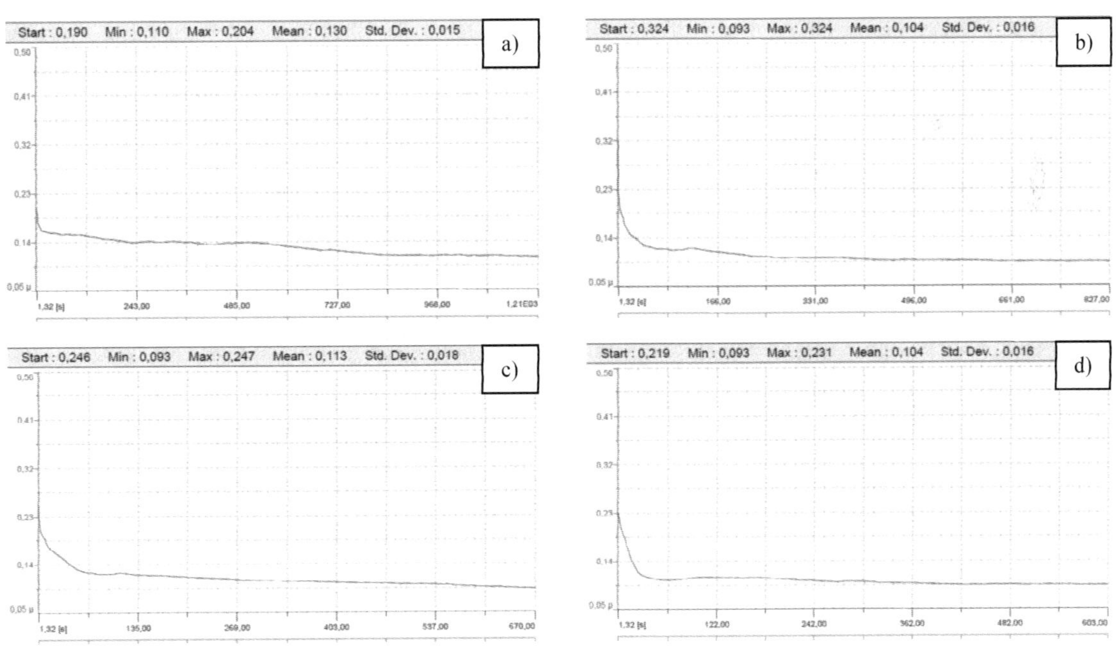

Figure 12. Coefficient of friction evolutions for the EN-GJS-400-15 cast iron at a load of 10N

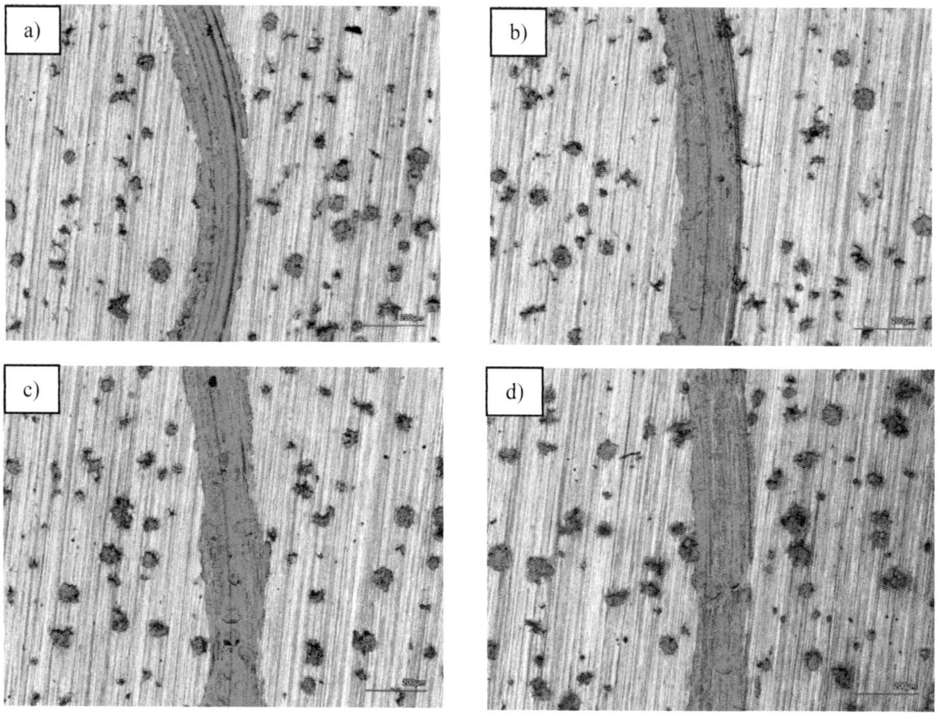

Figure 13. Wear tracks profiles for the EN-GJS-400-15 cast iron at a load of 7N

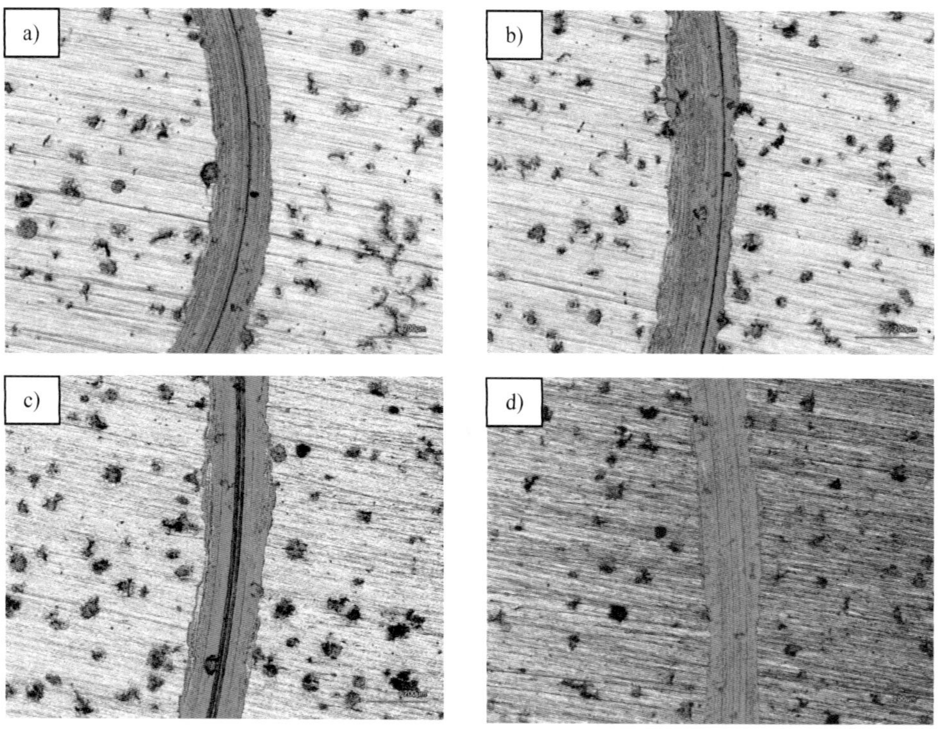

Figure 14. Wear tracks profiles for the EN-GJS-400-15 cast iron at a load of 10N

IOP Conf. Series: Materials Science and Engineering **477** (2019) 012060 doi:10.1088/1757-899X/477/1/012060

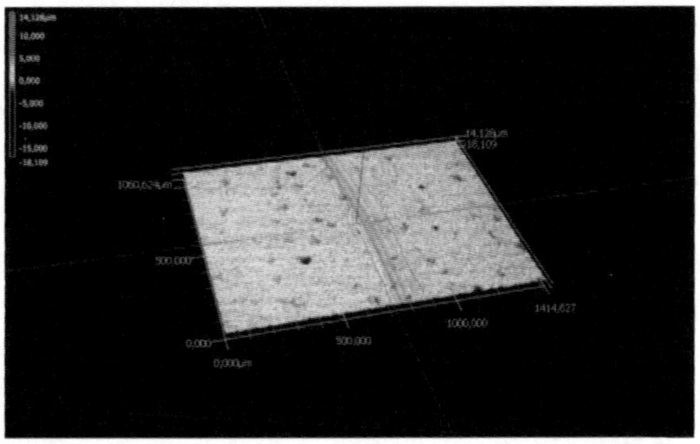

Figure 15. 3D wear track details for Face 2(4) of the EN-GJS-400-15 cast iron

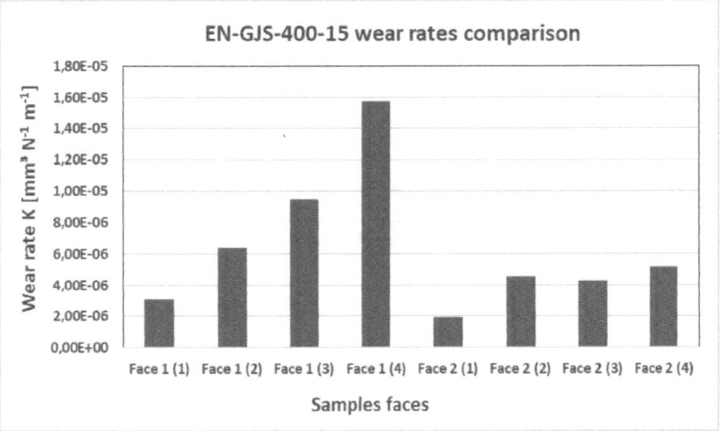

Figure 16. Comparison between the wear rates for the EN-GJS-400-15 cast iron

Table 9. Wear rate values for the EN-GJS-400-15 cast iron

Faces	Wear track depth, h (μm)	Wear track width, s (μm)	Cross section area, A (μm^2)	Volume loss, V (mm^3)	Wear rate, K (mm^{3}N^{-1}m^{-1})
Face 1 (1)	2.001	169.824	226.61	0.0021	$3.05 \cdot 10^{-6}$
Face 1 (2)	1.844	192.839	237.08	0.0045	$6.38 \cdot 10^{-6}$
Face 1 (3)	1.977	178.636	235.42	0.0067	$9.51 \cdot 10^{-6}$
Face 1 (4)	2.488	176.057	292.03	0.0110	$1.57 \cdot 10^{-5}$
Face 2 (1)	1.261	182.048	153.05	0.0014	$1.92 \cdot 10^{-6}$
Face 2 (2)	1.777	203.103	240.58	0.0045	$4.53 \cdot 10^{-6}$
Face 2 (3)	1.477	191.565	188.64	0.0053	$4.27 \cdot 10^{-6}$
Face 2 (4)	1.724	177.810	204.38	0.0077	$5.14 \cdot 10^{-6}$

It can be seen that all the coefficient of friction evolutions are quite similar with small values (between 0.085 and 0.130).

Figures 13 and 14 show that the EN-GJS-400-15 spheroidal cast iron presented a slightly abrasion wear. The wear rates values increase with increasing the radius and linear speed as it can be seen from Figure 16 and especially from the volume loss values (Table 9). The lowest wear rate toke places at a working speed of 400 rpm for the small radius, linear speed respectively distance.

4. Conclusions

The paper presented a comparison between the YSn83 antifriction alloy and EN-GJS-400-15 spheroidal cast iron regarding their cavitation erosion respectively dry sliding wear behavior.

After the cavitation erosion tests, YSn83 antifriction alloy has lost 637.5 mg material loss and show visible caverns until 300 μm depth. The EN-GJS-400-15 spheroidal cast iron lost only 13.33 mg material loss and show small pits until 100 μm.

For the dry sliding wear tests, in the case of the EN-GJS-400-15 spheroidal cast iron, the distance was 10 times more that in the case of the YSn83 antifriction alloy. After these tests, for the two tested materials the coefficient of friction was obtained respectively the wear rate was calculated. The YSn83 antifriction alloy show a severe abrasion wear and the EN-GJS-400-15 spheroidal cast iron show slightly abrasion wear. Regarding the tribometer parameters, by increasing the linear speed, the materials will lose more volume loss of material and the wear rate values will increase. At small linear speeds values, the wear rate is lowest.

From all the tests, the EN-GJS-400-15 spheroidal cast iron showed a superior cavitation erosion and dry sliding wear behaviour.

References

[1] Singh R, Tiwari S K and Mishra S K 2012 Cavitation Erosion in Hydraulic Turbine Components and Mitigation by Coatings: Current Status and Future Needs, *J Mater Eng Perform* **21** 1539-1551

[2] Frunzaverde D, Câmpian V, Nedelcu D, Gillich G R and Mărginean G 2010 Metallographic and numerical methods investigations about failure of a kaplan turbine runner blade, *WSEAS Transactions on Fluid Mechanics* **5**(3) 122-131

[3] Orłowicz A W, Mróz M, Tupaj M and Trytek A 2015 Materials Used in the Automotive Industry, *Archives of Mechanical Technology and Materials* **15**(2) 75-78

[4] Gelfi M, Gorini D, Pola A and La Vecchia G M 2016 Effect of Titanium on the Mechanical Properties and Microstructure of Gray Cast Iron for Automotive Applications, *J Mater Eng Perform* **25**(9) 3896-3903

[5] Nová I and Machuta J 2016 Monitoring of the diffusion processes during carburizing automotive steel parts, *Manufacturing Technology* **16**(1) 225-232

[6] Nedeloni M D et al. 2013 Research regarding the cavitation erosion resistance of the stainless steel with the 13% Cr and 4% Ni used to manufacture the components of Kaplan, Francis and Pelton hydraulic turbines, *Constanta Maritime University Annals* **19** 129-132

[7] Haţiegan C et al 2017 Vibration analysis of a hydro generator for different operating regimes, *IOP Conf. Ser.: Mater. Sci. Eng.* **163** 012030

[8] Frunzaverde D, Ciubotariu C R, Secosan E R, Campian C V and Fanica C 2016 Study on the Use of Elastomeric Coatings for Protection of Hydraulic Turbine Components Against Cavitation Erosion, *Mater Plast* **53**(3) 557-560

[9] Digulescu A et al 2016 Cavitating vortex characterization based on acoustic signal detection, *IOP Conf. Ser.: Earth Environ. Sci.* ***49*** *082009*

[10] Pola A, Gelfi M, Eleonora Depero L and Roberti R 2008 Study of annealing temperature effect on stress-corrosion cracking of aluminum brass heat-exchangers tubes by microdiffraction experiments, *Engineering Failure Analysis* **15**(1-2) 54-61

[11] Martínez J, Pereira A, Pérez J A and Mathia T 2015 Influence of machining of EN-GJL-250 and EN-GJS-400 cast irons on tribological behavior, *Archives of Mechanical Technology and Materials* **35** 12-22

[12] Palásti-Kovács B, Néder Z, Czifra A and Váradi K 2004 Microtopography Changes in Wear Process, *Acta Polytech Hung* **1**(1) 108-119

[13] Bąkowski H 2018 Wear mechanism of spheroidal cast iron piston ring-aluminum matrix composite cylinder liner contact, *Arch Metall Mater* **63** 481-490

[14] Nedeloni L, Korka Z I, Nedeloni M D and Pauliuc D 2017 Wear Resistance Assessment of

Fluoropolymer Coated Gears, *Annals of "Eftimie Murgu" University* **24**(1) 225-234
[15] Soiński M S and Derda A 2008 The influence of selected elements upon mechanical properties of ductile iron EN-GJS-500-7, *Arch of Mechanical Technology and Materials* **8**(3) 149-152
[16] Cîndea L et al 2016 The influence of thermal field in the electric arc welding of X60 carbon steel components in the CO2 environment, *Appl Therm Eng* **103** 1164-1175
[17] Foglio E, Lusuardi D, Pola A, La Vecchia G M and Gelfi M 2016 Fatigue design of heavy section ductile irons: Influence of chunky graphite, *Materials and Design* **111** 353-361
[18] Girelli L, Pola A, Gelfi M, Masotti M N and La Vecchia G M 2017 Performance optimization of high resistant white cast iron for severe working applications, *Metall Ital* **6** 5-10
[19] Janicki D 2018 Microstructure and Sliding Wear Behaviour of In-Situ TiC-Reinforced Composite Surface Layers Fabricated on Ductile Cast Iron by Laser Alloying, *Materials* **11**(75) 1-17
[20] Potoceanu N, Nedeloni M D, Chirus D and Florea D 2014 Cavitation erosion behavior of the antifriction alloy YSn83, *Materials Science Forum* **782** 257-262
[21] Lupinca C I and Marta C 2015 The Solidification Study of the Antifriction Sn-Sb Alloy, *Annals of "Eftimie Murgu" University of Resita* **22**(1) 277-284
[22] Dinescu I and Valentin I M 2013 Experimental research regarding the anti friction materials, *Review of the Air Force Academy* **1**(23) 41-44
[23] Dinescu I, Moldovan G and Mosoiu O 2014 *Experimental research regarding the manufacturing of the anti friction materials used in aeronautic constructions*, International Conference of Scientific Paper, Romania, May 22-24, pp 1-5
[24] Dojčinović M, Erić O, Rajnović D, Sidjanin L and Baloš S 2012 The morphology of ductile cast iron surface damaged by cavitation, *Metall Mater Eng* **18**(3) 165-176
[25] Chawla K, Saini N and Dhiman R 2013 Investigation of Tribological Behavior of Stainless Steel 304 and Grey Cast Iron Rotating Against EN32 Steel Using Pin on Disc Apparatus, *IOSR-JMCE* **9**(4) 18-22
[26] Lupinca C I, Nedeloni M D and Nedelcu D 2014 Gray Cast Iron Behavior in Cavitation Erosion, *Materials Science Forum* **782** 269-174
[27] Mitelea I, Bordeașu I, Pelle M and Crăciunescu C J 2015 Ultrasonic cavitation erosion of nodular cast iron with ferrite–pearlite microstructure, *Ultrason Sonochem* **23** 385-390
[28] Mitelea I, Bena T, Bordeasu I and Craciunescu C M 2018 Relationships Between Microstructure, Roughness Parameters and Ultrasonic Cavitation Erosion Behaviour of Nodular Cast Iron, EN-GJS-400-15, *Rev Chim* **69**(3) 611-617
[29] Ciubotariu C R, Frunzaverde D, Secosan E, Marginean G and Campian V 2016 Experimental Study Regarding the Cavitation and Corrosion Resistance of Stellite 6 and Self-Fluxing Remelted Coatings, *Stroj Vestn-J Mech E* **62**(3) 154-162
[30] Verma S, Dubey P, Selokar A W, Dwivedi D K and Chandra R 2017 Cavitation Erosion Behavior of Nitrogen Ion Implanted 13Cr4Ni Steel, *Trans Indian Inst Met* **70**(4) 957-965
[31] Gottardi G, Tocci M, Montesano L and Pola A 2018 Cavitation erosion behaviour of an innovative aluminium alloy for Hybrid Aluminium Forging, *Wear* **394-395** 1-10
[32] Kekes D, Psyllaki P and Vardavoulias M 2014 Wear Micro-Mechanisms of Composite WC-Co/Cr - NiCrFeBSiC Coatings. Part I: Dry Sliding, *Tribology in Industry* **36**(4) 361-374
[33] Kekes D, Psyllaki P, Vardavoulias M and Vekinisc G 2014 Wear Micro-Mechanisms of Composite WC-Co/Cr - NiCrFeBSiC Coatings. Part II: Cavitation Erosion, *Tribology in Industry* **36**(4) 375-383
[34] Bordeasu I et al 2017 A new concept for stainless steels ranking upon the resistance to cavitation erosion, *IOP Conf. Ser.: Mater. Sci. Eng.* **163** 012002
[35] Pola A, Montesano L, Gelfi M and La Vecchia G M 2016 Comparison of the sliding wear of a novel Zn alloy with that of two commercial Zn alloys against bearing steel and leaded brass, *Wear* **368-369** 445-452
[36] Pola A, Montesano L, Tocci M and La Vecchia G M 2017 Influence of Ultrasound Treatment

on Cavitation Erosion Resistance of AlSi7 Alloy, *Materials* **10**(3) 1-13
[37] Petrogalli C, Montesano L, Pola A, Gelfi M, Ghidini A and LaVecchia G M 2015 Improvement of Fatigue Resistance of a Tool Steel by Surface Treatments, *Procedia Engineer* **109** 154-61
[38] Nedelcu D, Nedeloni M D and Lupinca C I 2014 Cavitation erosion research on the X3CrNi13-4 stainless steel, *Materials Science Forum* **782** 263-268
[39] Hong S, Wu Y, Zhang J, Zheng Y, Zheng Y and Lin J 2016 Synergistic effect of ultrasonic cavitation erosion and corrosion of WC–CoCr and FeCrSiBMn coatings prepared by HVOF spraying, *Ultrason Sonochem* **31** 563-569
[40] Cao M, Zhao L, Wu L and Wang W 2018 Tribological Properties of New Cu-Al/MoS2 Solid Lubricant Coatings Using Magnetron Sputter Deposition, *Coatings* **8**(134) 1-16
[41] Kumar N and Kumar P 2016 Influence of machining parameters on surface roughness and dry friction, *Engineering Solid Mechanics* **4** 109-116
[42] Stepanova N, Zimogliadova T, Ognev A, Krivizhenko D, Maliutina Y and Zimogliadova O 2018 Effect of copper on the structure and antifriction properties of cast hypoeutectoid steel, *OP Conf. Ser.: Mater. Sci. Eng.* **286** 012024
[43] Paulin C, Chicet D, Paleu V, Benchea M, Lupescu Ş and Munteanu C 2017 Dry friction aspects of Ni-based self-fluxing flame sprayed coatings, *IOP Conf. Ser.: Mater. Sci. Eng.* **227** 012091
[44] Kazamer N et al. 2016 *Aspects concerning the wear and corrosion behavior of WC-CoCr coatings and respectively DLC/WC-CoCr systems*, NANOCON pp 383-389
[45] Cozza R C 2013 A study on friction coefficient and wear coefficient of coated systems submitted to micro-scale abrasion tests, *Surf Coat Tech* **215** 224-233
[46] Nedelcu D, Cojocaru V, Nedeloni M, Peris-Bendu F and Ghican A 2015 Failure analysis of a Ti-6Al-4V ultrasonic horn used in cavitation erosion tests, *Mechanika* **4** 272-276
[47] Cojocaru V, Campian C V and Frunzaverde D 2015 A comparative analysis of the methods used for testing the cavitation erosion resistance on the vibratory devices, *U.P.B. Sci. Bull., Series D* **74**(4) 257-262
[48] Nedeloni M D, Hatiegan C, Vasile O, Hamat C O, Fanica C and Gillich N 2015 Numerical Study Regarding the Influence of Material Components for a Booster - Ultrasonic Horn Assembly on the Natural Frequency, *RJAV* **12**(2) 155-160
[49] *** 2010 Standard Test Method for Cavitation Erosion Using Vibratory Apparatus ASTM G32-10
[50] *** 2000 Standard Test Method for Wear Testing with a Pin-on-Disk Apparatus ASTM G99
[51] Nedeloni M D and Nedelcu D 2012 Research through Direct and Indirect Cavitation Method for a Aluminum Specimen, *Annals of "Eftimie Murgu" University* **19**(1) 207-214
[52] Chirus D and Nedeloni M D 2013 Cavitation Erosion Research for AlSi12 Alloy Tested at Different Time Periods, *Annals of "Eftimie Murgu" University of Resita* **20**(1) 55-64
[53] Lupinca C I and Nedeloni M D 2014 Comparative study regarding the cavitation erosion behavior of Cu and Al alloys, *International Journal of Latest Research in Science and Technology* **3**(2) 95-99
[54] Hamat C O, Nedeloni M D, Haţiegan C, Ciubotariu R C and Pădureanu I 2015 Cavitation erosion research on C45 carbon steel. Part I: multiple tests of 180 minutes, *Annals of „Constantin Brâncuşi" University of Târgu Jiu* **3** 127-132
[55] Lupinca C I and Nedeloni M D 2014 The Study of Bronze Behaviour During Cavitation Erosion, *Indian Journal of Applied Research* **4**(5) 596-600
[56] Nedeloni M D, Cojocaru V, Ciubotariu R and Nedelcu D 2012 Cavitation Erosion Tests Performed by Indirect Vibratory Method on Stainless Steel Welded Samples with Hardened Surface, *Annals of "Eftimie Murgu" University of Resita* **19**(1) 215-226